International Conference on Robotics, Intelligent Automation and Control Technologies (RIACT 2020)

IOP Conference Series: Materials Science and Engineering
Volume 1012

Chennai, India
2 – 3 October 2020

Editors:

Arockia Selvakumar Arockia Doss
Daniel Schilberg

ISBN: 978-1-7138-4547-8
ISSN: 1757-8981

Printed from e-media with permission by:

Curran Associates, Inc.
57 Morehouse Lane
Red Hook, NY 12571

Some format issues inherent in the e-media version may also appear in this print version.

Institute of Physics
Dirac House, Temple Back
Bristol BS1 6BE UK

Phone: 44 1 17 929 7481
Fax: 44 1 17 920 0979

techtracking@iop.org

Additional copies of this publication are available from:

Curran Associates, Inc.
57 Morehouse Lane
Red Hook, NY 12571 USA
Phone: 845-758-0400
Fax: 845-758-2633
Email: curran@proceedings.com
Web: www.proceedings.com

TABLE OF CONTENTS

Author Index

IOP Conf. Series: Materials Science and Engineering 1012 (2021) 011001 doi:10.1088/1757-899X/1012/1/011001

PREFACE

This proceedings consists of the peer-reviewed papers from the 1st International Conference on Robotics, Intelligent Automation and Control Technologies (RIACT 2020), which was organized by the School of Mechanical Engineering at Vellore Institute of Technology, Chennai, India on October 2nd and 3rd, 2020. However, due to the COVID 19 pandemic the conference was held Online. The main objective of RIACT 2020 is to provide a platform for researchers and practitioners from both academic institutions and industries to meet and share cutting-edge developments in the areas of Robot Design, Development and Control, Rehabilitation Robots and Devices, Intelligent Automation and Fault Diagnosis, Agricultural Robots, Underwater Robots, Development of Drones, Control and Supervision Systems, Vision Systems, Actuators and Sensors, Mechatronics Systems, System Interface and Control, Sustainable Technologies and associated disciplines. This conference also provided an opportunity to exchange research evidence and innovative ideas and also the issues related to Robotics and Automation. This proceeding will include presentations on the latest research areas such as Agriculture Drones, Bio Mechanisms for rehabilitation applications and Deep learning in robotics. This conference had gathered an excellent group of keynote speakers from around the globe on different topics such as Robot Control Systems Using electroencephalography (EEG), electrooculography (EOG), and electromyography (EMG) signals, Future Robotics farming, Dynamic control of flexible manoeuvring systems, Intuitive Human-Robot Interaction in Teaching and for Care, Increase flexibility of industrial processes using advance 3D machine vision, Automation, Communications and Control for Demand Response: Potential Enablers for Decarbonisation, High-Performance Soft Wearable Robots for Human Augmentation: New Paradigm of Design and Control for Scientific Discovery, RPA: The 10 Points to Successful Projects, ST(R)EM generation: Alfa, Research and Education at the Convergence of Frontier Technologies, Warehouse Challenges and the Power of Automation.

RIACT 2020 was graced with the following eleven keynote speakers namely Professor M. Osman Tokhi from London South Bank University, London, UK; Professor Dr.-Ing. Daniel Schilberg from University of Applied Sciences Bochum, Germany; Professor Minoru Sasaki from Gifu University, Japan; Professor Vikram Kapila from NYU Tandon School of Engineering, USA; Professor Michael Short from Teesside University, UK; Dr. Asuman Suenbuel from SAP Labs, Silicon Valley, USA; Dr. Hao Su from City University of New York, City College, USA; Dr. Milena Y Krumova, from European Institute for Technologies, Education and Digitalisation, Bulgaria; Mr. Vincenzo Marchica from VINCIX Group, Italy; Mr. Stefano Tonello from IT+Robotics, Italy; ing. EJ (Erik) Pekkeriet from Agro Food Robotics, Wageningen University & Research, Netherlands. This year, RIACT 2020 had more than 350 participants from around the globe. A total of 203 research articles were submitted and based on the double-blind peer review process 68 full length papers were accepted to be published with IOP Conference Series: Materials Science and Engineering after presentations. The evaluation of all the papers was performed based on the reports from anonymous reviewers. We would like to take this opportunity to deeply express our heartfelt appreciation to Dr. G. Viswanathan, Founder and Chancellor, Vellore Institute of Technology; Shri Sankar Viswanathan, Vice President, Vellore Institute of Technology; Dr. Sekar Viswanathan, Vice President, Vellore Institute of Technology; Shri G V Selvam, Vice President, Vellore Institute of Technology; Ms. Kadhambari S. Viswananthan, Assistant Vice President, Vellore Institute of Technology; Dr. Sandhya Pentareddy, Executive Director, Vellore Institute of Technology. Our special thanks to all our Keynote Speakers, Session

IOP Conf. Series: Materials Science and Engineering 1012 (2021) 011001 doi:10.1088/1757-899X/1012/1/011001

Chairs, all the Committee Members and Reviewers involved in the technical evaluation of conference papers.

VIRTUAL DISCLAIMER FOR RIACT 2020: As it was uncertain about when COVID-19 pandemic and travel restrictions across the globe would end, the conference was conducted in virtual mode considering the safety of the participants. The conference was not postponed to a later period because the scientific community cannot share their research findings and there would have been a delay in publishing their research work. The virtual conference was organized by Vellore Institute of Technology, Chennai, India during October 2 & 3, 2020 using ZOOM: Video Conferencing platform.

The keynote speakers from across the globe delivered their talk for close to one hour which was based on the theme of the conference. A total of 14 parallel sessions were arranged for the paper presentation by the respective authors in the presence of 2 panel members (session chairs) for each session. Total of 15 minutes for presentation and 5 minutes for discussion was allotted for each paper. At the end of the keynote session, audience were provided 10 minutes to clarify their doubts with the speaker and a member from the organizing committee summarized and thanked the speaker. With regard to the paper presentation, the panel members (session chairs) and audience were given the opportunity to ask questions to the authors regarding their research findings for a duration of 5 minutes. RIACT 2020 had 355 participants and keynote speakers from USA, UK, Germany, Netherlands, Italy, Japan and Bulgaria. All the submitted papers were initially screened for plagiarism using TURNITIN software and only those papers which had less than 10% plagiarism were selected for presentation in the conference. Moreover, all the presented papers were double blind peer reviewed by two reviewers and the authors were instructed to submit the revised manuscript based on the reviewers comments. RIACT 2020 has been conducted successfully during October 2 & 3, 2020. It has been planned to organize the Second International Conference on Robotics, Intelligent Automation and Control Technologies during September 2021.

Best Regards,

Editors,

Dr Arockia Selvakumar Arockia Doss
Senior Associate Professor, Design and Automation Research Group, SMEC,
Vellore Institute of Technology, Chennai, Tamil Nadu, India.

Professor Dr.-Ing. Daniel Schilberg
Vice Dean, Mechatronics and Research, Head of Institute, Institute of Robotics and Mechatronics, University of Applied Sciences Bochum, Germany.

COMMITTEES

Conference Chair

Advisory Committee

Mr.R. Marimuthu, Consultant, TCS Limited, India

Dr. K. Annamalai, Professor, Design and Automation research Group, SMEC, VIT Chennai, TN, India

Dr. V. Sugumaran, Professor, Design and Automation research Group, SMEC, VIT Chennai, TN, India

Technical Committee

Dr Michael Short, Teesside University, UK

Dr Mark Ovinis, UTP, Malaysia

Dr Latifah Nurahmi, ITS Nopember, Indonesia

Dr Sasitharan Subramanian, ABB, Ludvika, Sweden

Dr. Richard L Jayaraj, CMHS, UAE

Dr Subhasis Bhaumik, IIEST, West Bengal, India

Dr Sudheer.A.P, NIT Calicut, India

Prof Ravi Prakash Tewari, MNNIT, Allahabad, India

Dr G B Krishnappa, VVCE, Karnataka, India

Dr. Dinakaran D, HITS, Tamil Nadu, India

Dr. S. Shankar, Kongu Engineering College, Tamil Nadu, India

Dr. I Thirunavukkarasu, MIT MANIPAL, India

Dr. Poongodi C, Kongu Engineering College, India

Dr. G. Satheesh Kumar, SSN College of Engineering, Tamil Nadu, India

Dr T Muthuramalingam, SRMIST, Tamil Nadu, India

Dr R Senthilnathan, SRMIST, Tamil Nadu, India

Dr G Shanmugasundar, Sri Sai Sairam Institute of Technology, TN, India

Dr. G. Ananthi, Thiyagarajar College of Engineering, Madurai, India

Dr P.D. Rathika, PSG College of Technology, TN, India

Dr. M. Sasikumar, Vellore Institute of Technology, Chennai, TN, India

Dr. V.G. Sridhar, Vellore Institute of Technology, Chennai, TN, India

Dr. I.S. Rajay Vedaraj, Vellore Institute of Technology, Vellore, TN, India

Dr. N. Maheswari, Vellore Institute of Technology, Chennai, TN, India

Dr. R. Ganesan, Vellore Institute of Technology, Chennai, TN, India

Dr. M. C. Lenin Babu, Vellore Institute of Technology, Chennai, TN, India

Dr. Sathis Kumar B, Vellore Institute of Technology, Chennai, TN, India

Dr. T.S.Pradeepkumar, Vellore Institute of Technology, Chennai, TN, India

Dr. P.S. Nithya Darisini, Vellore Institute of Technology, Chennai, TN, India

Dr. M.Sivagami, Vellore Institute of Technology, Chennai, TN, India

IOP Conf. Series: Materials Science and Engineering 1012 (2021) 011001 doi:10.1088/1757-899X/1012/1/011001

Dr. R.Manoj Kumar, Vellore Institute of Technology, Chennai, TN, India

Dr. R. Maheswari, Vellore Institute of Technology, Chennai, TN, India

Dr Jafferson J.M, Vellore Institute of Technology, Chennai, TN, India

Dr. Venugopal T, Vellore Institute of Technology, Chennai, TN, India

Dr. V.Berlin Hency, Vellore Institute of Technology, Chennai, TN, India

Dr Keyur B Joshi, Vellore Institute of Technology, Vellore, TN, India

Dr. Angalaeshwari S, Vellore Institute of Technology, Chennai, TN, India

Dr. T. Deepa, Vellore Institute of Technology, Chennai, TN, India

Dr. D.Subbuleksmi, Vellore Institute of Technology, Chennai, TN, India

Dr. Chinmaya Sahu, Vellore Institute of Technology, Vellore, TN, India

Dr. S. Radha,Vellore Institute of Technology, Chennai, TN, India

Dr. G. K. Revathi, Vellore Institute of Technology, Chennai, TN, India

Prof. Elango M, Vellore Institute of Technology, Chennai, TN, India

Dr. Joshuva Arockia Dhanraj, HITS, Tamil Nadu, India

Peer review declaration

All papers published in this volume of *IOP Conference Series: Materials Science and Engineering* have been peer reviewed through processes administered by the Editors. Reviews were conducted by expert referees to the professional and scientific standards expected of a proceedings journal published by IOP Publishing.

- **Type of peer review: Double-blind**
- **Conference submission management system: www.riact.co.in**
- **Number of submissions received: 203**
- **Number of submissions sent for review: 140**
- **Number of submissions accepted:68**
- **Acceptance Rate (Number of Submissions Accepted / Number of Submissions Received X 100):33.50**
- **Average number of reviews per paper: 2**
- **Total number of reviewers involved:71**
- **Any additional info on review process:** All papers had undergone plagiarism check and double-blind review by two reviewers. Based on the reviewer comments, revised manuscript were submitted by authors for final publication.
- **Contact person for queries:**

Dr. AROCKIA SELVAKUMAR AROCKIA DOSS
CONVENER – RIACT 2020
SENIOR ASSOCIATE PROFESSOR, DESIGN AND AUTOMATION RESEARCH GROUP,
SCHOOL OF MECHANICAL ENGINEERING (SMEC),
VIT CHENNAI, VANDALUR – KELAMBAKKAM ROAD,
MELAKOTTAIYUR, CHENNAI - 600 127.
Mobile: 9962681933
Email: arockia.selvakumar@vit.ac.in

Design and Development of In-pipe Inspection Robot for Various Pipe Sizes

Atul Gargade[1*] and Shantipal Ohol[1]

[1]Mechanical Engineering Department, College of Engineering Pune, India.

E-mail: gargadeaa14.mech@coep.ac.in

Abstract. In-pipe inspection robots are designed to pull out the human role from work load and risky working circumstances. In this paper an in-pipe inspection robot version 2 (IPIR version 2) is presented which composed of two driving leg systems, two supporting leg systems and a connecting body. Novelty of version 2 is its stability and diameter adaptability. Stability of version 2 is enhanced by adding two supporting leg systems in version 1 and diameter adaptability of version 2 is improved by optimizing its spring design. All major components of version 2 are designed safely. Solid modelling of all robot parts and its assembly is carried out in Solidworks 16. Mathematical modelling of version 2 is carried by Lagrange's method. A planetary geared DC motor with encoder (IG42E-104K) is used as prime mover of IPIR version 2. This robot has mainly employed aluminium as structural material. To verify the efficacy of driving mechanism, several experiments of version 2 are conducted in horizontal pipes, vertical pipes and couplings of 8 inches to 10 inches diameter range. This IPIR version 2 will be employed for offline visual checking of various pipe components like horizontal pipes, vertical pipes and couplings in water pipelines, gas pipelines and drain pipes etc.

Keywords: In-pipe inspection robot version 2 (IPIR version 2); driving mechanism; planetary geared DC motor; defects.

1. Introduction

From last couple of decades, robotics is becoming one of the most quickly expanding fields. As we know, variety of pipes is being used to build valuable lifelines like gas supply, water supply and sewerage systems in our modern community. Therefore, pipeline has become a crucial component of conveyance. In fact, pipeline has become an integrated module of dissimilar types of sectors, such as power plants, crude oil companies and suburban water supply, agriculture field and so on. But due to the years of usage of pipelines, defects like cracks, aging and corrosion takes place. Even blockages are also seen in the pipeline. Sometimes defects are taking place due to natural disasters and mechanical damages from unbiased observer. Therefore, it becomes very difficult to search such defects and their locations. Thus, maintenance plan of such pipelines should be scheduled. If we accomplish inspection activity by manual way then great amount of efforts, labours and time is required to dig out the pipes that are covered under the floor. The utmost reason of inventing in pipe inspection robot is to withdraw human being and come out with a fast and accurate inspection at low cost [1, 2].

Therefore, in-pipe robots are broadly classified into actively moving robots and passively moving robots. This classification is made on the basis of difference in operating source and controllability of steering mechanism. Active locomotion robots can be further sorted into six categories which are shown in figure 1. Pig type is a classic example of passively moving robots [3].

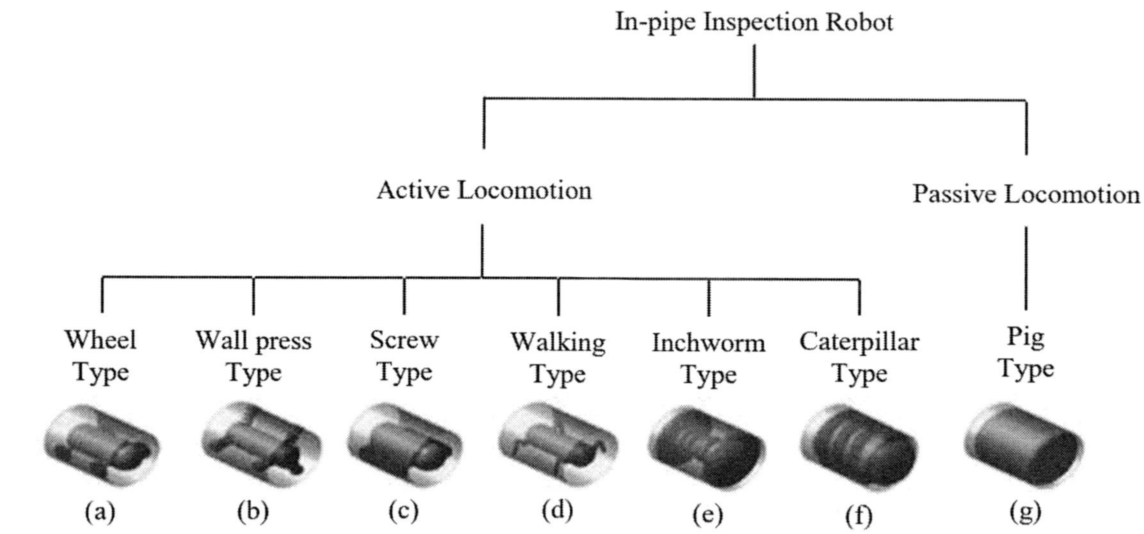

Figure 1. Sorting of in-pipe robots (a) Wheel type (b) Wall press type (c) Screw type (d) Walking type (e) Inchworm type (f) Caterpillar type (g) Pig type [4]

A wheel type robot shown in figure 1(a) is most preferred type of pipe inspection robot. Advantage of this type of robot is its very simple design. The wall press type robot shown in the figure 1(b) gives most tractive force as compared to other types. Due to high tractive force it can climb in vertical pipelines by pressing the wall by whatever mode they employ. In this type, design of robot is decided by the particular application. Figure 1(c) displays screw type (helical drive type) robot which follow the principle of screw. When it travel in the pipeline, it follows the helical path. Figure 1(d) depicts a walking type which is generally called as crawler robot. Advantage of this type of robot is that it can walk on any type of surface. Figure 1(e) shows inchworm type robot, which emulate motion of inchworm. Its movement through pipe is very slow like inchworm but it can be used for very small diameter pipelines. Figure 1(f) illustrates the robot with caterpillars. This type of robot comes up with additional grips than wheel type robot, which helps to reduce the slips inside the pipe. Figure 1(g) displays a pig type of in-pipe inspection robot. In this type fluid pressure is used to drive the robot passively inside the pipeline. Generally this type of robot is preferred for inspection of large diameter pipelines [5, 6].

This paper is arranged in VIII sections. Section II presents literature review of in-pipe inspection robot. Section III describes the construction and working of IPIR version 2. Section IV gives mechanical design and selection of version 2 components. Section V provides mathematical modelling of IPIR version 2. Section VI gives fabrication details and assembly of version 2. Section VII provides the result and discussion. Section VIII presents conclusion and future work.

2. Literature Review

Since long back, several researchers are working on robotic systems which can travel effectively and easily through the pipelines. Some of the most important and related pipe inspection robots are described in literature. Atsushi Kakogawa et al. [7] have presented screw drive in-pipe robot. This robot can travel across the vertically positioned bent pipes but still it required finding optimal values of arm length to reach to the pipe wall and the upper limit with minimum torque. Atul Gargade et al.

[8] have proposed a wall pressed wheel type in-pipe inspection robot. This robot can pass through horizontal and vertical pipes of 140mm to 200mm diameter range. It does not pass through bends and couplings. Muhammad Azri Abdul Wahed and Mohd Rizal Arshad [9] have presented wall pressed wheel type pipe inspection robot which can pass through pipes of 150mm to 230mm diameter range. That robot can pass through horizontal pipes and inclined pipes of a slope not more than 30 degree. It cannot pass through vertical pipes and other elements of pipeline. Hun-ok Lim and Taku Ohki [10] have developed wall pressed wheel type pipe inspection robot. That robot can pass through horizontal pipes, vertical pipes and 'T' joints. R. K. Jain et al. [11] have proposed a virtual prototype of in-pipe inspection robot. That robot has employed scissor mechanism for inspection of large diameter pipeline of 500mm to 1000mm diameter range. Ankit Nayak and S. Pradhan [12] have proposed an in-pipe inspection robot which is combination of screw type and wall pressed wheel type robot. They have provided mathematical treatment and demonstrated the efficacy of developed mathematical model. They have presented initial conceptual prototype of pipe inspection robot.

Majority of in-pipe inspection robots have used one of the basic types of mechanism directly whereas some robots have used combination of basic types. This paper presents a screw driven wall pressed wheel type in-pipe inspection robot version 2. This version is configuration of screw type and wall pressed wheel type robot. This presented configuration is providing better stability and diameter adaptability than existing screw driven wall pressed wheel type in-pipe inspection robot.

3. Construction and Working of IPIR version 2

3.1. Construction and Working of IPIR version 1

IPIR version 1 was consisting of forward leg system, backward leg system and a body. That version was able to pass through horizontal and inclined pipes of 10 inches diameter only. Therefore, its diameter adaptability inside different pipe sizes was poor. Weight of version 1 was 2.2 kg. Due to high weight its steerability in vertically pipe was very poor.

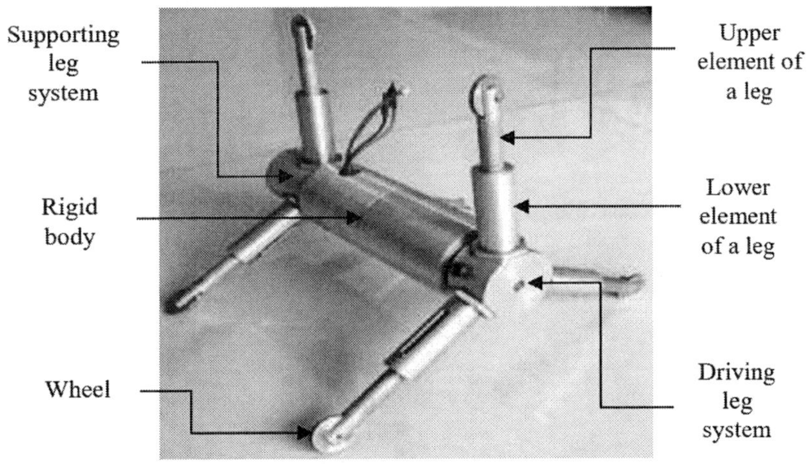

Figure 2. Working model of IPIR version 1 [13]

Therefore, IPIR version 2 has been developed to overcome the drawbacks of IPIR version 1. Stability of IPIR is improved in version 2 by adding two supporting leg systems. Spring design of version 1 is optimized to pass through pipes of 8 inches to 10 inches diameter range.

3.2. Construction and Working of IPIR version 2

In-pipe inspection robot version 2 mainly composed of a forward leg system, backward leg system, and a connecting body. Both the leg systems are mirror images of each other in construction. Each leg system consists of two sub leg systems. One is driving leg system with inclined wheels and another is supporting leg system with straight wheels. Driving leg systems of fore leg systems is mounted on the DC motor shaft whereas supporting leg system is fixed on a robot body. Both driving and supporting leg systems consists of three legs which are framed at an angle of 120 degree to one another to travel through pipes of 8 inches to 10 inches diameter range. Each leg comprised of a lower element, upper element, spring and a wheel which is shown in figure 3. To get better stability in vertical pipe wheels of both the leg systems are knurled.

Figure 3. Solid model of IPIR version 2

To get forward and backward motion of robot in spiral direction, wheels of driving leg systems are connected to top element at an angle of 15 degree. Wheels of supporting leg systems are kept at 90 degree to get rolling motion in to and fro direction. Springs are placed into lower element of each leg which does help to advance easily inside pipes of 8 inches to 10 inches diameter range. A planetary geared DC motor with encoder (IG42E-104K) is prime mover of IPIR version 2. In this robot, aluminium is used as structural material for fabrication of major parts.

3.3. Specification of IPIR version 2

Following table shows the specifications of IPIR version 2.

Table 1. Specifications of IPIR version 2

Sr. No.	Parameter	Dimension
1.	Length	300 mm
2.	Diameter (max)	250 mm
3.	Diameter (min)	200 mm
4.	Weight	2 kg
5.	Linear Velocity	113.5 mm/s

4. Mechanical Design and Selection of IPIR version 2 Components

Design of all major components and selection of some specific components of IPIR version 2 is shown in following table.

Table 2. Design of IPIR version 2 Components

Sr. No.	Parameter to Design	Parameters Considered for Design	Designed Parameters
1.	Linear velocity of robot	Helix angle, $\alpha = 15°$ Diameter, D = 270 mm	$V_H = 424$ mm/sec, $V_L = 113.5$ mm/sec
2.	Speed of motor	Helical Velocity, $V_H = 424$ mm/sec Radius of wheel, r = 135 mm	Motor speed, N = 30 rpm
3.	Spring force & Spring stiffness	Mass of robot considered for design, m = 4 kg Coefficient of friction, $\mu_s = 0.3$ Minimum compression of spring, $\delta_{min} = 15$ mm	Spring force , $[F_S]_{min} = 40.6793$ N Spring Stiffness, k = 2.7119 N/mm
4.	Spring wire diameter	Stiffness of Spring, k = 2.711 N/mm Outer diameter, $D_o = 16$ mm Free length, $L_f = 60$ mm	Spring wire diameter, d = 2.1 mm
5.	Power required	Weight of robot = 40 N Frictional force = 132 N Linear velocity = 113.5 mm/sec	Power required, $P_{required} = 20$ watt
6.	Selection of battery	Select Lithium-iron 12v & 2.5 Ah	Power provided, $P_{provided} = 30$ watt
7.	Design of motor shaft	Shear stress, $\tau_{max} = 90$ N/mm^2 Power, P = 20 watt	Motor shaft diameter, d = 7 mm

5. Mathematical Modeling of IPIR version 2

Dynamic Analysis of IPIR version 2 by Lagrange's Energy Method:

Here robot is performing general plane motion along x-direction.

∴ Its position vector can be written as,

$$\overline{R} = r\,\theta + x$$

∴ Velocity vector is,

$$\overline{v} = r\,\dot{\theta} + \dot{x}$$

∴ Acceleration vector is,

$$\overline{a} = r\,\ddot{\theta} + \ddot{x}$$

∴ The equation of driving force and driving torque in terms of generalize coordinates can written by Lagrange method as,

Equation of driving force is,

$$F\,(t) = \frac{d}{dt}\left(\frac{\partial L}{\partial \dot{x}}\right) - \frac{\partial L}{\partial x} \tag{1}$$

Equation of driving torque is,

$$T\,(t) = \frac{d}{dt}\left(\frac{\partial L}{\partial \dot{\theta}}\right) - \frac{\partial L}{\partial \theta} \tag{2}$$

As we know energy equation is,

$$T.E. = K.E. - P.E.$$

$$\therefore L = \left[\left(\frac{1}{2}\,mv^2 + \frac{1}{2}\,I\,\theta^2\right) - \left(\frac{1}{2}\,kx^2 + \frac{1}{2}\,k_t\,\theta^2\right)\right]$$

$$\therefore L = \left[\frac{1}{2}\,m\left((\dot{x} + r\dot{\theta})\sin\alpha\right)^2 + \frac{1}{2}\,I\left(\left(\frac{\dot{x} + r\dot{\theta}}{r}\right)\cos\alpha\right)^2 - \left(\frac{1}{2}\,kx^2 + \frac{1}{2}\,k_t\theta^2\right)\right] \tag{3}$$

By differentiating equation (3)

We get,

$$\frac{\partial L}{\partial x} = -kx$$

$$\frac{\partial L}{\partial \dot{x}} = m\,(\dot{x} + r\dot{\theta})\,\sin^2\alpha + I\left(\frac{\dot{x} + r\dot{\theta}}{r^2}\right)\cos^2\alpha$$

$$\frac{\partial L}{\partial \theta} = -k_t\theta$$

$$\frac{\partial L}{\partial \dot{\theta}} = mr\,(\dot{x} + r\dot{\theta})\,\sin^2\alpha + I\,\frac{(\dot{x} + r\dot{\theta})}{r}\,\cos^2\alpha$$

$$\frac{d}{dt}\left(\frac{\partial L}{\partial \dot{x}}\right) = m\,(\ddot{x} + r\ddot{\theta})\,\sin^2\alpha + \frac{I\cos^2\alpha}{r^2}\,(\ddot{x} + r\ddot{\theta})$$

$$\frac{d}{dt}\left(\frac{\partial L}{\partial \dot{\theta}}\right) = mr\,(\ddot{x} + r\ddot{\theta})\,\sin^2\alpha + \frac{I\cos^2\alpha}{r}\,(\ddot{x} + r\ddot{\theta})$$

∴ Equation (1) will become,

$$F\,(t) = m(\ddot{x} + r\ddot{\theta})\sin^2\alpha + \frac{I\cos^2\alpha}{r^2}\,(\ddot{x} + r\ddot{\theta}) + kx \tag{4}$$

∴ Equation (2) will become,

$$T\,(t) = mr\,(\ddot{x} + r\ddot{\theta})\sin^2\alpha + \frac{I\cos^2\alpha}{r}\,(\ddot{x} + r\ddot{\theta}) + k_t\theta \tag{5}$$

Where,

m = mass of robot (kg)

r = length of robot leg (mm)

α = helix angle of wheel (degree)

k = equivalent stiffness (N/mm)

k_t = torsion stiffness of wheel assembly (N.mm/rad)

I = inertia of robot (kg.mm^2)

θ = angular displacement of robot (rad)

$\dot{\theta}$ = angular velocity of robot (rad/s)

$\ddot{\theta}$ = angular acceleration of robot (rad/s^2)

x = linear displacement of robot (mm)

$\dot{x}$ = linear velocity of robot (mm/s)

$\ddot{x}$ = linear acceleration of robot (mm/s^2)

F (t) = driving force of robot (N)

T (t) = driving torque of robot (N.mm)

6. Fabrication and Assembly of IPIR version 2 Components

6.1. Fabrication Details of IPIR version 2 Components

All components of robot are fabricated by using lathe machine, milling machine and drilling machine. Hand tap is used for tapping operation. Due to lightweight and good strength of aluminum, it is used for fabrication of robot body and other parts of version 2.

Table 3. Fabrication Details of IPIR version 2 Components

Sr. No.	Component Name	Material Used	Machine Operations	Machines Used	Fabricated Component	Qty
1.	Rotor	Aluminium	Turning, facing, drilling & Tapping	Lathe, Drilling & Hand tap	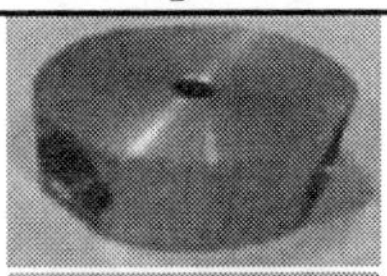	4
2.	Bottom element	Aluminium	Turning, threading, boring, facing & Slot	Lathe & Milling		12
3.	Top element	Aluminium	Turning, drilling, sleeting, facing & tapping	Lathe, Milling, Drilling & Hand tap		12
4.	Wheel	Mild steel	Turning, drilling, facing & Knurling	Lathe & Drilling	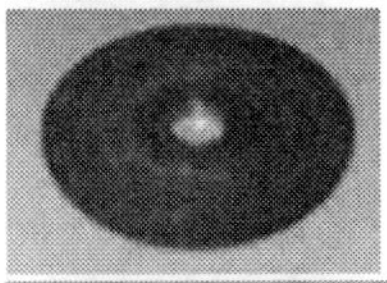	12
5.	Check nut	Aluminium	Turning, boring & threading	Lathe & Milling		12
6.	Spring	Stainless steel	Spring machine operation	Spring machine	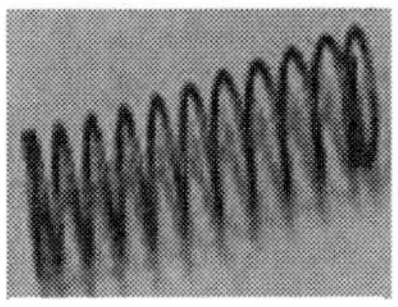	12

7.	Body	Aluminium	Turning, facing, boring & Treading	Lathe & Milling	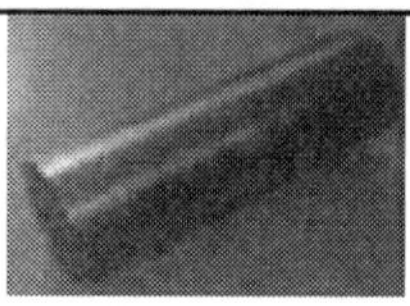	1

6.2. Working Prototype of IPIR version 2

Following figure shows fabricated model of IPIR version 2.

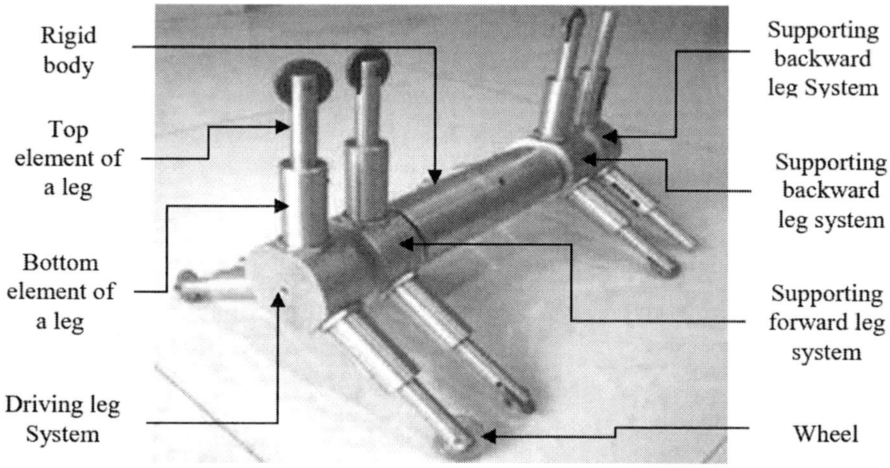

Figure 4. Working model of IPIR version 2

7. Result and Discussion

7.1. Result

To verify the efficacy of driving mechanism of version 2, its movement is conducted through horizontal pipes, vertical pipes and couplings of 8 inches and 10 inches diameter. All experiments are conducted for pipe elements of 900mm length.

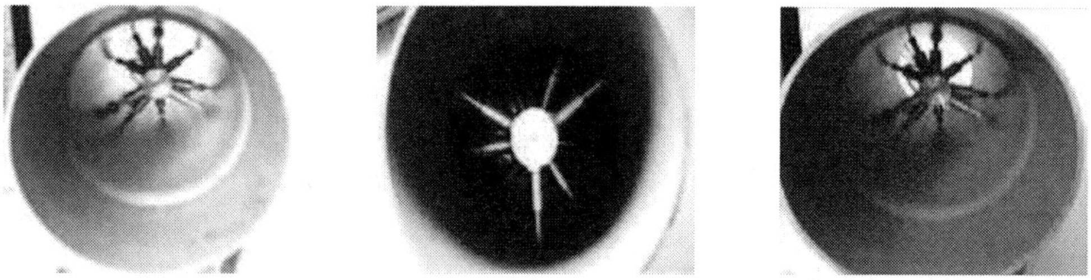

(a) Horizontal Pipe (b) Vertical Pipe (c) Coupling

Figure 5. Movement of IPIR version 2

7.1.1. Movement through 8 inches Diameter Pipe

Table 4. Experimental results in 8 inches diameter pipe

Pipe Position	Linear Displacement (mm)	Time (sec)	Linear Velocity (mm/s)
Horizontal forward	900	120.6	7.4626
Horizontal backward	900	73.2	12.2950
Vertical downward	900	60.3	14.9253
Vertical upward	900	183.9	4.8940
Coupling forward	900	122.8	7.3290
Coupling backward	900	75.1	11.9840

7.1.2. Movement through 10 inches Diameter Pipe

Table 5. Experimental results in 10 inches diameter pipe

Pipe Position	Linear Displacement (mm)	Time (sec)	Linear Velocity (mm/s)
Horizontal forward	900	36.9	24.3902
Horizontal backward	900	27.3	32.9670
Vertical downward	900	18.9	47.6190
Vertical upward	900	56.7	15.8730
Coupling forward	900	38.6	23.3161
Coupling backward	900	28.6	31.4685

It has been observed that least linear velocity of version 2 is found in vertically upward direction for 8 inches diameter pipe whereas higher linear velocity is found in vertically downward direction for 10 inches diameter pipe. Also velocity of robot is greater in reverse direction compared to forward direction.

7.2. Discussion

After real time experimentation of both the versions, their comparative study has been made which is given in following table.

Table 6. Comparison of IPIR versions

Performance Indicator	IPIR version 1	IPIR version 2
Weight	2.2 kg	2 kg
No. of actuators	01	01
Steerability	Poor	Good
Mobility	Horizontal pipes, Inclined pipes	Horizontal pipes, Vertical pipes & couplings
Size and shape adaptability	10 inches pipe	8 to 10 inches pipes

8. Conclusion and Future Work

IPIR version 2 have been developed which can travel easily through horizontal pipes, vertical pipes and couplings of 8 inches and 10 inches diameter. All major components of version 2 are designed carefully. Solid model of IPIR is created in Solidworks 16. Weight of IPIR is reduced in version 2. The efficacy of driving mechanism of version 2 has been verified by conducting its movement through horizontal pipes, vertical pipes and couplings of 8 inches and 10 inches diameter. On the basis of literature reviewed and experimental study of version 2, it has been concluded that IPIR version 2 is better in steerability and diameter adaptability than existing screw driven wall presses wheel type pipe inspection robots.

In future, design of robot will be optimized to pass through 45 degree and 90 degree bends. An ultrasonic sensor will be mounted on a robot body to detect the obstruction and point of obstruction inside the pipe. A CCD camera will be installed on forward leg system for real time in-pipe visibility.

References

[1] Atul Gargade and Shantipal Ohol. 2016. Design and Development of In-Pipe Inspection Robot. American International Journal of Research in Science, Technology, Engineering & Mathematics. USA. pp. 104-109.

[2] Atul Gargade and Shantipal Ohol. 2016. Development of In-pipe Inspection Robot. IOSR Journal of Mechanical and Civil Engineering (IOSR-JMCE). USA. pp. 64-72.

[3] Lei Shao, Yi Wang et al. 2015. A Review over State of the Art of In-Pipe Robot. Proceeding of 2015 IEEE International Conference on Mechatronics and Automation. China. pp. 2180-2185.

[4] Atul Gargade and Shantipal Ohol. 2020. A Review Over In-pipe Inspection Robot. Journal of SEYBOLD Report. pp. 1459-1468.

[5] Amit Shukla and Hamad Karki. 2016. Application of robotics in onshore oil and gas industry – A review part – I. Robotics and Autonomus systems. pp. 490-506.

[6] Iszmir Nazmi Ismail, Adzly Anuar et al. 2012. Development of In-pipe Inspection Robot: a Review. IEEE Conference on Sustainable Utilization and Development in Engineering and Technology. Malaysia. pp. 310-315.

[7] Atsushi Kakogawa, Taiki Nishimura and Shugen Ma. 2016. Designing arm length of a screw drive in-pipe robot for climbing vertically positioned bent pipes. Robotica. pp. 306-327.
[8] Atul Gargade, Dhanraj Tambuskar and Gajanan Thokal. 2013. Modelling and Analysis of Pipe Inspection Robot. International Journal of Emerging Technology and Advanced Engineering. pp. 120-126.
[9] Muhammad Azri Abdul Wahed and Mohd Rizal Arshad. 2017. Wall-press Type Pipe Inspection Robot. IEEE 2nd International Conference on Automatic Control and Intelligent Systems (I2CACIS 2017). Malaysia. pp. 185-190.
[10] Hun-ok Lim and Taku Ohki. 2009. Development of Pipe Inspection Robot. ICROS-SICE International Joint Conference. Japan. pp. 5717-5721.
[11] R. Jain, Abhijit Das and A. Mukherjee. 2019. Design Analysis of Novel Scissor Mechanism for Pipeline Inspection Robot (PIR). Proceedings of the 4th International Conference of Robotics Society of India at Indian Institute of Technology Madras. India. pp. 1-6.
[12] Ankit Nayak and S. Pradhan. 2014. Design of a New In-pipe Inspection Robot. 12th Global Congress on Manufacturing and Management (GCMM). pp. 2081-2091.
[13] Atul Gargade and Shantipal Ohol. 2017. Development of Actively Steerable In-pipe Inspection Robot for Various Sizes. Proceedings of the 3rd International Conference of Robotics Society of India at Indian Institute of Technology Delhi. India. pp. 1-5.

Design of an Anion Generator Equipped Fire-Bot

Himanshu Dubey[1], Pranav Suryadevara[2], Arockia Selvakumar Arockia Doss[1*]

[1] School of Mechanical Engineering, Vellore Institute of Technology, Chennai.
[2] School of Electronics and Communication Engineering, Vellore Institute of Technology

E-mail: arockia.selvakumar@vit.ac.in

Abstract- Fire accidents these days are becoming an alarming issue resulting in numerous deaths globally. In spite of having a robust system to put off the fire from spreading quickly, very less progress has been made in technology to rescue trapped people with only human efforts still taking place to rescue. While carrying out these manual rescue operations one major factor which hinders is the smoke, produced as a result of the fire, which greatly reduces the visibility thus making this task very risky. To address this issue, we have devised a negative ion generator, the anion (negative ion) generator is capable of removing a smoke and dirt particles from the air thereby increasing the visibility. This article presents a prototype where a fire extinguishing robot equipped with negative ion generator and can be deployed at an incident site.

Keywords-component: Fire Accidents, Negative Ion Generator, Fire Extinguishing, Smoke, Visibility.

1. Introduction

Fire accidents are generally the worst disasters that have a greater probability to happen in this modern-day world. It can happen quite easily and its impact ranges from a small house to a huge factory. It can be caused by very common things we observe from a common household gas cylinder blast, electrical transformer or current carrying wires failure to commercial chemicals and large-scale machinery failure. The emission of acid gases such as SO_x, N_xO_y, para-fluorocarbons (PFC) in untimely disasters lead to loss of human life and valuable assets. In order to treat such pollutants in a efficient and judicious manner usually three methods *i.e.* Dust precipitation via application of electrostatic charge, absorbing and adsorbing [1]. The anion generation method was found very promising in treating the air quality as negative ions react with positively charged particles which leads to energy transfer. As to attain oxidation of such species high electric field is taken into account to break the resistance of air. These ions react with any positively charged species present in air to form highly reactive charged cations which later oxidizes and leads to formation of harmless subdivisions [9]. Several anion generating devices or circuits are being developed in order to improve the air quality index which basically uses 5kV-7kV high voltage to produce negative ion [2]-[5],[7]. Few patents utilize an oscillator to convert input DC voltage to high- frequency AC voltage [4]-[5]. Ref. [6]-[8] uses Flyback converter, with a high ratio transformer, to transfer input voltage to a high DC voltage. Table 1 Shows health effects of some volatile organic compounds with threshold level value (TLV) in ppm [1].

Table 1. Volatile organic compounds with threshold level value

VOC	TLV	Formula	Effects
Acetone	10	CH_3COCH_3	Carcinogen
Benzene		C_6H_6	Carcinogen
Butyl Acetate		$CH_3COOC_4H_9$	Carcinogen
Carbon Tetrachloride	10	CCl_4	Probable carcinogen
Dichloroethane		CH_2ClCH_2Cl	Paralysis of nerve centers
EGM	5	$(CH_2OH)_2(C_2H_5)_2C$	Birth defects
Ethylbenzene		$C_6H_5C_2H_5$	Probable carcinogen
Formaldehyde		HCHO	Sore throat dizziness and headache
Methyl chloride		CH_3Cl	Carcinogen
Methyl ethyl ketone		$CH_3COC_2H_5$	Carcinogen
Perchloroethylene		C_2H_5Cl	Probable carcinogen
Phenol		C_6H_5OH	Headache, dizziness
Styrene		$C_6H_5CH{=}CH_2$	Probable carcinogen
Tetrachloroethylene		$CCl_2{=}CCl_2$	Probable heart and liver disease
Toluene		$C_6H_5CH_3$	Headache, dizziness
Trichloroethane	200	$CH_2ClCHCl_2$	Probable birth defects, heart problems
Trichloroethylene	50	$CHCl{=}CCl_2$	Liver and kidney disease, paralysis of nerve centers
Vinyl chloride	10	$CH_2{:}CHCl$	Carcinogen
Xylene		$C_6H_4(CH_3)_2$	Headache, dizziness

The prototype developed utilizes the application of high ratio Flyback transformer to attain high electric field that helps in ionizing the smoke pollutants and also helps in rapidly treating the pollutant particles along with toxic gases by passing the air through a pre-filter where particles of higher (Angstrom order) size will be accumulated and the air that passes receives energy from the highly condensed thus ionizing the atoms or molecules of the pollutant compounds. The ionized particles will be collected on the collector plates which would consist of opposite charge to that of the ionized air particles. Figure 1 Shows phases involved in treatment of polluted smoke.

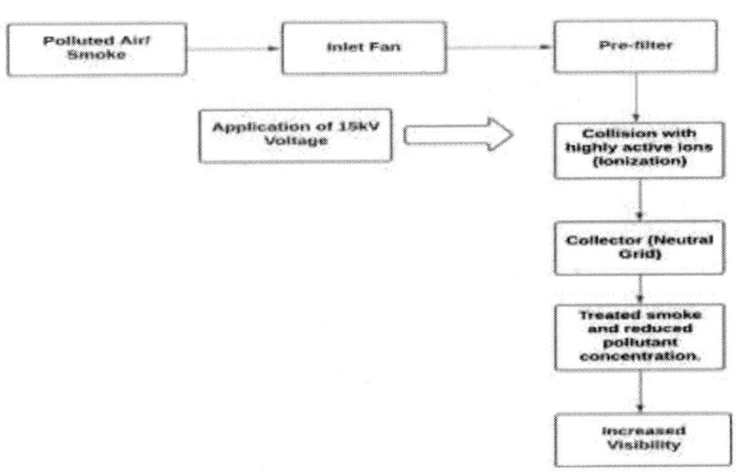

Figure1. Phases involved in treatment of polluted smoke

2. Conceptual design

2.1. Design

A novel Ion generator equipped mechanical layout was designed using CAD modelling Software (SolidWorks). The assembly comprises of treatment chamber of 210 x 120 x 122 mm as shown in **Figure 2** below.

Figure 2. Proposed 3D CAD model for smoke treatment box

In order to switch smoke, fans with 100 x 100 mm are set aside at Inlet and Outlet of box as shown in **Figure 3,** below

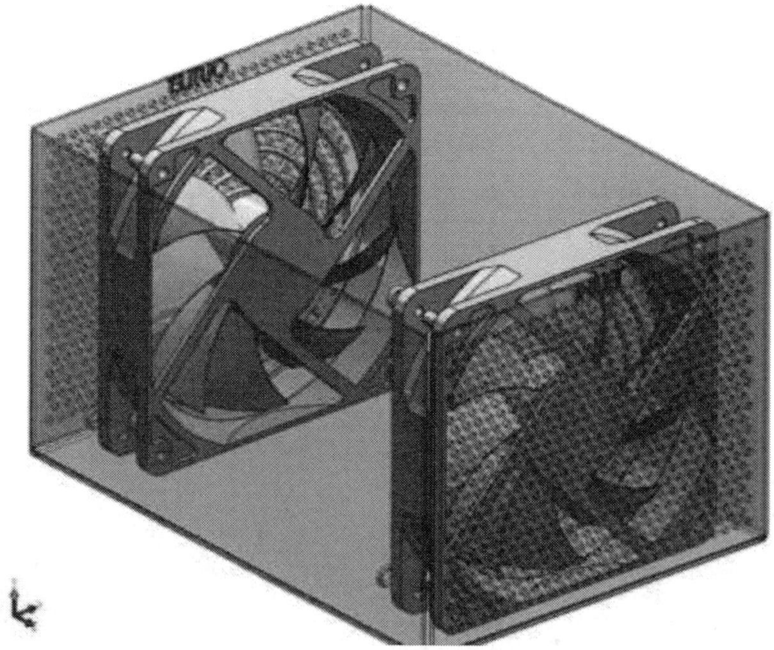

Figure 3. Assembly of fans to switch air between Inlet and Outlet

Figure 4 shows the entire assembly of proposed novel ion generator which is placed in along with other controllers to perform the task judiciously.

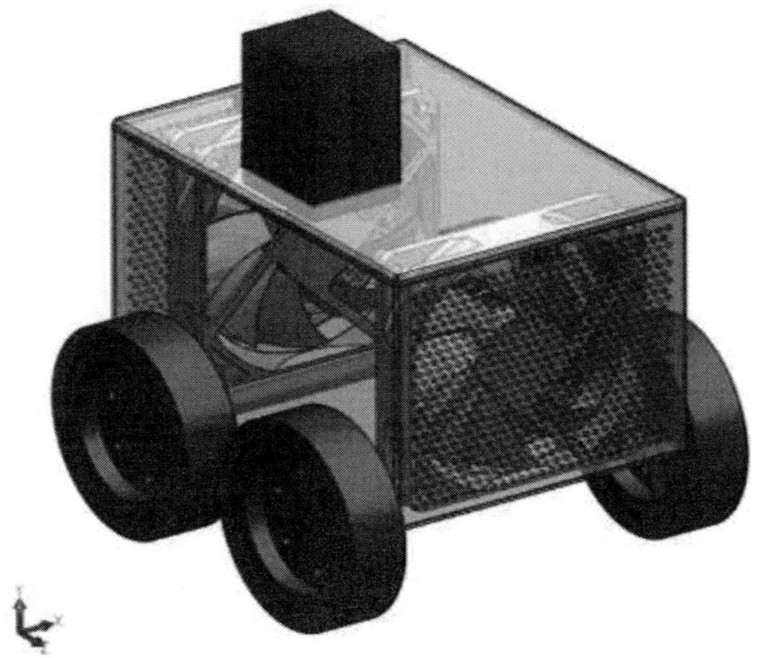

Figure 4. Anion Generator Fire-bot Assembly

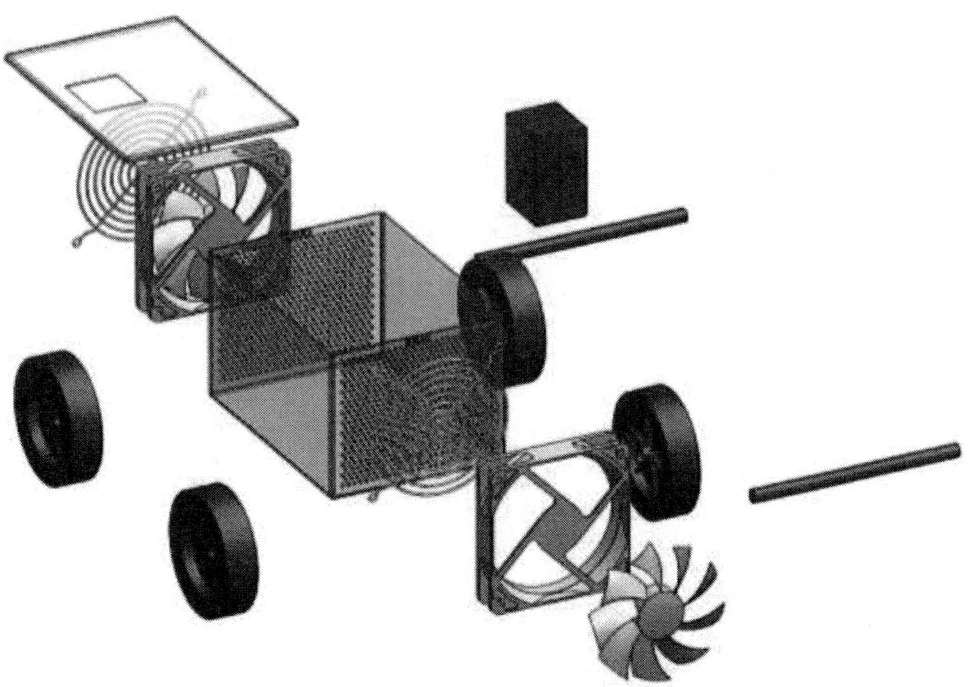

Figure 5. Exploded view of all the proposed Fire-bot CAD Model

2.2. Robot Chassis and Power Supply

The prototype uses a usual robot chassis with wheels. The power source used is a 9V battery which is used to power the Arduino board. Also it is used to power the DC motors which help in the motion of the robot; the circuit diagram is shown in **Figure 6.**

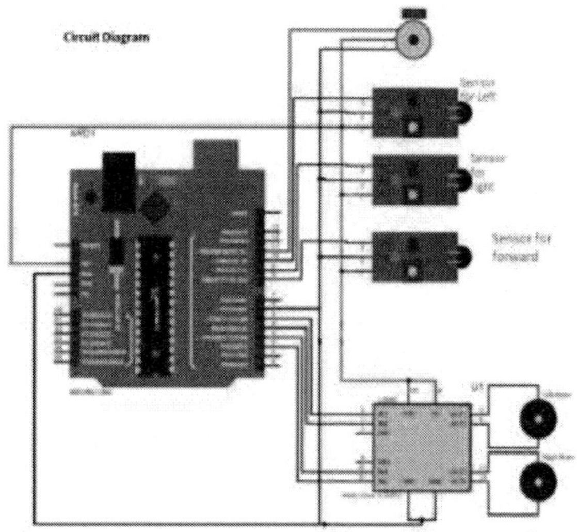

Figure 6. Circuit Diagram of the proposed control system

2.3. Anion Generator

The novel anion generator circuit was developed mainly comprises of High-Ratio Flyback transformer (FBT), Mosfet (IRF Z44N), Resistor (1.5k ohms), Copper winding, End pin and Input supply of 5V 2Amp. **Figure 7** shows the schematic of circuit diagram of anion generator designed in fusion 360 electrical. Moreover, to attain optimum potency of the device 9-9 turns of copper winding wire around metal core of FBT are used as feedback coil system. Also, high voltage end pins attract all the dirt particles to the neutral grid followed by ionization of harmful gases and also avoids all the severe effects as already mentioned in Table. 1

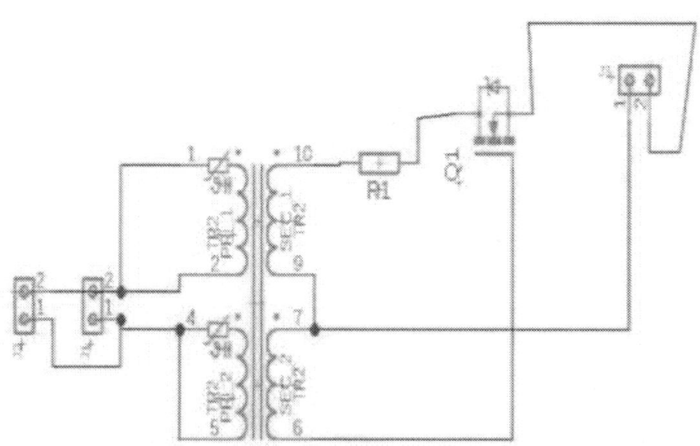

Figure 7. Schematic circuit diagram designed in Fusion 360 Electricals of an ion Generator

2.3.1. Component Selection

Flyback transformer (FBT)
A standard High-Ratio flyback of potency upto 32kV from Television motherboard is used for ion generation of High voltage. To attain optimum output end pins with maximum resistance were opt where as a result primary coil was taken into account. Where, one end pin of the active primary coil is connected to the neutral line which acts as a bin to the dust particles as it attracts all the pollutants over it. As Figure 3. Depicts the feedback coil of 9-9 turns over the core of FBT.

2.3.2. Mosfet
A standard Mosfet (i.e. IRF Z44N) is used in order to alter the flow.

2.3.3. Resistance
Resistor with resistance of 1.5K ohm were used to maintain appropriate flow of charge through Mosfet (IRF Z44N).

2.3.4. Heat Sink
To maintain even disperse of heat, Aluminum heat sink with 20x20 (in mm) was used accordingly.

2.3.5. Winding Wire
Insulated copper winding wire with 9-9 turns around the metal core to induce the charge flow rate via primary coil.

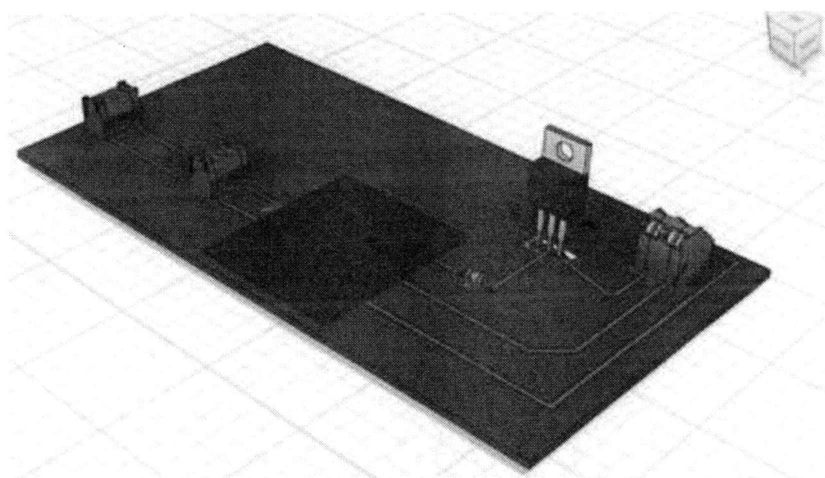

Figure 8. PCB circuit diagram for ion generator designed in Fusion 360 Electricals

2.4. Prototype development
The prototype developed comprises of an Arduino Uno equipped with an H-Bridge. The H-Bridge helps us in controlling the direction of the DC motors so that the fire-bot moves in the direction in we intend to move. Now we have actually made an input pin to facilitate the robot operation in two ways, they are automatic and the manual mode. This can be done by setting the pin either high (for automatic operation) or low (for manual operation). Now the robot developed can be operated manually by a person. The anion generator helps in removing the smoke and dust particles form the air surrounding so that there is an increased visibility which allows the fire Fighters to have an advantage in rescue operations they make. While considering the automatic mode of operation it is done with the help of IR sensors which detect the radiation emitted when heat is produced. Thus, the robot equipped with IR sensors in all directions can sense this radiation and move towards the place of burning to clear the smoke with the firemen trying to put the fire off. Figure 9,10 and 11 shows the various stages of prototype development of an ion generator equipped Fire-Bot.

Figure. 9. Prototype model of Robot chassis equipped with sensors and H-Bridge connection.

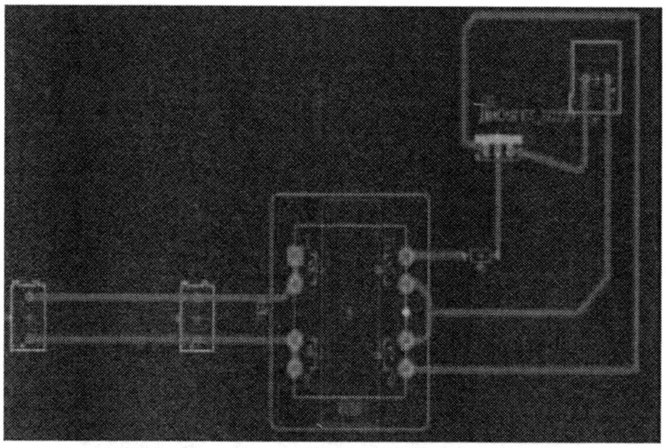

Figure. 10 Schematic circuit diagram of Fire-bot designed in Fusion 360 Electricals

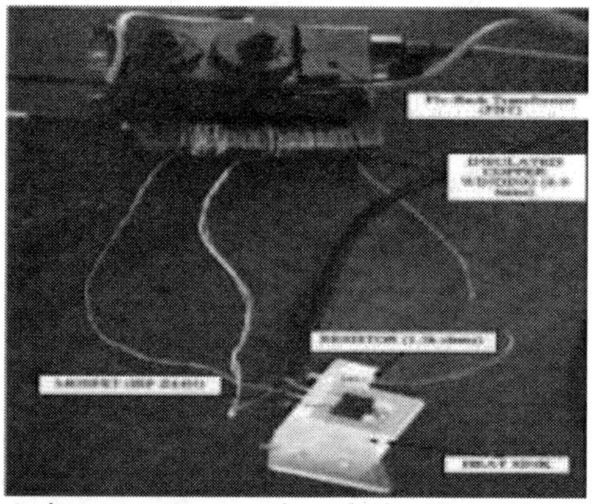

Figure. 11 Anion generator equipped Fire-Bot – Prototype Model

3. Experimental Studies

Later, the developed prototype as shown in Figure 11 was tested over trapped dry smoke of essence sticks in a jar at room temperature of 28^0 Celsius. Similarly, phases observed as shown in Figure. 12 &

13 before and after activity of prototype at different time intervals (T_0 & T_1) mentioned in table. 2 below.

Table. 2 Shows time taken by prototype to clear trapped smoke

S. No	Time Interval (in seconds)	Prototype State
1.	$T_0 = 0$ sec (Figure. 12)	Off
2.	$T_1 = 7$ sec (Figure. 13)	On

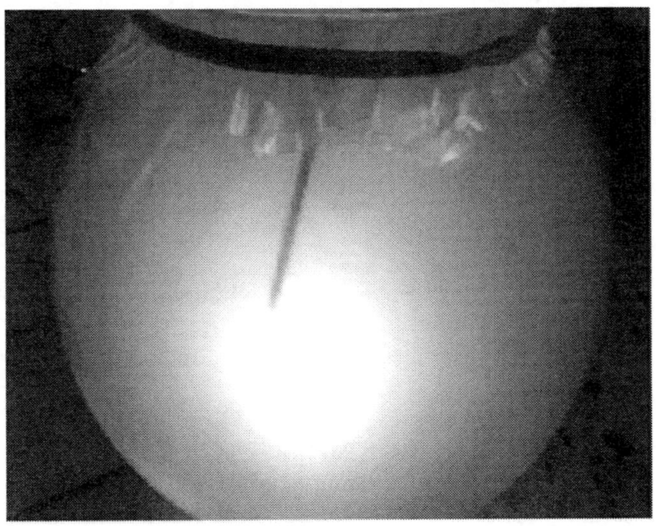

Figure. 12 Smoke trapped in glass jar at T_0=0 sec when device is Off

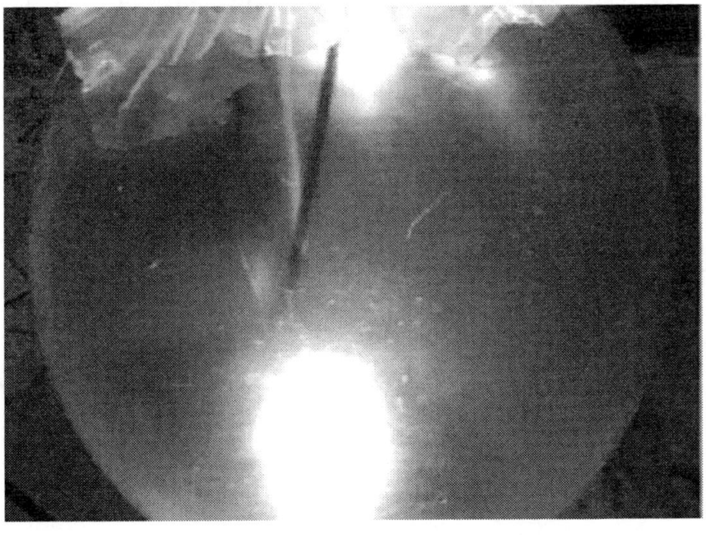

Figure. 13 Amount of essence stick smoke cleared at T_1= 7 sec when device is active

4. Summary and Future scope

An ion-generator designed and developed to deploy at fire incident site. Based on the development and experimental trials the developed fire-bot was found to be promising in terms of removing dust and soot particles of smoke up to great extent in short time span of 7 seconds. The developed prototype can also be equipped with a mini fire extinguisher to extinguish the fire.

The proposed design can be further developed by use of Aerial Robots instead of conventional ground-based ones. Also, further we could employ the use of IR imaging by using a thermal imaging camera (TIC), where we interpret thermal infrared radiation as visible light (VIBGYOR) thus areas of heat could be perceived through smoke or fire permeable barriers. This model can be further developed by employing the use of advancements in the areas of Artificial Intelligence (AI), Machine Learning (ML), and Deep Learning (DL). Via application of video monitoring trapped individuals can be rescued by guiding appropriate instruction and enhanced visibility as well. Also, we can use the aid of direction finders based on sound coming from the trapped people to be rescued.

References

[1] K. Urashima and Jen-Shih Chang, "Removal of volatile organic compounds from air streams And industrial flue gases by non-thermal plasma technology," in IEEE Transactions on Dielectrics and Electrical Insulation, vol. 7, no. 5, pp. 602-614, Oct. 2000, doi:10.1109/94.879356.

[2] Henri Maurice Pellin, "Negative Ionizer," US4102654.

[3] Charles Wilfred Topley, "Generation of Negative Ions," US3832554.

[4] Yoshinori Sekoguchi, Kenji Furuhashi, and Mamoru Morikawa, "Ion Generator and Air Conditioning Apparatus," US20040130271.

[5] H. Hsieh, "Negative Ions Generating Circuit Design with Decreasing High Frequency Noise and Apparatus Thereof," US20080191145.

[6] C. Li, Y. Du, and H. Li, "Negative Ion Air Purifier," US20060278080.

[7] C. H. Lin, C. M. Wang, and M. H. Hung, "High Voltage Driving Circuit with Negative Pulse Feature for Tubular Field Emission Lamp," International Review of Electrical Engineering, vol. 5, no. 4, July 2010, pp. 1430-1435.

[8] Müller, S., & Zahn, R.-J. (2007). Air Pollution Control by Non-Thermal Plasma. Contributions to Plasma Physics, 47(7), 520–529. doi:10.1002/ctpp.200710067.

[9] Chang, J.-S. (2001). recent development of plasma pollution control technology: a critical review. Science and Technology of Advanced Materials, 2(3-4), 571–576. doi:10.1016/s1468-6996(01)00139-5.

[10] C. Lai, M. Pan and M. Yang, "Study and development of a negative ion driving circuit with 8kV/4kV dual-output for air purifier," 2013 IEEE International Symposium on Consumer Electronics (ISCE), Hsinchu, 2013, pp. 39-40, doi: 10.1109/ISCE.2013.6570242.

[11] Seo KH, Mitchell BW, Holt PS, Gast RK. Bactericidal effects of negative air ions on airborne and surface Salmonella enteritidis from an artificially generated aerosol. J Food Prot. 2001;64(1):113-116. doi:10.4315/0362-028x-64.1.113

[12] Patil, Vijay & Patil, B P & Shimpi, NG & Jha, Ashish. (2012). Proposed Negative Ion Generator.

[13] Sawant, V., Meena, G. and Jadhav, D. (2012). Effect of Negative Air Ions on Fog and Smoke. Aerosol Air Qual. Res. 12: 1007-1015.

[14] Wu, C.C. and Lee, W.M. (2003). The Temporal Aerosol Size Distribution in Indoor Environment with Negative Electric Discharge. J. Aerosol Sci. 34: S999-S1000.

[15] Shiue, A., Hu, S.C. and Tu, M.L. (2011). Particles Removal by Negative ionic Air Purifier in Cleanroom. Aerosol Air Qual. Res. 11: 179-186

[16] Ki-Hyun Kim, Jan E. Szulejko, Pawan Kumar, Eilhann E. Kwon, Adedeji A. Adelodun, Police Anil Kumar Reddy, Air ionization as a control technology for off-gas emissions of volatile

Organic compounds, Environmental Pollution, Volume 225, 2017, Pages 729-743, ISSN0269-7491, https://doi.org/10.1016/j.envpol.2017.03.026.
[17] Černecký, J., Valentová, K., Pivarčiová, E., & Božek, P. (2015). Ionization Impact on the Air Cleaning Efficiency in the Interior, Measurement Science Review, 15(4), 156-166. doi: https://doi.org/10.1515/msr-2015-0023

Acknowledgments
The authors wish to thank Maulana Azad National Institute of Technology (MANIT), Bhopal for its effortless contribution and support.

Remote Controlled Pick and Place Robot

Meenakshi Prabhakar[1*], Valenteena Paulraj[1], Dhusyant Arumukam Karthi Kannappan[1], Joshuva Arockia Dhanraj[1], Deenadayalan Ganapathy[1]

[1]Centre for Automation & Robotics (ANRO), Department of Mechanical Engineering, Hindustan Institute of Technology and Science, Chennai 603103, Tamilnadu, India.

E-mail: vijiprabha2000@gmail.com

Abstract. In today's world sales and demand for products has grown exponentially. There is always a need of the products in a huge count in terms of production, handling of such a huge count of products is not possible without automation process but the instruments used of such automation process is of high cost which is not feasible for the small scale industries to implement such a high-cost machine for their use cases. To overcome the above problem of handling the products, a low-cost pick and place robot has been designed for handling the products, this robot is designed based on RRT:R configuration for the robot body, with this configuration this robot can handle products in complex positions. This robot is featured with the six DOF (degrees of freedom) for its movement. This robot is a tele-operated robot that can be controlled using a specialized mobile application developed using MIT APP INVENTOR 2 which makes the user use this robot without any special support for controlling, this robot gives a hassle-free experience for the user who is controlling the robot. By using this robot pick and place tasks can be simplified and the process can be automated and the rate of product handling can be increased.

Keywords: Pick and Place, Robotic Arm, Industrial Automation, Bluetooth Connectivity, Degrees Of Freedom.

1. Introduction

Pick and place robots automates the pick the place operation by picking an object and placing it in other different areas. Computerizing this procedure assists with expanding production rates. Pick and place robots handle dull assignments while opening up human laborers to concentrate on increasingly complex work. Many industries started using pick and place robots for moving operations. But small industries still use laborers for pick and place operation. To overcome this problem, the authors come up with a very potential pick and place robot with a very low budget. Developing this low budget pick and place robot can replace human labour and automate the small-scale industries [1]. This saves time and also improves productivity. Also, operations done by humans can cause damage to the objects, but using pick and place robot can avoid such damages. This project's main goal is to develop a pick and place robot that is embedded with a robotic arm having a gripper at its end. It is controlled through Bluetooth using an app developed in MIT app inventor 2. Robots are mainly used for doing highly dangerous jobs which couldn't be done by human beings. Thus, pick and place robots are also a type of robot which can solve this kind of problem. The robot does pick and place operation by receiving commands from the Bluetooth transmitter and receiver. It has six degrees of freedom. Pick and place robotic manipulator can be designed in many different ways according to the application it is going to be used. The total mechanism of this robot relay upon joints [2]. The joints are used to combine the two unbending continuous bodies in a robot which can be a revolving joint or linear joint. Joints are mainly used to characterize the integration of the arm as they choose the degree of freedom.

Kadam and Hwang [3] designed a flexible and easy "Pick and Place Robot" that can be controlled. The robot utilizes two Arduino microcontrollers associated with a master/slave arrangement. The master controls the automated arm while the slave controls the mechanical base. The arm uses servomotors to give movement in the necessary pivot. The base comprises of dc outfitted engines and followed wheels for shipping the mechanical arm with the griper. The aftereffect of this task is a small scale robot that gives pick and place operations that can be utilized in a few applications by changing the program of the regulator. The structure is intended to lift light loads. A characteristic and natural approach to communicate with a robot by utilizing signals and this concept was proposed by Latif et al., [4]. As a jump movement regulator gives a straightforward and compelling technique to precisely follow the development of human hands and fingers, can be utilized to give natural control of the robot. In this work, authors show a plan of an automated stage that is constrained by hand motions being followed by the jump movement regulator. The left hand is utilized for controlling stage versatility, utilizing motions to give unique movements, for instance, pushing ahead and moving in reverse and the correct hand is utilized for controlling the arm for pick and spot errands. The created robot can be utilized for far off observing of the troublesome climate and influence changes to it through a controller's arm. Likewise, it can self-sufficiently keep away from obstructions utilizing Ultrasonic sensors. In general, the planned framework model has demonstrated its possible abilities for climate observing and observation through pragmatic exhibits utilizing motions.

A prototype, that robot pick and placed, was developed and implemented by Abdulkareem et al., [5]. Here, they intended to use an automatic arm to pick and position items with a rotative base, elbow, and wrist motions. The different parts are bound together with joints, as are the joints in the body. The administrator shall use the cell phone key to select and position the object. The mechanical arm may be pushed by pressing different keys of the mobile device, for example, in the left or right direction, wherever. With far-off phone access, the authors have used the ATmega328P, DTMF and GSM Modules. Arduino IDE for control component programming and simulation security. For pick-up and place application, Ghadge et al.. [6] developed a robotic arm. An arma mechanical to pick and position products on customer's orders is developed using Arduino. It secures the collection and positioning of an object from source to goal. The responsive gripper in the arm does not impact the weight of the item. The robot is controlled and Bluetooth and Android mobile devices. Due to the customer's commands, the robot also travels. Four engines communicate with the miniature regulator at the end of the receiver. Two are also for the development of the engine and two remain for the arm. For robot power the blue control application is used.

Ongaro et al., [7] done a study on the control of untethered soft grippers for pick-and-place tasks. The undertaking is intended to build up a decision on the pick and place of a mechanical vehicle with a delicate getting gripper. For instance, it'll securely hand a bomb carefully to stay away from its blast while getting. The mechanical vehicle is golem application controlled for far off activity. All the sending end utilizing golem application gadgets, orders are shipped off the beneficiary to deal with the development of the golem either to move forward, in reverse and left or right, and so on. At the less than desirable end, four engines are interfaced with the microcontroller where two are for the body development. The golem application gadget transmitter goes about as a distance that has the upsides of sufficient change, while the recipient end Bluetooth gadget is taken care of to the microcontroller to drive DC engines through engine driver IC for essential work. From the above-mentioned, authors conclude that our robot is more unique because it is highly customizable. Authors have made the robot more user friendly by interfacing Bluetooth connectivity between the robot and the mobile app called servo. Each motor is controlled with a separate button for ensuring the movement of the desired link. Thus authors say our robot is unique from the other pick and place robot in this way. Once the circuit gets activated the process gets started by enabling the Bluetooth module. Then the degrees of the servos are initialized to rest position. Then it waits for the command. Each servo receives a separate command from each desired button respectively. The gripper operation is a toggle operation where the open and close operation is done through the same button [8]. The degrees of movement of each servo can be changed anytime by editing the code making it customizable according to our needs. Figure 1 shows the flow chart of the algorithm.

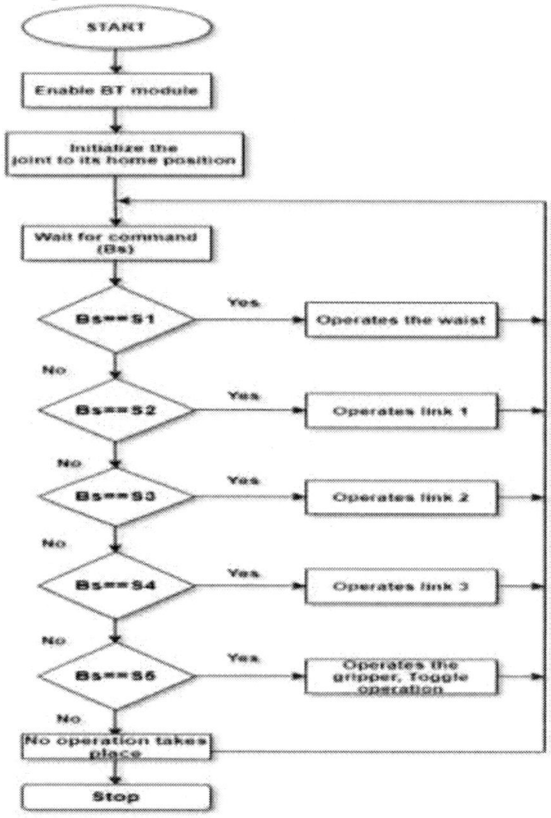

Figure 1. Flow chart algorithm

2. Design of Pick and Place Robot

The arm of the robot is designed with six degrees of freedom with the gripper. The manipulator is attached with 3 metal servos, one at the base, another one attaching the waist and link 1, and one attaching link 1 and link 2. The two micro servos are fixed near the link 3 and the gripper. The waist and link 4 are 360° rotatable. The other two links can go to and froth up to 180° and even more. The gripper uses a parallel jaw mechanism which is designed with two fingers. Figure 2 and Figure 3 shows the design of the arm, Figure 4 shows the design of the end effector (gripper) and Figure 5 to Figure 9 shows the 3D printed parts of the entire manipulator.

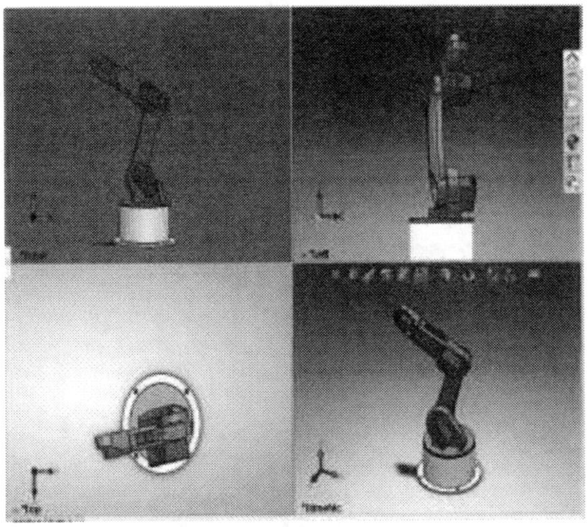

Figure 2. Design of the arm (top view)

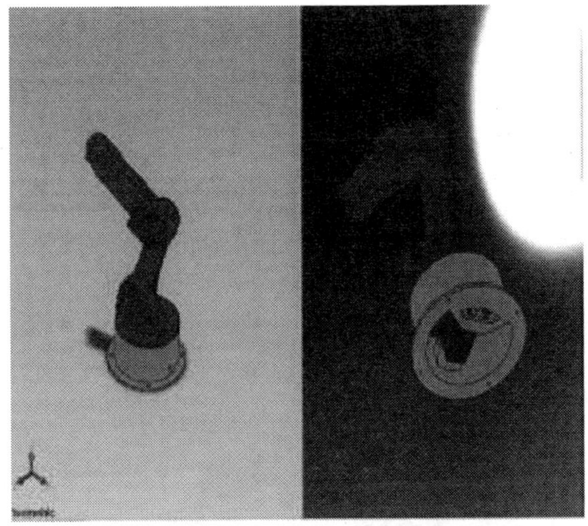

Figure 3. Design of the arm (side view)

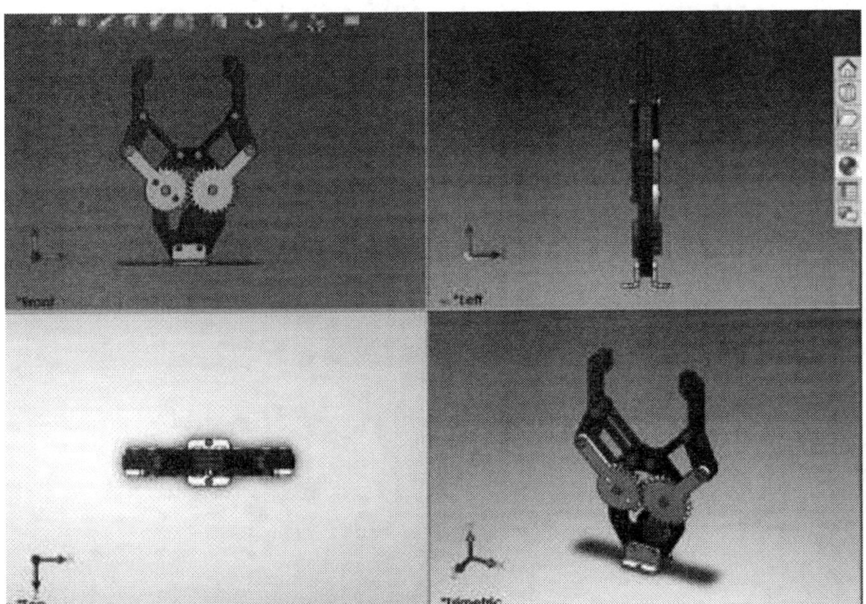

Figure 4. Design of the end effector (gripper)

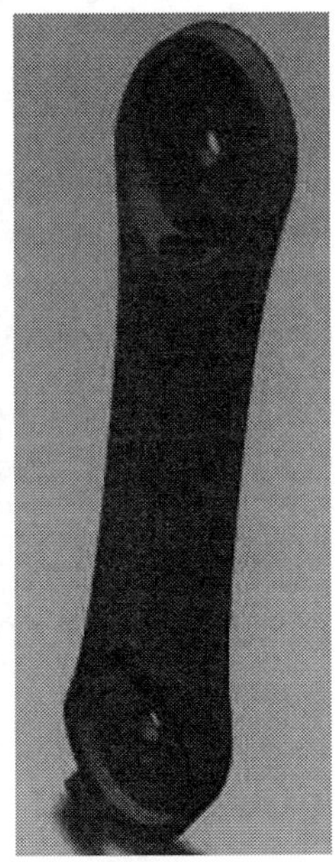

Figure 5. Link-1

3. Circuit Design and Simulation with App

Few essential components are needed for the design and simulation of the app. The Ardoino UNO is a Microcontroller Open Source Board built on and developed by Arduino [9], built on the ATmega328P microcontroller. The panel comes with collections of input / output (I/O) optical and analogue pins that can be interconnected with a number of boards and circuits. MG995 is a 180° rotary servomotor that produces an efficient 180° rotation. The servo is suitable for the construction of a robotic arm with a high motor wear. The servo is metal-conducted and can be mounted over a device such as a robot arm where motor running is immense. Strong yield strength compact and lightweight [10]. The servo is 180° (90 to each route) and operates just like the normal, but littler ones. To monitor these facilities, you can use any servo code, equipment, or library. It can be useful for those amateurs who need to move items without building a vital motor and appliance box, particularly because it fits in places that are small [11]. The HC-05 module is simple to use and is designed for the transparently configured wireless serial link by way of Bluetooth SPP (serial port protocol). HC-05 Bluetooth Module offers master-slave mode switching, which means that it does not receive or send data. An app called servo is developed through MIT inventor 2. The app will be operated by connecting to Bluetooth. The app is developed for operating the servo motors [12]. The degrees of rotation for the servos will be given through code. The process will start when the authors trigger the start button in the app. Thus authors can operate the servos by using the app invented. The authors have separate buttons for each servo.

Figure 6. Base

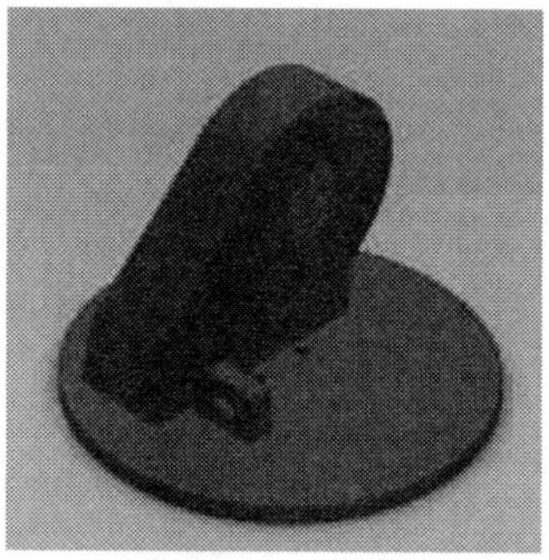

Figure 7. Waist

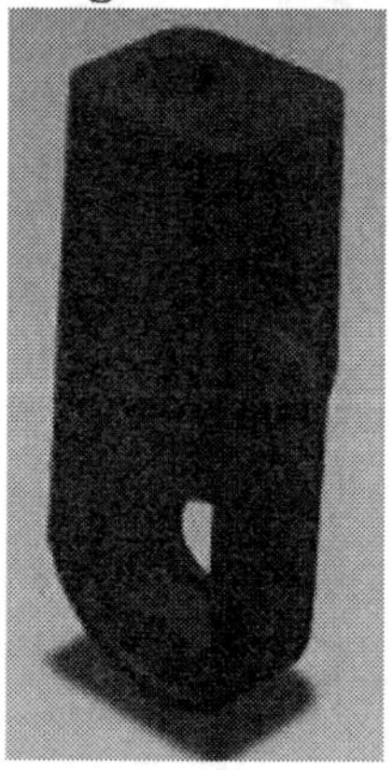

Figure 8. Link-2

Figure 9. Link-3

4. Calculations

The gripper force calculation is done for analyzing the grippe force. It is calculated using the following formula

$$\mu nfFg = w \qquad -(1)$$

Where μ = coefficient of friction of the finger contact surface against the part surface
nf = number of contacting fingers
Fg = gripper force
W = author sight of the part or object being gripped
$\mu = 0.75$
nf = 2
Fg = 0.75(2)
Fg = 50

1.5Fg = 50
Fg = 50/1.5
Fg = 33N

5. Results and Discussion

The block diagram shows the connection and relationship between every component. It represents the importance of the components. The Arduino UNO is the processor used here. The Bluetooth module is a receiver and transmitter [13]. It receives and transmits the signals from the app to the Arduino board which makes the arm move in the desired direction. The Arduino board controls the activity of the arm as per the commands received from the app. Figure 10, shows the code for the app.

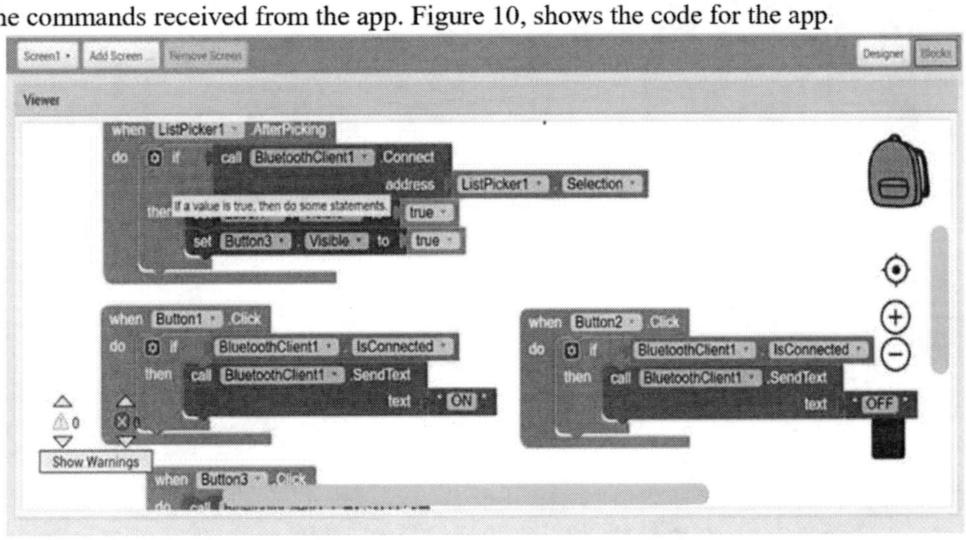

Figure 10. Code given for the app

The circuit diagram was designed and simulated in Proteus 8 Professional. As you can see, the processor is connected with each component of the circuit. The processor is connected with three servo motor MG995, two micro servo 9g, and a Bluetooth module HC 05. The power supply is given from the external source. The Bluetooth module is connected with an app developed by us [14]. The processor receives commands from the app as per the program which makes the arm move in a specified direction. Each servo motor is given a maximum and minimum degrees of movement in the coding, between which it can move [15]. Thus, the movement of the servo is controlled by the app using our mobile phones through which one can pick and place an object. Figure 11 shows the app developed in MIT APP INVENTOR, Figure 12 shows the block diagram of the interface with Arduino, and Figure 13 shows the Proteus circuit simulation. The circuit diagram was designed and simulated in Proteus 8 Professional. As you can see, the processor is connected with each component of the circuit. The processor is connected with three servo motor MG995, two micro servo 9g, and a Bluetooth module HC 05. The power supply is given from the external source. The Bluetooth module is connected with an app developed by us [14]. The processor receives commands from the app as per the program which makes the arm move in a specified direction. Each servo motor is given a maximum and minimum degrees of movement in the coding, between which it can move [15]. Thus, the movement of the servo is controlled by the app using our mobile phones through which one can pick and place an object. Figure 11 shows the app developed in MIT APP INVENTOR, Figure 12 shows the block diagram of the interface with Arduino, and Figure 13 shows the Proteus circuit simulation. We have designed a unique pick and place manipulator. Our design is 3D printed and assembled manually. The waist and link 4 are 360° rotatable. The other two links can go to and froth up to 180° and significantly more. The gripper utilizes a parallel jaw system which is designed with two fingers. It can lift a weight of 50g to 75g. The overall weight of the bot is 500g. The number of autonomous ways by which a unique framework can move, without violating any constraint forced on it, is called the number of degrees of freedom. The robot has six degrees of freedom. The authors calculated the gripper force to be 33N.

Figure 11. App developed in MIT App Inventor

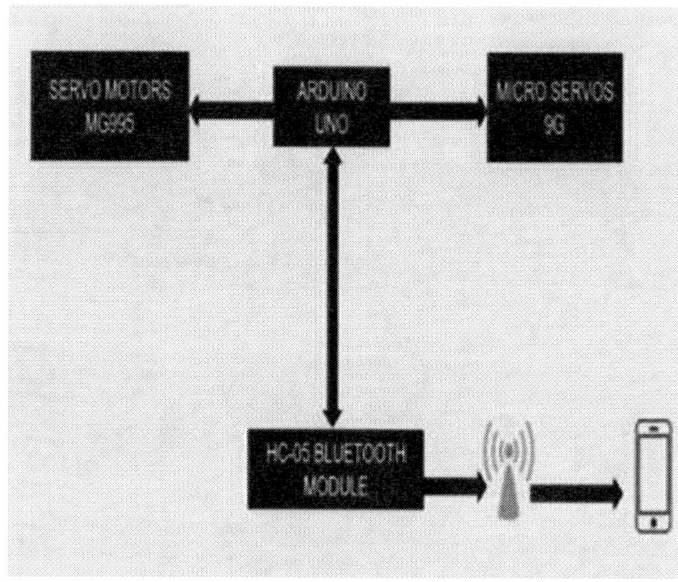

Figure 12. Block diagram of the interface with Arduino

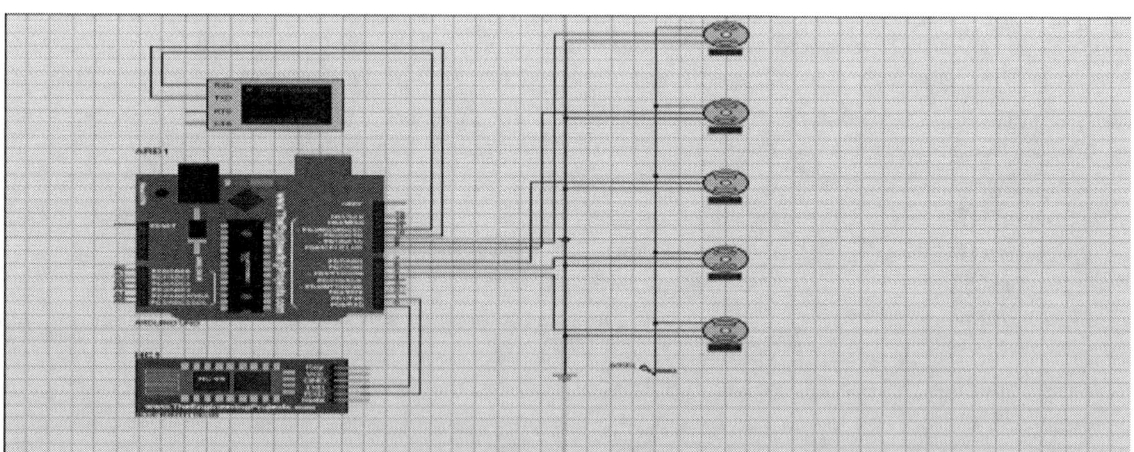

Figure 13. Proteus circuit simulation

6. Conclusion

As per considering the laws of a robot, a robotic arm that is highly customizable with an end effector is designed and developed. It is developed with an adjustable gripper. The gripper jaws move in an equal movement according to the gripper body. Utilized in and down, which through a mechanical linkage, powers the gripper jaws open and shut. The app is operated by connecting it to the Bluetooth module which is a serial port protocol. The circuit design is done and simulated in proteus. This robot can be fabricated with a very low budget. If this ideology is followed by all the industries, one could give way to the automation revolution "INDUSTRY 4.0". The user remains near the very edge of an innovative transformation that will alter in a general sense adjust how one live, work and identify with each other. In its scale, extension, and unpredictability, the change will be not normal for anything mankind has encountered previously.

REFERENCES

[1] Berscheid L, Meibner P, Kröger T 2020 Self-supervised Learning for Precise Pick-and-place without Object Model *IEEE Robotics and Automation Letters* **5** 4828-35.
[2] Dewi T, Nurmaini S, Risma P, Oktarina Y, Roriz M 2020 Inverse kinematic analysis of 4 DOF pick and place arm robot manipulator using fuzzy logic controller *International Journal of Electrical & Computer Engineering* **10** 2088-8708.
[3] Kadam AL and Hwang M 2020 Design and Implementation of Remote Controlled Robotic Arm Using GSM-Based Cell Phone for the Developing Countries *In Information Science and Applications* **621** 639-649.
[4] Latif AA, Nizamani MA, Shoro GM, Abassi F and Memon BR 2020 Design And Control Of Autonomous Robot Using Gesture Based Intuitive Interaction *International Journal of Advanced Computer Systems and Software Engineering* **1** 18-22.
[5] Abdulkareem A, Ladenegan O, Agbetuyi AF and Awosope CO 2019 Design and Implementation of a Prototype Remote-Controlled Pick and Place Robot *International Journal of Mechanical Engineering and Technology* **10** 1-12.
[6] Ghadge K, More S, Gaikwad P and Chillal S 2018 Robotic arm for pick and place application *International Journal of Mechanical Engineering and Technology* **9** 125-33.
[7] Ongaro F, Yoon C, Van Den Brink F, Abayazid M, Oh SH, Gracias DH and Misra S 2016 Control of untethered soft grippers for pick-and-place tasks *6th IEEE International Conference on Biomedical Robotics and Biomechatronics* 299-304.
[8] Raghul S, Mohankumar N 2020 Microcontroller Based ANN for Pick and Place Robot Coordinate Monitoring System *Lecture Notes in Electrical Engineering* **601** 340-348.
[9] Naga Mani Rahul T, Teja Reddy K and Viji Amutha Mary A 2019 Cost Effective Robotic System Using Arduino *Journal of Computational and Theoretical Nanoscience* **16** 3222-7.
[10] Sharma S, Sahai S, Joshi J and Hema N 2018 Stage-wise Development of a Remote Controlled Robotic Arm *Fifth International Conference on Parallel, Distributed and Grid Computing* **20** 364-367.
[11] Ishak MK, Roslan MI and Ishak KA 2018 Design of robotic arm controller based on Internet of Things (IoT) *Journal of Telecommunication, Electronic and Computer Engineering* **10** 5-8.
[12] Baby A, Augustine C, Thampi C, George M, AP A and Jose PC 2017 Pick and place robotic arm implementation using Arduino *IOSR Journal of Electrical and Electronics Engineering* **12** 38-41.
[13] Zhang J, Onaizah O, Middleton K, You L and Diller E 2017 Reliable grasping of three-dimensional untethered mobile magnetic microgripper for autonomous pick-and-place *IEEE Robotics and Automation Letters* **2** 835-40.
[14] Sawarkar MR, Raut TR, Nutan PN, Meshram SC and Pournima PT 2017 Pick and Place Robotic Arm Using Android Device *International Research Journal of Engineering and Technology* **4** 1706- 1709.
[15] Andhare P and Rawat S 2016 Pick and place industrial robot controller with computer vision *International Conference on Computing Communication Control and automation* 1-4.

Simulation of Grasp Localization Inferences for a Dual-Arm Collaborative Robot

Josin Hippolitus A[1*], R.Senthilnathan[1], Omar Malla[1]

[1]Department of Mechatronics Engineering, SRM Institute of Science and Technology, SRM Nagar, Kattankulathur, 603203, Kanchipuram, Chennai, TN, India.

E-mail : josinhippolitus@gmail.com

Abstract. Grasp Localization is an important problem in robotic manipulators which fundamentally aims at estimating a pose for the end effector which ensures that the object intended to be grasped does not slip off. This estimation is achieved through exteroceptive sensors used in robots such as vision systems which through features and learning-based approaches. One of the fundamental means to validate such an approach is by simulating the grasp routine for any given geometry and its probable grasp poses. This paper presents the details of a simulation work involving a dual-arm collaborative robot with each arm consisting of seven degrees of freedom. The dual-arm robot performs a grasp routine for a geometry that is imported from CAD software. To perform the simulation the various attributes of the robot are to be observed since the simulated robot is a mimic of a commercially available one.

Keywords: deep learning, grasp localization, collaborative robot

1. Introduction

Grasping is an important activity in manipulator robotics. It is the first step of a robot interacting with an object. The end-effector or the tool center point (TCP) frame, typically in a gripper scenario, must be guided to a suitable pose on the object to be grasped. When the geometry of the object is unknown, computer vision systems may be used to identify the right pose. The best grasp is achieved when the gripper successfully picks the object without any slip. To find the perfect grasp, grasp synthesis is performed, where deep learning is used [5]. For a given environment there can be any object with known geometry or unknown geometry. In model based approaches, the success rate of the grasp is very high, as the geometry of the object is known. Ever since a gripper is in use, analytical methods and expert knowledge in a certain application area has driven so far to execute a successful grasp. Nevertheless to say that they are most time consuming and least generalized. But when the robot is subjected to grasp a novel object, vision and touch based approaches are the more preferred methods to make the grasp successful. The cameras and tactile sensors help in providing the information about those novel objects subjected for grasp. Convolutional Neural Networks are employed to extract the features and identify the areas of interest for grasping using data driven methods. Here deep learning techniques are used for grasp localization.

2. Challenges, approaches for GL

In robotic grasp planning the major challenges observed are the tedious work on labelling the grasps by expert engineers and their biased nature on grasp localization. The biased nature on grasp localization is due to the nature of human beings' understanding of semantics and their experiences in the past. By automating the process of labelling, by reinforcement learning both challenges could be overcome.

A robotic gripper has to be aligned properly in an image plane according to the pose of the object. First the image plane coordinates has to map with the robot world coordinates as a part of grasp planning. The features of the novel object are marked during this process. Based on the detected features, the control system guided by a control algorithm will control the joints of the robotic manipulator to reach the object, also the end effector to orient itself to perform the grasp.

3. Deep learning for GL

The input data is an RGB-D image, which on using a convolutional neural network provides features of the objects present in the field of view. On further processing with deep neural networks, they yield multiple grasp configurations. The best grasp among them is found and is ranked as positive and negative grasps. Representation of RGB and depth image with rectangles is done by Ian Lenz et al., in 2014 [4]. Redmon et al. 2015 [7] introduced one-shot detection method which showcased a detection accuracy of 88% and detection speed of 76 milliseconds for an image. The decision making representational heuristics were obtained from the existing data by performing mathematical operations simultaneously and parallel. Normally, two-fingered grippers are used for grasping. But for challenging objects like a sphere, multi-fingered grippers are preferred [6].

4. DL approaches

The most common networks used for convolutional neural networks are AlexNet and ResNet. Among them ResNet has been identified as a better performing network than AlexNet. The ResNet has skip connections, which could skip unnecessary weights and biases to provide efficient results. Recent researches focus on PointNet, which takes the point cloud data directly and process the same. By doing so, the speed has been increased to a larger extent. Kumra et al. [8] implemented a much deeper network model and obtained detection accuracy upto 89.21%. Viereck et al. [2] Used (x,y,θ) and depth image as input and distance as output, which could tackle unexpected disturbances. Schwarz et al. [3] used two deep-learning approaches for object detection and semantic segmentation and one item model registration method; hence the system localized the requested item. Representation of RGB and Depth image with rectangles, Ian Lenz et al. [4] presented a method to apply structured regularization on the weights which is based on multimodal group regularization. Hence, multimodal inputs will be handled efficiently. He also used two deep networks to present a two-step cascaded system. In this method significant detections from the first were re-evaluated by the second. The first network is faster since it has fewer features and can effectively eliminate inefficient candidate grasps. The second is slower since it has much more features. But it finds its significance as it will only evaluate only significant detections.

LaserNet [9] is also a focus of research in recent times. LaserNet is the only probabilistic model, which predicts the probabilities of the dimensions of the bounding box. This has increased the detection performance as a whole. LaserNet works on the point cloud data obtained from LiDARs, to provide bounding boxes on the same. The predicted variance is used to determine an appropriate IoU (Intersection over Union) threshold for a pair of boxes. The likelihood of the box is used as its score [9].

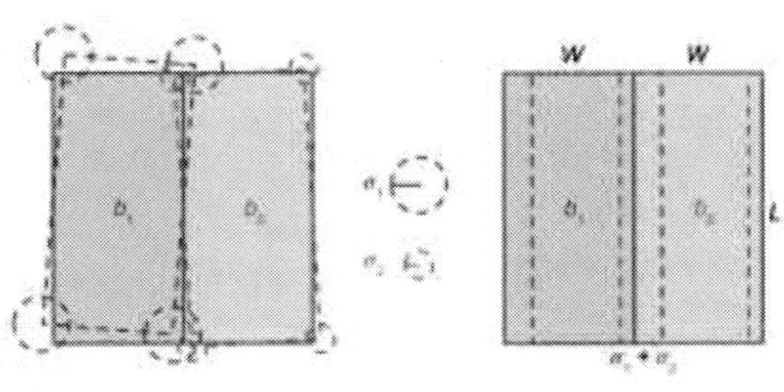

Figure 1. Illustration of the adaptive NMS technique [9]

During reinforcement learning of grasp localization, the success of the grasp is verified by few parameters; how successfully the gripper approaches and reaches the object in the desired pose, and how successfully it picks up the object without any slippage. When the gripper holds the object in the wrong location, then the possibility of the object falling is very high. Only when the object is being held in the right location, the grasp becomes successful. In reinforcement learning, the grasps as ranked, and the best ranked is selected based on a state of art method available and executed. The two-stage detection process of grasp rectangle and predicting the perfect grasp [4].

Johns et.al. [1] used a method that first predicts a score for every possible grasp pose. For the same, he created a grasp function. Within the scope of the paper no deep learning network is explicitly presented, though deep learning based grasp location is the current trend. The paper only highlights the simulation methodology after getting inferences from deep learning networks.

5. Importance of Simulation for GL

Reinforcement learning is a rigorous and time-consuming process by itself. Hence, we have to wait for a lot of time to find the result. But when we adopt simulation for doing the same, we could achieve quicker results. Mahler et.al [10] suggested using grasp simulation as one of the physics-based analyses to populate data when needed.

6. Simulation practices in grasp localization

There is also another handful number of researchers who used simulation for populating data. John et.al used 3D meshes for depth image and physics simulation [1]. Bohg et.al formed empirical methods using simulations on real robotic systems [2]. Amazon robotics also generated training data using simulations [11]. On the other hand, simulation practices for grasping are also a common practice nowadays [12]. [13] grasp is a simulation technique to compute grasp contacts using a robotic hand. They used Columbia Grasp Database. A soft finger contact model used GraspIt for simulating their robot [14]. K. Bousmalis *et al* used GraspGAN methodology [15] to obtain grasping performance equivalent to real-time grasping performance. The key to note that they obtained this without any real-world labels. OpenGRASP [16] is another toolkit used for robot grasping simulation which has extensibility and interoperability. [17] PyBullet is a more common simulator used for the simulation of grasping. In few grasp simulation projects [18] Q function estimation method is used for benchmarking the tasks.

In this project, grasp localization is simulated by the approach of the robot by generating simulated grasp pose for the end-effector. The simulated grasp pose is considered to be the ideal pose which would ensure that the object does not slip. The grasping simulation is performed by simultaneously applying the transformation which the end-effector is undergoing from time-to-time, to the object. This creates a visualization of the object being held by the robot and moved around.

7. 7-DOF Dual-arm Collaborative Manipulator

In this new era of collaborative work handling between machines and human beings, collaborative robots find its significance is not only material and tool handling, but also grasping, pick and place and many more serviceable arenas. Yi Ren et.al [19] presented a novel bio-mimetic object impedance control strategy for dual-arm co-operative 7 Degrees Of Freedom Humanoid manipulators. This manipulator exhibits human-like character, namely explicit compliance behavior during weak environment interaction and vice versa.

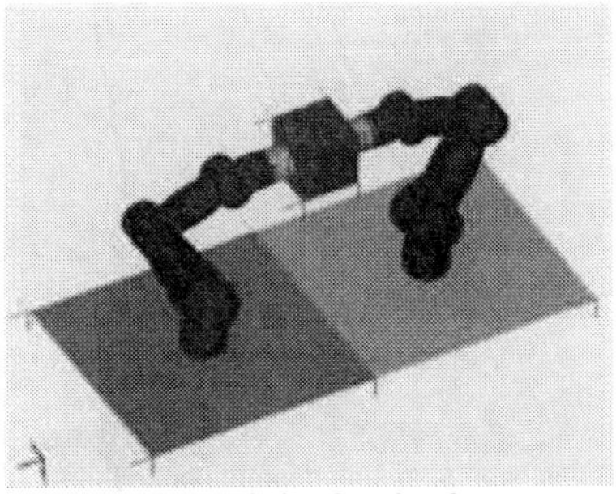

Figure 2. Virtual model of a dual-arm system [19]

Jun He [20] proposed a dual loop impedance control which exhibits adaptive stiffness for a dual-arm cooperative robot. This controller is used for grasping a given object in the environment.

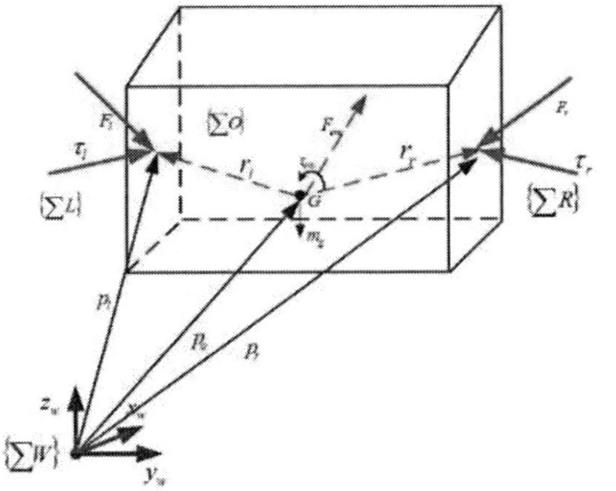

Figure 3. The coordinated manipulator grasping a common object [20]

In 2015 ABB has introduced YuMi, a 7 axis agile collaborative robot [21] which enabled people to work around and along with the robot.

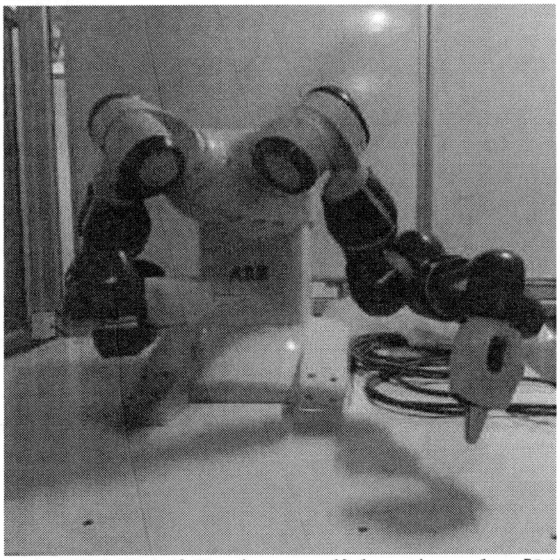

Figure 4. ABB YuMi Dual-arm collaborative robot [21]

IOP Conf. Series: Materials Science and Engineering 1012 (2021) 012004 doi:10.1088/1757-899X/1012/1/012004

8. Specification of the physical manipulator

The manipulator of ABB YuMi consists of a 7-axis configuration, which has reachability of 0.559 m and a payload capacity of 0.5 Kg. Its position repeatability of 0.02 mm makes it the best choice for grasp localization. Dudek et.al [23] have specified in his work about the dimensions of a manipulator, which is shown in **Figure 5. Figure 6** shows the simulated dual-arm YuMi robot.

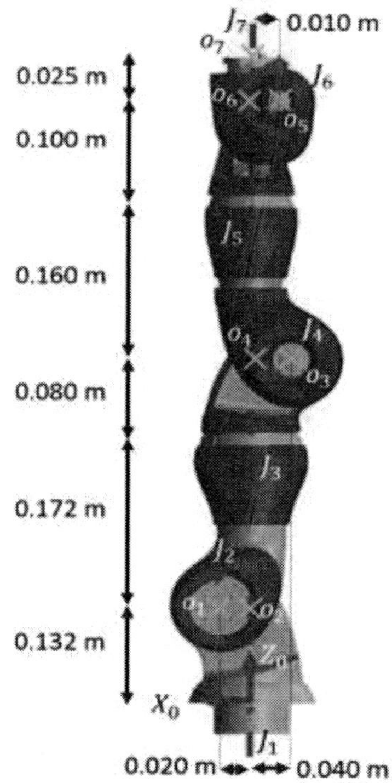

Figure 5. Specifications of the ABB YuMi dual-arm collaborative robot [23]

9. D-H parameters

D-H parameters are used to represent the architecture of a robot. Usually, they are provided with the robot itself by the manufacturer.

The D-H parameters of ABB YuMi is given here:

Table 1: D-H parameters of ABB YuMi [22]

Articulation	α_i (rad)	d_i(cm)	r_i(cm)	θ_i(rad)
1	0	0	16.6	q1-π
2	π/2	3	0	q2-π
3	π/2	3	25.15	q1
4	-π/2	4.05	0	q4-π/2
5	-π/2	4.05	26.5	q5+π
6	-π/2	2.7	0	q6-π
7	-π/2	2.7	3.6	q7+π

10. Results and Discussion

10.1. *Forward kinematics, transformation matrices*

Kinematic equations are used to calculate the position of the end effector from the joint parameter values. Forward kinematics uses these equations to calculate the position of the end effector to grasp the object in space using these joint parameter values. In other words, a set of joint angles are given and the position of the end effector has to be found.

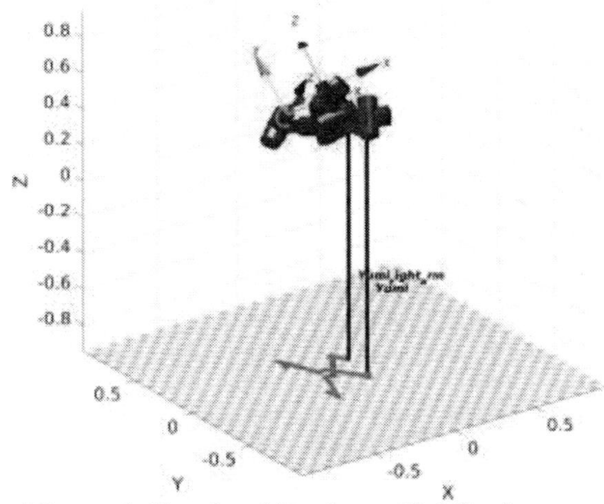

Figure 6. Simulated Dual-arm YuMi robot

10.2. Inverse kinematics, expressions for each joint angle

The joint parameters required to make the end effector move to a particular pose and rotation during the process of grasping can be determined by inverse kinematics. In other words, the position of the end effector is given and the joint parameters have to be found.

10.3 Workspace Analysis

10.3.1 Jacobian

When we have to deal with two different representations of the system, we use jacobian. They provide the relationship between joint velocities and end effector velocities of a robot manipulator. The robot moves from a resting position to the destination to grasp the object in space.

10.3.2 Singularities

As the robot manipulator moves to pick up the object in space, there is a possibility that two or more axes align to be collinearly resulting in undesirable motions and velocities. Wen et.al [24] states that singularities are the configuration the manipulability ellipsoid becomes degenerate.

10.3.3 Workspace plot

The workspace of the dual-arm YuMi robot simulated has been plotted and shown in **figure 7**.

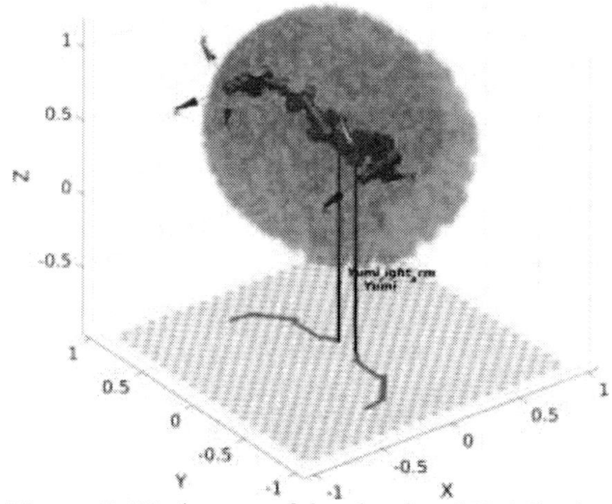

Figure 7. Workspace of the simulated YuMi robot

10.4 Trajectory Planning

10.4.1 Joint space planning

The joint space has been plotted below in figure 8 and figure 9. Figure 8 shows the joint space of the left arm and figure 9 shows the joint space of the right arm. In the left arm, the second joint and the sixth joint has the maximum displacement, which means that they rotate to the maximum extent. It is also identified that the first and fifth joints are at zero during the starting.

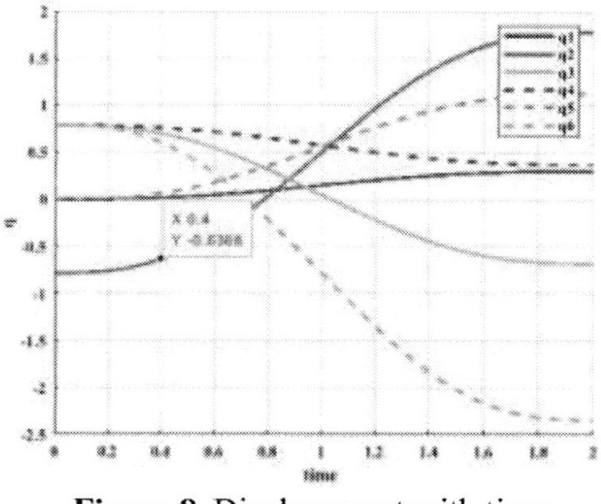

Figure 8. Displacement with time

Similarly, in the right arm also the second and sixth joint has maximum displacement. One significant point to note in the right arm is that the fifth joint moves from zero to negative, whereas in the left arm, it moves from zero to positive. The third arm moves from negative to positive in the right arm, whereas it is vice-versa for the left arm.

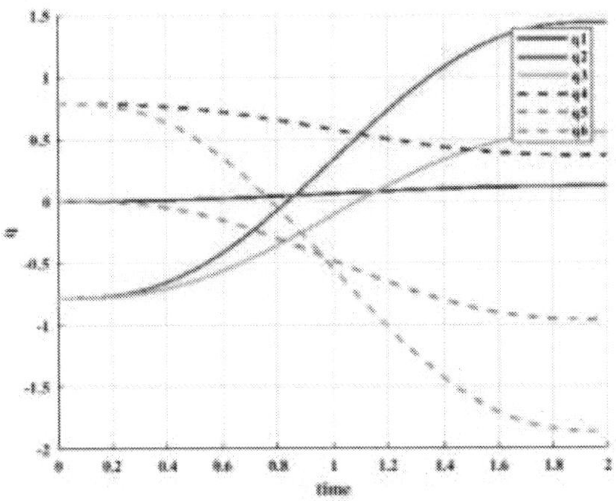

Figure 9. Displacement with time

As the displacement of the second and sixth joint is higher, so the velocity also. This is common in both arms. The velocity of the third joint oscillates in the negative, whereas the velocity of the fifth joint oscillates in the positive. This is vice-versa in the right arm. They are shown in **figure 10** and **figure 11**.

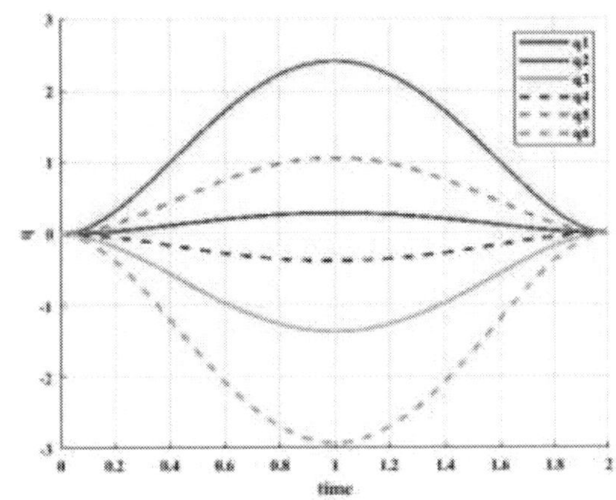

Figure 10. Velocity with time

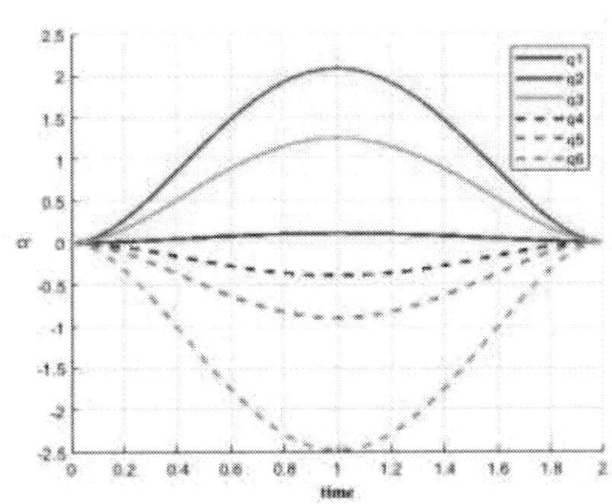

Figure 11. Velocity with time

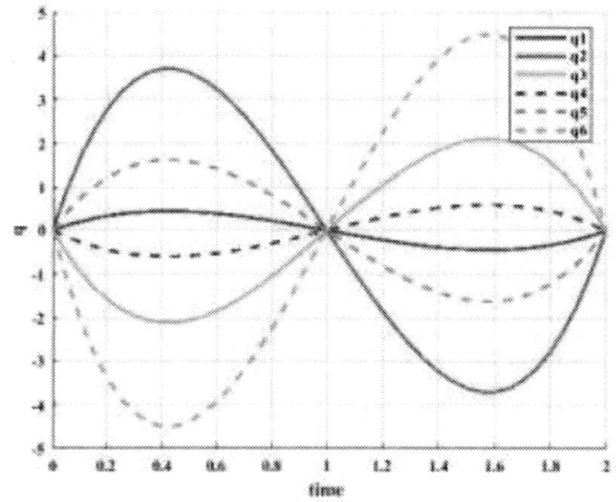

Figure 12. Acceleration with time

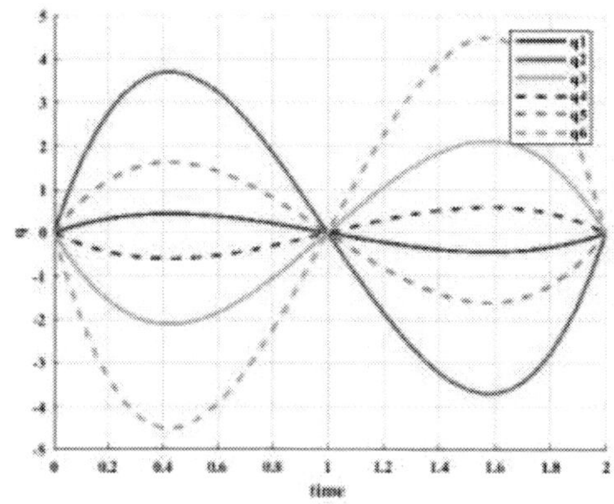

Figure 13. Acceleration with time

Figure 12 and **Figure 13** show the acceleration of the joint angles with time. For both arms, it is being identified that the sixth arm has maximum acceleration, whereas the first arm has minimum acceleration.

10.4.2 Cartesian space planning

When the robot joints are actuated the end-effector moves to reach the target and grasp it. The figures below, **figure 14** and **figure 15** depict the Cartesian space displacement of the left and right arm. In the left arm, the fourth joint deflects to the maximum in the Cartesian space, whereas in the right arm, the first joint moves up to the edge of Cartesian space.

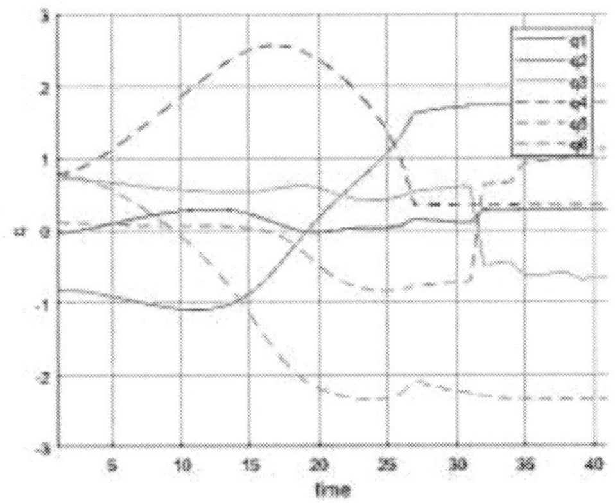

Figure 14. Displacement with time

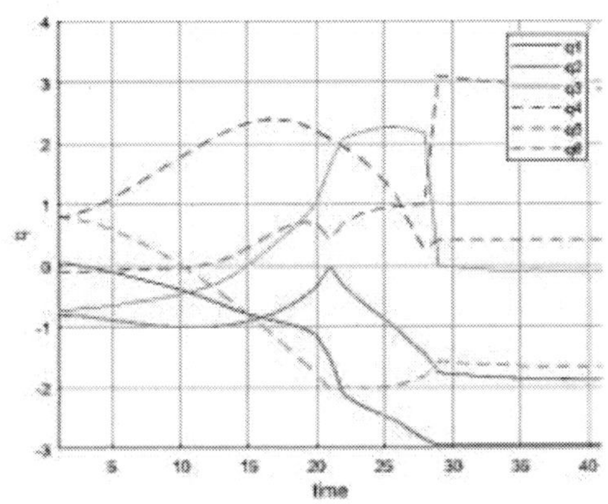

Figure 15. Displacement with time

10.5 CAD Import of Geometries

In order to simulate the grasp localization, the object needs to be imported into the robot simulation environment. The geometries of the objects either are extracted from any sensor such as time-of-flight cameras, LIDAR, etc or from CAD software for the visualization. In this paper the CAD import approach is presented. The CAD data of the object of interest is exported as an ASCII STL file. Most CAD software such as AUTOCAD, ProE, etc supports this feature. A third party tool [25] is utilized for converting the STL file into MATLAB workspace which may be subjected to some rotational transformations for pose presentation with reference to the robot's location. An arbitrary pose on the object was chosen based on manual observation of the geometry and target pose is created for the robot. A simulation step of the grasp is presented in **Figure 16** for a sample object.

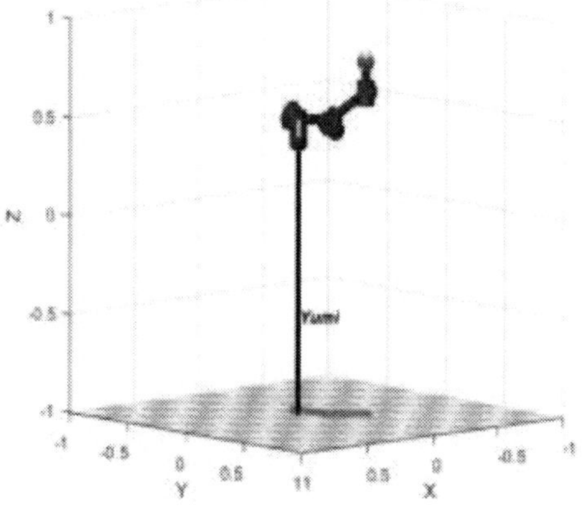

Figure 16. The simulation of the robot grasping the object in space is shown here

11. Conclusion

In the field of grasp localization, estimating the pose of the gripper for a successful grasp is a significant challenge. In this paper a grasp routine in seven degrees of freedom dual arm robot is simulated for the given object with known geometries which is imported from CAD software. This will also validate the feature based approaches. The paper discussed the details of the kinematics, workspace analysis, CAD import, planning of trajectories in Cartesian and joint space, the control strategy for executing the trajectory, and ultimately the grasping action. The main contribution of the paper is the simple methodology of simulating the grasp localization and grasping in a standard software platform thereby paving ways for various analysis of the grasp location problem. The results of simulation and the strategies adopted will be useful for technical community working in the field of grasp localization to quickly get started with grasp simulations using a redundant dual-arm manipulator.

References

[1] Johns, E.; Leutenegger, S.; Davison, A.J. Deep learning a grasp function for grasping under gripper pose uncertainty. In Proceedings of the 2016 IEEE/RSJ International Conference on Intelligent Robots and Systems (IROS), Daejeon, Korea, 9–14 October 2016; pp. 4461–4468.

[2] Viereck, U.; Pas, A.; Saenko, K.; Platt, R. Learning a visuomotor controller for real-world robotic grasping using simulated depth images. In Proceedings of the 1st Annual Conference on Robot Learning, Mountain View, California, USA, 13–15 November 2017; Volume 78, pp. 291–300

[3] Schwarz, M.; Milan, A.; Lenz, C.; Muñoz, A.; Periyasamy, A.S.; Schreiber, M.; Schüller, S.; Behnke, S. Nimbro Picking: Versatile Part Handling for Warehouse Automation. In Proceedings of the IEEE International Conference on Robotics and Automation (ICRA), Singapore, 29 May–3 June 2017

[4] Ian Lenz, Honglak Lee, and Ashutosh Saxena. Deep learning for detecting robotic grasps, Learning Methods in Robotic Grasp Detection, Multimodal Technologies and Interact. 2018, 2, 57. arXiv:1301.3592v6.

[5] Shehan Caldera; Alexander Rassau and Douglas Chai. Review of Deep Learning Methods in Robotic Grasp Detection, Multimodal Technologies and Interact. 2018, 2, 57.

[6] S. Carpin, S. Liu, J. Falco, K. Van Wyk, Multi-fingered robotic grasping: A primer, 2016, [online] Available

[7] Redmon, J.; Angelova, A. Real-time grasp detection using convolutional neural networks. In Proceedings of the 2015 IEEE International Conference on Robotics and Automation (ICRA), Seattle, WA, USA, 26–30 May 2015; pp. 1316–1322.

[8] Kumra, S.; Kanan, C. Robotic grasp detection using deep convolutional neural networks. In Proceedings of the 2017 IEEE/RSJ International Conference on Intelligent Robots and Systems (IROS), Vancouver, BC, Canada, 24–28 September 2017; pp. 769–776.

[9] *Gregory P. Meyer, Ankit Laddha, Eric Kee, Carlos Vallespi-Gonzalez, Carl K. Wellington*; Proceedings of the IEEE/CVF Conference on Computer Vision and Pattern Recognition (CVPR), 2019, pp. 12677-12686

[10] Mahler, J.; Liang, J.; Niyaz, S.; Laskey, M.; Doan, R.; Liu, X.; Ojea, J.A.; Goldberg, K. Dex-Net 2.0: Deep Learning to Plan Robust Grasps with Synthetic Point Clouds and Analytic Grasp Metrics. CoRR 2017, abs/1703.09312.

[11] Amazon Robotics. Amazon Picking Challenge 2016. https://www.amazonrobotics.com/#/picking challenge. Accessed: 2017-07-06

[12] Bohg, J.; Morales, A.; Asfour, T.; Kragic, D. Data-Driven Grasp Synthesis-A Survey. IEEE Transactions on Robotics 2014, 30, 289–309

[13] A. T. Miller and P. K. Allen, "Graspit! A versatile simulator for robotic grasping," in *IEEE Robotics & Automation Magazine*, vol. 11, no. 4, pp. 110-122, Dec. 2004, DOI: 10.1109/MRA.2004.1371616.

[14] H. Dang, J. Weisz, and P. K. Allen, "Blind grasping: Stable robotic grasping using tactile feedback and hand kinematics," *2011 IEEE International Conference on Robotics and Automation*, Shanghai, 2011, pp. 5917-5922, doi: 10.1109/ICRA.2011.5979679.

[15] K. Bousmalis *et al.*, "Using Simulation and Domain Adaptation to Improve Efficiency of Deep Robotic Grasping," *2018 IEEE International Conference on Robotics and Automation (ICRA)*, Brisbane, QLD, 2018, pp. 4243-4250, doi: 10.1109/ICRA.2018.8460875.

[16] León B. et al. (2010) OpenGRASP: A Toolkit for Robot Grasping Simulation. In: Ando N., Balakirsky S., Hemker T., Reggiani M., von Stryk O. (eds) Simulation, Modeling, and Programming for Autonomous Robots. SIMPAR 2010. Lecture Notes in Computer Science, vol 6472. Springer, Berlin, Heidelberg. https://doi.org/10.1007/978-3-642-17319-6_13

[17] E. Coumans and Y. Bai, *pybullet a python module for physics simulation games robotics and machine learning*, 2016-2017

[18] D. Quillen, E. Jang, O. Nachum, C. Finn, J. Ibarz and S. Levine, "Deep Reinforcement Learning for Vision-Based Robotic Grasping: A Simulated Comparative Evaluation of Off-Policy Methods," 2018 IEEE International Conference on Robotics and Automation (ICRA), Brisbane, QLD, 2018, pp. 6284-6291, doi: 10.1109/ICRA.2018.8461039.

[19] Yi Ren, Yechao Liu, Minghe Jin, Hong Liu, Biomimetic object impedance control for dual-arm cooperative 7-DOF manipulators, Robotics and Autonomous Systems, Volume 75, Part B, 2016, Pages 273-287, ISSN 0921-8890.

[20] J. He, M. Luo, and Q. Zhang, "Dual impedance control with variable object stiffness for the dual-arm cooperative manipulators," *2016 Asia-Pacific Conference on Intelligent Robot Systems (ACIRS)*, Tokyo, 2016, pp. 102-108, doi: 10.1109/ACIRS.2016.7556196.

[21] https://new.abb.com/products/robotics/industrial-robots/ irb-14000-yumi

[22] Meryem Taghbalout, Jean François Antoine, Gabriel Abba, Experimental Dynamic Identification of a YuMi Collaborative Robot,IFAC-PapersOnLine,Volume 52, Issue 13,2019,Pages 1168-1173,ISSN 2405-8963,https://doi.org/10.1016/j.ifacol.2019.11.354.

[23] Dudek and Jenkin, Computational Principles of Mobile Robotics and CS W4733 – Differential Drive Robots.(https://chess.eecs.berkeley.edu/eecs149/documentation/differentialDrive.pdf)

[24] 24. J. T. -. Wen and L. S. Wilfinger, "Kinematic manipulability of general constrained rigid multibody systems," in *IEEE Transactions on Robotics and Automation*, vol. 15, no. 3, pp. 558-567, June 1999, doi: 10.1109/70.768187.

[25] Don Riley (2020). CAD2MATDEMO.M (https://www.mathworks.com/matlabcentral/ fileexchange/3642-cad2matdemo-m), MATLAB Central File Exchange. Retrieved October 1, 2020.

Robust Design of Smart Lawn Cutter for Gardening Companion

Balaji A[1*], Shanthini E R[1], Rooshan M J[1], Sivaram V[1]

[1]Department of Mechatronics Engineering, Kongu Engineering College, Erode - 638060. Tamilnadu, India.

E-mail: bala2009mct@gmail.com

Abstract. This work discussed the fabricated model of android based grass cutter in agricultural field by implementing the new innovative concept with simpler construction and easy working. The cutting motor is employed to cut the grass in the field and the same is controlled by smartphone application. The components that are used to control the grass cutting level involve motor and gear arrangement. Below the gear arrangement, the cutting blade is fixed. And this arrangement will be simple and highly efficient for grass height adjustment system with up/down process motor. DC motor up and down motion is made using the rack and pinion gear setup. And then grass cutting motor on off control system will be achieved through android application. This application is connected to the Bluetooth communication for automatic the process. This work is motivated on preparation of the beneficial power basis for grass cutting process which is not reliant on the climates state and also prompt on its work. This setup is completely controlled using smart phone. In future, the automation of the device will play the most important role in world wide. In the meantime, technologies are merging with environmental awareness, consumers are looking for ways to contribute to the relief of their own carbon footprints.

Keywords: smart lawn cutter, arduino UNO, bluetooth wireless technology, wifi and zigbee modules

1. Introduction

In this paper, an idea is proposed to automate the lawn mower which is widely used to trimming the grass. It is made up with sharp linear blades operated with the support of motor enclosed on it. The high capacity battery is employed for the assistance of the linear blades. This project aims to trim the lawn which operates on an electrical source controlled by an Arduino UNO microcontroller interfaced with a Bluetooth module [1]. The Bluetooth module is paired with an android application in a smartphone. The instructions are controlled over by android application signal and the signal is transmitted via Bluetooth module to Arduino UNO [2]. The microcontroller board is equipped with different sets of digital and analog I/O pins that are linked with the mechanical setup [3]. To replace the traditional cutter by new innovative design smart cutter leads to provide the way to implement the smart device which reduces human fatigue as well as environmental pollution [4].

The design work is intensive on preparation of an helpful power source for grass cutting process which is not reliant on climatic condition and working nature [5]. This design is controlled using smart phone to trim the grass. Though it is a wireless controller the controlling range can be varied based on the module that have chosen [6]. The electric motor is the source which is used to rotate the blade to shape the grasses based on the regions and the places it can be actuated with different speeds. In upcoming years, the automation device in gardening is highly required in the worldwide lawns [7]. In the meantime, the technologies are integrating with environmental awareness and attract buyers, who are viewing the ways to contribute to the respite of their lawn. This new design will help both the consumer and environment [8]. The human made gardening convention faces various critical issues which in turn the requirement of automation in the same field [9]. The planned model consists of a linear designed blade which is operated substantially using the motor with the battery power supply [10].

In a conventional lawn mower, the heights of the blade cannot be accustomed spontaneously and the height alteration of the rotor blade is done by lifting the whole deck with the help of wheel supports and this process takes time of about at least 4 - 5 minutes to adjust the rotor [11]. This proposed work can control this process simply by pressing the button (up or down) in the android application. The motor runs spontaneously by increasing or decreasing the altitude of the blades with help of lead screw mechanism [7]. The lawn cutter with the revolving blades helps to cut the lawn at even long length to make it good looking. In the existing model of the grass cutter, IC engines were employed to rotate the blade which brings an environmental impact of pollution level [12]. The consumption of fuels and the emission of the fuels causes environmental hazardous. Nowadays, pollution is a main crisis in the worldwide issue, which is human intervention and will be proved in many places [13]. In case Gas powered lawn mowers if employed which also exhibits emission of gases leads environmental pollution [14]. The cost of the fuel is very high which will not provide the users to achieve the efficient in lawn maintenance. The maintenance of such a conventional engine is too difficult and tedious to handle for the human [15]. To avoid these drawbacks, an idea is proposed to rectify the drawbacks with innovative concept and design. The present lawn mowers can be replaced with the new idea and innovative technology leads to provide the pleasant lawn maintenance [15].

2. Frameworks of Smart Lawn Cutter

2.1 Android Application Development

Though many technological innovations are taking place in this world, the proposed solution is an automated device for trimming the grass evenly over long lengths. The linear motor is fixed and is joined in the blade which delivers the action of rotation to make the shaping of grass process through blades [14]. This work aims to trim the lawn which operates on an electrical source controlled by an Arduino UNO microcontroller interfaced with a Bluetooth module. The microcontroller is the controlling device has different sets of I/Os for interfacing various process achievement as well as for automation of different process [16]. The developed operating view of android application is shown in **Figure 1**.

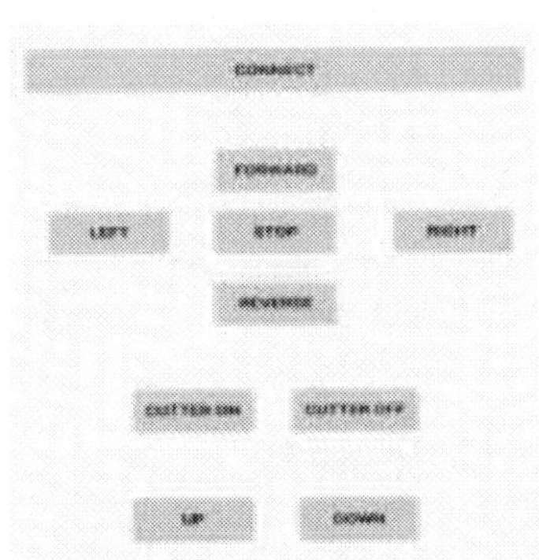

Figure 1. Android Application Operating View

The android mobile application is developed based on the required parameters and the same was paired with an android application on smartphone. Further, it is planned to pair with IOS platform for fully automation [17].

2.2 Technical Feasibility

The components required and the technologies adapted for the design are available at ease. Sufficient technology is available to carry out this project. Globally available software packages fulfil the requirements adapted for developing the project into a product for design verification [18]. The design which has to be implemented can be simulated through the available software and prototype can be developed according to the concept. The sample prototype represents the concept or the process which is going to be implemented. Based on the requirement of cutting level, lead screw mechanism can be adjusted with minimum of 4 cm and maximum of 25 cm range while cutting the grasses. After attaining the results from the prototype, the concept has been designed and fabricated into a product. The proposed system is technically feasible and makes the existing system incompetent [19].

2.3 Operational Feasibility

The operation of the entire system can be achieved using Arduino UNO and Bluetooth module relays. As per the programming instructions, the signals are transmitted from the Bluetooth module to the Arduino UNO microcontroller which in turn actuates the relay to turn ON the required motors.

2.4 Economic Feasibility

The components required for the fabrication of project are explained in below in the table 1.

Table 1. List of Components

S. No.	Components	Quantity
1.	Arduino UNO	1
2.	Bluetooth module	1
3.	DC Motor	4
4.	Wheel	3
5.	Chassis	1
6.	Battery	1
7.	Relay	6
8.	Capacitor	1
9.	Resistance	10
10.	Switch	1
11.	Blades	4
12.	Wires	As per required
13.	Other Peripherals	As per required

3. Principle of Operation of the proposed System

Robotics one of fastest growing field pave the way to get maximum output without the involvement of human efforts. An alternative is to be made to reduce the energy, wasted by the humans in mowing the lawns over different areas [20]. In this proposed project, human effort was reduced and also the safety was ensured while operating. The operation of the entire model can be achieved by using Arduino UNO, Bluetooth module relays which are used to actuate and send the signals. IOT helps to establish the fully automation device which is highly required in the gardening companion to eradicate the labour shortage in efficient manner.

This design is to trim the lawn evenly by using a smart lawn mower with android application, where Bluetooth module is connected to the smartphone. It can be adjusted as per the requirements, it has been designed with minimum of 4 cm and maximum of 25 cm range for height variation while cutting. To develop the communication between Arduino UNO and the android application using bluetooth module to implement the smart mower [21]. To implement the smart mower with all directions control based on the user requirement to cut the lawn in even manner. To achieve the smart lawn mower with control range of 20 meter between the mower and the communication device. To establish the smart lawn cutter can be worked in all climatic conditions [22].

4. CAD Model and Specifications

4.1 CAD Modelling

The below design **Figure 2** describes the modeling of smart lawn cutter. Initially the solid model of 3D is designed as per the planned dimensions and verified the model whether it should have stability in all aspects. This validation ensure the model and the design is widely suitable for fabrication. The whole setup is directly controlled through smart phone application which is easier to operate and flexible in accessibility. The material employed in the fabrication part is mild steel which is highly suitable for the lawn mower as disused by the various researchers. As mentioned in the solid works design, it is decided to have provision of three DC motors and one cutter motor can be placed to cut the grass in lawn. It is also confirmed that smart lawn cutter will cut the grass evenly in the lawn and it will provide neat look and pleasant environment.

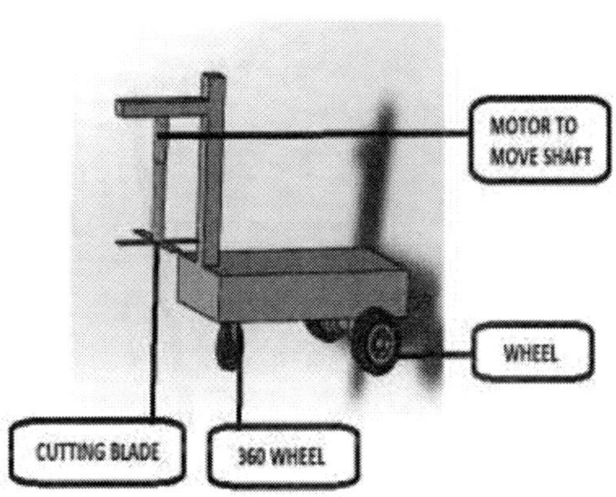

Figure 2. Solid works CAD Model

4.2 Design Specifications and Calculations

- Voltage, V= 12 V
- Current, I = 5 A
- Power, P = 250 w
- Speed, N = 250 rpm
- Distance of shaft from central axis= 4.5 cm
- Power, P = (2*pi*N*T)/60
- Torque, T = (250*60)/(2*pi*N)
  = 9.55 Nm
- Torque, T = L*d
- Load = 9.55 / 0.045
- Load = 212.22N (21.63 kg)
- Total load capacity = 21.63*4 = 86 kg

5. Experimentation

5.1 Mechanical Section

The fabricated model of smart grass cutter is shown in the **Figure 3.** The proposed lawn mower design is encompassed of smart controlled Bluetooth module with user friendly manner. The grass or the weeds can be cut by means of cutters rotating at high speed. Rotors are actuated using IOT (internet of things) concepts. The control system continuously monitors the motion and also the path to ensure the smooth operation cut in order to provide the shape and size in required manner. In this system, sensors will be placed to acquire the signals to eradicate the unwanted issues. No sensor is employed especially for cutting the grasses since it is controlled via Bluetooth modules through mobile application. Proximity sensor will be placed to avoid the damages of blades while the solid object intersects.

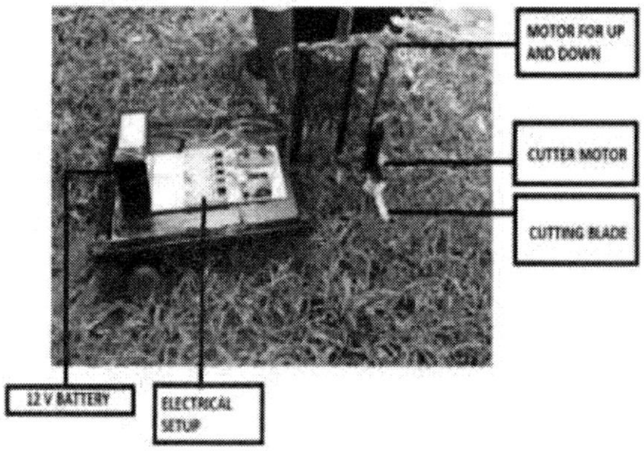

Figure 3. Fabricated Setup

The cutting area will be scanned and then the feedback will be sent to the controller by sensor and the programming will be controlled to create the required shape. The bot will receive these values and then start the process of moving. The system will constantly sending and receiving the data about the area to be removed and the surrounding environment. Blade rotating speed can also be controlled using the Smartphone with the help of an Android Application. The cutting Blade which is placed at the top of the mower is attached with the shaft and motor. Cutting Blade rotates at 2500 revolutions per minute .Blade will also be adjusted using the android application and it can be adjusted to a certain level. It is adjusted mainly because of the unwanted materials present in the lawn which will damage the cutting blade. The cutting area size will be accurately based on the capacity and efficiency of battery and motor respectively. In this fabrication work, it consumes around 50 Secs to complete the 5 x 5 feet area based on the experimentation. Proximity sensor will be placed in front of the device to avoid the obstacles. Any solid objects other than grass the device will detect and send the feedback to the system. Hence the rotating blade will stop when it interacts with solid objects.

5.2. Electrical Section

5.2.1 Electronic Circuit Design

In this section, all the required components are assembled in simulation software and verified. Software called PROTEUS is being used for the simulation. The design consists of proteus simulation and electric circuit.

5.2.2 Proteus Model

ATmega8 is the controller used for controlling the input and output parameters and used to follow the sequence of operation. The parameters such as motor on and off control are given to ports. To initiate the operation, push button start is given as input in the port 1.0. To move the motor forward the particular logic is created and that is saved in the form of hex file. Hex file is created while compiling the code in Arduino software likewise for backward also logic are given. For moving each and every direction there is some logic in coding and are implemented. This is the basic circuit to actuate the relay and Bluetooth module is used to make communicate with smart phone by using Tx and Rx mode and the below shown model **Figure 4** is general model of proteus software. Proteus software tool suite is used primarily for electronic design automation.

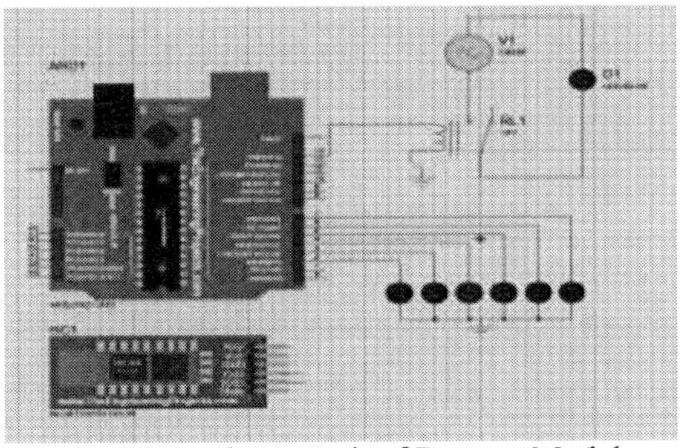

Figure 4. Photograph of Proteus Model

5.2.3 Electric Circuit

The electrical circuit consists of Arduino UNO as shown in Figure 5. A single-board microcontroller Arduino is built to make it easier to access interactive objects and their environment. An open source support of hardware features are built with ATMEL 8-bit AVR microcontroller device or a 32-bit ATMEL ARM device with a USB interface provision, which has 6 analog pins and 14 digital pins to transfer data. There are many third-party compatible versions found in the existing arduino boards. Arduino Uno R3 and Arduino Nano V3 are most commonly used for official purpose. The electromagnetic switch which works based on the principle of an electromagnetic fascination are called relays in which the magnetic field turns the relay armature for opening or closing the connections which in turn controls the direction of the motor. One contact type relays are used for small power applications whereas two contact type relays are used for high power applications. An LED is inbuilt which indicates the passage of every command given by user, it can be used in identifications of errors in an easy way and the resistor controls the over flow of supply. In the communication field, Bluetooth is the extensively used wireless device in most applications. Therefore, knowledge about a HC 05 Bluetooth is required in the interfacing process and data transmissions.

Mobile phones and personal communication devices are widely used to transmit and obtain data in the modern days using Bluetooth modules[23]. Host controller interface (HCI) is used to transmit or receive data from/to the host. The most popular HCIs are UART & USB. The possible best solutions are made by the Bluetooth module by using the Bluetooth embedded systems and the UART connections. Comparing to Wi-Fi and Zigbee, Bluetooth range is very less and operates at a frequency of 2.41 GHz and can be used in small ranges of applications. Capacitor is utilized to store energy which exist in excess by using the two parallel plates placed inside a dielectric medium.

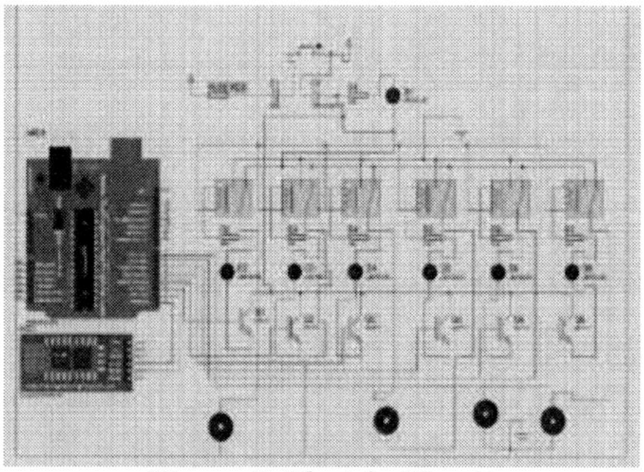

Figure 5. Electric Circuit

Relay circuit consists of relays on relay board to transfer signal. The rectifier circuit consists of a step down transformer step downs the 230V AC into 12V and fed as input to the motor. The entire electric section is shown in **Figure 5**, where the relay and rectifier circuits are developed on dot matrix board and the microcontroller are developed on the built on the controller development board.

6. Conclusion and Future Scope

The smart lawn cutter is designed and fabricated with innovative methodology to achieve the lawn trimming in even shapes. The fabricated machine is successfully integrated with an android application, where Bluetooth module is connected to the smartphone to perform various actions of lawn mower. The communication was developed between Arduino UNO and the android application using Bluetooth module to functionalize the smart mower in efficient way. The smart lawn mower was attained with control range of 20 meter between the mower and the communication device hence it depicts the effective model. The smart lawn cutter was established with working operations in all climatic conditions. In recent days, smartphone usage is highly increased at everywhere almost 80% of citizen owns smartphone on their own. In this project also, smartphone android application is employed for regulating the actions of smart cutter in all aspects which can be operated by anyone without instructive knowledge. Automatic control has been established with the help of IOT.

References

[1] Gupta K 2020 SOLAR GRASS CUTTER *Global Journal For Research Analysis (GJRA)* **8**

[2] Bhalodi T, Bhujbal N, Doshi K, Goregaonkar R and Jagtap S 2020 Environmental Friendly Solar Grass Cutter *International Journal of Research in Engineering, Science and Management* **3** 177-80

[3] HARITHA M, SRINIVASULU M and SIVA B SOLAR GRASS CUTTER WITH AUTOMATIC TRACKING

[4] More S M, Pujari R K, Jadhav S S, Kalli K R and Mali D P Fully Automatic Solar Grass Cutter

[5] Kulhariya S K, Tomer A S, Krishan K, Yadav N K and Mishra R K 2020 Performance evaluation of solar operated bidirectional grass cutter *Journal of Pharmacognosy and Phytochemistry* **9** 329-34

[6] Khillare S R, Morey D P, Ghoti B A, Thorat S S, Pimple S D and Sharma M O 2020 A Review of Fully Automated Grass Cutter Using Solar Power *International Journal of Research in Engineering, Science and Management* **3** 108-11

[7] Daniyan I, Balogun V, Adeodu A, Oladapo B, Peter J K and Mpofu K 2020 Development and Performance Evaluation of a Robot for Lawn Mowing *Procedia Manufacturing* **49** 42-8
[8] Afan M, Muzaffar M, Supervision A and Rabbani M A Grass Cutter Machine
[9] Ibrahim B, Brahmaiah V S and Sharma P 2020 Design of smart autonomous remote monitored solar powered lawnmower robot *Materials Today: Proceedings*
[10] Shah A D, Mujawar S J, Sutar P R and Prasad S R SOLAR POWERED INTELLIGENT GRASS CUTTER ROBOT
[11] Kalburgi S, Nair V G and Guruprasad K 2020 *Intelligent Systems, Technologies and Applications*: Springer) pp 213-20
[12] Asha N, Monica J, Saraswathi R, Rahul R and Ravikiran R 2017 Smart Grass Cutter *Perspectives in Communication, Embedded-systems and Signal-processing-PiCES* **1** 97-9
[13] Seiferth B 2020 *Development of a system for selective pasture care by an autonomous mobile machine*: Springer) pp 7-13
[14] Khan D M, Mumtaz Z, Saleem M, Ilyas Z, Ma Q and Ghaffar S 2019 Solar Powered Automatic Pattern Design Grass Cutting Robot System Using Arduino
[15] Nautiyal M, L Grabow G, L Miller G and L Huffman R 2010 Evaluation of Two Smart Irrigation Technologies in Cary, North Carolina. In: *2010 Pittsburgh, Pennsylvania, June 20 - June 23, 2010,* (St. Joseph, MI: ASABE)
[16] Singh M, Padhan A, Saurabh S, Priyesh K, Rai M and Naveen T A Review and Comparative Analysis of Solar, Electric and Gasoline Lawnmowers: An Extensive Study
[17] Green P, Cheydleur G, Li K, Deng R and Sanchez R 2018 Free smartphone apps to support human factors/ergonomics work *Ergonomics in Design* **26** 4-16
[18] Pane M, Ehkan P, Saleh K and Irwanto M 2018 A Mobile Surveillance Robot Over The Wifi Network Using Atmega 8
[19] HASHIYANA V, HAIDUWA T, SURESH N, SHISHIIVENI T and MUTALYA A 2020 A Prototype of an Android Application Controlled Lawnmower. In: *2020 IST-Africa Conference (IST-Africa)*: IEEE) pp 1-5
[20] Siregar I 2020 Design tools multifunction equipment. In: *IOP Conference Series: Materials Science and Engineering*: IOP Publishing) p 012022
[21] Vesali F, Omid M, Kaleita A and Mobli H 2015 Development of an android app to estimate chlorophyll content of corn leaves based on contact imaging *Computers and Electronics in Agriculture* **116** 211-20
[22] Robinson H, MacDonald B and Broadbent E 2014 The role of healthcare robots for older people at home: A review *International Journal of Social Robotics* **6** 575-91
[23] Zachiotis G A, Andrikopoulos G, Gornez R, Nakamura K and Nikolakopoulos G 2018 A survey on the application trends of home service robotics. In: *2018 IEEE International Conference on Robotics and Biomimetics (ROBIO)*: IEEE) pp 1999-2006

Design and Analysis of Universal Gripper for Robotics Applications

Balaji A[1*], Mithil J[1], Jason Gousanal J[1]

[1]Department of Mechatronics Engineering, Kongu Engineering College, Erode – 638060. Tamilnadu India.

E-mail:bala2009mct@gmail.com

Abstract: A universal gripper is a component that consists of fine dust material enclosed in fine rubbery membrane for holding and assembly. The gripper is especially applied for quickly gripping and discharging large number of objects having complex geometries in wide applications. The objects which are in various shapes and different geometries are difficult to grasp which confirms that it is not suitable for apply in various pick and place operations. In this project, a low cost modular gripper is implemented with suitable mechanical design, structural analysis and real time analysis. In this association, the designed gripper can hold many parts and place it at once while sustaining relative distance and relative coordination at their state. The universal gripper acts as a vacuum tube to grip the object it needs to relocate. This universal gripper can hold various objects such as eggs, grocery items, industrial components, heavy objects etc., therefore, a selected experimental setup has been developed with the composition of shapeless gripper and perform simulation in holding of the different items. Results of the tests and experiments shows the proposed Universal Gripper are validated with feasibility options.

Keywords: *Gripper, Granules, Silicon Rubber, Vacuum Chamber, Gripping Mechanism*

1. Introduction

The effective end user, an end effector or gripper is the momentous module in the robotic manipulator. It assists the robot's end effector in the effective manner and allows the robot to deploy objects [1]. In recent times, the end effector pay more attention in diverse tasks in various fields such as automation, medical and food processing industries [2]. In manipulators, end effector has major role in grasping actions i.e. to pick and place the object from one location to another location respectively which is a tedious process for all grippers for different geometries [3]. In order to overcome this problem, soft grippers have been manufactured for gripping indulgent kind of geometric objects Soft grippers used in industries are designed to grip the objects which are having minimum tension [4].

Universal grippers are proposed to replace the soft grippers which are able to grip irregular shape of geometric object [5]. It will be used for minimum as well as maximum tension objects, even that object having minimum thickness. This kind of grippers are recently identified by the researches and analysis of its parameters for effective utilization in industries like Food processing industries is done [6]. Universal grippers are standard grippers which may be used in industrial manipulators for gripping industrial objects. The design, modeling and fabrication of Universal Gripper is analyzed to reduce complication while handling soft materials [7].

To execute different operations of end effectors, the tools such as spray guns, hammer can be utilized which are similar to tools used for industrial robots. End effectors are usually designed for the task that has to be carried out as a mechanized material handling which involves the linkage of work piece with the required places. The connected region is the place where the potential of the machine is fully utilized. There are two variety of components available for the wrist attachment: namely grippers and tools [8].

Using the gripper as interface is fundamentally a point-to-point process in which the working of machine is carried out. The construction of manipulator consists of several job-related classifications it might be essential to exchange the end effectors by the instrument itself in the industrial operations. For example, in on-site construction process, grippers are mostly employed in handling the glass based materials [8]. Grippers are those which are used for holding an object and placing an object in the work space. The path defining for the end effector movement is basically a point to point movement process [9, 10]. The construction of manipulator purely based on the end effector utilization in the required application of industrial process[11]. For example, in on-site construction process, grippers are mostly employed in handling the glass based materials [12]. This work introduces the detailed idea about universal gripper and its basic mechanism. In fact it describes the design of universal gripper using solid works based on the selection of rubber and granular materials [11]. The process also explains the working of universal gripper and the geometric gripping method with granular material as coffee powder and thermocol balls [13]. From this work, the analysis of gripper material and comparison of stress, strain and displacement for various varying loads can be established. The main novelty involved in the work is an effective as well as exclusive universal gripper is introduced with different granular materials for industry handling applications.

2. Materials and Methods

For various applications, the grippers are classified as follows: Mechanical grippers, Magnetic grippers, Vacuum grippers. Using grippers for mechanical clamping of objects with the required mechanisms is the common tasks in various places, where hydraulic or pneumatic operated devices are used to grip the objects. The different types of mechanism available in the industry are parallel jaws, finger crossed mechanisms, contraction and retraction type, Anthropomorphic type of hands etc., among these the parallel jaws mechanism grippers are widely applied to hold the object using the direction controlled surfaces [14]. These devices have one or more moving fingers where the different objects can hold but it is not suitable for soft materials. Finger based grippers will hold the component in the tip of the jaws. Anthropomorphic types require three finger where the object with definite structure can be picked and placed. This type of gripper works as same as the human hand but it is available in the research laboratory for further execution. However the end effector with provision of varieties of objects handling type are highly demanded in the industry to compensate for action of different objects [15]. Electromagnetism is the basic principle employed in magnetic grippers to hold the ferrous components. This type of gripper can be suitable for the magnetic materials not applicable for variety of objects [16].

The most common gripper named vacuum grippers where the suction cups is positioned to do the grasping actions of defined bodies, where the vacuum is produced by a vacuum ejector through the pump. It is mainly used for roofs, interior lightings and outer wall assembly, door, and windows assembly. Vacuum grippers are used for a wide range of applications from cases like soft flexible packaging to soft material handling operations, but most commonly seen handling objects are smooth, flat, and solid. This kind of gripper uses soft rubber, plastic, or elastic suction (vacuum) cups for picking objects so that there is no damage to the product being handled [17]. The soft material handling of irregular shape and size are difficult in the existing scenario. The main principle behind universal gripper is jamming the object and will transform the shape associated with the object using the granular materials. The rubber pouch filled with granular materials is the holder part of the gripper. The rubber material bag is filled with a sand like porous material [13]. The compressed air is allowed to the rubber bag to soften its action. Then the ball is pushed to the opposite direction of the target. The transition will made on the target based on the geometry by compressing the ball. This method leads the attention gripping forces to hold the objects. The present study utilizes a real time experimental setup to analysis the gripping force variation in the universal gripper with respect to various granular materials. The flow chart as shown in figure illustrates the process involved in the analysis. The methodology of the project is designed as follows Figure.1.

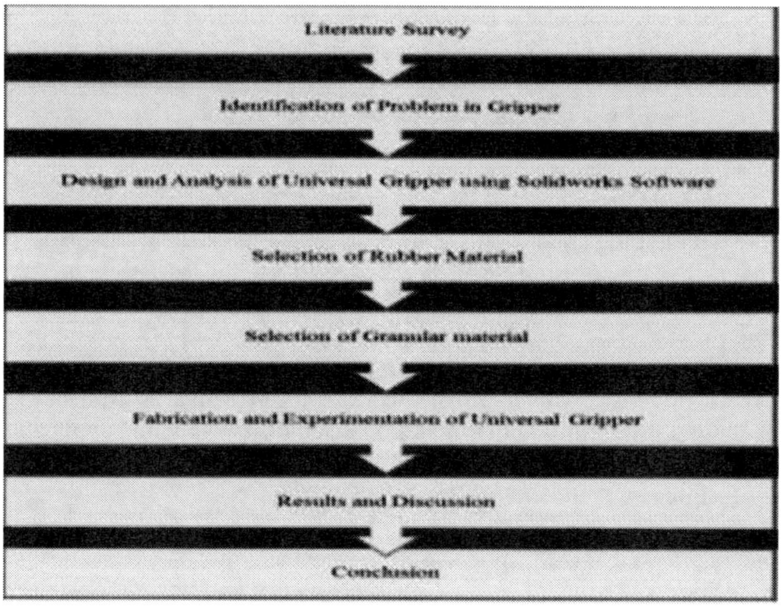

Figure 1. Methodology of the Work

2.1 Selection of Vacuum Cup

In order to handle the workpiece safely the force required to produce by suction cups is calculated theoretically. The atmospheric conditions and the application of the gripper is taken into consideration.

The factors required for selection of cups are

- ✓ Application
- ✓ Material
- ✓ Surface
- ✓ Shape
- ✓ Interior volume
- ✓ Materials
- ✓ Environment

The comparison of vacuum pressure and atmospheric pressure are discussed in table 1.

Table 1. Comparison of Vacuum vs Atmospheric Pressure

% of vacuum	mm of Hg	Bar	Psi
10	76.92	0.1	1.47
20	153.85	0.2	2.84
30	230.77	0.3	4.41
40	307.69	0.4	5.88
50	384.62	0.5	7.35
60	461.54	0.6	8.82
70	538.46	0.7	10.29
80	615.38	0.8	11.76
90	692.31	0.9	13.23
100	769.23	1	14.70

2.2 Design of Gripper Using Solid works

The gripper is designed using solid works software shown in Fig. 2. and the stress strain analysis is done for various loads. The holder is designed with ABS plastic material and the gripper bladder is designed using natural rubber and latex rubber. The various analysis were carried out to exhibit the result of the design & analysis of gripper in all aspects.

Figure 2. SolidWorks Design of Gripper

3. Experimentation

Universal gripper is working based on the process of "congestion". When a granulated resource like coffee powder is compressed into the rubber material bag forms, it becomes very rigid as shown in the figure. The contact between porous particle decreases as the pressure increases. This helps to hold the work piece. This is observed during the handling of eggs and soft materials. A vacuum crammed bag of coffee granules is made tighten as much as for transition and the stamp ruins unaffected. When replacing coarse coffee grains with fine grains of powder will have chance to break the bag while applying pressure. It was also tested with fine porous materials such as nuts, cereals, rice, etc. The experimentation was carried out by using the industrial manipulator ABB IRB 1410 for testing the universal gripper model with vacuum cup as shown in Figure. 3. The pressure is controlled using the pressure regulator which is attached with FRL unit of the compressor line for ABB IRB1410 end effector.

A latex rubber material bag is filled with coffee grounds as shown in figure and attached to an air hose. On applying minimum pressure the material inside the bag gets positioned accordingly. On pressing the material bag against the workpiece, the porous materials move and form a particular structure. The compression of grounds hold the object and the rubber material bag also helps to holding the object. The rubber holder is filled with coarse coffee ground grains which is attached to the wrist part of the manipulator. The funnel is inserted into the other end. Then minimum amount of porous material is poured into the material bag. The bag is partially inflated on removing the funnel, permits the grounds to come down to the bottom of the bag.

Figure 3. Experimentation of Universal Gripper

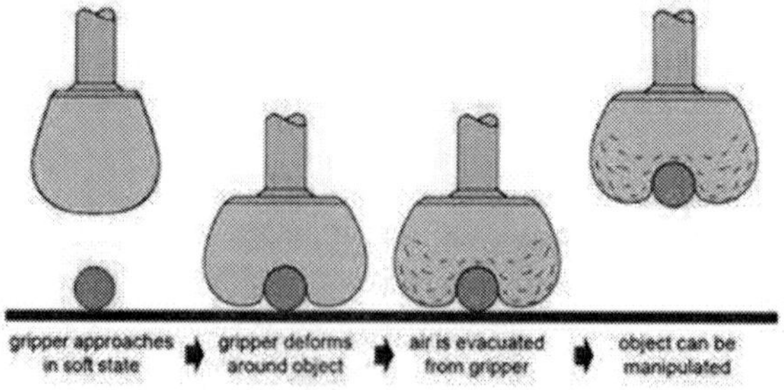

Figure 4. Form of Gripper While Handling Objects[13]

Figure 5. a) Stress at 10 N/m^2

Figure 5. b) Strain at 10 N/m^2

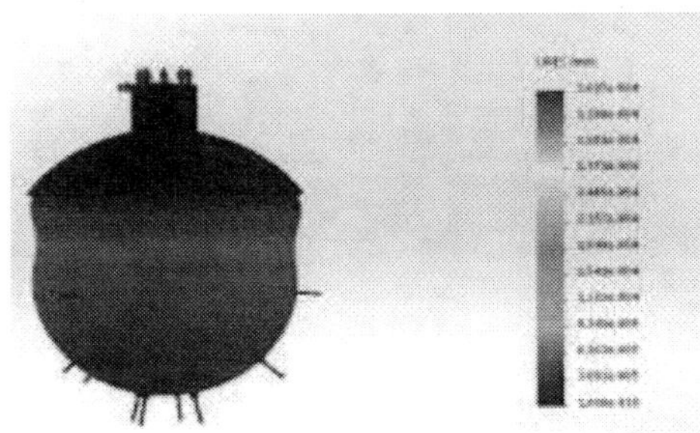

Figure 5. c) Displacement at 10 N/m^2

Figure 5. d) Displcament at 20 N/m^2

Figure 5. Simulation of Gripper

Stress, strain and displacement analysis were made to check the stability of the gripper and its deformation on different loading conditions such as 10 N/m^2, 20 N/m^2, 30 N/m^2 as shoun in Figure 5.. After the air is backed out slowly the porous material will remain trapped inside the material bag. The process is further repeated on adding coffee granules after inserting the funnel. The size of the funnel is checked periodically where the rubber bag stick with an inch of the funnel to perform the actions. If the coffee grounds in the rubber material bag is enough then the rubber material bag is removed from the tube and the coffee grounds stay inside the rubber material bag. A rough filter in the form of small piece of fabric is placed to avoid the granules from falling out. A fabric that is good is used for moving air in and out. Then inflate the rubber partially. The grains are then loosened and it gets positioned accordingly. Then gently press down the rubber material bag on the surface of the object. Now the air is sucked out from the rubber material bag to holding the object. With maximum pressing on the object will deform the rubber material bag as shown in Figure 4.

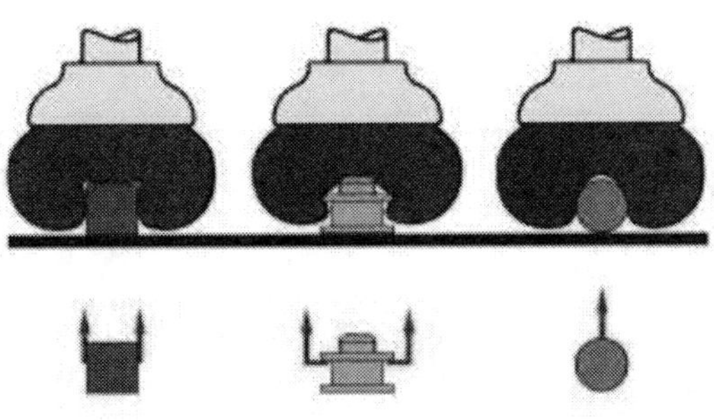

Figure 6. Gripping Mechanism[13]

The rubber material bag shrinking itself when applying the suction to hold the object. The rubber material deteriorates the object when releasing the suction. This is the principle followed in the pick and place applications as shown in Figure 6. If the vacuum is compacted, the object will be tapped firmly with the vacuum bag. The placing can be achieved through the gradual releasing of suction pressure. Then the object will be placed and detached in the gripper. If the air is blown forcefully get into the rubber material bag then the gripper will forcefully eject the object as shown in Figure. 7 (a) & (b) which is examples of gripper actions while handling water bottle and bottle cap.

Figure 7(a). Gripper handling water bottle

Figure 7(b). Gripper handling bottle cap

4. Conclusion

In this paper, the universal gripper was designed and fabricated which will be used to hold and move various materials. The vacuum induced gripper with jamming and unjamming role in the transition enables the actions of handling various geometrics of objects. It has been identified that the performance will get increased while using suitable granular and rubber materials has been chosen. Reliability also exposed by this method of gripping actions. The gripper with air locked construction provide the required potential for different environment of material handling. The future of robotics is depend on the composition of end effector with flexible handling options as well as more accessible objects. Regardless with shape, the universal gripper cab be used to hold different objects. This innovation pays the more attention in various domains such as food processing, defense industry, material handling, home and commercial places etc. The major task of the designed gripper is achieved in the grasping various objects. Thus the productivity, cost and performance of the industry is effectively decreased.

References

[1] Tahir A M, Zoppi M and Naselli G A 2018 PASCAV Gripper: a Pneumatically Actuated Soft Cubical Vacuum Gripper. In: *2018 International Conference on Reconfigurable Mechanisms and Robots (ReMAR)*: IEEE) pp 1-6

[2] Tai K, El-Sayed A-R, Shahriari M, Biglarbegian M and Mahmud S 2016 State of the art robotic grippers and applications *Robotics* **5** 11

[3] Manti M, Hassan T, Passetti G, D'Elia N, Laschi C and Cianchetti M 2015 A bioinspired soft robotic gripper for adaptable and effective grasping *Soft Robotics* **2** 107-16
[4] Zhong G, Hou Y and Dou W 2019 A soft pneumatic dexterous gripper with convertible grasping modes *International Journal of Mechanical Sciences* **153** 445-56
[5] Lin K-Y and Gupta S K 2017 Soft fingers with controllable compliance to enable realization of low cost grippers. In: *Conference on Biomimetic and Biohybrid Systems*: Springer) pp 544-50
[6] Bogue R 2016 Flexible and soft robotic grippers: the key to new markets? *Industrial Robot: An International Journal*
[7] Park W, Seo S and Bae J 2018 Development of a Hybrid Gripper with Soft Material and Rigid Structures. In: *2018 IEEE/RSJ International Conference on Intelligent Robots and Systems (IROS)*: IEEE) pp 5930-5
[8] Hao Y, Wang T, Ren Z, Gong Z, Wang H, Yang X, Guan S and Wen L 2017 Modeling and experiments of a soft robotic gripper in amphibious environments *International Journal of Advanced Robotic Systems* **14** 1729881417707148
[9] Piltan F, Sulaiman N, Soltani S, Marhaban M and Ramli R 2011 An adaptive sliding surface slope adjustment in PD sliding mode fuzzy control for robot manipulator *International Journal of Control and Automation* **4** 65-76
[10] Yang G J and Choi B W 2013 Smooth trajectory planning along Bezier curve for mobile robots with velocity constraints *International Journal of Control and Automation* **6** 225-34
[11] Fantoni G, Capiferri S and Tilli J 2014 Method for supporting the selection of robot grippers *Procedia CIRP* **21** 330-5
[12] Mehmet S G, John E and Wan Nurshazwani Wan Z 1970 Vision based Object Recognition of E-Puck Mobile Robot for Warehouse Application *International Journal of Integrated Engineering* **6**
[13] Brown E, Rodenberg N, Amend J, Mozeika A, Steltz E, Zakin M R, Lipson H and Jaeger H M 2010 Universal robotic gripper based on the jamming of granular material *Proceedings of the National Academy of Sciences* **107** 18809-14
[14] Hisyam Abdul R, Nur Amirah Mohamad H, Mohd Hafiz A J Z, Dirman H, Babul Salam Ksm Kader I, Muhamad Saiful H and Siti Nor Zawani A 2020 Force Control for One Degree of Freedom Haptic Device using PID Controller *International Journal of Integrated Engineering* **12**
[15] Amend J R, Brown E, Rodenberg N, Jaeger H M and Lipson H 2012 A positive pressure universal gripper based on the jamming of granular material *IEEE transactions on robotics* **28** 341-50
[16] Xu Q, Kan J, Chen S and Yan S 2014 Fuzzy PID based trajectory tracking control of mobile robot and its simulation in Simulink *International journal of control and automation* **7** 233-44
[17] Wiseman Y 2017 Safety mechanism for SkyTran tracks *International Journal of Control and Automation* **10** 51-60

Hydrodynamic Modeling and Design of Robotic Fish using Slender Body Theory

Palmani Duraisamy[1], **Manigandan Nagarajan Santhanakrishnan**[1*]

[1]Department of Electronics and Instrumentation Engineering, SASTRA Deemed University, Thanjavur, Tamil Nadu-613401, India.

E-mail : manigandanns@eie.sastra.edu

Abstract. Design of bio-inspired robotic fish has been an important area of research for applications such as water quality monitoring, aquatic animal behavioral study and underwater exploration. The modeling of robotic fish requires high dimensional analysis, which increases the design complexity. To address this complexity, approximate models are preferred and frequently used in the design of robotic fishes. The accuracy of these models mainly relies on various hydrodynamic parameter identification of which drag coefficient estimation is highly challenging. This paper utilizes a simple deceleration motion model to estimate the drag coefficient effectively so that the linearization error is reduced. The non-linearity in the deceleration model is simplified and mapped as a linear model to predict the drag coefficient. The estimated drag coefficient is used for dynamic modeling of robotic fish using slender body theory. The dynamic model is validated through experimental setup using accelerometer and visual sensor data. The maximum swimming speed of 0.32 ms-1 is achieved at 1.2 Hz caudal fin oscillations.

Keywords: robotic fish, dynamic modeling, deceleration model, drag estimation, speed tracking.

1. Introduction

Underwater vehicles propelled by natural methods (bioinspired) has been an interesting area of research which, provides greater performance in terms of maneuverability, stealth, noise, and energy efficiency compared to other engineered designs. Among various bioinspired designs, robotic fish is the most remarkable one to achieve agile motions. Dynamic models of robotic fish are essential to achieve desired motion response often necessary in many real-time applications. Generally, dynamic model utilizes Newtonian [1] or Lagrangian mechanics [2] and Lighthill's models [3-4] frequently for modeling the kinematics and hydrodynamics of the robotic fish. Lagrangian models require higher order dimensional analysis, which increases the system complexity. On the other hand, Newtonian models are simple, computationally efficient and frequently used for modeling a robotic fish [5]. Among different approaches of Lighthill's model, slender body theory [6] effectively models the hydrodynamics of a swimming fish, if the dimensions and oscillations of the fish are small compared to its physical length.

The parameters such as caudal fin thrust and drag coefficient are necessary for modeling a robotic fish. Without a physical model, its highly challenging to determine these parameters. The drag coefficients for objects with simple geometry can be directly calculated using numerical computations, but for arbitrary shapes it is estimated using experimental approaches [7-9]. Numerical method uses CFD like tools [10], whereas on the other hand analytical methods uses experimental design for its estimation. In model-based estimation [1,3], the robotic fish swims at a constant speed so that the thrust generated by robotic fish is exactly balanced by drag force. The overall drag coefficient is estimated using least mean square method through graphical approach, which demands high-speed motion capture camera for velocity and displacement measurement. In constant speed methods [11-15], a robotic fish is towed at different speeds using a motion control setup. The towing force is measured using force sensor and the slope of velocity versus squared speed plot gives us the drag coefficient. The requirement of suitable force sensors and motion control unit complicates the system design. Wain Leung Chan *et al* [7] presented an experiment that simultaneously estimates drag coefficient and added mass, which also requires a high-resolution optical encoder setup and timing precision unit for displacement measurement. The proposed work introduces a simple deceleration motion model of robotic fish for drag coefficient estimation. The principle behind the experiment is inverse velocity-time relation to estimate the drag coefficient to mass ratio. The least mean square fit of experimental data estimates drag coefficient to mass ratio. Next, the thrust generated by a robotic fish is modeled using slender body approach by considering caudal fin oscillations. The Newtonian dynamic model relates the thrust and drag to estimate the speed of a robotic fish at given instant. The closed form control system of robotic fish requires this speed estimation to follow the predefined trajectories to achieve the desired motion.

2. Modeling of Robotic Fish

2.1. Kinematic Model

The proposed robotic fish design is inspired from carangiform swimmers. Its oscillation is confined to rear on-third of the body to produce swimming thrust. These swimmers are efficient in terms of speed and maneuverability because of its streamlined structure [4]. The free body diagram of proposed robotic fish and its equivalent kinematic model is shown in figure 1. Oscillation of tail is modeled [3] using the expression as follows,

$$h(x,t) = x\tan(\theta(t)) \tag{1}$$

where, x is the length of the oscillating body of the robot, $\theta(t)$ is oscillation angle sinusoidally varies with respect to time.

$$\theta(t) = \theta_m \sin(2\pi f t) \tag{2}$$

where, θ_m is oscillation amplitude and f is the frequency of oscillation.
with the use of slender body theory, the thrust generated by carangiform robotic fish can be expressed as follows,

$$F_T(t) = \frac{\rho\, A(l)}{2}\left[\left(\frac{\partial h(x,t)}{\partial t}\right)^2 - v^2\left(\frac{\partial h(x,t)}{\partial x}\right)^2\right]_{x=l} \tag{3}$$

where, $F_T(t)$ is instantaneous thrust force, ρ is density of water, $A(l)$ is area of the circle calculated using overall dimension of tail as a diameter, v is horizontal speed of robotic fish along *-x* direction, and l is length of the tail. The squares of the derivative values are averaged for one cycle of oscillation and it is given by,

$$G_1(\theta_m, f) = \frac{\rho\, A(l)}{2}\,\frac{1}{T}\int_0^T \left(\frac{\partial h(x,t)}{\partial t}\right)^2 dt \tag{4}$$

$$G_2(\theta_m, f, v) = \frac{\rho\, A(l)}{2}\, v^2\, \frac{1}{T}\int_0^T \left(\frac{\partial h(x,t)}{\partial x}\right)^2 dt \tag{5}$$

Therefore, the thrust generated by the robotic fish is a function of amplitude of oscillation (θ_m) and frequency of oscillation (f).

$$F_T(t) = G_1(\theta_m, f) - G_2(\theta_m, f, v) \tag{6}$$

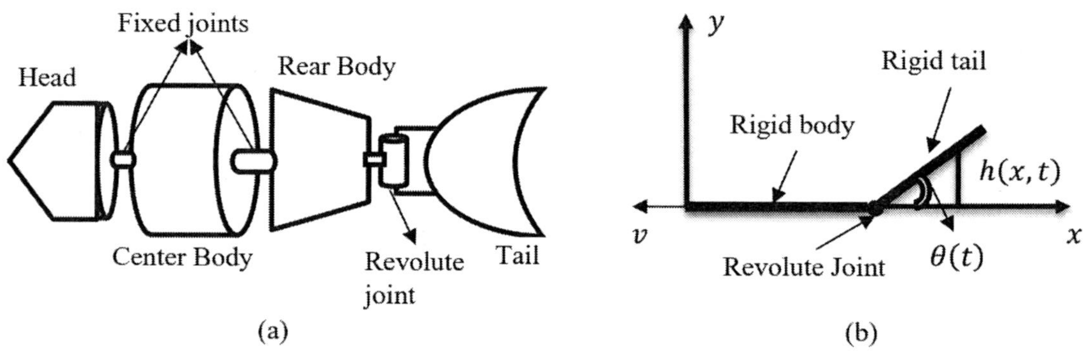

Figure 1. (a) Free body diagram of proposed robotic fish. (b) Equivalent kinematic model adapted from [3].

2.2. *Dynamic Modeling*

Dynamic modeling of robotic fish follows the Newton's law of motion,

$$m\,\dot{v} = F_T(t) - F_D(t) \tag{7}$$

where, m is the mass of the physical robotic fish, $\dot{v}$ is derivative of velocity, and $F_D(t)$ is horizontal drag force acting opposite to the thrust ($F_T(t)$). Drag force acting on the robotic fish is expressed as follows,

$$F_D(t) = \frac{1}{2}\,\rho\,A\,C_d\,v^2(t) \tag{8}$$

where, A is surface area of the robotic fish, C_d is drag coefficient of robot body and v is speed of the robotic fish.

An estimation of drag coefficient utilizes decelerated motion of a robotic fish. Under undisturbed decelerating motion, the generated thrust is zero and speed of the robotic fish decreases gradually during which at any instant the force of inertia ($m\,\dot{v}$) counter balances the body drag ($F_D(t)$) [8].

$$m\,\dot{v} = -F_D(t) = -\,\frac{1}{2}\,\rho\,A\,C_d\,v^2(t) \tag{9}$$

The non-linear differential equation (10) can be transformed into linear equation as follows,

$$\frac{\partial v(t)}{\partial x} = -k\,v^2(t) \tag{10}$$

with the drag coefficient to mass ratio,

$$k = \frac{\rho\,A\,C_d}{2m} \tag{11}$$

the equation (11) is integrated and simplified as follows,

$$x(t) = k\,t + x_0 \tag{12}$$

where, $x(t) = \frac{1}{v(t)}$ is reciprocal function of speed and $\frac{1}{x_0} = v_0$ is the speed at time t=0 (initial speed of the robotic fish). The reciprocal function is plotted using decelerated robotic fish data. The slope of the function estimates the overall drag coefficient. The drag coefficient C_d is computed from the overall drag coefficient by substituting area, body mass and density of water. The slender body theory assumes that this drag coefficient is constant throughout the motion.

3. Physical Prototype Design

Propulsion mechanism of a real fish utilizes the tail part effectively to generate required thrust for motion. The proposed design of robotic fish utilizes the same principle and generates required thrust using tail oscillations. The prototype was built with four discrete parts namely head, body, lateral body and tail to ensure the streamlined structure. Each of this part was fabricated using MakerBot 3D printer using polylactic acid (PLA) filament to provide moderate rigidness in the structure. Parts of the robotic fish are connected by fixed joints except the tail. The tail is attached by revolute joint using servomotor to produce oscillatory motion. The complete prototype design is shown in figure 2. The prototype is designed such a way that the center of mass point is fixed at the body center of the robotic fish so that its aligned with respect to the body axis horizontally. Any corrections in the center of mass was compensated by adding weights inside of the robotic parts so as to balance the buoyancy force such that more than 95% of body parts are immersed into water and hence, ensures surface swimming. A high torque dynamixel servomotor AX-12 is used to produce oscillation motion of the tail, which has an inbuilt closed loop control system to reach the desired position accurately at rated speed. Prototype is completely sealed using waterproof skin made of silicone material. The rotary motion of the servomotor was programmed using Matlab software and interfaced via CM-530 controller to produce oscillation pattern of the tail. The physical parameters of robotic fish prototype are detailed in table 1.

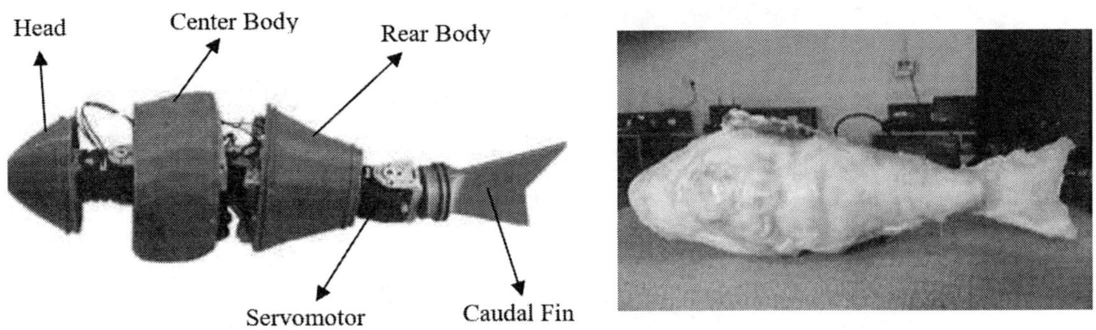

(a) (b)

Figure 2. (a) A 3D printed robotic fish prototype. (b) Water proof seal using silicone material.

Table 1. Details of physical parameters of robotic fish.

Symbols	Parameters	Value with units
L	Length of the robotic fish	0.59 m
l	Length of the tail	0.16 m
C	Span of the tail	0.096 m
A	Surface area	0.153 m^2
m	Mass	5.5 kg
ρ	Density of fluid	998.2071 kg m^{-3}

4. Experimental Setup

The robotic fish swimming trials were conducted in a circular tank of 1.8 m diameter. An overhead camera is fixed above the circular tank and an inertial measurement unit (IMU) sensor is fixed on the head of the robotic fish. The accelerometer data from the IMU sensor is recorded and is used for calculating speed of the swimming robotic fish. The overhead camera captures the motion of robotic fish at a speed of 25 frames per second. The decelerated motion of robotic fish is generated using a setup built inhouse as shown in figure 3. Robotic fish is towed using a string connected to servomotor. The servomotor is excited by Lego Mindstorms EV3 controller which is programmed to generate variable acceleration, so that it can produce constant speed during initial periods. After achieving constant speed of the robotic fish, the servomotor is stopped, so that the deceleration motion can be established. The decelerated robotic fish motion is captured using overhead camera for different initial speeds. The Kinovea software is used for calculating the speed of the robotic fish at different instants of time. The experimental trial of cruise swimming motion is shown in figure 4.

Figure 3. A setup to generate constant speed for decelerated motion capture of robotic fish.

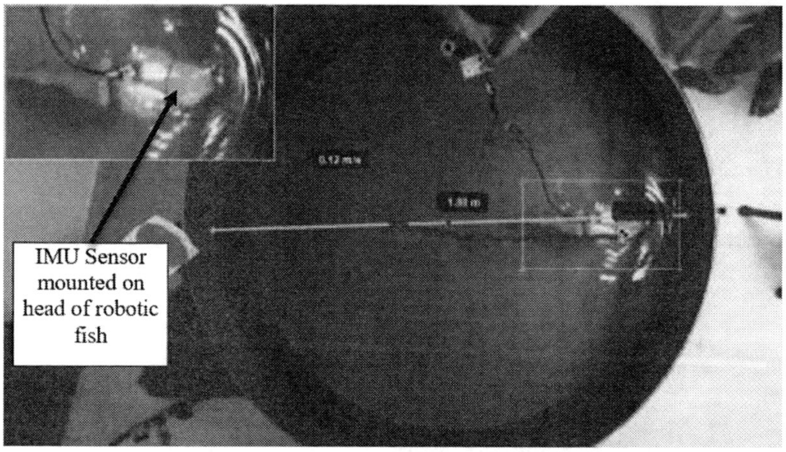

Figure 4. Top left: swimming robotic fish integrated with IMU sensor, center: experimental trial in a circular tank of 1.8 m diameter, where the red line shows trajectory of swimming plotted using Kinovea software.

5. Results and Discussion

The speed of decelerated robotic fish motion is mapped to reciprocal function. The resultant reciprocal speed data is plotted and the slope of this line is estimated using least mean square method as shown in figure 5. Using this slope (k), the drag coefficient (C_d) is computed as 3.1032 using equation (11). Next, a simulation is performed to calculate speed of the robotic fish using the estimated drag coefficient. The swimming trials of robotic fish was conducted at different frequencies of oscillations. The speed of the robotic fish is tracked using overhead camera and accelerometer data from IMU sensor. Speed of robotic fish is plotted based on the simulation data, camera, and accelerometer data as shown in figure 6. The results of speed profile for different frequencies also produces similar relation with simulation and experiments. Due to the precise estimation of drag coefficient, it is seen that the real-time swimming speed of the robotic fish closely follows simulation results.

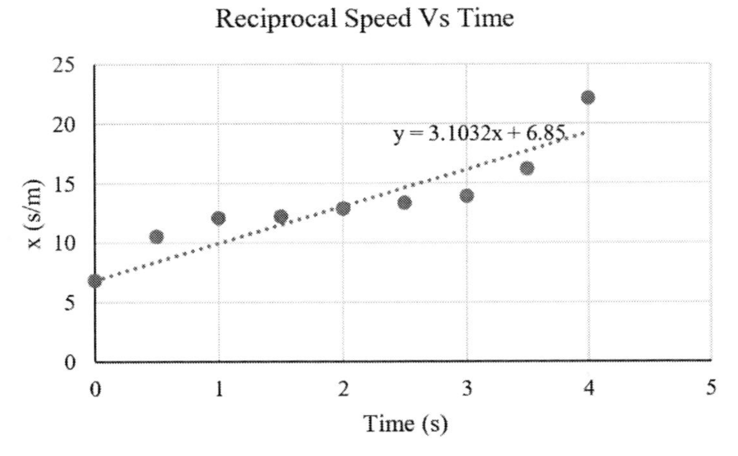

Figure 5. Least square fit of reciprocal velocity versus time plot using decelerated robotic fish motion data.

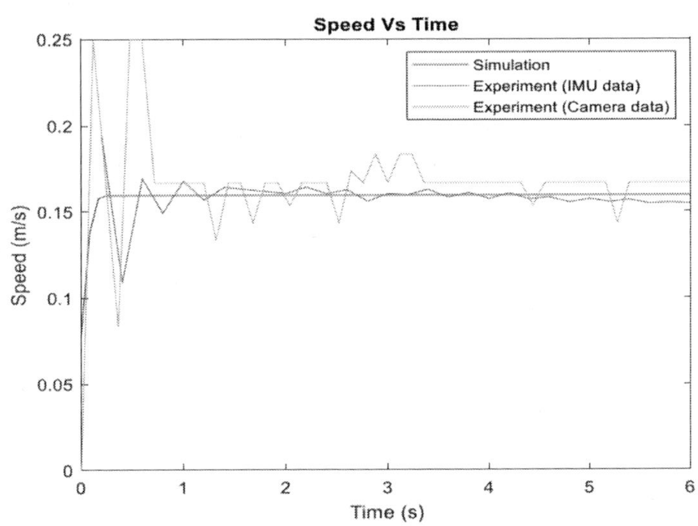

Figure 6. Speed profile of robotic fish using simulated dynamic model and its experimental validation for a tail oscillation frequency of 1.2 Hz.

6. Conclusion and Future Scope

The model-based validation of speed profile for a robotic fish is accomplished in this work. An efficient estimation of drag coefficient was established by using deceleration motion model. Proposed method estimated the drag coefficient, which is closer to the experimental value, hence the accuracy of the dynamic model improved significantly. The dynamic model of the robotic fish was validated through simulation and experiments using camera and IMU sensor. In the future works, speed profile of proposed robotic fish can be utilized for the design of closed loop controllers in order to improve the speed tracking performance. This work may also be extended for two-dimensional maneuvering applications.

References

[1] Saurab Verma and Jian-Xin Xu 2018 Analytic modelling for precise speed tracking of multilink robotic fish IEEE Transactions on Industrial Electronics 65 5665–72
[2] Ozmen Koca G, Bal C, Korkmaz D, Bingol M C, Ay M, Akpolat Z H and Yetkin S 2018 Three-dimensional modeling of a robotic fish based on real carp locomotion Appl. Sci. 8 180
[3] Li X, Ren Q and Xu J X 2016 Precise Speed tracking control of a robotic fish via iterative learning control IEEE Transactions on Industrial Electronics 63 2221–28
[4] McMasters RL, Grey CP, Sollock JM, Mukherjee R, Benard A and Diaz AR 2008 Comparing the mathematical models of lighthill to the performance of a biomimetic fish Bioinsp. Biomim. 3, 247–69
[5] Palmani Duraisamy, Rakesh Kumar Sidharthan and Manigandan Nagarajan Santhanakrishnan 2019 Design, modeling, and control of biomimetic fish robot: a review J Bionic Eng. 16 967–93.
[6] Lighthill M J 1960 Note on the swimming of slender fish Journal of Fluid Mechanics 9 305
[7] Wai Leung Chan and Taesam Kang 2011 Simultaneous determination of drag coefficient and added mass IEEE Journal of Oceanic Engineering 36 422–30
[8] Bilo D and Nachtigall W 1980 A simple method to determine drag coefficients in aquatic animals Journal of Experimental Biology 87 357–59
[9] Wai Leung Chan, Taesam Kang, Young Jae Lee, Sang Kyung Sung and Kwang Joon Yoon 2007 Swimming study on an ostraciiform fish robot International Conference on Control, Automation and Systems (Seoul) (South Korea/IEEE) 700–05
[10] Amit Tyagi and Debabrata Sen 2006 Calculation of transverse hydrodynamic coefficients using computational fluid dynamic approach Ocean. Eng. 33 789–809
[11] Li Wen, Guanhao Wu, Jianhong Liang and Jinlan Li 2010 Hydrodynamic experimental investigation on efficient swimming of robotic fish using self-propelled method Int. J. Offshore Polar Eng. 20 167–174
[12] Li Wen, Tianmiao Wang, Guanhao Wu, Jianhong Liang and Chaolei Wang 2012 Novel method for the modeling and control investigation of efficient swimming for robotic fish IEEE Trans. Ind. Electron. 59 3176–88
[13] Li Wen, Tianmiao Wang, Guanhao Wu and Jianhong Liang 2013 Quantitative thrust efficiency of a self-propulsive robotic fish: Experimental method and hydrodynamic investigation IEEE/ASME Trans.Mechatronics 18 1027–38
[14] Saurab Verma and Jian-Xin Xu 2017 Data-assisted modeling and speed control of a robotic fish IEEE Trans. Ind. Electron. 64 4150–57
[15] AC Fernandes and FPS Mineiro 2007 Assessment of hydrodynamic properties of bodies with complex shapes J. Appl. Ocean Res. 29 155–66

Acknowledgement

The authors would like to thank Dr. S. Rakesh Kumar, Assistant Professor, SASTRA Deemed University for his kind and helpful suggestions on our experimental apparatus and implementations.

IOP Conf. Series: Materials Science and Engineering 1012 (2021) 012008 doi:10.1088/1757-899X/1012/1/012008

Line following robots on factory floors: Significance and Simulation study using CoppeliaSim

Saharsh Oswal[1*] and Saravanakumar D[1]

[1]School of Mechanical Engineering, Vellore Institute of Technology, Chennai, India

E-mail:saharshrajoswal@gmail.com

Abstract. Line following robots has been around for many years now. Research and development of these robots have been keenly studied and examined for many industrial applications. Line following robots can provide maximum effectiveness to in-house transportation of jobs inside a factory, which is traditionally done by forklifts and different types of cranes. These traditional methods are high maintenance and extremely unsafe for the employees. This paper presents the design, assembly and dynamic simulation of a line following robot integrated with a proximity sensor for collision avoidance and vision sensors for line tracking. The model is designed using Autodesk Inventor 2018 and the assembly and simulation is carried out using CoppeliaSim software. This simulation study intends to provide significance of the implementation of such robots on a factory floor, which can cover a path of distance 10 meters in approximately 8 seconds.

1. Introduction

Logistics and supply chain management has become a crucial part of the entire manufacturing operational timeline. But, we often tend to forget the logistics dealt within the factory where products are manufactured, tested and packaged. Traditionally, jib cranes, conveyor belts, overhead cranes and forklifts are used to transfer material or products between different production and machining operations. Most of the aforementioned methods are regulated and inspected manually. Currently, with Industry 4.0 at its peak, autonomous technologies are used in every aspect of product development, logistics and management. In modern manufacturing, the ultimate aim is to reduce the cycle time and enable the possibility of manufacturing more complex and customizable products [1]. Along with these parameters, the operational efficiency has to not be compromised so as to ensure minimal financial loss and ill-usage of raw material & energy.

Most defects in manufacturing are caused due to processing errors carried out by individuals who fail to follow operating schedules and instructions, which generally happen when companies fail to fix an established set of rules or "kaizens". Most manufacturing processes consist of more than one operations in the assembly line which makes the transfer of unfinished jobs difficult, complex and even unsafe due to high temperatures and heavy payloads. Owing to the concerns and risks arising due to manual errors and delay between two or more operations, the methodology of "lights-out manufacturing" has been entrenched.

Although now, most production processes are being carried out autonomously using CNC operated machines, SCADA controlled systems using G-codes and PLC programmed panels. The performance gets compromised when one specific job has to be transferred between two or more manufacturing and machining processes. This in-house multi-variable logistic problem has given a kick-start to autonomous mobile robots which help in picking and delivering jobs on factory floors. Conventionally, manually operated cranes (jib or overhead), conveyors or forklifts or even combination of these machines are used for the transit of jobs on a factory floor. The biggest concern while embedding automation in manufacturing is the risk of quality control and the pressure on the equipment. Not only do these run operational risks, but also leads to complicated labor management, schedule delays and increased accidents which also threatens safety issues of the employees.

The basic requisites in the motion planning of mobile robots are path plans, localization and collision avoidance [2]. The path planning tends to get very complicated in a dynamic or static environments such as markets or shopping malls, and also automated factories and warehouses, where autonomous mobile robots can be conveniently used for in-house logistics. The navigation and localization of the auto-vehicle have to be pre-determined, regulated and tested to ensure minimal errors. We are well aware that some robots can perform tasks such as milling, welding and turning with higher precision and speed as compared to humans. Also, robots can withstand high temperatures and survive in conditions where humans are unable to. Similarly, mobile robots have proven to be far more efficient than human-operated cranes and forklifts so as provide lesser time-delays and high safety measures.

Line following robots have been a part of many industrial and commercial applications for a long time. The idea of using this kind of an approach for motional dynamics is due to the simple construction and programming. These robots can be configured conveniently for any tracks inside a factory, commercial space and even hospitals [3]. The usage of line following robots on factory floors is an integral part of automating the material handling and unfinished goods. The hardware of the robot consists of the robot body, motors, sensors and control boards while the software includes the program code, controller design, sensor parameters and method design [4].

The low-cost wheeled mobile robot (as shown in figure 1) which is proposed in this paper serves as a good example on which manufacturers can understand the significance of autonomous path following robots in factories.

Figure 1. Mobile robot for factory floors

Line following robots can be used in various environments such as hospitals, commercial spaces and also, factory floors. These robots can be conveniently used to substitute manual workforce involving the transportation of goods in factories and warehouses. Optimizing manufacturing process has been the core requirement with the increase in population, demand and customizable priorities [5].

The objective of this paper is to provide a comprehensive as well as a simulation study on a state of art mobile robot which tracks and follows a pre-defined path so as to provide transit amidst the walls and machinery on the factory floor. It intends to give aid in the better perception of the path planning robot technologies and also guide manufacturers to combine and optimize techniques for better line following executions in both single and multi-robot environments.

2. Limitations

Logistics Implementation of any techniques in a conventional space takes time to develop and get accustomed to. Even though implementing line following robots in factories is beneficial in safety, productivity and finances, there are several challenges to be faced.

- Skill gap: The employees at most factories are used to the forklift and crane system of transportation of "work-in-progress' items. Implementing such robots can cause a gap in pre-existing skills, which might take some time to settle in.
- Factory plan adjustments: The Manufacturing Resource Plan (MRP) has to be altered integrating the line following action of the robot. Besides, the layout has to be changed throughout the enterprise so as to promote data standardization and interoperability.
- Disruptive technology: Most blue collars have found comfort in using forklifts and cranes for in-house logistics, changing the mindset and familiarizing the workers with this technology can be a difficult task. Not only humans but even other machines might also have to be customized in accordance with the line following tracks.

3. Robot Design

The robot used for the simulation is a differential drive robot, which is basically a mobile robot with wheels placed on both sides of the main base. The movement of the entire robot body is based on these two separately placed wheels. In addition to these wheels, a castor wheel is added to the center-back of the robot body for static and dynamic balance, which orients itself in accordance with the other two wheels [6]. The robot steers itself autonomously with the help of necessary sensors and motors. The parts are modelled individually in Autodesk Inventor 2018.

The main base of the robot is a tray-type design so as to provide better assistance to carry unfinished jobs from one machine to another. Many items in a factory are either heated to high temperatures, or too fragile or maybe even radioactive. Hence this robot offers assistance and control over most logistical features on the factory floor, which cannot be carried out by humans.

The dimensions of this robot are assigned in accordance with the average payload for most components manufactured (as shown in figure 2 and 3). The differential wheels are set as 200mm in diameter while the castor wheel is set as 80mm in diameter. The robot is designed with a base of 1000mm by 750mm. The total height of the model base is 600mm, which gives it an appropriate structure to travel over a line in a dynamic environment.

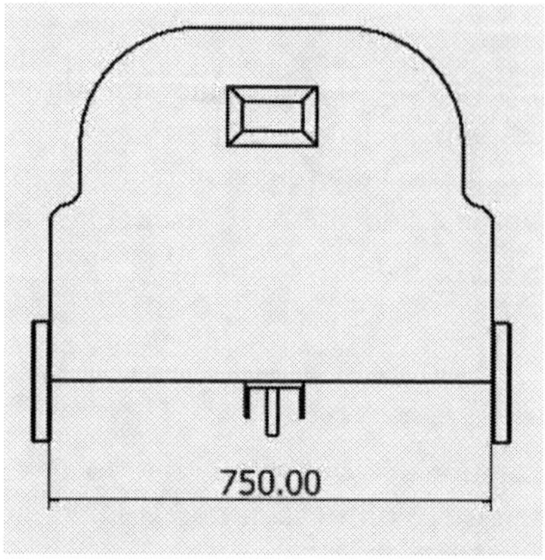

Figure 2. Front view of the robot

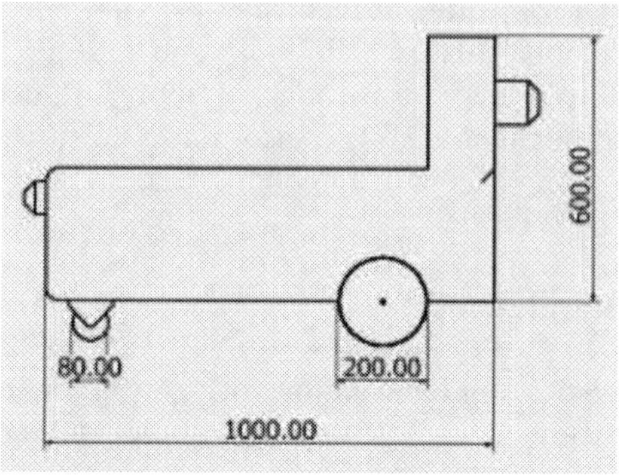

Figure 3. Side view of the robot

The dynamic assembly of this robot is done in CoppeliaSim Edu as it allows us to add the necessary joints, sensors and the necessary scaling factors (as shown in figure 4). The same software is also used for the simulation of the line following action in a factory environment, to be demonstrated later in the paper. CoppeliaSim is a simulator software specifically for robotics and automated systems, in which, every object of the system can be individually controlled via an embedded script written in many programming languages. The script for this robot is written in Lua programming language.

Figure 4. Assembly in CoppeliaSim for dynamic simulation.

CoppeliaSim allows importing the '.stl' files to its scene and orient all parts as per the robot design. All parts are converted to convex shapes and made invisible using the layer properties. The entire robot is set as collidable, measurable, detectable and renderable. Revolute joints are added to the side wheels, castor holder and the wheel, with the motors, enabled only for the differential wheels. A cone-type proximity sensor and three orthographic vision sensors are added to the front and base of the robot respectively. A non-threaded child script (program) is linked to the model base of the robot body.

After the assembly is completed, the objects are arranged in a particular hierarchy for the simulation (as shown in figure 5). Only the imported parts are visible and the convex shapes and sensors are set to invisible using the layer option, as they are used for only dynamic simulation.

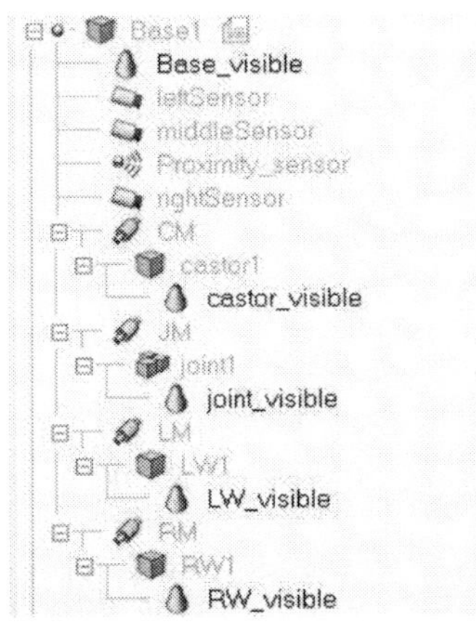

Figure 5. Hierarchy of robot objects. The sensors and motors have been set to invisible.

The mass of the entire robot body is set as 218.3kgs which is optimum for a line following mobile robot carrying a heavy payload. The corresponding principle moments of inertia in the x, y and z-direction are formulated as shown in table 1. Lesser the moment of inertia, lesser the force required for a body to perform the non-translational motion, which works in favor of a line following mobile robot.

Table 1. Principle moments of inertia of the robot

Direction	Moment of Inertia (kg m^2)
I_{xx}	19.288
I_{yy}	17.217
I_{zz}	9.439

4. Program Instructions

CoppeliaSim allows the dynamic assembly of the robot integrating features such as sensors, joints, motors and even embedded scripts so as to link the code to the robot. Other robots simulating software such as Webots, Gazebo and RobotStudio do not allow engineers for remote monitoring or safety double-checking. CoppeliaSim supports a number of programming languages such as C/C++, Python, Java, Lua, Matlab or Urbi and also allow integrating robots designed by other third-party companies [7]. This simulator not only facilitates the design of a factory environment and the mobile robot but also permits the program to execute in accordance to the robot features.

The code for the robot is written in Lua programming language. Lua is a powerful, fast, lightweight language designed to support procedural programming and it is embedded in CoppeliaSim software. A non-threaded child script is used here which follows a precise execution order as specified by the user. The code consists of three main parts:

- Initialization: Here the objects are retrieved individually using 'sim.getObjectHandle' and values are assigned to certain variables such as the speed of the robot. The User Interface is also designed here using 'simUI.create', which in this case consisted of options to stop the robot, resume the robot, decrease and also increase the speed of the robot [8].
- Actuation: This part of the code deals with the execution and implementation of various sensors and motors initialized before. The varying speeds for the wheels are computed as per the path detected. On basis of the principles of the line following robot, the code is generated. The parameters for the sensors and motors are specified and called.

- Other functions: These are basically the functions associated with the UI initialized before.

5. Sensors

The sensors play the most vital role in the implementation of the line following action of the robot. A single cone-shaped proximity sensor and 3 different vision sensors are used for this particular simulation. The proximity (ultrasonic) sensor allows the robot not to collide with an obstacle in case the path has been blocked or obstructed. The vision sensors are used to detect and differentiate the color of the path from the surroundings [9]. Moreover, as sensors are usually light-weight and reliable, consequently not causing any alterations to the weight, inertia and torque of the robot.

The proximity sensor is set as invisible for the simulation. The radius of the cone is 5 meters and the range is set to 1 meter, with an angle of 90 degrees (as shown in figure 6). For the collision simulation, all objects have to be set to detectable, so that they are on par with this ultrasonic sensor. If an object is detected by this sensor, both the left and right wheels move backwards at different speeds, hence in a curve. This helps the robot to re-trace itself onto the path.

An ultrasonic sensor can be used as a proximity sensor which can regulate the distance of an obstacle by emitting ultrasonic waves. It converts the reflect sound wave into an electric signal and this time is used to calculate the distance between the robot and the object. The sensor has two main parts, namely the transmitter and the receiver [10]. The transmitter emits the sound wave and the receiver is where the sound waves travel back to. These sensors can easily detect the surface of an obstacle and perform collision avoidance algorithm in order to facilitate maximum safety in the factory environment.

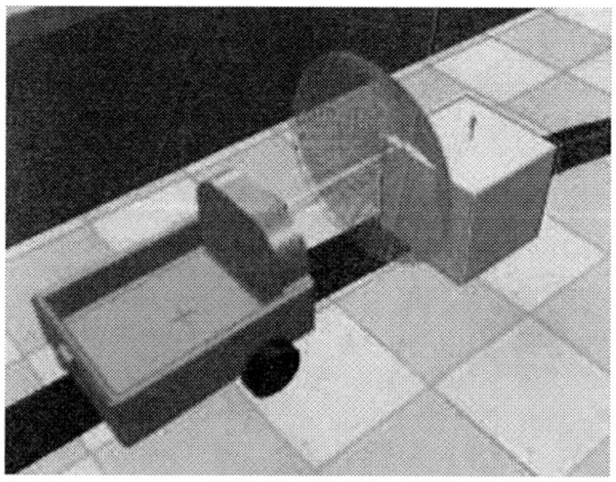

Figure 6. Proximity sensor detecting obstacle

The vision sensors (as shown in figure 7) are the most important components of the robot architecture as they help with the detection of the path, and in turn, allow the robot to move forward. All three sensors act as cameras placed 200 mm apart from each other. The orthographic nature with the resolution of 1x1 helps to detect the color beneath. The far clipping plane is 0.15 meters which associates itself with the code guiding this robot to move straight, turn right or left accordingly [11].

Vision sensors are used in this application just as a camera so as to return 1x1 images of the surface underneath the robot. The line to be followed has to be segmented from the factory floor using a set of three such sensors, working simultaneously. Infrared sensors can be used to transmit and receive IR rays which are in-turn collected by the photodiodes based on the Planck's radiation law, Stephan Boltzmann law, and Wien's Displacement law [12]. As IR sensors reflect back when a white surface is detected, it provides the necessary inputs needed for the line following action to take place. These sensors are used to inspect the track to be followed using a combination of both image acquisition and image processing.

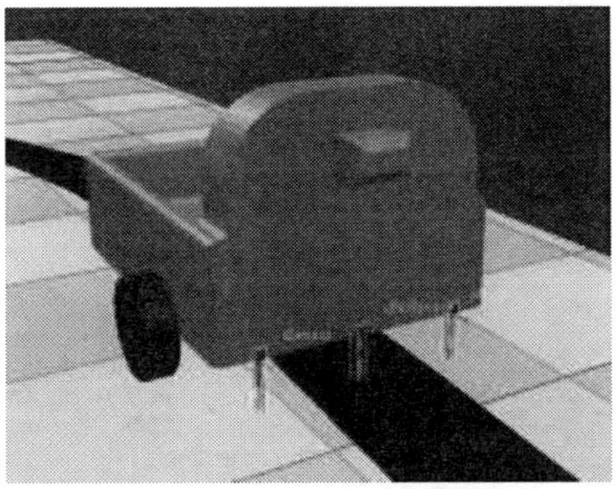

Figure 7. Left, middle and right vision sensors

The vision sensors are placed 20 cm apart from each other keeping the line following operation under consideration. The track is 25cm in width which allows the middle sensor to almost always stay with the edges of the track. As soon as the left or right sensor detects a black image i.e. the line underneath, the middle sensor will detect the factory floor. This allows the steering action to take place almost instantly when the middle IR sensor detects a non-black image [13]. Most line following operations have a narrower track to be followed, but as this takes place inside a factory, maximum safety and minimum delay has to be ensured. These configurations allow the robot to follow the line at sharp turns and even U-turns as the steering delay is almost negligible.

The combination of these two types of sensors not only help in avoiding collision with factory equipment but also navigate the robot throughout the track. These sensors detect the necessary information and corresponding processors convert and analyses the raw data into valuable information which in turn, guides the mechanical action of the robot. The combination of this signal processing technology and dynamic configuration of motors enables the robot to manoeuvre over a pre-defined line in a factory environment.

6. Simulation and Results

The simulation for this robot is carried out using CoppeliaSim. The software allows us to check the dynamic properties of the robot and assign different actuation codes to every component. The basic principles used for this simulation:

- If the middle vision sensor detects a black image, both the wheels move forward with the same speed (v).
- If the left sensor detects a black image, the left wheel moves with a decreased speed (0.2v), hence steering the robot to the left.
- Similarly, if the right sensor detects a black image, the right wheel moves with a decreased speed (0.2v), hence steering the robot to the right.

Therefore, this enables the robot to take travel over a certain "track", turning left or right and even take U-turns. The vision sensors allow the robot to travel by differentiating the line color and its surroundings. This difference between black and non-black pixels is detected and processed to accede the motion of the robot.

For the simulation, a factory environment is created (as shown in figure 8) consisting of a black track for the robot to trace and follow, a stacking and storage area and grey blocks to portray machinery. A closed path is created with a horizontal segment including both left and right turns. A floating view (a display screen) is also added and associated with the middle visual sensor of the robot, so as to check the image being detected.

The main base of the robot is a tray-type design so as to provide better assistance to carry unfinished jobs from one machine to another. Many items in a factory are either heated to high temperatures, or too fragile or maybe even radioactive. Hence this robot offers assistance and control over most logistical features on the factory floor, which cannot be carried out by humans.

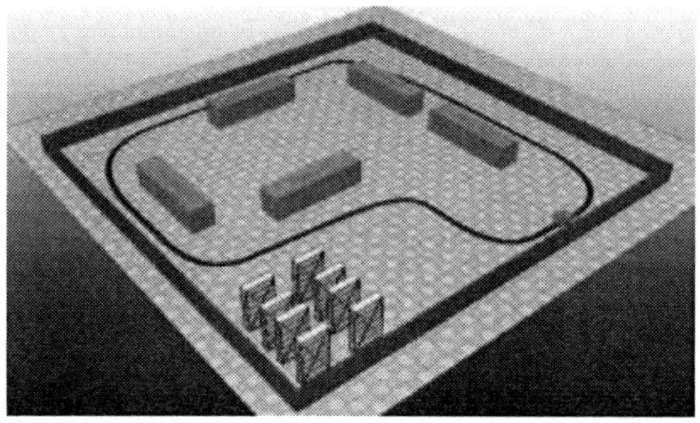

Figure 8. Factory scene set in CoppeliaSim

The robot moves along the track designed on the factory floor with varying speed due to the curves and turns. The linear and angular velocity varies in accordance with the path being followed. All sensors and motors work simultaneously which enables the robot to navigate itself and travel efficiently in the environment.

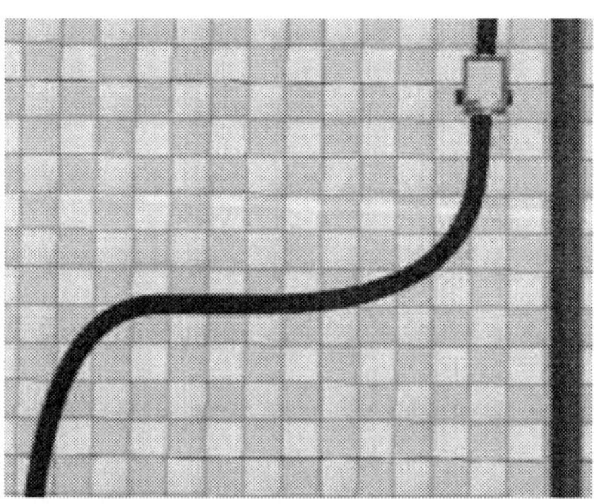

Figure 9. Top view of path to be taken

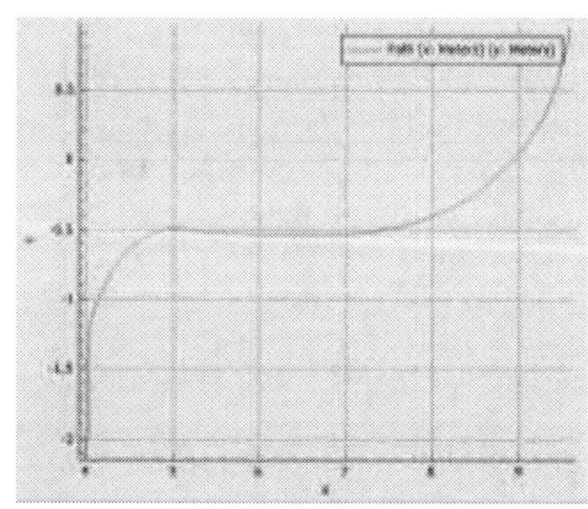

Figure 10. Path curve formulated on an X-Y graph

An X-Y graph is added to trace the path being taken by the robot. The X-axis indicate the absolute x-positions of the robot model base and similarly the Y-axis indicate the absolute y-positions of the robot model base. The points generated are linked so as to form a curve on the graph. This graph determines the path being followed by the robot in real-time and helps in the monitoring and verification of the line following action. Figures 9 and 11 are top views of the pre-defined line on the factory floor, the robot has to detect the line and follow, once the code is run. The sensors detect the line and enable the motors to move along the line as shown in the graphs (figure 10 and 12).

For the first case (figure 9), the speed of the robot is set as high as 1.5 m/s, taking a total time of 75 seconds, covering a path 90 meters in length. The line followed has 2 turns, one towards the right and another to the left. The velocity varies all throughout, due to the proximity between the middle sensor and the line to be detected underneath. As the robot reaches the first turn, the right sensor detects a black image which executes the actuation code so as to steer the robot to the left by decreasing the speed of the right differential wheel. Correspondingly, during the next turn, the left sensor detects a black image, hence the left wheel decreases it speed to obtain a left steering motion. As no obstacles are detected by the embedded proximity sensor, no disruption is anticipated in the kinematic motion. The simulation runs as per the code provided in the child script. The robot can be stopped at any time using the user interface, and the speed can also be increased or decreased so as to support the payload.

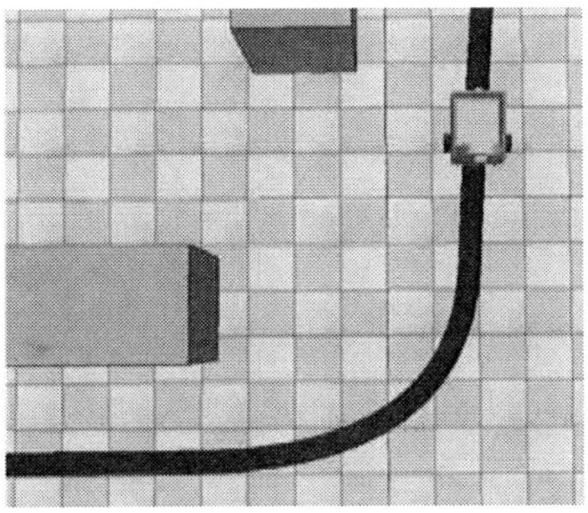

Figure 11. Top view of path to be taken

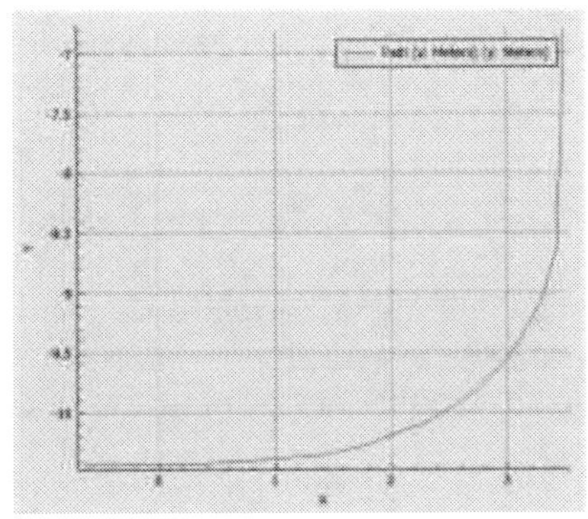

Figure 12. Path curve formulated on an X-Y graph

In the second case (figure 11), with the same speed, this robot covers a path of 80 meters in 62 seconds. As noticed, this path consists of a steep turn to the left, which corresponds to a black image being detected by the left sensor, leading to a decrease in speed in the left wheel, causing a left steering motion. Even though the robot passes along different machinery, it is at a safe distance from any damage or else the proximity sensor would detect and source a feedback motion. This path takes less time than the path in figure 9 as it consists of only one turn. The resultant speed during a turn is considerably less in case of a differential drive line following robot. On taking the average for the entire path in the environment, this mobile robot manoeuvres over a pre-defined path with an average of 1.25 meters per second. The decrease in speed is due to the varying speeds of specific motors at the turns in the designed path.

The main aim of this simulation is to facilitate the robot to travel over the pre-defined path on the factory floor. The model base moves as per the data collected from the sensors and the actuation followed by the corresponding motors. This paper signifies the possibility of line following robots on a factory floor by assembling a state of the art robot with necessary components and a code which helps the simulation to succeed.

7. Conclusion and Scope

The advancement in manufacturing technology and automation has set its mark in the era of Industry 4.0. This paper describes all the instructions and steps to be followed to design, assemble and simulate a line following robot in a factory environment. On simulating, the robot travels over the pre-defined path which promises to be beneficial in most industrial scenarios. This method of transportation is far more efficient and superior than traditional methods. It requires minimal maintenance, almost no workforce and no hazards to safety and security.

Forklifts require fuel and tank replacements, which increases maintenance costs and duties. Cranes take up space in a manufacturing layout and overhead cranes are extremely unsafe and heavy-duty.

Using line following robots overcomes all these issues and delivers promising results for the transportation of "work-in-progress" goods on factory floors.

These robots can be an essential aspect to optimize a factory operation and the manufacturing quality can be enhanced to a greater level. However, safety regulations regarding the same are extremely necessary and should be evaluated keeping all constraints in mind. According to the US Department of Labor, there were only 38 robot-related accidents in the United States between 1987 and 2016 [14]. Despite the possibility of these accidents is low, safety measures should be taken for the protection of the employees and the machinery involved. Laser scanners can also be used to guarantee collision avoidance between the robots and workers on the factory floor. All mobile robots should be assessed individually and all dangers should be checked regularly.

The sensors and actuators need multiple tests and re-planning which ensure the efficiency of the robot to perform the operation [15]. The simulation study ensures that this following robot with differential wheels, a castor joint, a proximity sensor and 3 vision sensors can operate over a globalized path in a factory environment. Considering suitable turns and shifts in the path, the robot if found to cover a distance of 10 meters in 8 seconds, when the speed is set as 1.5 m/s.

Optimizing a facility's configuration increases the manufacturing outcome, quality and the supply output. Automated machinery, Implementation of Industrial IoT and cloud-based design and the usage of autonomous robots can contribute to an enterprise's journey towards quality assurance, vigorous inspection and extensive production. The assistance provided by these robots can help manufacturers to contrive optimized factory and assembly layouts, with incremented operational efficiency.

References

[1] "How replacing forklifts with AMRs increases safety", Mobile Industrial Robots (https://www.mobile-industrial-robots.com/en/insights/amr-safety/how-replacing-forklifts-with-amrs-increases-safety/)

[2] S. Sedhumadhavan, E. Niranjana, "An Analysis of Path Planning for Autonomous Motorized Robots", International Journal of Advance Research, Ideas and Innovations in Tech6nology (Page 1234-1257, Volume 3, Issue 6)

[3] Beatriz Arruda Asfora, "Embedded Computer Vision System Applied to a Four-Legged Line Follower Robot", 23rd ABCM International Congress of Mechanical Engineering, December 6-11, 2015

[4] Shomitro Kumar Ghosh, Toheen Bhuiyan, Raihan Chowdhury, Ismail Jabiullah, Daffodil International University, "A Line Follower Robot with Obstacle Detection by Ultrasonic", A Steganographic Apps-based Patient"s Information Encryption-Decryption.

[5] Buniyamin N., Wan Ngah W.A.J., Sariff N., Mohamad Z, "A Simple Local Path Planning Algorithm for Autonomous Mobile Robots", International Journal of Systems Applications, Engineering & Development (Page 151-159, Issue 2, Volume 5, 2011)

[6] S Samuel Abhishek, K.Veladri, "Trajectory Planning of a Mobile Robot", National Conference on Technological Advancements in Mechanical Engineering (Page 296-302)

[7] "List of Robotic Simulation Softwares", Ricky Ruijiao Li, University of Essex (https://sites.google.com/site/ruijiaoli/resources/roboticssimulationsoftwarelist)

[8] Regular API function list by CoppeliaSim (https://www.coppeliarobotics.com/helpFiles/en/apiFunctionListAlphabetical.htm)

[9] "Line following robot using Arduino", Saddam, Circuit Digest. (https://circuitdigest.com/microcontroller-projects/line-follower-robot-using-arduino)

[10] "Sensors", Fierce Electronics (https://www.fierceelectronics.com/sensors/what-ultrasonic-sensor)

[11] M. S. Islam & M. A. Rahman, "Design and Fabrication of Line Follower Robot", Asian Journal of Applied Science and Engineering, Page 27-32, Volume 2, No 2 (2013)
[12] Jagruti Chaudhari, Asmita Desai, S Gavarskar, MGM's Jawaharlal Nehru Engineering College, "Line Following Robot Using Arduino for Hospitals", 2nd International Conference on Intelligent Communication and Computational Techniques (ICCT), September 2019.
[13] "Vision sensors and their modern uses", AZO Sensors (https://www.azosensors.com/article.aspx?ArticleID=1129)
[14] "The Importance of Robot Safety on Your Factory Floor", Genesis Systems, IPG Photonics (https://www.genesis-systems.com/blog/importance-robot-safety-factory-floor)
[15] Yitao Ding and Ulrike Thomas "Collision Avoidance with Proximity Servoing for Redundant Serial Robot Manipulators" IEEE Systems Journal PP(99): 1-11, September 2018

Comparative Need Analysis of Industrial Robot Calibration Methodologies

Sumedh N Chiwande[1*], Shantipal S Ohol[1]

[1]Department of Mechanical Engineering, College of Engineering Pune, Pune, Maharashtra, India

Email:chiwandesn13.mechphd@coep.ac.in

Abstract-The paper focuses on a study of literature about available Industrial robot calibration technique, technical relevance of its utility and summaries findings & reasons behind specific features of each type of method. Comparative analysis helps to identify and assess the present knowledge to identify the future research areas for developing the research interest on the particular needs. The study focuses on advantages and disadvantages of different measuring systems used for calibration. It is observed from literature survey that the research focus about the calibration technique has moved towards the micro level classification of an industrial robot calibration process. The use of micro analysis, benchmarking and standardization of a calibration process requires an additional research. Therefore various recent contributions on industrial robot calibration techniques have been analysed for establishing a commonly acceptable methodology.

Keywords: Industrial robot, repeatability accuracy, robot calibration.

1. Introduction

An industrial robot is a robot system used for production, manufacturing and service industries. Industrial robots are programmable, automated and able to move on three or more axes in space. There are many applications of industrial robots with repetitive actions, which include pick and place, manufacturing goods inspection, welding, and painting, labelling, and testing. With these repetitive requirements robots need to be verified and calibrated with a certain interval of time through standard verification systems.

There is a direct impact of the effectiveness of a robot when it is calibrated. Repeatability is one of the main measurable characteristics in terms of evaluated path, position and orientation called as pose. These are the parameters to be ensured for achieving every time the robot has to be set up and the motion of the end effector.

1.1 Technical Parameters Associated with Calibrations

Performance parameters are effectively considered while planning calibrating a robot. Repeatability is most significant amongst all and hence it is referred to as priority. It is the ability of a robot to visit the said position again and again through a taught path [1]. The second and the most significant parameter is robot's accuracy, attributed to its links, inertial join compliance, on the other hand impact of robot

performance parameter, interaction of parameters and their responses are still a source of statistical investigation [2].
All the robot manufacturers have been mentioned in their information site about the robot specification data, but the robot user does not know the optimal conditions under which it has to be operated and are acceptable. For example, the specification mentioned on industrial robot ABB 1520ID is listed as ±0.35mm repeatability, 1000mm/s speed and 4kg payload [3]. However through this information it is not understood by the user about the mentioned repeatability at the known speed and stated payload capacity. With such information in industrial robot process parameters, few or no standard is present in any robot manufacturing industries. So it is not clear for the robot user to understand the effects or the significance of the parameters in some reference is unimportant. Therefore it is very much important to examine the interface among the various robot parameters, and determine optimum conditions under which a given mix of values can be obtained. All these optimal parameters considerations lead to improved design and deployment of industrial robots.
However, there is an unavailability of guidelines or standards to develop the in-house calibration procedures and to use specific meteorological gauges [4].

2. Industrial Robot

In the actual robot system, the robot is relatively difficult to program. Not only it is difficult to determine how the robot is to perform correctly its function, but also it is difficult to get the robot to achieve a desired location. Following are the performance parameters of Industrial Robot.

1. Degrees of freedom – DoF is equal to the number of axes a robot has.
2. Working envelope – Space or volume where a robot can reach.
3. Kinematics –Determines robot's possible motions.
4. Payload – Weight carrying capacity of a robot.
5. Speed – Ability of a robot to position the end effector in space.
6. Acceleration – Capacity of the arm to accelerate quickly.
7. Accuracy – It is degree of closeness of the actual value to the set value or target value.
8. Repeatability –Capacity of a robot to visit a taught position again and again.

As a result, robots are used in applications which reduce the amount of programming effort. For example, they are used in applications which have small part counts. Because the programs are hard to change, the robot is used for relatively medium production quantities. In this way one programme can be used for more profit. Basically, the reason robots do not achieve desired positions is because they are operated in an open loop manner. It is difficult to measure the true position of the robot tool, but it is easy to measure positions of joint actuators.

2.1 INDUSTRIAL NORMS FOR ROBOT CALIBRATION

The ISO-9283 standards describe the detailed methods and verifying the following performance parameters of an industrial robot [4]:

1. Pose accuracy and repeatability: The differences which occur between a command and attained pose, and the variation in the attained pose for repetitive visits to a command pose.
2. Multi-directional pose accuracy variation: The deviation between the different mean attained poses achieved when visiting the same command pose n times from three orthogonal directions.
3. Distance accuracy and Repeatability: The variations in the distances for the repetitive movements between the two poses.
4. Position stabilization time: The robot performance which can measure how fast a robot can stop at the taught pose.
5. Position overshoot: At the attained poses it makes smooth and accurate stops.

6. Drift of pose characteristics: Variations at pose accuracy over the precise time interval.
7. Exchangeability: The deviation of the barycenter when different robots of the same type are exchanged under the same environmental conditions, mechanical mounting and the use of the same task programme.
8. Path accuracy and path repeatability: Independent of the shape of the command path.
9. Path accuracy on reorientation: To record the influence of three-directional orientation alterations on a linear path in a simple way.
10. Cornering deviations: The maximum deviation from the taught path after the robot edge point of the corner.
11. Path velocity characteristics: it is a measure as the performance characteristics of an industrial robot with respect to the path velocity.
12. Minimum posing time: The time taken for a robot to stabilise at the attained pose.
13. Static compliance: The maximum amount of displacement per unit of applied load.
14. Weaving deviations: A combination of one or motions superimposed on a path.

However the robot metrology has been studied for more than twenty years. Since then a basic theoretical knowledge has same. Latest technology builds better robots having better repeatability with minimal source of errors. Meteorological instruments are also getting better because of micro technology and are less expensive. Therefore most of the small and medium scale industries started adopting the ISO-9283 standard. However, the position performance of a robots remain a single measure specifies as "position performance and related test methods according to ISO-9283". As mentions in this standards, the average unidirectional position pose accuracy and pose repeatability at five different poses which is obtained from thirty number of measurement cycle according to ISO-9283: 1998 [4].

3. Recent Observations for Industrial Robot Calibration

Performance parameters for industrial robots have been developed in various research centres across the world with respect to the heavy industrial needs. However, there is no universally common platform that has been adopted. Researchers are still developing the commonly accepted procedure and instrumentation for examining the repeatability of industrial robots.

In response to the researcher in the field of robot metrological methods are still improving the repeatability and accuracy results. A measurement method in robot metrology ranges from Dial indicator, Theodolite, laser tracker, Two dimensional machine vision, camera calibration, portable coordinate measuring machines and others.

Table No. 1 Comparative Analysis of Research in The Field of Robot Calibration

Sr. No.	Author, Country &Year	Journal	Objective	Methodology
1	G. Boye A .Olabia, J. Palosa, France, 2017[5]	Procedia Manufacturing 11 (2017) 132 – 140	Calibration method based on Tool Centre Point (TCP) is applied in this research paper to a collaborative robot. The author used the hand guiding element is applied to a setup a human robot interaction for find the responses.	A systematic methodology followed by descriptive analysis is applied to a standard sphere, and then compare all robot's position and orientation using the single reference.
2	M. Alia,	Procedia Computer	Self calibration technique of an articulated industrial robot has	The research methodology is adopted in the present study to

	MSimic, France, 2017 [6]	Science 112 (2017) 1601–1610	been reviewed in this paper and proposed the original method, which help to reduce the production, maintenance down times & cost.	implement using the ultrasonic triangulation based, online autonomous calibration for industrial robots.
3	Z Usman,U K,2016[7]	IJM&QE 2016, 204-214	Research has been started to examine the impact of robot movements on the measurements of robot accuracy.	To provide a methodology to divide experiments for investigating and validating using the various geometric and surface parameters of holes.
4	S. Makris, G.Chryssolouris Greece,2014[8]	RCIM 32 (2015) 65–71.	In this paper, the author proposed an algorithm to define path which leads to desire pose of an end effector of industrial robot manipulator.	Model algorithm parameters and grid parameters are verifying and validate to reduced the process time for better understanding of desired pose of a robot manipulator.
5	M. Svaco, Croatia,2014 [9]	Procedia Engineering 69 (2014) 459 – 463.	A stereovision system is mounting on an industrial robot end effector for calibrating point in space using a non-contact method.	The D-H parameters are used to modify model parameters which contain joint encoder offset values that are used as a input to the robot controller.
6	Albert Nubiola, Canada, 2014 [10]	Precision Engineering 38 (2014) 472–480.	This system comprises a telescoping ball-bar and two custom-made fixtures for the 6D measurement.	A novel 6-D measurement system is proposed in this paper, which comprises of a single telescoping ball bar. Two fixtures, one at robot base and the other at the end effector are attached respectively.
7	Yongjin (James) Kwon, Hong S Korea, 2014 [11]	Robotics and CIM 30 (2014) 451–459	This study emphasizes on two critical process parameters, process capability & accuracy of vision calibration is discussed in this article from the remote site of a robot.	The vision guided robot system is directly proportional to the process capability of a robot system. In normal condition higher the precision assembly higher will be the positioning accuracies but there is a imperfect mathematical conversion of coordinate.
8	G Du, P Zhang China,2013[12]	Robotics & CIM 29 (2013)484–492	In this method the author has demonstrated the camera is used for a calibration board which is attached at the end effector of a robot from where the camera can see it.	The author reviewed this technique, to notice the corners with reference to the images of the calibration board are discussed. The position and orientation of the robot of the robot can be estimated from the identified corners. When a robot grabs the objects with different poses, there is a change in kinematic parameters.

9	J Yin, Y Gao China,2012[13]	Energy Procedia 14 (2011) 977-982	The major contribution of this research paper is the utilization of measurement with respect to the robot end effector to compensate for errors.	The deviations in the parameters of the industrial robot have been corrected with the help of nominal parameters. This will lead to the development of better corrective measures in calibration.
10	G. Duelen, K. Schroer (1991) [14]	RCIM 08 (1991) 223-231	The authors are mention here about the absolute Cartesian pose accuracy with same magnitude as robot's repeatability. It is directly benefited for calibrating the robot.	In this methodology the calibration procedure determine kinematic and gear parameters along with the static elasticity effects. Few suggestions have been noted for the robot design which leads to maintain the better repeatability of the industrial robot.
11	G. Legnani, M Tiboni (2014) [15]	Mechanism & Machine Theory,73 (2014)25-48	Wire sensors are used for measuring robot pose system. Effect of some parameters on measuring system on robot calibration has been studies in this article.	A low cost measurement system and the study of performance parameters with relevant outcome is discussed. Also a simulation tools, and the measurement system is used to optimise the experimental phase.
12	Y.Meng, (2007) [16]	RCIM, 23 (2007)436-446	Low cost calibration apparatus with serial robot manipulator is used in this research paper.	The major advantage of this methodology is that using the image sequence of a calibrating object & robot base length that are required.
13	Y. Zeng, W. Tian, W. Liao (2016) [17]	RCIM, 42(2016) 113-120	In this article the authors proposed the method for the without modifying, classification or modification of kinematic parameters	Error compensation methods were applied for positional error of robot's TCP. Kringing method applied to determine positional errors of industrial robot TCP.
14	G. Alici, B. Shirinzadeh [18]	Mech & m/c Theory,40 (2005)879-906	A systematic approach is discussed to estimate the robot manipulating Cartesian positioning errors using analytical functions.	Errors are identified and corrected using first order inverse kinematic model. A laser based dynamic system is used for generating positioned data.
15	G. Du, P. Zhang (2015) [19]	RCIM,31 (2015)91-100	Online robot programming with the help of hybrid sensor for the measurement of position and orientation.	Inertia measuring unit and position sensor are used in an online robot self-calibration method.
16	H. Zhuang, J. Yan, (1998) [20]	JRS (1998) Issue 7	A Stewart platform is used to find out the accuracy of a robot. Experimental studies validated the usefulness of the method.	The methods are effective for kinematic calibration of Stewart platforms. This approached may be applied to different types of parallel manipulator.
17	I.Park,B. Lee, (2012) [21]	IEEE/ASME ToM (2012) 1059-1067	The authors proposed the technique to identify entire kinematic parameters errors of industrial robot manipulator.	The proposed technique is verified with seven degree of freedom humanoid manipulator arm by simulation with experimental verification.
18	G.Du, H. Shao (2016) [22]	RCIM (2016)39-48	Online self-calibration method is discussed in this article.	GOOGLE GRB3016 robot used to verify and validate a methods for getting convenience, precision and efficiency.

19	O. Barawid, K.Ishii (2008) [23]	IFAC Proceeding Volumes (2008)1596-1601	A calibration method has been developed to identify the exact pose of the two-dimensional laser scanner, mounted on robot's head.	The system is runs on the autonomous guidance with the developed calibration setup to measure the accuracy improvement.

4. Comparative Analysis of Available Robot Calibration Methodologies

The accuracy of the measurements made by 3D using collaborative robot systems is dependent on tool centre point and robot [5].Setup is expensive, process is time consuming and it is a challenge for robot calibration, due to expensive and complexity in the calibration process.

In this article the self calibration procedure integrated with following four steps, need of data collection, estimation of joints poses, identification of kinematic parameters and assessment robot accuracy [6].It is observed that the manipulator's working conditions are comparatively easy. However, it is costly and very much complex. The proposed method in this article paper is more accurate and required less time.
The study of an industrial robot for consider to measuring various geometric and surface parameters of holes has been discussed in this paper [7].

As summarised in [8], the algorithm develop has moves slowly towards the required configuration once identified and evaluate a number of industrial robot configurations.
To minimise the computational time a robot grid configuration has to reduce the search pace. This study focuses on an experienced robot programmer to generate automatically a robot desired path from intermediate point to the goal position.

The study outlines the problem of reaching the taught position, from the robot base the first three joints or an axis if considered, the best results are obtained. Robots absolute positioning errors are measured by considering spatial coordinate of the sphere centre keeping at different pose of industrial robot configurations. The maximum positioning error in the proximity of calibration points was decreased to 35% after calibration procedure [9]. This helps in identifying the optimising procedure of offset values that has minimizing the errors.

The mean absolute positioning error improved by 54.8% when it is validated and verified with the help of the FARO laser tracker and nearly 10000 robot configurations. This calibration process required relatively less workspace, it is manual, less skill required but it takes less time to perform. Laser tracker produces significantly better results.

5. Recent Findings in Robot Calibration

The existing review studies published papers and articles till to date. It provides a good initiative for the researchers with limitations.
Facts observed concerning the robot industry are

a) The review articles from the last thirty years were limited, and the studies are limited.
b) Standardised Industrial robot calibration system are not provided by ISO standard.
c) Improvement strategies for repeatability in all the research papers are limited.

In the present study, it has been observed in a literature review is to analyze and evaluate a present situation and the research inclination in the specialised area of industrial robots calibration methods.
The reference papers have been analysed to bridge a research gap, which raises the questions are
1. Should it focus on fundamental work aiming at significance of robot geometric errors compared to non-geometric errors?

2. Should standard calibration quality be related to the repeatability of the calibration instrumentation and the method of calibration?
3. What would be the optimal ways in which robot measurement can be selected?
4. How is observability of robot kinematic error parameters related to the selection of calibration configurations and method?

Requirements for possibilities of practical implementation of uniform calibration system:
1. Robot calibration improvisation for faster, efficient & inexpensive calibration is needed.
2. The calibration load needs to be decided either by manufacturer or by end user.
3. Designing a robot to accommodate on-line calibration capability.
4. Use of advanced technology to make robot calibration more simple and user friendly.

With regards to the implementation issues, the "fast and cheap" guideline automatically rules out expensive instrumentation such as Laser Tracker system, Camera Calibration, Coordinate Measuring Machine or Theodolite. From the calibration cost viewpoint the use of above instrumentation is an extremely costly and time-consuming process.

6. A Robot Calibration System

There are many factors to consider during the calibration solutions. For the actual industrial robot, measurements are necessary to determine the best fit set of parameters. Tools positions with respect to the joint positions are required for measurement. Since the fit is as good as the original data collections. That is the significance of getting the valid set of data.

It is well known that the data used in calibration require sufficient joint exercise. This means the joints must move significantly. These must include repeatability, setup-time, type of measurement, contact vs. noncontact, required software, and controller interface and cost. However, set-up time and cost is the major decision driver. Many calibration systems are having high cost and less significance in the robotic field, many production industries and researchers still not used these concepts.

Industry trends currently heading towards macro to micro analysis. Depending upon the application, i.e. for many large scale applications such as pick and place, welding operation etc this approach has worked well. But at the same time it's costlier for smaller applications.

6.1 Brief discussion on Various Calibration Measuring Techniques

Currently industrial robots are still practice programming using teach pendant. Human operators are required to guide the robot to the desired locations. The motions are recorded in the controller and that can be edited easily if required, within the robotic language and played back to perform the repetitive tasks. The robot calibration phase is the most significant step for any industrial robot.

Most of the calibration measurement instruments are available such as dial indicator, external ball bar system, camera calibration, laser tracker systems, theodolite, or coordinate measuring machines are too slow or overly expensive or both.

A flowchart as shown in Fig. 1, is simply a graphical representation of steps for selection of calibration methods. It shows steps in sequential order and is used in presenting the flow processes selection. Cameras calibration and camera vision systems have a standard automation component.

IOP Conf. Series: Materials Science and Engineering 1012 (2021) 012009 doi:10.1088/1757-899X/1012/1/012009

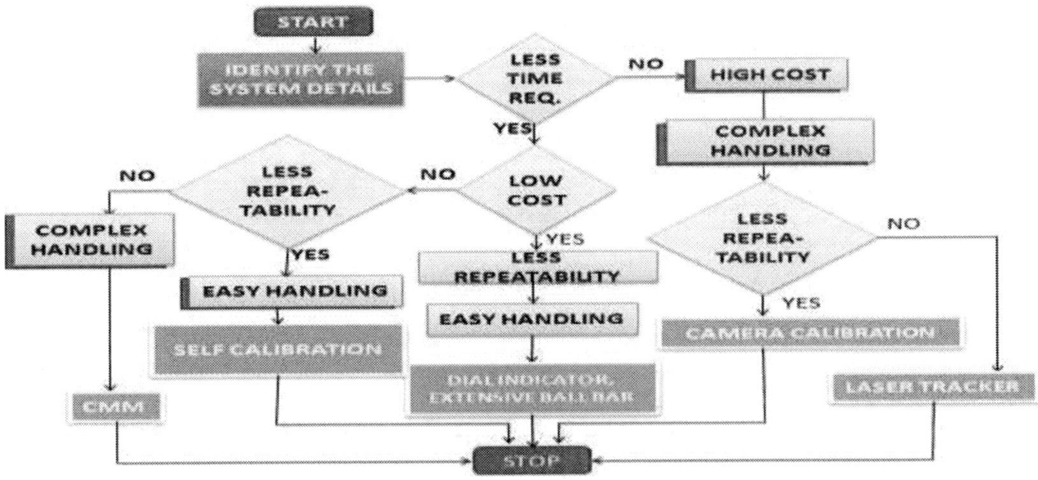

Figure 1. Flow Chart for Selection of Calibration Method

Integrated multi-camera systems of a robot cell are used for inspection, part identification, and real time inspection of assembly accuracy.

Few cameras may fix in the robot workspace area and the others may fix to the moving end effector to assist the online component alignment. This system requires additional hardware and the additional software.

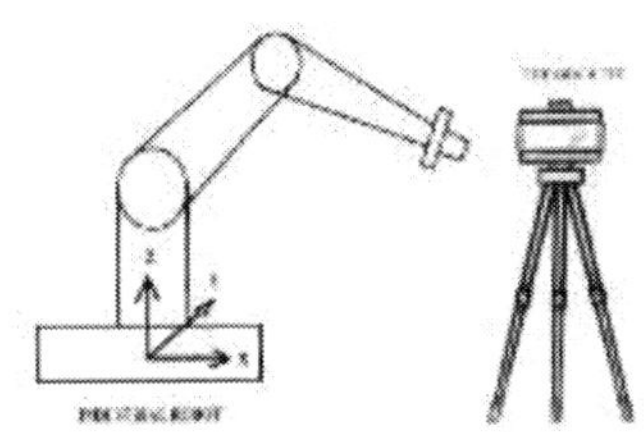

2.a Theodolite measurement setup

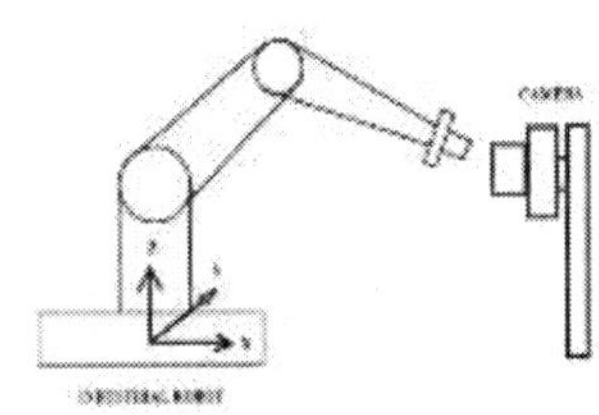

2.b Camera Calibration setup

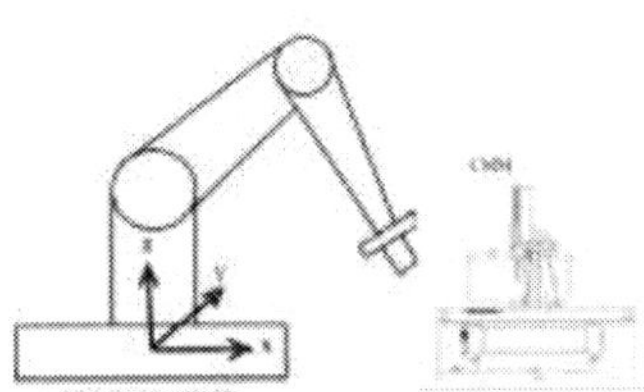

2.c CMM Calibration setup

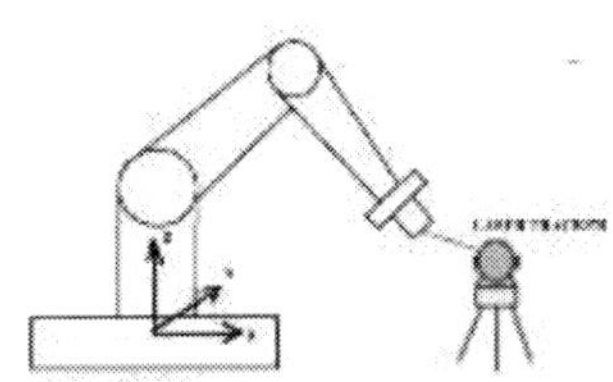

2.d Laser Tracker Setup

Figure 2. Commonly Used Robot Calibration Methods

In calibration of robot, a theodolite is used for automatic partial pose measurement system. It required a rotational platform measurement from the camera system for finding out the spherical illuminate target. After calibration, the system can continuously track the target with sufficient speed, and accuracy. Since this approach requires dedicated system hardware and only a partial pose can be measured.

A frame is mounted on tool coordinator at the end effector centre. After driving each axis to its reach limit, the Coordinate Measuring Machine (CMM) position indicator has set to zero. As CMM is used for pose measurement in industrial robot calibration.

Key equipment for calibration is measurement systems. Measurement equipment and calibration systems must be appropriately matched for reliable identification of model parameters. The above table 1 presents measurement systems. Non-contact measurement systems with respect to their use in robot calibration.

Static measurement based on theodolites and triangulation principles. Another way is dynamic approach which is based on a laser tracking interferometer used to measure distance and the direction to determine position and orientation.

7. Conclusions

After critically analysing all facts related to existing robot calibration methods, their comparative analysis, making a ready reference flowchart for selection of suitable method following conclusions have been summarised as below:

1. This study highlights the importance of low cost industrial robot calibration that can be used as a check on robot manufacturing quality and maintenance needs. The calibration procedure must be industry usable. This means that the procedures should be automated. The goal is to develop low cost equipment that can easily be used on the shop floor for incidental and periodic robot calibration.

2. Based on the review of reference article, it is observed that the measuring system is having an important issue for finding a robot's process parameters.

3. It is observed that some proposed standards for better understanding of testing and various specification of a robot process performance with reference to the industry. Testing should be carried out on various point to point and continuous path in the work space.

4. Finally the data produced from a calibration procedure is useful not only for improving robot accuracy, but also for analysing the quality in which a robot, or several robots of a single type, is manufactured. Because the calibration data is very extensive, a system is being developed to manage and analyse them based on the needs of the calibration procedure and the needs for quality control.

References

[1] B.W. Mooring, Z.S. Roth, R. D. Morris, Fundamentals of Manipulator Calibration, Wiley, New York, (1991), 301-305
[2] B. Greenway. Tutorial robot accuracy, Industrial Robot: An International Journal, 27(4) (2000), 257- 265.
[3] ABB ROBOTICS, Product specification – IRB1520 Robot documentation M2012, rev J, RW5.13, (2012), pp11-15, 12-17.
[4] ISO 9283 Manipulating Industrial Robots-Performance criteria and Related Test Methods International Standards Organization (1998).
[5] G. Boye, A. Olabia, J. Palosa, 3D metrology using a collaborative robot with a laser triangulation Sensor. Procedia Manufacturing 11, (2017), 132-140.
[6] M. Alia, M. Simic, F. Imad. Calibration method for articulated industrial robots. Procedia Computer Science 112, (2017), 1601–1610.
[7] Z. Usman, P. Radmeh, An investigation of highly accurate and precise robotic hole measurements using non-contact devices. International Journal of Metrology and Quality engineering 2016, 204-214.

[8] K. Kaltsoukalas, S. Makris, G.Chryssolouris. On generating the motion of industrial robot manipulators. Robotics and Computer-Integrated Manufacturing 32(2015) 65–71.
[9] M. Svaco, B. Sekorana, F. Suligoj, B. Jerbic. Calibration of an Industrial Robot using a Stereo Vision System. Procedia Engineering 69 (2014) 459 – 463.
[10] A. Nubiola,A.Ilian. Absolute robot calibration with a single telescoping ballbar. Precision Engineering 38 (2014) 472–480.
[11] Y. Kwon, J. Hong. Integrated remote control of the process capability and the accuracy of vision calibration. Robotics & CIM 30 (2014) 451–459.
[12] G. Du, P. Zhang. Online robot calibration based on vision measurement. Robotics and Computer-Integrated Mfg 29 (2013) 484 – 492.
[13] J. Yin, Y. Gao. Pose Accuracy Calibration of a Serial Five DOF Robot. Energy Procedia 14 (2011) 977-982.
[14] Z. Chen, L. Zhuo, K. Sun, C. Zhang. Extrinsic Calibration of a camera and a Laser Range finder using point to line constraint. Procedia Engineering 29 (2011) 4348-4352.
[15] G. Duelen, K Schroer. Robot calibration—Method and results. RCIM 8(1991)223-231
[16] G. Legnani, M Tiboni. Optimal design and application of a low-cost wire-sensor system for the kinematic calibration of industrial manipulators. Mechanism and Machine Theory 73 (2014) 25-48
[17] Y. Meng, H Zhuang. Autonomous robot calibration using vision technology. Robotics and Computer-Integrated Manufacturing 23 (2007) 436-446
[18] Y. Zeng, W. Tian, W. Liao. Positional error similarity analysis for error compensation of industrial robots. Robotics and Computer-Integrated Manufacturing 42(2016)113-120
[19] G. Alici, B. Shirinzadeh. A systematic technique to estimate positioning errors for robot accuracy improvement using laser interferometry based sensing. Mechanism and Machine Theory 40(2005) 879-906
[20] G. Du, P. Zhang. Online robot calibration based on hybrid sensors using Kalman Filters. Robotics and Computer-Integrated Manufacturing 31(2015) 91-100
[21] H. Zhuang, J. Yan. Calibration of Stewart platforms and other parallel manipulators by minimizing inverse kinematic residuals. Journal of Robotic Systems 15(1998)7
[22] I. Park, B. Lee. Laser-Based Kinematic Calibration of Robot Manipulator Using Differential Kinematics. IEEE/ASME Transactions on Mechatronics 17 (2012) 1059-1067
[23] G. Du, H. Shao. An online method for serial robot self-calibration with CMAC and UKF. Robotics and Computer-Integrated Manufacturing 42 (2016)39-48
[24] O. Barawid, K. Ishii.Calibration Method for 2-Dimensional Laser Scanner Attached on a Robot Vehicle. Proceeding IFAC (2008)1596-1601

Acknowledgments

The authors wish to express sincere gratitude and support from Dr B B Ahuja, Director, College of Engineering Pune, Pune.

Industrial robot performance analysis using low cost set- up

K Chachane[1*], S Ohol[2] and S Chiwande[2]

[1]Department of production Engineering and Industrial Management, College of Engineering pune, Pune, Maharashtra, India

[2]Department of Mechanical Engineering, College of Engineering Pune, Pune, Maharashtra, India

Email: kbchachane@gmail.com

Abstract— Industrial robot calibration is technique used to increase industrial robot repeatability and accuracy. In our project, we have analyzed various calibration techniques of articulated industrial robots and proposed an original method to analyze Industrial robot performance analysis. The proposed calibration set-up technique is using a mechanical setup of Linear Variable differential Transformer (LVDT) probe and inductive proximity sensor. The Statistical process control (SPC) software is using to monitor the real time data of the process and extracting the data for statistical analysis. The proposed Industrial robot calibration technique is applying at ABB Six Degrees of Freedom industrial robot and analyzing the performance gets affected by various Robot parameter and Internal/External factors. The calibration method is proposed to evaluate the repeatability of an articulated industrial robot using component of a Statistical Analysis course.

Keywords: Robot calibration, Robot repeatability, parameter identification, Statistical Analysis.

1. Introduction

Industrial robot calibration is the method of calibrating Industrial robot to achieve better accuracy and repeatability during its task. Industrial robot calibration is an important and very first task we used to do while installation of Industrial robot. In various Automation industries, to reduce maintenance cost and shut down time, Calibration technique comes into the picture. In the Industrial and Automation industries, various Calibration methods are present with various different and unique techniques. So depend upon the requirement of various Industrial robot performances and application, one choose calibration method to calibrate the industrial robot to achieve desired output. In our proposed calibration set-up, we have analyzed the change in repeatability with various industrial robot factors and variables. To analyze how robot speed affecting the robot accuracy and repeatability. We have used ABB IRB 1520ID robot. We have performed experimentation at two different robot speed i.e. 150mm/s and 100mm/s with 30 no. of repetitions.

2. Robot Calibration

Industrial robot calibration is the method of calibrating Industrial robot to achieve better accuracy and repeatability during its task.

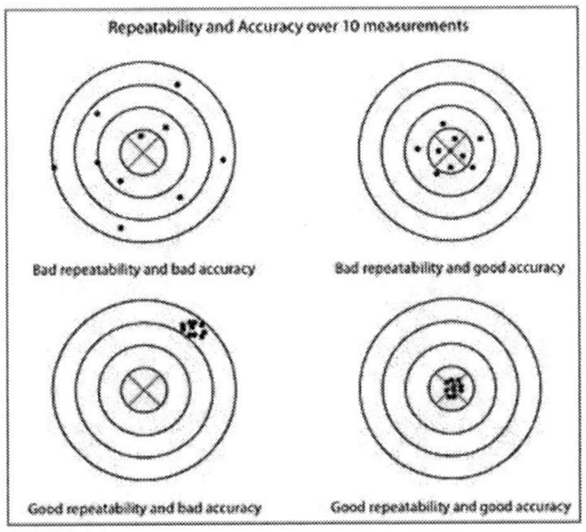

Figure1. Repeatability and accuracy over 10 measurements

3. Methodology

The low cost calibration set-up method has the following method of the project, once the experimental design has setup. The robotics programming at different angular/linear position, speed and zone will be done for no. of execution of a cycle. For each cycle, the real time data is being monitored on DAT-SPC in graphical charts form and after each cycle the data gets collected in excel format for statistical analysis. By applying the statistical process control method, the repeatability of the robot will be calculated.

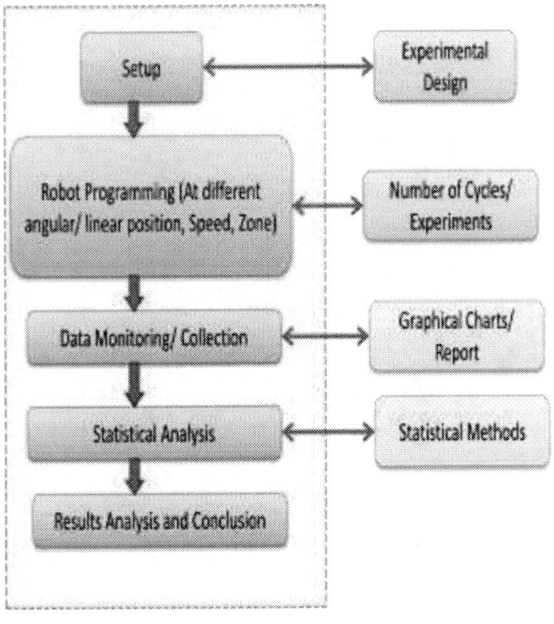

Figure2. Step by step execution of the task for project work

IOP Conf. Series: Materials Science and Engineering 1012 (2021) 012010 doi:10.1088/1757-899X/1012/1/012010

4. Experimental Setup

The Robot programming was run thirty times (n=30) for each experimentation trials of controllable external/internal factors.

- For Robot calibration testing, this is the mechanical arrangement consisting of 5 LVDTs such as 2 in X-direction (X+ X-), 2 in Y- direction (Y+ Y-) and one in Z-direction.
- Rectangular shaped prism tool connected to the end-effectors of robot.
- Inductive proximity sensor is used to count the number of cycles.

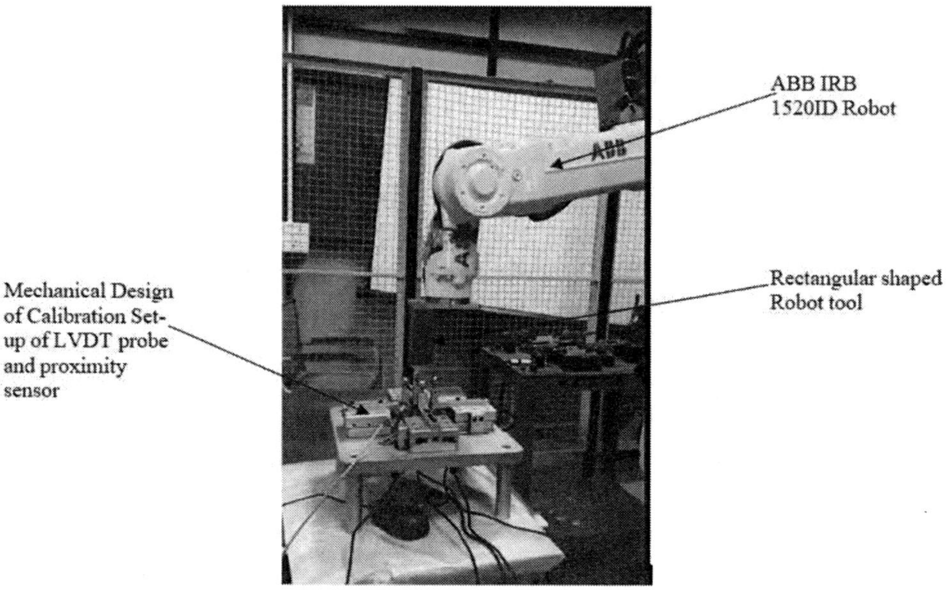

Figure3. Experimental Setup of ABB IRB 1520ID Robot and Calibration setup

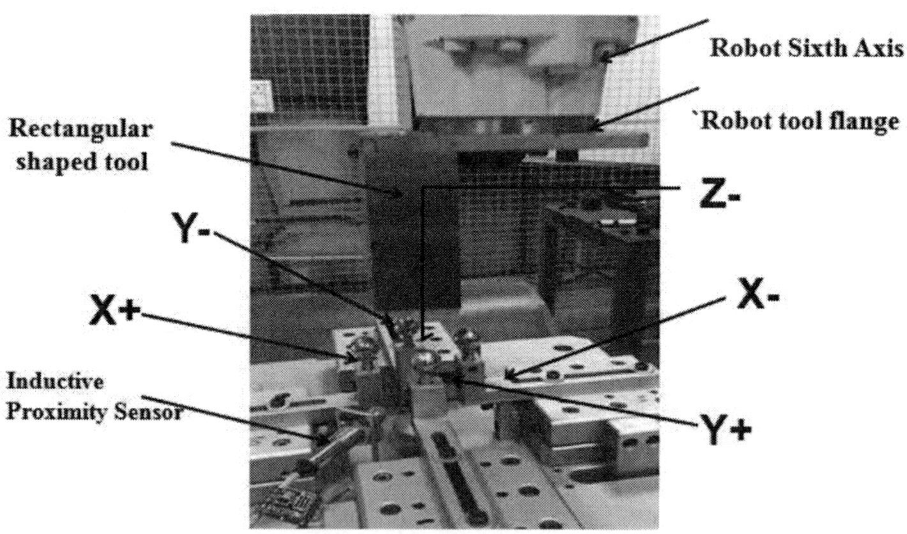

Figure4. Experimental Setup 5 LVDT probe of Industrial robot calibration method

5. Planning set-up for the experimental trial

The step by step procedure has been planned during the planning stage of an experiment. As the output of this planning stage, the external factors and environmental factors were fixed out. The final output of the whole experimentation will be the repeatability of the industrial robot. The repeatability will be determined by carrying out various experimentation trials and doing the mathematical calculations. The factors and parameters to be observed during the experimentation trials were chosen based on the literature review survey and by the feedback from Robot faculties from the college premises. The various controlling parameters differed in the experiment trials are presented in Table 1. Robot speed, payload, orientation, leveling of a calibration setup, working volume, and the presence of intermediate points are the five controlled factors, each having a range of operating points, which are listed in Table 1.

Table1. External factors determined after the planning procedure

External Factors	Range	Measurement Accuracy	Setting	Predicted Effects
Robot Speed (S)	10mm/s-3000mm/s	Based on Industrial Robot Controller	100 mm/s - 3000 mm/s	Significant
Payload (P)	0 – 4 kg	Scale unit accuracy	1kg- 4kg	Highly Significant
Height of adjustable table	Approx. 3ft		3 points throughout its length	Significant
Working Volume (W)	0-1000mm	Discrete Variable	5 points within work envelope	Highly Significant
Robot Movement (M)	Discrete	Discrete Variable	Linear, Angular	Questionable Significance

After external parameters were selected, then considered environmental factor which should be same throughout experimentation trials.

Table2. Experimental factors determined after the planning procedure

Experimental Factor	Desired Experimental Level	Measurement Accuracy	How to compensate	Predicted Effects
Environmental Factor	20 ° C – 25° C	Ambient Temperature (Based on the Room temp.)	Keeping the controller door open as it gets heated more	Could affect robot performance

6. Results

The data is then extracted from DAT SPC statistical software and exported into excel sheet to do the statistical analysis

Table3. The table data shows the step by step method to calculate the repeatability for the sample of 30 size of LVDT probe Robot speed at 150mm/s

No. of Cycle	Port1	Port2	Port3	Port4			Repeatability (mm)
	X+(X1)	X-(X2)	Y+(Y1)	Y-(Y2)	$(X1\text{-}X2)^2$	$(Y1\text{-}Y2)^2$	$\sqrt[2]{(X1-X2)^2+(Y1-Y2)^2}$
1	-0.2634	-0.078	0.001	0.026	0.034373	0.000625	0.187
2	-0.2635	-0.079	0.002	0.027	0.03404	0.000625	0.186
3	-0.2635	-0.079	0.002	0.027	0.03404	0.000625	0.186
4	-0.2635	-0.079	0.002	0.026	0.03404	0.000576	0.186
5	-0.0385	-0.077	0.014	0.075	0.001482	0.003721	0.072
6	-0.0385	-0.077	0.013	0.075	0.001482	0.003844	0.073
7	-0.0386	-0.078	0.014	0.075	0.001552	0.003721	0.073
8	-0.0387	-0.077	0.015	0.076	0.001467	0.003721	0.072
9	-0.0386	-0.079	0.014	0.075	0.001632	0.003721	0.073
10	-0.0385	-0.077	0.014	0.075	0.001482	0.003721	0.072
11	-0.0388	0.085	0.017	0.0877	0.015326	0.004998	0.143
12	-0.0385	0.085	0.017	0.0877	0.015252	0.004998	0.142
13	-0.03	0.088	0.018	0.089	0.013924	0.005041	0.138
14	-0.03	0.085	0.018	0.088	0.013225	0.0049	0.135
15	-0.04	0.086	0.018	0.088	0.015876	0.0049	0.144
16	-0.04	0.088	0.018	0.089	0.016384	0.005041	0.146
17	0.001	0.12	0.022	0.13	0.014161	0.011664	0.161
18	0.001	0.12	0.022	0.13	0.014161	0.011664	0.161
19	0.12	0.15	0.021	0.25	0.0009	0.052441	0.231
20	0.12	0.15	0.021	0.24	0.0009	0.047961	0.221
21	0.11	0.15	0.035	0.25	0.0016	0.046225	0.219
22	0.11	0.19	0.035	0.24	0.0064	0.042025	0.22
23	0.002	0.25	0.03	0.28	0.061504	0.0625	0.352
24	0.002	0.25	0.04	0.28	0.061504	0.0576	0.345
25	1.002	0.34	0.041	0.31	0.438244	0.072361	0.715
26	1.001	0.35	0.042	0.3	0.423801	0.066564	0.7
27	0.99	0.22	0.035	0.44	0.5929	0.164025	0.87
28	0.99	0.22	0.035	0.45	0.5929	0.172225	0.875
29	1.002	0.34	0.041	0.31	0.438244	0.072361	0.715
30	1.001	0.35	0.042	0.3	0.423801	0.066564	0.7

Table4. The table data shows the step by step method to calculate the repeatability for the sample of 30 size of LVDT probe at Robot speed at 100mm/s

No. of Cycle	Port1	Port2	Port3	Port4			Repeatability (mm)
	X+(X1)	X-(X2)	Y+(Y1)	Y-(Y2)	$(X1-X2)^2$	$(Y1-Y2)^2$	$\sqrt[2]{(X1-X2)^2+(Y1-Y2)^2}$
1	-0.2634	-0.079	0.001	0.024	0.034003	0.000529	0.186
2	-0.2635	-0.079	0.002	0.025	0.03404	0.000529	0.186
3	-0.2635	-0.078	0.003	0.026	0.03441	0.000529	0.187
4	-0.2635	-0.079	0.002	0.026	0.03404	0.000576	0.186
5	-0.0385	-0.077	0.014	0.074	0.001482	0.0036	0.071
6	-0.0386	-0.075	0.013	0.075	0.001325	0.003844	0.072
7	-0.0387	-0.771	0.0149	0.0876	0.536263	0.005285	0.736
8	-0.0388	-0.772	0.0148	0.0877	0.537582	0.005314	0.737
9	-0.0385	-0.077	0.013	0.075	0.001482	0.003844	0.073
10	-0.0385	-0.076	0.014	0.075	0.001406	0.003721	0.072
11	-0.0387	-0.771	0.0148	0.0877	0.536263	0.005314	0.736
12	-0.0384	-0.771	0.0149	0.0877	0.536703	0.0053	0.736
13	-0.03	-0.772	0.0148	0.089	0.550564	0.005506	0.746
14	-0.03	-0.772	0.0149	0.0891	0.550564	0.005506	0.746
15	-0.035	0.085	0.015	0.088	0.0144	0.005329	0.14
16	-0.036	0.086	0.0151	0.08	0.014884	0.004212	0.138
17	-0.038	0.085	0.015	0.087	0.015129	0.005184	0.143
18	-0.039	0.088	0.0151	0.09	0.016129	0.00561	0.147
19	-0.031	0.088	0.021	0.0889	0.014161	0.00461	0.137
20	-0.031	0.088	0.021	0.0889	0.014161	0.00461	0.137
21	0.11	0.19	0.035	0.25	0.0064	0.046225	0.229
22	0.12	0.18	0.035	0.24	0.0036	0.042025	0.214
23	0.002	0.25	0.041	0.287	0.061504	0.060516	0.349
24	0.002	0.24	0.04	0.28	0.056644	0.0576	0.338
25	1.0011	0.35	0.041	0.3	0.423931	0.067081	0.701
26	1.001	0.35	0.042	0.4	0.423801	0.128164	0.743
27	0.99	0.22	0.036	0.44	0.5929	0.163216	0.87
28	0.97	0.24	0.035	0.43	0.5329	0.156025	0.83
29	0.003	0.25	0.04	0.28	0.061009	0.0576	0.344
30	1.002	0.347	0.041	0.3	0.429025	0.067081	0.704

7. Calculations and Discussion

7.1 To Calculate the Repeatability

Repeatability= $\sqrt[2]{(X1-X2)^2+(Y1-Y2)^2}$

Where X1 is displacement of LVDT probe in X+ direction
X2 is displacement of LVDT probe in X- direction
Y1 is displacement of LVDT probe in Y+ direction
Y2 is displacement of LVDT probe in Y- direction

7.2 To calculate mean

IOP Conf. Series: Materials Science and Engineering 1012 (2021) 012010 doi:10.1088/1757-899X/1012/1/012010

Mean= $\frac{X1+X2+X3+\cdots+Xn}{n}$

Where X1, X2….Xn is each value from sample of experimentation trial
n is size of population

7.3 To calculate Standard deviation

Standard deviation(S) = $\sum_{i=1}^{n}\frac{(Xi-\dot{X})^2}{n-1}$

Where S is population standard deviation
N is the size of the population
Xi is each value from the population
Ẋ is population mean

7.4 To calculate Relative Standard deviation (RSD)

Relative Standard deviation (RSD) = $\frac{S}{\dot{X}}$

Where S is population standard deviation
Ẋ is population mean

Table5. The table data shows the value of Mean, Standard Deviation and Relative Standard Deviation for the sample of 30 size of LVDT probe at Robot speed 150mm/s

Statistical term	Value
Standard Deviation (S) (mm)	0.2554
Mean (X) (mm)	0.2837
Relative Standard Deviation (RSD)	0.9003
%RSD	90.03%

Table6. The table data shows the value of Mean, Standard Deviation and Relative Standard Deviation for the sample of 30 size of LVDT probe at Robot speed 100mm/s

Statistical term	Value
Standard Deviation (S) (mm)	0.2927
Mean (X) (mm)	0.3878
Relative Standard Deviation (RSD)	0.7547
%RSD	75.47%

7.5 Graphs plotted on minitab 17 software

Histogram Plot of repeatability for each sample size (N=30) at robot speed 150mm/s plotted on Minitab 17 software, The graph itself shows that the value of repeatability is increasing throughout the sample size and we are not getting good repeatability.

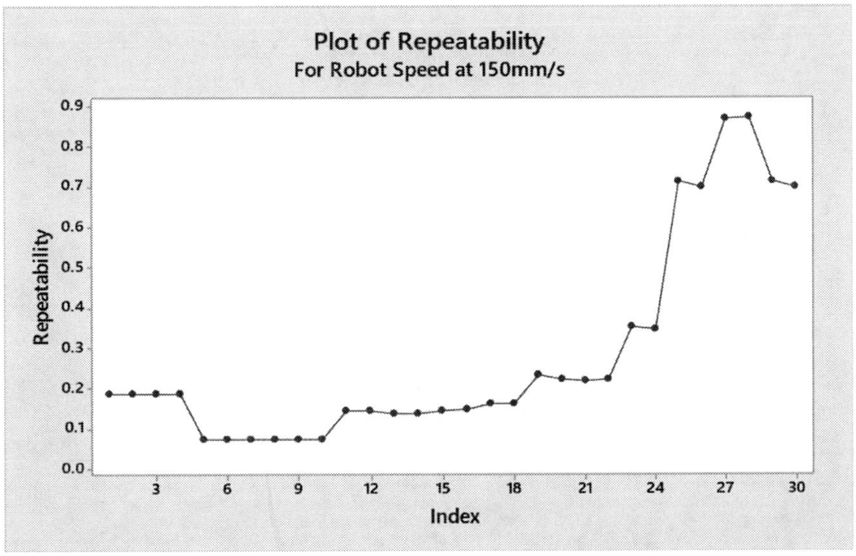

Figure5. Plot of repeatability for each sample size (N=30) for robot speed at 150mm/s

Histogram Plot of repeatability for each sample size (N=30) at robot speed 100mm/s plotted on Minitab 17 software, The graph itself shows that the value of repeatability is increasing and decreasing throughout the sample size but repeatability is better as compare to the repeatability at 150mm/s.

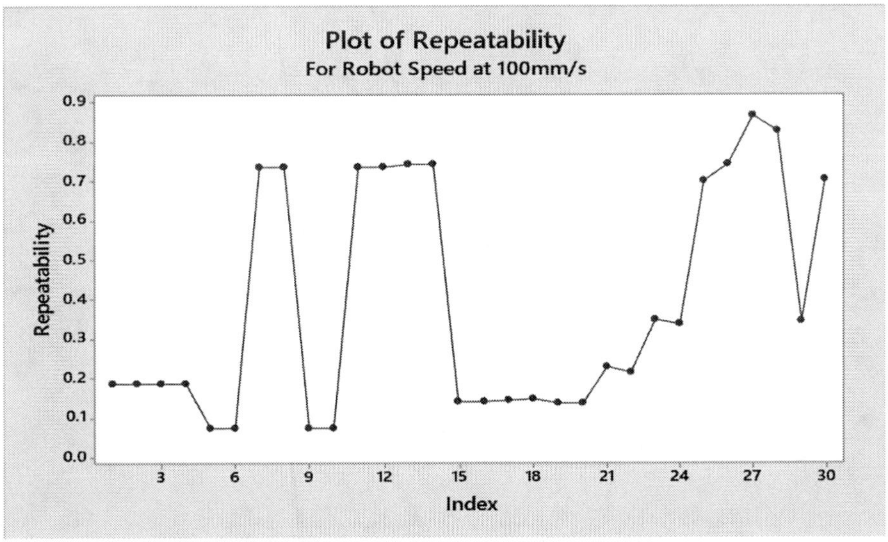

Figure6. Plot of repeatability for each sample size (N=30) for robot speed at 100mm/s

Line graph plotted on Minitab 17 Software, to compare the repeatability at robot Speed 150mm/s & 100mm/s, where repeatability is better at Robot speed 100mm/s as compared to the repeatability at 150mm/s.

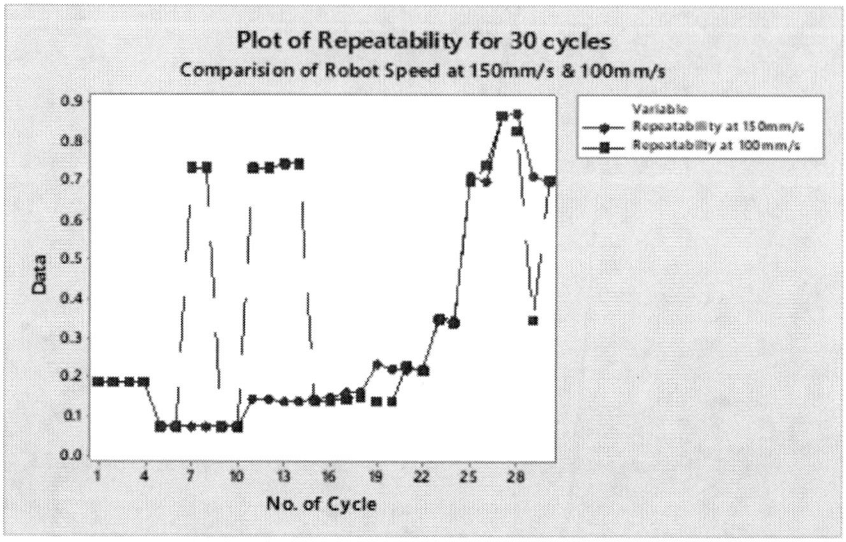

Figure7. Line Plot for comparison of repeatability for each sample size (N=30) for robot speed at 100mm/s & 150mm/s

Line plot of Mean, Standard Deviation and % Relative standard deviation are plotted on Minitab 17 software at robot speed 100mm/s & 150mm/s .The graph itself shows that the value of Relative Standard deviation at robot speed 100mm/s is less (%RSD= 75.47%) as compare at robot speed at 150mm/s (%RSD= 90.03%), which shows that repeatability is better at 100mm/s as compare to 150mm/s.

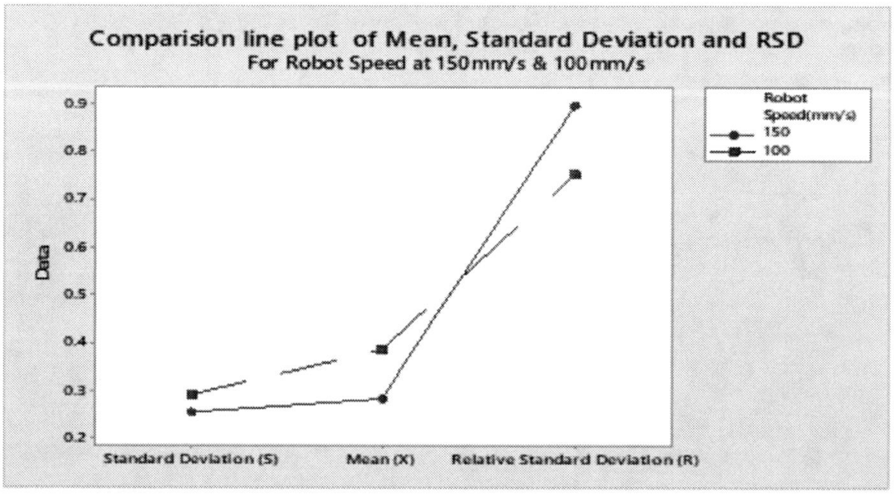

Figure8. Comparison Line plot of Mean, Standard Deviation and % Relative standard deviation at robot speed 150mm/s & 100mm/s

8. Conclusion

Robot calibration system capable of improving the robot position repeatability has been presented and improvement measure for the pose accuracy of the robot manipulator by using LVDT sensors are part of our Position Sensor range and it works by converting the rectilinear motion of an object into a corresponding electrical signal. LVDT sensors are well known for being more precise, they can measure movements which are smaller than 0.1mm as its has least count of 0.1mm.
Robotics programming has been done for calibration testing at different orientation (Linear/ angular), speed and zone and have observed that as the speed of the robot gets decreased from 150mm/s to 100mm/s the readings were close to master point means accuracy getting improved.
The repeatability has been calculated at Robot speed 150mm/s & 100mm/s. The Mean, Standard Deviation and % Relative standard deviation are 0.2837mm, 0.2554mm & 90.03% respectively at robot speed 150mm/s. The of Mean, Standard Deviation and % Relative standard deviation are 0.3878 mm, 0.2927mm & 75.47% respectively at robot speed 100mm/s. The value of Relative Standard deviation at robot speed 100mm/s is less (%RSD= 75.47%) as compare at robot speed at 150mm/s (%RSD= 90.03%), which shows that repeatability is better at 100mm/s as compare to 150mm/s.

References

[1] Jose Mauricio, S. T. Motta and R. S. Mcmaster 1999, Modeling, Optimizing and Simulating Robot Calibration with Accuracy Improvement, Journal of the Brazilian Society of Mechanical Science Print version ISSN 0100-7386
[2] Gianni Campion, Paolo Fiorini and Sandra Martelli, May 2002, Robot Calibration using Mobile Camera, Proceedings of the IEEE International Conference on Robotics & Automation Washington
[3] Chandra Sekhar Gatla, Ron Lumia, John Wood and Greg Starr, 6, December 2007, An Automated Method to Calibrate Industrial Robots Using a Virtual Closed Kinematic Chain, IEEE Transactions on robotics, vol. 23
[4] Yu Sun and John M. Hollerbach, May 19-23, 2008, Active Robot Calibration Algorithm, IEEE International Conference on Robotics and Automation Pasadena, CA, USA
[5] Emanuele Lubrano, Mohamed Bouri and Reymond Clavel, May 9-13, 2011 ,Ultra-high-precision Industrial Robots Calibration, IEEE International Conference on Robotics and Automation Shanghai International Conference Center, Shanghai, China
[6] Xiongzi Li and Biao Zhang, 2011, Toward General Industrial Robot Cell Calibration, IEEE 5th International Conference on Robotics, Automation and Mechatronics (RAM)
[7] Xiaoping Zhang, Wenyu Yang, Xuegang Cheng and Wei Zhu, 2012, A Practical Calibration Method for Spray Painting Robot in Factory, Springer-Verlag Berlin Heidelberg, C.-Y. Su, S. Rakheja, H. Liu (Eds.)
[8] Alexandr Klimchik, Yier Wu, Stephane Caro, Benoit Furet and Anatol Pashkevich, 2013 Advanced robot calibration using partial pose measurements, Project ANR COROUSSO, Project ANR ROBOTEX
[9] Erick Nieves and Ning Xi, May 31 - June 7, 2014, Robot Control System for Multi-position Alignment Used to Automate an Industrial Robot Calibration Approach, IEEE International Conference on Robotics & Automation (ICRA) Hong Kong Convention and Exhibition Center, Hong Kong, China
[10] Marwan Alia and Milan Simicb, 2017, Calibration method for articulated industrial robots, International Conference on Knowledge Based and Intelligent Information and Engineering System, Melbourne, Australia

[11] Yuye Caia, Hao Gu, Cheng Lib and Huashan Liua, 2018, Easy industrial robot cell coordinates calibration with touch panel, Department of Information Science and Technology, Donghua University, Shanghai, China, Corporate Research Center, ABB, Robotics and Computer–Integrated Manufacturing, Shanghai, China

[12] Zexiao Xie, Pengfei Zong, Peng Yao and Ping Ren, 2019, Calibration of 6-DOF industrial robots based on line structured light, Department of Automation, College of Engineering, Ocean Optik- International Journal for Light and Electron Optics 183, University of China, Qingdao, China

[13] Indian Standard, Manipulating Industrial Robots — Performance Criteria and Related Test Methods, Bureau Of Indian Standards, 2005

IOP Conf. Series: Materials Science and Engineering 1012 (2021) 012011 doi:10.1088/1757-899X/1012/1/012011

Wearable Non-invasive Health Monitoring Device for Elderly using IOT

B Sumathy[1*], S Kavimullai[1], S Shushmithaa[1], S Sai Anusha[1]

[1]Instrumentation and Control Engineering Department, Sri Sairam Engineering College, Chennai – 600044.

E-mail:bsumathy.ice@sairam.edu.in

Abstract. Patients and old age people suffering from Chronic, systemic diseases like heart diseases, asthma diseases, Alzheimer and dementia, Kidney diseases etc. needs regular health monitoring and extra care during any emergency. During the world of pandemic, those people suffered a lot and needed more attention from expert/ specialists. Due to the difficulty in consulting the expert specialists in person, wearable technology plays an important role, which evolved for ease of use and advancements for monitoring patient's health status. So, a wearable health monitoring system using IOT is proposed to monitor the regular health parameters periodically. The proposed system is a single integrated device consists of sensors for measuring vital physical parameters like pulse rate, respiratory rate, and temperature. Those data and initial predictions are sent via IOT cloud platform and expert opinions are received for further action as remote monitoring. If any critical changes are found, the same has been transmitted to doctor end and closest people of contact. If temperature changes are found below or above the standard value about 97 to 99 F, along with any respiratory difficulties with abnormal values are brought immediately to the notice of the specialist, due to the recent treats in covid. The sensor values are monitored time to time and connected to Arduino with GSM module for alert message. The proposed patient health monitoring system based on IOT helps the doctors and family members to keep track of the patient's health. The covid pandemic period also made us realize to monitor even normal healthy person, which is the need of the hour. Also, the unit is wearable, small in the form of belt/collar, light weight and cheap. The aim of the study is fulfilled, to prognosticate the possibility of unidentified or untreated health effects and monitor the patient health.

Keywords: IOT, HealthCare, Heartbeat sensor, Temperature sensor, Respiratory sensor.

1. Introduction

We live in a world where technology plays a vital role in monitoring human health with an easily accessible wearable technology. This is possible using IOT technology. IOT is the platform which connects various devices, transfer data with a wireless network without human interaction and it is embedded with sensors, actuators etc. Health is the crucial need for the body [1,2]. One of the important factors for superior life is health [1]. Many of us especially oldest people are failed to be monitored and to maintain their health regularly. Thus, there is a non-existence of good medical maintenance between patients, which leads to systemic diseases like heart diseases, attacks, and death rates in old age patients. Also, during the pandemic situation, it is necessary to monitor body temperature continuously which help in reduces the spread of covid- 19 like viruses and helps to protect all from diseases.

In traditional methods, hospitals and doctors plays a major role in health monitoring [5]. But it requires a lot of time for early booking, appointment, waiting for the doctor's consultation and checkup [5]. Regular visits to the hospital are mandatory even for minor problems. By using the latest technology efficiently, this kind of scenarios could be avoided. IOT enables the doctors and healthcare experts to be more connected with the patients and helps in monitoring them regularly from a distant. The main motive of patient health monitoring is to give alert well in advance or critical worsening and it can be achieved by the proposed system.

So, a "Wearable Non-invasive Health Monitoring Device for Elderly using IOT", is proposed to monitor the health status of patients regularly. The system is a single integrated device consists of sensors to monitor the health periodically. Also, the system is proposed where the permitted personal can ingress these data stored using any kind of IOT based platform and based on these rates received, the diseases are identified by the doctors and experts from a distance. Patients are suffering from heart diseases, asthma diseases, Alzheimer and dementia, kidney diseases etc. needs more care, regular checkups and need more consultation from expert/specialists. This health monitoring system is in the form of belt/collar and it monitors the heart rate (pulse rate), temperature and respiratory rate regularly, if any critical changes is been monitored, it directly informs to the doctor and other closest people of the patient. The proposed system is useful for many patients especially during covid-19 period, patients in quarantine ward for emergency needs. Also, this proposed system would meet the demands of proper health monitoring system and SAVE LIVES.

The flow of the paper is mentioned below. Introduction is the first part, followed by literature survey explaining various patient monitoring systems and devices. Part 3 explained about the method adapted that has included the parts, requirements, and specifications. Discussion of the results obtained in narrated in the next part, followed by conclusion and future works.

2. Literature Survey

In [1], the authors proposed the health monitoring customized system where the system monitors the heart rate in terms of number of pulses and temperature of the persons along with some environmental parameters like gas level in the room via sensors. The data is transmitted through wireless networks which makes an alert to the attendees and healthcare expert, by maintaining a single database of patients, apart from personalization of critical health related criteria. The authenticated medical expert collects the data, analyze and take decisions virtually using the data, without meeting in person. Deepak et. al, [2] proposed a monitoring system using ARM controller and GPRS modem. The system is associated with three sensors which are temperature, pulse rate. These sensor data which are attached to the patient body is sent to the ARM7 controller, and the data is transmitted with the help of Bluetooth module to Smart phone application and cloud with the help of GPRS module. In [3], the authors presented a healthcare monitoring system by IOT platform by utilising My Signals with wireless network system LoRa (Low power long range). Electrocardiogram signals, heartbeat pulse signals, SPO2 level signals, and temperature signals are measured and integrated with LoRa and My Signals. The performance and sensor effectiveness are checked, and analyzed. The vital parameters data are analyzed both by analytical and statistical methods.

In [4], an IOT based health supervisory system is proposed and the data are transmitted via cloud platform. The above device is made as a wearable device and is shortly called as WISE. The authors captured the real time data from the patients using the vital bio signal sensors. The device is made as a integrated device which measures the pulse rate, BP and temperature etc. In [5], a system is developed that measures the heart rate and body temperature using the LM35 and pulse sensor, respectively. These sensors are integrated to the uno board for the control of signals and channelling via wireless networks. In paper [6], the authors proposed a design for the finding pulse rate and functioning of heart. The device is basically a warning kind of prototype and if the values of parameter is beyond the limitation, divergent values of output of Arduino Uno goes to alerting system and further action takes

IOP Conf. Series: Materials Science and Engineering 1012 (2021) 012011 doi:10.1088/1757-899X/1012/1/012011

place. The proposed design has features like reliable, easy assessable and robust. In [7], the authors aimed to develop the sensor-based healthcare observation and analysing system, without mesh of wiring systems which collects the real time data of bio signals of the patient under investigation. The various parts used are sensors for acquisition of bio signals, signal conditioning units, controlling parts and relevant application software programmed using JAVA. Heartbeat rate, and other vital physiological information is been continuously checked, and further given as output for doctor's decision. Also, the data are stored for future reference and save life on time.

In [8], the authors reviewed the various latest technologies in wearable biosensors which allows health care providers to predetermine and decide the patient's vital bio signals traits after the treatments or therapeutics. The challenges faced by healthcare professional are reviewed. In [9], the authors proposed a model for basic health observation unit which is not connected with wire. This model sends the short communication message to the attendee and medical expert. The authors divided the prototype into different divisions as data processing, signal acquisition, and the last division is transmission, receiving divisions. Graphical User Interface is devised for all basic information in the first division. The basic data is entered and checked for vital good signs and there is any deviation in optimum parameters, the same is transferred for further guidance to medical expert. In [10], Prakash Patilet. Et. al, proposed a device which could continuously observe the biological body signals. The observed signals are compared with the optimum or threshold values of signal. When there is much difference between the two, alerting system comes into effort. The warning signal is sent wireless to healthcare personal and neighbours of close contact. In [11], an innovative patient health-care monitoring system is proposed which integrates the wearable devices and application to monitor the patient critical physiological data. Sensors and the Bluetooth chip are been attached to E-health shield which in turn connected to the Arduino Uno and is utilized only for small path transmission and reception between the Arduino and mobile device.

3. Proposed Methodology

Figure 1. shows the implementation of the system. The core objective of the unit is to prognosticate the unidentified and untreated health issues by easily wearable health monitoring system mainly for the elderly patients. This system differs from other wearable health monitoring systems by including all the three sensors-temperature sensor, respiratory sensor, and heartbeat sensor. The sensors embedded in the wearable belt will sense the heart rate (pulse rate), temperature rate, respiratory rate of the patient. The LCD will display the output for reference and the buzzer will intimate as an emergency alarm for checking the health of the person.

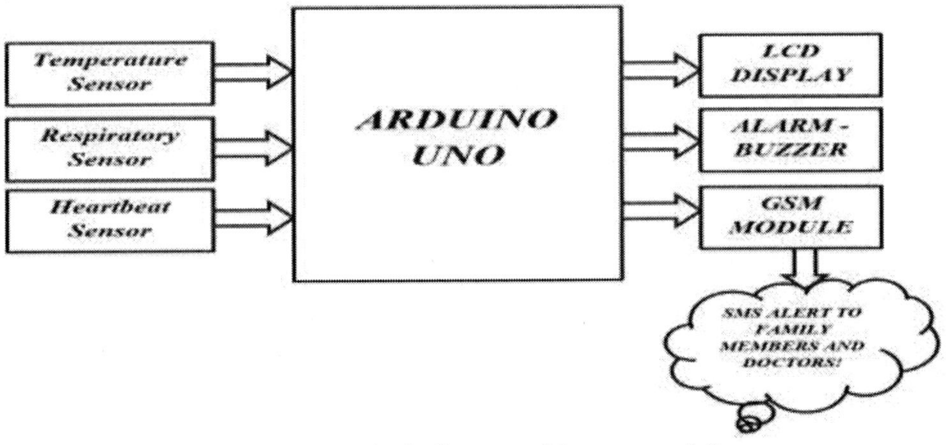

Figure 1. Block diagram of the proposed System

The sensors connected to the microcontroller of the Arduino provide the output data of the patients. An optimum estimate value is fixed at the initial stage, when it exceeds the limit a warning report data will be transmitted to patient's family members and doctors for emergency situations using GSM module and IOT cloud. Based on the health status, the doctor can decide the condition of the patient and appropriate medications will be provided. The system will predict the possibilities of getting affected by heart attacks, corona virus, asthma etc. and will help in maintaining regular health check-up and reduces death rate.

The proposed method is followed by hardware and software descriptions. The following section explains about the hardware components, specifications, and software descriptions in the proposed system. The important components are temperature, heartbeat and pulse sensor, GSM module, Arduino Controller.

3.1. Temperature Sensor

The LM35 temperature sensor (**Figure 2**) can be used to measure the body temperature more easily. The range of temperature is between (-55°C to 150°C), 0.1°C resolution precise. It is chosen for its cheaper cost and for easier availability. The LM35 does not have a temperature sensitive resistor like a thermistor. The temperature sensor works based on diodes; with the help of diodes the temperature changes the voltage at a known rate. The sensor will measure small changes and will provide an output of an analog voltage between 4 and 20V DC based on the measurements. The three pins of the sensor are connected to the GROUND, power supply (+5V) to Vs and any analog pin on the Arduino.

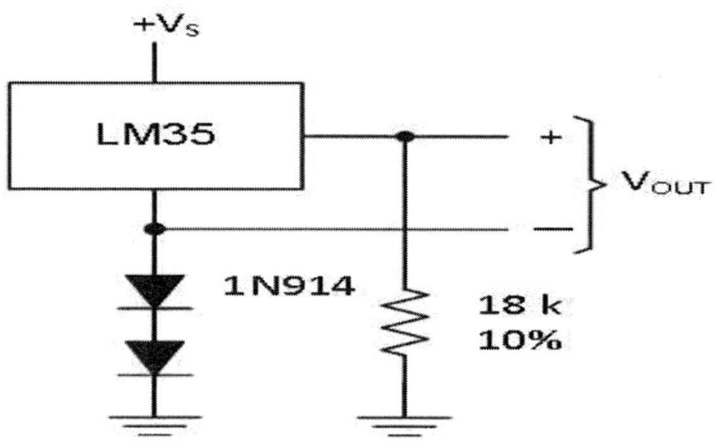

Figure 2. Circuit Diagram of Temperature Sensor

3.2. Heartbeat Sensor

The HeartBeat Sensor (**Figure 3**) is used to measure the rate of heartbeat. To find the heart rate (Pulse rate), the person under investigation's finger is kept in the heartbeat sensor it measures the heart rate (pulse rate) and coverts the data as an analog output. Heartbeat sensor is based on the principle of light modulation. When the IR signals are allowed to fall on the finger vein due to blood flow, light amplification getting changed. According to the principle of heartbeat sensor, the changes in the intensity of light passing through the organ will measure the changes in the volume of blood in an organ. IR LED will be used as the source of light in heartbeat sensor and any Photo diode will be used as a detector. The sensor has its own clip where we can insert our finger and this sensor also has three pins for connecting VCC, GND and the data. The pulses will be read by the Arduino, it will measure the heart rate and display it in LCD. Heartbeat sensor output is transmitted to the Arduino with the output pin, LCD is connected to Arduino in the 4-bit mode.

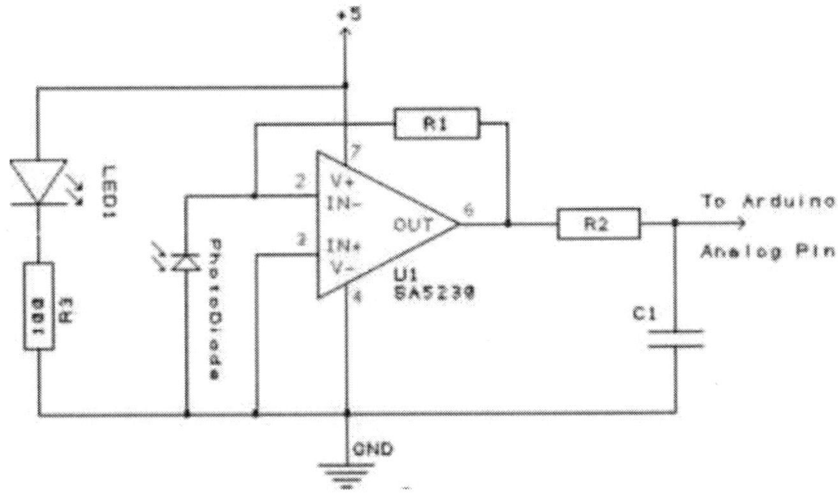

Figure 3. Heartbeat sensor circuit diagram

3.3. Respiratory type of sensor

Respiratory type of sensor is integrated into the system and number of breaths per minute is counted and noted. If the patient is exposed to high pollution it alerts and helps in prevent attacks. Thus, the respiratory sensor is immensely helpful in monitoring patients with asthma, chronic bronchitis etc. The sensor is available in affordable cost and comfortable for elderly having long sleep sometimes. It will be easily wearable for the elderly. The wearable device is made as a belt around the chest of the patient which will monitor their respiratory rate.

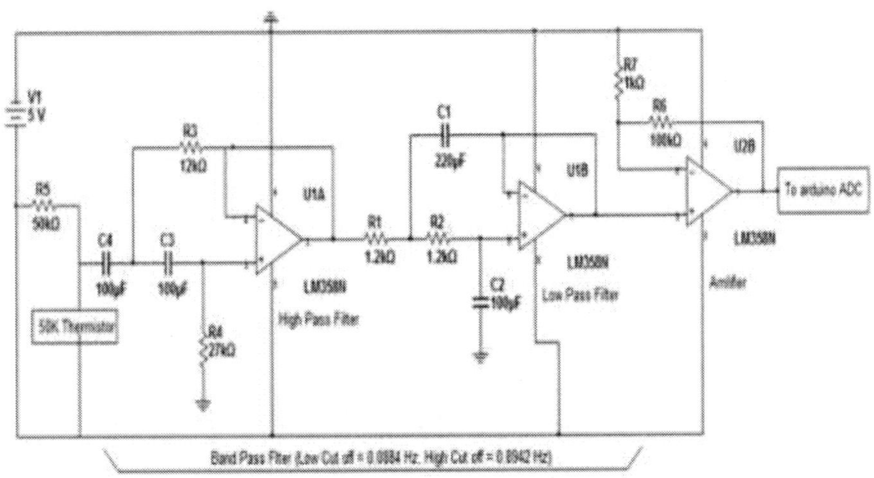

Figure 4. Respiratory sensor circuit diagram

The breath rate sensor is provided with 50 k thermistor in a voltage divider biasing circuit as shown in **Figure 4.** The value of change of resistance in the thermistor is monitored which is a function of inhale or exhale of oxygen/air. Also, the output is an ac signal in which dc components are eliminated by using the band pass filter. The range that be measured using this sensor is between 5 to 50 breaths/min. The output is suitably signal conditioned and send to the Arduino controller for further control action. The sensing element induced by inhaling and exhaling is measured by displacement variations. The sensor has elastic strap which can be adjusted according to the length and needs of the person. It determines chest contraction, and the output must be in a waveform.

3.4. GSM Module

GSM is an open technology used as communication bridge for transmitting voice and data service operating at different frequency bands (**Figure 5**). GSM is developed using the technique of time division multiple access (TDMA) for communication. The input will be divided in to 8 timeslots for transmitting and receipting data. TDMA technique is used to avoid the traffic over multiple data at the same time. The data rate transmission is between 64 kbps to 120 Mbps. GSM module consists of mobile system, base station subsystem, network subsystem. The system will instruct the user via SMS and meets the needs and requirements of the user. Security alert is also achieved with end to end encryption of data. The user can call at anytime from anywhere which is the most efficient featureand simply GSM is a mobile phone which need SIM card to identify themselves for a network. GSM module that works Via GPS module to send the location information and output processed data as message to the family members or to the registered health centre.

Figure 5. GSM Module

3.5. ARDUINO UNO

Arduino Uno consists of multiple digital and analog input/output pins. Arduino is used to connect the sensors; thus, it is a microcontroller board with 14 pins. The controller is used to collect the data from the sensors, process the integrated data and generate the output through the LCD display to alert the present health condition of the elderly. The communication with a computer can be done by Arduino, another microcontrollers or Arduino board. The ATmega328P micro-controller are provided with the serial communicator, crystal oscillator, USB port, and voltage controller that is used to control the voltage level of the microcontroller. The serial pins 0 (Rx) and 1 (Tx) are used for the transmission and reception of data. Arduino IDE when connected to board using the external USB wire is meant for programming. The input to the controller may be connected using various sensors and actuators. The reference analog value may be in between 0 and 5V. Uno controller module is shown in **Figure 6.**

Figure 6. ARDUINO Module

Overall Hardware Specifications
The overall hardware specifications is shwon in the Table 1.

Table 1. Hardware Specifications

Module	Item	Specifications
Controller	Operating voltage, digital pins, Flash memory	5V, 14, 32KB
Temperature Sensor	Temperature range, power, output impedance	-55C to 150C, 4-30V, 0.1W for 1mA load
Heartbeat sensor	LED, Gain, Power	100, 3.3 V
Respiratory	Bandwidth, Consumption	0 to 15Hz, 4mA
LCD	Power, Display	5V, plasma display
Buzzer	Rated Voltage, Operating Voltage Rated Current, Resonant	6V, 4 to 8V, <30mA, -2300Hz
GSM module	Voltage, Low Power Consumption, Operating	5V, 1.5mA, -40C to 85C

3.6. Flow Chart

Figure 7 represents the working flow of the human health monitoring system. The results would be analyzed for any critical conditions and immediate action would be taken accordingly. This system also helps in predicting the health of the patient well in advance to avoid any emergency conditions. Thus, remote monitoring and alerting system in the new era is achieved with the development of wireless and wearable technology.

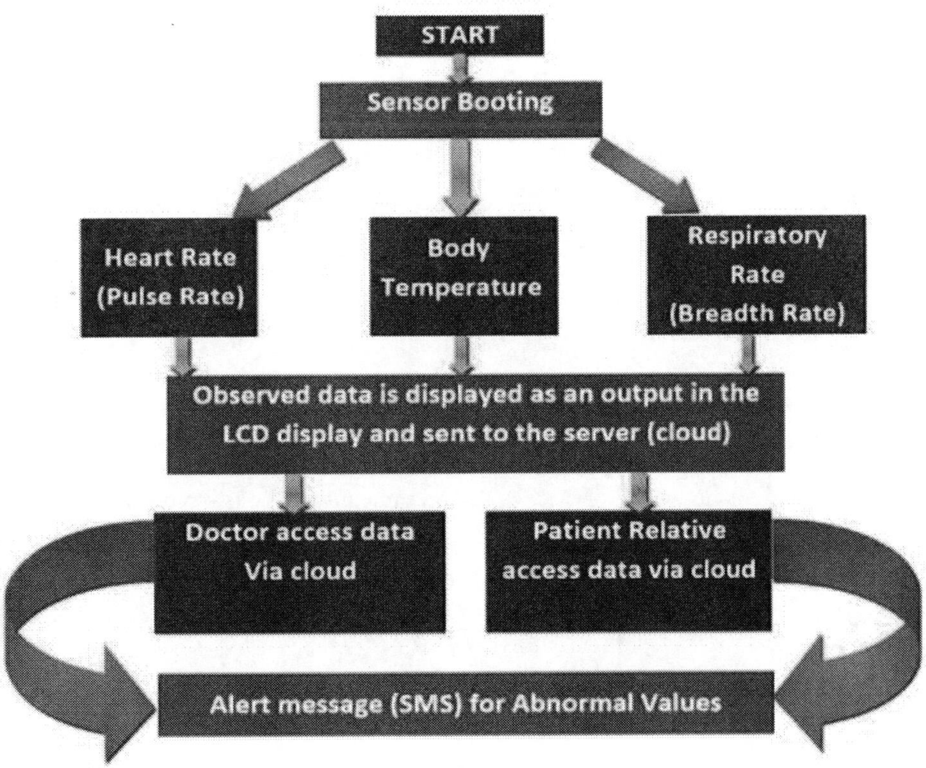

Figure 7. Flow chart of the Proposed System

IOP Conf. Series: Materials Science and Engineering 1012 (2021) 012011 doi:10.1088/1757-899X/1012/1/012011

3.7. IOT Platform

The IOT Platform is used to transfer the collected, processed data from the controller to the cloud through any internet source. The data reaches to the doctor's and other closest people of the patient. Based on the processed data speed the IOT hardware system will collect the information of the data and respond to the instructions. The following activities done using the IOT software as data collection, data processing, storage, and evaluation of the instructions. In data collection, the data given by the Arduino using the sensors which is processed in the microcontroller and the data is stored and programmed. Also, the device integration ensures that all the components of IOT are well integrated. In IOT based healthcare, patient health condition prone to any chronic disease is evaluated and predicted. With the help of GSM, transmission of collected information reaches to the destination i.e. any health centre or people of close contact. An immediate warning for any emergency is also incorporated. Google Cloud IOT platform is used and using HTTP's the analysis of the real time data with security to the patient is done.

4. Results and Discussion

The output from temperature, heartbeat and respiratory sensor are tested individually and integrated output for the patient is collected and transmitted to the health center and people of close contact. The overall control is by the Arduino controller and data transmission using IOT cloud platform. The output from the temperature sensor is tabulated below in Fahrenheit (Table 2) and Celsius (Table 3).

Table 2. Temperature sensor output (Degree Fahrenheit)

Test	Normal Value	Tested Value	Result
Patient 1	97°F to 99°F	100.4°F	Abnormal
Patient 2	97°F to 99°F	95°F	Abnormal
Patient 3	97°F to 99°F	98°F	Normal

Table 3. Temperature sensor output (Degree Celsius)

Body Temperature	Status
Less than 36°C	LOW - Unstable
36°C TO 38°C	NORMAL - Healthy
38°C and above	HIGH - Abnormal

Graph of Temperature Sensor

The output temperature values are tabulated, and the corresponding graph is shown in **Figure 8**. If the temperature goes below 95°F (36°C) the person will be in the state of hypothermia and the healthy person temperature range is between 36°C to 38°C and unhealthy if goes beyond 38°C.

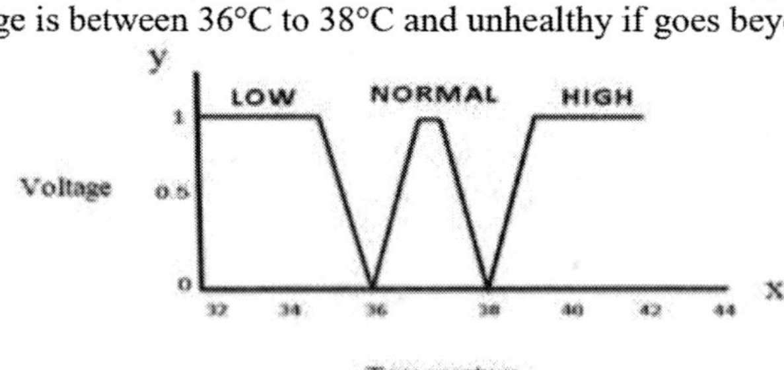

Figure 8. Temperature Graph

The output from the heartbeat sensor is tabulated below (Table 5 – Reference & Table 5). If the pulse rate is below 60 or above 100 BPM the person must go for medical advice. If the output of heartbeat sensor is in between 60 to 100 Beats Per Minute, the person health is normal.

Table 4. Heartbeat sensor output (Reference Table)

Pulse Rate	Status
Less than 60 BPM	LOW - Unstable
60 BPM To 100 BPM	NORMAL - Healthy
More than 100 BPM	HIGH - Abnormal

Table 5. Heartbeat sensor output

Test	Normal Value	Tested Value	Result
Patient 1	60 to 100 beats	106 beats	Abnormal
Patient 2	60 to 100 beats	80 beats	Normal

Graph of Heartbeat Sensor

The output of heartbeat sensor values is tabulated, and the corresponding graph is shown in **Figure 9.** The graph also explicit the normal and abnormal values in terms of Beats Per Minute.

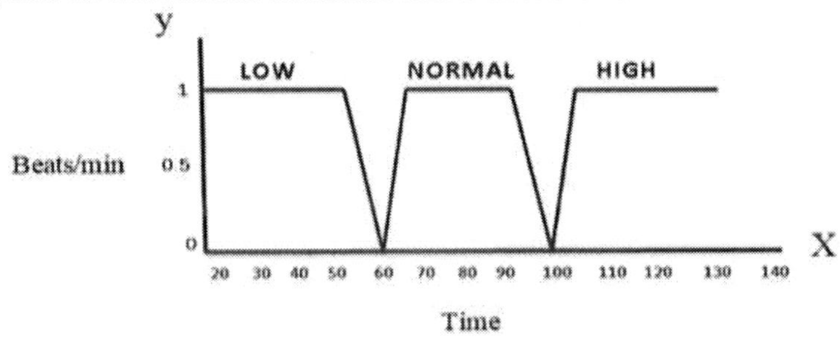

Figure 9. Heartbeat sensor Output Graph

The output from the respiratory sensor while resting condition is tabulated below (Table 6- Reference, Table 7). If the person breathes at a rate of value below 12 BPM or above 25 BPM, the person condition is abnormal and needs a medical advice. If the output of respiratory sensor is in between 12 to 25 Breadths Per Minute, the person health is normal.

Table 6. Respiratory sensor output (Reference Table)

Breathing Rate (Breaths Per Minute)	Status
Less than 12	LOW
12 to 25	NORMAL
More than 25	HIGH

Table 7. Respiratory sensor output (Breaths Per Minute)

Test	Normal Value	Tested Value	Result
Patient 1	12 to 25	23	Normal
Patient 2	12 to 25	10	Abnormal
Patient 3	12 to 25	25	Abnormal

Graph of Respiratory Sensor

The output of respiratory sensor values is tabulated, and the corresponding graph is shown in **Figure 10.**

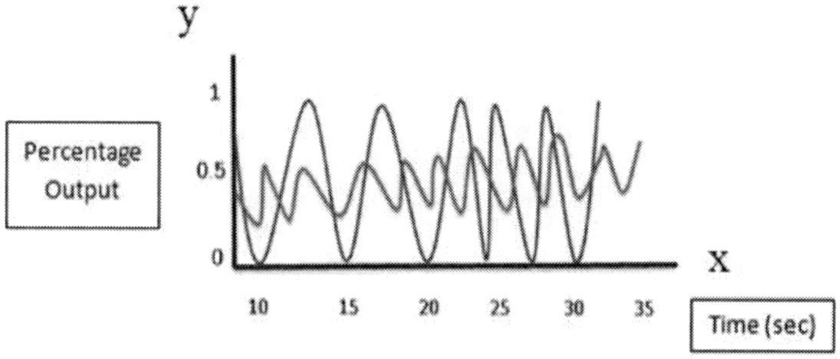

Figure 10. Respiratory sensor Output Graph

The graph explicit the normal and abnormal values in terms of percentage output with respect to time. Red colour denotes the abnormal pattern and blue colour shows the normal pattern for healthy persons.

In **Figure 11**, the overall monitoring circuit for pulse rate and temperature measurement is shown. The LM35 temperature sensor and Heart rate grove sensor is connected to the controller Arduino. The sensors have three pins for connecting VCC, GND, data and the microcontroller will process the data to be sent to the patient's closest members and to the doctor using the GSM module and cloud server. The data is collected and processed using the controller. The doctor can view the health of patient by using unique ID of the patients. The temperature sensor will give an output based on the voltage across the terminals. Contact type temperature sensors are used to detect the temperature. The principle of temperature sensor is when the temperature exceeds the threshold value it gives HIGH output, when the temperature drops below than the value, it gives LOW output. The heartbeat sensor will give an analog output when the patient places his/her finger on the clip attached with the sensor. The principle of this sensor is that it gives an output based on the light modulation and calculate the pulse rate based on intensity change. The respiratory sensor will be attached to belt/collars which will sense the breathing rate, BPM. Glow of white LED indicates for low heart rate and yellow indicates high values. Also, the body temperature by red LED is abnormal high value and green indicates the lowest value.

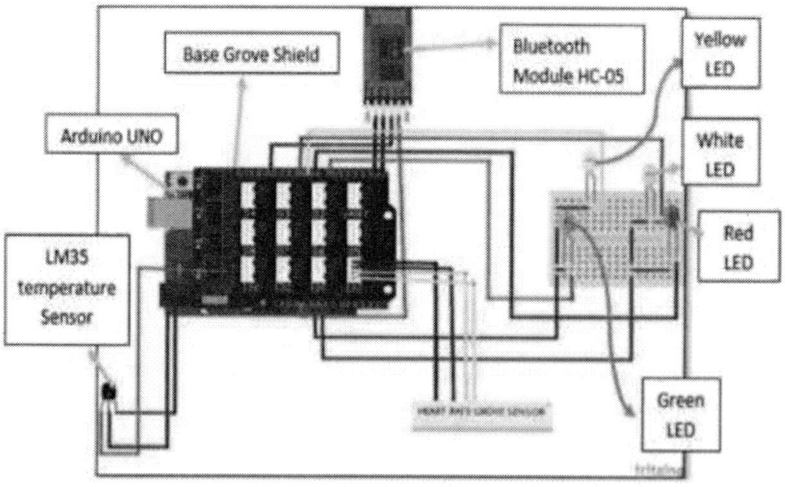

Figure 11. Overall monitoring circuit for pulse rate and temperature measurement

Figure 12. depicts the connection diagram of respiratory sensor with Arduino controller.

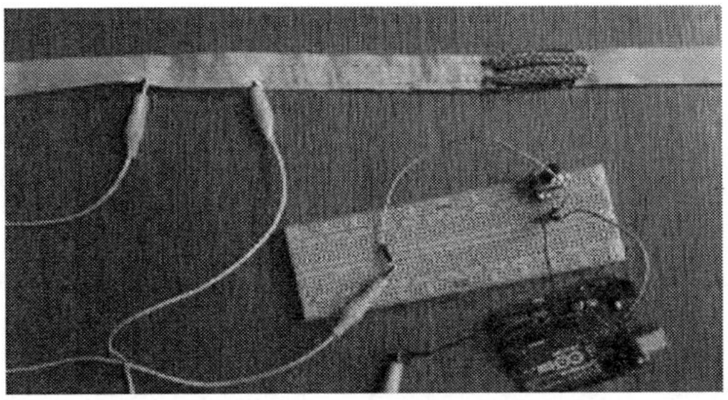

Figure 12. Interfacing of Arduino with Respiratory sensor

After performing the sensor operations, the data will be collected and stored and transmitted to the hospital end for further actions. Any emergency, treatment would be given accordingly. The SMS alert message for a sample is shown in the **Figure 13.** Currently the alert message is giving the output of temperature, heartbeat and respiratory sensor output depending on the reference values of each vital parameters. The output is transmitted to the patient's contacts and emergency calls. For any abnormal values, the patient health would be monitored in contact mode.

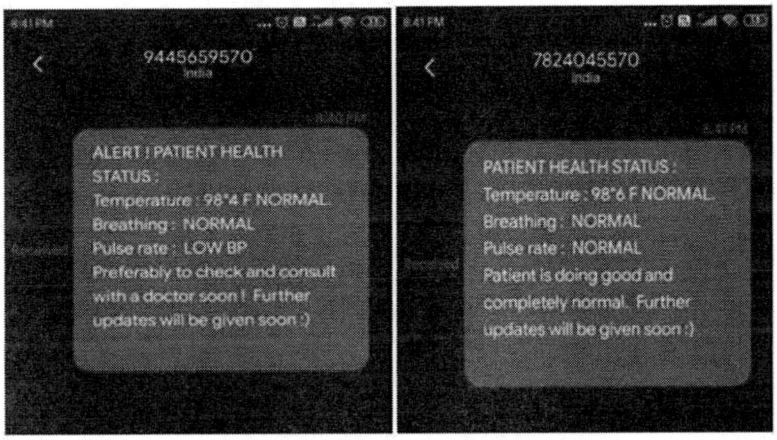

Figure 13. SMS Alert Message Output

5. Conclusion and Future works

Smart health monitoring system for elderly is proposed and implemented to monitor the basic vital body parameters like body temperature, heart rate and some measures of patient's condition as respiratory rate. The devised system checks the vital parameters and signs of the person under consideration and further processed using the integration unit of sensors, GSM wireless network system. The patient's wireless health monitoring reduces the time consumption in gathering the patient's data, more accurate than the manual system. The transmitted data is stored, analysed with proper analytic tool, and based on the report, the prediction of progress of diseases are done. This wearable device, in turn avoid the risky situation of patients with systemic organ diseases like heart, kidney and respiratory diseases. By using IOT devices combined with cloud environment, the database is shared to the hospital for further treatment and intensive care. Patients could be benefited by sitting in home itself, the basic tests and regular check-ups can be done and could avoid visiting the clinic frequently especially during this Covid period. Positively, the death rate of patient's could be prevented, and treatment can be introduced at the right time to the needy. As a future work, this wearable device could be further enhanced with more features like reduction in size, more vital parameters and with less cost.

References

[1] Md. Milon Islam, Ashikur Rahaman and Md. Rashedul Islam. 2020. Development of Smart Healthcare Monitoring System in IOT Environment *Springer Nature Singapore Pte Ltd.*
[2] Deepak N, Rajendra Prasad C H and Sanjay Kumar Ch. 2018. Patient Health Monitoring using IOT *International Journal of Innovative Technology and Exploring Engineering* **8**.
[3] Islam M S, Islam M T, Ali F. Almutairi, GankokBeng, Amid N M N. 2019. Monitoring of the Human Body Signal through the Internet of Things (IoT) Based LoRa Wireless Network System *Journal of Applied Sciences* **9** (9).
[4] Jie Wan, Munassar A A H. AI-awlaqi, Ming Song Li, Michael O'Grady, Xiang Gu and Jin Wang. 2018. Wearable IoT enabled real-time health monitoring system *EURASIP Journal on Wireless Communications and Networking*.
[5] Shivleela Patil, Dr. Sanjay Pardeshi. 2018. Health Monitoring system using IoT *International Research Journal of Engineering and Technology* **05** (04).
[6] Hari Kishore K, Surendra Nath K V, Hari Krishna K V N, Pavan Kumar D, Manikanta V and Fazal Noor Basha. 2019. IOT Based Smart Health Monitoring Alert Device *International Journal of Innovative Technology and Exploring Engineering* **8**(6).
[7] Senthamilarasi C, Jansi Rani J, Vidhya B and Aritha H. 2018. A Smart Patient Health monitoring system using IOT *International Journal of Pure and Applied Mathematics* **119** (16) 59-70.
[8] KyeonghyeGuk, Gaon Han, Jaewoo Lim, Keunwon Jeong, Taejoon Kang, Eun- Kyung Lim and Juyeon Jung. 2019. Evolution of Wearable Devices with Real-Time Disease Monitoring for Personalized Healthcare. *Nanomaterials*.
[9] Priya B, Rajendran S, Bala R and Gobbi R. 2009. Remote Wireless Health Monitoring Systems *IEEE Conference on Innovative Technologies in Intelligent Systems and Industrial Applications (CITISIA)*.
[10] Prakash H Patil, Pratyush Singh, Swatee Biradar and Prasad Rane. 2013. Design and Implementation of Wireless Patient Health Monitoring System *International Journal of Engineering Research and Technology* **2** (6).
[11] Abou-EINour A, Tarique M, Foyez D, Hassan L and Rashed A. 2019. Location Aware Health Monitoring System for Emergency Cases *International Journal of Scientific and Technology Research* **8** (12).
[12] Mary Livinsa Z, Mary Valantina G, Godwin Premi M S and Merlin Sheeba G. 2019. E Health Monitoring Systems in Smart Environments *Journal of Pharmaceutical Sciences and Research* **11**(9) 3130- 3132.
[13] Asad Khaliq, Addullah Awan M, Umair Saleh M, Waseem Abbas M, Husnul Maab and Salman Khan. 2014. Real Time Wireless Health Monitoring System *IOSR Journal of Electrical and Electronics Engineering* **9** (3) 31-38.
[14] Nur Hudha Wijaya, Fajar Ahmad Fauzi, Elsayed T Helmy, Phong Thanh Nguyen and Rachmad Andri Atmoko. 2020. The Design of Heart Rate Detector and Body Temperature Measurement Device Using ATMega16 *Journal of Robotics and Control* **1** (2) 40-43.
[15] Michael Chu, Thao Nguyen, Vaibhav Pandey, Yongxiao Zhou, Hoang N. Pham, Ronen Bar-Yoseph, Shlomit Radom-Aizik, Ramesh Jain, Dan M. Cooper and Michelle Khine. 2019. Respiration rate and volume measurements using wearable strain sensors *NPJ Digital Medicine*, **2** (8).

Modelling and Analysis of Shoulder Traction Surgery Kit

K Aruna prabha [1*], S Swetha [1], Priyadarsini Morampudi[1], CH Naveen Reddy[1]

[1]Department of Mechanical Engineering, VNR Vignana Jyothi Institute of Engineering and Technology, Telangana, India

E-mail: arunaprabhakolluri@gmail.com

Abstract- The shoulder traction surgery kit supports and improves efficiency in performing arthroscopic shoulder surgeries. Shoulder Traction Kit provides a convenient method to apply traction and position the patient's arm for shoulder procedures. It provides a proper vision within the shoulder joint and provides easy access for sub-acromial space. This paper presents a traction based shoulder kit aiding surgeries where the design and outlay requirements are reviewed and collected from research papers and suggestions taken from medics. This project initially deals with a simple design and model of the shoulder traction kit which is reliable and cost effective. A design which is simple to handle and easy to maintain has been the main moto of this project. Initial equipment outlay is modelled using software CATIA V5, SOLID WORKS and analysis is done on ANSYS. The model is simple, with few mechanisms involved and flexible which is easy to use. Based on analysis results and observations it is found that the designed shoulder traction kit can withstand 8 Kg loads with lesser deformations, stress within limits without any failure.

Keywords: Traction, Shoulder kit, Modelled, CATIA V5, ANSYS.

1. INTRODUCTION

In ortho medics traction force is applied to attain a convenient and accessible position for a shoulder arthroscopy. Traction is a pull force applied to the two objects to keep an interface to perform actions. In ortho field this phenomenon is applied to construct devices using pulleys and loads to aid a fractured or injured bone/ joint in order to cure or restore or aid for surgery. The tension created by traction suspends and stretches muscles and bones for certain amount of time [1]. A tapered/ dislocated joint is fixed by incessantly and steadily pulling the bone by using gravitational force when 7-8Kg weights are suspended onto pulley device [2]. An economical and reliable traction kit is built which stretches and suspends the shoulder by using supports cables, pulleys and loads. Point tractions with required inclination are obtained by assigning loads onto three point traction chain [3]. Generally, shoulder arthroscopy is carried by appropriately positioning the subject in lateral decubitus or beach chair depending on the conditions and the effectiveness. This gives an idea for design and developing a traction kit useful for surgeons [4]. Shoulder positioning such as lateral decubitus is desirable as it avoids difficulties and improves efficiency of surgical procedures. This benefit surgical group taking into account of patient's safety [5]. Grace Benson[6] developed a traction based unit containing a hook belt, a sleeve strap 25° inclination maintained suspended by the pulleys a scale with load suspended. This gave an idea for many orthopedic doctors for setup and making use of traction in performing complex shoulder surgeries. The shoulder to be operated is subjected to traction to gain joint recess by using arrangement which includes A frame hooks holders etc where one end of rope is in tension and pulls the shoulder the other end is pulled by traction making a lateral decubitus position[7] as shown in figure 1. The importance of LD position is emphasized as it gives much room for set up of surgical equipments with better image and operative environment [8]. Neurological and cardio complications and damages cannot be ruled out in LD position caused by traction hence it is advisable that the weights should be restricted below 8kg to prevent tearing and injuries[9][10][11]. Ravi Kumar Ray et al [12] developed a traction kit which is useful for 2 positions in shoulder surgeries which is much economical and reliable. This paper intends to present a solid model of shoulder traction kit which is easy to handle and use. The basic design provides us with a excellent mechanism for the sake of arthroscopic shoulder surgery but it fails at several aspects which include durability, easy handling, safety etc. Improved version of this design presented in the paper provides the user to use the equipment with greater balance, improved accuracy while holding the patients hand firmly against the equipment at required angle. The user can comfortably use as it has height and level adjustments, simple nut and bolt mechanisms with greater life span. Traction device for arthroscopy of shoulder and wrist possesses simple structure and easy to operate characteristics. The system of traction device for arthroscopy of shoulder and wrist can attain maximum traction (pull force) when simply rotate the tension wheel. The pull force of this traction can be able to adjust, which can achieve change the size of the traction force effect through adjust the weight of Farmar. The traction can adjust traction force, angles and height, etc. Thus, it is very convenient to operate. The figure 2 shows the fully designed shoulder traction surgery kit.

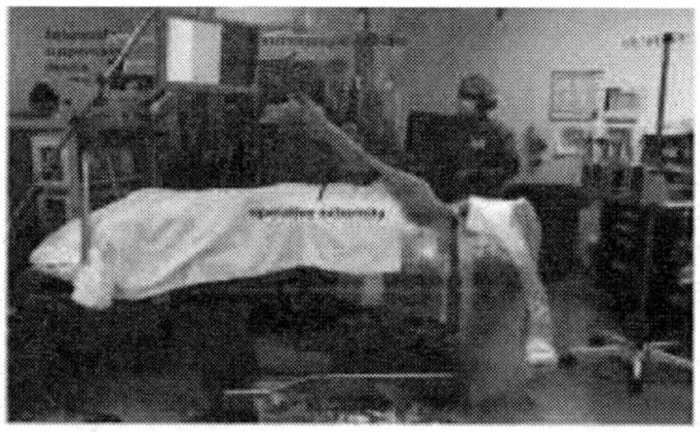

Figure 1: Lateral Decubitus Position for Shoulder Arthroscopy [7]

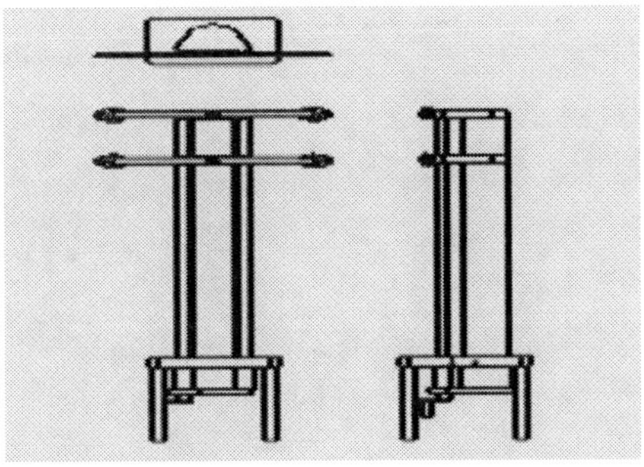

Figure 2 The Design View of Shoulder Traction Kit

2. MODELING

Initially a basic model of the shoulder traction kit has been made to get a realistic view of the design. The images below are the replica of the idea behind the project which has been done using CATIA V5 software. The design shows few unique parts and mechanical motions which delivers most important mechanism. This design has got the following parts namely angle plug, adjustable rod, gears, top cap, sliding cap, handle, guide rod, pulley, base jig etc The base components or parts which have been designed to be assembled are below.

2.1. Base Jig & Central Threaded Rod

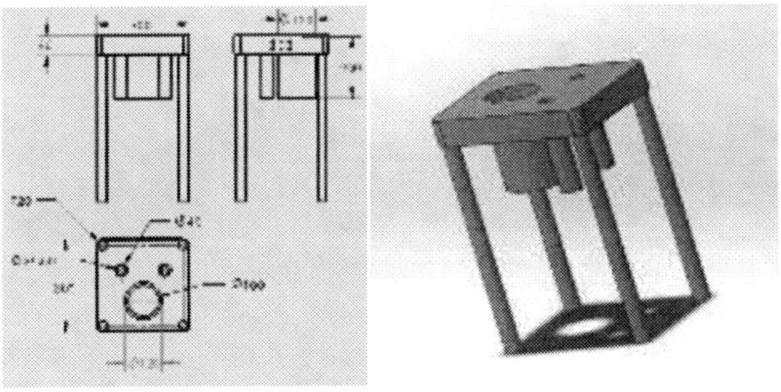

Figure 3 Base Jig

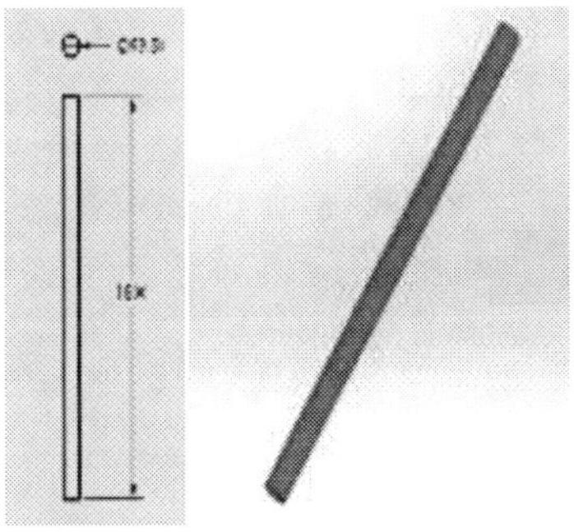

Figure 4 Central Threaded Rod

The above figure 3 represents the main base of the whole mechanism. It supports the 2 guide rods and the central threaded rod. The 2 holes of 25.5 mm diameter are to hold the 2 guide rods and the 100 mm diameter slot is to hold the central threaded rod. The above figure 4 represents the central threaded rod of diameter 99.50 mm – M12, which is 1800 mm long as per the requirements of the customer. Its position is fixed and any kinds of movements are restricted.

2.2. *Guide Rod& Sliding Cap*

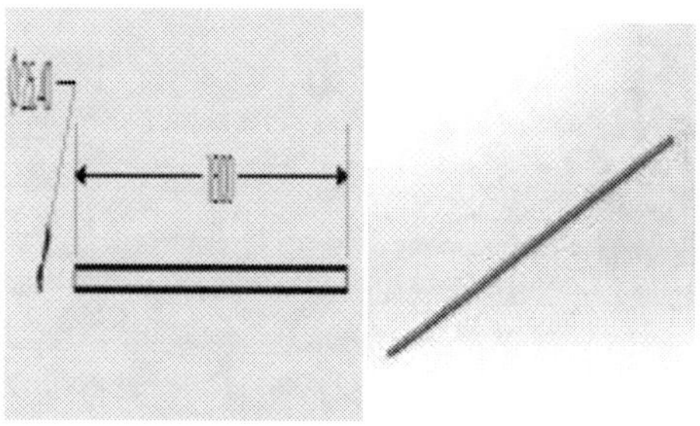

Figure 5 Base Jig

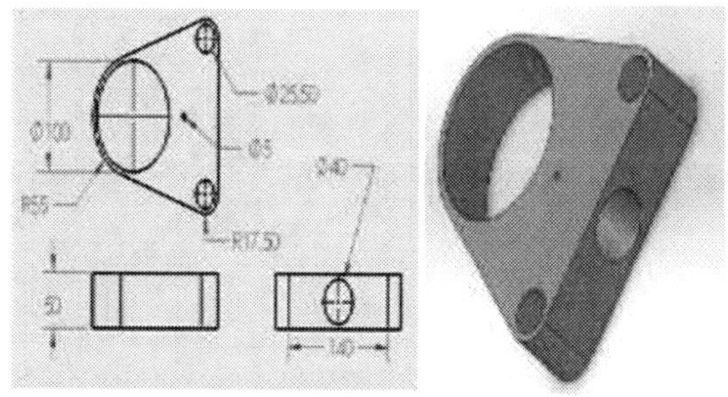

Figure 6 Sliding Cap

As represented in the above figure 5 the guide rod is 1800 mm long and 25.40mm diameter, this allows and also restricts the movement of the sliding cap, vertically. The above figure 6 shows the dimension of the sliding cap which has a 100 mm diameter hole which has internal threading to allow its movement along the vertical axis of the central threaded rod. The front view shows there is also another slot of 40mm diameter which is to facilitate the placement of the angle plug. The two similar holes of diameter 17.50 mm are to allow the passage of the guide rods and hence the sliding cap attains stability. Also, it has a 5mm diameter hole which aligns with the angle plug slots to fix upon a required angular position.

2.3. *Top Cap & Horizontal Hand*

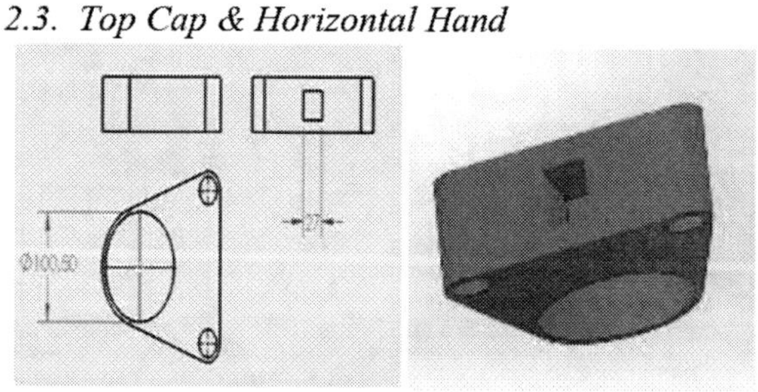

Figure 7 Top cap

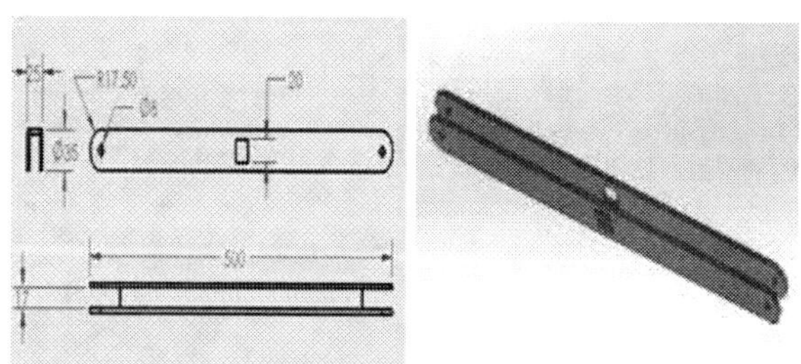

Figure 8 Horizontal Hand

The top cap shown in figure 7 is very similar to the sliding cap and also has a central hole of 100.5 mm diameter and the side view has a square slot unlike the sliding cap, the square slot has been chosen just to restrict the movement of the plug which mates the caps and the horizontal hands. The figure 8 shown above is the horizontal hand which is in a U-shape when seen from the side view. There are 2 holes of diameter 8 mm present at either ends to facilitate the entrance of the pin that locks the pulley and the horizontal hand. There is also a square slot in the centre of the rod to facilitate the entrance of the plugs depending on its position.

2.4. *Pulley and Angle Plug*

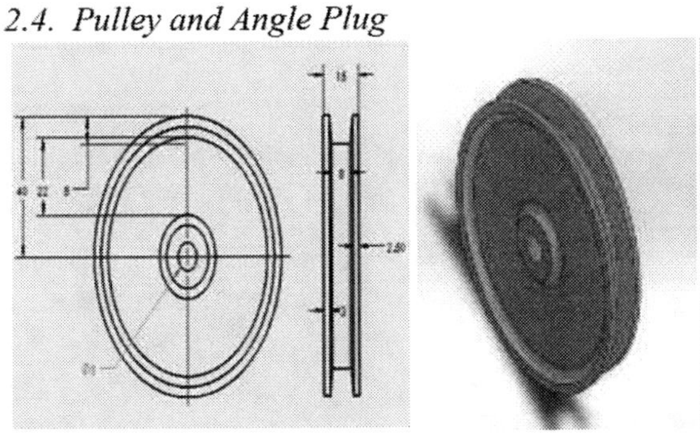

Figure 9 Pulley

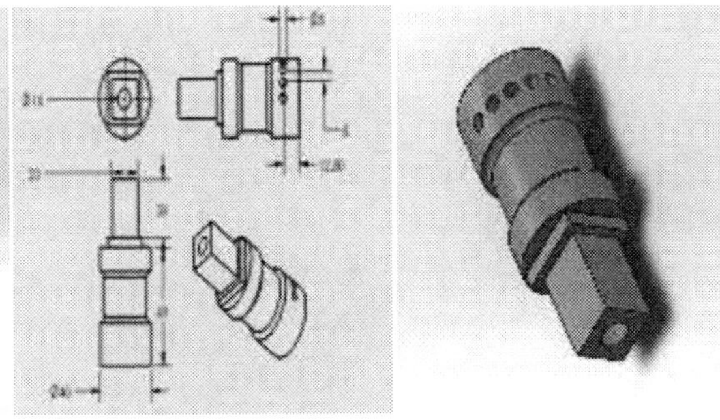

Figure 10 Angle Plug

The pulley in figure 9 is of 80 mm diameter with a central slot of 8mm diameter to allow the passage of a pin and hold it together with the horizontal hand for4 pulleys. The angle plug as shown in the figure 10 has been designed to mate/hold together, both the handle and the sliding cap. The one end of the plug has holes of 5mm diameter, these are present to facilitate the locking at require angle , based on the requirements of the horizontal hand. There is slot in the sliding cap which aligns with these slots and a pin is to be placed when they align to lock in the angle required. The whole 60mm length of the plug fits into the sliding cap leaving the 30mm part visible which has a central hole for the screw to fit in and tighten the locking further.

2.5. *Static Plug& Handle*

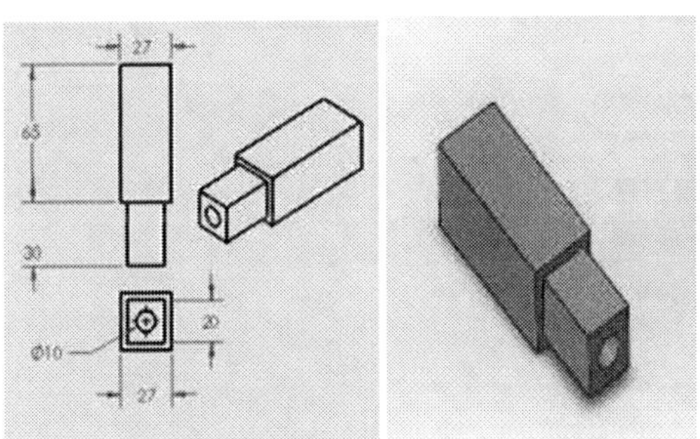

Figure 11 Static Plug

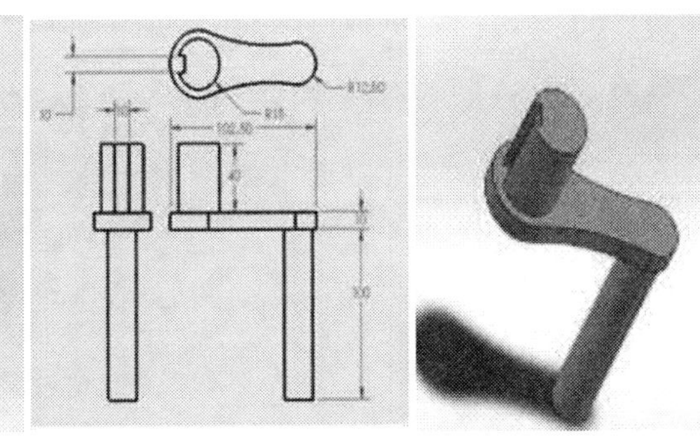

Figure 12 Handle

The static plug in figure 11 is used to mate the 2nd horizontal hand with the top cap which is in static position. The bigger square end sits inside the slot allotted in the top cap and the other end is set in the horizontal rod. A mating hole of 10mm is put down and the plug is tightened in the horizontal rod with the help of screw. The handle in figure 12 goes in the slot of one of the guide rods and is freely movable and rotatable, the key slot on the handle of 10 mm mates with the smaller gear.

2.6. *Small Gear & Big Gear*

The smaller gear in figure 13 mates with the handle using a key and slot of 10x10 mm. The gear was designed according to the distance between the 2 centre points of the locations. The pressure angle taken = 14.5^0, Module taken = 1, Pitch circle diameter = 84 mm, Bore = 30 mm.

The bigger gear in figure 14 is to be welded to the end of the central threaded rod. This meshes with the smaller gear which can give rotary motion to central threaded rod.

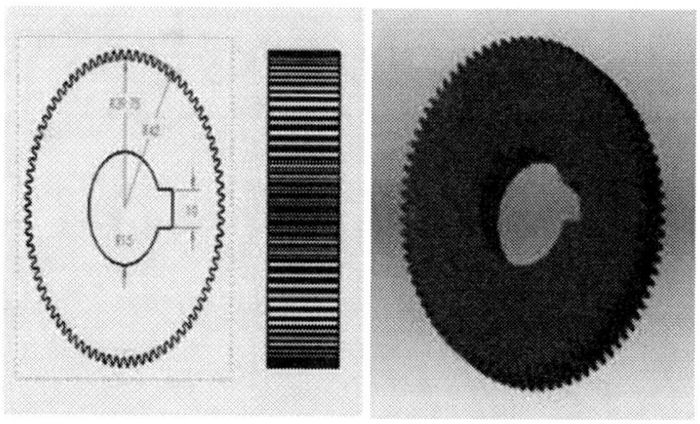

Figure13 Small Gear

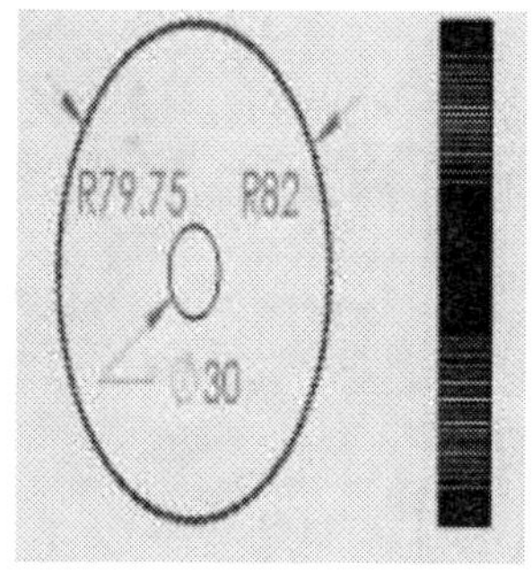

Figure 14 Big Gear

3. Assembled Model

The overall assembly of the above mentioned parts adds up to the shoulder traction kit as shown in figure 15. The working principle is that the handle is rotated to move the threaded rod which provides motion to the sliding cap so that the height can be adjusted. The sliding cap holds the horizontal hand using the angle jig, the angle of the horizontal hand can be adjusted and the cord is placed on the pulley. The second horizontal rod is used based on the requirement of the operation. A total of up to 8 kg loads can be placed on either of the hands through the pulleys.

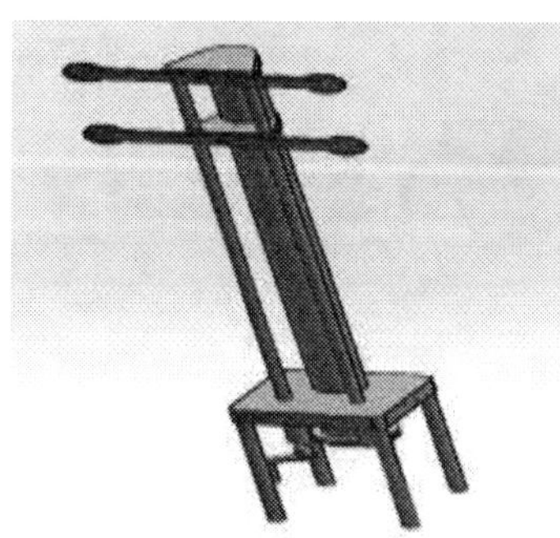

Figure 15 Shoulder Traction Kit

4. ANALYSIS

The analysis is done on ANSYS software. The model is subjected to maximum load conditions of 8 kg on the pulleys keeping base fixed subjected to boundary conditions as shown in figure 16. The resultant output analyzed are equivalent stress, factor of safety, total deformation etc are within permissible limits which validate the model. Figure 17 and 18 are stresses and total deformations obtained in ANSYS.

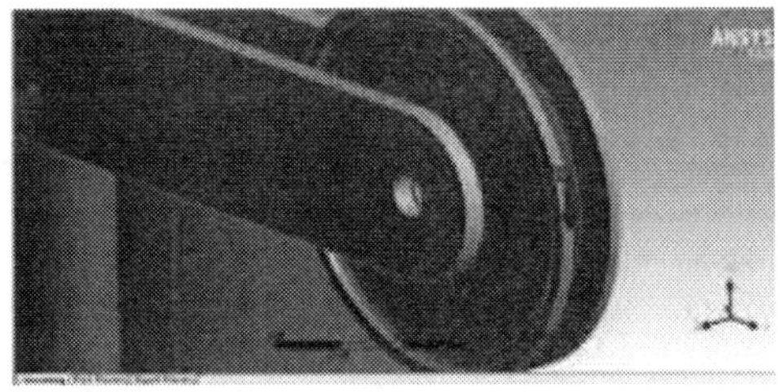

Figure 16 Load Applications on Pulleys

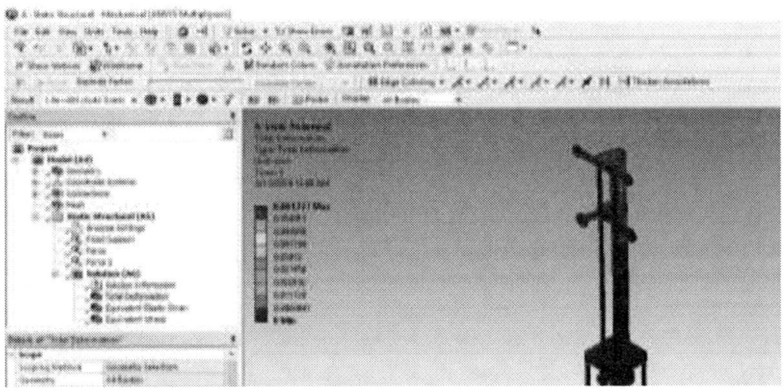

Figure 17 Total Deformation

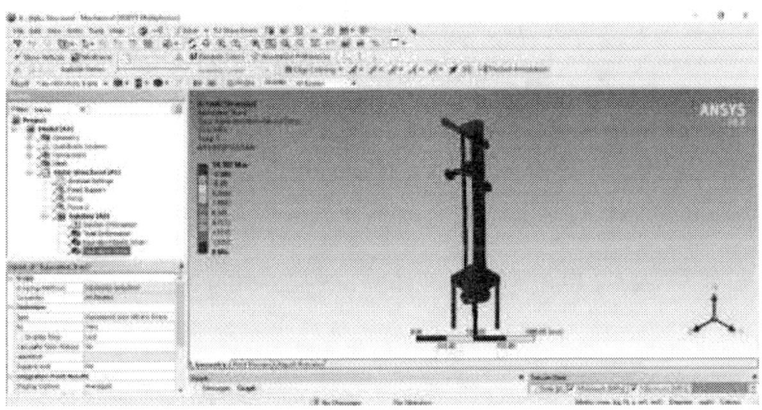

Figure 18 Stress Induced

5. DISCUSSION & CONCLUSIONS:

The developed design and model of shoulder traction kit aids arthroscopic shoulder surgeries with simple mechanisms involved. It also assures that the equipment has greater balance, improved accuracy, providing greater comfort to user as height can be adjusted by rotating the handle. The visualization of the equipment model is done by CATIA V5, SOLIDWORKS software and the reliability of the product is checked using ANSYS software. The design shows unique parts which involves few mechanical motions and mechanisms making it easier to use. The model when analyzed in ANSYS and subjected to max load conditions of 8 kg, the equivalent stress, strain, factor of safety, total deformation are found to be within limits. The product designed covers all the demerits and is specially made to cut down the price of the equipment to a great level and which works equally well. Scope for the project is very high since the product if produced is very much affordable and reliable for the customer.

References

[1] Christina M. Peruto, MD, Michael G. Ciccotti,MD, Steven B. Cohen, MD, Shoulder Arthroscopy Positioning: Lateral Decubitus vs. Beach Chair, A Concise Review, Arthroscopy - Journal of Arthroscopic and Related Surgery,Volume 25, Issue 8, August 2009, pp 891-896.

[2] Karel D. Capek, Ramon Zapata Sirvent, Ted T. Huang, Management Of Contractural Deformities Involving The Shoulder (Axilla), Elbow, Hip And Knee Joints In Burned Patients, Total Burn Care (Fifth Edition),2017, pp 573-588.e1

[3] Murat Songur, M.D. and ErcanSahin, M.D., A Simple Traction Assembly for Shoulder Arthroscopy in Lateral Decubitus Position: A Cost- Effective Alternative, Arthrosc Tech, 2015 August, Volume 4, Issue 4, pp e341–e343.

[4] Jorge Rojas, Filippo Familiari, Alexander Bitzer, Uma Srikumaran, Rocco Papalia and Edward G. McFarland, Patient Positioning in Shoulder Arthroscopy: Which is Best?, Joints, Volume 7, Issue 2, 2019 June, pp 46–55.

[5] Selby R. Philadelphia, Setup and patient positioning, PA: Saunders/Elsevier; 2010, pp 53–58.

[6] Grace Benson, Mechanical Traction for Stiff Shoulders, Physical Therapy, Volume 15, Issue 5, September 1935, pp 191–194.

[7] Alexander H. Jinnah, M.D., Sandeep Mannava, M.D., Ph.D., Johannes F. Plate, M.D., Ph.D., Austin V. Stone, M.D., Ph.D., and Michael T. Freehill, M.D. Basic Shoulder Arthroscopy: Lateral Decubitus Patient Positioning, Arthroscopy Techniques, Volume 5, Issue No 5, October 2016, pp e1069-e1075.

[8] X. Li, J.K. Eichinger, T. Hartshorn, H. Zhou, E.G. Matzkin, J.P. Warner, A Comparison Of The Lateral Decubitus And Beach-Chair Positions For Shoulder Surgery: Advantages And Complications, J AmAcad OrthopSurg,Volume 23, 2015, pp 18-28.

[9] W.D. Stanish, D.C. Peterson, Shoulder Arthroscopy And Nerve Injury: Pitfalls And Prevention, Arthroscopy, 11 (1995), pp 458-466.

[10] S.C. Weber, J.S. Abrams, W.M. Nottage, Complications Associated With Arthroscopic Shoulder Surgery, Arthroscopy, Volume 18, 2002, pp 88-95.

[11] D.J. Knight, R.P. Mahajan, Patient Positioning in Anaesthesia, Contin Educ Anaesth Crit Care Pain, Volume 4, 2004, pp 160-163.

[12] Ravi Kumar Ray, Helen Aung, Steven Andrew Corbett, Cost-Effective Traction for Arthroscopic Shoulder Surgery, Open Journal of Orthopedics, Volume 4, Issue No.5, May 2014, pp 130-136.

Development of lower body exoskeleton, mathematical modeling and video analysis of its prototype for obtaining customized joint actuation

S. S. Ohol[1], **K. D. Kalantri**[1*]

[1]Department of Mechanical Engineering, College of Engineering Pune (COEP), India

E-mail: kalantrikd18.mech@coep.ac.in

Abstract – Exoskeleton technology have shown its importance in various fields of application such as military, medical, industrial and commercial. But wide use of this technology is limited due high cost and Customized application according to user. If the exoskeleton has different gait cycle than unique natural gait cycle of user, it will lead to the injuries. Customization of exoskeleton gait cycle can overcome this challenge. Customized gait cycle according to user means control unit should provide output in form of position and torque in accordance to natural gait cycle of unique user. In this paper, A methodology is proposed and validated to obtain customized gait cycle for exoskeleton using video analysis and MATLAB simulation. Validation of Methodology is performed on normal human walking gait cycle. Video analysis performed in HALEX (Human Assistive Lower Limb Exoskeleton)is considered as basic method to conduct video analysis of gait cycle for unique user and data obtained from their experiment considered as base data for position vs time analysis. Later data obtained from video analysis is used as input data to obtain results fromMATLABSimscape-Simulink simulations using genetic algorithm. Above method is applied on biped robot to obtain results for normal human gait analysis. Simulations gives results in both position and torque wrt time. Torque values are compared with values obtained from mathematical model and position values are compared with values obtain from Video gait analysis. Validation of methodology gave satisfactory results and later suggested method can be used to obtain resulted torque and gait cycle for exoskeleton.

Keywords—Biped Robot, MATLAB-Simscape, genetic algorithm, exoskeleton, Normal Human Gait.

1. INTRODUCTION

From human to robot now back to human that is the trend in every industry. Yes, that's right combination of human with robot overcomes many limitations of not only human but also robots. This is the primary need of robotic exoskeleton technology. Humans have limitations like risk of injury, fatigue due to repeated type of work and lack of strength for challenging task while robot's high initial cost and cost of robot increases with increase in flexibility of control. But these limitations can be resolved if human dexterity of control and strength of machine is combined. This led to innovation and evolution of robotic exoskeleton in various fields of application.

In recent years Robotic Exoskeleton became a serious part Robotic industry. Application of exoskeleton is found in field of military, medical, industrial and civilian.

Exoskeleton technology has shown its importance in various fields of application such as military, medical, industrial and commercial. But wide use of this technology is limited due high cost and Customized application according to user. For exoskeleton to obtain human like gait motion links must have predefined position with respect to (wrt) time and to obtained predefined position of links wrt actuator should provide predefined torque wrt time. In this paper, methodology is proposed and validated to obtain customized gait cycle for exoskeleton using video analysis and MATLAB simulation Validation of Methodology is performed on normal human walking gait cycle. Video analysis performed in HALEX (Human Assistive Lower Limb Exoskeleton) is considered as basic method to conduct video analysis of gait cycle for unique user and data obtained from their experiment considered as base data for position vs time analysis. Later data obtained from video analysis is used as input data to obtain results from MATLAB Simscape-Simulink simulations using genetic algorithm.

Methodology to obtain customized gait pattern according to user have two steps. Step one is video analysis using video analysis software and step two is using data obtain from video analysis as input for genetic algorithm to perform simulation to obtain torque for actual gait pattern with exoskeleton.

To validate above methodology experiments are performed are in four steps, first step is validation video analysis method of gait cycle using with normal gait pattern available. Second steps are generation of mathematical model of human body for walking to obtain torque required at various phases of walking. Third step is simulation of biped robot with gait cycle input from step one to obtain torque and position of joint wrt time. At the end comparison of results of simulation with video analysis to validate the gait cycle and mathematical model to validate torque required at various phases.

2. LITERATURE REVIEW

Need of customized Gait cycle for exoskeleton – skeleton have shown their versatile use in medical field for gait rehabilitation and augmentation to strength of paraplegic patients [4]. A driven gait orthosis (DGO) has been developed that can move the legs of a patient in a physiological way on the moving treadmill the orthosis is adjustable in size so different patients can use it. Actuators at the knee and hip joints are controlled by an edge controller. With the DGO the legs of patients with different degrees of paresis and spasticity could be trained [5].Human Gait Cycle is dependent on various physical parameters of users like height, weight, gender, ages, lifestyle etc. and customization gait cycle for exoskeleton means obtaining gait cycle same as users not general gait cycle which one for all. The main concern was reaching the target angle within the allowable error quickly enough to track the trajectory (customized gait) during walking. If accurate control is not used, which would have aided in eliminating steady state error due to gravity and other factors because a control unit would have also worked against any sort of resistance from the patient? Because many paraplegic patients also have muscle spasticity, their muscles will fight the motion of the machine forcing the integrator term to wind up. Therefore, by overpowering the spasticity to move the joint, the machine could seriously injure the patient[3].This type of injury can also affect normal user using exoskeleton as unique gait of user is altered by overpower exoskeleton To eliminate this risk, it is a need to develop a method to obtain customized gait from particular user. In this paper video analysis is used to study gait cycle of user and using MATLAB simulation to obtain torque related to gait cycle of user.

3. Step I : Study of Gait Cycle

Study of normal human gait using video analysis to obtain angle position vs time graph throughout the normal gait cycle

Human walking is common yet most important form of human movement. Gait analysis measures, analyze and assesses the bio mechanical features that are associated with the walking task. Therefore gait analyses have application all considerable fields of locomotion.

Dartfish is a video solution provider from Switzerland. Dartfish provides online and offline platform to view, analyze and edit video for individuals and corporate.

The variation of joint angles of subject (age 21) walking on a treadmill were recorded using a video camera fitted on a stand of three legs in a sagittal plane while subject walked on a motorized treadmill one, the treadmill is often used in exoskeleton development programs because it allows standards and controlled conditions and it needs small space. These video was analyzed for joint angles on Dartfish Pro, video analysis software.

Analysis of Hip Angle and knee angle is performed on 25 frames per second settings.

The basic requirement video analysis method to be used is that it should give unique gait cycle for unique user which may different from average gait cycle mass users. This step is very important step to obtain customized output gait cycle for exoskeleton according users. Such type of video analysis can provide gait cycle accommodating different physical parameters like height, weight, gender, age etc. this analysis is performed on Dartfish sports analysis software.Fig. 6 and Fig. 8 show measurement of angle of knee and Hip wrt time on treadmill setup. Fig. 7 and Fig. 9 show the result with highlighted single cycle in Gait.

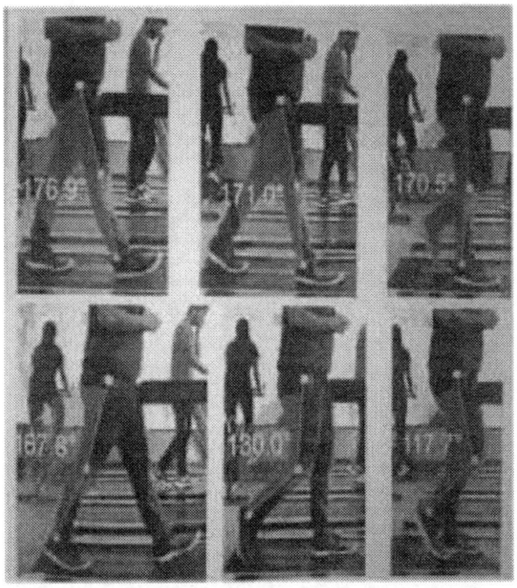

Fig. 1 Measurement of angle of knee wrt time [1]

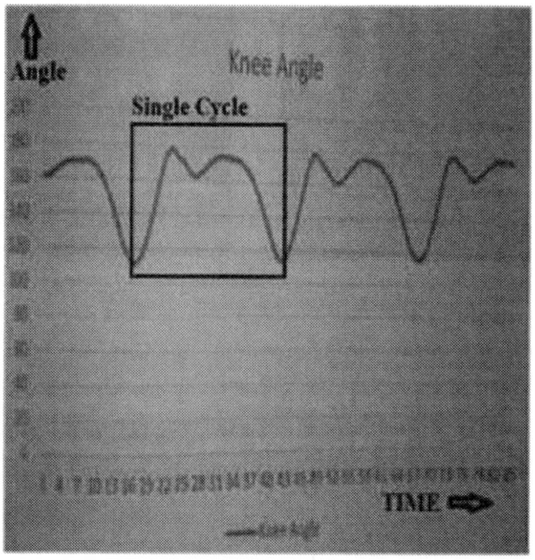

Fig. 2 Results of Knee angle (Single cycle in Gait is highlighted by rectangle) [1]

From Fig. 2 shows that single cycle is highlighted by rectangle and it is repeating the same pattern wrt time. Therefore angular position at knee joint angle ranges from 115 to 180 degrees i.e. 65 degrees.

Fig. 3 Measurement of angle of hip wrt time [1]

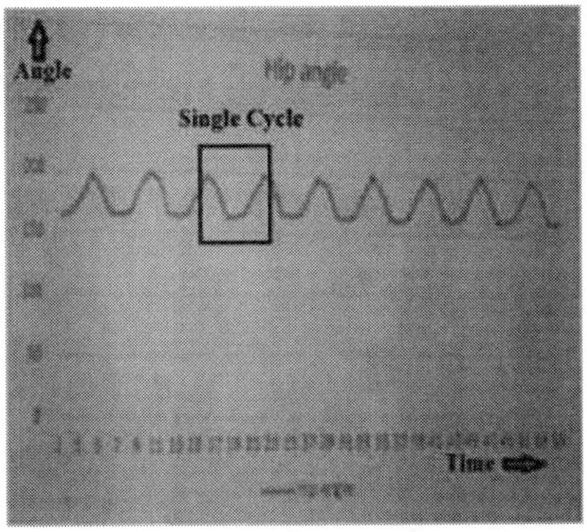

Fig. 4 Results of Hip angle (Single cycle in Gait is highlighted by rectangle) [1]

From Fig.4 shows that single cycle is highlighted by rectangle and it is repeating the same pattern wrt time. Therefore angular position at hip joint angle ranges from 160 to 200 degrees i.e. 40 degrees.

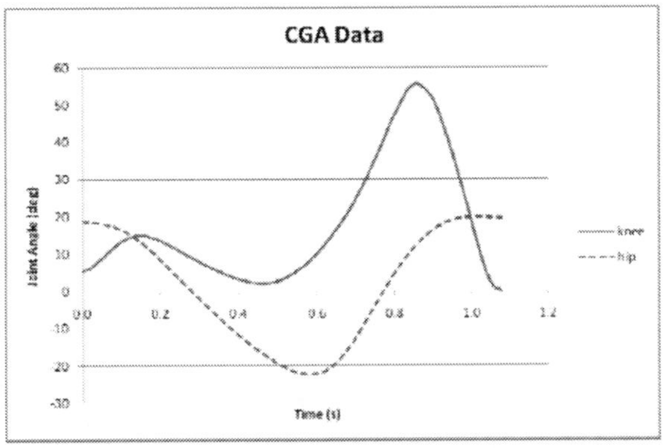

Fig. 5 The CGA Trajectories Used For the hip and knee angle shows the range of the joint angle used to achieve walking. [3]

Fig. 5 is reference gait cycle for average user. This average normal human gait cycle is reference to validate video analysis process to obtain unique gait for unique user. From fig. 5 shows that both knee and hip angular position are shown single graph. Due to different reference angle for measurement of angular position graphs are inverted and starts near zero degree. Therefore angular position at knee joint rangesfrom 0to 55 degrees i.e. 55 degrees and similarly at hip joint ranges from -22to 19 degrees i.e. 41 degrees

4. Step II : Calculations

Calculation of torque required for human walking at various stages of normal human gait using Mathematical Model

To generate mathematical model of human body to obtain torque required at various phases of gait following data of human body weight and height is used from Table 1 and Table 2 respectively.

Table. 1 Human Body Weight Distribution [2]

Weight Distribution of Body Parts		
Segment	**Average %**	**Average (kg)**
Total (M)	100	80
Head (Mh)	6.81	5.45
Trunk (Mt)	43.02	34.4
Total Arm (Ma)	4.715	3.77
Upper Arm	2.63	2.10
Forearm	1.5	1.2
Hand	0.585	0.47
Total Leg	20.37	11.58
Thigh (Mth)	14.47	11.58
Shank (Msh)	4.57	3.66
Foot (Mfo)	1.33	1.06

Table. 2 Human Body Height Distributions

Height Distribution of Body Parts		
Segment	**Average %**	**Average (cm)**
Total (H)	100	180
Total Leg	56.2	101.16
Thigh (L1)	28.1	50.58
Shank (L2)	28.1	50.58
Foot (L3)	14.05	25.29
Total Arm	39.3	70.74
Head	15.7	28.26
Torso	28.1	50.58

For generation of mathematical model for normal human walking subject of height 180 cm and weight 80kg is considered. From video analysis it was noted that average gait cycle time for normal human walking is 1second and stride length is 60 cm. Torque require at crucial phases in gait cycle are calculated.it was observed in modelling that torque required at ankle throughout gait cycle in negligible.

Fig. 6 shows stance and swing phases of normal gait cycle of human along with 8 sub phases. To obtain torque at these sub phases following formulas are derived.

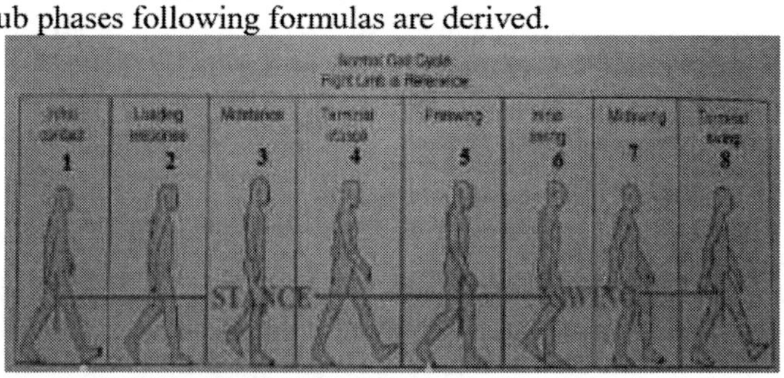

Fig. 6 Phases of Normal Human Gait Cycle [6]

For stance phase -

Human walking requires less torque than maximum torque capacity human joint. Human walking is only possible if sufficient friction is available between ground and human foot. Thus, friction plays very crucial role in calculation of torque phase. In stance phase both horizontal and vertical forces are acting on joints.

Average step length = 60 cm = 0.6 m
average duration of one gait cycle = 1 sec
s= u*t + 0.5*a*t*t
0.6 = 0 + 0.5*a*1*1
a = 1.2 m/s*s
Total mass of human body = M = 80 Kg
F = M*a = 80*1.2 = 96 N
Therefore Inertia force = I = F = 96 N
Friction coefficient = u = 0.5
Normal force = N = M*g = 80*9.81 = 784.8 N
Maximum Friction force = Fmc = N*u = 0.5*784.8 = 392 N
Fc = actual friction force required.

Force required for human walking obtain from ground i.e. friction force. Therefore required friction force (Fc) should be less maximum available friction force.

After calculating torque required at various sub phases, mid stance is the critical sub phase with highest torque required within stance phase.

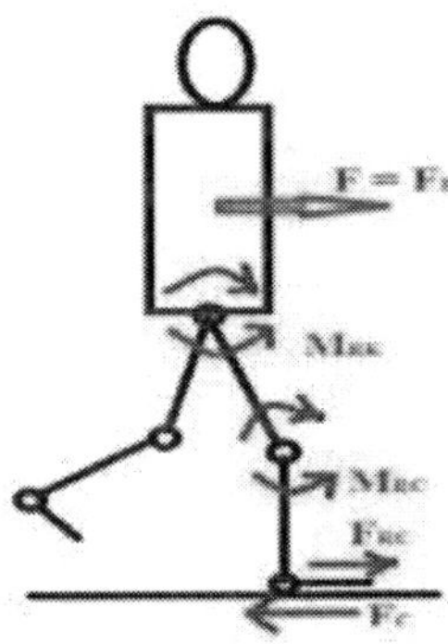

Fig. 7 mid stance sub phase of Normal Human Gait Cycle

Frc = reaction of friction on human body = Fc
Mrc = reaction moment of Fc at knee
Mk = torque required at knee b
Mrc = reaction moment of Fc at knee
Mk = torque required at knee by actuator
Mrk = reaction moment due to torque at knee at hip
F = force required for walking at defined speed
Mh = F* 0.3 N m = 96 * 0.3 = 28.8 N m
Mk = Mh + F*0.8 = 28.8 + 96*0.8 = 105.6 N m
Mk = Mrc = Frc * 0.5 = 105.6 N m
Frc = 212 N
Frc = Fc = 212 <Fmc = 392 N

Thus, Mh = 28.8 N m (at hip)
Mk = 105.6 N m (at Knee)

For swing phase -

For swing phase legs perform pendulum free swing motion so equation requires calculating torque for swing phase are different from stance phase.

After calculating torque required at various sub phases, terminal swing is the critical phase with highest torque required within swing phase.

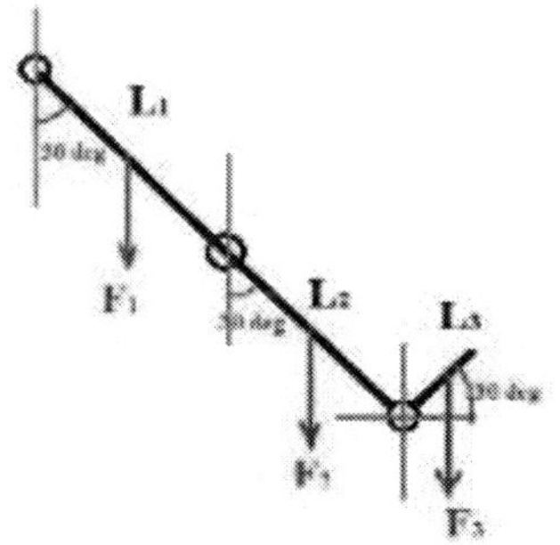

Fig. 8 Terminal swing sub phase of Normal Human Gait Cycle

Moment equation at hip

Mo1 = L1*0.5*sin(30)*Mth*g
+Msh*g*[L1+L2*0.5]*sin(30)+Mfo*g*[(L1+L2)*sin(30)+l3*cos(30)*0.5]

Mo1=50.58*0.5*0.5*11.58*9.81+3.66*9.81*[50.58+50.58*0.5]*0.5+
1.06*9.81*[(50.58+50.58)*0.5+25.29*0.86*0.5]

Mo1=1436.5+1362+639 N cm

Mo1 = 34.4 N m

Moment equation at Knee

Foot is not in contact with ground normal force and
friction forces are not considered in moment calculation.

Mo2=Msh*g*L2*0.5*sin(30)+Mfo*g*[L2*sin(30)+L3*0.5*cos(30)]

Mo2 = 454 + 376 N cm

Mo2 = 8.3 N m

Thus,

Mh = 34.4 N m (at hip)

Mk = 8.3 N m (at Knee)

5. BIPED ROBOT SYSTEM

This is simple stick and brick body version of Biped robot walker system. MATLAB Simulation using Simscape is performed on this system to compare results with video analysis and calculated torque. Genetic algorithm with 100 generation and 100 population size is used i.e. total 10000 experiments are performed on this biped robot system to learn human like walking. For genetic algorithm converge on results according to initial condition provided at the beginning of the experiments [7].

For Biped robot system,

Length of thigh is 40 cm.

Length of shank is 40 cm.

Density is 1000kg/ m^3 (Human body density is 980 kg/m^3)
Total weight is 5.588 kg.

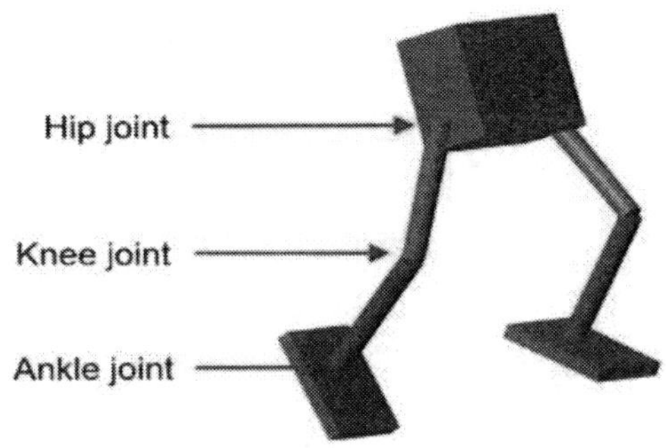

Fig. 9 Biped Robot System using simple Stick and brick model [7].

For normal human walking 6 DOF are required for human i.e. Ankle Roll, Ankle pitch, Knee, Hip Roll, Hip Pitch and Hip Yaw. Above biped robot model have same DOF as normal human.

6. RESULTS & DISCUSSION

Results of MATLAB Simulation of Biped Robot.

a. Hip roll

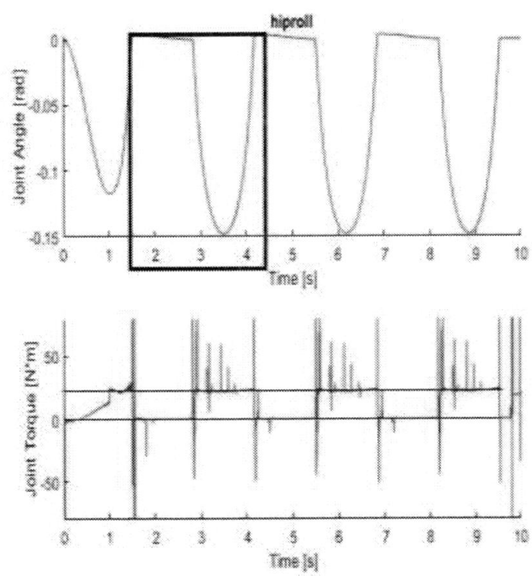

Fig. 10 Joint angle and joint torque wrt time for Hip roll

From fig. 10 shows the angular position and torque required at hip roll. In angle vs time it is observed single cycle in gait as highlighted in rectangle and variation of total angle throughout the cycle is -0.15 rad to 0 rad i.e.range is 8.6 degrees. Range and Max joint torque is 0 Nm to 20 Nm and 20 Nm respectively.

b. Hip pitch

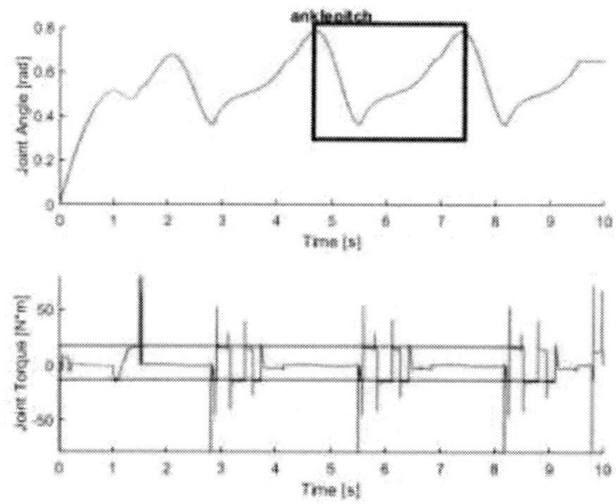

Fig. 11 Joint angle and joint torque wrt time for Hip pitch

From fig. 11 shows the angular position and torque required at hip pitch. In angle vs time it is observed single cycle in gait as highlighted in rectangle and variation of total angle throughout the cycle is 0.3 rad to 0.8 rad i.e.range is 28.6 degrees. Range and Max joint torque is -15 Nm to 20 Nm and 20 Nm respectively.

c. Hip Yaw

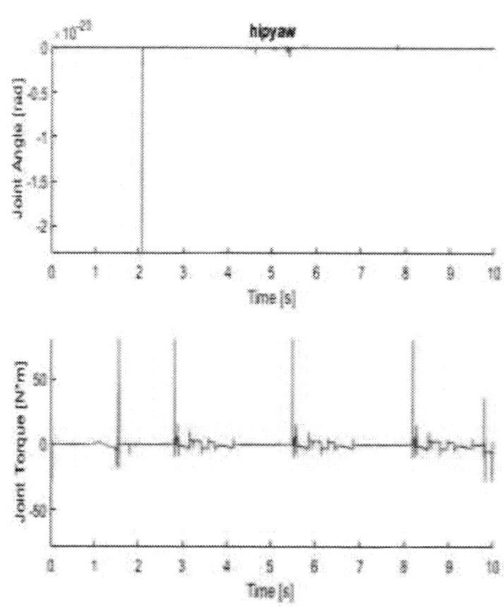

Fig. 12 Joint angle and joint torque wrt time for hip yaw

IOP Conf. Series: Materials Science and Engineering 1012 (2021) 012013 doi:10.1088/1757-899X/1012/1/012013

From fig. 12 shows the angular position and torque required at hip yaw. In angle vs time it is observed that joint angle range and torque required is negligible. Therefore contribution of hip yaw in gait cycle torque requires for normal human walking is negligible.

d. Knee

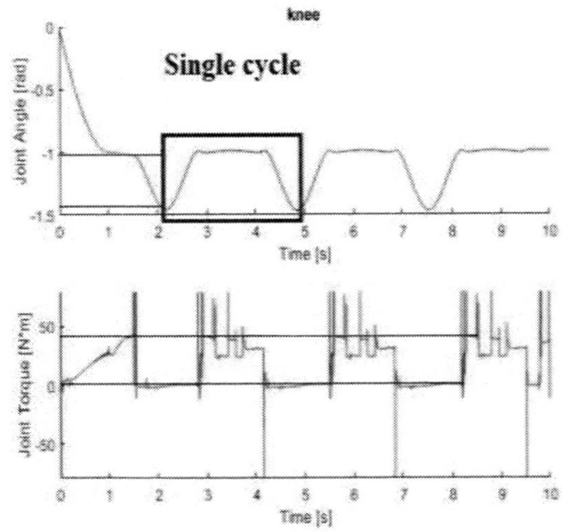

Fig. 13 Joint angle and joint torque wrt time for Knee

From fig. 13 shows the angular position and torque required at Knee joint. In angle vs time it is observed single cycle in gait as highlighted in rectangle and variation of total angle throughout the cycle is -1 rad to -1.5 rad i.e.range is 28.6 degrees. Range and Max joint torque is 0 Nm to 45 Nm and 45 Nm respectively.

e. Ankle Pitch

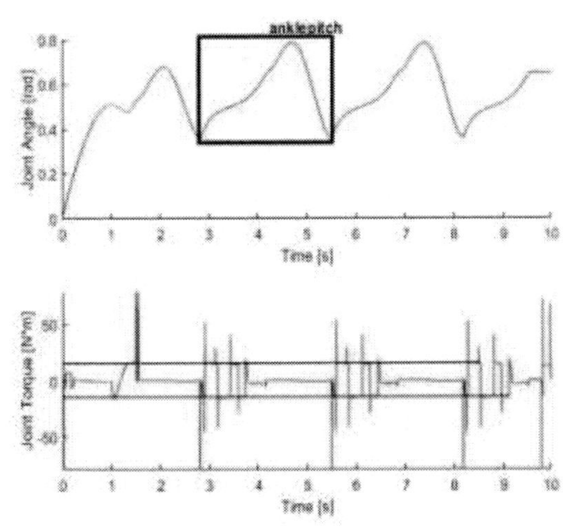

Fig. 14 Joint angle and joint torque wrt time for Ankle pitch

From fig. 14 shows angular position and torque required at ankle pitch. In angle vs time it is observed single cycle in gait as highlighted in rectangle and variation of total angle throughout the cycle is 0.4 rad to 0.8 rad i.e.range is 22.9 degrees. Range and Max joint torque is -10 Nm to 20 Nm and 20 Nm respectively.

f. Ankle Roll

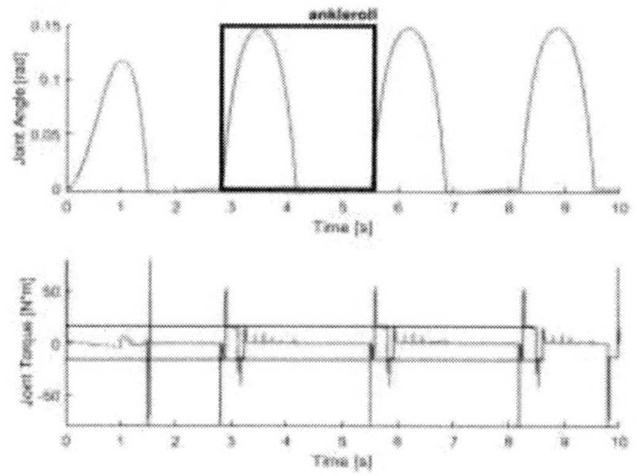

Fig. 15 Joint angle and joint torque wrt time for Ankle roll

From fig. 15 shows angular position and torque required at ankle roll. In angle vs time it is observed single cycle in gait as highlighted in rectangle and variation of total angle throughout the cycle is 0 rad to 0.15 rad i.e.range is 8.6 degrees. Range and Max joint torque is -10 Nm to 20 Nm and 20 Nm respectively.

7. SUMMARY OF RESULTS

1. MATLAB simulation using sim scape and genetic algorithm is able to teach biped robot to walk human like gait cycle.

2.Results obtain from video analysis of human gait using treadmill setup (Fig. 2 and Fig. 4) matches with reference gait cycle (Fig. 5). From Hip Angle of gait cycle using video analysis matches exactly to reference gait in terms of pattern and range. Also Knee angle of gait cycle using video analysis matches the pattern only inverted (due different reference angle for measurement) and rage is 10 degrees less than reference gait cycle.

3.Gait cycle can be different for different person according there physical parameters like weight, height, gender, age etc. So gait cycle obtain from video analysis is different from overall average gait cycle for unique test subject.

4. Gait cycle obtained from biped robot system for hip pitch angle matches for pattern with gait of video analysis gait cycle. But range of angle is 12 degrees smaller

5. Gait cycle obtained from biped robot system for knee angle matches for initial and terminal phases of gait cycle to gait cycle obtain from video analysis but biped robot gait is flattened for mid swing phase. Thus range of angle varies for almost 30 degrees.

6. FromMathematicalmodeling of human walking maximum torque is required at knee during mid stance during Stance phase is 105.6 Nm. Similarly Maximum torque required for biped robot system is at mid stance phase and equal to 45 Nm.

7. It can be observed from mathematical modeling and biped robot simulation that Torque required at hip joint is less than knee joint throughout the gait cycle.

8. from mathematical modeling it was calculated Torque at required at ankle is negligible but simulation shown torque required at ankle is significant.

8. CONCLUSIONS

1. Results obtain from video analysis and simulation of biped robot for position vs time graph i.e. Gait cycle are as following.

   For knee –
   From Fig. 2 and Fig. 13 we can observe that range of gait cycle of Video analysis and biped robot simulation is 65 deg and 28.6 degree. Pattern of simulated results matches with video analysis 50 %.

   For hip –
   From Fig. 4 and Fig. 11 we can observe that range of gait cycle of Video analysis and biped robot simulation is 40 deg and 28.6 degree. Pattern of simulated results matches with video analysis 70 %.

   From above result we can conclude that gait cycle obtained is fairly matching for unique user.
2. Results obtain from Mathematical Modeling and simulation of biped robot for Torque vs time graph are as following.

   For knee –
   From mathematical modeling max torque in Stance and swing phase are 105.6 Nm and 8.3 Nm respectively and from simulation of biped robot in stance and swing phase are 50 Nm and negligible.

   For hip –
   From mathematical modeling max torque in Stance and swing phase are 28.8 Nm and 34.4 Nm respectively and from simulation of biped robot in stance and swing phase is almost equal and around 20 Nm.

   From above result we can conclude that Torque required for mathematical model and simulation follows pattern.
3. MATLAB Simulation using Simscape with genetic algorithm can be used to teach human like walking to Model of human body with and without exoskeleton.
4. Torque results obtain from simulation are in correspondence to torque results obtain from mathematical modeling of normal human walking.
5. Overall above methodology to generate customized gait for exoskeleton using Video analysis of test subject and MATLAB simulation to generate torque require at joint actuator can be used.

Still there scope in refining proposed methodology, above method can be used on different exoskeletons to further check out this method to obtain customized gait cycle for exoskeleton. Simulation results can be improved by using Bio Mechanical simulation of human body i.e. more human like biped robot instead simple stick and brick diagram so that actual human like weight distribution in robot can be obtain. Optimization results can be improved by designing more detail cost function and allowing more iteration to learn.

IOP Conf. Series: Materials Science and Engineering 1012 (2021) 012013 doi:10.1088/1757-899X/1012/1/012013

References

[1] Y. M. Pirjade, A. U. Kotkar, N. M. Patwardhan, D. R. Londhe, T. P. Shelke and S. S. Ohol, Nov 2019 " Human Assistive Lower Limb Exoskeleton" Asian Journal of Convergence in Technology, AJCT, vol. 5, no. 2, ISSN No. : 2350-1146, I.F-5.11.

[2] Paolo de Leva (1996) Adjustments to Zatsiorsky-Seluyanov's Segment Inertia Parameters . Journal of Biomechanics 29 (9), pp. 1223-1230.

[3] Katherine A. Strausser et al. , 2010 "prototype medical exoskeleton for paraplegic mobility: first experimental results" Proceedings of the ASME 2010 Dynamic Systems and Control Conference, Cambridge, Massachusetts, USA.

[4] Gabi Zeilig et al 2012 "Safety and tolerance of the ReWalk™ exoskeleton suit for ambulation by people with complete spinal cord injury: A pilot study"The Journal of Spinal Cord Medicine 2012 VOL. 35 NO. 2.

[5] GeryColombo et al, November/December 2000 "Treadmill training of paraplegic patients using a robotic orthosis"Journal of Rehabilitation Research and Development Vol . 37 No . 6

[6] HeikkiUustal, "Study of Noraml Gait" MD Medical Director, Prosthetic/Orthotic Team JFK-Johnson Rehab InstituteEdison, NJ.

[7] (2020) The mathworks website. [online] https://www.mathworks.com/help/reinforcement-learning/ug/train-biped-robot-to-walk-using-reinforcement-learning-agents.html;jsessionid=fc0f8c0aa4292db7f004f094bba6

[8] (2020) The Global Optimization toolbox from Mathworkswebsite.[online] https://www.mathworks.com/products/global-optimization.html

[9] Prabhakar Naik, Jayant Unde, Bhushan Darekar and S. S. Ohol, 5th July 2019 "Pneumatic Artificial Muscle Powered Exoskeleton" International Conference on Advances in Robotics (AIR) 2019 at Indian Institute of Technology, Madras (IIT Madras), India. Article No.: 33 Pages 1–7

[10] Walsh et al., (2007) "A quasi-passive leg exoskeleton for load-carrying augmentation" International Journal of Humanoid Robotics Vol. 4, No. 3 (2007) 487–506.

[11] Adam Zoss, H. Kazerooni, Andrew Chu, 2005 "The Mechanical Design of the Berkeley Lower Extremity Exoskeleton (BLEEX)" IEEE/RSJ International Conference on Intelligent Robots and Systems, Edmonton, Alta., Canada.

[12] Y. M. Pirjade, D. R. Londhe, A. U. Kotkar, T. P. Shelke and S. S. Ohol, February2020 "Design and Fabrication of a Low-cost Human Body Lower Limb Exoskeleton" ICMRE-2020 in Barcelona, Spain.

ACKNOWLEDGMENT

Authors are thankful to College of Engineering Pune (COEP), Production Department, Robot study circle (RSC COEP), Robotics and Automation Laboratory. Authors would like to thank Dr. B. B. Ahuja, Director, COEP for his constant guidance and support.

Inverse kinematic analysis of a fixture for fine alignment of central axis of top and bottom frame using simulation

Aglawe P S [1*], **Ohol S S**[1]

[1]Department of Mechanical Engineering, College of Engineering Pune – 411005, Maharashtra, India

E-mail: pankajsaglawe@gmail.com

Abstract - In the field of robotics, the design and development of Stewart Platform is currently the widely discussed topic because of its complexity as well as the usefulness in many fields like military operations, biomedical, Stealth operations, etc. This research work focuses on designing of fixture using rotary actuators and Simulation of Stewart Platform in MATLAB. Fixture consists of controlling of different degrees of freedom of the module, precise linear as well as rotary motion and motor position controller for every degree of freedom and enhanced real time control of each module. Inverse kinematic analysis of a fixture using Sim-mechanics of MATLAB is carried out. This fixture has a unique design considering space constraints and weight constraints. Above research work for automatic alignment of parallel robot with a payload of @5kg can be useful for many applications.

Keywords: Simscape. PS-Simulink, Stewart Platform, MATLAB Sim-Mechanics

1. Introduction

The Stewart platform (SP) has been a popular research topic in robotics since its first appearance on the scientific agenda in 1965 in the renowned work by Stewart. Many publications concerning its kinematics, dynamics, work space estimation, path planning and force sensing applications have been published since the time of Stewart's original publication.

Stewart [3] "describes a mechanism, which has six degrees of freedom (DOF) controlled in any combination by six motors each having a ground abutment". Stewart platform which is parallel link manipulator is used as a flight simulator to train the helicopter pilots.

Lee ,Shah and Denavit and Hartenberg [1,2] developed Kinematic equations for parallel manipulator . Artificial Neural Network (ANN) Algorithm for solution of inverse kinematics of hexapod is devised by many researchers [4-7].The authors [8] briefly studied various methods used for deformity correction for automatic bone alignment. Morishita and Tojo [9] used fuzzy logic method for inverse kinematic analysis of multi joint robot manipulator. Many researchers simulated series robots using MATLAB driven inverse kinematic analysis to obtain desired trajectory [14, 15, 16]. Inverse and forward kinematic solution of robot manipulator is also described [10, 11, 13]. Chiddarwar and Babu proposed Artificial neural network model for inverse kinematics of serial manipulator[12].Author proposed adapted PSO algorithm for inverse kinematic solution to medical robot which is used for fracture treatment .Author tested on medical robot and results are compared with made classical PSO

[16]. Arockia and Arul performed kinematic analysis of tri-Glide and tripod parallel manipulator and experimental results are verified using ADAMS [17].
This research work presents an approach to understand and realize motion of real and it through several electric actuators. Designing of fixture using rotary actuators and observations of simulation of fixture in MATLAB has been discussed in details. The fixture movement is obtained by controlling of different degrees of freedom of the module, precise linear as well as rotary motion. The motor position controller for every degree of freedom and enhanced real time control of each module achieves synchronised motion for desired movements.
Solid model is imported in MATLAB function block, using inverse kinematic equation, desired rotation of motor is carried out which is useful in desired motion of top plate. Using Simulink-PS converter block physical signal which is used as input to plot the graph. This signal was using Scope Block and the graph of position, velocity, acceleration & torque vs time is obtained for 3 simulations of top plate namely A) linear motion along X, Y, Z axis. B) Yaw and C) along an inclined plane.

2. Inverse Kinematics for the Customised Stewart plat-form (fixture).

The purpose of inverse kinematics is to determine six leg lengths with respect to the position and the orientation of the moving frame.

Kinematics:

The precise locations of the actuator frames Bi and top platform frames Pi were obtained through CAD software and later through external measurements using Coordinate Measuring Machine (CMM).

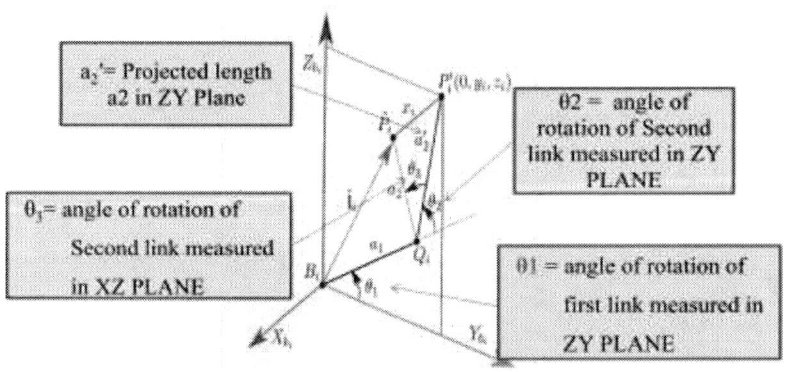

Figure 1: Kinematic diagram of an isolated 3-R manipulator

The position vector of the ith joint on the top platform with respect to the ith actuator frame is given by

$$\boldsymbol{L_i = t + R_o.(p_i) - b_i} \quad (1)$$

Where R_o is the combined rotation matrix of the moving frame with respect to the bottom fix frame. O(X, Y, Z).

$$R_o = R_z(\emptyset).R_y(\theta).R_x(\psi) \quad (2)$$

$$\theta_3 = \sin^{-1}\frac{x_i}{a_2} \; and \; a_2' = a_2 \cos\theta_3 \quad (3)$$

$$\theta_2 = \cos^{-1}\left[\frac{y_i^2+z_i^2-(a_1^2+a_2^{'2})}{2a_1a_2^{'}}\right] \quad (4)$$

$$\varphi = \tan^{-1}\frac{a_2^{'}\sin\theta_2}{a_1+a_2^{'}\cos\theta_2} \quad (5)$$

$$y_i = A[\cos\varphi\cos\theta_1 - \sin\varphi\sin\theta_1] = A\cos(\theta_1+\varphi) \quad (6)$$

$$\theta_1 = \cos^{-1}(\frac{y_i}{A} - \varphi) \quad (7)$$

Using equation (7) the final solution for joint angle of the ith actuator was obtained. On successive evaluations of the top frame locations Pi for any given pose of the moving platform and base frame locations Bi, the joint angles of of all the active joint actuators for I =1, 2, 3, 4, 5, and 6 were obtained

3. Design Constraints

3.1. Space constraints

4. The robot should not exceed the dimension 300mm × 300mm × 300mm and should not be less than the dimension 50mm × 50mm ×50mm.

These dimension limits are selected considering the following points:

1. All kinds of degrees of motion those are possible by a fixture. To acquire this dimension constraints are necessary.
2. It should be easy to move the top plate from one place to another.
3. It should be easy to handle (operate).

3.2 Weight Constraints

The robot module should not be more than 5kg. This weight limit is considering the following points:

1. It should be moderately portable.
2. Its motors should sustain its load

Table 1. Weight of components of Customized Stewart Platform (fixture)

Top Platform	Weight in kg
Motors	1
Crank with pin	0.9
Connecting Rod	0.3
Bottom Platform	0.4
Top Platform	1.1

4. Basic Structure of the Robot

The robot is basically motor operated robot with a cluster of modules. The main components that will be used for making this robot are Servo motors for each degree of freedom in motion..

Some other design considerations are:

1. It should be able to communicate with other modules without using external input device.
2. It should be environment friendly and also able to avoid danger and obstacle.

The actual model and the designed model have components of same mass and the centre of gravity is located at the same location of actual model and designed model. It has same link lengths and same spherical joints.

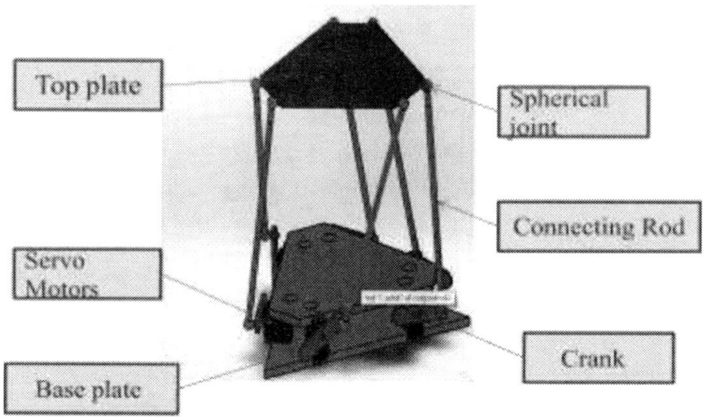

Figure 2: solid model of Customized Stewart platform for MATLAB Simulink (Sim-Mechanics)

5. Simechanics

5.1. MATLAB function block.

MATLAB function block helps in getting the desired output by processing the inputs given by the user by interpreting the equations written in the block. The equations of inverse kinematics are coded in the MATLAB function block and produced the desired output of rotation of the motors. This output was then used to get the desired motion of the platform.

Interfacing the output given by the MATLAB function block with the input of the platform.

This output was given as the actuation to the revolute joint between the motor and the crank.

The output was converted in the Physical signal by using Simulink-PS converter block and then was provided as input. The output from the revolute joint was converted into Simulink signal using PS-Simulink converter block.

This signal was then traced using Scope Block and the graphs of Position, Velocity, Acceleration, actuator Torque against Time was obtained.

The Simulation was then run in the Mechanics explorer.

Following 3 simulations are carried out:

1-Top plate moving in linear motion along X, Y, Z axis.

2-Top plate rotating and moving in linear direction(yaw)

3-Top plate moving along an inclined plane.

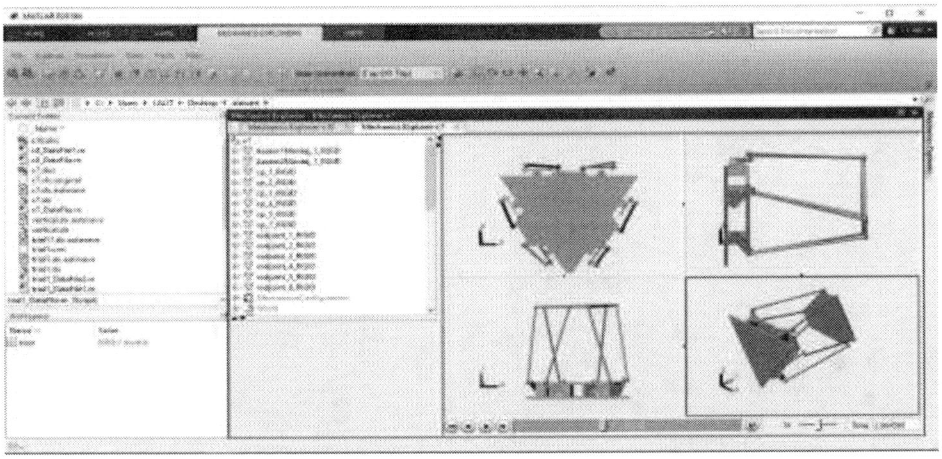

Figure 3: Top plate moving in linear motion along X, Y, Z axis.

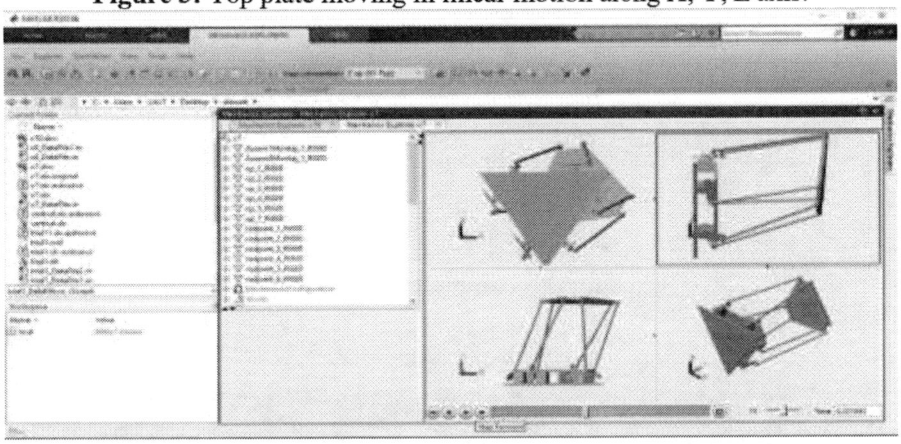

Figure 4: Top plate rotating and moving in linear direction(yaw)

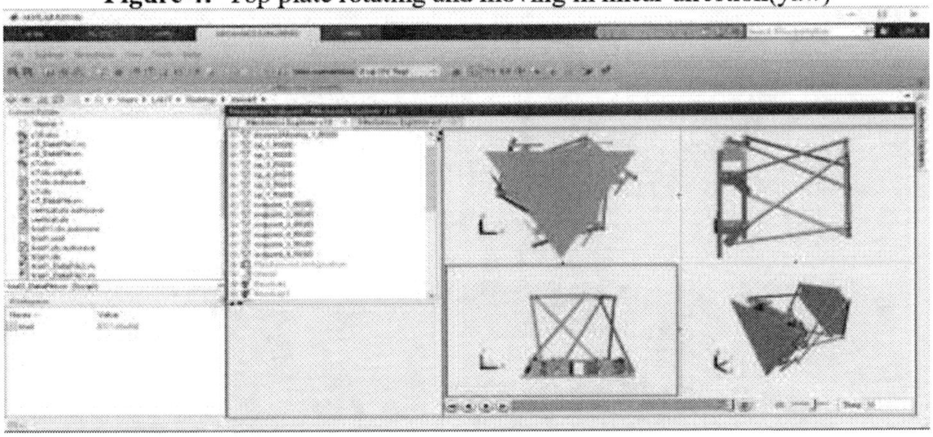

Figure 5: Top plate moving along an inclined plane

5.2. Plant model of Stewart Platform using SimMechanics

Simulink is a graphical programming environment developed by Mathwork.. It is widely used in automatic control and digital signal processing for multidomain simulation and Model-Based Design.

IOP Conf. Series: Materials Science and Engineering 1012 (2021) 012014 doi:10.1088/1757-899X/1012/1/012014

Using MATLAB Simulink [18] plant model of fixture (customised Stewart Platform) created as shown in Figure 6.

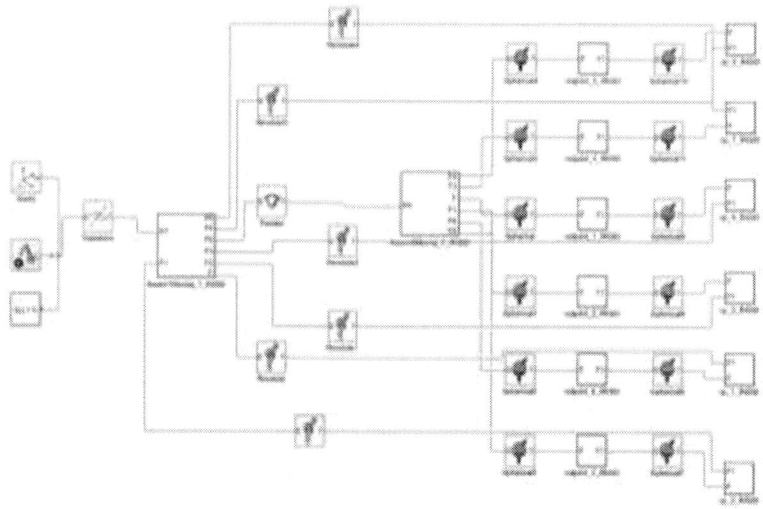

Figure 6: Plant model of Stewart Platform using SimMechanics

6. Analysis of motion of Stewart's Platform

Graphs of position, velocity acceleration and torque versus time are plot. They are as follows Mat lab Simulation results- (GRAPHS)

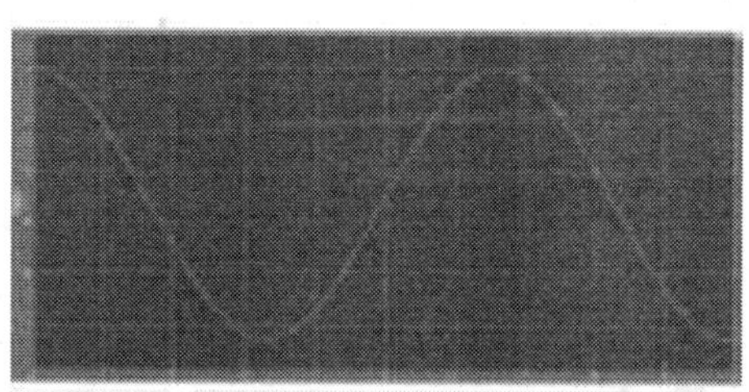

Graph 1: Angular displacement of Customized Stewart platform(fixture)

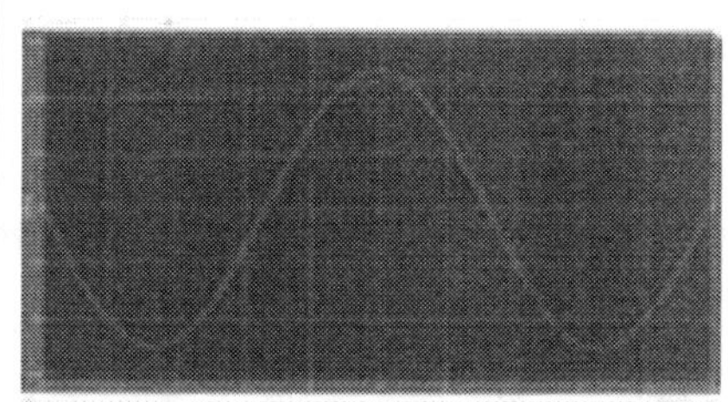

Graph 2: Angular velocity of Customized Stewart platform(fixture)

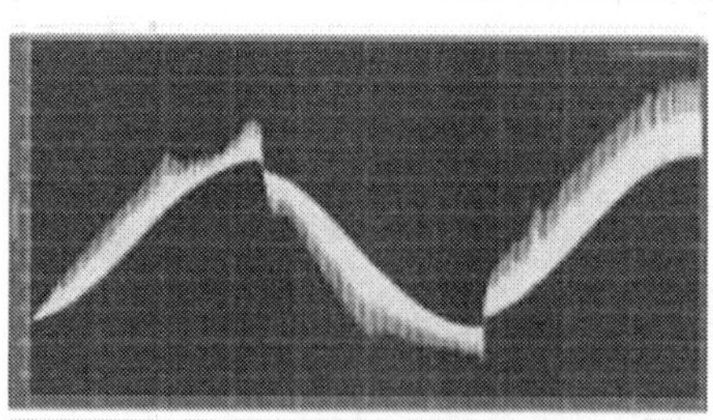

Graph 3: Angular acceleration of Customized Stewart platform(fixture)

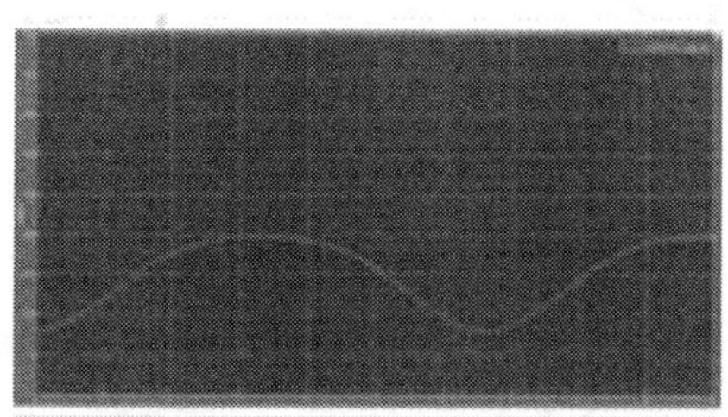

Graph 4: Angular torque of Customized Stewart platform (fixture)

IOP Conf. Series: Materials Science and Engineering 1012 (2021) 012014 doi:10.1088/1757-899X/1012/1/012014

Graph	Analysis of graph
Graph 1: Angular displacement Customised Stewart platform (fixture)	It is a shifted sine curve. The maximum displacement of shaft from its zero position is **1 radian**,in clockwise direction.
Graph 2: Angular velocity of Customised Stewart platform (fixture)	It is sine curve. The maximum velocity of shaft from its zero position is **.5 rad/sec** .
Graph 3: Angular acceleration of Customised Stewart platform (fixture)	The maximum acceleration of shaft from its zero position is **.8 rad/sec** ^{2}
Graph 4: Angular torque of the Customised Stewart platform	The maximum torque for shaft from its zero position is **.03Nm**.

The same curve is traced by all the motors when alternate motors moving in clockwise and remaining alternate in counter-clockwise direction. Above graphs are obtained for Top plate moving in linear motion along X, Y, Z axis.

Graph 5: Angular displacement of Customised Stewart platform(fixture)

Graph 6: Angular velocity of Customised Stewart platform(fixture)

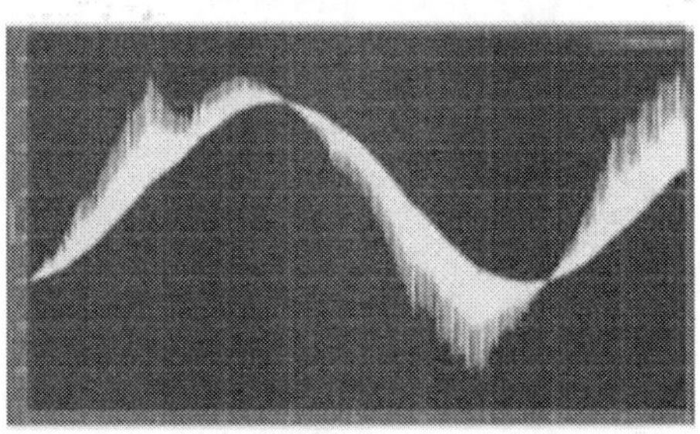
Graph 7: Angular acceleration of Customised Stewart platform (fixture)

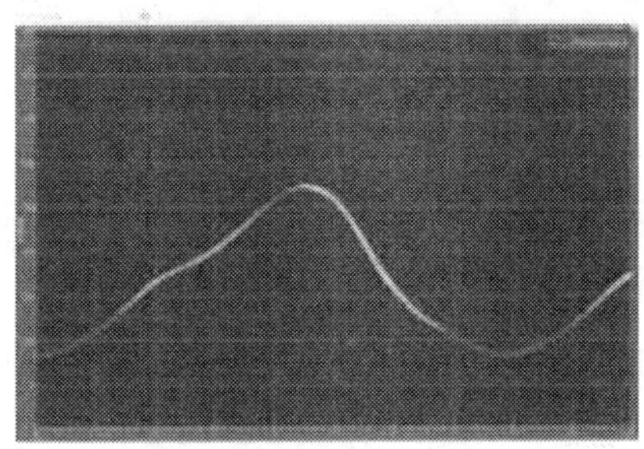
Graph 8: Angular torque of Customised Stewart platform (fixture)

Graph	**Analysis of graph**
Graph 5: Angular displacement of Customised Stewart platform(fixture)	It is a shifted sine curve. The maximum displacement of shaft from its zero position is 1.2radian,in clockwise direction.
Graph 6: Angular velocity of Customised Stewart platform(fixture)	It is sine curve. The maximum velocity of shaft from its zero position is **.5 rad/sec**
Graph 7: Angular acceleration of Customised Stewart platform(fixture)	The maximum acceleration of shaft from its zero position is **.6 rad/sec^2**
Graph 8: Angular torque of the Customised Stewart platform(fixture)	The maximum torque for shaft from its zero position is **.045 Nm**.

The same curve is traced by all the motors when alternate motors moving in clockwise and remaining alternate in counter-clockwise direction.Above graphs are obtained for Top plate rotating and moving in linear direction (yaw).

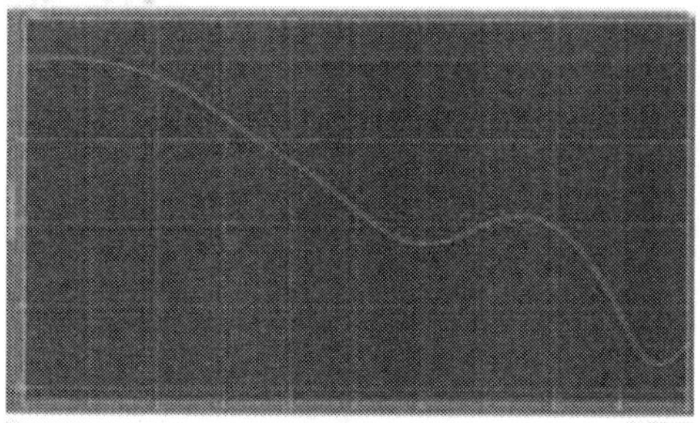
Graph 9: Angular displacement of Customised Stewart platform(fixture)

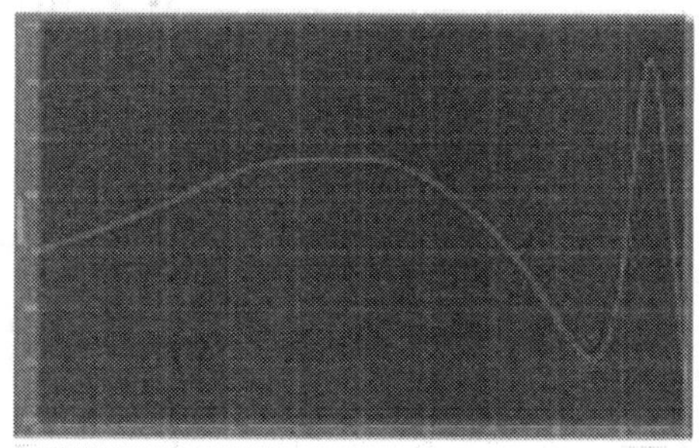
Graph 10: Angular acceleration of Customised Stewart platform(fixture)

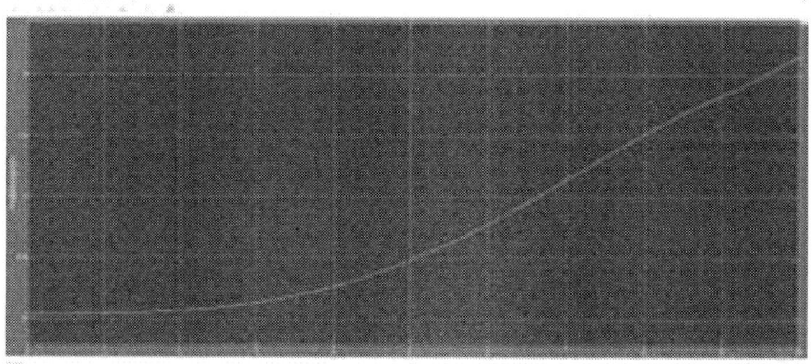
Graph 11: Angular displacement of Customised Stewart platform(fixture)

Graph	Analysis of graph
Graph 9: Angular displacement of Customised Stewart platform (fixture)	The Top Platform is slowly coming downwards.
Graph 10: Angular acceleration Customised Stewart platform (fixture)	The maximum acceleration of shaft from its zero position is **0.175 rad/sec ^{2}** .
Graph 11: Angular displacement of Customised Stewart platform (fixture)	The displacement of shaft from its zero position is increasing. The same curve is traced by all the motors when alternate motors moving in clockwise and remaining alternate in counter-clockwise direction.

7. Results and Discussions

This paper aims at designing of fixator using rotary actuators and Simulation of fixator in MATLAB.

It consists of controlling of different degrees of freedom of the module, precise linear as well as serpentine motion and motor position controller for every degree of freedom and enhanced real time control of each module in assistance with feedback from proximity/visual sensory system.

To obtain linear movement in x, y, z axis in both direction (positive & negative direction), sine wave is used as an input for angular displacement and angular velocity. [9] Amplitude of displacement is 0.5 rad. Therefore, maximum velocity amplitude is 0.5 rad/sec.

Software mathematical operation on given input to obtained required output i.e. acceleration & torque. From graph no. 3& graph no.4 it is observed that maximum angular acceleration is+0.8 rad /sec^2 .Also from graph no.4 maximum angular torque is+0.03 N-m & minimum angular torque is-0.03 N-m. Therefore, the optimum torque is 0.025Nm.

This paper focuses designing of fixture using rotary actuators and Simulation of Stewart Platform in MATLAB, fixture consists of controlling of different degrees of freedom of the module, precise linear as well as rotary motion and motor position controller for every degree of freedom and enhanced real time control of each module for Inverse kinematic analysis of a fixture Sim-mechanics of MATLAB.

8. Conclusions

In this work, modified 6-DOF parallel manipulator with rotary actuator which is designed for effective scheduled micro movement of all possible alignments, has been discussed. This work, based on inverse kinematics, will provide a baseline for determining the six leg lengths corresponding to the position and the orientation of the moving platform. The simulation of Stewart platform and designing of the rotary actuator assisted fixture is also discussed. Simulink module of the MATLAB software is used for completing simulation while estimating maximum angular displacement, velocity along angular acceleration and torque. The maximum values obtained for displacement, velocity and acceleration using simulation are 0.5 rad, 0.5 rad/sec and 0.8 rad/sec^2. The maximum and minimum values of angular torque are in the range of ± 0.03 N-m.

Therefore this simulation predicting the working range of fixture can be helpful for various applications related to fine alignment in all possible directions.

References

[1] Lee K M,and Shah D K .1998. Kinematic analysis of a three-degree-of-freedom in-parallel actuated manipulator. *IEEE Journal of Robotics and Automation*; **4**:354-60

[2] Denavit J and Hartenberg R S. 1955. A kinematic notation for lower-pair mechanisms based on matrices. *Transaction of ASME Journal of Applied Mechanics*; **22**: 215-21.

[3] Hostens I, Anthonis J and H Ramon.2005. New design for a 6 dof vibration simulator with improved reliability and performance. *Mechanical Systems and Signal Processing*; **19, 1, (105)**:(2005).

[4] Cheng Y, Ren G and Liang Dai S.2004. The multi-body system modeling of the Gough–Stewart platform for vibration control. *Journal of Sound and Vibration*; **271, (599)**:3-5.

[5] Navvabi H and Markazi A H D.2018. New AFSMC Method for Nonlinear System with State-dependent Uncertainty: Application to Hexapod Robot Position Control. *Journal of Intelligent & Robotic Systems*. **10.1007/s10846-018-0850-4**.

[6] Djukic D J, Zhukov Yu A, Korotkov E B, Moroz A V and Slobodzyan N S.2018. Hexapod digital control using the inverse dynamics and it implementation on the radiation-resistant arm-microcontroller. *Issues of radio electronics*;**7**:103-10.

[7] Zhukov Y A, Korotkov E B, Moroz A V, Zhukova V V and Abramov A M.2018. A control of

hexapod based on the neural networks solution for forward kinematics. *IOP Conference Series: Materials Science and Engineering;* **441:**1757-899.

[8] Aglawe P S, and Ohol S S.2020 .Deformity correction methods for fracture bone alignment: an overview.*International Conference on Mechanical Engineering for Sustainable Development (ICMESD-2020);* **Feb 17-18. 2020**:466-71.

[9] Morishita T and Tojo O.2012.Integer inverse kinematics method using Fuzzy logic. *Intelligent Service Robotics*,**6**,101-08.

[10] Wenjun L, Yufeng L, Tingli Y, Zhixing S and Meitao F.2010.Numerical study on inverse kinematic analysis of 5R serial robot. *International Forum on Information Technology and Applications (IFITA).*

[11] Jha P and Biswal B B.2013.Inverse Kinematic Solution of Redundant Robot Manipulator using Artificial Neural Network.*Proceedings of International Conference on Advances in Mechanical Engineering ;***May 29-31**, 2013, COEP, Pune.

[12] Chiddarwar S S and Babu N R.2010 .Comparison of RBF and MLP neural networks to solve inverse kinematic problem for 6R serial robot by a fusion approach. *Engineering Applications of Artificial Intelligence*; **23**: 1083–92.

[13] Jha P,Biswal B B, and Sahu O P.2015. Inverse kinematic solution of robot manipulator using hybrid neural network. *International Journal of Materials Science and Engineering;***3**: 31-38

[14] Wei W, Xin Z, Han Li-li, Min W and Zhong You-bo. 2018 .Inverse kinematics analysis of 6 – DOF Stewart platform based on homogeneous coordinate transformation.*Ferroelectrics*;***522***:108-21.

[15] Zhukov Y A, Korotkov A B, Moroz A V, Zhukova V V and Abramov A M.2018.Adaptive neural network control of hexapod for aerospace application.IOP Conf. Series: *Materials Science and Engineering* ;441.doi:10.1088/1757-899X/441/1/012001 IOP Publishing.

[16] Sancaktar I, Tuna B,Ulutas M.2018. Inverse kinematics application on medical robot using adapted PSO method. *Engineering Science and Technology, an International Journal*; **21**: 1006–10.

[17] Arockia S A and Arul K M.2014. Experimental investigation on position analysis of 3 – DOF parallel manipulators. *Procedia Engineering*; **97** :1126 – 34.

[18] MathWorks. Simulink..[Online]. Available: https://en.wikipedia.org/wiki/Simulink

IOP Conf. Series: Materials Science and Engineering 1012 (2021) 012015 doi:10.1088/1757-899X/1012/1/012015

Design, development and analysis of human exoskeleton for enhancing human capabilities

R V Vibhute[1*], S S Yeole[1], S D Waghmare[1], T A Tonde[1], S S Ohol[1]

[1]Department of Mechanical Engineering, College of Engineering, Pune, India

E-mail: vibhuterv16.mech@coep.ac.in

Abstract. The study of the exoskeleton is widely recognized as the current challenge for robotic research. The exoskeleton helps us understand the complex gait cycles of the human locomotion. This paper aims to design a lightweight and easy to use powered dynamic exoskeleton with a higher weight carrying capabilities and produce gait cycles which imitate human locomotion. In this paper, we have proposed some novel mechanisms to overcome the limitations of heavy load-carrying capacities of the soldiers and also help with gait rehabilitation of patients with spinal cord injury, stroke, or neurological disorder. The exoskeleton is divided into the upper body and the lower body. The upper arm and the forearm are actuated by DC Geared motors and thus the upper limb can trace any path in the parasagittal plane. The thigh and calf segments are actuated by DC Geared motors to mimic the human gait cycle. The torques required to achieve this motion are discussed here.

1. Introduction

An exoskeleton is a wearable device powered by a system of actuators that enhance limb movement with increased strength and endurance. The design aims to control the movement of legs and hands to assist in carrying heavy loads by signaling the motors. A powered exoskeleton can prove to be a vital device to those who have lost control of their limbs.

A rigid, powered and mobile exoskeleton is designed for a person weighing a maximum of 100 kg and height 175 cm. The upper body of exoskeleton can lift a maximum of 5 kg of weight and the lower body of the exoskeleton can carry a maximum load of 40 kg. Gearboxes have been designed to increase the torque concentration of the actuators.

This exoskeleton design is inspired by HALEX (Human Assistive Lower Limb Exoskeleton) which is a lower limb exoskeleton developed for assisting humans in gait locomotion. [1]

2. Anatomy study of human body

2.1. Weight Measurement

The weight of human body parts is in a fixed proportion. The percentage of weight of the upper arm and forearm in human body weight is 2.63 and 1.5 respectively and that of thigh, shank, and the foot is 14.47, 4.57, and 1.33 respectively. [2]

2.2. Length measurement

A human has a fixed proportion of length for various parts. According to the study, the mean value of the forearm and upper arm is chosen to be 280 mm and 320 mm respectively. Similarly, the mean value of hip to knee and knee to ankle length is chosen to be 620 mm and 260 mm respectively. [3]

IOP Conf. Series: Materials Science and Engineering 1012 (2021) 012015 doi:10.1088/1757-899X/1012/1/012015

2.3. Degrees of freedom

In a human body, every joint has different degrees of freedom. The designed DOFs of the exoskeleton should be aligned with the natural DOFs of the human body to prevent undue stresses on human joints. Considering natural DOFs and the complexity of the design, 1 active DOF is provided each at the shoulder and the elbow for extension/flexion. Also, 2 DOFs (active DOF for extension/flexion and passive DOF for abduction/adduction) are provided at the hip and 1 DOF is provided at the knee (active) and ankle (passive) each.

3. Upper Body Exoskeleton

3.1. Torque Calculations

The calculations for torque are necessary to obtain the required drive. Motor selection and gear designing are dependent on the torque value at each joint. Considering,

L_h = *Length of hand* L_f = *Length of fore arm* L_u = *Length of upper arm*
W_h = *Weight of hand* W_f = *Weight of fore arm* W_u = *Weight of upper arm*
W_{fl} = *Weight of forearm link* W_{ul} = *Weight of upper arm link*
W_{em} = *Weight of elbow motor* W_{sm} = *Weight of shoulder motor*
W = *Load on hand*

Maximum torque at the elbow can be calculated as,

$$T_{\text{elbow}} = (W + W_h) * \left(\frac{L_h}{2} + L_\text{f}\right) + (W_\text{f} + W_\text{fl}) * \left(\frac{L_f}{2}\right) \quad (1)$$

Maximum torque at the shoulder can be calculated as,

$$T_{\text{shoulder}} = (W + W_h) * \left(\frac{L_\text{h}}{2} + L_\text{f} + L_\text{u}\right) + (W_\text{f} + W_\text{fl}) * \left(\frac{L_\text{f}}{2} + L_\text{u}\right) + (W_\text{u} + W_\text{ul}) * (L_\text{u}/2) + (W_\text{em} * L_\text{u}) \quad (2)$$

3.2. Analysis

The trajectories that every human joint trace is different. Some of the human joints can be easily reproduced using various robotic joints like revolute and prismatic, whereas some human joints have complex motion which makes it difficult to reproduce using simple robotic joints.

The analysis of the upper arm and forearm assembly is obtained for various trajectories using regression analysis.

3.2.1. Vertical Line Trajectory. A vertical line of length of 500mm at a distance of 400mm from the shoulder is chosen as the trajectory for analysis.

Table 1. Equations for Vertical Trajectory

	Equation for Elbow	Equation for Shoulder
Angular Displacement θ	$-0.0541t^2 - 0.8952t + 136.49$	$0.0287t^2 + 0.2599t - 62.973$
Angular Velocity θ'	$-0.067\,t + 1.7175$	$0.0574\,t + 0.2599$
Angular Acceleration θ''	-0.067	0.0574

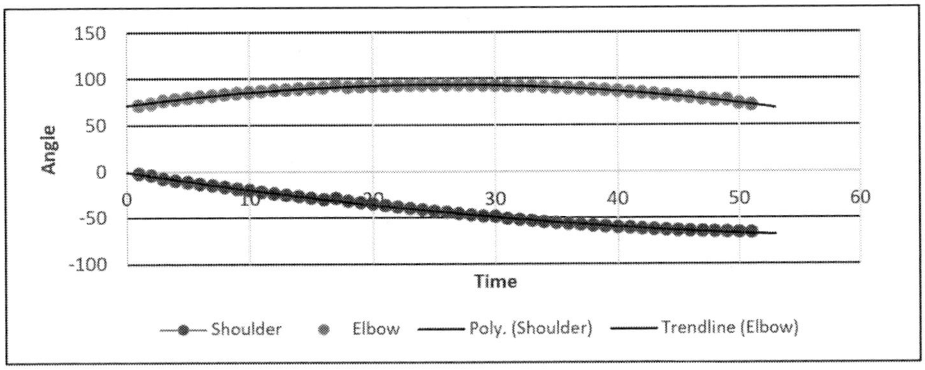

Figure 1. Regression Analysis of Vertical Trajectory

3.2.2. Horizontal Line Trajectory. A horizontal line of length of 380mm starting from 200mm from the shoulder and collinear with the shoulder is chosen as the trajectory for analysis.

3.2.3. Circular Trajectory. A circle of diameter 350mm, center at 350mm horizontally apart from the shoulder is chosen as the trajectory for analysis. The trajectory starts from uppermost point of the circle in anti-clockwise direction.

Table 2. Equations for Horizontal Trajectory

	Equation for Elbow	Equation for Shoulder
Angular Displacement θ	$-0.0335t^2 + 1.7175t + 71.028$	$0.015t^2 - 2.0768t - 0.7122$
Angular Velocity θ'	$-0.067\,t + 1.7175$	$0.03\,t - 2.0768$
Angular Acceleration θ''	-0.067	0.03

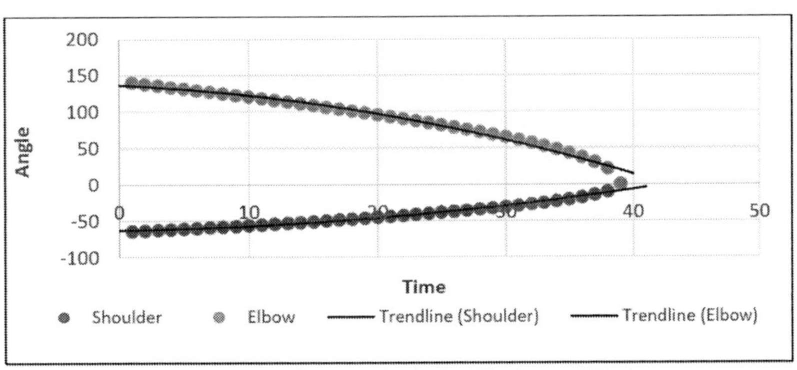

Figure 2. Regression Analysis for Horizontal Trajectory

Table 3. Equations for Circular Trajectory

	Equation for Elbow	Equation for Shoulder
Angular Displacement θ	$-0.0397t^3 + 1.8929t^2 - 21.808t + 124.12$	$0.0242t^3 - 0.8524t^2 + 4.5579t - 20.915$
Angular Velocity θ'	$-0.1191t^2 + 3.7858\,t - 21.808$	$0.0726t^2 - 1.7048\,t + 4.5579$
Angular Acceleration θ''	$-0.2382t + 3.7858$	$0.1452t - 1.7048$

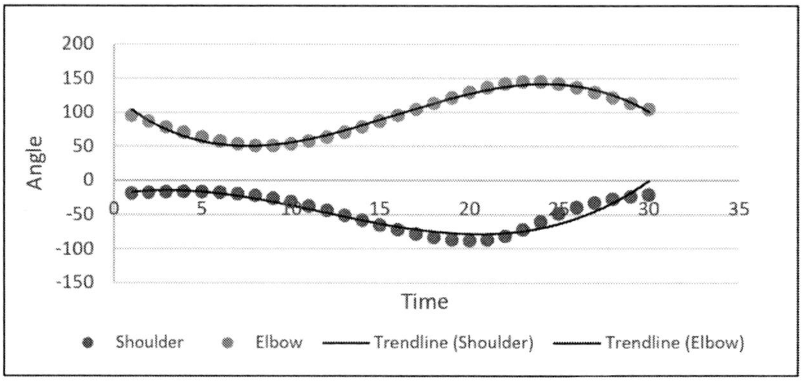

Figure 3. Regression Analysis for Circular Trajectory

4. Lower body exoskeleton

4.1. Gait analysis

An ideal kinematics appearance of the human gait cycle for walking centers of physiotherapy, orthopedics, and development of exoskeleton as a reference data for the kinematic joint parameter is the purpose of this study [4]. A person walking on a treadmill and the legs moving in parasagittal plane was video recorded. The angles of the hip(θ_1), knee (θ_2), and ankle(θ_3) of the right leg at equal intervals of time were measured by analyzing each frame in this video.

By quadratic regression,

$$(\theta_1) = -0.1915t^2 + 2.0252t + 172.8344 \quad (3)$$

$$(\theta_2) = 4.1370t^2 - 11.3538t + 161.3970 \quad (4)$$

$$(\theta_3) = -0.0168t^2 + 3.0867t + 41.3745 \quad (5)$$

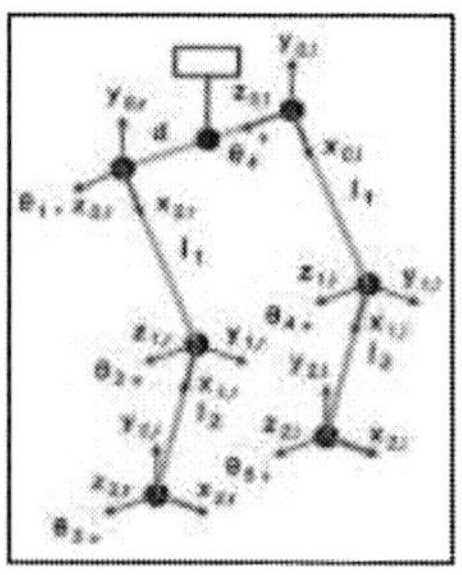

Figure 4. Distribution of reference axes on the lower limb exoskeleton [6]

4.2. Dynamic analysis

Using data from the proposed mechanisms and from Minchala *et al* [5], torque values for lower limb exoskeleton were obtained by using the following formulae,

$$\tau = \frac{\mathrm{d}}{\mathrm{dt}}\left(\frac{\partial L}{\partial \dot{q}_i}\right) - \frac{\partial L}{\partial q_i} \tag{6}$$

$$L = K - P \tag{7}$$

Where i represents the DOF, q_i represents the generalized coordinates (articular coordinates θ_i), τ represents torque applied to joint i. Also, K, P and L represent the kinetic energy of the system, potential energy of the system and Lagrangian function respectively.

The torque values for the lower limb exoskeleton were acquired by using the data from the propound mechanics and from Minchala et al. [6]

5. Design of exoskeleton

5.1. Upper arm module

In a human body, the shoulder or the Glenohumeral joint permits three degrees of freedom. It can perform abduction/adduction, extension/flexion and internal/external rotation. There is minimal translational motion in this joint.

An exoskeleton must incorporate the changes during the entire motion. The exoskeleton, if designed for all three degrees of freedom would become quite bulky and difficult to carry. Hence the degree of freedom is restricted to one. The exoskeleton is made to perform extension/flexion. Power is transmitted using DC motor coupled with a gearbox and miter bevel gears are used to reduce lateral dimension.

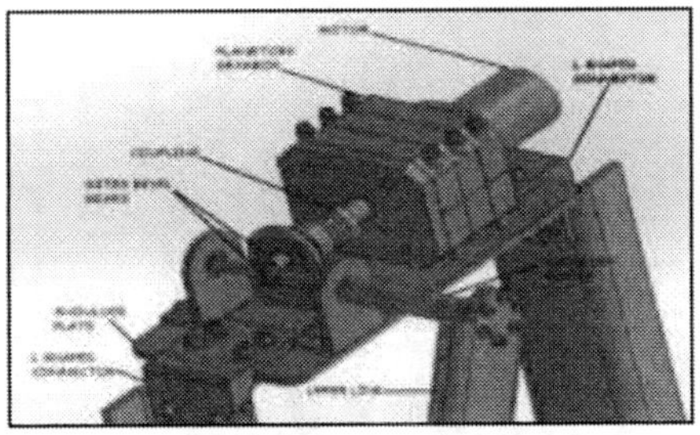

Figure 5. Shoulder module

5.2. Forearm module

The elbow joint permits two degrees of freedom which are extension/flexion and pronation/supination. The extension/flexion movement occurs in the parasagittal plane around a coronal axis.

The forearm exoskeleton is designed to serve one degree of freedom. Only extension/flexion is achieved using a DC Motor coupled with a gearbox and miter bevel gears are used to reduce lateral dimension.

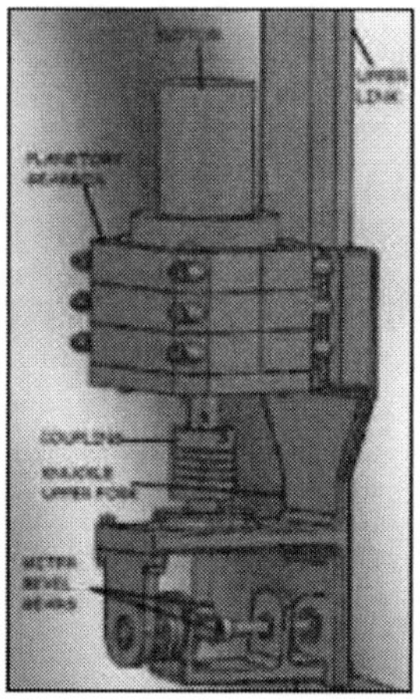

Figure 6. Fore arm module

5.3. Hip module

The rotational movement of the human hip is possible because of the ball and socket hip joint. It allows movement in 3 planes namely parasagittal, coronal and transverse. The hip module of this exoskeleton allows rotational movement in the parasagittal and coronal planes. The hip flexion/extension in parasagittal plane is active DOF and is controlled by a DC motor whereas the hip abduction/adduction in coronal plane is passive DOF.

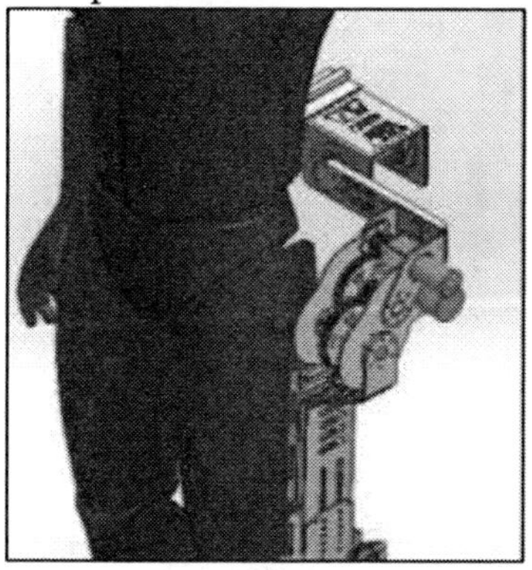

Figure 7. *Hip module and thigh segment*

5.4. Knee module

As the torque required at the knee joint is 21.68Nm, the gearbox provided a reduction of 5:1. Hence, the weight of knee and tibia part is kept optimum.

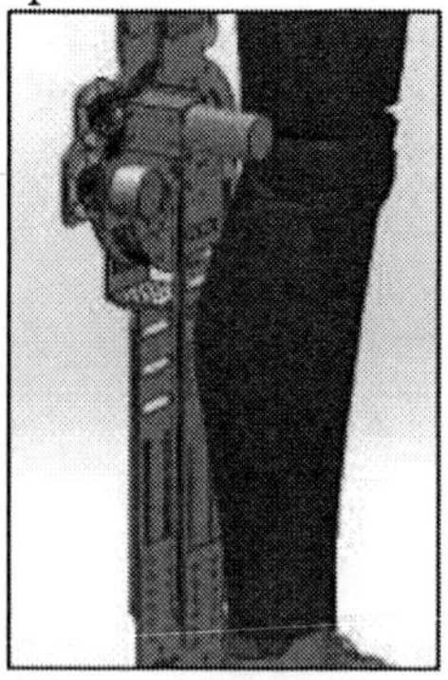

Figure 8. Knee module and calf segment

5.5. Ankle module
It should be compliant with the motion of wearer's foot. It should be light weight so as to not pose any high torque requirement for upper joint modules. It should assist the wearer during the gait locomotion.

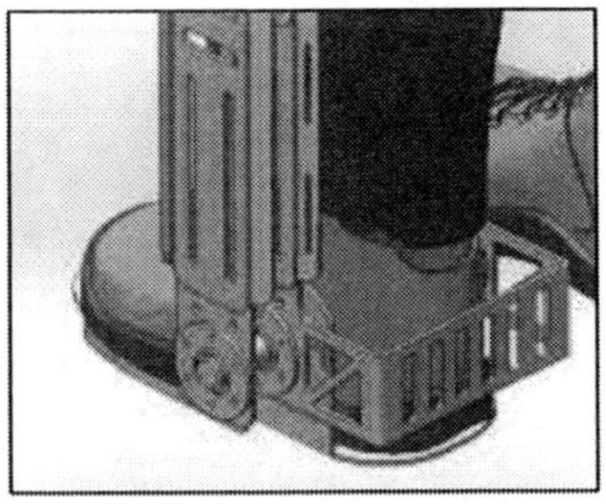

Figure 9. Ankle module

5.6. Standard Motor
DC Geared RS 550 (12 V) Motor is used. Peak specifications of motor are 7.86 oz.-in/A, i.e. 0.055 Nm/A torque, 17100 RPM and 76% efficiency. For 5A current, torque will be 0.275 N-m.

5.6.1 Motor alignment for upper limbs. The motor at the shoulder joint has to be perpendicular to the arm links for power transmission. However, this protrudes out of the assembly by increasing the width of the exoskeleton. To avoid this, the motor is located parallel to the shoulder clavicle. The power transmission is achieved using a pair of mitre bevel gears. This reduces the lateral dimensions of the exoskeleton and avoids protrusions. For this alignment, Mitre gears with 18 teeth and 0.6 modules are used with both the motors.

5.7. Gearbox
RS 550 motor does not provide sufficient torque hence gearboxes were self-designed. Planetary Gearboxes of 0.6 modules are used to obtain reduction in upper limbs whereas Compound Gearboxes of 1.25 modules is used in lower limbs.

Table 4. Specification of self-developed gearboxes

Joint	Torque(N-m)	Design Reduction Ratio	Gearbox Stages
Elbow	26.06	125:1	3 – 5:1, 5:1, 5:1
Shoulder	57.9	216:1	3 – 6:1, 6:1, 6:1
Hip	47.53	9:1	2
Knee	21.68	5:1	1

6. Conclusions and Future Scope
A human power amplification exoskeleton will be available to the soldiers, rescue workers, trekkers or the physically disabled. The exoskeleton will be easy to operate and will assist the user for carrying the load up to a maximum of 5 kg by the upper limb and 40 kg by the lower limb. The exoskeleton can bear the static human weight of 100 kg. The self-weight of the full-body exoskeleton is 20 kg.

By using composite materials, the self-weight of the exoskeleton can be further reduced. The use of composite materials can help in increasing strength. The load-carrying capacity can be increased by incorporating higher torque motors. Strain Wave Gearing or Harmonic gearing, which provides maximum output torque with minimum size and weight, can be implemented. Electro-myographic sensors can be used to predict real-time intentions of the user and assist accordingly.

References

[1] Y M Pirjade, A U Kotkar, N M Patwardhan, D R Londhe, T P Shelke, and S S Ohol 2019 *Human Assistive Lower Limb Exoskeleton* (Asian Journal of Convergence in Technology) Vol. 5

[2] De Leva P 1996 *Adjustments to Zatsiorsky-Seluyanov's segment inertia parameters* (U.S.A.: Bloomington)

[3] Gordon Claire C. *et al* 1989 *Anthropometric Survey of U.S. Personnel: Summary Statistics Interim Report*

[4] Sadiq Jafer Abbass and Ghaidaa Abdulrahman 2014 *Kinematic analysis of human gait cycle* Vol.16 No.2, pp.208-222 (Iraq: Nahrain University)

[5] Minchala, Ismael Palacio Kenneth Astudillo Darwin Vázquez-Rodas Andrés 2017 *Mechatronic Design of a Lower Limb Exoskeleton* (Ecuador: University of Cuenca)

Acknowledgement

We express our sincere gratitude towards Dr S S Ohol for guiding us. We thank the Robot Study Circle, the Mechanical Engineering Department, and the Production Engineering Department, College of Engineering, Pune for providing state of the art facilities which have proved to be invaluable for the completion of this research.

Design and Development of an Autonomous Quadruped Robot

Sandeep B [1], **Tamil Selvan P** [1*]

[1] School of Mechanical Engineering, Vellore Institute of Technology, Chennai

E-mail: tamilselvan.p@vit.ac.in

Abstract This research paper contains the work done on the '*Design and development of an Autonomous quadruped robot*' – a system engineered to complete tasks like carrying weaponry and materials for military applications. The primary aim of the project was to design a quadruped that works successfully in harsh conditions and irregular terrain were human intervention is impossible or hard. Initially, individual components of the design were designed and modifications were made according to the requirements. The development was carried out in five major steps including frame assignment, DH parameter analysis, Kinematic and dynamic analysis, Multibody simulation, Stability analysis and execution of gaits. The average time taken for the device to complete on cycle was found to be 10 seconds in crawl gait and around 8 seconds in trot gait. Further tests on the final device also directed a consistent results in its functionality and time taken for the quadruped to complete one gait cycle.

As such, the performance of the quadruped is found to be successful and also open to further modification for better performance and varied usage requirements since the research on autonomous quadruped is at its early stages of development.

Keywords Autonomous quadruped, gait patterns and implementation, DH parameters, Multibody simulation, Stability criterion, SSM,

.

1. Introduction

Locomotion on legs is one of the foremost and reliable type of movement patterns around us in the environment. Proof of first use of movement among biologically apt beings are found around 585 million years ago. Nature helped the evolution of legs as a source of movement than use of wheels because more than half of the world cannot be travelled by use of wheels. Mammals exhibit "erect" locomotion posture (ELP)[14]. Humans show ELP in their movement too. Here, limbs are held right under the torso. Designing a robot with inspiration from their biological counterparts can lead us to provide very critical ethical and logical functions for the robots. Also, it has been seen from various studies that artificial intelligence is more efficiently developed in bio-inspired compliant designs. Here, the focus is on quadruped robots, which belongs to a broader class of robotic system with legs. Quadruped robots have four legs with each leg consisting of two or more degrees of freedom.

The first design for a legged robot system was developed by Chebyshev in 1870[1]. It was a simple 4 link system with a dual axis leg motion. Further improvements on the legged robotics systems came with the development of mechanical horse by researcher Rygg in 1893[1].This was a purely mechanical model with a pedal to help in movement. Further along 1965, Ralph Mosher along with

General Electric developed a Walking Tuck. It was an electro mechanical model which could follow mechanical motion with electrical inputs, but was huge in size[1]. MH Raibert [5] in 1986, with his study on Legged Robots that Balance, opened the doors to new developments in the field of robots. This was further supported by an extensive study on Dynamic effects of static stability in 1998 [25] by Gonzalez de Santos, Jimenez and Armada. The further improvements on quadruped robots came with the incorporation of biological inspiration to robotic design [11,15,20]. This boom was followed by Tekken[13], Boston dynamics spot mini[27] and MITs Cheetah cub[29] with impeccable stability and smooth gait patterns. Further development lead to BigDog[23] which was discontinued due to its loud noise of hydraulics and heavy weight. In this research, the focus is on developing a quadruped robot from initial design to final gait stability stage which is highly stable in maneuverability and can carry substantial amount of load while also being economically feasible.

2. Design of Quadruped Robot

The overall design process of the robot is compliantly based on the interdisciplinary design approach. This approach lends its focus on CAD, kinematic and dynamic models, MSAand FEA. This is a existing engineering approach to do research for R&D requirements for modelling of a prototype (shown in Figure 2).

In this approach a CAD drawing is designed based on the quadruped model according to the given criteria or robot requirement. The designed quadruped is then validated using multibody dynamics (MBD) mechanism. The simulation parameters are set such that it is in close proximity to the real-world conditions. For this the material properties and characteristics are set according to the real robot and also the conditions of the environment are taken into consideration.

The step by step process is as follows (See Fig 1) :

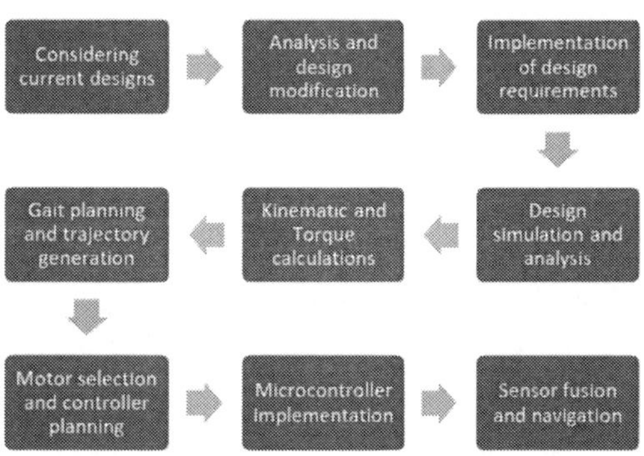

Figure 1 Development approach

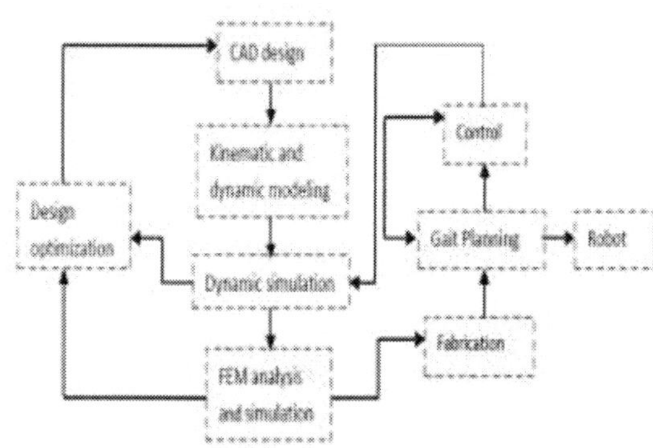

Figure 2 Design Approach

The overall simulation is based on kinematic and dynamic analysis. The trajectory and gait planning is done considering the feedback and stability of the quadruped.

To understand the biological inspiration behind the quadruped robot, consider the locomotion and limb arrangement of a dog. A dog has mainly two bones in their hind legs, Femur and Tibla which helps in the movement of legs.Here, each leg of a dog gives 3 DOF and contains mainly 2 parts ignoring the feet as in doesn't help in the degrees of movement. This 3 DOF movement of the legs help it traverse in all terrains in all direction with proper gait patterns and also provides us with a better agility and maneuverability compared to other quadruped designs.This led us to developing a similar design for the quadruped with 4 legs with 2 main segments providing 3 degrees of freedom.

The structure of leg design in dogs by the nature was rather complex and hard to achieve with perfection, which directed us to developing a similar, simpler structure with the capability and structural nature to follow the maneuverability requirement of the robot. Considering the leg of a dog in robotic terms a structure in which all 4 legs are symmetrical is developed in nature where, the upper hip joint provides the robot with the ability to turn in the lateral plain, helping in direction change. Whereas the lower hip and knee joints helps in locomotion of the robots according to gait patterns.

The basic design of this mammal inspired quadruped robot consists of 4 limbs to hold the torso and assist in its movement. Each leg is a rigid 2 link mechanism with each of it offering 3 DOF. Hence, the overall degrees of freedom of the robot is 12. All the legs consists of 3 DC motors with one motor acting on the upper hip for lateral movements in the XY plain, and other 2 motors in the lower hip and knee joints respectively. The total weight of the robot is around 60 kg with each leg carrying a wait of approximately 10 kg. The structure of the robot is developed in aluminum with a modular and flexible design consisting of no permanent joints. Each motor of the robot is controlled using 12 slave micro controllers controlled by a master. The upper hip helps in roll movement and lower hip and knee helps in the pitch of the robot respectively.

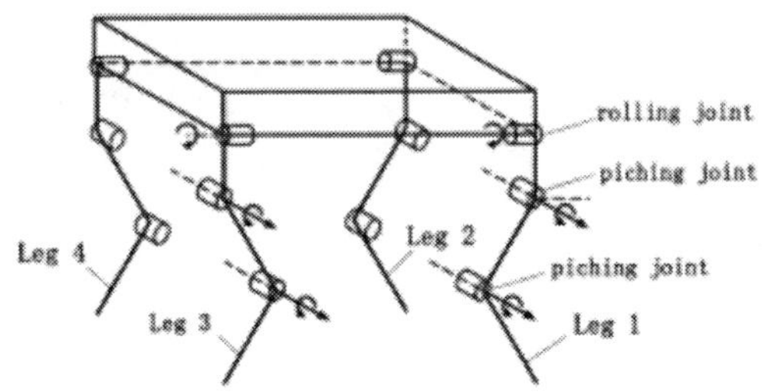

Figure 3 Joint types (with ref. to Quadruped locomotion by Gonzalez de Santos)

The torso of the robot helps in stabilizing the movements with each motor housed in its on casing. The casing works as a support to carry additional axial load during the lateral movement. The four legs provide a stable state at its initial homing position where the legs support the entire body at static state. The stability is maintained at motion using stability criterions. The lower hip and knee motors helps in vertical movement with knee motor carrying the maximum load and hence requiring maximum torque, whereas the upper hip joint helps in horizontal rotation so as to help change the direction of the robot.

Table 1. Design description of quadruped

S. No.	Characteristics	Value
1	M_{robot}	20kg (torso) + 10kg (each leg)
2	M_{motor}+ M_{extra}	3.5 kg (each)
5	Width	360 mm
6	Length	720 mm
7	Length of link 1	80 mm
8	Length of link 2	400mm
9	Length of link 3	400mm
10	Centroid coordinate	(-360,-180,-720)mm
12	Maximum speed Vmax	4 m/s
13	Gait	Trot , crawl

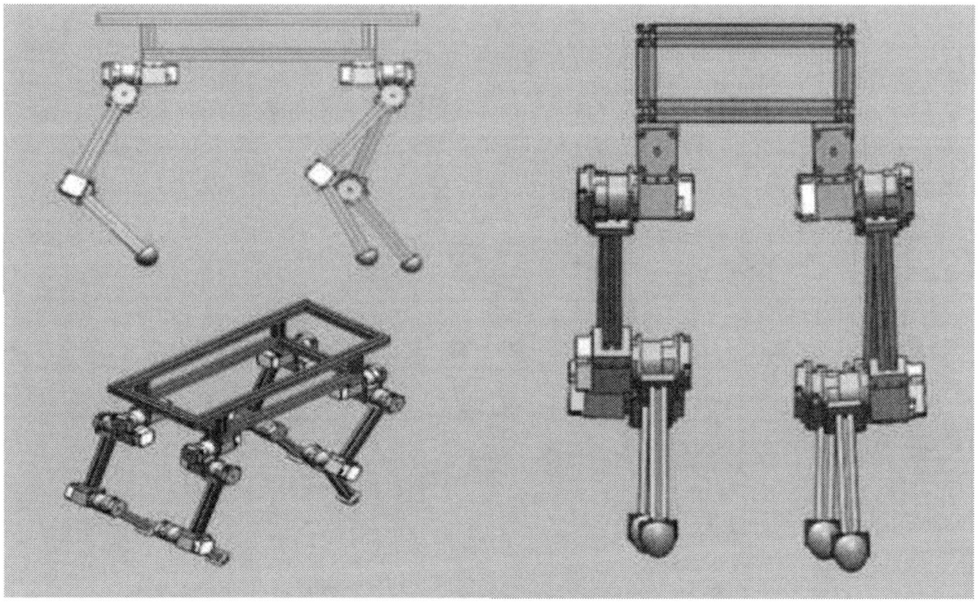

Figure 4 CAD Model of Quadruped

2.1 Torque Calculation

For calculation of motor torque, the quadruped is considered to be statically stable with all the wieght concentrated towards the centre of the body. The entire robot is considered to be in a maximum torque condition with reaction force acting vertically opposite to the ground surface.

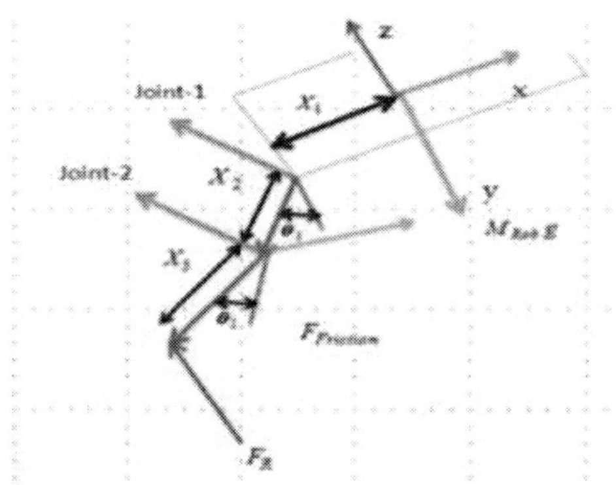

Figure 5 Free body diagram

$$W_{rob} = M_{rob}g \quad (1)$$

Considering the torque at joint1

Here, only the friction acting on the body is taken into consideration, hence the net torque at joint1 will be

$$\tau_1 = F_{Friction} * X_2 \quad (2)$$

$$F_{Friction} = \mu F_R \quad (3)$$

$$\text{Net torque at joint1} = \tau_1/\eta \quad (4)$$

$$\eta = \text{Motor efficiency} * \text{gear efficiency}$$

Now, Considering torque at joint2

Only the reaction force from ground is taken into picture, so the overall net torque at joint2 can be given as,

IOP Conf. Series: Materials Science and Engineering 1012 (2021) 012016 doi:10.1088/1757-899X/1012/1/012016

$$\tau_2 = F_r \sin(\theta_2) * X_3 \quad (5)$$

$$\text{Net torque at joint2} = \tau_2/\eta \quad (6)$$

η = Motor efficiency * gear efficiency

Here,

$X_2, X_3 = 40$ cm

$\mu = 0.25$ (flat surface)

From MBD simulation, The maximum approximate reaction force $F_R = 60N$,

From equation (3.2)

$\tau_1 = (0.25*60) * 40 = 600$ N-cm

ie, From (3.4)

Net torque $\tau'_1 = 600 / 61.88 = 9.69$ Kgcm

Similarly, Considering equation ,
when $\theta_2 = 45^o$

$\tau_2 = 60 * 0.71 * 40 = 1704$ kgcm

Net torque $\tau'_2 = 1704 / 61.88 = 27.53$ Kgcm

at $\theta_2 = 90^o$

$\tau_2 = 60 * 40 = 2400$ kgcm

Net torque $\tau'_2 = 2400 / 61.88 = 38.74$ Kgcm

2.2 Motor Selection

The maximum torque requirement of the robot is at its knee and is around 38 kgcm. Hence the motor is coupled with a planetary gearhead to give sufficient output torque. The gear head offers a gear ratio of 230:1 so as to be able to produce a high torque for joint torque requirements.

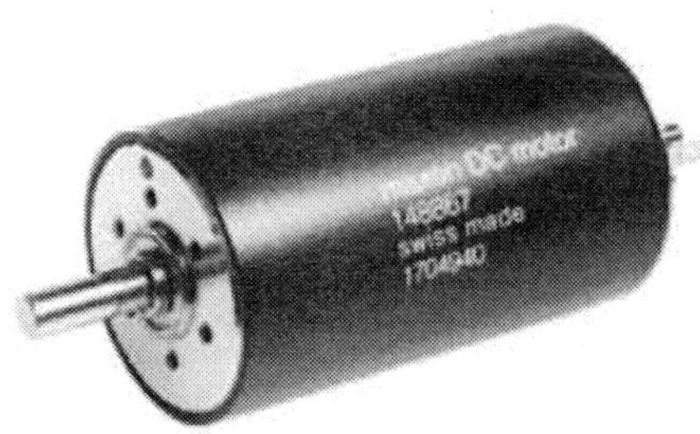

Figure 6 Maxon RE 40 motor

Figure 7 Planetary gearhead GP 52C

(ref : Maxon.com)

Table 2 Specifications of motor

Nominal Voltage	V	24
Nominal speed	rpm	6940
Nominal torque	Kgcm	1.804
Nominal current	mA	6
Stall torque	Kgcm	24.677
Max. efficiency	%	91
Terminal resistance	ohm	0.299
Torque constant	Kgcm/A	0.30
Weight	kg	0.48

IOP Conf. Series: Materials Science and Engineering 1012 (2021) 012016 doi:10.1088/1757-899X/1012/1/012016

Table 3 Specifications of gearhead

Reduction		230:1
Max. motor shaft diameter	m	0.008
Max. continuous torque	kgcm	305.94
Max. efficiency	%	68
Weight	kg	0.92

3. Kinematic analysis

Frame assignment of the robot is based on an arbitrary reference in both joint and cartesian space. The torso of the robot is defined to be in a global reference frame with each legs defined in its on local reference frames. Each joint of the leg is taken in separate frames and hence referenced to the DH parameters. The convention is taken such that the Z axis is pointed in the direction of the shaft ie, outwards and X and Y axis as mutually perpendicular to the direction of Z.

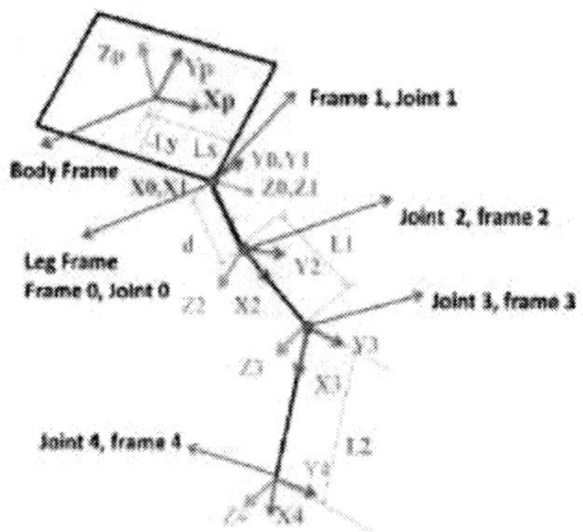

Figure 8 Frame assignment of the quadruped

DH parameters of the quadruped robot is defined with reference to the frame assignment and geometrical configuration considering the movement parameters of the quadruped.

Table 4 DH Parameters

S.No	***Angle of Twist (α_{i-1})***	***Length of link (a_{i-1})***	***Link offset (d_i)***	***Joint angle (θ_i)***
1	*0*	*0*	*0*	θ_1
2	$\pi/2$	$-l_1$	*0*	θ_2
3	π	$-l_2$	*0*	θ_3

3.1. Forward kinematics

The forward kinematics of the quadruped is calculated using the DH parameters in the above table. The total transformation is calculated by summing the transformations from individual frames. The end effector frame is then added to the final transformation.

Considering the transformation from 0th frame to 1st frame,

$$ {}^{0}_{1}T= \begin{bmatrix} \cos(\theta_1) & -\sin(\theta_1) & 0 & 0 \\ \sin(\theta_1) & \cos(\theta_1) & 0 & 0 \\ 0 & 0 & 1 & 0 \\ 0 & 0 & 0 & 1 \end{bmatrix} \quad (7) $$

Transformation from frame 1 to 2,

$$ {}^{1}_{2}T= \begin{bmatrix} \cos(\theta_2) & -\sin(\theta_2) & 0 & l_1 \\ 0 & 0 & -1 & 0 \\ \sin(\theta_2) & \cos(\theta_2) & 0 & 0 \\ 0 & 0 & 0 & 1 \end{bmatrix} \quad (8) $$

Transformation from frame 2 to 3,

$$ {}^{2}_{3}T= \begin{bmatrix} 1 & 0 & 0 & -l_2 \\ 0 & \cos(\theta_3) & -\sin(\theta_3) & 0 \\ 0 & \sin(\theta_3) & \cos(\theta_3) & 0 \\ 0 & 0 & 0 & 1 \end{bmatrix} \quad (9) $$

Hence, the transformation from fame 0 to 3 implies,

$$ {}^{0}_{3}T= \begin{bmatrix} C_{23}C_1 & S_{23}C_1 & -S_1 & -C_1(l_1+l_2C_2) \\ C_{23}S_1 & S_{23}C_1 & C_1 & -S_1(l_1+l_2C_2) \\ S_{23} & -C_{23} & 0 & -l_2S_2 \\ 0 & 0 & 0 & 1 \end{bmatrix} $$

Now, considering the end effector and completing the final transformation

$$ {}^{3}_{4}T= \begin{bmatrix} 1 & 0 & 0 & -l_2 \\ 0 & 1 & 0 & 0 \\ 0 & 0 & 1 & 0 \\ 0 & 0 & 0 & 1 \end{bmatrix} \quad (10) $$

$$ {}^{0}_{4}T = [{}^{0}_{1}T\,{}^{1}_{2}T\,{}^{2}_{3}T\,{}^{3}_{4}T] $$

$$
{}^{0}_{4}T = \begin{bmatrix} C_{23}C_1 & S_{23}C_1 & -S_1 & -C_1(l_1+l_2C_2+l_3C_{23}) \\ C_{23}S_1 & S_{23}C_1 & C_1 & -S_1(l_1+l_2C_2+l_3C_{23}) \\ S_{23} & -C_{23} & 0 & -l_2S_2 - l_3C_{23} \\ 0 & 0 & 0 & 1 \end{bmatrix}
$$

Similarly, transformation for the remaining legs are found to be same as they are similar in nature and their frame assignment are taken to be same. This also helps in reducing the complexity of the calculation

$$P_x = x = -cos(\theta_1)\ (l_1 + l_2cos(\theta_2) + l_3cos(\theta_2)cos(\theta_3)) \tag{11}$$
$$P_y = y = -sin(\theta_1)\ (l_1 + l_2cos(\theta_2) + l_3cos(\theta_2)cos(\theta_3)) \tag{12}$$
$$P_z = z = -l_2\ sin(\theta_2) - l_3cos(\theta_2)cos(\theta_3) \tag{13}$$

3.2. Inverse kinematics

For calculation of inverse kinematics of the leg frame, it should be noted that each end position has more than a one unique way of achieving success and hence they have more than one result. That is in each case there are multiple sets of joint angles for the end effector to reach the desired position. But in a legged robot, all these solutions may not be equally desirable due to factors like say stabili ty and motion type. This is when inverse kinematics comes into picture.

There are various methods used for the calculation of inverse kinematics, like graphical, algebraic and pseudo-inverse methods.Now, consider the algebraic method for inverse kinematic cacluation.
Assuming a new 4*4 matrix that is similar to a transformaton from frame 0 to frame 4

$$
{}^{0}_{4}T = \begin{bmatrix} r_{11} & r_{12} & r_{13} & x \\ r_{21} & r_{22} & r_{23} & y \\ r_{31} & r_{32} & r_{33} & z \\ 0 & 0 & 0 & 1 \end{bmatrix} \tag{14}
$$

By multiplying the inverse of first transformation matrix with transformation equation,

$$ {}^{0}_{1}T^{-1}\,{}^{0}_{4}T = [{}^{1}_{2}T\,{}^{2}_{3}T\,{}^{3}_{4}T] \tag{15}$$

By comparing the value on both sides, three sets of independent equations wit r_{ij} are obtained,

$$xcos(\theta_1) + ysin(\theta_1) = (L_2\cos(\theta_2 + \theta_3) + (L_1\cos(\theta_2) + d)$$
$$xsin(\theta_1) - ycos(\theta_1) = 0$$
$$Z = L_2\sin(\theta_2 + \theta_3) + L_1\sin(\theta_2)$$

Solving these equations, joint angles of leg 1 are found,

$$\theta_1 = tan^{-1}\ (y/x) \tag{16}$$
$$xcos(\theta_1) + ysin(\theta_1) - d = k \tag{17}$$

By taking the square of equation *16* and *17* and adding them, the third joint angle is obtained,

$$\theta_3 = cos^{-1} \{(z^2 + k^2 - l_1^2 - l_2^2)/(2l_1l_2)\} \quad (18)$$
$$P_1 = l_1 + l_2cos(\theta_3) \quad (19)$$
$$P_2 = l_2cos(\theta_3) \quad (20)$$

By multiplying the equation *19* and *20* with position equation *17* and *15* second joint angle is found,

$$\theta_2 = cos^{-1}\{(k*P_1*z*P_2) / (P_1^2 + P_2^2)\} \quad (20)$$

Hence, leading to all the required solutions for the joint angles. But, this joint angles are with the respect to the leg frame where the frame of reference is with respect to the leg. Now, considering the torso as global frame of reference, the velocity analysis is carried out.

3.3. Jacobian Velocity analysis

The jacobian associates with the angular velocity of the robot body and helps in defining the linear velocity at points of the defined modular structure. So, considering joint 0 as the global point of reference the angular velocity of the arbitrary link [i] is calculated.

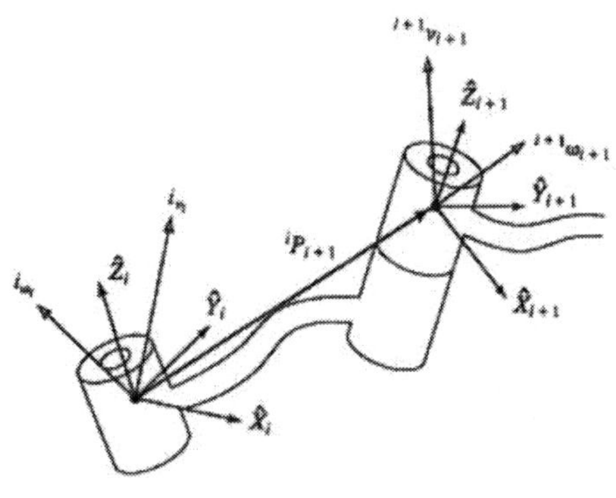

Figure 9 Velocity vector of referencing link (courtesy Quadruped locomotion by Gonzalez de Santos)

Therefore, angular and linear velocity can be given as,

$$^{i+1}_{i+1}\omega = {}^{i+1}_{i}R^{i}_{i}\omega + \dot{\theta}_{i+1}{}^{i+1}_{i+1}Z \quad (21)$$

$$^{i+1}_{i+1}V = {}^{i+1}_{i}R(^{i}_{i}V + {}^{i}_{i}\omega \; {}_{i+1}{}^{i}P) \quad (22)$$

Eq.1 (courtesy Quadruped locomotion by Gonzalez de Santos)

Now, differentiating the position equation with respect to the transformation matrix, the transformation of velocity from 0 to 4 can be defined as,

$$ {}^{0}_{4}V = \begin{bmatrix} S_1(l_1+l_2C_2+l_3C_{23}) & -C_1l_2S_2 \\ -C_1(l_1+l_2C_2+l_3C_{23}) & S_1l_2S_2 \\ 0 & -l_2S_2+l_3C_{23} \end{bmatrix} \quad (23) $$

3.3.1. linear and angular velocity from leg frame
From equation *21* and *22*,
For Link 0

$$ {}^{0}_{0}\omega = [0] \quad (24) $$
$$ {}^{0}_{0}v = [0] \quad (25) $$

For Link 1

$$ {}^{1}_{1}\omega = \begin{bmatrix} 0 \\ 0 \\ \dot{\theta}_1 \end{bmatrix} \quad (26) $$

$$ {}^{1}_{1}v = \begin{bmatrix} 0 \\ 0 \\ 0 \end{bmatrix} \quad (27) $$

For Link 2

$$ {}^{2}_{2}\omega = \begin{bmatrix} sin\theta_2\dot{\theta}_1 \\ cos\theta_2\dot{\theta}_1 \\ \dot{\theta}_2 \end{bmatrix} \quad (28) $$

$$ {}^{2}_{2}v = \begin{bmatrix} 0 \\ 0 \\ -d\,\dot{\theta}_1 \end{bmatrix} \quad (29) $$

For Link 3

$$ {}^{3}_{3}\omega = \begin{bmatrix} sin(\theta_2+\theta_3)\dot{\theta}_1 \\ cos(\theta_2+\theta_3)\dot{\theta}_1 \\ \dot{\theta}_2+\dot{\theta}_3 \end{bmatrix} \quad (30) $$

$$ {}^{3}_{3}v = \begin{bmatrix} l_1sin(\theta_3)\,\dot{\theta}_2 \\ l_1cos(\theta_3)\dot{\theta}_2 \\ -l_1cos(\theta_3)\dot{\theta}_2 + -d\dot{\theta}_2 \end{bmatrix} \quad (31) $$

Since the end effector is in the end of the frame let us consider the change to be negligible.
Now, comparing it with the body frame,

3.3.2. Linear and angular velocities from the torso
There will not be any change in linear velocities even if the change is considered from the torso. The only evident change will be seen in the joint velocities from the body. This can be calculated by a multiplication factor of rotation from the body frame. Hence,

$$ {}^{3}_{3}\mathrm{v} = \begin{bmatrix} l_1 sin(\theta_3)\,\dot{\theta}_2 \\ l_1 cos(\theta_3)\dot{\theta}_2 \\ -l_1 cos(\theta_3)\dot{\theta}_2 + -d\dot{\theta}_2 \end{bmatrix} $$

This is further multiplied by the end effector and in turn multiplied by the rotation matrix so as to receive the final joint velocity matrix

$$ {}^{4}_{4}\mathrm{Jv} = \begin{bmatrix} 0 & P_1 & l_2\cos(\theta_{23}) \\ -k\cos(\theta_1) & -P_2\sin(\theta_1) & -P_3\sin(\theta_1) \\ -k\sin(\theta_1) & -P_2\cos(\theta_1) & -P_3\cos(\theta_1) \end{bmatrix} \begin{bmatrix} \dot{\theta}_1 \\ \dot{\theta}_2 \\ \dot{\theta}_3 \end{bmatrix} \quad (32) $$

4. Trajectory planning of quadruped robot

Leg movement is basically described by the motion of a robot based on its trajectory. Here, trajectory is the time history of the position, its velocity and acceleration of each degree of freedom. When simulation is done without a proper trajectory planning, the robot movement is found to be random and irregular and the robot lacks balance. This is due to the sudden change in velocity in the cartesian space.

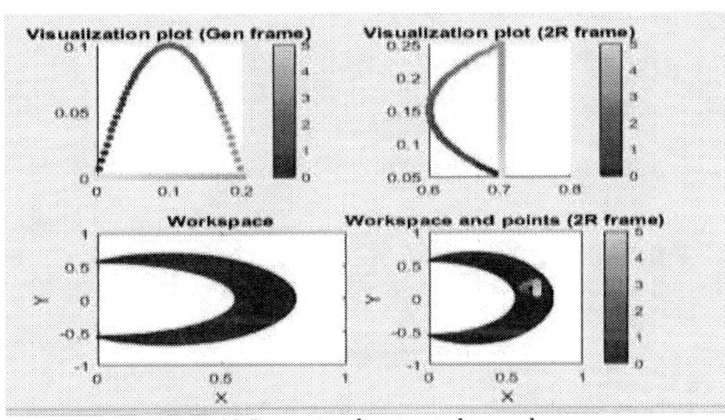

Figure 10 Point visualization in leg workspace

Trajectory planning is necessary for kinematic based motion analysis. Hence, the trajectory is generated with the help of probable via points so as to maintain smooth profile. Since the quadruped has 3 degrees of freedom. The robot shows a variation in all 3 axes of the cartesian space.

Considering the values of a function and its respective derivatives at two of the points, there is exactly one function of third degree that has the same four values, this is called a Cubic Hermite spline. There are two common ways of using the following method.

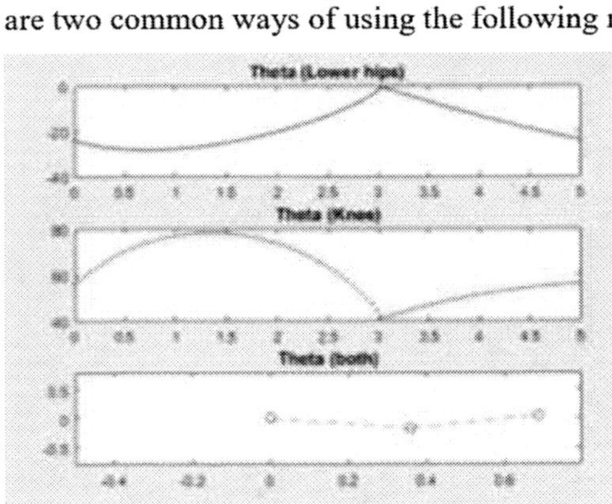

Figure 11 Joint angle generation

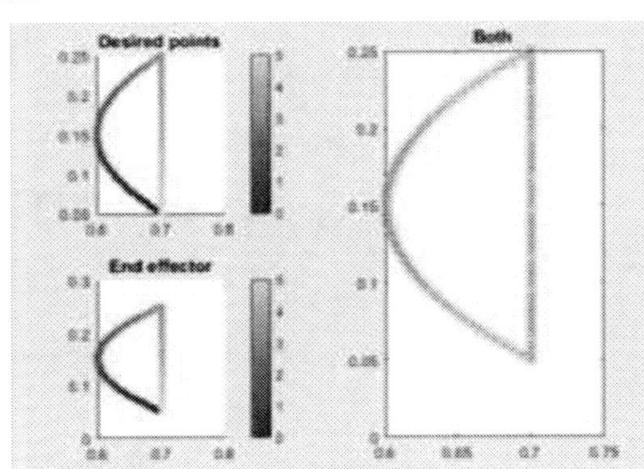

Figure12 Trajectory generation

$$f(x) = ax^3 + bx^2 + cx + d \qquad (33)$$

If the value of the said function is known at various data points, cubic interpolating the curve helps in approximating the function and its value by a continuously differentiable function, which is piecewise cubic. So, since the via points are already defined by the inverse kinematic function, using those points in MATLAB to validate a cubic spline and hence a cubic profiling of the robot motion is generated.

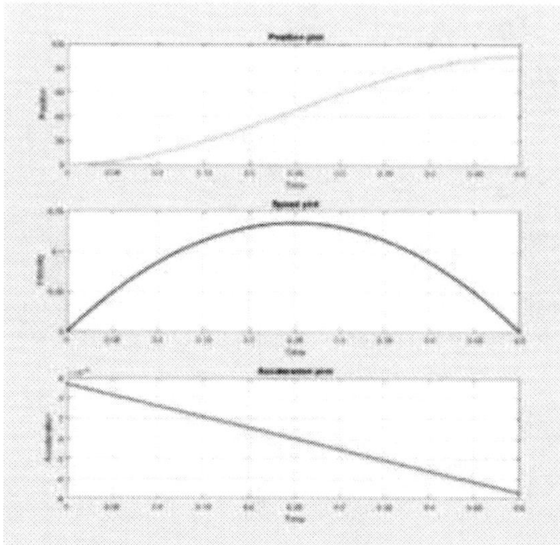

Figure 13 cubic profiling curves

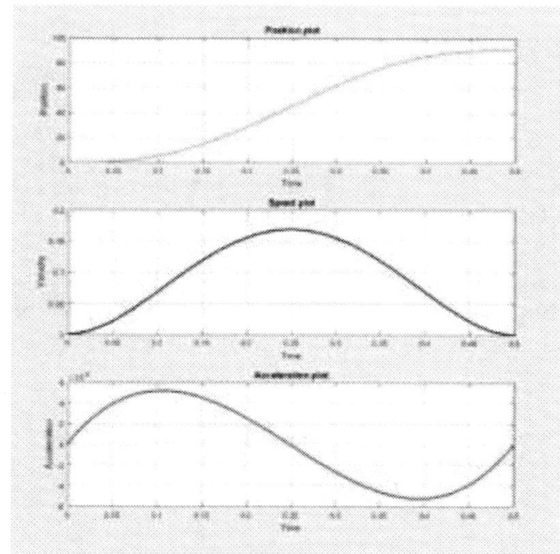

Figure 14 Fifth degree profiling curve

The given cubic polynomial is further intrapolated to the fifth degree with the help of MATLAB. The curve is then further smoothened to generate a fifth degree curve. This curve follows a softer curve where even the critical points in the third degree curve is further intrapolated to return smoother curve points

5. Gait Planning and stability analysis

A centroid based analysis of a quadruped is done considering a foot polygon. In a four legged robot, the foot polygon is basically referred in rectangular and triangular bases This section helps in finding a computational method to find the centroid of the foot polygons formed by the robotic leg.

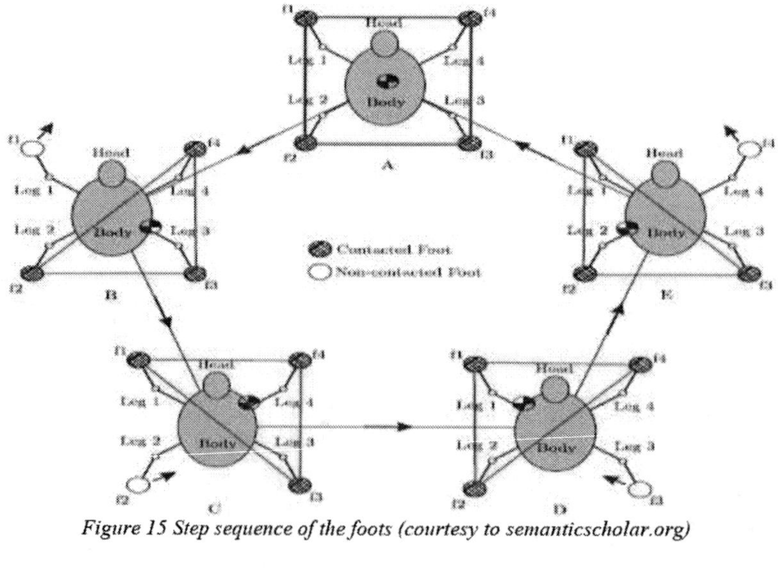

Figure 15 Step sequence of the foots (courtesy to semanticscholar.org)

During a quadruped walking, a triangular foot configuration is made the moment the first leg is lifted of the ground. For example, figure 5.3(d) shows the triangular foot of a polygon and its centroid coordinate ($Cg(xg(t), yg(t))$) position can be represented by,

$$x_g(t) = \{x_1(t) + x_2(t) + x_3(t)\}/3 \quad (34)$$
$$y_g(t) = \{y_1(t) + y_2(t) + y_3(t)\}/3 \quad (35)$$

where, $xi(t)(i= 1, 2, 3)$ and $yi(t)(i= 1, 2, 3)$ are the positions of each foot in the x- and y-direction correspondingly. At the straight state of the quadruped, a rectangular foot pattern similar to Fig. 17 is made. The centroid point of rectangular polygon can be found using the equation,

$$x_g(t) = 1/(s_1(t) - s_2(t)) \{s_1(t)x_{g1}(t) - s_2(t)x_{g3}(t)+y_{g3}(t) - y_{g1}(t)\} \quad (36)$$

$$yg(t) = \begin{cases} s_1(t)\{x_g(t) - x_{g1}(t)\} + y_{g1}(t), \text{ or} \\ s_2(t)\{x_g(t) - x_{g3}(t)\} + y_{g3}(t) \end{cases} \quad (37)$$

where $xi(t)(i= 1, 2, 3, 4)$ and $yi(t)(i= 1, 2, 3, 4)$ are the positions of each end effector duly. The parameter $s1(t)$ shows the slope character of the line passing through the centre point of the triangle $f1f2f3$, $Cg1(t)$, and also the centre of triangle $f1f3f4$, $Cg2(t)$. The point $s2(t)$ represents the slope of the line segment travelling along the centroid of triangle $f1f2f4$, $Cg3(t)$, also the centroid of triangle $f2f3f4$, $Cg4(t)$. See the step configuration in Fig 15.

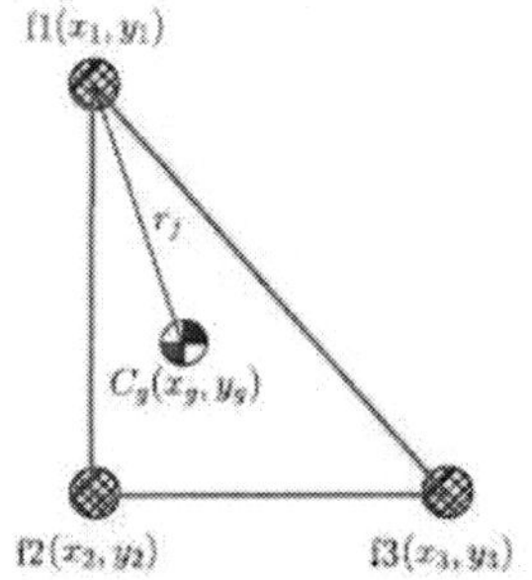

Figure 16 Triangular foot polygon

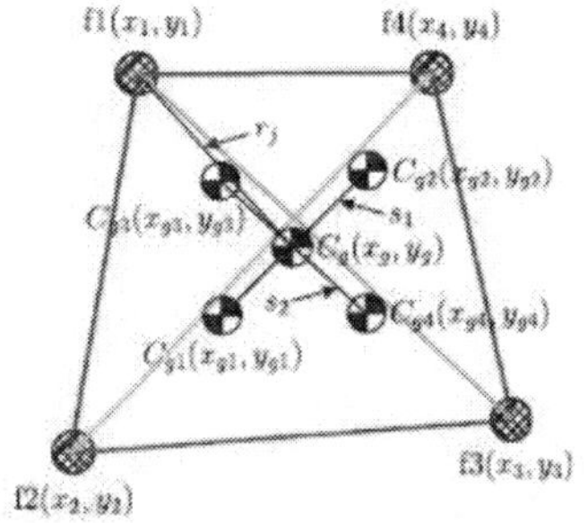

Figure 17 Rectangular foot polygon

The centroid of of all the foot polygons formed during the motion of the robot was noted and the walking balance is analyzed. Also, the balance of the quadruped robot during motion is found by using a simple index of its movement. Through various simulations, it is confirmed that the movement of a quad robot on legs requires a waddling in its motion to walk with is stable balance.

5.1 Walking balance index

Even if the overall center of mass in directed towards the centre, the balance of the configuration maybe slightly different depending on the shape of the foot polygon. It is usually considered that, as the polygon is more spacious so is the balance leading to a increased walking balance. Hence, this helps in generation of a simple performance index.

$$I_B = min(r_j(t)), j = 1, \cdots, n \quad (38)$$

Here, $\min(rj(t))$ is the minimum value that can be attained by $rj(t)$. The n is 3 for a triangular polygon and 4 in the case of a rectangular foot polygon . That is, the total distance between the proper centroid and the vertex 'j' of a polygon in (Fig.16 and Fig.17) , $rj(t)$, is calculated by

$$r_j(t) = \sqrt{(\{x_g(t) - x_j(t)\}^2 + \{y_g(t) - y_j(t)\}^2)}. \quad (39)$$

Practically, the value $I_B$in (38) represents the lowest distance between the centroid of the support polygon and position of each foot in a walking balance. It is then infered that the case with a rather large I_B in study by referring to many walking patterns with better potential of smotth walking. The walking starts with the initial step of the quadruped as follows,

Table 5 Initial position of four legs

Foot	*X (m)*	*Y (m)*	*Further remarks*
f1	*0.2*	*0.5*	*Rectangular posture as shown in figure 17.*
f2	*0.2*	*0.2*	
f3	*0.4*	*0.2*	
f4	*0.4*	*0.2*	

After going to the initial homing position, there are multiple sequence of steps so as to maintain a stable dynamic and static posture. The steps are followed in such a fashion with reference to the centroid based balancing criteria so as to make the robot traverse in a stable position.

Table 6 Example cases for step by step walking

Case	*Order of footsteps*	*Further remarks*
1	*f1 f2 f3 f4 f1*	*See Figure 15*
2	*f1 f2 f4 f3 f1*	
3	*f1 f3 f4 f2 f1*	
4	*f1 f3 f2 f4 f1*	
5	*f1 f4 f2 f3 f1*	
6	*f1 f4 f3 f2 f1*	

On managing the step criteria with respect to the timing parameters a stable walking method is obtained. The walking is maintained such that the robot is stable and without tendency to slack during motion. The motion parameters for uneven and irregular terrain is obtained similarly. (refer Figure 18)

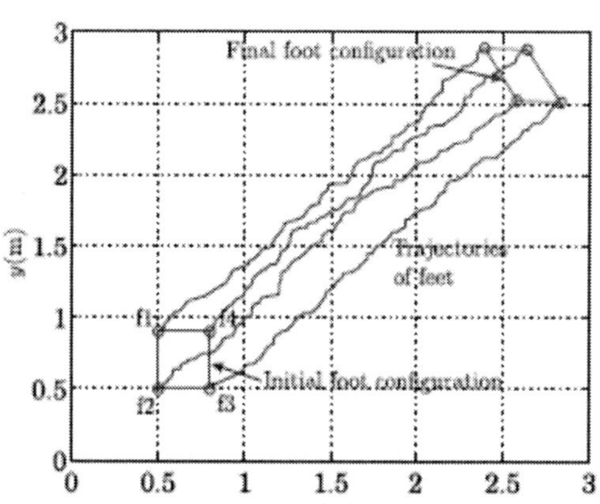

Figure 18 Reference foot trajectory for walking

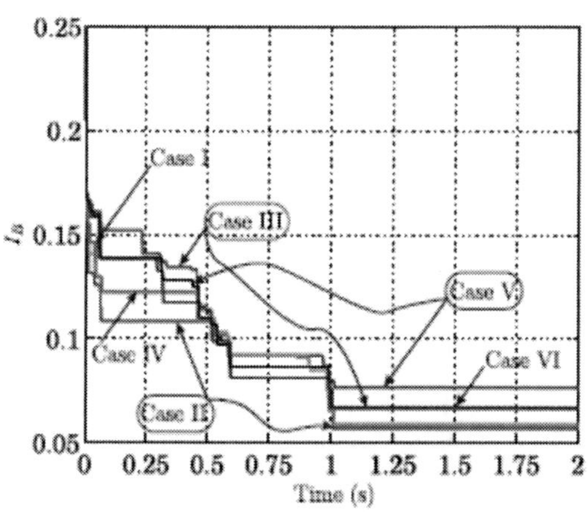

Figure 19 Balance index

(courtesy Centroid based analysis of quadruped robot walking balance)

As the robot starts its motion, all the designed centroid trajectories are being moved in another form. This implies that the centroid position trajectory of the bot depends on the order of foot movement during motion. In theory, the wriggling trajectories directs our attention to the idea that the four-legged walking robot walks with a waddle like movement similar to a duck during movement. Considering the facts, it is not a strange behaviour, and such a wriggle like motion is naturally occurring for the structural balance of the robot. This is because it is necessary for the quadruped to adjust its motion during walking so as to prevent itself from falling while in motion. But these motions are rather

unstable and undesirable. That is why sudden changes in centroid trajectories are not allowed for dextrous walking and such characteristics requires the need for an excessive walking. Furthermore, it is noted from the trend of that the walking behaviour of the robot that the movement can be more balanced by properly scheduling the entire walking order.
Next, to understand what style helps in better walking balance, look at the performance index defined in (18) for the walking task. Fig. 19 shows the indices of walking balance for all the legs as assigned in Table 6. From Fig.19, it is to be confirmed that walking style case 3 defined in Table 6 maintains the best balance is the front part of the quadruped, but Case 5 is better than the others for the rear. Also, the overall walking balance of the Case 2 is found to be the lowest of all for the walking task. In a four-legged task, the idea for the walking motion of the robot is to be controlled by removing any unwanted waddle like motion. One major reason for this is that waddling tends to disturb the overall control performance of the quadruped. In theory, if the COM of the quadruped vehicle is to be directed to the centroid of foot polygon in all the steps during motion, it is really easy to guarantee the overall walking balance in each step taken during the process. So, in order to identify a characteristic of waddling in a quad robot movement, it is useful to check the trajectory of the centroid of the foot polygon formed in every step of its motion.

6. Experimental Results

6.1 Multibody simulation of the quadruped leg

Leg of the quadruped robot shows 3 DOF which includes one roll and two pitch motion. This simulation helps in the calculation of joint torque and joint motion for proper selection of actuators. The variation in torque and joint motion is observed.This simulation helps in understanding the leg ratio requirements to achieve required output torque and also joint angles for the projection.

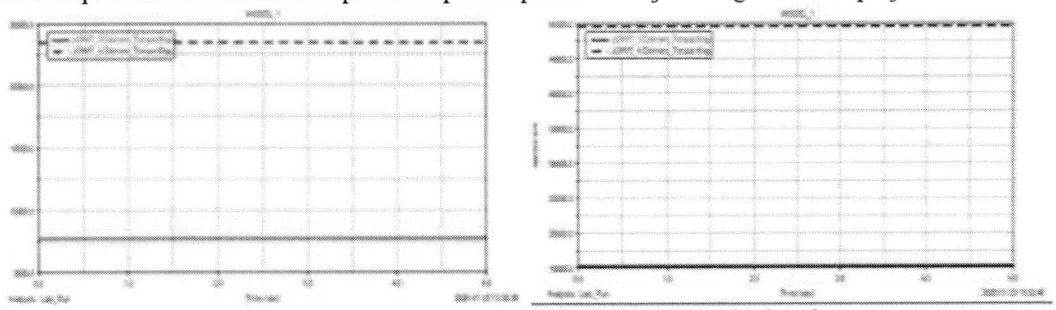

Figure 20 Torque requirement at medium and maximum load condition

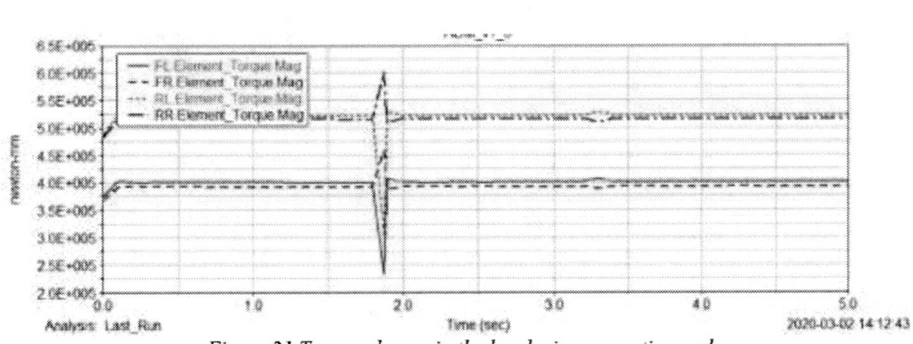

Figure 21 Torque change in the leg during one action cycle

6.2 Multi-body simulation of the robot

This section shows the whole-body simulation of the quadruped robot. This simulation was carried out in an Adams – MATLAB embedded atmosphere. The quadruped simulation carries out a proper analysis of the robot motion. The joint characteristics are observed along with motion analysis for smooth motion planning. The simulation also helps in observing the stability criterion and making necessary changes to make the quadruped stable. Crawl and Trot gaits were implemented in simulation and the crawl gait is found to be the most stable and also requires the most torque for its movements

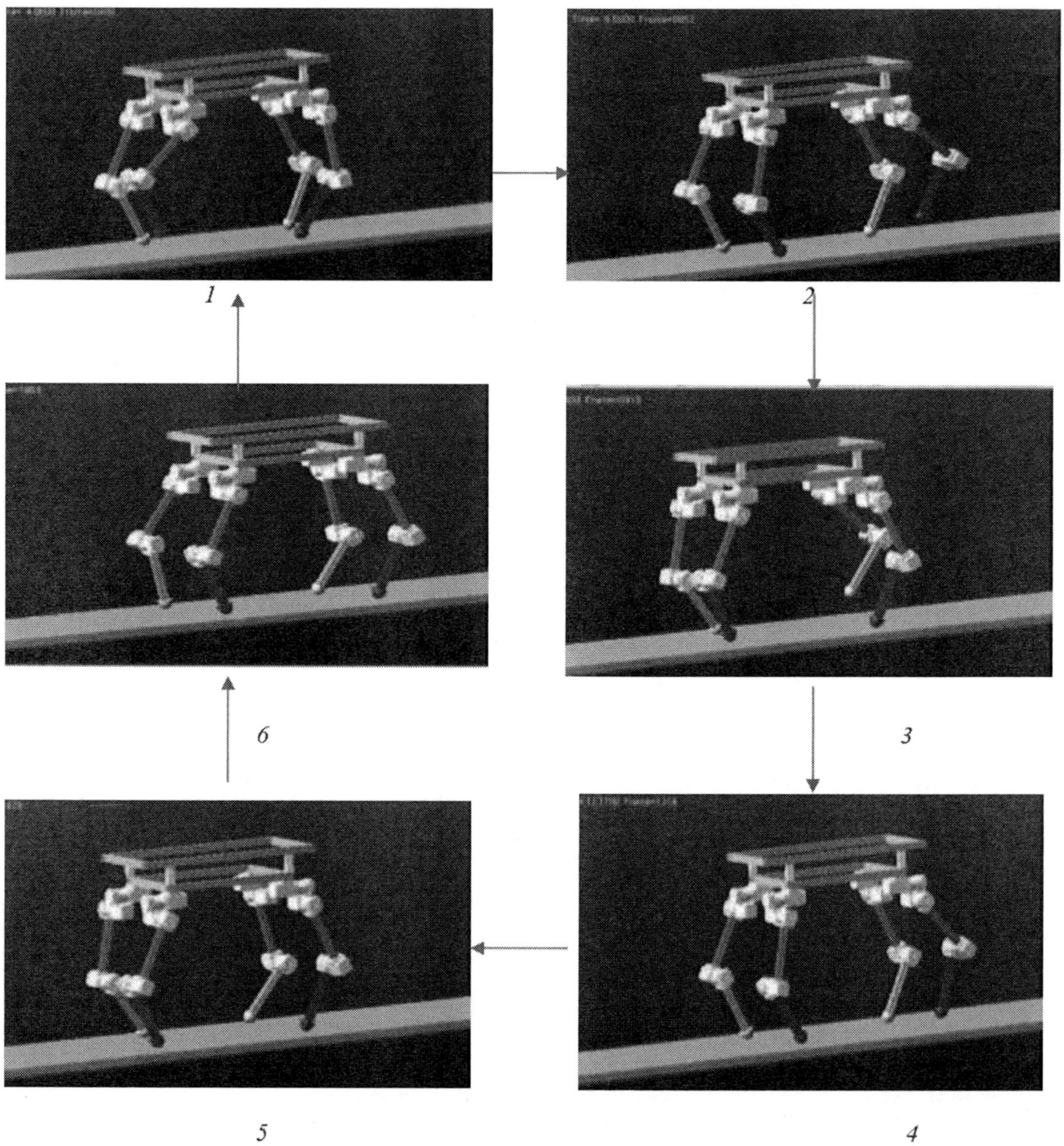

Figure 22 Gait simulation

The simulation cycle returned the following results ,

CRAWL GAIT :

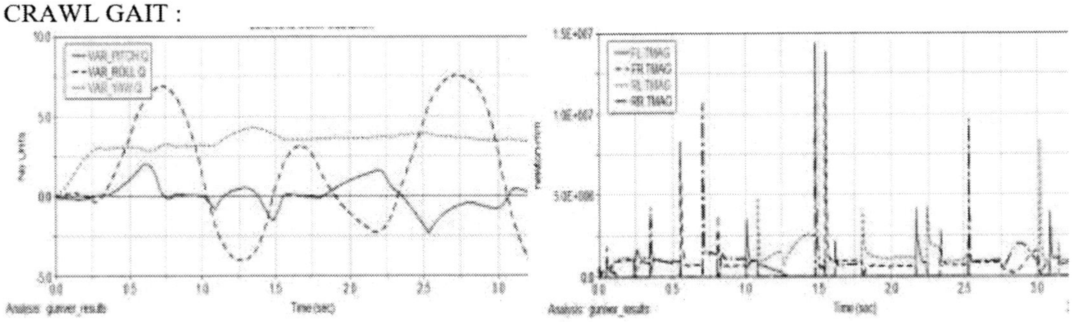

Figure 23(a) Roll, pitch and yaw acting on the quadruped (b) Torque

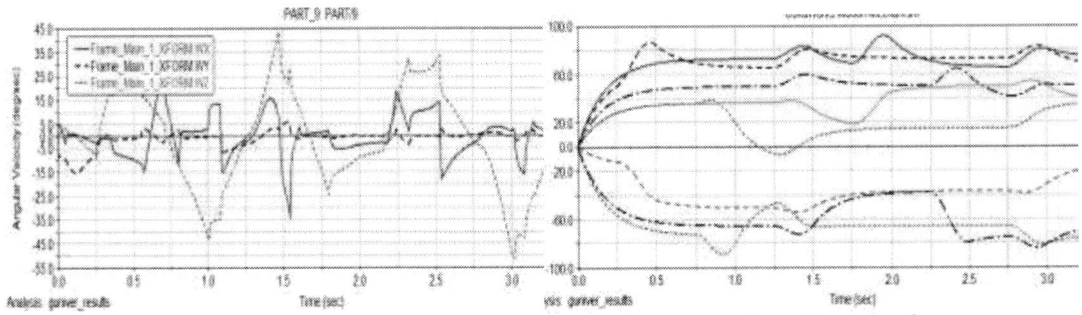

Figure 24(a) Angular velocity variation in x, y and z axis during crawl gait (b) angular change

TROT GAIT :

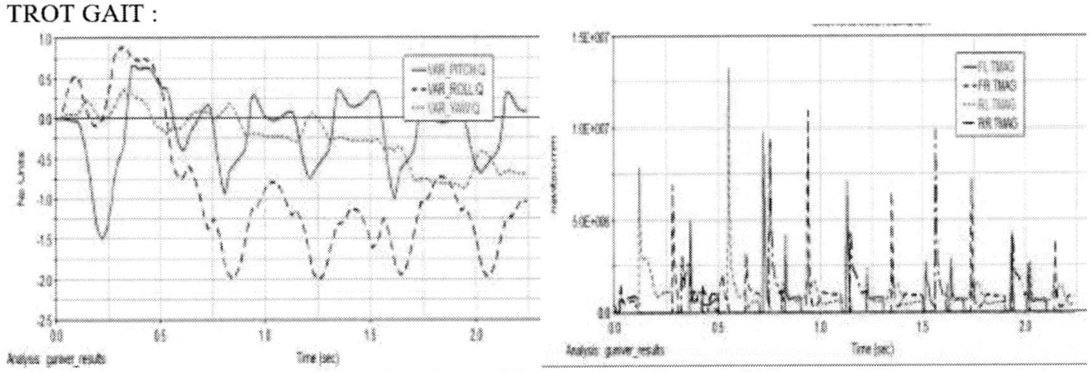

Figure 25(a) Roll, Pitch and Yaw of the robot during Trot gait (b) Torque variation

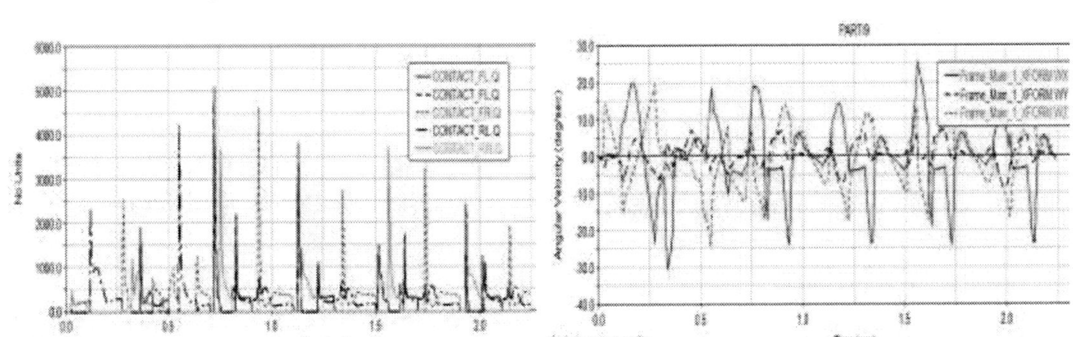

Figure 26(a) Contact force on each leg during trot gait (b) Angular variation

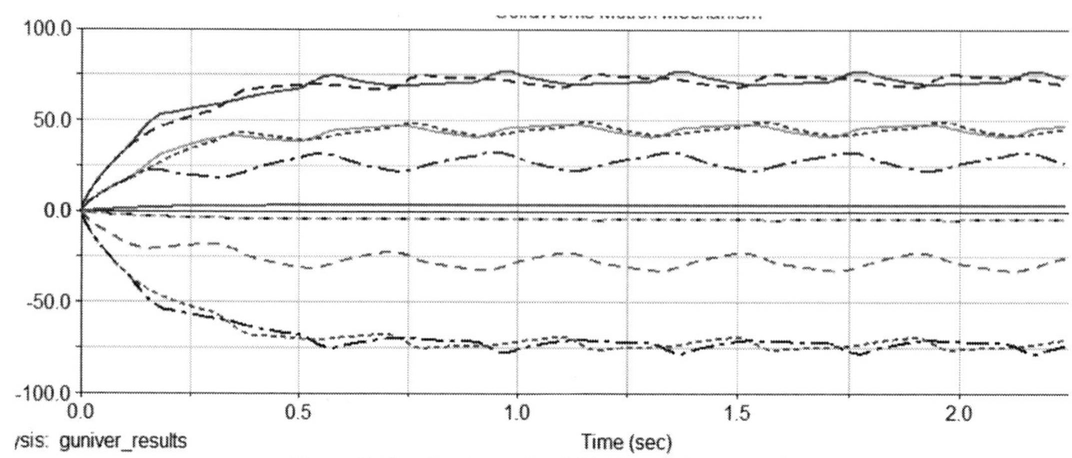

Figure 27 Angular change in all 12 joints during trot gait

7. Overview of Control Architecture

The motion control of quadruped is based on a 3 Channel Input capture (IC) method consisting of slave and master controls. The controllers are connected serially to the PC with the master controller controlling the entire structure through the CAN bus.

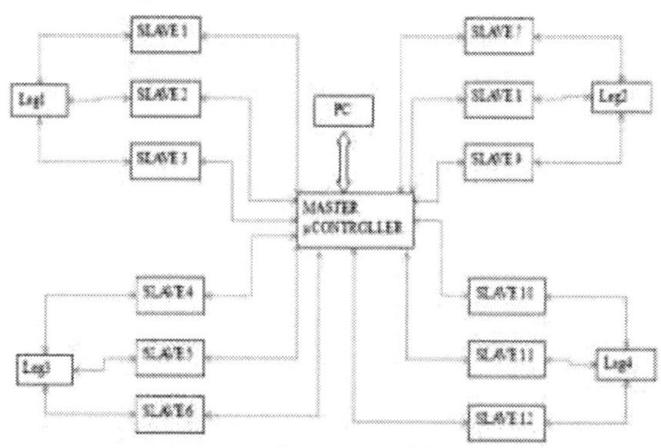

Figure 28 Control structure of the quadruped

A well-defined single axis motion control is tested, were the slave card is connected to the motor at joints and master card for the connection of all the slave cards and interface to the host computer. The master and slave cards are connected through CAN bus and a distribution control is used to split one control into 12 different signals according to movement requirements. The host communication is carried out through a RS232 serial bus. The motor is tested in various frequency ranges of PWM and the movement speed is considered according to the stability of motion.

Motion will be controlled and tested in both open loop and close loop circuits. In the open loop motion control, the control will be based on PWM signal and interrupt timer and in the closed loop control, it will be focused on the error percentage of servo loop. The motion will follow a particular motion trajectory according to requirement of movement smoothness (cubic, quadratic or fifth degree). The closed loop control will direct to us the positional feedback of motor with each instantaneous change.

8. Conclusion

Steps were taken to make the system rugged and rough with an ability to traverse in all terrains. The entire system is based on a closed PID controller loop with one master controller which transfers information to 12 slaves so as to control the joint movement. The micro controller used is PIC18F4011 micro-controller. The P-control algorithm is what helps us in achieving the control of speed an point

of the robot. The basic concept of servo loop is to calculate the total error and subtract it incrementally whereby achieving the point value of the PWM duty cycle. The error can be found by removing the final point from the given initial or target point. This helps in calculating the final position whereby the multiplying the error with the proportionality gives us the absolute value of its absolute position. This duty cycle directs us in generating the PWM which is then directed to the motor as its input. A feedback control loop is used to overcome the absence of an absolute encoder.
The quadruped was controlled using proper gait patterns with the help of kinematic analysis. The gaits were extrapolated using trajectory parameters to achieve a smooth profile and hence completing the proper gait patterns and stabilizing the overall movement. The research is hence concluded to be successful in terms of functionality and open to further necessary improvements.

References

[1] Pablo Gonzalez de Santos, Elena Garcia and Joaquin Estremera ''Quadrupedal Locomotion An Introduction to the Control of Four legged Robots'' by Springer-Verlag London Limited 2006
[2] F.E. Fish, P.B. Frappell, R. V. Baudinette, P.M. Macfarlane, ''Energetics of Terrestrial Locomotion of the Platypus OrnithorhynchusAnatinus', The Journal of Experimental Biology ,2001, 204, pp.797–803.
[3] R.W. Blob, A.A. Biewener ''Mechanics of limb bone loading during terrestrial locomotion in the green iguana (iguana iguana) and American alligator (Alligator Mississippiensis)'', The Journal of Experimental Biology,(2001), 204, pp.1099–1122.
[4] J.M. Parrish, ''The Origin of Crocodilian Locomotion, Paleobiology'', 1987, Vol. 13, No. 4, pp. 396-414.
[5] M.H. Raibert ''Legged Robots That Balance'' MIT Press, 1986.
[6] Stable Stair-Climbing of a Quadruped Robot. https://arxiv.org/pdf/1809.02891
[7] J. F. Vincent, O. A. Bogatyreva, N. R. Bogatyrev, A. Bowyer, A. K. Pahl, ''Biomimetics: its practice and theory''. Journal of the Royal Society Interface, 2006, 3(9), 471-482.
[8] A.E. Rawlings, P.J. Bramble, S.S. Staniland, ''Innovation through imitation: biomimetic, bioinspired and biokleptic research''. Soft Matter, 2012, 8, 6675-6679.
[9] B. Webb, T. Consilvio, ''Biorobotics'' 2001 Mit Press.
[10] Stoeter, Sascha A., Ian T. Burt, and N. Papanikolopouslos. "Scout robot motion model." In Robotics and Automation, 2003. Proceedings.2003.
[11] Harada, Kanako, Sheila Russo, Tommaso Ranzani, Arianna Menciassi, and Paolo Dario. "Design of Scout Robot as a robotic module for symbiotic multi-robot organisms." In Micro-Nano Mechatronics and Human Science (MHS), 2011 International Symposium on, 2011 pp. 511-513.
[12] Kimura, H., Fukuoka, Y., and Cohen, A. H. (2007). Adaptive dynamic walking of a quadruped robot on natural ground based on biological concepts.International Conference on, 2003, vol. 1, pp. 90-95
[13] Fukuoka, Y., and H. Kimura. "Dynamic locomotion of a biomorphic quadruped 'Tekken'robot using various gaits: walk, trot, free-gait and bound." Applied Bionics and Biomechanics 6, 2009, no. 1,pp 63-71.
[14] Li, Yibin, Bin Li, JiuhongRuan, and Xuewen Rong. "Research of mammal bionic quadruped robots: a review." In Robotics, Automation and Mechatronics (RAM), 2011 IEEE Conference on, 2011 pp. 166-171.
[15] Zhou, Xiaodong, and Shusheng Bi. "A survey of bio-inspired compliant legged robot designs." Bioinspiration & biomimetics 7, 2012: no. 4.
[16] Chen, Xianbao, Feng Gao, Chenkun Qi, and Xianchao Zhao. "Spring parameters design to increase the loading capability of a hydraulic quadruped robot." In Advanced Mechatronic Systems (ICAMechS), 2013 International Conference on, 2013, pp. 535-540.
[17] Estremera, Joaquin, and Kenneth J. Waldron. "Leg Thrust Control for Stabilization of Dynamic Gaits in a Quadruped Robot." In Romansy 16, 2006, pp. 213-220. Springer Vienna.
[18] Vivian Allen, Ruth M. Elsey, Nicola Jones, Jordon Wright and John R. Hutchinson, "Functional specialization and ontogenetic scaling of limb anatomy in Alligator mississippiensis, Journal of anatomy, 2009, pp 423-445

[19] T. McGeer ''Passive dynamic walking''. Int. J. Rob. Res. 9, 2 (March 1990), pp. 62-82. [31] Anand Kumar Mishra, "Design, Simulation, Fabrication and Planning of Bio- Inspired Quadruped Robot", 2014 May

[20] Li, Yibin, Bin Li, JiuhongRuan, and Xuewen Rong. "Research of mammal bionic quadruped robots: a review." In Robotics, Automation and Mechatronics (RAM), 2011 IEEE Conference on, 2011, pp. 166-171. IEEE, 2011.

International Journal of Robotics Research, 26:475–490.

[21] Q. Huang, K. Yokoi, S. Kajita, K. Kaneko, H. Arai, N. Koyachi, K. Tanie, "Planning walking patterns for a biped robot," Robotics and Automation, IEEE Transactions on , vol.17, 2001, no.3, pp.280,289.

[22] Mori, Makoto, and Shigeo Hirose. "Development of active cord mechanism ACM-R3 with agile 3D mobility." In Intelligent Robots and Systems, 2001. Proceedings. 2001 IEEE/RSJ International Conference on, 2001, vol. 3, pp. 1552-1557.

[23]Boston Dynamics Corp. (2008). Bigdog overview. (Online) http://www.bostondynamics.com/img/ BigDogOverview.pdf.

[24] JJ craig ''Introduction to Robotics Mechanics and Control'' 3rd edition, 2004, by Pearson publication

[25] Gonzalez de Santos, P., Jimenez, M., and Armada, M. ''Dynamic effects in statically stable walking machines'' Journal of Intelligent and Robotic Systems, 1998, 23(1), pp 71–85.

[26] A. Crespi and A.J. Ijspeert: AmphiBot II: an amphibious snake robot that crawls and swims using a central pattern generator, Proceedings of the 9th International Conference on Climbing and Walking Robots, 2006, pp 19-27.

[27]Boston Dynamics Corp. (2010). http://www.bostondynamics.com.

[28] Harada, Kanako, Sheila Russo, Tommaso Ranzani, Arianna Menciassi, and Paolo Dario. "Design of Scout Robot as a robotic module for symbiotic multi-robot organisms." In Micro-NanoMechatronics and Human Science (MHS), 2011 International Symposium on, 2011, pp. 511-513.

[29] Drenner, Andrew, Ian Burt, Tom Dahlin, Bradley Kratochvil, Colin McMillen, Brad Nelson, Nikolaos Papanikolopoulos et al. "Mobility enhancements to the scout robot platform." In Robotics and Automation, 2002. Proceedings. ICRA'02. IEEE International Conference on, vol. 1, 2002, pp. 1069-1074.

[30] Walking Truck - Wikipedia. https://en.wikipedia.org/wiki/Walking_truck

Gyro-Stabilized Camera Control in drones for Military Applications

J Bharath Singh[1*], B Jaison[2]

[1]Department of Electronics and Instrumentation Engineering, R.M.K. Engineering College, Tamil Nadu
[2]Department of Computer Science Engineering, R.M.K. Engineering College, Tamil Nadu

E.mail: j.bharathsingh@gmail.com

Abstract. The usage of drones or unmanned aerial vehicle in military applications is increasing. The input or the control of drones is becoming a challenging task for the developers. One of the input recognition methods is the gesture recognition. Pilot's head movement using a gyroscope is used to control the angle of an on- board FPV (First Person Video) Camera. MPU 6050 Sensor collects the displaced values from three axis and sent it as an input for the processor which reads the possible values. The values are then compared and processed and given to the output. The output is controlled by servomotor which actuates mechanical moves based on MPU sensor. This whole setup can be connected through wireless RC (Radio Controller) transmitter and receiver. This motion controlled mechanical arms can be used for various applications like FPV (First Person View) in drone enabled surveillances, Military, Search and Rescue etc.

1. Introduction
As we know many recognition methods are used for human computer interaction. Vision recognition, Speech recognition, gesture recognition, Movement recognition are the commonly used recognition methods used in latest technology. These recognition methods have grown to a tremendous level due to rapid development of hardware and software. This gesture recognition has got a good attention recently. There are two types of gestures. In static gestures the posture of the body or hand denotes a sign whereas the dynamic gestures the movement of the body or hand conveys some message. The main advantages of the gesture recognition are accurate, high stability and time saving. Gesture recognition are commonly used in mobile phones unlocking, Domestic and Industrial automation, Automobiles, Security services, Transportation services and Robots design.

2. Drones
A Drone is commonly an Air supervisor without a human on board. It is commonly called as an unmanned aerial vehicle. It look like an aircraft. The unmanned Aerial Vehicle system will have a ground based controller, a communication protocol and a UAV. The unmanned aerial vehicle can be operated in various angle of degrees by using a remote control. Either a human or computer programming can be used for operating the remote.
A human can do manually whereas the automatic operation can be achieved by programming. When the missions are secret and dangerous, using a UAV will be best method when comparing to manned aircraft. The drones are not only used only for defense applications, it can be also used in surveillance, Agricultural (Spraying pesticides from air), Photography, transportation to human non access areas, forest fire controlling and domestic gaming purposes. UAVs work on the battery power and depending on the work the battery power can be increased. The battery selection is an important criterion because

the UAV's have to carry the battery with them. Heavy batteries will decrease the UAV's efficiency. UAVs operated using remote control is often called as Remotely Piloted vehicles (RPV).
UAVs has major role in Target and decoy, combat, logistics, research and development, Civil, agriculture and aerial photography. The major components of the drone are Body platform, Battery, Computing components, Sensors, Actuators, software and a hardware software interface board. The use of drones is keep on increasing to reduce the risks to the humans.

3. Gyro-Stabilized Camera
A small shake or error in the image will be very large after amplification. Digital stabilization is possible after the image is taken but for large amplification images, digital stabilization is not practically possible. This can be overcome by the Gyro-stabilized camera systems which uses either Fiber optic gyroscope or MEMS (Micro Electro Mechanical systems to track any movement in the camera base.[10]
If the gyroscope detects any movement, it gives an input to the processor and the camera will be tilted in the counter opposite direction to avoid the fall. The main applications of these gyro stabilized camera are mobile phones to play games, marine, vehicle and mast tower. The recent advancement provides a maximum zoom ability of the camera and the sensitivity is increased to improve accuracy.

4. Literature Survey
A various control algorithms are available for the stabilization of the drones. The commonly used algorithms are PID, Linear Quadratic Regulator (LQR), Sliding mode, Back stepping, Feedback linearization, Adaptive, Robust, Optimal, L1, H∞, Fuzzy logic and Artificial neutral networks. Each algorithm has its own advantages and disadvantages. After the comparison of many algorithms, the author confirmed PID control algorithm as the best one in terms of accuracy and analysis. [1]
The usage of unmanned aerial vehicle is common in military applications. When the history is analyzed, the role of UAV in the military applications is more. The author studied the history of drones and presented a detailed report on the usage. [2]
The main applications of MPU6050 sensor is the detection of signals from the gestures. The angles of the drones are calculated and the position is balanced. The response from the sensor is fed to the processor for further calculations and analysis. [3]
Kui Liu in "Dual-sensor approaches for Real time robust hand gesture recognition" proposed two effective methods. In the first method, Images from a left and right stereo cameras setup are merged to form a pair. The second method actively uses the Kinect depth camera and a inertial sensor to find the acceleration and angular velocity within the framework of a hidden Markov model [4].

5. Problem Identification

5.1. Existing Model

For running drones and to control the cameras fixed Joystick and the potentiometers are needed. These potentiometers controls camera only when the measurement is fed into it manually. The other existing model involves a handheld device either wired or wireless called Teach Pendant. The existing systems is not providing an accurate results.[7]

5.2. Problem Identification

In existing model for controlling the drone and the camera attached to it, two persons are needed. Since it is controlled using joystick and potentiometers it is not time efficient and accuracy is the major disadvantage. For turning the camera to a particular angle, the angle's information should be fed to the potentiometer independently. As the observation from the later, this increases the complexity and reduces the efficiency of the system.

6. Proposed Methodology
The proposed methodology and the block diagram is given in figure 1. Here the gestures are sensed by the sensor and it is fed to the Arduino Nano. The processor will give commands according to the gestures signal received to the drones. Here the drones and the cameras are completely controlled by the processor. The drone flies according to the PWM signals received.

IOP Conf. Series: Materials Science and Engineering 1012 (2021) 012017 doi:10.1088/1757-899X/1012/1/012017

The conventional transmitter and receiver is communicated via Arduino. The wearable is packed with two sensors (Accelerometer and gyro sensor) commonly known as MPMU 6050 sensors (gyroscope). Gyroscope is used for measuring or maintaining orientation and angular velocity of the camera.
The other concept deals with the hand gesture which helps to turn the camera to the desired direction. In the below block diagram, the Gyroscopic sensor can be attached to our wrist or head and then it can be used to operate a drone in real time. At first the drone is made to move and the camera starts recording. When the hand is moved to an angle, the sensor collects the displaced values from the three axes(X, Y and Z) and the collected value is fed to the Arduino Nano.

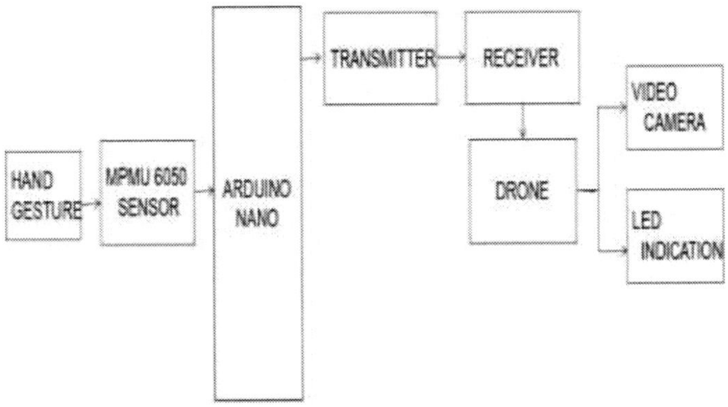

Figure 1: Schematic block of proposed system

After the samples are processed in Arduino, it is fed to the transmitter. Since it is wireless communication the transmitter transmits the signal (i.e. Movement of the hand) to the receiver end. Once the receiver end receives the drone changes its direction according to the angle the hand is moved. Gun is attached to the drone and hence their movement is simultaneous since a gun cannot be used in the prototype instead of the gun laser light is used.
The drone and the gun can be controlled by the same person. And the gun can be triggered in need. The sensor (gyroscopic) can be attached to our hand and then it is used to operate the drone in real time. When the hand is moved to a particular angle, the equivalent displaced places say (X, Y, and Z) is given to the sensor. The sensor calculates the angle from these displaced values and it is fed back to the transmitter which is used to operate the drone. When the sensor is moved, the camera and the gun present in the drone will also move to the direction. The videos can be recorded using the camera via an application called Plug and Play.
The respective angle of movement can also be calculated and a precise rotation of the drone can be done by programming. The working module can be made automatic by defining the angle of displacement of the drones. Making automatic by writing codes will give a precise and accurate working model. Getting angles of displacement can be done by using codes in Arduino. getAngleX, getAngleY, getAngleZ are the program snippets for finding the angle of displacement. The values corresponding to the angle of deflection is listed in Table 1.

angleX:-18.17	angleY-21.29	angleZ-1.64
angleX:-18.17	angleY-21.27	angleZ-1.63
angleX:-18.16	angleY-21.30	angleZ-1.63
angleX:-18.15	angleY-21.32	angleZ-1.64
angleX:-18.16	angleY-21.28	angleZ-1.65
angleX:-18.17	angleY-21.27	angleZ-1.65

7. Interference and Analysis

7.1. Normal Operating mode

Figure 2 shows the working of the model in normal operating mode.

- Step 1: When the sensor is placed still and when the transmitter is made ON
- Step 2: When the camera is placed in the normal position
- Step 3: The camera takes a photo of the normal position in an application called Plug and Play

Figure 2: Normal Operating mode

7.2. When the sensor is moved upwards

Figure 3 shows the working of the model when the sensor is moved upwards.

- Step 1: The sensor is rotated upwards.
- Step 2: The camera is also rotated to the position the sensor rotates.
- Step 3: The camera captures the rotated view.

Figure 3: Sensor moved upwards

7.3. When the sensor is moved downwards

Figure 4 shows the working of the model when the sensor is moved upwards.

- Step 1: The sensor is rotated downwards
- Step 2: The camera is also turned to the position the sensor rotates.
- Step 3: The camera captures the rotated view.

Figure 4: Sensor moved downwards

The Laser beam and the Camera will turn accordingly to the respective direction by the sensor values and the snapshot taken from the camera can be monitored. There are many applications available in website for connecting the camera using Wi-Fi to the device and it can record for further purpose. The Wi-Fi can be password protected for enhanced security

8. Drawbacks

The new form of recognition like Gesture interfaces provides a lot of advantages. These Gesture based interfaces provide the user with a new form of interaction. This kind of input also gives new issues that are not relevant to the traditional input. Correctly recognizing the gestures, the developer needs a system. Learning and remembering are accurately not possible.

Designing a system that recognizes accurately and quickly the gestures is difficult task. The other difficult task for the developer is to provide a manual for quick and easy learning of these gestures. And the sensor senses the hand movement and moves the particular object respectively. But the disadvantage is whenever a hand is moved unintentionally the particular object also starts to move.

9. Conclusion and Future scope

The challenges in the usage of unmanned aerial vehicle in military applications have been studied and a fair solution is proposed. Two major disadvantages like the accurate positioning and the real time streaming of the place are discussed in this paper. With the help of the sensors and the real time camera, the solutions for the disadvantages are provided. The usage of very low cost processor in the system made this work, a reliable and a cheap control mechanism is designed for the applications. The outcome of the idea is to provide accurate positioning in unmanned vehicles. It helps in the usage of unmanned aerial vehicles in military can save lives. It is also specifically designed to provide visual picture in places where human interaction is less possible.

By placing the fire extinguisher below the drones instead of gun, it can be used to let go of fire in places where the human interaction is not at all possible.

References

[1] Andrew Zulu, Samuel John, 2014, A Review of Control Algorithms for Autonomous Quadrotors, Open Journal of Applied Sciences, 4, 547-556

[2] Cyprian Aleksander Kozera, April 2018, Military Use of Unmanned Aerial Vehicles – A Historical Study, Safety & Defense 4(1):17-21 DOI: 10.37105/sd.4

[3] A Yudhana1, J Rahmawan1 and C U P Negara1, 2017, Flex sensors and MPU6050 sensors responses on smart glove for sign language translation, IOP Conference Series: Materials Science and Engineering, Volume 403.

[4] K. Pfeil Kui Liu in, 2015 March, Dual-sensor approaches for Real time robust hand gesture recognition" Ph.D thesis

[5] Rafiqul Zaman Khan., 2012 August, Hand Gesture Recognition: A Literature Review, International Journal of Artificial Intelligence & Applications 3(4):161-174,

[6] Hand Mantecón T, del-Blanco CR, Jaureguizar F, García N, 2019,A real-time gesture recognition system using near-infrared imagery. PLoS ONE 14(10): e0223320. https://doi.org/10.1371/journal.pone.0223320

[7] Ernesto Paiva; Juan Soto; Julio Salinas; William Ipanaqué, Oct. 2016, Modeling, simulation and implementation of a modified PID controller for stabilizing a quadcopter, Publisher: IEEE

[8] C. Georgoulas, A. Raza, J. Guettler et al.,2014, Home environment interaction via service Robots and the leap motion controller, in the International Symposium on Automation and Robotics in Construction(International Association for Automation & Robotics in Construction, IAARC, 2014, 174–193)

[9] A. Sanna, F. Lamberti, G. Paravati et al., 2013, A kinect-based natural interface for quadrotor control, Entertainment Comput. 4(3), 179–186.

[10] MIRMOJTABA GHARIBI,, 2016, RAOUF BOUTABA and STEVEN L. WASLANDER Internet of Drones, , IEEE

[11] Liu, 2015, Dual-sensor approaches for real-time robust hand gesture recognition," Ph.D. (The University of Texas at Dallas, 2015).

[12] L. E. Potter, J. Araullo, and L. Carter,2013, The leap motion controller: A view on sign language, in Proceedings of the 25th Australian Computer-Human Interaction Conference: Augmentation, Application, Innovation, Collaboration (ACM, 2013), pp. 164–175.

[13] J. Nagi, A. Giusti, G. A. Di Caro, and L. M. Gambardella, 2014,HRI in the sky: Controlling UAVs using face poses and hand gestures, in 8th ACM/IEEE International Conference on Human-Robot Interaction (HRI) (HRI, 2014), pp. 1–2.

[14] K. Pfeil, L. K. Seng, and J. Laviola, 2013, Exploring 3D gesture metaphors for interaction with unmanned aerial vehicles," in International Conference on Intelligent User Interfaces(ACM, NY, 2013), pp.

[15] Olha Sushchenko and Andriy Goncharenko, 2016, Design of Robust Systems for Stabilization of Unmanned Aerial Vehicle Equipment, International Journal of Aerospace Engineering, Volume 2016 , Article ID 6054081.

IOP Conf. Series: Materials Science and Engineering 1012 (2021) 012018 doi:10.1088/1757-899X/1012/1/012018

Emulation of Multi-Rotor Functionalities in Fixed Wing Airplane for Farm Based Land Surveillance with Controlled Fluid Dispersion System

Aagashram Neelakandan[1*], Chellakrishna V[1], Ajay Antuvan M[1], Avinash M[1] and Ranjith Kumar R[1]

[1] Department of Mechanical Engineering, St. Joseph's Institute of Technology, Chennai, Tamil Nadu, India

E-mail: aagashram@gmail.com

Abstract- The World is emerging with Technologies in every discipline, and India plays a vital role in the export of agricultural products. Since it has a huge amount of manifestation, it requires a lot of manpower. The Lack of skilled labour can also cause depletion of crops production rate. So the use of Technologies in agriculture reduces manpower drastically and increases efficiency. The production rate of crops in agriculture is based on various parameters like temperature, humidity, rain, etc., which are natural factors and are not controlled by farmers. The field of agriculture also depends on some factors like pests, disease, fertilizers, etc., which can be controlled by giving proper treatment to crops. Even though the use of pesticides can protect the crop, it may also have harmful effects on human health. So usage of technologies like quadcopter is an efficient process. But as it involves a huge amount of Initial and Operational cost, the quadcopters are replaced with Fixed Wing Drones. This paper depicts about the design and development of Fixed Wing Airplane known as AGRIWING, which combines all the desirable features of a conventional agricultural fixed wing such as mapping of farm, gathering heat signatures, etc., with the added capabilities of the autonomous fluid dispensing system, as in the Drone Spraying System with the advantages of Fixed Wing Airplane including reduced cost, improved flight time, robustness and more.

1. Introduction

India plays a major role in the field of agriculture. It requires a lot of manpower to increase the crop production rate. So there is a need to employ various technologies which can maintain the production rate with less manpower and thus resulting in increased efficiency. There is a high need for these systems and they form an integral part for Precision farming. In this rising population, the need to produce food as equally important as the need to produce quality foods, because the use of large number of pesticides and chemicals can damage the crops. The World Health Organization (WHO) estimates that more than three lakh people die from the toxic exposure of pesticides per year across the world [1]. Various pesticides used in agriculture causes health effects like skin allergies, asthma, and it also leads to a serious effect on fetal growth [2]. Some pesticides even contain carcinogenic substances. Many side effects of using these pesticides mostly occurs in developing countries because they need to fulfil their rising demand for food and it majorly affects the growing generations like children, Teenager and even adults by reducing the visual ability and resulting in lots of neurological problems. To reduce these effects, the amount of pesticides and other chemicals that are used for cultivation are to be controlled. To achieve this, Precision farming is desired and for implementing this concept various technologies are required. Various technologies which has been currently employed for this operation are the Unmanned Aerial Vehicles (UAV),which comprises of a multirotor with sensors for guidance and mapping along with a dispersion system which can be used for fertilizers, pesticides and even seeds. But all these systems incurs huge cost and it is not well suited for developing countries like India and it is probably one of the major reasons which explains why India is not a pioneer in the use of Agriculture technologies and various researches are done to rectify the issue.

2. Related Works

There have been many works previously done in the field of spraying pesticide for plants. One such work describes the use of Hexacopter UAV for pesticide spraying for crop protection [3]. There is also another work which states the design of blimp with a quadcopter for automatic aerial spraying system [4] and there is another paper which makes use of the Atmega 328 for controlling the drone and sprayer system [5]. But the problem with all these systems is that it uses a form of multirotor system and high-end sensors which is not only expensive and but also difficult to obtain.

3. Solution

Technological development on autonomous flight system has been developed to a level, when there is sudden change in Mission planning, decision problem involving obstacles decision [17] are no longer a hurdle. The reason that it is been considered as a problem because most of the solutions provided by others authors, who try to eliminate human from the process by incorporating more technologies, thus increasing the overall cost of the system [20]. This may be suitable for a farming system in developed nations where the majority of the farms are big and owned by a company, producing a specific type of crops. But this is not the case in India where most of the farmers own a relatively small strip of land and cultivate seasonal crops, which indeed depends on the monsoon of that year. Also in most of the farms, surveillance mapping is done by either a fixed wing aircraft or a multirotor but pesticide spraying is done only by a multirotor [14]. So all the works done earlier are harder to replicate and economically not viable to replace the damaged part. To eliminate these problems and to create a solution which can emulate the function of an agricultural multirotor, we have designed a Fixed wing UAV which is named as AGRIWING, whose purpose is to act as Farm Surveillance system along with a Pesticide sprayer system. The design choice mainly relies upon the fact that cost including capital cost, running cost and maintenance cost, for a multirotor is more when compared to a Fixed Wing Aircraft.

4. Proposed Methodology

For reducing the use of pesticides in crop production, it has to be sprayed in the areas where it is actually needed because increasing the time of spraying increases the rate at which it gets mixed with air and thus increases chances of falling on the wrong plants which affects the production rate. Therefore our design needs to be robust with minimal parts which can detect the areas needed for pesticides and can spray the right amount of fluid on the areas detected in the previous step.

4.1. Design of Fixed Wing

Our design of fixed wing is based on a Blunt Nose Ft Versa with modification for various attachments. The reason that the Blunt nose design was chosen because it has an extra compartment for the storage of the other electronic components and also provides protection for the flight controller (Omnibus F2) and Global Positioning System (GPS) module. The Figure. 1 shows the actual AGRIWING without the winglets and sprayer system. Also availability of the flat base gives us the place to mount mapping and spraying modules in a custom designed enclosure which can speed up the insertion of the module in the flight. The module in Figure. 2 shows the quick insertion attached to flight bottom which is designed for the mapping and spraying system containing the Raspberry Pi Zero board, Pi Camera and an extendable microphone array. All these parts are separately powered, removing the dependency of the battery in plane and thus ensuring longer flight time. Also the Figure. 2 shows the Mapping and

Spraying module that has a nozzle with a valve, a liquid storage section and a pump. The flow is controlled by a Raspberry Pi that adjusts the valve which is actuated by a servo motor connected to the valve. All these parts can be removed, replaced and easily upgraded.

Figure 1. The black and orange colour gives better contrast to sky and makes it easy to spot

Figure 2. 3D model of quick insertion design and Mapping and Spraying System

4.2. Crop Monitoring Systems

Monitoring of the crops is the important step because it determines the areas of the field which needs to be sprayed with pesticide and also determines the amount of spraying needed for that particular area. Plants are monitored based on two ways, one is Vegetation Index and another is through the detection of pests in the field.

4.2.1. Vegetation Index. A Vegetation Index (VI) is a method used to analyze the vegetation properties of the plants by enhancing the terrestrial photosynthetic activity and canopy structural variations [6]. There are various types of Vegetation Index used for agriculture. But some crop require a multispectral camera [19] which is expensive and the returns for that investment is considerable only for large farms. But in India, most of the farm land are small and the rate of return is low. Therefore, other Vegetation index like the Visible Atmospheric Resistant Index (VARI) [7] and Green Leaf Index (GLI) [8] are used because both of them can be measured by RGB spectrum camera. The formula for Indices are,

$$VARI = (C_{GREEN} - C_{RED}) * (C_{GREEN} + C_{RED} - C_{BLUE})^{-1} \quad (1)$$

$$GLI = (2*C_{GREEN} - C_{RED} - C_{BLUE}) * (2*C_{GREEN} + C_{RED} + C_{BLUE})^{-1} \quad (2)$$

where *Cxxx* is the color reflected by the plant. Using (1) and (2), plant growth can be monitored and areas where growth is depleted due to external factors are marked and thus allowing the system to spray fluids in only those affected areas.

4.2.2. Detection of Pests. A Vegetation Index can give details of the plant growth and health, indicating whether the plant is growing or depleting. But it doesn't give details about the type of pests affecting the plant. This identification of pests is important because it enables the system to localize the spraying of different types of pesticides for various types of pests because spraying all kinds of pesticides for unaffected areas affects the vegetation and also increases the cost. For detection of pests, the audio is recorded and it is processed after the flight. The recorded audio is compared with the insect's sound to create a heat map. By using both mapping techniques the areas of pesticide spraying is decided and it is stored in Raspberry Pi whilst it can be overwritten by the user if they want to.

4.3. Control Systems for Mapping and Spraying

Control system for sprayer uses information from the heat maps formed by the combined monitoring systems.

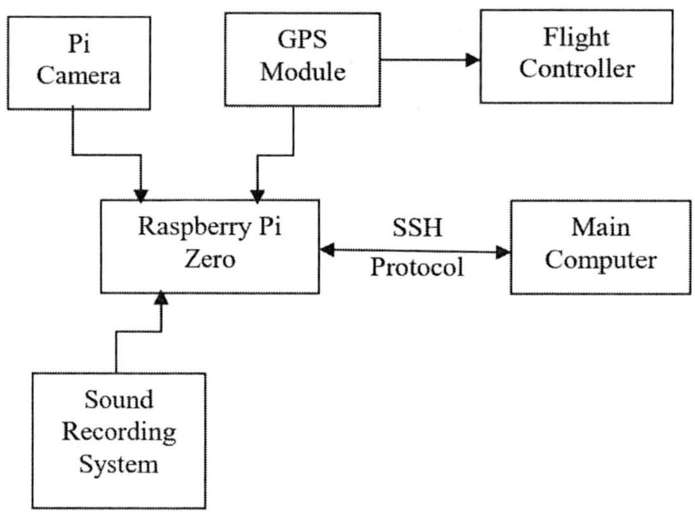

Figure 3. Block Diagram of Crop Monitoring System

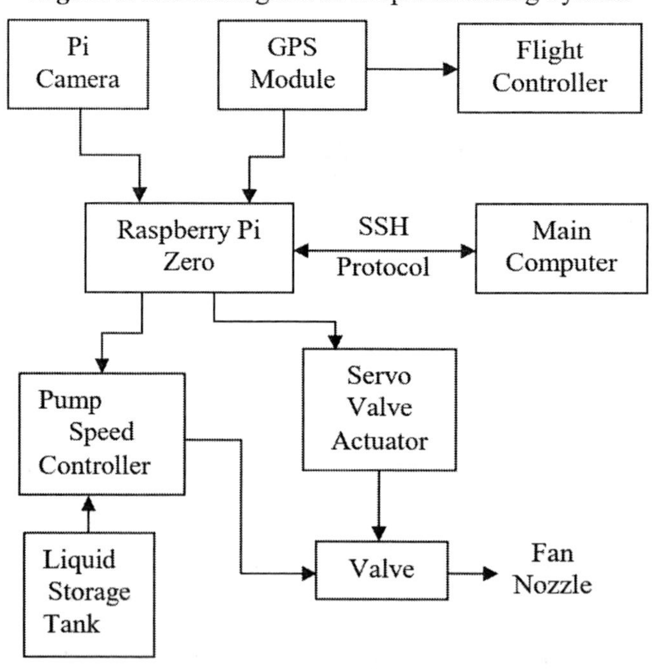

Figure 4. Block Diagram showing Liquid Spraying System

Flight paths are needed to be designed in a way such that multiple flights can take place simultaneously for large farms [13]. Since flight path determination for the AGRIWING is same as that for Drone, existing method is adopted. In Figure. 3 the block diagram shows the monitoring systems which has the Raspberry Pi Zero which processes the data and acts as storage system. Once the Flight Mission is planned and the route is set using Flight Controller Software, the Raspberry Pi takes images using a Pi camera for land patches and embeds that image with GPS data from the GPS Module. In addition to that the Microphone data is also stored along with the agricultural land images. After completing the mission, the flight lands and all the data metrics is viewed using SSH in the main computer field.

During the manual inspection, the user can change the amount of liquid spraying or pesticide spraying by colouring the heat maps appropriately. After inspection, the spraying mode sets the same mission route and Raspberry Pi will spray the plants based on the Heat maps which is generated during the mapping stage. Based on the data values stored in the heat maps in their respective GPS locations, the spraying system can vary the amount of liquid sprayed and this can be done by using two controls methods. The Figure. 4 shows the sprayer system consisting of a pump controlled by the Raspberry Pi, that takes liquid from the storage tank and pumps it to the valve which is actuated by a servo motor. By manipulating the valve position and flow rate of the pump using Raspberry Pi, the amount of liquid sprayed is controlled. Finally, the liquid is sprayed using a fan nozzle as it provides more coverage and better spread even in the presence of drift wind [3]. Spraying system is designed based on the modified and simplified version of Variable Spraying Actuator [9].

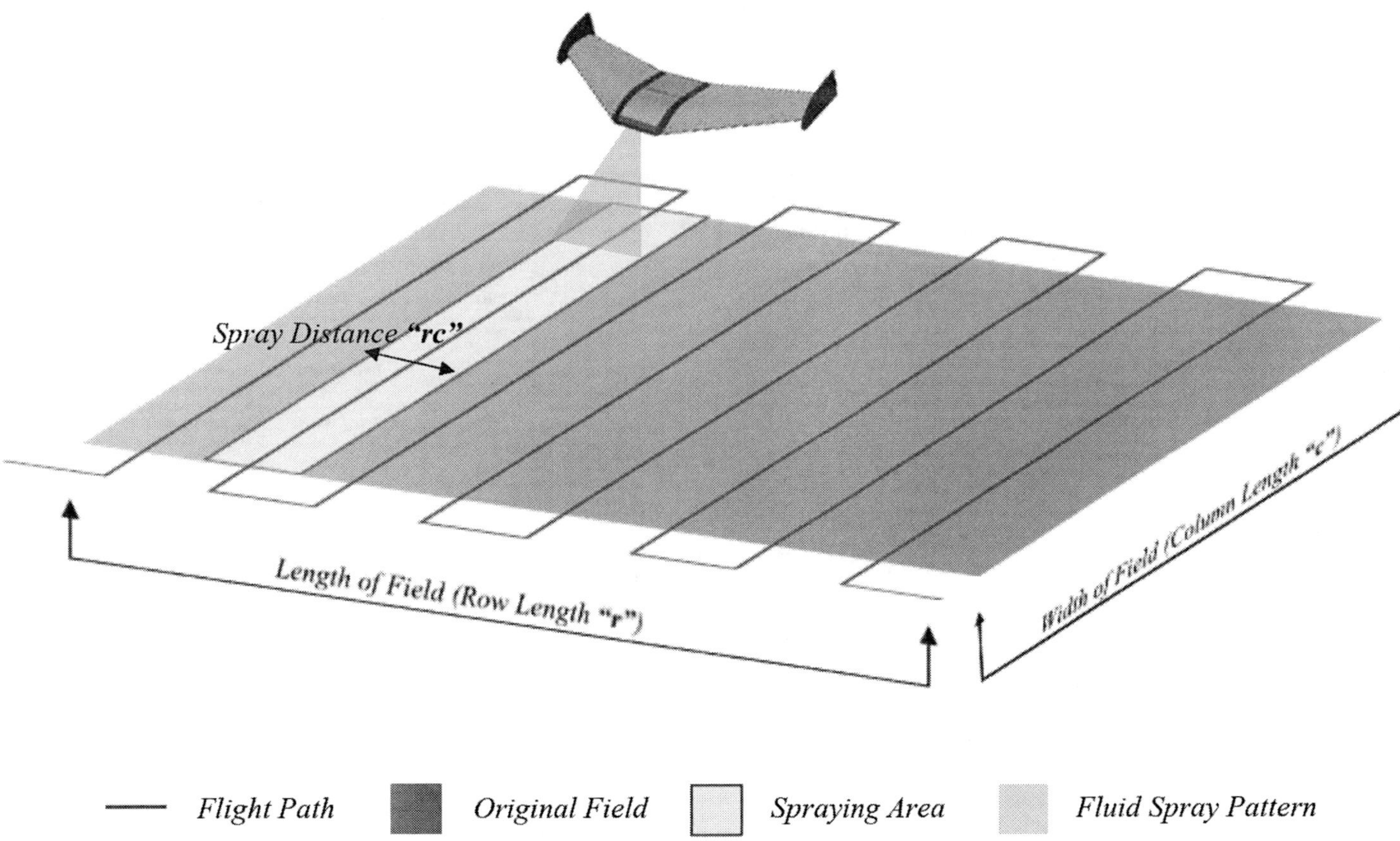

Figure 5. Illustration showing Spray Control System

4.4 Valve Position Control System

The Servo Valve Control System, also known as Valve Position Control System, whose main purpose is to spray the fluid based on the spray pattern measured by Vegetation Index. Servo Valve Control is built up on the image processing to calculate the movement of servo in correspondence to plants health measured using vegetation index [10]. Servo valve control provides the output signal for the servo to alter the position which controls the amount of pesticide to be sprayed. Servo Valve Control gets the input form the Modified VARI (MVARI) image from farmers, in which farmer marks the areas of affected crops based on his expertise. This input is passed on to the Servo Control's subsystems. The overall working of the AGRIWING and parameters used for calculation is shown in Figure 5. This input is unrolled based on the flight path and effective length distance which can be covered by spray nozzle in single pass.

4.4.1 Unrolling of the MVARI Image. The input MVARI is unrolled based on the flight path and effective length which can be covered by spray nozzle in one pass which is Spray Distance also known as row count **"rc"**. For getting optimal Spray Distance ***(rc)*** without overlapping the spray, the flight path and length of the Field ***(r)*** is used. The spray distance is calculated using the following formula,

$$\textit{Spray Distance, } rc = r/nop \quad (3)$$

Where, *nop* or *Number of Passes* denotes the number of times the plane crosses the field before reaching its end point, which is based on the factor of turning radius of plane. Based on this value, the MVARI is unrolled into a strip which proceeds to the normalizing of the strip image.

4.4.2 Normalizing Unrolled Image. The Unrolled MVARI image strip consists of ***rc*** number of Pixels in each row (along the flight path) and individual pixels are made up of 24 Bit Colour data, with 8 bit for Red, Green and Blue each. These data has to be converted into a form of an image showing the amount pesticide required based on the MVARI image which can be turned into a Signal for Valve Servo. As each pixel is having 3 Colour channel, each colour channel is assigned a coefficient which is based on the notation from VARI image, i.e., Red pixels should have more coefficient value as it needs more amount of pesticides. Due to this process, 3D Matrix (3 Channel) is linearly transformed into 2D Matrix (1 Channel). The normalizing formula used is,

$$Norm = (R_{COEFF} * C_{RED} / 255) + (G_{COEFF} * C_{GREEN} / 255) + (B_{COEFF} * C_{BLUE} / 255) \quad (4)$$

Where *Cyyy* is the colour value in *yyy* channel. The coefficients are chosen in a way that give optimal angle based on the sunlight falling on the field at the time of measurement.

4.4.3 Valve Opening Angle or Servo Signal. From the normalized image strip the average sum of pixel values at each row is calculated and it is mapped to the Servo Valve operating range of angles from 0° to 180° degrees. Dimensionality reduction of 2D Matrix to 1D array is done by the formula,

$$SA(i) = \sum_{i=0}^{l} \Sigma_{j=0}^{rc} N(i,j) * 180 \quad (5)$$

Where, *l* is the length of flight path and *SA* is an array containing Servo angles in degrees. From equation 5, the angles are calculated and these angles are plotted on the graph in Y-Axis with respect to time for completing the flight in x axis as shown in the Figure 6. Due to Light disturbance and cloud movements there will be light differences in field which affects the processing of VARI image and thus introduces noise in the Servo angle output signal. Some of these noises can also be caused by wind [11]. So in order to reduce noise, the Kalman filter is used, which reduces the external noise without actually losing the data [18]. This filtering application is limited due to its highly computational intensive process and for small farms it provides less advantage, as lighting changes across the field is minimum. The overall process flow chart is shown in Figure 6, where a specific

IOP Conf. Series: Materials Science and Engineering 1012 (2021) 012018 doi:10.1088/1757-899X/1012/1/012018

image is converted into VARI image which is then Modified, Unrolled and converted to normalized image from which the servo signal is generated and filtered using Kalman filter.

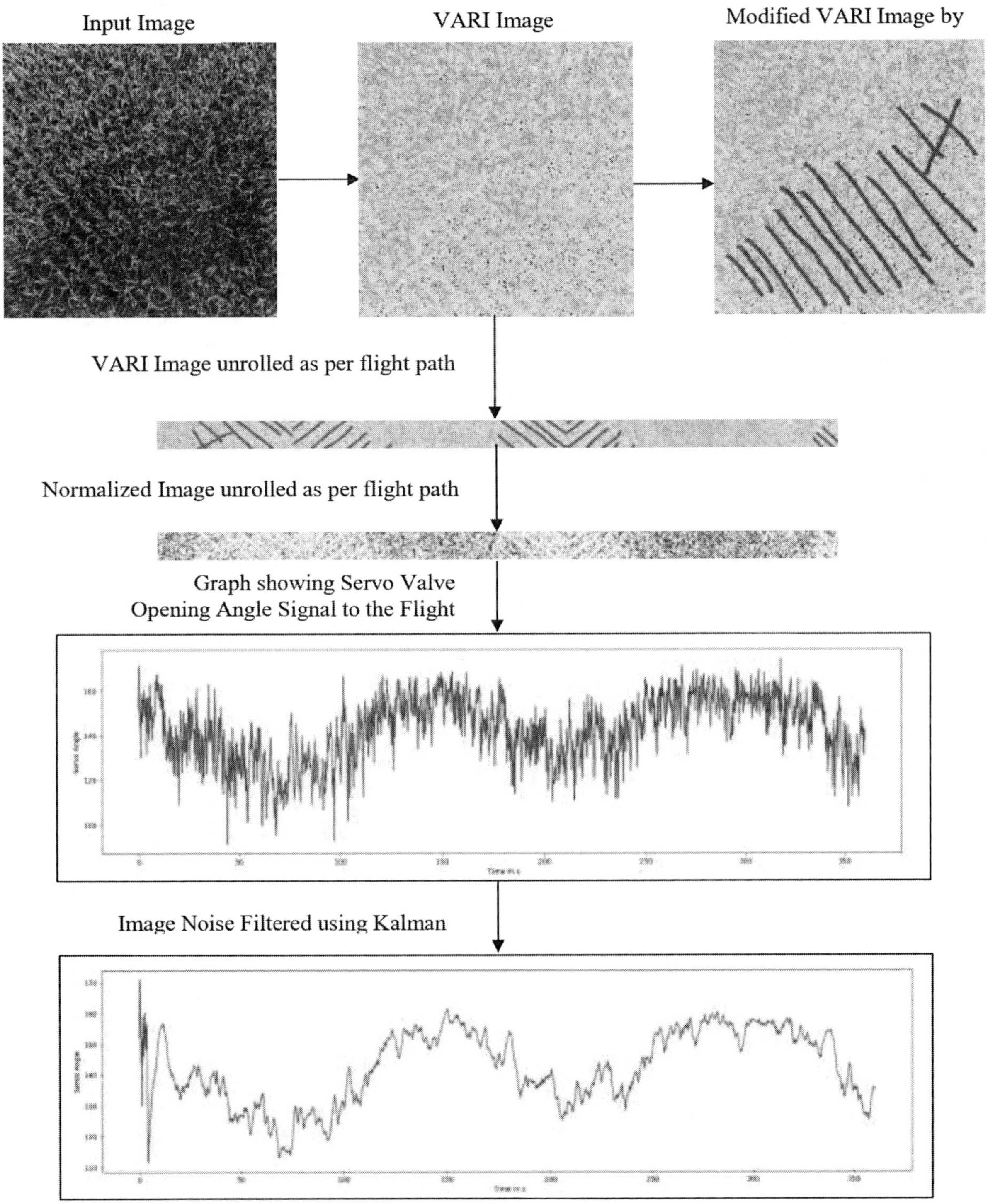

Figure 6. Process Flowchart of Valve position Control System

5. Conclusion

The Fixed Wing drone system with selective spraying method was developed and tested for agricultural purposes like Vegetation Mapping and Pesticide Spraying. Based on the development and experimental investigation the following points are concluded,

- The method of spraying is designed for the Fixed wing, which allows it to spray the right amount of pesticide to affected plants while travelling in the normal flight path.
- Mapping of the Agricultural Field is done using RGB Camera, which calculates VARI index and based on that value, amount of pesticides to be sprayed is calculated.
- The developed system can fly at a maximum speed of 60 km/hr in no load condition, which allows it to map the field faster than the conventional drones.
- Due to the emulation of multirotor functionalities in fixed wing, the development cost, running cost and maintenance cost of this system is cheaper than the existing multirotor system [21].

The future scope of our system is to detect multiple crop vegetation and use machine learning to make more accurate pest detection and to increase the wide usage of the system in other terrestrial areas like river banks [12]. Using a custom flight controller powered by smartphones [16] or the one built on Raspberry Pi Zero [15] can further reduce the cost. Also, in future we will develop additional modules like seeding system and weed removal system.

References

[1] Anon WHO | Poisoning Prevention and Management
[2] Anon *TRAINING FOR THE HEALTH SECTOR*
[3] Zhang Y L, Lian Q and Zhang W 2017 Design and test of a six-rotor unmanned aerial vehicle (UAV) electrostatic spraying system for crop protection *Int. J. Agric. Biol. Eng.* **10** 68–76
[4] . P D P R H V 2014 Development of Automated Aerial Pesticide Sprayer *Int. J. Res. Eng. Technol.* **03** 856–61
[5] Shivaji C P, Tanaji J K and Satish N A 2017 Agriculture Drone for Spraying **2** 34–6
[6] sayer D 2018 Overview of the radiometric and biophysical performance of the MODIS vegetation indices *New Perspect.* **26** 49–83
[7] Weiss M J, Hofeldt A J, Behrens M and Fisher K 1997 Comparing RGB-Based Vegetation Indices With NDVI For Drone Based Agricultural Sensing *Retina* **17** 105–8
[8] Eng L S, Ismail R, Hashim W and Baharum A 2019 The use of VARI, GLI, And VIgreen formulas in detecting vegetation in aerial images *Int. J. Technol.* **10** 1385–94
[9] Ling W, Du C, Mengchao Z, Yu W, Ze Y and Shumao W 2018 CFD Simulation of Low-attitude Droplets Deposition Characteristics for UAV based on Multi-feature Fusion *IFAC-PapersOnLine* **51** 648–53
[10] Wang L, Lan Y, Yue X, Ling K, Cen Z, Cheng Z, Liu Y and Wang J 2019 Vision-based adaptive variable rate spraying approach for unmanned aerial vehicles *Int. J. Agric. Biol. Eng.* **12** 18–26
[11] Yao W, Lan Y, Wen S, Zhang H, Zhang Y, Wang J and Xie C 2019 Evaluation of droplet deposition and effect of variable-rate application by a manned helicopter with AG-NAV Guía system *Int. J. Agric. Biol. Eng.* **12** 172–8
[12] Kooistra L, Salas E A L, Clevers J G P W, Wehrens R, Leuven R S E W, Nienhuis P H and Buydens L M C 2004 Exploring field vegetation reflectance as an indicator of soil contamination in river floodplains *Environ. Pollut.* **127** 281–90
[13] Anand K and R. G 2019 An Autonomous UAV for Pesticide Spraying *Int. J. Trend Sci. Res. Dev.* **Volume-3** 986–90
[14] Mogili U R and Deepak B B V L 2018 Review on Application of Drone Systems in Precision Agriculture *Procedia Comput. Sci.* **133** 502–9
[15] Wang S, Zhen Z, Zheng F and Wang X 2015 Design of autonomous flight control system for small-scale UAV *2014 IEEE Chinese Guid. Navig. Control Conf. CGNCC 2014* 1885–8
[16] Masoud M, Jaradat Y, Farhoud M and Almdallaleh A 2016 On Implementing a Low Cost

Quadcopter Drone with Smartphone Control
[17] Chandler P R and Pachter M 1998 Research issues in autonomous control of tactical UAVs *Proc. Am. Control Conf.* **1** 394–8
[18] Cis P R, Des Q U E, Classiques C, Que D E S, Thodes D E S M, Erreurs D E C D, Signal L E, Partir D S I R, Bruit D E M, Ans E S, Version U N E and Algorithme M E L 2000 Application of Kalman filtering to noise reduction on microsensor signals Application de filtres de Kalman à la réduction de bruit 443–50
[19] Meivel S M, Professor A, Maguteeswaran R, Gandhiraj N B and Srinivasan G 2016 Quadcopter UAV Based Fertilizer and Pesticide Spraying System *©International Acad. Res. J. Eng. Sci.* **1** 2414–6242
[20] Valerdi R 2005 Cost metrics for unmanned aerial vehicles *Collect. Tech. Pap. - InfoTech Aerosp. Adv. Contemp. Aerosp. Technol. Their Integr.* **3** 1753–8
[21] Anon DJI announces $15,000 agricultural drone designed to spray crops - The Verge

Optimization of quadcopter frame using generative design and comparison with DJI F450 drone frame

Jerrin Bright[1], R Suryaprakash[1], S Akash[1], A Giridharan[1*]

[1] School of Mechanical Engineering, Vellore Institute of Technology Chennai, Tamilnadu, India

E-mail: giridharan.abimannan@vit.ac.in

Abstract: This research accentuates to explore designing the drone frame using Generative design tools. A quadcopter is designed using Autodesk generative design embedded in Fusion 360. The simulation results such as static stress-strain, modal frequency and displacement results of additive manufactured quadcopter are compared with a DJI flame wheel F450 drone frame. The generative designed frame has minimum displacement compared to traditional designed drone frame. It is observed that generative designing technique along with additive manufactured frames yields better frames with improved resistance to fracture and minimum displacement compared to traditional designed DJI flame wheel F450 drone frame.

1. Introduction

The term "drone" characteristically stated to any unpiloted aircraft that can carry out a striking range of tasks, vacillating from military operations to package delivery. It can be as large as an aircraft or as small as the palm of your hand. Drones are classified based on the following factors- size, range, abilities, weight. According to aerial platforms, drones are of four types- single rotor, multi-rotor, fixed wing and fixed wing hybrid. This research is fixated on multi rotor drones, chiefly quadcopters. Few common classifications of quadcopters based on its frame size include true-X, square, hybrid X, H and stretched X. Generative design is an iterative design progression that proposes promising answers to a design issue. Design techniques such as topology optimization, lattice optimization etc., are focused on improving pre-existing models, contrasting generative design emphases solely on creating new possible designs with the limited constraints or parameters.

Generative designing is an iterative design exploration process, pioneered by Autodesk, uses artificial intelligence [AI] algorithms and high computing power of cloud to generate numerous design options emerging from one CAD design, done by defining certain parameters. Principally, by setting our advanced goals and its constraints, we will autonomously be able to generative thousands of design outcomes satisfying the desired requirements. The designer will not be able to evaluate all the design outcomes generated without cognitive catastrophe [1]. Few of the input parameters include the Design space consists of preserve and obstacle geometry; the design conditions include structural constraints and structural loads; design criteria include objective and manufacturing technique; and the material used for workpiece. Traditional optimization techniques focus on one variable whereas generative design is aimed at providing different design solutions and these design variants helps in meeting the demands for various other practical applications which will logically be difficult to describe mathematically [2].

Design problems in various technical fields are solved using generative design tools. Several researches have contributed in using generative design to obtain reasonable solution. Recently, the MaRS Innovation District in Toronto, Canada built an office environment, in which the floor plan was designed using generative designing. Similarly, Airbus uses generative designing to reimagine the central panels for the A320 Aircraft. Likewise, there has been a lot of research/ projects going on in these domains starting from the advances in generative designing [3] to the modeling and analysis of Quadcopter F450 Frame [4]. Harish Doraiswamy [5] proposed a catalogue-based framework where conceptual building design options of the interactive exploration can be enabled based on the adjustable view preferences. Authors compared different frames designed in generative designing to that of conventionally designed frames. The variability of the designs must be carefully controlled to advance various stimulating designs, such that the design generated will be comprehensible, unpredictable and most importantly desirable [6].

The conventionally designed DJI Flame Wheel F450 frame is compared with the design obtained from Generative designing drone frame. The Autodesk's Generative Design, hosted in Fusion 360 software is used to create model using the base parameters and structural constraints set by the user before generation is initiated. Simulation results were then compared including static stress-strain analysis, displacement and model frequency tests performed on the frames obtained from generative designing along with the results of DJI Flame Wheel F450 frame.

2. Dynamic Modeling and Calculations

The aerodynamic lift can be explained with the help of the Bernoulli's Conservation of Energy. This theory is also called as "Longer path" or "Equal transit theory". The equation for calculating lift (L) is given as,

$$L = \frac{1}{2} \times \rho \times V^2 \times A \times C_L \qquad (1)$$

Where, ρ is air density, (~1.2 kg/m^3)
V is velocity, m/s
A is surface area of blades, m^2
C_L is coefficient of lift

Table 1. Consists of all the components along with its weight

Component	Number of individual components	Weight of individual component (in gm)	Net Weight of individual component (in gm)
RS2205S Motor	4	28.8	115.2
Pixhawk controller [a]	1	73	73
M8N GPS	1	32	32
5'' Propeller	4	4	16
BLHeli ESC	4	14	56
FC Hub PDB	1	8.5	8.5
Camera	1	12	12
Telemetry	1	15	15
iA6b Receiver	1	14.6	14.6
Lipo Battery 3S	1	200	200

[a] Including the controller and anti-vibration stand.

The components and its weight used for modeling are given in table 1. The total weight of the component obtained is 542.3 g. Therefore, considering the frame weight to be 250 g approximately and by adding 20% buffer on component mass, the total load [M] got to be 1000 g. G is the gravitational constant calculated by the gravitational effects of the law of universal gravitation, which is 9.8 m/s^2 precisely.

$$M = 1000\ g = 1\ kg \tag{2}$$

$$F = M \times G = 10\ N \tag{3}$$

Therefore, for a quadcopter the thrust of each motor is equivalent to 10 N. Propeller design is one of the most imperative criteria to be studied to analyze the performance of the quadcopters, which is expressed as

$$\eta = \frac{P_{out}}{P_{in}} \tag{4}$$

where, η is the efficiency of the propeller

P_{out} is the output power, that's the ability of the propeller to yield the given thrust (N) at the given airspeed (m/s).

P_{in} is the input power, that's the power given by the motor. In other words, the input power of the propeller is the output power of the motor.

Generally, a 2-blade propeller yields two pressure pulses and a 3-blade propeller will yield three pressure pulses per revolt. Therefore, a 3-blade propeller will be intrinsically smoother and quieter than a 2-blade propeller, making 3-blade propeller comparatively efficient than a 2-blade propeller.

3. Generative designing and selection of quadcopter frames

The human computer interaction targeting the generative design process can be divided into three dynamic phases- **Define; Generate** and **Explore**. Define is principally the phase we find our problem and convey it to the computer via fixing the goals and constraints. Generate is the place where the computer will look into the design and constraints defined by the user and works through those options in various different angles. Explore is where the computer will show the user, the options the computer devised using the conditions defined in the Explore phase. Explore phase is a decision making and interaction phase between the computer and the user. The generative design process is done as said with the help of AI algorithms including convolutional neural networks, Generative adversarial networks [GANs] including Conditional GANs [CGANs], Conditional Wasserstein GANs [CWGANs], Boundary Equilibrium GANs [BEGANs] and so on. Ivan Sosnovik [7] familiarized convolutional encoder-decoder architecture and stated the issues with image segmentation tasks. Similarly, Hongbo Sun [8] explained and studied four different typical exploration approaches of Reinforcement Learning [RL] converging on the development of Solid isotropic microstructure with penalization [SIMP] and Bi-directional evolutionary optimization [BESO] for structure-based topology optimization difficulties. Sangeun [9] proposed a framework consisting of iterative design exploration and design evaluation parts. YI Ji-jun [10] developed an element self-governing nodal density method for topology optimization, which is a part of generative designing.

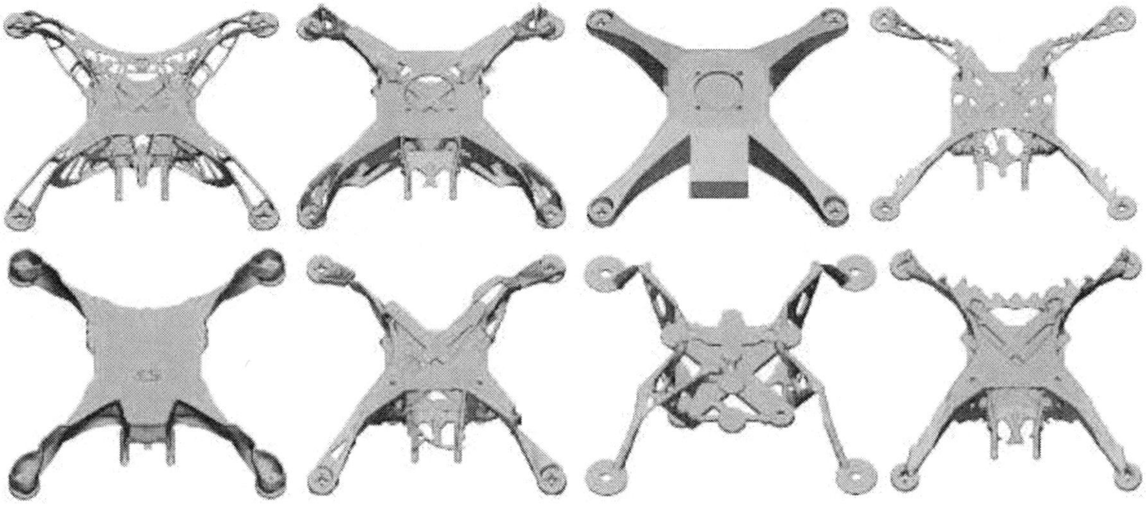

Figure 1. The various outcomes created when generative design was applied.

A quadcopters model is created using the specifications given in table 1 and generative design is applied to the model. In generative design, after various iterations, 74 designs where unconventionally generated and out of which 8 design outcomes have considered being the most suitable and pertinent based on our requirements and those 8 designs are shown in figure 1. The stress-strain, vibration, total weight of the frame, minimum factor of safety, etc. factors are considered to select the best outcome. In every iteration, the edifice of the CAD model will be tested and then, the AI algorithm will learn from the testing done in each and every step and will apply the changes at the subsequent iterations thus producing optimistic design which ultimately meets the desired expectations of the user [11]. Most of the CAD systems rely on the core constraints and measure to paradigm the geometry of the CAD model, though the procedural aspect of the constructional geometry is hidden from the user [12].

One key advantage to be noticed in these designs is that generative design slowly chips away material from a huge block, which implies that however complex the geometry of the CAD design, singularity of the design is maintained, in other words, a design is produced as a single part. The iterative process starts with an optimal block mass considering the hole and obstacle boundary implemented while fixing the design parameters. The Algorithm is designed in such a way that the removal of mass is done in a streamlined way ensuring maximum materialistic properties. The cloud-based system simultaneously processes multiple designs with unique topologies and properties. The design after generative designing checks if the structure is as per the requirements specified as goal via the parameter and structural constraints failing which the design will be discarded [13]. Figure 2 depicts the most effectual iterative design attained among various other design outcomes; hence these designs are used to compare with the DJI F450 frame considering factors such as processing status, study, visual similarity, material, piece part cost, fully burdened cost, volume, mass, maximum Von-Mises stress, minimum factor of safety and maximum displacement global [14].

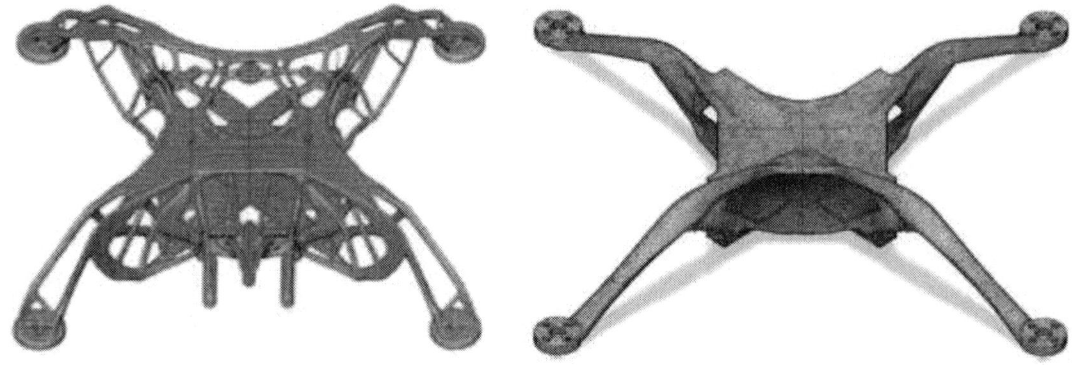

(a) Frame 1 (b) Frame 2

Figure 2 Optimal design outcome obtained from generative design

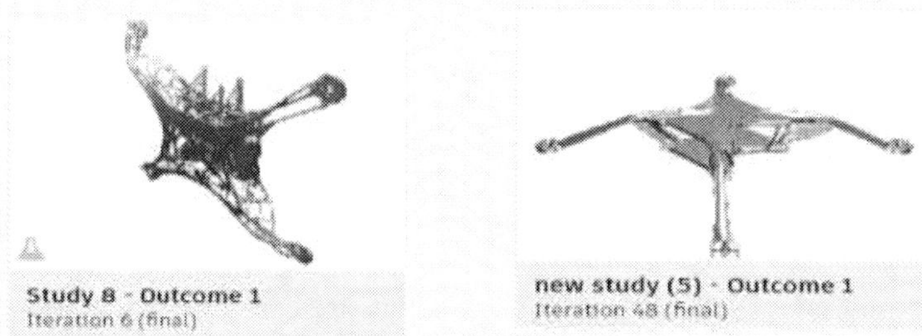

Properties	Study 8 - Outcome 1 Iteration 6 (final)	new study (5) - Outcome 1 Iteration 48 (final)
Status	Completed	Converged
Material	ABS Plastic	ABS Plastic
Orientation	-	X+
Manufacturing method	Unrestricted	Additive
Visual similarity	Group 4	Unique
Production volume (pcs.)	2500	2500
Piece part cost		
Range (USD)	303 - 624	272 - 524
Median (USD)	418	348
Fully burdened cost		
Range (USD)	303 - 624	272 - 524
Median (USD)	418	348
Volume (mm^3)	214,162.43	260,586.4
Mass (kg)	0.227	0.276
Max von Mises stress (MPa)	1.5	0.1
Factor of safety limit	2	2
Min factor of safety	13.3	137.63
Max displacement global (mm)	6.22	0.01

(a) Frame1 (b) Frame2

Figure 3. Properties logged when Generative design was implemented

In this study, the quadcopter frames designed are 450mm in size, also termed as the wheelbase which refers to the diagonal motor-motor distance. The size is particularly chosen, as the DJI frame is of the same wheelbase, which brings in a legit comparison between design specifically in terms of weight and load capabilities. Figure 3 depicts the snapshot of properties logged during generative designing. The user is facilitated in the execution of all the steps by a simplified GUI that proposes the steps in a chronological order [15]. For engineering design optimization and exploration, supervised learning is very operative making mapping out viable regions quite simpler [16]. After finishing the initial design space exploration, the node-geometry was refined for additive manufacturing using multi-stage structural optimization [17].

Table 2. Comparison of generative design frame with DJI F450 drone frame

Factors	Frame 1	Frame 2	DJI F450 frame
Weight of the frame (g)	227	267	330
Minimum factor of Safety	13.3	133	3.301
Manufacturing Method	Additive Manufacturing [b]	Additive Manufacturing [b]	Advanced manufacturing
Maximum von Mises Stress (MPa)	1.5	17.11	21.33
Maximum Displacement global (mm)	6.22	0.01	4.016
Material used	ABS Plastic [a]	ABS Plastic [a]	Polyamide Nylon
Filament spent (mass, length)	2 g, 0.7 m [b]	1 g, 0.21 m [b]	Not available

[a] Choosing the material is most crucial part of design as it withstand the weight of components exerted on the frame, resistance against impact etc. ABS material is chosen, due to its high rigidity, impact resistance, and tensile strength. Dark color ABS will be prefered as other light colors will degrade over time.

[b] The filament used while 3D printing is obtained from CURA software, where the length and weight of the filament used includes the supports structures generated.

[c] Additive manufacturing technique retrieves data from the CAD file converting it into STL file which will then be sliced in softwares like CURA generating G-Codes used for 3d printing. Slicing will ultimately encompass the information of each and every layer to be printed in form of triangles [18]

4. Analysis on generative design and DJI F450 drone frame - comparison

In this section, the influence of factors given in table 2 is analyzed. Von Mises criterion is principled on von Mises-Hencky theory, which is also known as shearing energy theory or the maximum distortion energy theory [19]. Theory emphasizes that when Von Mises Stress equals or exceeds the effort limit, a ductile material shows weaknesses in some parts of its structure [20].

Table 3 CAD design, obstacles and preserved boundaries of generative designs

	Frame 1	**Frame 2**
CAD Design	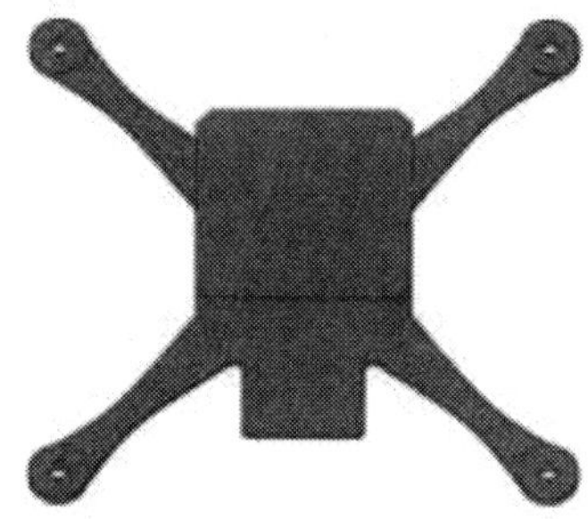	
Obstacle and preserved boundaries		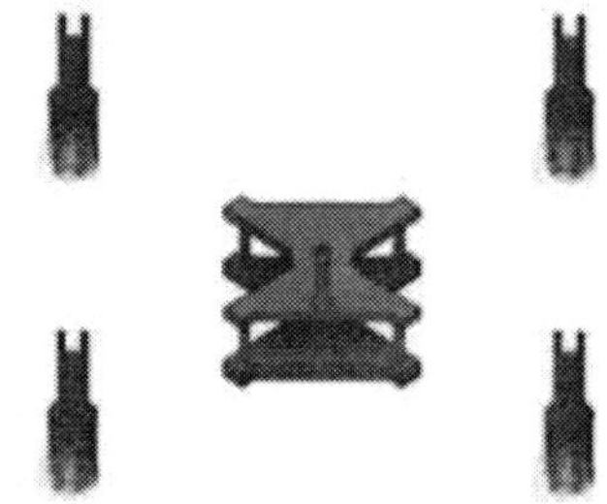

Figure 4. The schematic representation of DJI F450 drone frame [21]

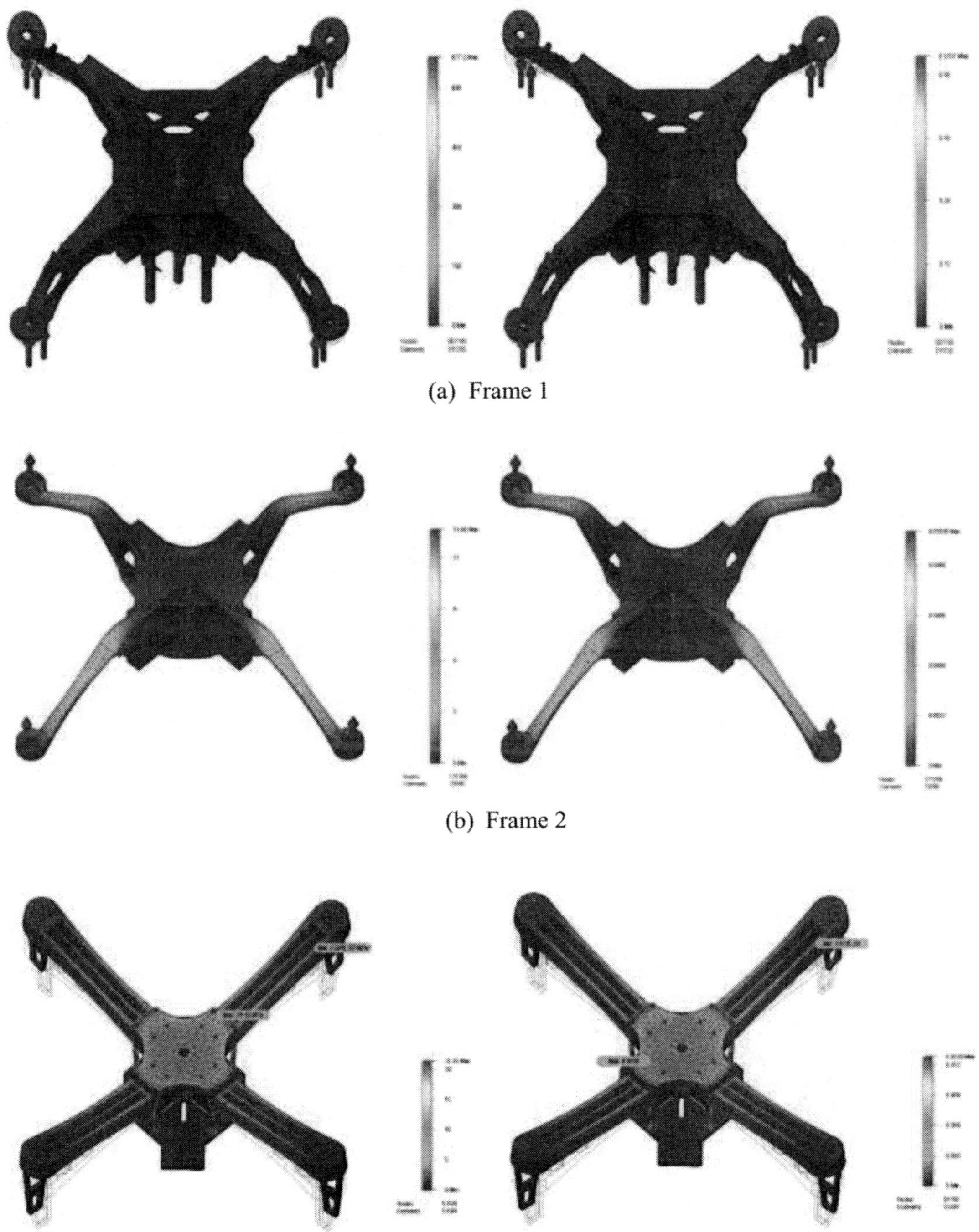

(a) Frame 1
(b) Frame 2
(c) DJI F450 frame

Figure 5. Comparison of stress-strain analysis between different frames

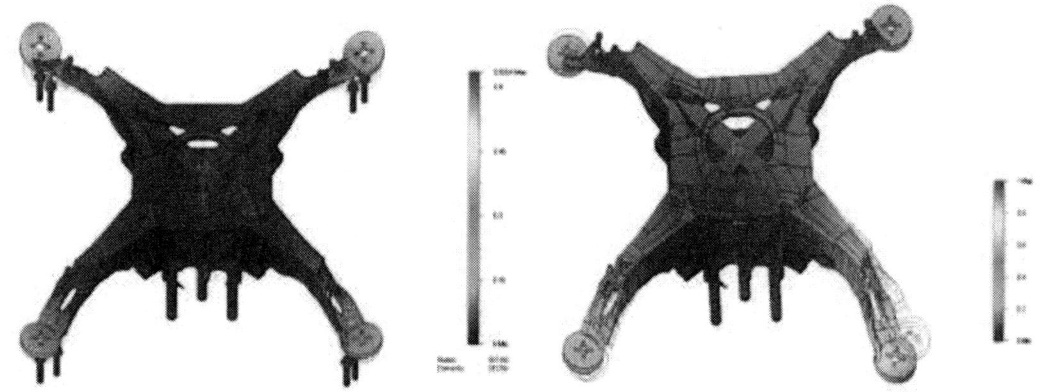

(a) Frame 1

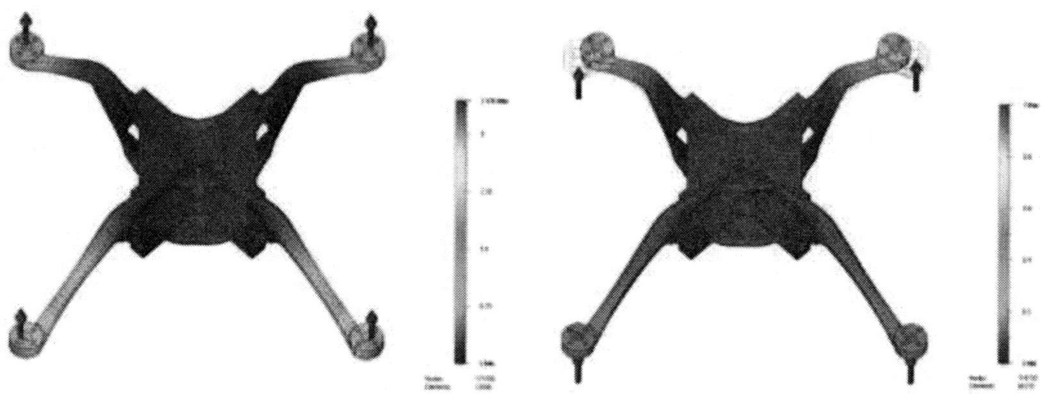

(b) Frame 2

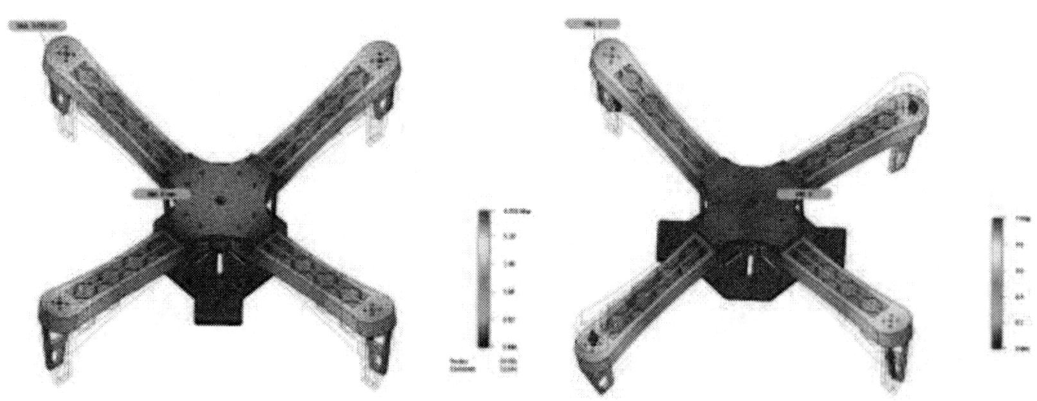

(c) DJI F450 frame

Figure 6. Comparison of displacement and modular frequency between different frames

By studying the designs of frame 1, frame 2 and DJI frame that's figure 4, various shift in criterions has been observed and reflected in the following contexts. One factor to reminisce is that all the three designs where in safe range, that's the framework is safe to apply the load stated while simulating and thus the frame will not fail in static load conditions as per the structural load constraints in designing.

The traditionally designed figure 4 has severe stress points in the standoff region caused due to the axial loading at the end of the drone arm. The critical point on the standoff on repeated loading may fail in due course of time. While the generatively designed frames have no critical points of failures, it can ultimately withhold repetitive and cyclic loads. Generative design automatically iterates the mass distribution in such a way that there are no/less critical points of failure while decreasing maximum mass from the design while optimising its mechanical properties. The symmetric nature is also optimizing the centre of mass to be exactly in the centre of the frame which is a major factor to be considered while designing a drone. The major mass is distributed in the standoff section connecting the centre region and drone arm, this is majorly observed in all the designs computed throughout the experiment. The mass in the standoff section is to counter the turning moment caused by the thrust and the load couple. And also, additional overhanging structures were also created to counter the stress in that region. Another perceptible factor is the trend of the mass distribution in the standoff region, mass of the frame and factor of safety, more the mass in the region increases the factor of safety of the frame but in turn increasing the mass of the frame. More the number of overhangs less is the mass in the standoff region, less is the factor of safety when compared to a generatively designed drone frame with a solid mass in the standoff structure. The traditional F450 frame has a lot of mass reduction features incorporated as triangles has multiple critical points of failure despite the fact the mass is significantly greater than the generatively designed drone frames.

The overall performance of the drone frame designed via generative designing method has three notable factors healthier than the traditionally designed F450 drone frame. Starting with the factor of safety, Frame 2 has a factor of safety of almost 40 times that of figure 4. From the data and material parameters it can be inferred that the probability of failure of figure 4, that's the DJI frame is higher than the other drone frames designed using generative design. Secondly, Von Mises stress governs if a material given will yield or fracture in time. The Von mises stress value of figure 4 is 11.4 times greater than frame1 thus concluding the resistance of yield or fracture is more for frame 1 when compared to a traditionally designed drone frame. Finally, the maximum displacement under the load condition for frame 2 is just 0.01 mm which is approximately 400 times as compared to figure 4 which indeed proves the ability to resist deformation of the generatively designed drone frame. There is also reduction of weight when compared to the DJI frame. The weight of frame 1 is 227gms the wispiest of the 3 designs, followed by frame 2 with 267gms and then the DJI frame with 330gms. Thus, are numerous data logged and studied via various experiments and analysis.

5. Conclusion

In brief, generative design is a beautiful symbiosis between man and computer that imitates natural worlds evolutionary progression with the help of the computing power of the cloud to provide thousands of solutions to one engineering problem rather than applying them retrospectively. As a result of this computational processes, innumerable remarkable diverse structures that looks very organic in design are created. An immaculate analysis is done, between the traditional designing method and generative designing technique, comparing with two generative designed drone frames [Frame1 and Frame2] alongside a DJI Flame Wheel F450 frame [figure 4]. Thus, from the analysis and simulation data logged, it is proven with the help of these frames that the outcomes attained by generative design technique is much better in many aspects when compared to other traditional designing.

References

[1] Buelow PV 2002 Using evolutionary algorithms to aid designers of architectural structures 315-336
[2] McKnight, M. 2017 Generative Design: What it is? How is it being used? Why it's a game changer KNE Eng., **2** 176–181
[3] Wu, Jun & Qian, Xiaoping & Wang, Michael 2019 Advances in generative design. Computer-Aided Design **116** 102733
[4] Ahmed, Md & Zafar, Mohd & Mohanta, Jagadish. 2020 Modeling and Analysis of Quadcopter F450 Frame 196-201
[5] Harish Doraiswamy, Nivan Ferreira, Marcos Lage, Huy Vo, Luc Wilson, Heidi Werner, Muchan Park, and Cláudio Silva 2015 Topology-based catalogue exploration framework for identifying view-enhanced tower designs **34** 1-13
[6] Janssen P. A 2006 generative evolutionary design method Digital Creativity **17** 49–63
[7] Sosnovik, I.; Oseledets, I 2019 Neural networks for topology optimization **34** 215–223
[8] Sun, H.; Ma, L 2020 Generative Design by Using Exploration Approaches of Reinforcement Learning in Density-Based Structural Topology Optimization **4** 10
[9] Oh, Sangeun et al. 2019 Deep Generative Design: Integration of Topology Optimization and Generative Models. Journal of Mechanical Design **141** 1
[10] Yi, Jijun & Zeng, Tao & Rong, Jian 2014. Topology Optimization for Continua Considering Global Displacement Constraint. Strojniski Vestnik. **60** 43-50
[11] Bashu Aman 2020 Generative Design for Performance Enhancement, Weight Reduction, and its Industrial Implications
[12] Sivam krish 2011 A practical generative design method. Computer-Aided Design. **43** 88-100.
[13] Vlah, D., Žavbi, R. and Vukašinović, N 2020 Evaluation of topology optimization and generative design tools as support for conceptual design **1** 451–460
[14] Conner Sharpe, Tyler Wiest, Pingfeng Wang, and Carolyn Conner Seepersad 2019 A comparative evaluation of supervised machine learning classification techniques for engineering design applications. Journal of Mechanical Design **141** 1-38
[15] M D Stroia et al 2020 Case studies for automotive components using CAD and CAE Techniques **1426** 012047
[16] Buonamici, F., Carfagni, M., Furferi, R., Volpe, Y. and Governi, L 2020 Generative Design: An Explorative Study **18** 144-155
[17]] Mühlbauer, Manuel 2020 Smart Structures -A Generative Design Framework for Aesthetic Guidance in Structural Node Design Application of Typogenetic Design for Custom-Optimisation of Structural Nodes
[18] Kaufui V Wong, Aldo Hernandez 2012 A Review of Additive Mnaufacturing 10.5402/2012/208760.
[19] Shahroz Khan, Muhammad Junaid Awan 2018 A Generative Design Technique for Exploring Shape Variations **38** 712-724
[20] M D Stroia, D E Mosteanu, I Virca, E Raduca, C Popescu, C Haţiegan 2020 Case studies for automotive components using CAD and CAE techniques **1426** 012047
[21] Burggräf, P., Pérez Martínez, A.R., Roth, H. et al. 2019 Quadrotors in factory applications: design and implementation of the quadrotor's P-PID cascade control system **1** 722

Heuristic Algorithm for Cooperative Motion Planning & Controlling of Unmanned Surface Vehicles

Tirumalapudi Raviteja[1],Rajay Vedaraj.I.S[1*], R Siva Krishnam Raju[1], M Yugandhar[1]

[1]School of Mechanical Engineering, Vellore Institute of Technology, Vellore, Tamilnadu, India

E-mail: `rajay@vit.ac.in`

Abstract. From the past decades autonomous technology is growing significantly in all applications. In Marine applications, Unmanned Surface Vehicles (USV's) is emerging research technology. For this research mobile robotics motion, path planning technologies are adopted and extended. The objective of this paper finds out the shortest route and Collision free path between the two points at the dynamic environment in moving time using the proposed modified A* algorithm. The simulation of the proposed algorithm is using python - OpenCV.

1. Introduction

In automation technology is all over the world using engineering and science applications like automotive, robotics, software development, Medical and agricultural, etc. Initially, this technology was used remote controller with a human operator for controlling with the machine [1][2]. In WW-II time USV's are used for the removal of explosive things in naval mines. Later this technology is upgrading to Semi-autonomous and fully autonomous systems [3]. Present USVs are using for environment monitoring, research, commercial, surveillance, inspection and navigation, and mapping in naval operations [4][5]. USV's are classified from Submersible vehicles [6] and they are Autonomous underwater vehicles, Autonomous underwater gliders, and remotely operated vehicles. An overview of the simple functionality architecture of a USV is outlined in Fig 1.

Different types of USV's prototypes are explained in Table 1. Most of the USV's are using research and navy applications of surveillance of the environment and inspection only.[7]

2. Motion Path Planning Algorithms

In mobile robotics for motion path planning developed few algorithms based on robot application, local and global environment, and moving positions. USV's also the same as adopting A* algorithm and modifies for the shortest route and obstacle avoidance.

IOP Conf. Series: Materials Science and Engineering 1012 (2021) 012020 doi:10.1088/1757-899X/1012/1/012020

Table 1. Different Prototypes of USV's.

S. No	Prototype Name	Manufacturing	Applications	Description
1	NOMARS[8]	DARPA ,US& 2020	Navy	Smaller watercraft operating
2	Large Unmanned Surface Vehicle [9][10]	US& 2020	Navy	Defense operations
3	MV Yara Birkeland	France[11][12] & 2019	Commercial	Port&.Industrial Operations
4	Sea Hunter [13]	DARPA , USA& 2018	Research	Detect diesel-electric submarines
5	USV Maxlimer [14]	British& 2018	Commercial	Crossed North sea and awarded Ocean Discovery X Prize
6	UAPS20 Unmanned Autopilot System [15]	Italy& 2016	Research	Autopilot system
7	Buffalo Automation [16][17][18]	New York& 2015	Commercial	Hazard identification and avoidance
8	Liquid Robotics [19][20]	US& 2013	Research	Guinness World Record awarded by a travelled longest journey by an autonomous
9	Scout (autonomous boat)[21]	Spain& 2013	Research	Research
10	Piranha Unmanned Surface Vessel [22]	Washington& 2010	Research	For manufacturing, this USV'S composite material was replaced by lightweight carbon-nanotube for less weight
11	ASW Continuous Trail Unmanned Vessel [23]	DARPA , US& 2010	Research	Detect diesel-electric submarines
12	Unmanned surface vehicle Piraya [24]	Sweden& 2009	Navy	Single operator control multiple USV's Pirayas
13	USV RSV (Marine Tech)	France& 2008	Research	Research
14	Fleet-class unmanned surface vessel [25]	US& 2008	Navy	Anti-submarine mission
15	Silver Marlin	Israel[26]& 2006	Navy	surveillance and reconnaissance
16	Protector USV [27]	Israel& 2005	Navy	surveillance and reconnaissance
17	Spartan Scout [28]	US& 2001	Research	Research
18	FL-boat [29]	German& 1911	Navy	Remote control

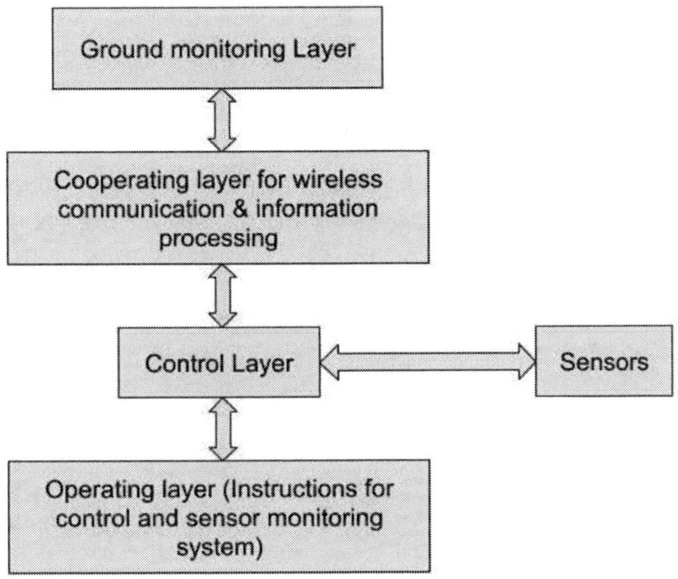

Figure 1. Simple functionality architecture of a USV's

A few USV's algorithms are shown in Fig 2. But basic and successful algorithms are Dijkstras and A* only.

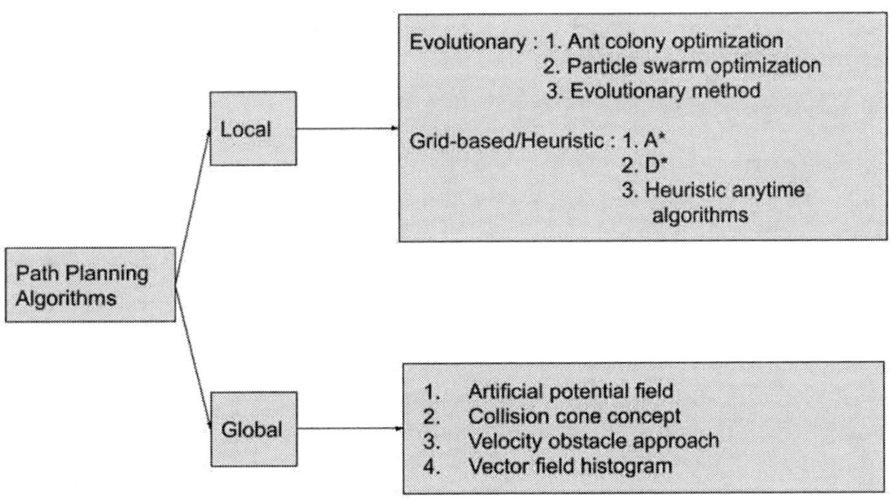

Figure 2. Classifications of Path Planning algorithms

- Ant Colony Optimization (ACO): It was first presented by Marco Dorigo during the 90's in his Ph.D proposal. This calculation is presented dependent on the rummaging conduct of a subterranean insect for looking for a way between their state and source food. At first, it was utilized to tackle the notable mobile sales rep issue. After ward, it is utilized for taking care of various hard improvement issues [31].

- Particle Swarm Optimization (PSO): The PSO calculation look through the space of a target work by modifying the directions of individual operators, called particles, as the piecewise ways shaped by positional vectors in a semi stochastic way. The development of an amassing molecule comprises of two significant parts: a stochastic segment and a deterministic segment. Every molecule is pulled in toward the situation of the current worldwide best and its own best area ever, while simultaneously it tends to move haphazardly..
- Evolutionary algorithms (EA): It frequently performs well approximating answers for a wide range of issues since they in a perfect world don't make any suspicion about the basic wellness scene. Methods from transformative calculations applied to the displaying of natural development are commonly restricted to investigations of micro evolutionary cycles and arranging models dependent on cell measures. In most genuine utilizations of EAs, computational intricacy is a forbidding element.
- A* Algorithm: A* is a calculation that is broadly utilized in path finding. The calculation productively plots a walk able way between various hubs, or focuses, on the diagram. A non-proficient approach to discover a way on a guide with numerous obstructions, path finding from directs A toward B can be troublesome. A robot, for example, without getting a lot other direction, will proceed until it experiences a hindrance, as in the way discovering guide to one side underneath. In any case, the A* calculation brings a heuristic into a standard diagram looking through calculation, basically preparing at each progression so a more ideal choice is made. With A*, a robot would rather discover a way in a manner like the graph on the privilege beneath.

$$f(n) = g(n) + h(n)$$

  Where,
  f(n) = total estimated cost of path through node n
  g(n) = cost so far to reach node n
  h(n) = estimated cost from n to goal
- Dijkstra's Algorithm: Dijkstra's Algorithm works by visiting vertices in the chart beginning with the item's beginning stage. It at that point over and again analyses the nearest not-yet-inspected vertex, adding its vertices to the arrangement of vertices to be inspected. It extends outwards from the beginning stage until it arrives at the objective. Dijkstra's Algorithm is ensured to locate a most brief way from the beginning stage to the objective, as long as none of the edges have a negative expense.
- Heuristics: Utilizing a decent heuristic is significant in deciding the presentation of A*. The estimation of h (n) would in a perfect world equivalent the specific expense of arriving at the objective. This is, nonetheless, unrealistic on the grounds that we don't have the foggiest idea about the way. We can, nonetheless, pick a technique that will give us the specific worth a portion of the time, for example, when going in an orderly fashion without any obstructions. This will bring about an ideal exhibition of A* in such a case.
- Artificial Potential Field: A potential field is any physical field that complies with Laplace's condition. Some normal instances of potential fields incorporate electrical, attractive, and gravitational fields. A potential field calculation utilizes the counterfeit possible field to control a robot around in a specific space. For USV's, we believe a space to be partitioned into a network of cells with hindrances and an objective hub. The calculation doles out a fake expected field to each point on the planet utilizing the potential field capacities which will be portrayed further in the clarification. The robot reproduces from the most noteworthy potential to the least potential. Here, the objective hub has the most reduced potential while the beginning hub will have the greatest potential. Henceforth, we can say that the USV's moves from most minimal to the most noteworthy potential.

- Vector Field Histogram: Another method snag shirking technique for versatile robots has been created and executed. This strategy, named the vector field histogram (VFH), licenses the recognition of obscure snags and dodges impacts while at the same time guiding the versatile robot toward the objective. The VFH technique utilizes a two-dimensional Cartesian histogram network as a world model. This world model is refreshed consistently with run information inspected by ready range sensors. The VFH strategy in this manner utilizes a two-stage information decrease measure so as to process the ideal control orders for the vehicle.
- Velocity Obstacle approach: It is the arrangement of all speeds of a robot that will bring about a crash with one more robots at some second in time, expecting that the other robot keeps up its present speed. On the off chance that the robot picks a speed inside the speed obstruction, at that point the two robots will in the long run impact, in the event that it picks a speed outside the speed impediment, such a crash is ensured not to occur. This calculation for robot impact evasion has been more than once rediscovered and distributed under various names: in 1989 as a moving board approach, in 1993 it was first presented as the "velocity obstacle", in 1998 as impact cones, and in 2009 as illegal speed maps. A similar calculation has been utilized in oceanic port route since in any event 1903

Way arranging approaches for USV's can be partitioned into nearby and worldwide methodologies Grid-based way organizers are fit for discovering answers for the reasonable issues and fulfilling targets like ideal and goal. Network based worldwide way organizers have been built up that takes care of such complex way arranging issue with a generous low computational time. Such heuristic methodologies have been widely utilized in mechanical way anticipating the most recent twenty years and have been the best option of way organizers contrasted with nearby methodologies.

2.1. A* Algorithm

A* is a graph search and path planning algorithm for optimization of the shortest distance with weights in many fields. In robotics, A* algorithm is used for planning [32]and re-planning [30] for the shortest route from the starting point to the goal point. Same as adopted in this USV's system also find out the optimal path from kalaimagal nagar point to Broken Bridge Point of Adayar River shown in Fig 3.

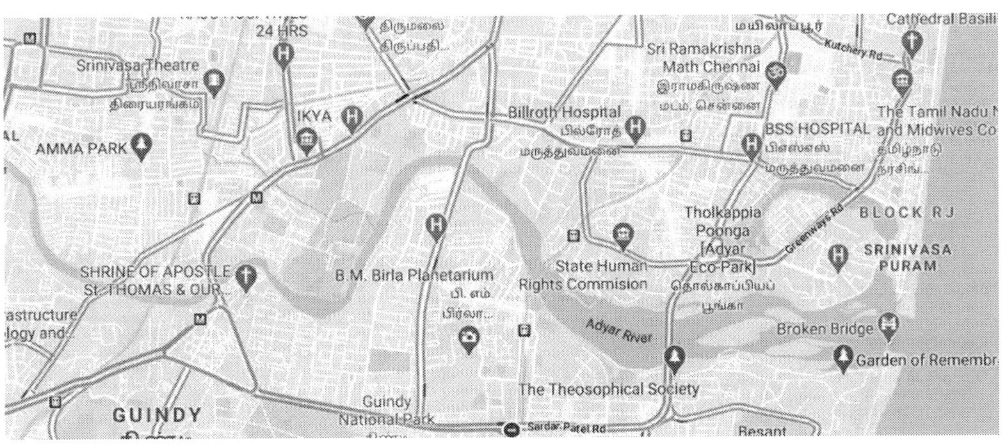

Figure 3. Map of Adayar River (Source: Google Maps)

Algorithm 1: A* algorithm

function A*:
$C(x_I) = 0$
$Q \cdot \text{ins ert}\,(x_I, \quad C\,(x_I) + h\,(x_I))$
visited $(x_I) = x_I$
while $Q \neq \emptyset$ x = Q. pop()
if x = xG
return SUCCESS
for u ∈ U(x)
x=f(x,u)
if not visited (x') or
$C\,(x') > C(x) + d\,(x, x')$
visited $(x') = x$
$C\,(x') = C(x) + d\,(x, x')$
Q.insert $(x', C\,(x') + h\,(x'))$
return FAILURE

3. Methodology

It implements the proposed motion path planning algorithm into 3 layers. In the initial layer, the path planning algorithm planned the optimal path information of the surrounding environment using the sensors and measure using the following equations.

Global positioning systems (GPS) devices are received signals from the satellite and deform the route from the start position to the destination position.

$$P^N = f\,(p + \delta p, O^n)$$

depth sensor is used for measure the depth of the current state place,

$$\mathrm{P_{ds}N} = \mathrm{Z} + \delta\mathrm{d(noise)}$$

$$\text{Doppler velocity } \log \mathrm{V^B} = \mathrm{V_1} + \mathrm{b_v} + \delta_\mathrm{v}$$
$$\text{Fiber gyro optic effect } w^m = w + b_e(\text{gyro}) + b_f(\text{bias}) + \delta_f(\text{noise})$$

The middle layer is path planning, it works as detecting the object and keeping a distance from the objects for the collision. It always maintained path smoothness for control of the USV in all conditions. The last task layer is the UAV reaching the destination position and controlling the USV safely is shown in Fig 4.

4. Mathematical Model of USV

The basics and mathematical equations are used for deriving the dynamic modelling of USV. Using these equations estimate the current position and optimal path of the Vehicle,

$$\text{Coordinate fixed frame} = \begin{pmatrix} O_N \\ X_N \\ Y_N \\ Z_N \end{pmatrix} \&\text{Body frame of the UAV} = \begin{pmatrix} O_A \\ X_A \\ Y_A \\ Z_A \end{pmatrix}$$

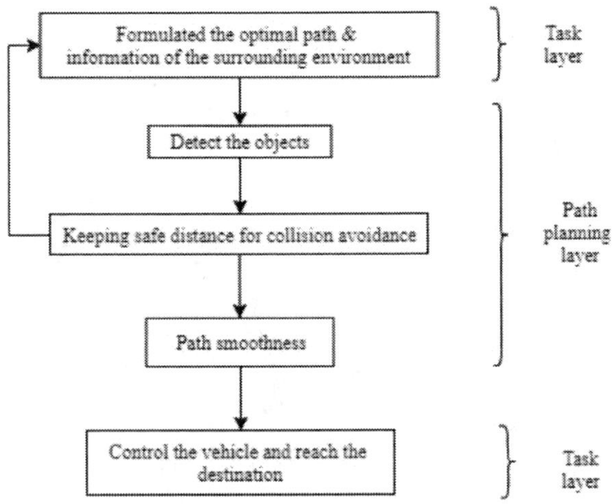

Figure 4. Proposed Path planning methodology

Orientation of the frame using Euler angles $= \begin{pmatrix} \varphi \\ \theta \\ \psi \end{pmatrix}$

The governing equation of the dynamics of the USV body is

$$M_m\dot{v} + C(v)v + D(v)v + g_n(\eta) = T(v, v)$$

Velocity angles of the body frame $= \begin{bmatrix} v & v & \omega \end{bmatrix}$

Mass moment Inertia of the vehicle $\mathrm{M_m}\dot{v}$

$= \begin{bmatrix} \mathrm{diag}\{\mathrm{m,m,m}\} & 0 \\ 0 & \mathrm{diag}\,\{\mathrm{I_X, I_Y, I_Z}\} \end{bmatrix} - \mathrm{diag}\,\{\mathrm{X_u, Y_v, Z_w, K_p, M_q, N_r}\}$

Centripetals and Coriolis forces C(v)

$= \begin{bmatrix} \mathrm{S}\,([\mathrm{m_p m_q m_r}]) & 0 \\ 0 & \mathrm{s}\,(\mathrm{I_X p I_Y q I_z r}) \end{bmatrix} + \begin{bmatrix} 0 & \mathrm{S}\,([\mathrm{X_v V Y_U U Z_\omega W}]) \\ \mathrm{S}\,([\mathrm{X_v V Y_v U Z_\omega W}]) & \mathrm{S}\,([\mathrm{K_p}p\mathrm{M_q}q\mathrm{N_r}r]) \end{bmatrix}$

Hydrodynami Damping force

$D(v) = -\,\mathrm{diag}\,(\mathrm{X_u|v|} \quad \mathrm{Y_v}|v| \quad \mathrm{Z_\omega}|\omega| \quad \mathrm{K_p|p|} \quad \mathrm{M_q|q|} \quad \mathrm{N_r|r|})$

Gravitational and Buoyancy force $(\mathrm{g_n n}) = \begin{pmatrix} \mathrm{W^B + B^B} \\ \mathrm{r^B * B^B} \end{pmatrix}$

As per the above expressions derived the mathematic model of the USV and some of the terms as assumed as deserve for this model.

5. Results

A reproduction result is introduced in this subsection to exhibit the presentation of a USV's arrangement independently exploring in a useful oceanic climate. More in particular, the essential point of this test is to check a way arranging calculations capacity to manage impact shirking necessity.

The simulation results of the geographical view of Adayar River generated path is shown in Fig 5 and element results are presented in Table 2. In conclusion, this approach is suitable for the real path planning of USVs.

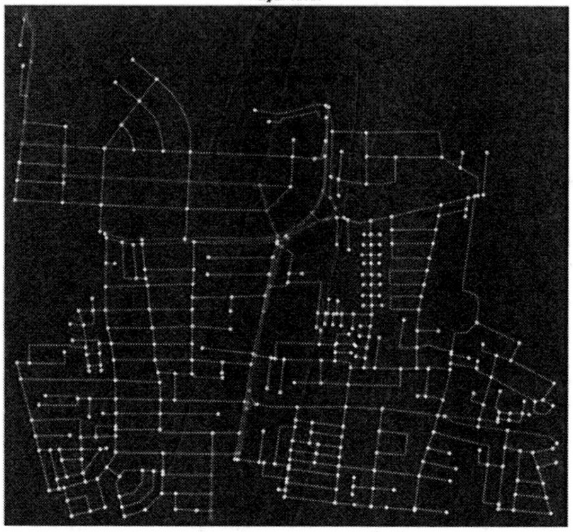

Figure 5. Output Graphical outline view of Adayar River

Table 2. Result Elements of the USV.

S No	Element	Result
1	Motion Speed	Quick Fast
2	Inertia	Small
3	Major drawback	Wind disturbing
4	Response	Less
5	Sensors	State estimation, Environment perception, Situation awareness
6	Applications	Ocean and atmosphere exploration, Remote sensing, agricultural, etc.,

6. Conclusion

Heuristic A* algorithm detained for the shortest motion path planning of USVs in a dynamic environment. The generated way is safer, and keeping a safe distance from the dynamic obstacles was simulated using python-OpenCV. In future work, this work extends in very conjugation paths with the above-modified heuristic A* algorithm. Conclusion, it is obvious that such an ideal methodology is reasonable for worldwide way arranging of USV's. In future work, it is intended to expand the work being developed of a way adherent methodology working in formation with proposed approach for a responsive way arranging in situations including close experiences.

Another expansion of the current work lies in considering heading edge requirement for USV's, in such cases, where way length is a higher priority than computational time. This The A* calculation have been effectively actualized on a pragmatic USV's arrangement stage to feature the presentation of self-governing route. The outcomes have demonstrated that by utilizing such a framework, the vehicle can self-navigate and securely explore a pragmatic sea climate. Heuristic A* algorithm detained for the shortest motion path planning of USVs in a dynamic environment. The generated way is safer, and keeping a safe distance from the dynamic obstacles was simulated using python-OpenCV. In future work, this work extends in very conjugation paths with the above-modified heuristic A* algorithm.

References

[1] Waymo 2017, "Remote Assistance for Autonomous Vehicles in Predetermined Situations," Us9720410B2, **12**.
[2] T. Raviteja and I. S. Rajay Vedaraj(2020), "AN INTRODUCTION of AUTONOMOUS VEHICLES and A BRIEF SURVEY," J. Crit. Rev.**7**(13) .doi: 10.31838/jcr.07.13.33
[3] J. E. Manley(2008), "Unmanned surface vehicles, 15 years of development," Ocean.1–4, doi: 10.1109/OCEANS.2008.5152052.
[4] M. Caccia, M. Bibuli, R. Bono, and G. Bruzzone , "Basic navigation, guidance and control of an Unmanned Surface Vehicle," Auton. Robots, vol. 25, no. 4, pp. 349–365, 2008,
[5] R. jian Yan, S. Pang, H. bing Sun, and Y. jie Pang (2010), "Development and missions of unmanned surface vehicle," J. Mar. Sci. Appl.9(4) ,451–457.
[6] S. Negahdaripour and X. Xu (2002), "Mosaic-based positioning and improved motion-estimation methods for automatic navigation of submersible vehicles," IEEE J. Ocean. Eng., 27(1), 79–99.
[7] J. Larson, M. Bruch, and J. Ebken (2006), "Autonomous navigation and obstacle avoidance for unmanned surface vehicles," Unmanned Syst. Technol. **7**(9)
[8] D.B.Larter,https://www.defensenews.com/naval/2020/01/21/darpas-latest-mad-science experiment-a-ship-designed-completely-without-humans/.
[9] D.Larter,https://www.defensenews.com/naval/2019/12/21/the-us-navy-gets-its-large-unmanned surface-vessels-in-2020-with-strings-attached/.
[10] R. O'Rourke (2019), "Navy Large Unmanned Surface and Undersea Vehicles: Background and Issues for Congress," 27.
[11] World's First Autonomous Ship to Launch in 2018.
[12] Savvides,www.freightwaves.com/news/revolution-for-inland-shipping-depends-on-thesuccess of-the-yara-birkeland.
[13] J. Vincent, www.theverge.com/2016/4/8/11391840/us-navy-autonomous-ship-sea-hunter christened.
[14] en.marinetech.fr/Marine-survey-USV-a380/
[15] www.sielnet.com/index.php/products/usv.
[16] phys.org/news/2015-10-yacht.html.
[17] techweek.com/buffalo-automation-ai-startup-boat/
[18] www.plymouthherald.co.uk/news/plymouth-news/robot-water-taxis-trialled-tech-4049999.
[19] T. Woody, qz.com/64819/we-welcome-our-new-solar-powered-ocean-going-robot-overlords/
[20] A. Park, www.sbs.com.au/news/wave-powered-robot-arrives-in-australia.
[21] www.huffpost.com/entry/scout-robotic-solar-boat-n-3575669.
[22] newatlas.com/nanotube-reinforced-carbon-fiber-piranha-usv/14321/.
[23] www.darpa.mil/news-events/2018-01-30a
[24] saab.com/naval/Submarines-and-Surface-Ships.
[25] news.usni.org/2016/04/07/
[26] www.elbitsystems.com/elbitmain/area-in2.asp?parent=10\& num=92\& num2=92.
[27] www.defense-update.com/products/p/protector.htm.
[28] www.news.navy.mil/search/display.asp/story/id=10964.
[29] muse.jhu.edu/article/173293/summary.
[30] M. Dakulovi and I. Petrovi (2011), "Two-way D* algorithm for path planning and replanning," Rob.Auton. Syst., **59**(5) ,329–342. doi: 10.1016/j.robot.2011.02.007.
[31] G. Xia, Z. Han, B. Zhao, and Y. Yang (2019), "Unmanned surface vessel path planning based on quantum ant colony algorithm," Harbin Gongcheng Daxue Xuebao/Journal Harbin Eng. Univ.**40**(7),1263–1268.
[32] R. T. Tirumalapudi and R. Vedaraj (2020), Mobile Robots Obstacle Avoidance and Shortest Path Planning Using Algorithms-A Review, TEST Eng. Manag, **83**,22434–22443.

Design of Dual Copter for Surveillance Applications

Balaji A[1*], Mithil J[1], Jason Gousanal J[1]

[1]Department of Mechatronics Engineering, Kongu Engineering College, Erode - 638060. Tamilnadu, India.

E-mail : bala2009mct@gmail.com

Abstract. In this paper, an innovative design model of autonomous aerial vehicle is designed and fabricated with dual rotor setup. Nowadays, innovative unmanned aerial vehicles (UAVs) are widely required for surveillance, monitoring, data collection, reconnaissance and generally in places where humans face hazardous. In this model and design of UAVs , aero dynamic principles are adopted through reasonable design considerations and assumptions. Most of the researchers and academicians are currently pursuing research on AVs with various structures, wing designs, fabrication methods, controllers and materials to increase the efficeincy as well as for various applications. In this proposed work, an effort is made to design and fabricate the dual copter with wide moving ranges for surveillance applications in military intelligence system. This proposed design will be suited for all environmental conditions to collect the data as well as monitoring the system. Dual copter chassis are fabricated using 3D printer after the validation and analysis the proposed design using Solidworks CAD tool. The chassis are fabricated with the provision of space where the required sensors, actuators and controllers can be assembled. Lithium polymer batteries are employed to achieve the better performance for copter. The fabricated setup is integrated with radio frequency controller to achieve the flight control through RF transmitter. Hence the performance of the fabricated device is validated and analyzed through various performance measures of dual copter.

Keywords: Brushless motor, PWM (Pulse Width Modulation), Radio Frequency Controller, Servo Motor

1. Introduction

A quadcopter, also known as quadrotor helicopter or quadrotor which are operated and controlled by four rotors. Quadcopters are commonly made up with two pairs of fixed propellers one is clockwise and another one is anticlockwise (ACW) [1]. The desired total thurst can be attained by the regulation of rotor speed and also the rotor can be located in both lateral and longitudunal directions in the space by controlling rotor speed with various range. The desired torque and turning force can be achieved by rotor controller based on the applications [2]. The control issues due to the induced torque can be avaoided through counter rotor rotations where the comparatively short blades are required which is simpler to construct [3]. This kind of air vehicles are considered as the first victorious vehicle among various model which have vertical take-off landing (VTOL) provisions [4].Two decades ago before 2000s, the electronics advancements endorsed the manufacturing of inexpensive low-weight air vehicles with global postioning systems, cameras, apropriate controllers, acclerometers [5]. A Quadcopter with four self-governing rotors requires swash plate mechanism which will lightening the quadcopter for flying [6]. The swash plate mechanism was required to allow the helicopter in order to achieve the required degrees of freedom (DOF), in the meantime the control can be obtained by incorporating two more rotors [7]. The development of quadcopter are stabilized in recent days by controlling the four self-governing rotors without integration of electrical devices which are tedious and impossible [8]. Quadcopter control is basically a tedious one and intersect more problems while operating for different environmental conditions [9]. The quadcopters are opearted with six DOF (three linear movements and three rotational movements) and self-governed inputs of four i.e. speed of rotor and the performance achieved by coupled motion of rotational and translational [10]. Because of complicated aerodynamic effects, the results may be highly nonlinear [11]. Quadcopter has own damping system to prevent the little friction or motion with the provision to stop moving and remain stable which is not achievable in ground vehicles [12]. This results in the quadcopters emerging with flexible configuration which becomes very popular for small aerial vehicles. Its minimum size and maneuverability, provides the provision for quadcopters in any environments. **Figure. 1.** shows the design of quadcopter used for surveillance purposes. In the recent research environments, quadcopters have received more attention among the researchers to achieve the various innovation techniques of rotors for different applications [13]. The controller design requires more consciousness while selecting the quadcopters based on the required applications [11]. There are some uncertainties in the quadcopters such as degradation of actuators, environment disturbances and improper time delays in communication [14]. The above disturbances are amplified in the proposed design if it lost the its control in unbalanced situation [15]. Hence an effort is proposed to overcome the above mentioned drawback and this concept will functionalize the qaudcopter movement in easier as well as achievable [16].

However, earlier forms of prototypes grieved from poor performance, and later the same type of premodeled design highly demands too much pilot interaction i.e. workload due to the unsatisfied augmented stability as well as limited ability of controller. [17]. These innovations have modern microprocessors that offer more electronic integration at minimum cost which are completely autonomous control [18]. These advantages made the quadcopters feasible for commercial, military, and even hobbyist purposes [19].

The advantages of the proposed dual copter are

- Increase the flying time
- Increase the efficiency of the copter

Figure. 1. Quad Copter

To overcome the drawbacks in the quadcopter it has been planned to design and fabricate the dual copter for surveillance applications in the military intelligence system and to implement the sound design and model with uniqueness in all aspects for uninhibited missions, fast and high operations of dual copter. Also will validate and optimize the proposed design using CAE tools to ensure the well-designed model of the dual copter.

2. Proposed System

The proposed work will favor to terrestrial as well as aerial moving ranges for surveillance applications in the military intelligence system with best design of system. The proposed work will find the better accomplishment in the surveillance system with wide range and high load carrying capacity. The novelty of the work is the dual copter covers wide range of distance and high load carrying capacity compare to other drones currently exhibits. The propeller is designed in the enclosed manner except the top and bottom by the way air never leaks on sideward hence a cumulative amount of thrust is produced in this system it paves to higher energy efficiency. The centre of gravity and centre of pressure is aligned perfectly it ensure that the system is aerodynamically stable for manoeuvre and control. This proposed design will be suited for all environmental conditions. The advantages of the proposed model are

- Easily maneuver
- Reduce the usage of propeller blade number
- Enclosed cylindrical-shaped frame
- Increasing flying time

3. Methodology and Design

The details of stages involve for the development of dual copter was shown in the **Figure. 2.** The design is the most important stage for any product. In this work, the design and analysis of model were performed using solid works software. Initially the part drawings has to be completed with the specified dimension as shown in **Figure. 3. (a), (b) & (c)** which represents front view, top view and side view of the dual copter design respectively. The reason for choosing solid works over AutoCAD is the fact that the ease with working in the former one. The dimensions in the Figure. are mentioned in mm.

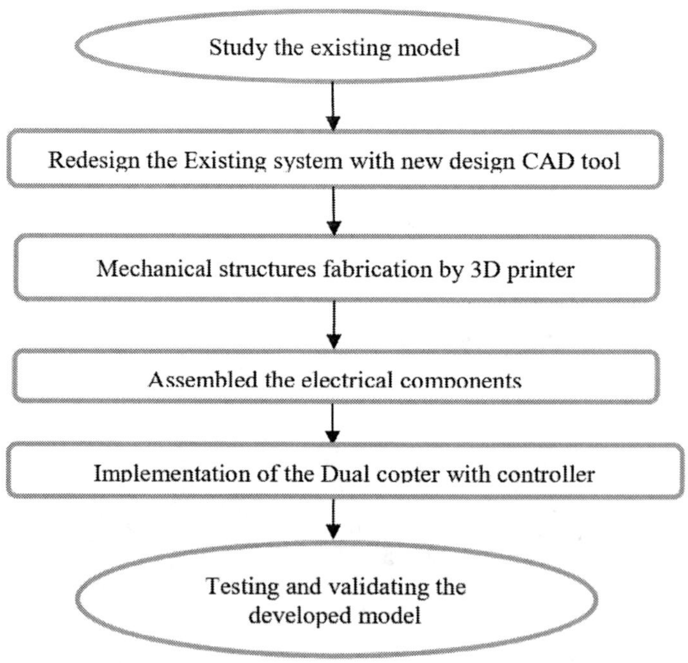

Figure. 2. Methodology

Since the project require the outer frame with less weight ratio, the design is ensured to achieve the smooth functioning of flying by using the simulation package. It plays a major role at the time of lifting the dual copter and the motor should be fixed in a correct position to prevent the error happening in the propeller blades.

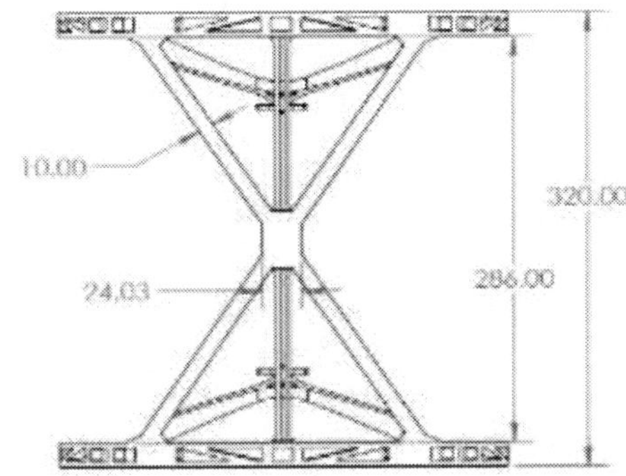

Figure. 3 (a) Front View of Dual Copter model

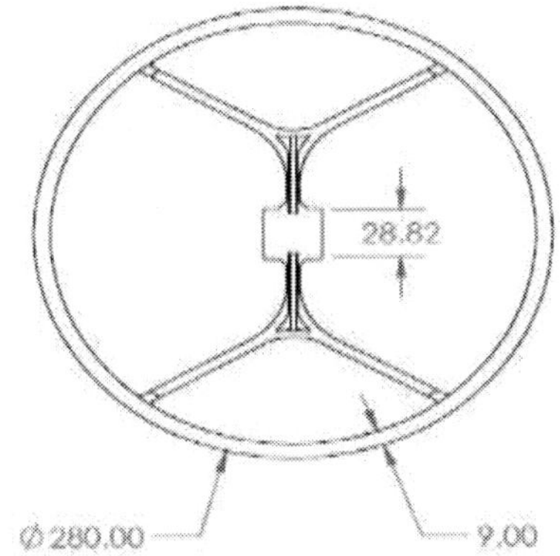

Figure. 3 (b) Top View of Dual Copter model

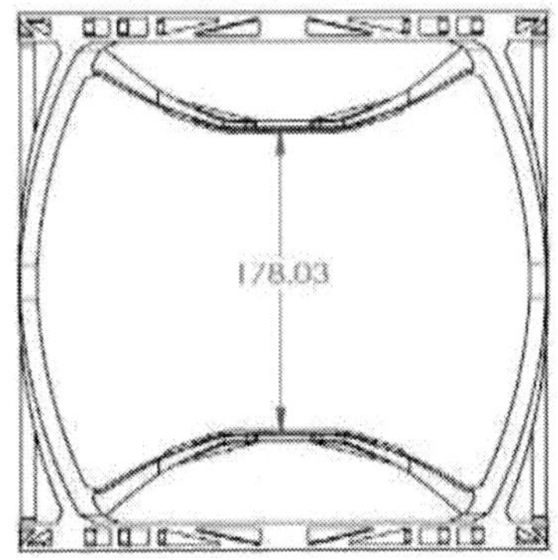

Figure. 3 (c) Side View of Dual Copter model

Figure.3. Solid Model of Dual Copter

The radio controller has been selected with proper range of control to actuate the rotor as required for the betterment of flying and landing. The discharge rate of battery should be maintained to make the proper functioning as well as operating the copter [20].Technical success of the system relies upon the work expected from the mechanical and electrical section. The fabrication of the project contains validated mechanical structure and electrical connections. The integration of the motor and other electronic components was made with the mechanical frame for execution of the proposed model. Then the model can be tested physically to verify the working efficiency.

4. Electrical Section

Transmitter transmits the radio frequency signals which will be received by the receiver at the rate of different inputs. The received signals will transfers to both the motors in the form of PWM (Pulse Width Modulation) signals to control the rotors [21]. But the brushless motor has three different poles which are actuated by PWM signals when it tends to rotate [22]. The three poles are activated with alternating signals to control the speed of brushless motor for performing the actions. Normally servo motor is actuated by the interpret of direct PWM signal. Both servo and brushless motor are controlled by the PWM signals. There are two servo motors are implemented for pitch, roll, yaw motion controls, this motions are achieved by air flippers which is attached to the shaft of servo motors [23]. Angular position of the servo motor is attained by the respective PWM input signals, for example 0.5 PWM to servo is 90°, 1 PWM to servo is 180°. And these PWM signals of brushless and servo motors are directly controlled by our radio control joystick. The electrical section of the project is modelled as shown in **Figure. 4.**

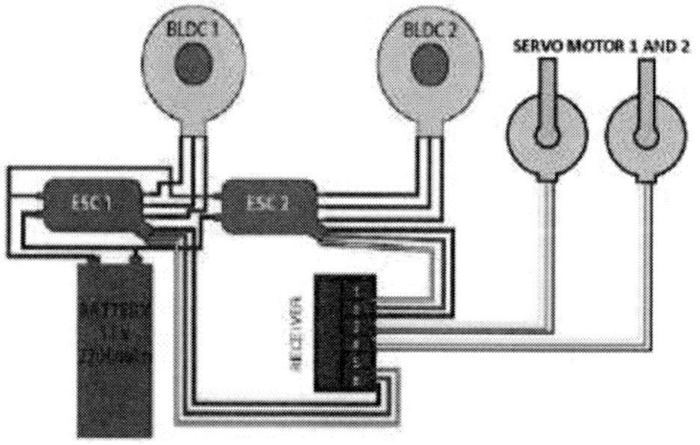

Figure. 4 Electrical Section of the model

5. Mechanical Section

The frame is constructed by additive manufacturing method with printing material called ABS (Acrylonitrile butadiene styrene). Normally ABS and Nylon is preferred for this kind of application. ABS is best suited for the frame because of its high strength and less weight-density ratio. 3D printer named CUBEPRO is used to fabricate the mechanical frame. The additive layer thickness was adopted as 200μm. Honey-comb structure is preferred for its strength and reliability. This frame tends to withstand load, sudden shocks and vibrations of motor occurs at the time of fly. Frame can be easily fabricated because of the availability of ABS material as plenty in the market and economically beneficial. Frame is in cylindrical shape and concentrated air flow produced by the two propellers which are thrusts downward by the way the drone lift upwards vertically. This airflow passes through flippers of the servo motor. The air is diverted by the flippers to attain the pitch, roll, yaw motions. The prototype of the fabricated model has been shown in the **Figure. 6.**

IOP Conf. Series: Materials Science and Engineering 1012 (2021) 012021 doi:10.1088/1757-899X/1012/1/012021

6. Experimentation

The experimentation was carried out with the required specifications which has been disseminated as follows. The assembly of the mechanical and electrical section was made to make the intelligent dual copter and analyzed the performance of the model. The **Figure. 5** shows the block diagram of the working layout. In this work, two sections were fabricated and connected to achieve the conduction. The model is the excitation of dynamics and longitudinal motion as well as lateral motion.

Brushless Motor:

$$Thrut\ (Kg) = \left((2.83 \times 10 - 12) \times (RPM^2) \times (DIA^4) \times \left(\left(((AIR\ DENSITY) \times 23.936) \div 29.92\right) \times CF\right) \div 2.2\right)$$

Thrust = 600 g

Where,

- RPM is mentioned in kV, it is multiplied with battery voltage,
- DIA, diameter for the total length of propellers,
- AIR DENSITY can be calculated separately,
- CF changes based on prop type but should just be "1" for the calculations.
- RPM = (Motor KV Racing * Battery Voltage)
- Speed = 10000 RPM

Battery:

- Voltage = 11 (V)
- Energy = Voltage * current * time = 24200 (J)

Servo Motor:

- Rotation = 180 degree
- Torque = 1.8 kgcm

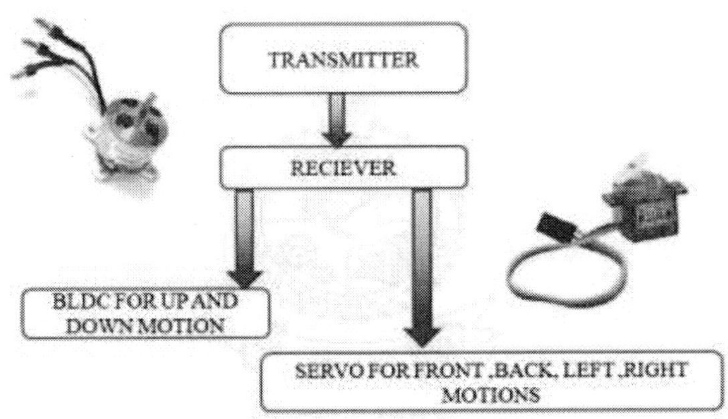

Figure. 5 Block diagram of the working layout

Electromagnetic waves is used in the process for controlling the dual copter with various level. Transmitter emits the signal and the receiver absorbs the signal in the form of PWD signals at the rate of 2.4 GHz. Then these signals are transferred to brushless and servo motors to control the up and down movement of the dual copter. Servo motor is attached with a RADAR to change the movement of left and right positions in the dual copter [19]. The radio frequency controller has the control range of 500m. It can be varied with the configuration of channel in the transmitter. Position of motor in the respective place must be accurate during the fly time [24]. The function of the dual copter is validated with various actions to accomplish the goal. The fabricated model with all components is represented in the **Figure. 6.**

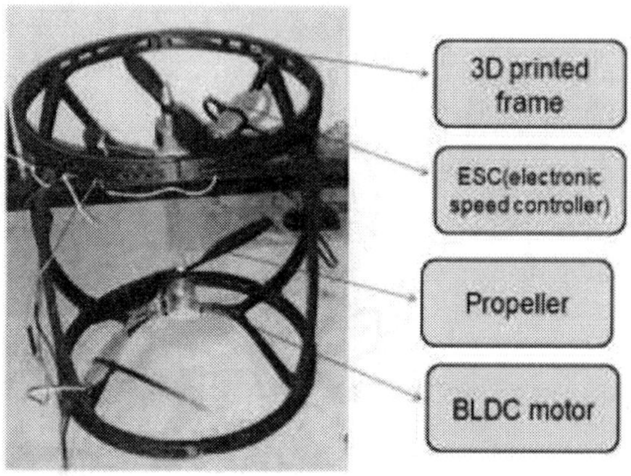

Figure. 6 Dual copter- Prototype

7. Conclusion and Future Scope

Thus the design and fabrication of prototype of dual copter was successfully developed with intelligence system. The developed system shows that high bandwidth of control, wide range and high load carrying capacity with adaption of any environmental conditions. Thus, the greater accuracy and control over the process has been achieved through this design. The model of the proposed design are validated and optimized using CAE tools to ensure the well-designed dual copter.

The advancement of this model plays vital role in both aerial and terrestrial. GPS autonomous tracking and obstacle avoiding UAV with accurate PID controls can be implemented in future with the developed model [25]. It can be used to program the controller using Arduino board to perform the system integration. Then, synchronize the Zig bee module with mobile phone in the fabricated system to implement the IOT concepts.

References

[1] Mohammadi M and Shahri A M 2013 Adaptive nonlinear stabilization control for a quadrotor UAV: Theory, simulation and experimentation *Journal of Intelligent & Robotic Systems* **72** 105-22

[2] Yu Y, Ding X and Zhu J J 2013 Attitude tracking control of a quadrotor UAV in the exponential coordinates *Journal of the Franklin Institute* **350** 2044-68

[3] Hernández Y L, Frías O G, Lozada-Castillo N and Juárez A L 2019 Control Algorithm for Taking off and Landing Manoeuvres of Quadrotors in Open Navigation Environments *International Journal of Control, Automation and Systems* **17** 2331-42

[4] Bnhamdoon O A A, Hanif N H H M and Akmeliawati R 2020 Identification of a quadcopter autopilot system via Box–Jenkins structure *International Journal of Dynamics and Control* 1-16

[5] Rabah M, Rohan A, Han Y-J and Kim S-H 2018 Design of fuzzy-PID controller for quadcopter trajectory-tracking *International Journal of Fuzzy Logic and Intelligent Systems* **18** 204-13

[6] Carlton Z, Wei W and Cohen K 2020 *Multi-Rotor Platform-based UAV Systems*: Elsevier) pp 129-52

[7] Bolandi H, Rezaei M, Mohsenipour R, Nemati H and Smailzadeh S M 2013 Attitude control of a quadrotor with optimized PID controller

[8] Khatoon S, Gupta D and Das L 2014 PID & LQR control for a quadrotor: Modeling and simulation. In: *2014 International Conference on Advances in Computing, Communications and Informatics (ICACCI)*: IEEE) pp 796-802

[9] Sugimoto Dimitrova R, Gehrig M, Brescianini D and Scaramuzza D 2019 Towards Low-Latency High-Bandwidth Control of Quadrotors using Event Cameras *arXiv* arXiv: 1911.04553
[10] Reyes-Valeria E, Enriquez-Caldera R, Camacho-Lara S and Guichard J 2013 LQR control for a quadrotor using unit quaternions: Modeling and simulation. In: *CONIELECOMP 2013, 23rd International Conference on Electronics, Communications and Computing*: IEEE) pp 172-8
[11] Deskovski S, Sazdovski V and Gacovski Z 2011 Advanced guidance laws and navigation systems for UAV: theoretical and practical findings in a developing country *Proceedings of COSY* 333-44
[12] Liu Z-J, Lv Q and Wang D-l 2010 Modeling And Simulation Control Of a Small-sized Quadrotor Helicopter [J] *Computer Simulation* **7**
[13] Bouabdallah S and Siegwart R 2007 Full control of a quadrotor. In: *2007 IEEE/RSJ International Conference on Intelligent Robots and Systems*: Ieee) pp 153-8
[14] Meyer J, Sendobry A, Kohlbrecher S, Klingauf U and Von Stryk O 2012 Comprehensive simulation of quadrotor uavs using ros and gazebo. In: *International conference on simulation, modeling, and programming for autonomous robots*: Springer) pp 400-11
[15] Fernando H, De Silva A, De Zoysa M, Dilshan K and Munasinghe S 2013 Modelling, simulation and implementation of a quadrotor UAV. In: *2013 IEEE 8th International Conference on Industrial and Information Systems*: IEEE) pp 207-12
[16] Rezaei M 2014 Dynamic modeling, simulation and control of a quadrotor using MEMS sensors' experimental data *Modares Mechanical Engineering* **14** 176-84
[17] Dikmen İ C, Arisoy A and Temeltas H 2009 Attitude control of a quadrotor. In: *2009 4th International Conference on Recent Advances in Space Technologies*: IEEE) pp 722-7
[18] Amlashi N J S, Rezaei M, Bolandi H and Sedigh A K 2014 Robust second order sliding mode control for a quadrotor considering motor dynamics *International Journal of Control Theory and Computer Modeling (IJCTCM)* **4** 9-25
[19] McNamee P and Barrett-Gonzalez R M 2020 Modeling and Simulation of Quadcopter Dynamics in Steady Maneuvers. In: *AIAA Scitech 2020 Forum,* p 1133
[20] Jun L and Yuntang L 2012 Modeling and PID control for a quadrotor [J] *Journal of Liaoning Technical University (Natural Science)* **1**
[21] Gong X, Hou Z-C, Zhao C-J, Bai Y and Tian Y-T 2012 Adaptive backstepping sliding mode trajectory tracking control for a quad-rotor *International Journal of Automation and Computing* **9** 555-60
[22] Wang L, He C and Zhu P 2014 Adaptive sliding mode control for quadrotor aerial robot with I type configuration *International Journal of Automation and Control Engineering (IJACE)* **3** 20-6
[23] Li J, Qi X and Han S 2013 Attitude decoupling control for quadrotor aircraft based on active disturbance rejection control technique *Electronics Optics & Control* **20** 44-8
[24] Amouri A, Mahfoudi C and Zaatri A 2019 Dynamic Modeling of a Spatial Cable-Driven Continuum Robot Using Euler-Lagrange Method *International Journal of Engineering and Technology Innovation* **10** 60
[25] Lee S-h, Kang S H and Kim Y 2011 Trajectory tracking control of quadrotor uav. In: *2011 11th International Conference on Control, Automation and Systems*: IEEE) pp 281-5

Modelling and Simulation of Autonomous Indoor Robotic Wastebin in Webots for Waste Management in Smart Buildings

Manju Mohan[1], RM. Kuppan Chetty[1*], K. Mohammed Azeem[1], P. Vishal[1], B. Poornasai[1] and Vijayram Sriram[1]

[1]Centre for Automation and Robotics (ANRO), School of Mechanical Sciences, Hindustan Institute of Technology and Science, Padur, Chennai – 603103.

*Corresponding Author : rmkuppan@gmail.com

Abstract. Waste bins are being used in an indoor environment by every common man for over many decades. Proper disposal of the waste and the waste collection system in an indoor environment is an important aspect of the cleanliness of the environment, the outlook of the environment, and as well as considering the health aspects of the people. An improper waste collection system in an indoor environment leads to a foul smell and unhygienic condition, thus inherently results in environmental pollution. Another crucial challenge is devising a safe navigation methodology for the smart waste bin in the indoor environment and efficiently model and simulate its performance. Therefore, in this paper, Mobile Smart Waste Bin with a smart waste collection system is proposed using the technical advancements in robotic technology with the navigation techniques. A simple line following navigation technique is employed in such a way that, when the waste bin gets filled with trash, the corresponding bin autonomously moves to the common collection point. This is done to analyse the integration of the physical and intrinsic characteristics of the system through simulation studies and the results are reported. The performance of the proposed indoor robotic dustbin bin is modelled with the underlying physics as well as the environment in webots virtual simulator. The effectiveness of the system is analysed through simulation studies in webots simulator and the results are visualized and reported. Thus, the designed mobile smart bin and proposed waste collection system will be a better autonomous solution compared to existing ones in the indoor environment of Smart Buildings.

Keywords: smart waste bin, line follower, navigation technique, smart buildings, webots.

1. Introduction

The development of smart buildings can be accomplished with the implementation of various indoor robotic technologies which makes the day to day activities much easier with less manpower and maximum utilization of technological advancements. The use of robotic technologies is very significant in the development of smart buildings. Advancement in robotic technologies is the key part of the smart buildings irrespective of various needs and applications [1]. In places such as apartment, multi-story buildings, in large scale industries, or institutions, there occurs a large accumulation of wastes at various time intervals. The applicability of a proper waste collection system is very significant. Therefore, solid waste collection in an indoor environment possesses a major challenge in most places. [2] Usage of Waste bins indoors are part of our lives for over many decades and mostly the waste collection system is not properly maintained due to improper monitoring and irregular disposals which leads to the foul smell and unhygienic condition which inherently results in environmental pollution.

The existing method of waste collection is an unwieldy and intricate process, which involves full human effort; valuable time; cost and hazardous for human life and is not well-matched with the advancement in robotics technology for providing smartness in waste bins which is much safer considering the health aspects of the people. One of the smart ways of waste collection and management is the use of smart waste bins interlinked through an IoT cloud-based network that can be monitored and controlled remotely. Hence, the concerned persons will be notified about each bin status and thereby achieving the removal of wastes in a proper way. The design of such a smart waste bin with real-time monitoring is presented in [3] where the level of the bin is monitored using the capacitance. An ultrasonic sensor is used to open and close the lid of the bin whenever the persons are nearby the bin.

Such smart bins are connected to the IoT cloud, where the bin status is communicated and monitored through the app. However, these systems are suitable only for large-scale external bins, and when comes to the application in smart buildings for indoor waste management it is inevitable to incorporate the advancement of robotic technologies to effectively manage the solid wastes in a well-organized manner. The design of a smart waste bin for smart waste management is represented [4] in which the system consists of a load cell sensor to measure the weight of the waste collected and an ultrasonic sensor is used to measure the level of waste inside the bin. This system also uses GSM and Bluetooth for the transmission of data required for waste management. An Android application is used to indicate a particular worker.

An intelligent waste alert and prediction system by employing a machine learning technique has been discussed in [5] which explains the steps involved to develop a smart waste bin which notifies the person in charge to guide the garbage trucks to collect garbage only from the cites where the bin is filled. The machine learning concept is used to gather data about waste generation in a particular area. Smart waste management for a green environment was discussed in [6] which briefs on a smart waste management system that measures the level of garbage collected by the bin using an ultrasonic sensor. GPS equipped waste collection trucks are used to collect the wastes. This system is based on the Bootstrap platform. Hence the smartness of the bin could be enhanced by incorporating the mobility to the waste bin which can safely navigate to the common waste collection point using autonomous navigation techniques similar to a mobile robot, making it a robotic waste bins placed in different locations.

These robotic waste bins could be connected to the common cloud and its operations could be monitored through a centralized server. The major challenge in such work is to apply simple and safe navigation techniques for the waste bins and to effectively model and simulate them to enhance its performance and applications. Modeling and Simulation is a two-part problem used to study the behavioral characteristics of the system to effectively measure its performance in applications intended for. As the smart robotic waste bin is proposed in this work, it is necessary to model the waste with the underlying physics, as well as the environment and as well as perform simulation studies to measure the performance of the applied navigational capabilities. Modeling and simulation of the motion of a Wheeled Mobile Robot are proposed in [7, 8] where a traction model was used for the Wheeled Mobile Robot. This model has employed the dynamic equations of motion to mimic the motion of the mobile robot. The simulation results portray that the controllers do not consider the wheel slip and give a poor tracking performance for the mobile robot.

Numerous modeling and simulation tools are found in the literature such as Matlab, Scilab, RoS, Microsoft RDS, V-REP, MRSim, RobotSim, etc., where the behavior of the system can be analyzed. However, to create a model of the environment and robot system as the replica of the real one along with the simulation capability in which the system performance can be visualized in a better manner is of much interest. One such tool is the Webots [9] developed by Cyberbotics Limited is an open-source virtual mobile robotics simulation software which provides an interactive prototyping environment for modeling, programming, and simulating mobile robots. The provided robot libraries enable to transfer of the control programs to the available real mobile robots. Each robot can be equipped with a large number of available sensors and actuators. Programming can be done with various development environments, can simulate them, and can transfer the resulting programs onto the real robots.

Navigation methods of simulated mobile robots in the Webots environment are discussed in [10] with the method of metric navigation which is based on the occupancy grid. During the navigation process, the robot has to continuously calculate the occupancy values of its neighborhood and has to insert the result into the grid. Exploration was the major parameter for the simulation of the robot in the webots platform. Collaborative multi-robot mapping and exploration in webots were discussed in [11] with an exploration algorithm which uses two-dimensional, grid-based, probabilistic road map and a potential field model. A potential field-based approach is employed which is based on the attraction of cells. The magnitude of attraction of the cells is dependent on the information stored in the map. As no previous map information is available, the path planning approach employed in this model is closed-loop and time-critical.

Trajectory planning and collision avoidance algorithm for mobile robots is presented in [12] using the webots simulator environment with techniques for line following and collision avoidance of the designed mobile robot. This technique uses infrared sensors. The virtual assembly is tested on multiple scenarios to analyze the robot's ability to follow a path in a straight line as well as in congested curves; to detect obstacles, and to navigate around them to avoid collisions. The development of a robot capable of avoiding obstacle's as well as following a line is discussed in [13], which employs the integration of a fuzzy logic system into this mobile robot in both the static and dynamic environment was presented in [13]. In this model, collision avoidance and line following algorithm to navigate in static and dynamic environments with the integration of fuzzy logic fusion is explained. The obstacle avoidance model for the e-puck robot by using the fuzzy logic controller was modeled and simulated by using webots.

Eight IR sensors were employed to feed the input and the output of the motor speed was used to construct the Fuzzy Logic rules. The robot's performance was observed in terms of distance and time when the robot navigated through simple, average, and complex obstacle environments [14]. Navigation of simulated mobile robots in the webots is presented in [15]. Various robotic navigation algorithms were described with aid of robot simulation programs and it shows the advantages and disadvantages of those different navigation paradigms. However, considering the application of the proposed robotic waste bins, a simple Line Following method is chosen as the navigation method in this work and it is applied in the waste management system.

It is observed from the literature, neither of the above-mentioned manuscripts has addressed the integration of robotic technologies along with the IOT and waste management systems. Here falls our novelty. The novelty of this manuscript lies in integrating the robotic technologies for waste management and testing the behavior of such systems in a virtual environment where the exact systems as well as the environment can be modeled and simulated using a virtual simulator, which could be useful for an easy understanding of this complex system visually. Here falls our second novelty. From the literature found and discussed, we were able to understand that there are numerous efforts made in modeling mobile robots in various simulation platforms, incorporating various navigation techniques. But ultimately, there is no application of an autonomous indoor based robotic dustbin incorporating any one of the navigation techniques present.

This and the Swachh Bharat initiative of the Government of India inspired us and helped us to formulate the main objective of this paper which is modeling and simulation of autonomous indoor based robotic dustbin using a line following based navigation technique in Webots. The proposed system intends to reduce human effort, time, and cost to an extent. It also preserves the immaculateness of the environment promoting the healthy living of people. The modeling and the parameters of the navigation technique of the proposed robotic dustbin are discussed in the methodology section. The simulation carried out for the proposed system in the Webots platform is discussed in the simulation studies section.

2. Methodology

Inspired from the literature and as a preliminary study, modeling and simulation of a mobile smart waste bin in webots is presented in this section. This section also describes the use of different sensors, actuators, and controllers associated with the modeling of the mobile smart bin.

2.1. *Design Perspective and Background*

In this proposed model, the line follower technique is prioritized as the navigation technique because, in an indoor environment, by the usage of physical markers, it is easy to implement this technique. Certain navigation techniques are already addressed in the literature; such techniques will increase the implementation cost and computational complexity. But our focus is on incorporating mobility to a smart waste bin that doesn't require much complexity in navigation. In an indoor environment, the line follower based navigation system for the waste bin is easy to implement since it can be en-routed to a specific path as per the convenience and it will be cost-effective.

2.2. *System Parameters*

Figure 1. below shows the block diagram of the indoor robotic waste bin. This figure depicts the integration of the input units (sensors 1, 2&3) and the output unit (motor driver) with the microcontroller. It also depicts the inter-relation between the microcontroller and the WIFI module and the cloud server. Here the smartness is achieved by integration of Infra-Red sensor, ultrasonic sensor, color sensor, Arduino Mega, DC motors, and NodeMCU to form a completely autonomous system.

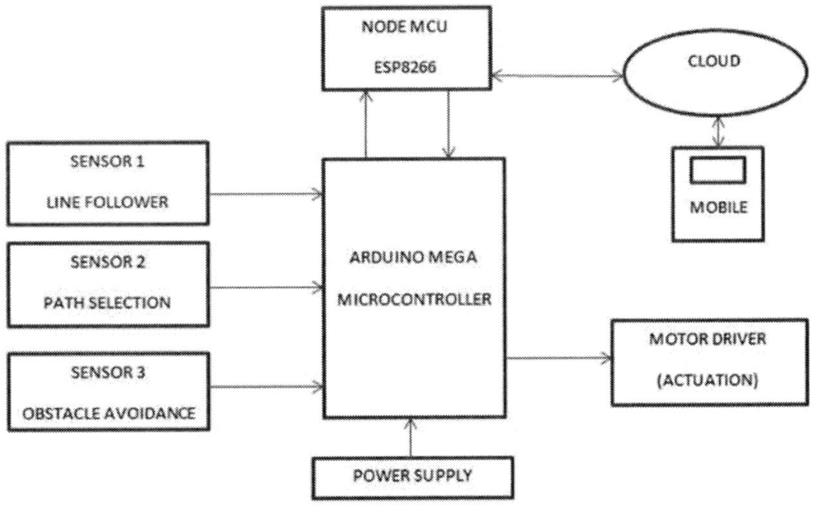

Figure 1. Block diagram of the proposed indoor robotic waste bin.

The line follower Navigation of the dustbin is established by detecting the black line on a white surface or white line on the black surface using the Infra-Red sensor. In addition to this, the path selection is also established by using color sensors. The sensors are connected with the controller and the actuation is controlled accordingly. When the dustbin reaches a junction during its navigation, the color sensor detects the color tags and the controller decides to navigate in a particular direction as per the command. Obstacle avoidance of the dustbin is also incorporated by having an ultrasonic sensor, which stops the actuation of the bin whenever it perceives an obstacle nearby to the bin. A power supply unit provides the necessary power for the operation of the dustbin. Once the dustbin reaches the common collection point, another dustbin returns to the initial position of the previous dustbin by following the same path.

Figure 2. below shows the software flowchart of the indoor robotic waste bin. This flowchart depicts the flow of the process sequentially. It depicts the algorithms of 3 different functions, i.e., path detection, path selection, and obstacle avoidance. For line follower navigation, 3 array Infra-Red sensors are used in the robot for following the line. When the 3 array IR sensor detects the black or white line, the signal is sent to Arduino mega which further controls the actuation of the robot. 3 array Infra-Red sensors are used because it is accurate, it requires lesser calibration and it was enough because the robot is not designed to navigate at higher speeds for the designed application. The robot follows the path to a common collection point.

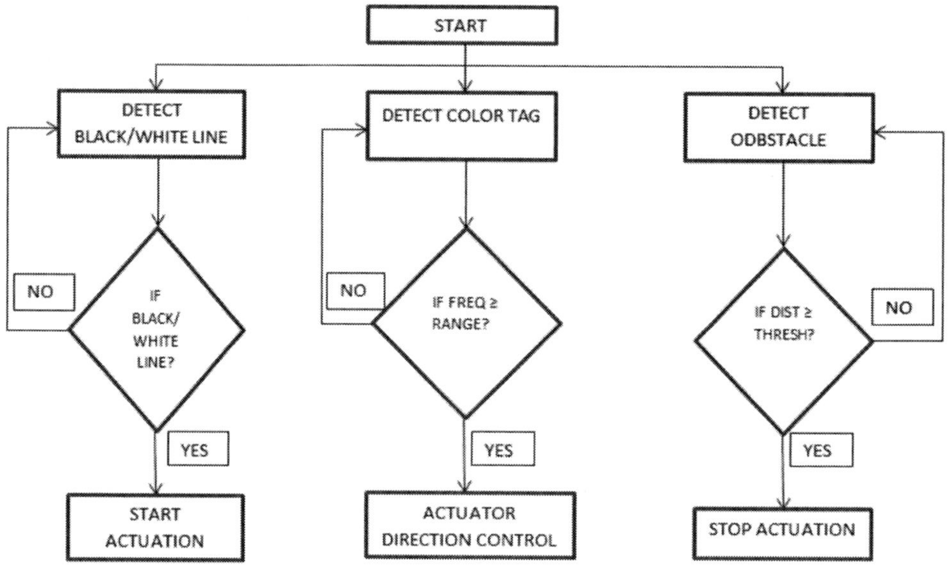

Figure 2. Software flowchart of the proposed indoor robotic waste bin.

TCS3200 Color Sensor is used to detect the colored tags placed at the junctions of the path of the line follower robot. It is used for the path selection of the robot to turn left or right or to go straight and to take alternative directions if any obstacle is present in the path. When the robot detects a particular color, the signal is sent to Arduino mega which controls the actuation of the motors. An ultrasonic sensor with a frequency of 40 kHz and the time of flight method is used for obstacle avoidance during the navigation of the robot when the obstacle falls inside the threshold value. If it encounters an obstacle, it stops and waits till the obstacle is cleared and the robot continues the navigation. If the obstacle remains stable at the position, then after a delay the robot detects an alternative path. Brushless DC motors are used as the actuator for navigation purposes. An L298N H-BRIDGE is used as the motor driver.

A rechargeable Lithium Polymer (LiPo) battery of 2200mAh, 11.1V is used as a power supply to the entire system. The entire electronic system is mounted within a weatherproof casing and fixed in the trash bin. An Atmega2560+NodeMCU ESP8266 microcontroller with 32 Mb (megabits) of flash memory, working on the Arduino development environment is used as a controller to integrate the sensors, actuators, and communication devices to provide the necessary actions required by the mobile smart bin. The advantage of using this controller is that all of the modules can work together or each separately, eliminating the interface issues.

NodeMCU is used as the communication device which connects the system with the IoT cloud by broadcasting the position and the status of the bin along with the bin ID. Further, it also broadcasts the signal for actuation. When the robot reaches a common collection point, Arduino mega sends a signal to the NodeMCU which communicates to the control server through the cloud, it sends a signal to the Arduino mega of another robot placed near the common collection point. This robot navigates through the same path as the previous robot and it reaches the initial position of the previous robot.

2.3. Environment Modelling

Webots is an open-source robot simulator developed by Cyberbotics Limited. This virtual platform provides an expeditious prototyping environment to the user for modeling, programming, and simulating mobile robots. The in-built robot libraries enable the user to test their hypothesis virtually as well as allows the transfer of the programs developed to the available physical robots. Each robot can be equipped with a large number of sensors and actuators. The user can define various properties of the model like shape, size, color, texture, etc. with ease. This and the acceptability of various programming paradigms make this platform more flexible and user friendly to use.

Webots Simulation Software (2020a version 1) was used to implement the working algorithm of the indoor robotic dustbin in a virtual simulated world. This virtual simulation world is known as an environment in Webots. These environments are modeled by using the scene tree; which allows creating & adjusting various features in the environment such as the background, size, simulation world time, luminescence of the simulation world, and other settings required for the indoor robotic dustbin.

Figure 3. below shows the Webots simulation environment depicting various elements. This figure depicts a simulation environment where two mobile robots, i.e., Bin A&B are present at the starting point and collection point respectively. A black line on a white surface that has two different loop paths has been shown. Four different static obstacles (obstacles 1, 2, 3 &4) and one dynamic obstacle (black colored mobile robot) are shown.

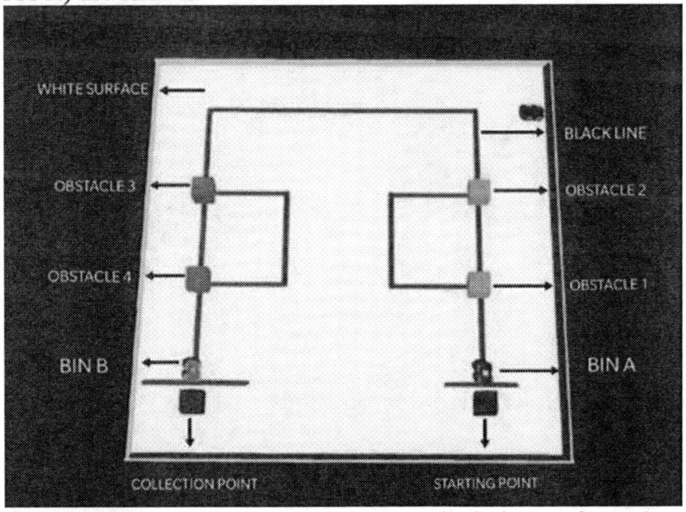

Figure 3. Webots simulation environment depicting various elements.

The basic setting required for the environment is to have four walls acting as a boundary for the simulation environment, in which each wall is four cross four meters in dimensions (4m x 4m). Four square-shaped obstacles were used to represent the presence of obstacles in the real-world. The boundary wall and static obstacles are created by using the add node feature in the scene tree of the simulation world.

The appearance and Geometry of those objects can be tinkered with using the corresponding sub-nodes present in the parent node of the corresponding objects. After this, add the 'bounding Object' node with 'Static obstacle'. (I.e. the square-shaped obstacles) and add 'Physics' to make sure that it acts as an obstacle during the simulation. The physics node implemented in the obstacles allows giving density, mass, Centre of mass, etc. for that obstacle. And since the indoor robotic dustbin follows a line to navigate, the floor is converted to a floor that has a black colored line etched on a white colored surface. The black line is etched on the floor as per the requirement of the indoor robotic dustbin's navigation around the simulated indoor virtual world.

2.4. Robot Modelling

Figure 4. below shows the four-wheeled Robots in skid steer locomotion with sensors modeled in the Webots environment. The red colored mobile robot has an emitter module at the top of its body and two distance sensors at the front face of the body. The blue colored mobile robot has a receiver module at the top of its body and two distance sensors at the front face of the body.

The robot is made of four wheels and a body above which the dustbin is placed. For simulation purposes, the dustbin is not added to the robot. The mobile robot was modeled to have a cuboid-shaped box as a base is that, the other shaped bases would not provide the required support needed for holding the bin placed on it. And for simplicity, the dustbin structure is not added into this robot figure and as a result, only the base of the robot is implemented in the virtual Webots platform. The reason for doing this is that, if the intelligence that is provided to the robot works well with the base alone, then it obviously will work perfectly with a bin placed on it. This robot was created in Webots simulation software without importing the pre-existing robot modules. The Robot comprises four wheels, two IR sensors, two distance sensors, and an emitter-receiver module to navigate around this indoor world.

Figure 4. 4 wheeled Robots with sensors modeled in the Webots simulator environment.

Figure 5. below shows the top view of the robot with the position of wheels, sensors, and emitter and receiver modules. It depicts the distance sensors placed at the left and right positions of the front face of the robot body and the angular range it covers.

The Red colour lines represent the direction in which the distance sensors emit its wave for measurement of obstacle in front of it. The Blue arrow indicates the front part of the four-wheel robot. The graph-like grid present behind the top-view of the robot spans a dimension of 0.2 x 0.3 meters. The text labels mentioned corresponds to the name of the corresponding devices. The point to note is that the coordinate system represents the robot body and its wheels both of which are oriented in the same direction. The x-vector (in red) defines the body of the robot, the y-vector (in green) defines the top of the robot, and the z-vector (in blue) defines the forward-face of the robot.

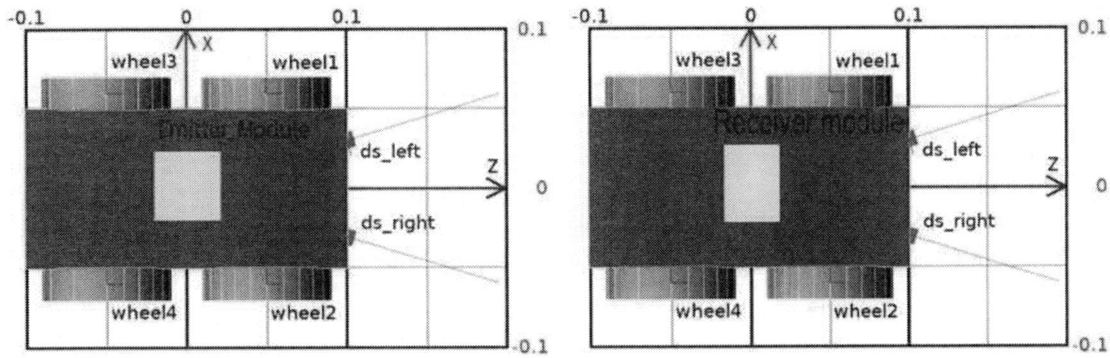

Figure 5. Top view of the mobile robots.

2.5. Sensor Modelling

The various sensor used in the four-wheel robot was modeled and designed using the various functions available in Webots instead of using the readily-available Proto sensor nodes. The sensors were designed since the already available sensors were not suitable and customizable according to the needs of the four-wheel robot. The sensors used in each of the two robots are included under the Children node of each bot. The point to note is that is both the bots are not identical to the sensors on-board. The primary robot (Red-colored bot) has an emitter module that communicates with the secondary robot (Blue colored bot) which is equipped with the Receiver module. The distance sensor used to detect the obstacles is a generic one in comparison to the IR sensor which can be also used to calculate distance instead of detecting black color. Both the bots are kept at the same communication channel so that the information emitted by the primary bot is received by the secondary bot.

The range at which the signal is emitted is limited to one meter. The rate at which the sensor emits a signal and refreshes is per the Time step value given in the controller. The necessary libraries and header files are included in the controller as it helps us to interlock the components in the scene tree, sensors in the bot, and the value received by the signal's interaction with the environment with the controller.

2.6. Navigational Algorithm

The intelligence or commands to be obeyed by the mobile robot at any given instant are provided by the controller which comprises of the code with which the mobile robot functions as per the user's wish. The developed controller allows the mobile robot to detect and follow a black line on a white surface, then it also allows the mobile robot to avoid obstacles present and finally allows the mobile robot to send a signal to the corresponding entity on whether the mobile robot has reached its destination or not.

The pseudo code for the above-given functions is given below:

Let Ri be the robots in the system, where i = number of robotic dustbins in the system
Let Rsi be the status of the robot i
Let Rsi = 01, When the bin is full;
    00, when the bin is empty
    11, Empty bin in the dump yard
Let dSL and dSR be the Left and Right distance sensor
Let IRL and IRR be the Left and Right IR sensor used to detect the line
Let WL and WR be the Right and Left Wheels of the Robot Ri.
STEP 1: Check Rsi; when Rsi =00, Position the bin in its location when empty

When Rsi = 01, i.e. bin is full; Execute the Navigation algorithm, to move the robot to follow the line and reach the dump yard.

STEP 2:
When dSL and dSR = 0; i.e. not triggered provided that
IRL and IRR =0
Drive Robot Ri with WL and WR Clockwise with velocity VL = VR =V
When IRR =1 and IRL = 0
Drive Robot Ri with WL in clockwise and WR Counter-Clockwise with velocity VL =V and VR =-V
When IRR =0 and IRL = 1
Drive Robot Ri with WL in Counter-Clockwise and WR in Clockwise with velocity VL =-V and VR =V
STEP 3: When dSL or dSR = 1; i.e. anyone detects obstacle
Drive Robot Ri with WL in Counterclockwise and WR Clockwise with velocity VL =V and VR =-V
Repeat STEP 2 and STEP 3, until when both dSL and dSR = 0 and IRL and IRR =1
Stop the Wheels, by WL=WR= 0.
Communicate from Emitter to Receiver that the target position is reached.

3. Simulation Studies

A single robot was modeled and it was incorporated with the various in-built proto-sensors present. But the downside of using proto-nodes / proto sensors is that the sensor parameters present in these pre-built nodes or sensors cannot be tweaked to suit the application. This subjected to create the nodes and sensors. Then implementing a color sensor was tried but the problem with the color sensor is that its accuracy as well as precision was not up to the mark. The output obtained via the color sensor had a very low-resolution value and often deviated a lot due to the amount of light or shadows employed in the simulated environment. This subjected to model a distance sensor. Its initial readings were very promising and the response time was way faster compared to the color sensor which was tested previously. The accuracy and precision of this IR-Based Distance sensor was more efficient.

This made us to implement the IR-based distance sensor into the mobile robot. With these findings, the mobile robot was fine-tuned to perfection. Then the second mobile robot was modeled with the same above-mentioned sensors. The biggest challenge faced in simulation studies is the mode of communication to be incorporated between the robots present so that it can transmit information between one and another. For the communication purpose, emitter-receiver module was used. There is also a supervisor node which allows modeling leader-follower based applications, but the application was way off than the leader-follower application. The emitter-receiver module was capable of transmitting data at a very high range within the simulation world. This and the presence of low latency between the transmission of data and the corresponding action taken by the corresponding robot made us incorporate this into the robot.

In the static environment and with static obstacles the robot's performance was very well. Then the same was tested with a dynamic obstacle (the robot in black color in figure 6). Some of the constraints of the mobile robot were that it cannot be accelerated beyond the value of 10. I.e. 10 rad/s. The maximum acceleration can be set as per the user's wish but since the mobile robot was modeled to be deployed indoors where the speed is not much required; the maximum speed was set to be 10rad/s in the simulation environment.

3.1. Simulation Results and Discussions

When the simulation begins, the primary bot (red) travels along the black line while avoiding obstacles by taking a short detour and then joining back to the primary line track to be followed. And if the bot is subjected to face a dynamic obstacle it automatically treats it as a static obstacle and avoids it as well. And once the bot reaches the common collection point, the emitter module present in the primary bot sends a signal to the secondary bot. Once the receiver bot receives this signal, it travels through the same line by avoiding obstacles to the point where the primary bot was initially present. Since the primary bot was able to handle the dynamic obstacle more precisely and without any collision it can be concluded that this proposed technique can be implemented in the real world. By this, it can be observed that this same methodology can be applied in an indoor environment to navigate the various filled smart dustbins to a common collection point and replacing those dustbins with the secondary dustbins can be implemented autonomously without human intervention.

Experimental investigations were not done in this work. Simulation results were only reported in this work. The simulation studies were done in Webots virtual simulator. It was approached in this manner because; analysing the effectiveness of the system through simulations before implementing it as a prototype model would give a better analysis. By this approach, the integration of the physical and intellectual characteristics of the system through simulation studies can be studied before implementing it as prototype model.

Figure 6. below shows the list of the overall simulation carried out in the Webots simulation platform. It shows the simulation results carried out at various time instant. At instant 1, bin A starts navigating from the starting point. At instant 2, bin A avoids the static obstacle 1. At instant 3, bin A en-routes through an alternate path, i.e., loop 1. At instant 4, bin A avoids static obstacle 2. At instant 5, bin A avoids a dynamic moving obstacle. At instant 6, bin A follows the path towards the destination. At instant 7, bin A avoids the static obstacle 3. At instant 8, bin A avoids the static obstacle 4. At instant 9, bin A reaches the collection point. At instant 10, bin B follows the same path and avoids the obstacles in between. At instant 11, bin B en-routes towards the initial position. At instant12, bin B reaches the starting point.

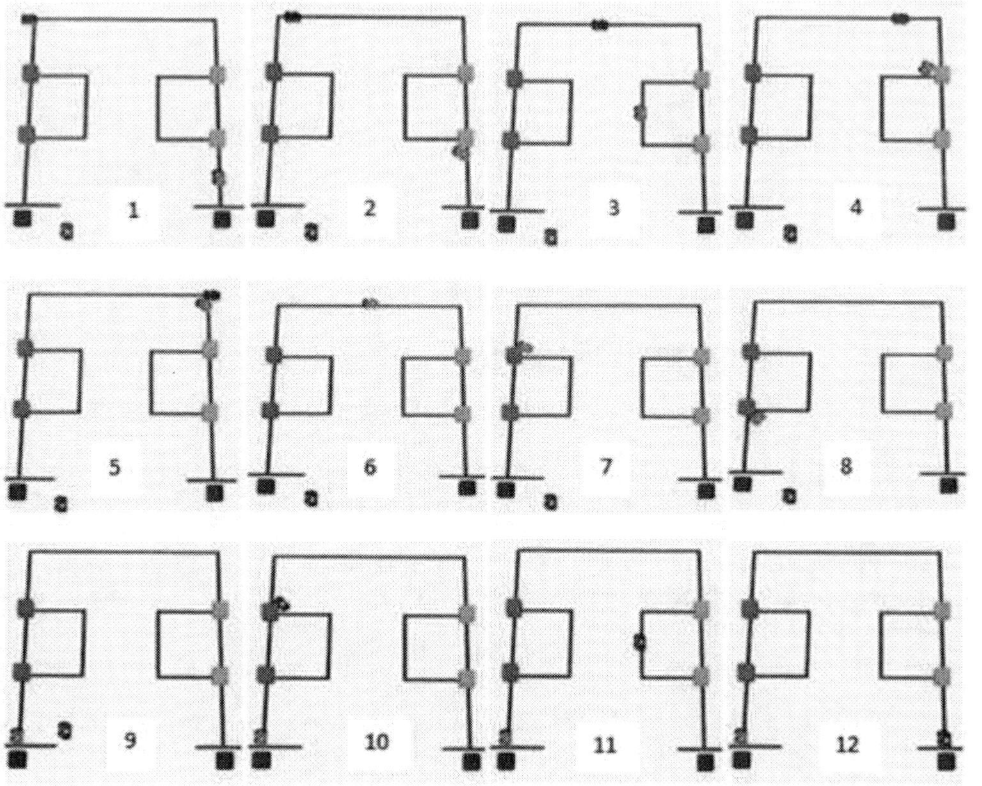

Figure 6. Simulation results of the proposed system.

4. Conclusions

In most of the buildings globally poses a challenge on effective solid waste collection and maintenance of the waste bins. In this work, an indoor robotic waste bin with the line follower navigation technique is modelled and simulated in webots virtual simulator. A smart waste collection system incorporating the indoor robotic smart bins, where the smart bin has the mobility to move to the common collection point by localizing itself in the environment, is simulated and presented in this work. Simulation investigations were carried out where it can be concluded that the integration of the physical and intellectual characteristics of the system through simulation studies can be studied before implementing it as experimental prototype model. The simulation results prove that the proposed indoor robotic smart bin model can be implemented in real time efficiently. This system could find an application in smart buildings where the waste collection could be practiced more smartly and efficiently.

Our future work is to investigate the performance of the proposed robotic waste collection system in our Institution. **Figure 7**. below shows the architecture of the waste collection system using robotic smart waste bins and navigation techniques. The figure depicts the autonomous navigation of the robotic dust bin through a specified path to the collection point and the autonomous navigation of a secondary robotic dustbin replacing the previous one. The communication between the two robotic dustbins takes place through the cloud server. It is integrated with the line follower navigation system, path selection system, and obstacle avoidance and communication protocols to monitor the position of the robotic smart bins enabling the effective waste collection system, which is planned to be implemented in our Hindustan Institute of Technology and Science campus.

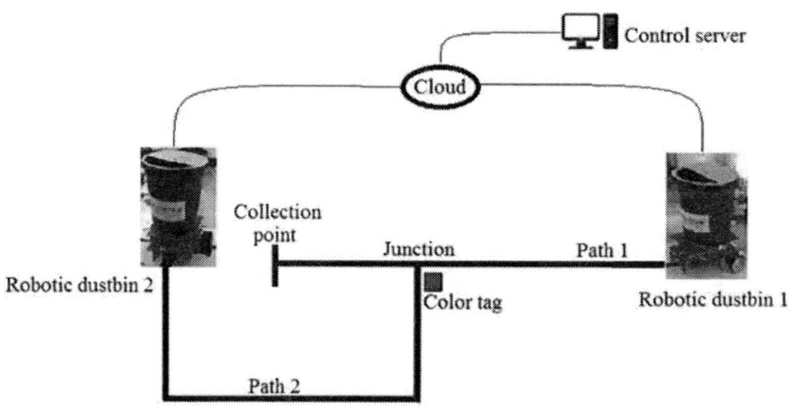

Figure 7. The architecture of the proposed waste collection system.

Therefore, the proposed system could be considered as a robotic smart waste bin system, where the bins could mobilize, localize its location; communicate its status to the cloud. During the bins are serviced at the waste collection point, spare bins could automatically replace the serviced bins enhancing the effectiveness of the system. This concept is proposed due to the evolution of smart buildings and smart homes in recent times and it could be a potential technical advancement in the area of waste collection systems and IoT.

References

[1] Shujaatullah Khan, Srinivasan Venkat and Kayalvizhi Jayavel 2017 Autonomous Smart Waste Collection System Using the Internet of Things In. *Ieee International Conference on Power, Control, Signals, and Instrumentation Engineering* 1181-86.

[2] Kellow Pardini, Joel J. P. C. Rodrigues, Sergei A. Kozlov, Neeraj Kumar & Vasco Furtado 2019 IoT-Based Solid Waste Management Solutions: A Survey *Journal of Sensor and Actuator Networks* **8**(1), 5.

[3] Manju Mohan, RM. Kuppan Chetty, Vijayram Sriram, Azeem, P. Vishal, and G. Pranav 2019 IoT Enabled Smart Waste Bin with Real-Time Monitoring for efficient waste management in Metropolitan Cities In. *International Journal of Advanced Science and Convergence* **1**(3) 13-19
[4] Aksan Surya Wijaya, Zahir Zainuddin, Muhammad Niswar 2017 Design a Smart Waste Bin for Smart Waste Management 5th *International Conference on Instrumentation, Control, and Automation* Yogyakarta, Indonesia 62-66.
[5] Cyril Joe Baby, Harvir Singh, Archit Srivastava, Ritwik Dhawan and P. Mahalakshmi 2017 Smart Bin: An Intelligent Waste Alert and Prediction System Using Machine Learning Approach *Ieee Wispnet conference* 771-774.
[6] Teh Pan Feil , Shahreen Kasim, Rohayanti Hassan, Mohd Norasri Ismail, Mohd Zaki Mohd Salikon Husni Ruslai, Kamaruzzaman Jahidin, Mohammad Syafwan Arshad 2017 SWM: Smart Waste Management for Green Environment *Ieee 6th ICT International Student Project Conference* 5.
[7] Orazio Miglino, Henrik Hautop, Stefano Nolfi. Evolving Mobile Robots in Simulated and Real Environments *Artificial life* **2**(4) 417-434.
[8] Balakrishna, R., & Ghosal, A. 1995 Modeling of slip for wheeled mobile robots *Ieee Transactions on Robotics and Automation* **11**(1) 126–132.
[9] Michel, O. 2004 Cyberbotics Ltd. Webots™: Professional Mobile Robot Simulation *International Journal of Advanced Robotic Systems* **1**(1) 5.
[10] Richárd Szabó Navigation of simulated mobile robots in the Webots environment *Periodica Polytechnic* Ser. El Eng* **47**(1-2) 149-163.
[11] Scott, A. F., & Yu, C 2009 Cooperative multi-agent mapping and exploration in Webots® 4th *International Conference on Autonomous Robots and Agents* Wellington, New Zealand 56-61.
[12] Almasri, M. M., Alajlan, A. M., & Elleithy, K. M. 2016 Trajectory Planning and Collision Avoidance Algorithm for Mobile Robotics System *Ieee Sensors Journal* **16**(12) 5021–28.
[13] Siti Hajar Ashikin Mohammad, Muhammad Akmal Jeffril, Nohaidda Sariff 2013 Mobile Robot Obstacle Avoidance by Using Fuzzy Logic Technique *Ieee 3rd International Conference on System Engineering and Technology* Shah Alam, Malaysia 331-335.
[14] M. Izzuddin Ibrahim, Nohaidda Sariffl , Juliana Johari, Norlida Buniyamin 2014 Mobile Robot Obstacle Avoidance in Various Type of Static Environments Using Fuzzy Logic Approach *2nd International Conference on Electrical, Electronics and System Engineering* 83-88.
[15] Michael Dooley, Paolo Pirjanian, Nikolai Romanov 2009 Application of localization, positioning, and navigation systems for robotic enabled mobile products. In. *Wipo Ip Portal.* WO/2009/132317.

Aerodynamic Analysis of Hybrid Drone

S Krishnaraj[1*], R Senthil Kumar[1], A Gokula Krishnan[2], G Ganeshkumar[2], M Mohan[2], M Nirmal[2]

[1]Department of Mechanical Engineering, Sri Sairam Engineering College, West Tambaram, Chennai - 600044

E-mail: krishnaraj.mech@sairam.edu.in

Abstract. Drones are Unmanned Aerial Vehicles (UAV's) or Remotely Piloted Vehicles (RPV's) whose application ranges from photography to defence. There are two major classifications of drone, one, a fixed wing; and another, a quadcopter. A hybrid drone is a kind of drone which has a mix of features of the two kinds of drone. It has quad-rotors for Vertical Take-Off Landing (VTOL), and a fixed wing for cruise. Aerodynamic analysis for a hybrid drone is done using Computational Fluid Dynamics (CFD) as a tool. A virtual wind tunnel test is done by means of this analysis. It is done in order to examine the various technical edges of the hybrid drone over the rest of the types. Firstly, the main entity of the drone, the wing, of 3 different profiles(aero-foils), is analysed for better lift coefficient as it is the major contributor for the thrust during cruise (for 0 deg angle of attack). The velocity contours are viewed to interpret the aerodynamic advantage of each wing. Then the whole vehicle is analysed after inculcating the wing. The various aerodynamic parameters such as the lift force, lift coefficient, drag force and drag coefficient of the hybrid drone are determined from the analysis. A hybrid drone with optimum lift has been designed from the inference.

Keywords: Unmanned Aerial Vehicles (UAV's) or Remotely Piloted Vehicles (RPV's), Vertical Take-Off Landing (VTOL), Computational Fluid Dynamics (CFD), angle of attack, aerodynamics, lift, drag.

1. Introduction

UAVs (Unmanned Aerial Vehicles) or Remotely Piloted Vehicles, are aircrafts without human pilot or personnel operating it onboard. It is either controlled by an operator by the virtue of a wireless remote, or a fully autonomous flight with controllers on-board with algorithms for navigation.

There are three main classification of drones, namely;

- Quad (or multi)-rotor drone
- Fixed wing drone
- Single-rotor drone

1.1. Problem Encountered

Drones are employed in various applications, but there are notable and practical limitations for the same. In medical field, the transportation of organs, blood, etc, in various geographies of the world, employ drones. The major problem here is that fixed wing drones are used, owing to the advantage of having good stability and range, whereas having the limitation that the drone cannot be landed or hovered in any location.

Moreover, in defence field, quadcopters are employed in surveillance, and emergency situations such as, rescue operations during natural calamities as they possess VTOL (Vertical Take-off Landing) feature. But due to unstable nature of this type of drone and lower operation range, its application is less effective and less efficient too.

1.2. Quad (or multi)-Rotor Drone

It's a type of drone that has rotors on the corners of its structure (frame) as shown in **figure 1**., which are coupled with propellers, that produce lift and the thrust required for the flight. It has VTOL (Vertical Take-Off Landing) feature in it. The disadvantage part of it is that it has lower range and lesser stability during cruise.

Figure 1. Quad-rotor drone; Succo; "Spying quadcopter: espionage and observation"; 2014; Pixabay https://commons.wikimedia.org/wiki/File:Spying_quadcopter.jpg

1.3. Fixed Wing Drone

It's similar to a normal flight in working but is e-propelled most of the times. Unlike the other types, it has wings to produce lift as shown in **figure 2.** and a propeller with horizontal axis of rotation to move forward. The major disadvantage is the requirement of run way for take-off and landing.

Figure 2. Fixed-wing drone; Ssgt Reynaldo Ramon; "IAI Heron 1 UAV in flight"; 2003; Wikimedia Commons; https://commons.wikimedia.org/wiki/File:IAI_Heron_1_in_flight_2.JPEG

1.4. Single-Rotor Drone

Its functionality is similar to that of a usual helicopter, with a single central rotor with vertical axis of rotation as depicted in **figure 3.** Again, the disadvantages of low stability and complex operation makes it a less preferred model.

IOP Conf. Series: Materials Science and Engineering 1012 (2021) 012023 doi:10.1088/1757-899X/1012/1/012023

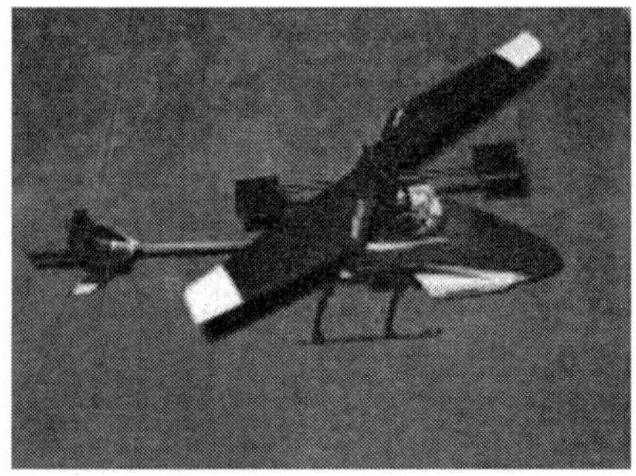

Figure 3. Single-rotor drone; Wesha; "E-flite Blade mSR (Micro Single Rotor) remotely controlled model helicopter"; 2010; Wikimedia Commons; https://commons.wikimedia.org/wiki/File:E-flite_Blade_mSR_Micro_Single_Rotor_Model_Helicopter_3.jpg

1.5. Proposed solution

In order to overcome the disadvantages of quadcopter and fixed wing drone, a hybrid drone is proposed as shown in **figure 4.** A hybrid drone is a combination of a quad-copter and a fixed wing drone. It takes the advantage of VTOL from quad and a stable cruise journey from fixed wing. This type of drone is still under R&D and study.

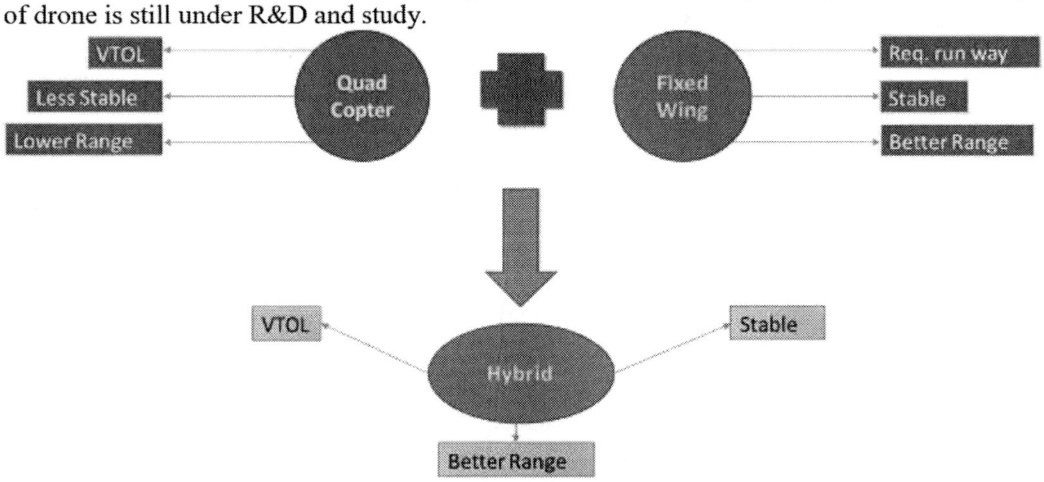

Figure 4. Hybrid drone Explanation.

1.6. CFD Deployment

Computational Fluid Dynamics (CFD) is used to predict the nature of fluid flow. Here, it is deployed to find the aerodynamic parameters of the wing and the hybrid drone, such as the lift force, drag force, lift coefficient, and drag coefficient. Using CFD as a tool, the flow characteristics, velocity contours are obtained.

1.7. Governing Equations

i. Continuity Equation:

$$\frac{\partial \rho}{\partial t} + \nabla .(\rho V) = 0$$

ii. General Transport Equation:

$$\frac{\partial (\rho \emptyset)}{\partial t} + \nabla .(\rho V \emptyset) = \nabla .(D \nabla \emptyset) + S$$

iii. X-Momentum Equation:

$$\frac{\partial(\rho u)}{\partial t} + \nabla.(\rho V u) = \nabla.(\mu \nabla u) - \frac{\partial p}{\partial x} + S_u$$

iv. Energy Equation:

$$\frac{\partial(\rho h)}{\partial t} + \nabla.(\rho V h) = \nabla.\left(\frac{k}{C_p}\nabla h\right) - \frac{\partial p}{\partial x} + S_h$$

v. Species Transport Equation:

$$\frac{\partial(\rho Y_i)}{\partial t} + \nabla.(\rho V Y_i) = \nabla.(D_i \nabla Y_i) - \frac{\partial p}{\partial x} + R_i$$

2. Literature Review

Ugur C. Yayli et al., says about optimization of fixed wing aircraft in order to reduce the take-off/landing distance. The lift, drag, velocity(air) and various parameters are observed by changing the air foil used for wing, tail configuration, fuselage design. This is similar to our project, except for the fact that the analysis is performed on fixed wing [1]. Pooneh Aref et al., discusses about the aircraft wing's aerodynamic characteristics with and without propellers using CFD as a tool. The local lift, drag coefficients are found considering four cases at a time, without propeller, with propeller (inboard)(ccw), (cw) and with propeller (outboard)(ccw). The variation in the parameters are plotted in a graph for comparison and the best case is found [2]. Dhwanil Shukla et al., says about the aerodynamic analysis of a multi-copter. Specifically, he discusses about the consequences pertaining to the interaction between two adjacent propellers and its effect on performance. Blade–vortex interaction is stated as the reason for the hinderance of performance [3]. K Sreelakshmi et al., says about aerodynamic analysis of UAV and finding the lift and drag forces for a fixed wing drone. The results in the form of plots are made using Ansys software [4]. Piotr Kardasz et al., discusses about the limitation of commercial drones being dependent on battery for power. Also discusses about the hinderance due to battery failure, breakdown, obstacle collision, etc. He also adds upon to the research on batteries and various other new power sources for drones [5]. Chun Fui Liew et al., summarizes the different types of UAVs and their features. Specifically, it gives a brief note on research activities pertaining to UAVs from the year 2001 [6]. Hazim Shakhatreh et al., discusses the various challenges relating to UAVs such as battery-related problems, collision and other scopes of research in UAVs. It also briefs upon the civil infrastructure development and the related challenges [7].

Oscar Vestlund says, the battery powered UAVs limits the powered time and poses the need to design a UAV with extended flight time. This study investigates the possibility of designing a large VTOL UAV which is capable of a higher range. Design with high lift to drag ratio is achieved and the best VTOL type is suggested based on the aerodynamic and structural analysis [8]. Özgür Dündar et al., discusses the Fixed Wing Drone and the Hybrid drone in terms of their endurance capabilities. It comprises the power and energy requirements to achieve the maximum endurance during cruise, climbing and landing and the best type is suggested for the prominent application [9]. Ventura Diaz et al., discusses about CFD analysis of three different multirotor drones. The analysis of drones is done with different rotors and for various weather conditions. A comparative study is done for various configurations [10]. K Christodoulou et al., discusses finding the aerodynamic parameters of a propeller, which is incorporated in a drone, made from scratch. It is found through CFD analysis using three different turbulence models, and the appropriate design for the propeller is selected from the results [11]. Vasile Prisacariu et al., says about numerical analysis of a UAV wing for various angles of attack and stresses the requirement of wind tunnel experiments to confirm the results of the CFD analysis done [12]. C T Lao et al., discusses the effects of ground-effect on the lift, performance, and various aerodynamic parameters of the wing. He adds that it improves lift, performance and Oswald efficiency, while producing an inconsistent moment coefficient [13]. Khuntia SK et al., says about the optimization of design of a wing for better efficiency using the outcomes of CFD analysis of the same. It is achieved through analysis of iterative designs of wings using ANSYS software. In addition, the strength of the wing is also analysed using the same software and by comparison it is found to be advantageous than the traditional wing in load carrying capacity and various other parameters [14]. Zhe Ning discusses the aerodynamic parameters pertaining to the wake structure of the propellers. In addition, a couple of bio-mimicked propellers have been analysed for aerodynamic performance and acoustic advantages [15].

3. Parts of a hybrid drone

3.1. Frame

It is the main part of the drone, to which all the other components are joined. It is selected based on the size, load, and various dynamic requirements. There are many types of frames, namely,

1. X-type frame
2. H-type frame
3. Z-type frame
4. Plus-type frame

3.2. Wing

Wing is the major contributor for lift of the drone. In a hybrid drone, the wing plays a major role in counteracting for the weight and propelling the drone during the cruise. The wing contains ailerons, flaps to control the lift parameters to greater extent.

The basic unit of a wing is the aero-foil, from which it is extruded. The aerodynamic parameters of the aerofoil are the most important in determining the type of the wing. The main features that determine the properties of wing are:

- Aero-foil:
  - ➢ Aero-foil is the basic entity of a wing, which forms the cross-section of the same.
- Wing Planform:
  - ➢ It refers to the shape of the wing from top-view after its being extruded.
  - ➢ Classified as Straight wing, Swept wing, and Delta wing.

3.3. Fuselage

In an UAV, it is the part which houses all the hardware parts of the drone, such as battery, ESC, flight controller board, receiver, etc. It is basically the covering of the hardware components of the UAV (drone). It adds to the major weight of the drone as it contains the hardware components.

3.4. Tail

It is the component that helps in controlling the direction of the UAV. This is again an aero-foil, which is extruded to form the tail.

4. Design calculation for drone

During **Cruise flight**, the following conditions are considered:

Steady Flight + Level Flight

$(\frac{dV}{dt} = 0)$ $(\gamma = 0)$

(Constant Vel) (No inclination w.r.t horizontal)

Now, the following values are considered for finding the required parameters,

Weight (Structure, payload, battery) = 1kg

Velocity of wind = 10m/s

Chordal length = 0.2m

Kinematic Viscosity=1.5*e-5 m^2/s

Density of Fluid medium(air)= 1.225 kg/m^3

4.1. Lift - Coefficient Calculation

Now, the Reynold's Number is found for further calculations:

Reynold's Number = $\boldsymbol{Re} = \frac{V*l}{v}$

= 10*0.2/(1.5*e-5)

= 1,33,333.

(*Here 'l' is the Chordal length of aero foil in 'm'*)

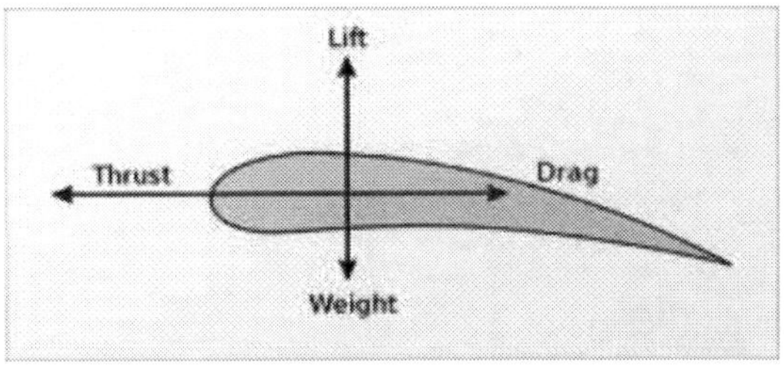

Figure 5. Aero-foil Demonstration (Free Body Diagram); Adapted from Grahamuk; "Graphic of aerofoil cross section"; 2013; Wikimedia Commons https://commons.wikimedia.org/wiki/File:Aerofoil.svg

From the **figure 5.**, it is clear that

=>Thrust = Drag (1)

=>Lift = Weight (2)

From Eq. (2), we get, W=L;

$$\Rightarrow \boldsymbol{W} = \frac{1}{2}\boldsymbol{\rho} * \boldsymbol{S} * \boldsymbol{V}^2 * \boldsymbol{C_L} \ \left(since\ L = \frac{1}{2}\rho * S * V^2 * C_L\right) \quad (3)$$

Rewriting the above, we get:

$$\Rightarrow \boldsymbol{C_L} = \frac{2*W}{\rho*S*V^2} \quad (4)$$

Now, to find **Area(S)**, *take Aspect Ratio (for low speed drones) = 12.5*

w.k.t., **A.R(Aspect Ratio)** = $\frac{b^2}{S} \Rightarrow 12.5 = \frac{b^2}{b*C}$

$\Rightarrow \boldsymbol{b} = 12.5 * 0.2 = \mathbf{2.5m}$

Now, **S(Area)= b*C** = 2.5* 0.2 = $\mathbf{0.5m^2}$,[for **rectangular planform**]

where, *b-span of the aerofoil in m,*

C-Chordal length or characteristic length in m

Substitute 'S' in Eq. (4), we get:

∴ $(\mathbf{C_L})_{\mathbf{req}}$= 2* (1*10/0.5) / (1.225*10^2) = **0.3265**

4.2. Minimum Thrust Calculation

Consider, Eq. (1) ÷ Eq. (2):

$$=> \frac{T}{W} = \frac{D}{L}$$

=>T=D*W/L = W/(L/D)

Now,

$$\mathbf{=>T_{min} = W/(L/D)_{max}} \quad (5)$$

Note:

$$L = \rho * \frac{AV^2}{2} * C_L \quad (6)$$

$$D = \rho * \frac{AV^2}{2} * C_d \quad (7)$$

Where,
A= Area in m^2,
V= Velocity in m/s,
ρ= Density in Fluid in kg/m³.
C_L= Lift Coefficient.
C_d= Drag Coefficient.
Putting Eq. (6) and (7) in Eq. (5), we get,

$$\mathbf{T_{min} = W/ (C_L/C_d)_{max}} \tag{8}$$

To find maximum value of C_L/C_d, consider,

$$\frac{d((C_L/C_d)}{dC_L} = 0$$

we know that,

$$C_d = C_{d_0} + KC_L^2 \tag{9}$$

Where,
C_{d_0} = Drag Coefficient, when α=0
$K = \frac{1}{\pi eAR}$
e = Oswald efficiency
AR = Aspect ratio = $\frac{b^2}{S}$
b = Span length
S = Area

Again,

$$\Rightarrow \frac{d((C_L/C_d))}{dC_L} = 0$$

$$\Rightarrow \frac{C_{d_0} + KC_L^2 - C_L(2KC_L)}{C_d^2} = 0$$

$$\Rightarrow C_{d_0} = KC_L^2$$

$$\therefore C_L = \sqrt{\frac{C_{d_0}}{K}} \tag{10}$$

Now, substitute Eq. (10) in Eq. (9)
Therefore, we get,

$$\therefore C_d = 2C_{d_0} \tag{11}$$

Substitute Eq. (10) and Eq. (11) in $\left(\frac{C_L}{C_d}\right)_{maximum}$

$$\Rightarrow \left(\frac{C_L}{C_d}\right)_{maximum} = \frac{\sqrt{\frac{C_{d_0}}{K}}}{2C_{d_0}}$$

$$\therefore \left(\frac{C_L}{C_d}\right)_{maximum} = \frac{1}{\sqrt{4KC_{d_0}}} \tag{12}$$

Substitute Eq. (12) in Eq. (8):
We get,

$$\mathbf{T_{min} = W * \sqrt{4KC_{d_0}}} \tag{13}$$

Here, k=1/ (π*1*12.5) =.0254, sub. in Eq. (13)

$$\Rightarrow T_{min} = 1*10* \sqrt{4 * 0.0254 * 0.02} = 0.45 \text{ N}$$

5. Design of wing

Many standard aero-foils are available in NACA library as stated in the previous chapter. There are two kinds of aero-foils, symmetric and asymmetric. For the analysis one of symmetric type and two of asymmetric type are chosen. Here, 3 aerofoils, namely, NACA 0012 (symmetric), NACA 6409 (asymmetric), and NACA 6412 (asymmetric), are considered for comparison and the best aero-foil with optimum Cl to Cd ratio is taken for designing the hybrid drone.

5.1. NACA0012
The various characteristics of the aero-foil are (as depicted in **figure 6.** and **figure 7.**):

Max. camber = 0%
Max. camber position = 0%
Thickness = 12%

Figure 6. NACA0012 Profile.

Figure 7. NACA0012 wing.

5.2. NACA6409
The various characteristics of the aero-foil are (as depicted in **figure 8.** and **figure 9.**):

Max. camber = 6%
Max. camber position = 40%
Thickness = 9%

Figure 8. NACA6409 Profile.

Figure 9. NACA6409 wing.

5.3. NACA6412
The various characteristics of the aero-foil are (as depicted in **figure 10.** and **figure 11.**):

Max. camber = 6%
Max. camber position = 40%
Thickness = 12%

Figure 10. NACA6412 Profile.

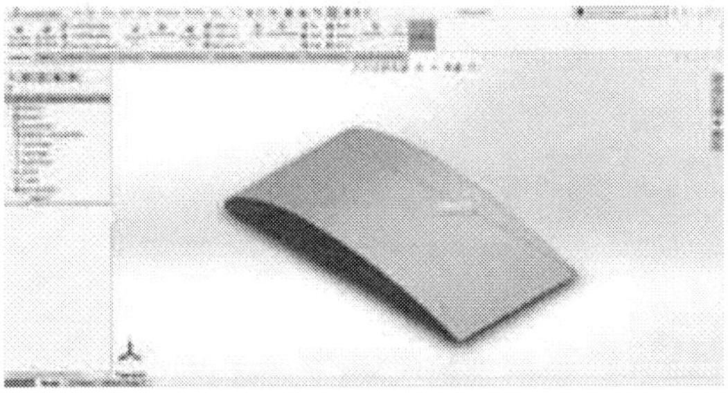

Figure 11. NACA6412 wing.

6. Aerodynamic analysis of wing
The aerodynamic analysis is done to select the wing with best lift coefficient to drag coefficient ratio, and optimum lift coefficient for the specific condition considered. The analysis is done using

Converge CFD software for the above initial and boundary conditions. The initial conditions specify the values of physical variables during the start of the simulation, i.e., when time, t=0 s. The boundary conditions specify the values of the physical variables in the inlet, outlet, and various walls. The initial and boundary conditions are as follows:

Reynolds Number (Re) = 1.33e+05
Fluid Medium = Air
Kinematic Viscosity (ν) = 1.52e-05 m^2/s
Angle of Attack (α) = 0 deg
Temperature (T) = 20 deg Celsius
Velocity (V) = 10 m/s
Pressure (P) = 1atm (101325 Pa)

The grid independence test has been done, considering the system constraints, it is found that 0.01mX0.01m mesh is optimal with Adaptive Mesh Refinement around the aero-foil surface of 5 levels of embedding. The *k-ω SST* turbulence model is selected as it is external flow and it is advantageous as it switches between the equations it solves for lower Y+ value (in Viscous sub-layer) and higher Y+ value (turbulence layer).

6.1. NACA0012

Figure 12. below shows that the velocity distribution is symmetrical about the chord, since the aerofoil is symmetric. This is the cause of the negligible lift coefficient as the forces in the top and the bottom cancel each other. **Figure 13.** shows that the lift coefficient is very much lower (0.019~0) than drag coefficient (0.157), since it is a Symmetric Aerofoil. This value very much lesser than the required value of lift coefficient ((Cl)req=0.33). **Figure 14.** shows the lift to drag coefficient ratio. The Cl/Cd ratio is an important factor for finding the aerodynamic parameters of an aero-foil. The value converges to 0.125 for this case, which is very negligible.

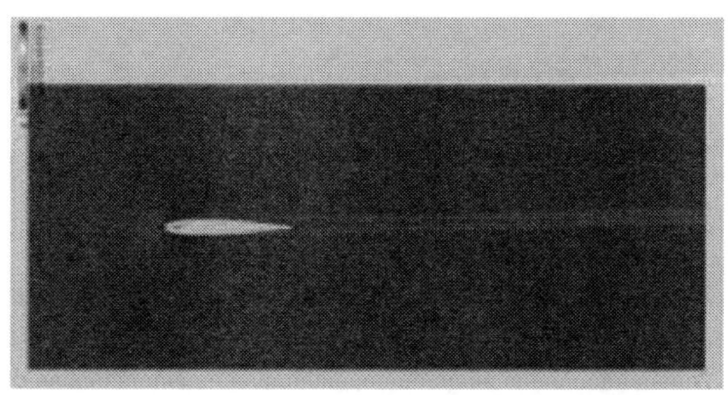

Figure 12. NACA0012 Velocity contour.

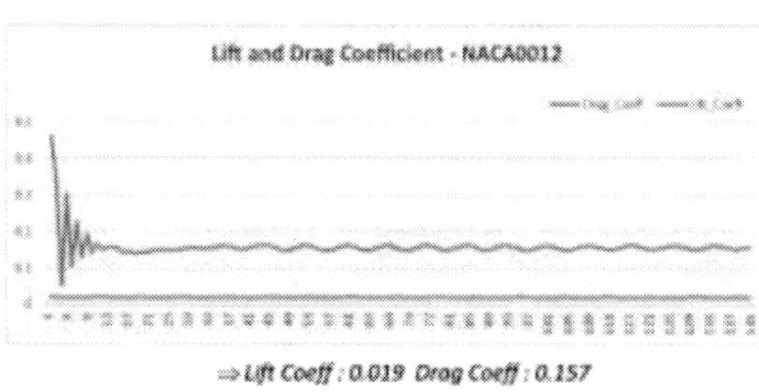

Figure 13. NACA0012 Lift and Drag Coefficient.

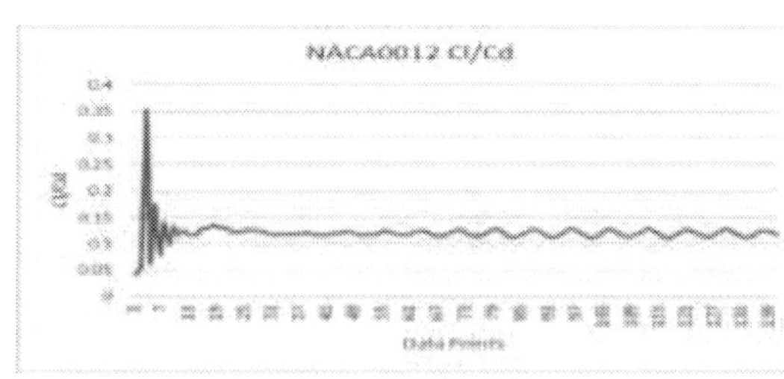

Figure 14. NACA0012 Cl/Cd ratio.

6.2. NACA6409

Figure 15. below shows that the velocity distribution is not symmetrical about the chord, since the aerofoil is asymmetric. The velocity below the aero-foil is lower as compared to upper side. Hence the pressure below the aero-foil is far higher than the upper side, which offers higher lift. **Figure 16.** shows that the lift coefficient is very much higher (1.27) than drag coefficient (0.032). This value is amply greater than the required value ((Cl)req=0.33), which is recommendable. **Figure 17.** shows the lift to drag coefficient ratio. The value converges to 38.9 for this case, which is optimal.

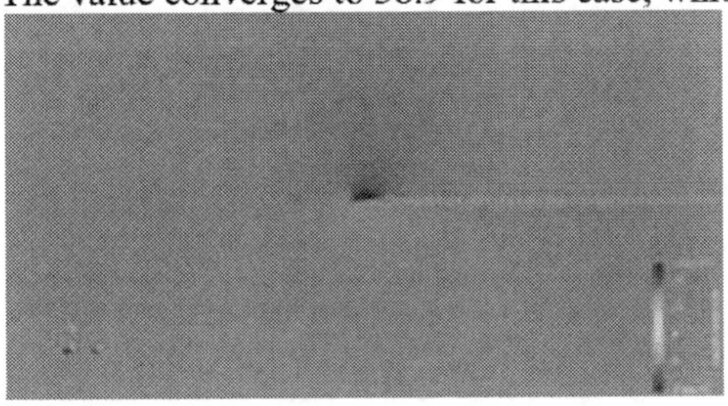

Figure 15. NACA6409 Velocity contour.

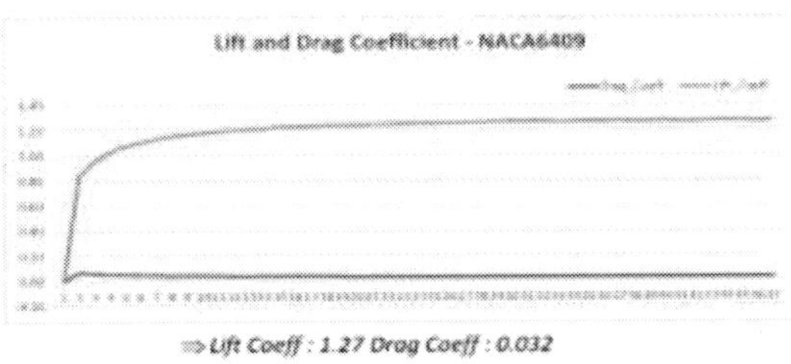

Figure 16. NACA6409 Lift and Drag Coefficient.

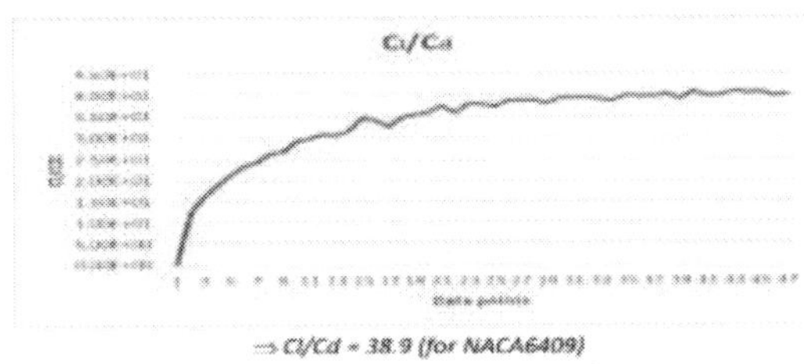

Figure 17. NACA6409 Cl/Cd ratio.

6.3. NACA6412

Figure 18. below shows that the velocity distribution is not symmetrical about the chord, since the aerofoil is asymmetric. This shows that the velocity in the leeward side of the aerofoil is considerably lower than that in the upper and lower regions. This causes higher drag coefficient. **Figure 19.** shows that the lift coefficient is lower (0.079) than drag coefficient (0.562). This value much lesser than the required value of lift coefficient ((Cl)req=0.33). **Figure 20.** shows the lift to drag coefficient ratio. The average value of the trend converges to 0.14 for this case, which is negligible.

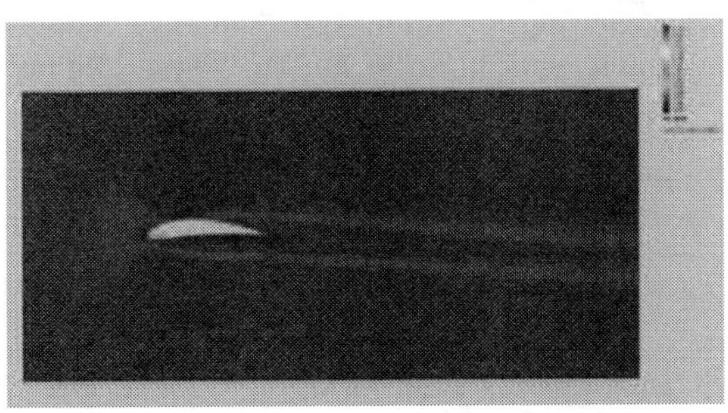

Figure 18. NACA6412 Velocity contour.

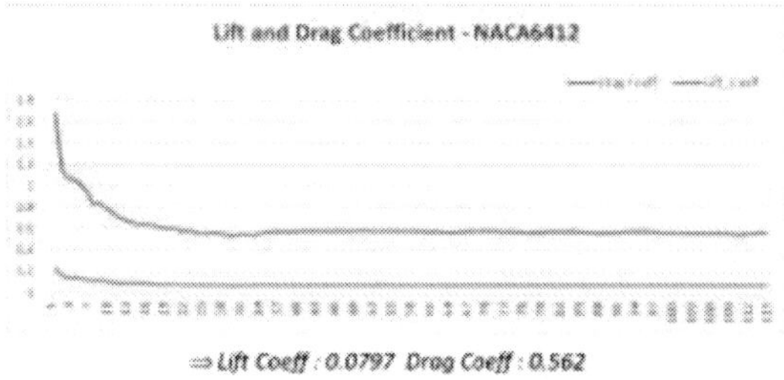

Figure 19. NACA6412 Lift and Drag Coefficient.

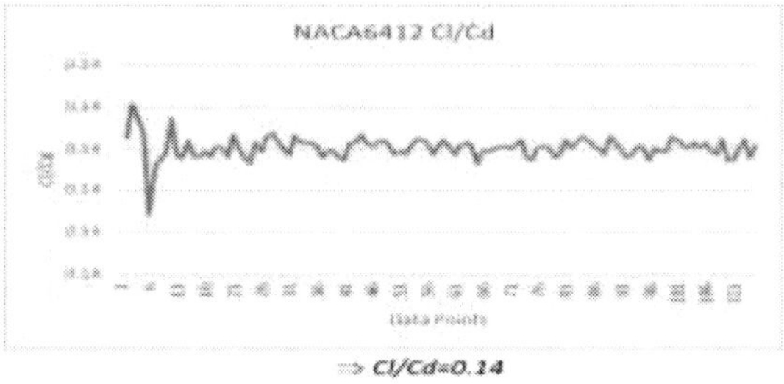

Figure 20. NACA6412 Cl/Cd ratio.

6.4. Comparison of Simulation

Here 1-NACA0012, 2-NACA6409, 3-NACA6412. **Table 1.** shows that the Cl/Cd ratio is optimal for Aero-foil-2, i.e., NACA6409. The plot, **figure 21.** shows that the lift coefficient is higher and drag coefficient is lower for Aero-foil 2, i.e., NACA6409, so it is recommended for the wing. The plot (**figure 22.**) shows that the Cl/Cd ratio is optimal for Aero-foil-2, i.e., NACA6409, so it is selected for the wing.

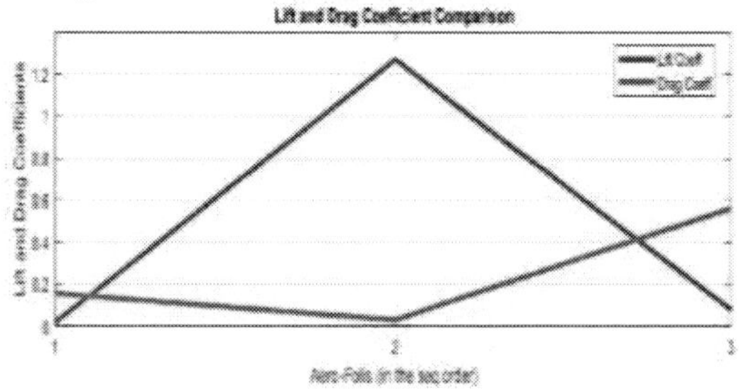

Figure 21. Comparison of lift and drag coefficients.

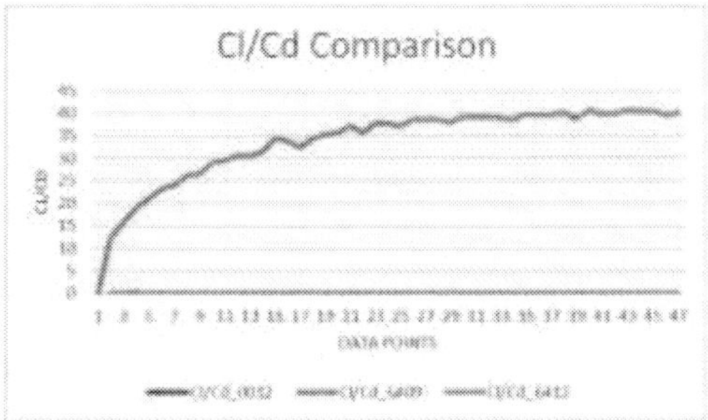

Figure 22. Comparison of Cl/Cd ratios.

Table 1. Comparison of Aerodynamic parameters.

Sl. No.	Aero-Foil	Lift Coefficient	Drag Coefficient	Cl/Cd
1	NACA 0012	0.019	0.15	0.125
2	NACA6409	1.27	0.032	38.9
3	NACA6412	0.0797	0.56	0.144

7. Design of hybrid drone

The wing is incorporated in the drone design, the various views of the same are as shown below. **Figure 23.** shows the front view of the drone and exposes the frontal area of the same. **Figure 24.** shows the top view of the drone, describing the planform of the wing and the rotor positioning. **Figure 25.** shows the side view of the drone, describing the fuselage design in detail. **Figure 26.** shows the dimensions of the drone, conforming to the one, when it would be fabricated. **Figure 27.** shows the isometric view of the drone in a rendered fashion.

Figure 23. Front View.

IOP Conf. Series: Materials Science and Engineering 1012 (2021) 012023 doi:10.1088/1757-899X/1012/1/012023

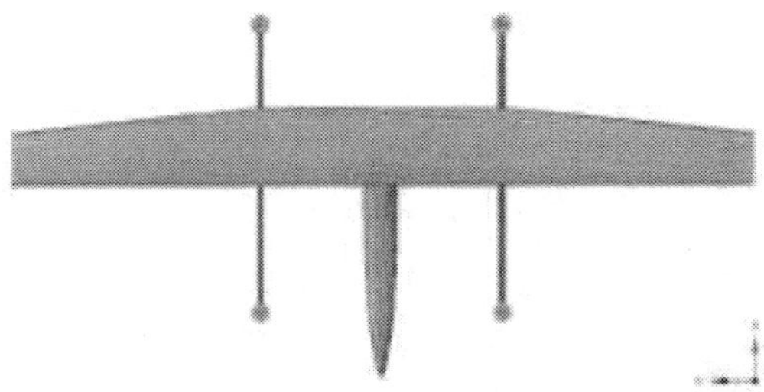

Figure 24. Top View.

Figure 25. Side View

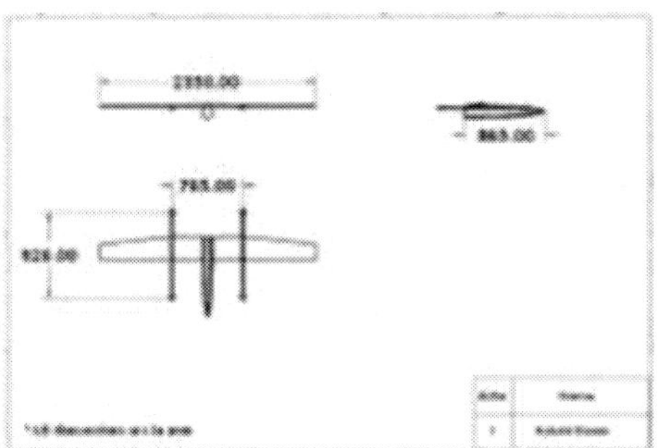

Figure 26. Dimensions of the drone.

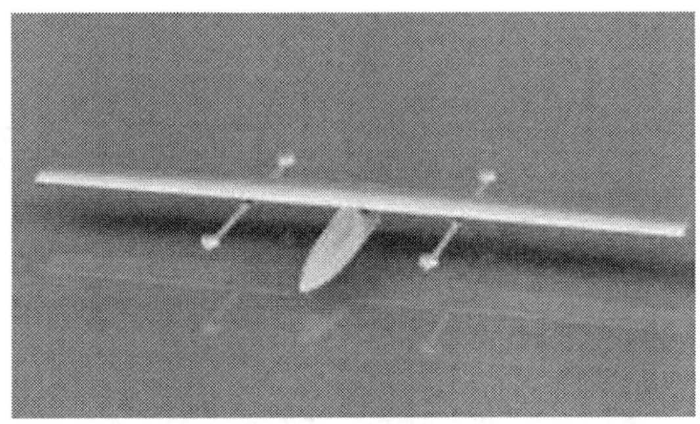

Figure 27. Isometric View.

8. CFD analysis of hybrid drone

Figure 28. and **Figure 29.** show that the velocity pattern is similar to that of the wing, which plays a major role in the aerodynamic parameters of the drone. The velocity in the leeward side is much lower (~0), which accounts for the drag. This low velocity region, which causes the drag, is due to the flow separation accorded to the geometry of the drone. There are various other parameters which do affect the drag formation. They are:

- Wake formation in the leeward side.
- Higher pressure gradients in the leeward side of the geometry, i.e. drastic variation of pressure due to partial vacuum in the critical region and high pressure in the region away from it.

As the flight is considered for cruise operation only the above parameters are taken into concern. The parameters like angle of attack, etc., are not taken into consideration due to the nature of application. Thus, the analysis on the drone body has been done.

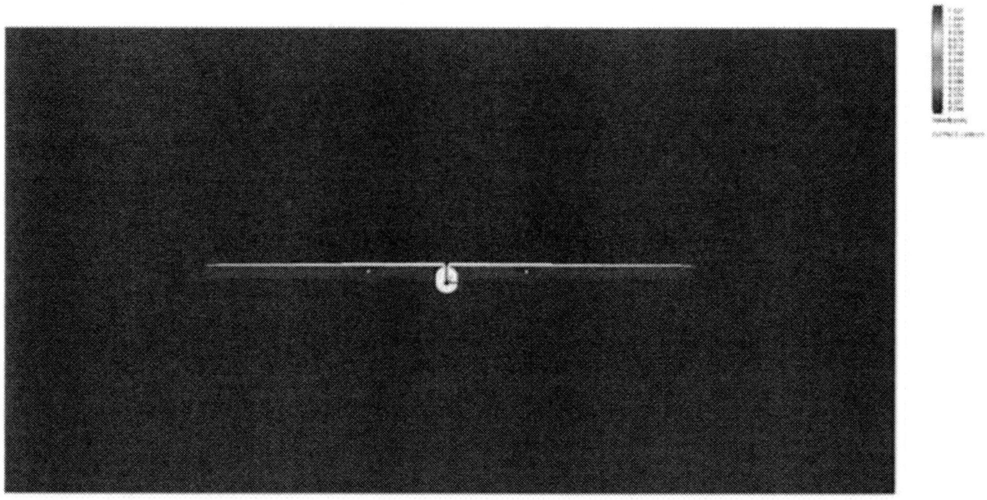

Figure 28. Velocity contour front view.

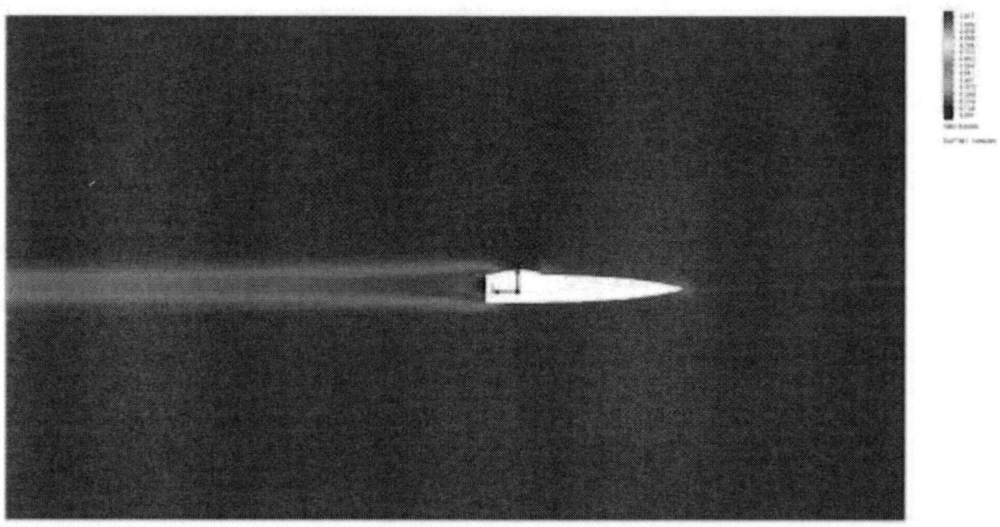

Figure 29. Velocity contour side view.

9. Conclusion

The aerodynamic analysis of various wings, the major contributor for the lift of the hybrid drone, was carried out and the solutions converged (reached steady state). The best lift to drag coefficient ratio (38.9) offering wing, i.e., NACA6409, was selected for the drone for operation during the cruise condition and incorporated on the same. The hybrid drone was analysed and the results converged to an optimum value. As the results were favourable, a prototype model was made by 3D printing using PLA (Poly Lactic Acid) as material. A new type of drone with hybrid characteristics has been designed and tested for aerodynamic properties, and the advantages were recorded.

The project has a wide scope as it involves an upcoming technology in it. The future scope would be to develop a test-prototype to perform wind tunnel test and ascertain the simulation results. Then develop an original model for the hybrid drone and test its practical functionality. After tuning its dynamic functionalities, the drone would be further developed for applications such as air-ambulance service, surveillance, defence applications, etc.

IOP Conf. Series: Materials Science and Engineering 1012 (2021) 012023 doi:10.1088/1757-899X/1012/1/012023

References

[1] Ugur Yayli, C., Cihan Kimet, Anday Duru, Ozgur Cetir, Ugur Torun, Ahmet Aydogan, C., Sanjeevikumar Padmanaban and Ahmet Ertas, H. 2017. Design optimization of a fixed wing Aircraft *Advances in Aircraft and Spacecraft Science* **4(1)** 65–80.

[2] Pooneh Aref, Mehdi Ghoreyshi, Adam Jirasek, Matthew Satchell, J., and Keith Bergeron. 2018. Computational Study of Propeller–Wing Aerodynamic Interaction Aerospace 5(3) 79.

[3] Dhwanil Shukla and Narayanan Komerath. 2018. Multirotor Drone Aerodynamic Interaction Investigation Drones 2(4) 43.

[4] Sreelakshmi, K., and Kota Jagadeeswar, K. S. R. 2018. IOP Conf. Ser.: Mater. Sci. Eng. 455 012044.

[5] Piotr Kardasz, Jacek Doskocz, Mateusz Hejduk, Paweł Wiejkut and Hubert Zarzycki. 2016. Drones and possibilities of their using Journal of Civil & Environmental Engineering 6(3) 233.

[6] Chun Fui Liew, Danielle DeLatte, Naoya Takeishi and Takehisa Yairi. 2017. arXiv:1711.10085v2.

[7] Hazim Shakhatreh, Ahmad Sawalmeh, Ala Al-Fuqaha, Zuochao Dou, Eyad Almaita, Issa Khalil, Noor Shamsiah Othman, Abdallah Khreishah and Mohsen Guizani. 2018. arXiv:1805.00881v1.

[8] Oscar Vestlund. 2018. Aerodynamics and Structure of a Large UAV Independent thesis Basic level. Mälardalen University, Sweden.

[9] Özgür Dündar, Mesut Bilici and Tarık Ünler. 2020. Design and performance analyses of a fixed wing battery VTOL UAV Engineering Science and Technology 23(5) 1182-93.

[10] Ventura Diaz, Patricia Yoon and Seokkwan. 2018. High-fidelity computational aerodynamics of multi-rotor unmanned aerial vehicles AIAA SciTech Forum.

[11] Christodoulou, K., Vozinidis, M., Karanatsios, A., Karipidis, E., Katsanevakis, F., and Vlahostergios, Z. 2019. Aerodynamic Analysis of a Quadcopter Drone Propeller with the Use of Computational Fluid Dynamics Chemical Engineering Transactions 76 181-186.

[12] Vasile Prisacariu, Henri Coandă and Mihai Viteazul. 2016. CFD Analysis of UAV Flying Wing INCAS BULLETIN 8(3) 65-72.

[13] Lao, C. T., and Wong, E. T. T. 2017. IOP Conf. Ser.: Mater. Sci. Eng. 370 012006.

[14] Khuntia, S. K., and Ahuja, A. S. 2018. Optimal Design and CFD Analysis of Wing of a Small-Scale UAV to Obtain Maximum Efficiency Journal of Aeronautics & Aerospace Engineering 7(1) 207.

[15] Zhe Ning. 2018. Experimental investigations on the aerodynamic and aeroacoustic characteristics of small UAS propellers Graduate Theses and Dissertations Iowa State University, US.

Design and Analysis of Coaxial Ducted Propulsion Systems of Unmanned Aerial Vehicles

Billa L R Omkar[1*], Arockia Selvakumar Arockia Doss[1], T. Christo Michael[1]

[1]School of Mechanical Engineering, VIT Chennai, Chennai, Tamilnadu – 600127

E-maill: omkar.billa@gmail.com

Abstract. Drone Technology is one of the major areas of research in numerous aspects and purposes. Less flight time and more power consumption has become the major problem in flying drones and UAVs. This project aims in analysing the thrust and other parameters of Coaxial propellers by ducting them uniquely and comparing them with conventionally ducted propellers. Ducting increases the thrust and in this, two propellers were individually ducted. Increase in the efficiency will allow the drones to fly effectively with adequate power consumption. The experiment results were tabulated and compared with each different case such as hover and forward flight configurations. In this Coaxial system one propeller rotates in Clockwise whereas the other in Counter-clockwise direction.

Keywords: Coaxial propellers, Drone, Duct, Efficiency, Thrust, UAV.

1. Introduction:

Coaxial ducted propellers can increase the efficiency and decreases power consumption of Unmanned Aerial Vehicles (UAVs). Ducts increases additional weight but also increases the efficiency of the UAV. Less flight time, thrust and more power consumption have become the major problems in drone technology. The stack of coaxial propellers provides more thrust without increasing UAV's power consumption [1]. Coaxial propellers do not operate on motor's maximum power [2]. In hover flight case, pusher configuration increases 2 to 4% efficiency. Using multi-blade propellers such as tri-blade decreases the noise and because of lower RPM it also reduces the risk [3]. The swirl generated by the puller motor (front motor) let the aft rotor to benefit from additional tangential velocity [4]. Also, propellers which are ducted have a better efficiency than non-ducted one. This is because to generate maximum thrust with minimum disturbance, the resulting airflow trajectory moves downward [5]. Increase in battery and payload problems occurs when UAV is developed [6]. This problem can be addressed by ducting propellers which can increase thrust by 94% [7]. Among square and circular ducts, circular ducted system was found to be 4.5% more efficient than the square one [8]. This is why circular ducts were opted for these experiments. Lift of the propeller increases with increase in the propeller blade pitch [9]. Thus, ducting propellers which are coaxial increases thrust and efficiency. The application of ducted propellers will be an advantage only in the case of low speed and high loading [10].

Based on the literature survey, to achieve more thrust and more flight time with less power consumption, ducting is necessary. In this research, coaxial propellers with Separate ducts, Common duct and no duct were tested and compared its result with each other. The experiments were done in two configurations, which are, forward and Hover flight configurations. Both the propellers rotate in the opposite directions. All the experiments were done in static air conditions.

IOP Conf. Series: Materials Science and Engineering 1012 (2021) 012024 doi:10.1088/1757-899X/1012/1/012024

2. Conceptual design:

Based on the literature survey [1][2][3][5][8][9], the following conceptual designs were made. Designing of a Thrust measuring stand was made as shown in the Fig 1. This stand was used to measure the thrust using a weighing machine by connecting the duct setup on the top end of it. The stand pole is 90 degrees to the other pole which will be rested on weighing machine as shown in Fig 1.

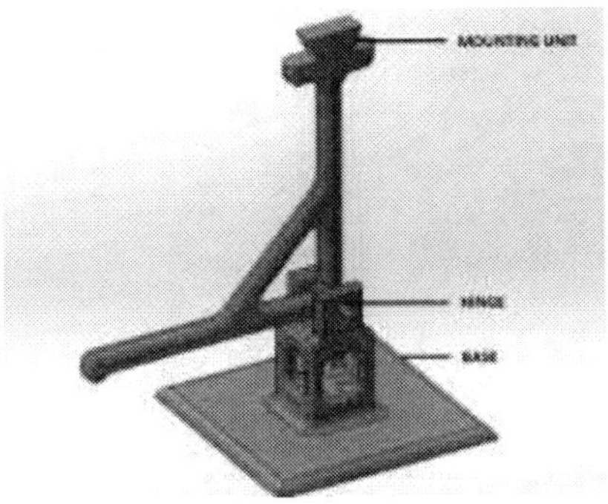

Fig 1. Thrust measuring stand

To conduct the experiment on Co-axial propeller system with no duct, a CAD model was designed as shown in Fig 2. This model was made according to the dimensions of the BLDC motors and the propellers. For the experiment on Co-axial propeller system with a common duct, a duct was designed as shown in Fig 3. In this setup, both BLDC motors were ducted using one duct and its dimensions are shown in Fig 3. Propeller of motor 1 (M1) (puller) is 26.9mm from the top of the duct and propeller of M2 (pusher) is 10 mm distance from the bottom of the duct. A CAD model was designed also for the experiment on Co-axial propellers using two separate ducts as shown in Fig 4, 5. As shown, each duct in this setup has 72.81mm height and 167.26mm, 139mm of inlet and outlet diameter whereas the distance between propellers maintained constant in all the experiments.

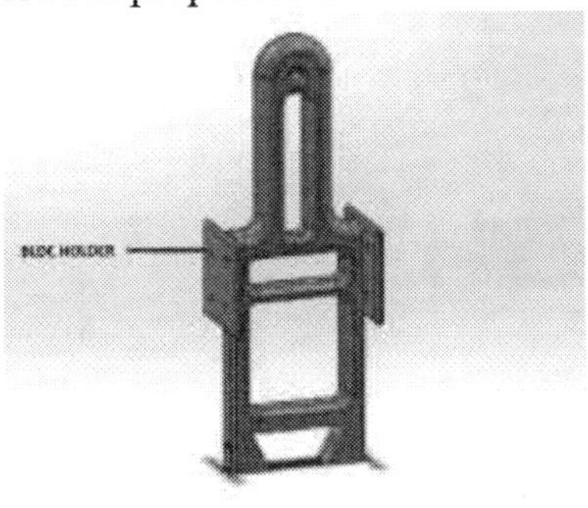

Fig 2. Design for Co-axial propellers with no ducts

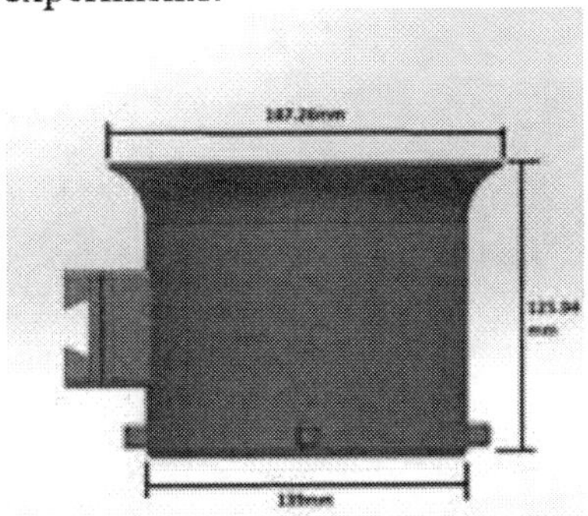

Fig 3. Common duct design

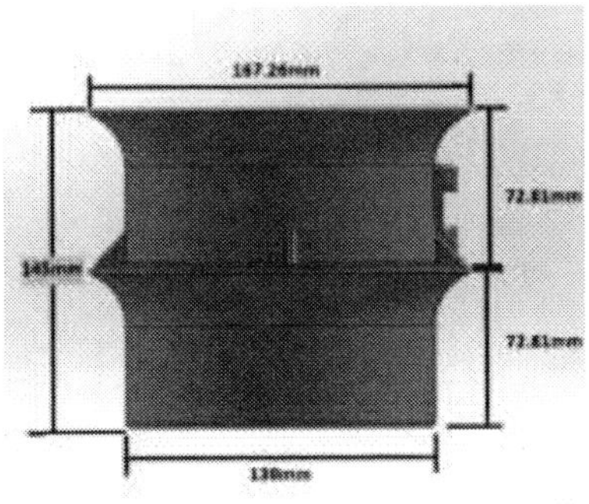

Fig 4. Separate ducts design (side view)

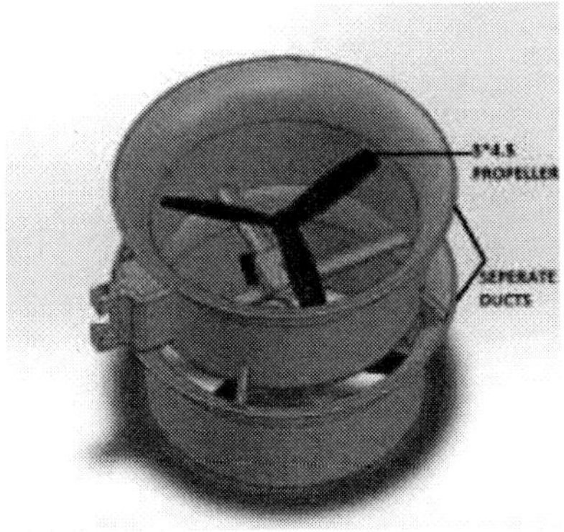

Fig 5. Separate ducts design

IOP Conf. Series: Materials Science and Engineering 1012 (2021) 012024 doi:10.1088/1757-899X/1012/1/012024

3. Development of prototype:

Cylindrical duct geometry was properly designed based on the adequate parameters and needs. All the designs were 3D printed using PLA filament and these ducts are durable, light weight and made as shown in Fig 6,7,8. Usage of ducts changes the flight characteristics depending on whether it is a forward flight or hover flight. Two tri-blade propellers which are 5 inches and have 4.5 inches fixed pitch, 2 BLDC motors of 2300KV along with 30A speed controller and 4 cell LiPo battery of 1000mAh, 30C/60C were used. For a propeller if the pitch angle increases, power loading ratios increases [11]. To control the speed of the BLDC motors, digital multi servo tester was used. Anemometer, Tachometer and Multimeter were using to check wind speed, RPM of the propellers and power.

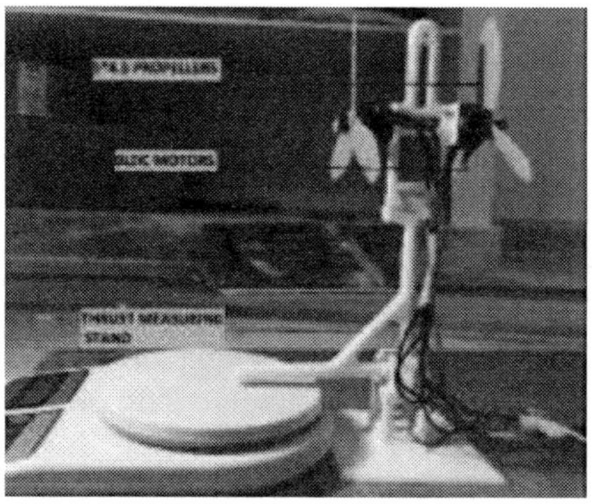

Fig 6. No duct experiment

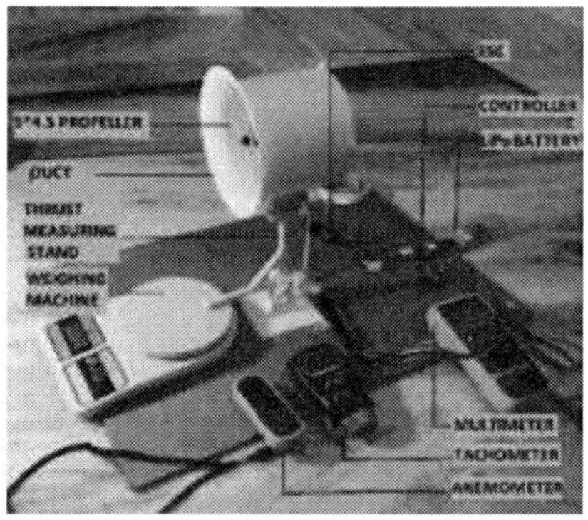

Fig 7. Common duct experiment

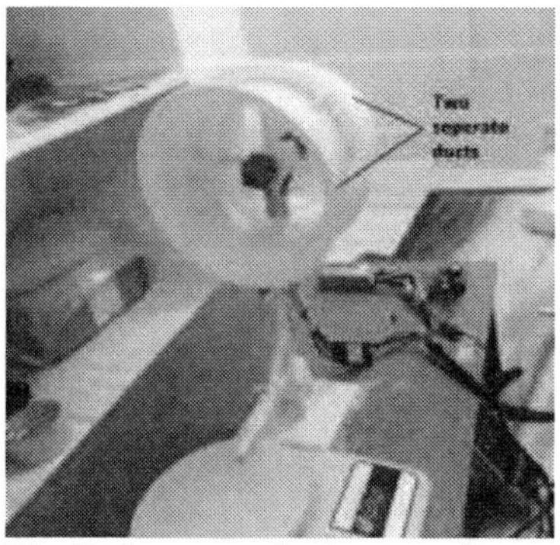

Fig 8. Separate ducts experiment

Two propellers were fixed properly to the two BLDC motors and both the motors were connected to an ESC. M1 acts as puller/tractor and M2 acts as pusher motor. ESC was connected to the motors M1, M2 and a servo controller which was used to control the BLDC motors speed and also connected to a lithium polymer battery (LiPo). Entire unit was mounted to the stand as shown in the Fig 9.

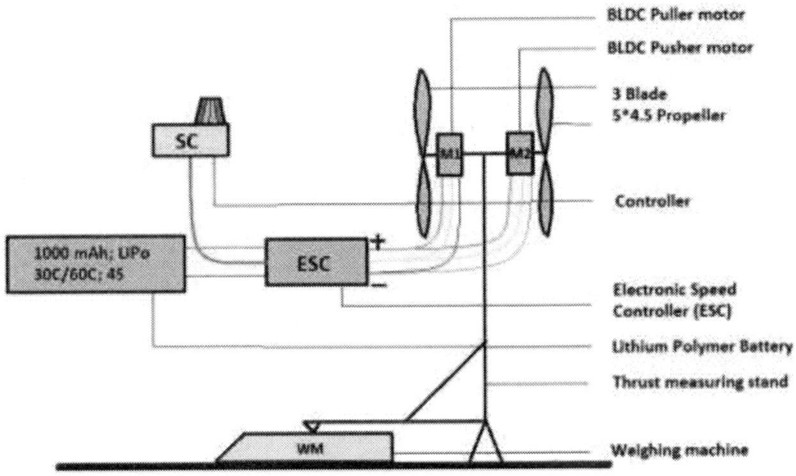

Fig 9. Schematic representation of the experimental setup.

4. Experimental studies:

Two polycarbonate propellers rotate in opposite direction are 5 inches whereas the top rotor turning CCW (puller/tractor) and the bottom is CW (pusher) and the clearance between propellers and duct inner wall is 3mm. Duct upper radius is 167.26mm and the height of the duct is 125.94mm. This duct includes two propellers and the distance between two propellers is 90mm whereas duct wall thickness is 3mm. Diameter (D) of propeller is 127mm and height (H) between two propellers is 90mm. It is proven that by comparing upper and lower rotor in torque balance setup, lower rotor operates at lower efficiency compared to the upper rotor. [12]

H/D value of this setup is 0.7, whereas large manned aircraft use a 0.1 H/D ratio [13]. The weight of the duct used in common duct experiment is 130 grams and can be reduced, were as the weight of duct used in separate duct experiment is 125 grams. Weight of the motor along with propeller is 36 grams. In this, both propellers rotate in opposite directions whereas it is proven that coaxial propeller system has the same efficiency whether the propellers rotate in same or opposite direction [14]. Drones are very vital for mankind and even in military. Drones have vital role in military, by helping in detection of buried bombs and bombers [15]. Testing of coaxial propellers with no duct, common duct and separate ducts were done. Several experiments were carried out on various cases and had tested in both forward and hover flight conditions. The following experiments were done in the static air condition.

4.1. Experiment without a duct:

In this, BLDC motors were aligned coaxially with 90mm distance between each other. Both rotate in the opposite direction. BLDC motors were operated using a controller and had tested at varies power inputs and noted the thrust generated by this setup.

4.1.1. Results:

Result shows that at 107.2 watt, thrust generated was 500g and is represented in the graph shown in Fig 10. There was a uniform increase in the thrust by increase in the current. 50g of thrust generated at 10 watt of power supply.

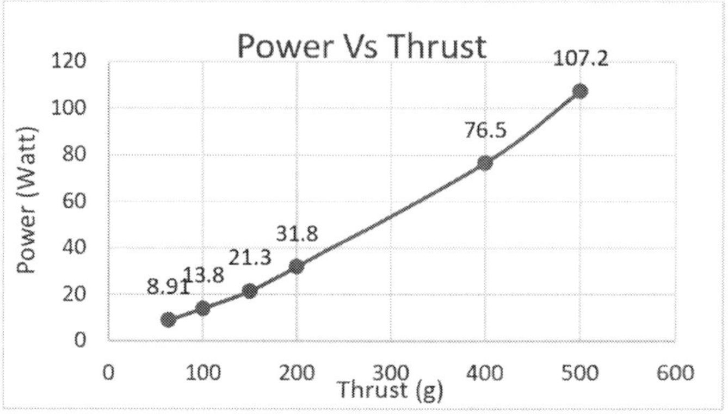

Fig 10. Power vs Thrust Graph

4.2. Experiment with Common duct:

In this, BLDC motors were aligned coaxially and both the propellers were ducted using one common duct. This duct is mounted to the thrust measuring stand which measures thrust using a weighing machine.

4.2.1. Forward flight:

Entire duct setup was aligned in forward flight configuration and tested. In order to measure current, rpm and wind speed; multimeter, tachometer and anemometer were used. In forward flight experiment, the coaxial rotor needs less power compared to the equivalent solidity single rotor upto moderate advance ratios [16]. Same tri-blade propellers, motors, battery, ESC, servo tester were used like before in this and the following experiments and tested. Only duct varies between experiments.

4.2.2. Results:

At 7.6 watt of power input, 65g of thrust was generated and at 66.43 watt power input, 500g of thrust was generated. Results shows that the thrust is increased compared with the results of no duct experiment. Power vs Thrust graph in Fig 11 shows the detailed results of this experiment.

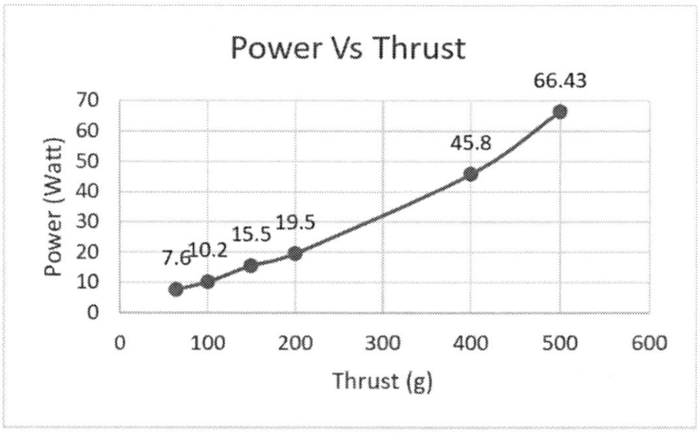

Fig 11. Power vs Thrust Graph

4.3. Hover flight:

Experiment was carried out in hover flight configuration with the same setup. In hover flight configuration, duct will have tendency to move up by forcing the wind down. Thrust was measured by using two scales and tested properly.

4.3.1 Results:
At 42.9 watt of power, this flight configuration has generated 300g of thrust and it varied with respect to power as shown in the graph which is in Fig 12. Whereas in forward flight condition gravitational force is at 90 degrees to the force of thrust. Where as in forward flight configuration, to generate 300g of thrust, 32 watt power was consumed. Increase in power consumption is because of the gravitational force acting opposite to the direction of flight.

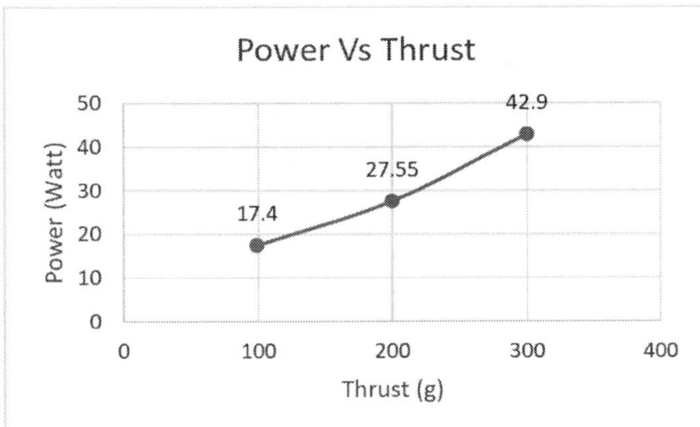

Fig 12. Power vs Thrust Graph

4.4 Experiment with Separate ducts:
In this, BLDC motors were aligned coaxially and both BLDC motors were ducted separately using two ducts. Propeller of M1 is 26.9mm from the top of its duct and propeller of M2 26.9mm distance from the top of its duct. This duct weighs 125 grams and can be reduced using lightweight materials.

4.4.1 Forward flight:
Entire duct setup was aligned in forward flight configuration and tested. In order to measure current, rpm and wind speed; Multimeter, tachometer and anemometer were used. Same components from previous experiments were used except a new duct design.

4.4.2 Result
From the obtained results, at 9.71 watt power input, 55g of thrust was generated and at 72.85 watt power input, 500g of thrust was generated. Thrust is more than the thrust generated by no duct setup and less than common duct setup. Detailed data of the experiment is graphically represented in Fig 13.

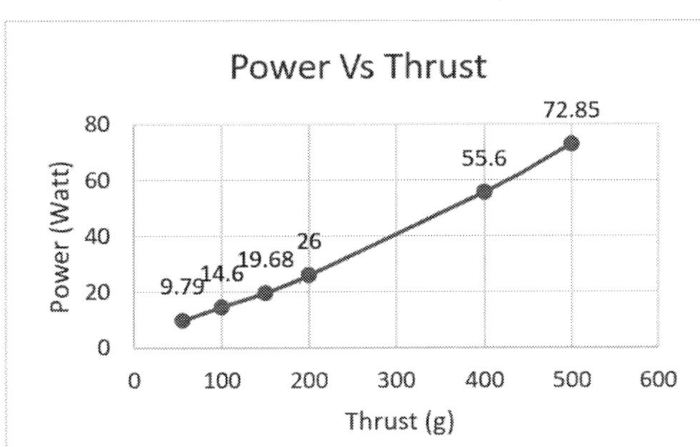

Fig 13. Power vs Thrust Graph

4.5 Hover flight

Entire setup was mounted as shown in Fig 14 to measure hover flight characteristics of the setup. Used another weighing scale from the top of the duct to measure its thrust accurately.

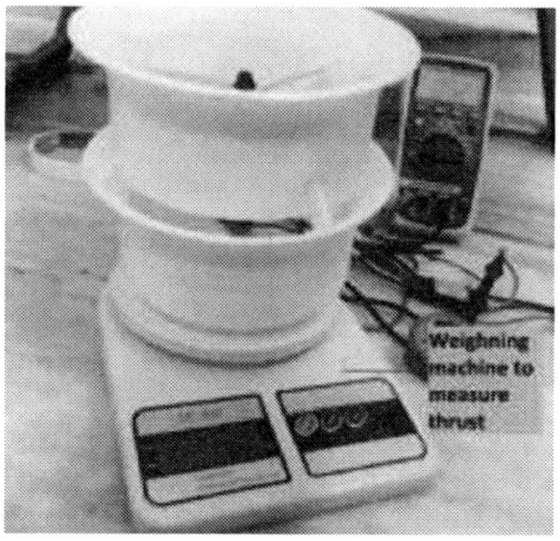

Fig 14. Hover flight setup

4.5.1 Result

Result shows that at 15.4 watt power input, 100g of thrust was generated and at 57 watt of power input, 300g of thrust was generated. Thrust is less compared to the thrust generated by common duct setup. It was tested at different power inputs and results are shown in Fig 15.

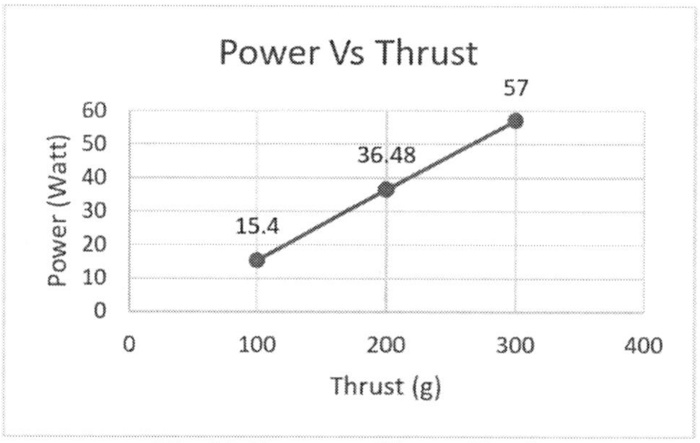

Fig 15. Power vs Thrust Graph

5. Comparative study:

Comparing power and thrust from the results of No duct, Common duct and Separate duct shows that Common duct has less power consumption compared to the setup with no duct. Whereas a Separate duct setup result shows that the thrust is slightly less compared to the Common duct setup. As shown in Fig 16, P v T graph, common duct can generate more thrust with less power consumption than both the other cases.

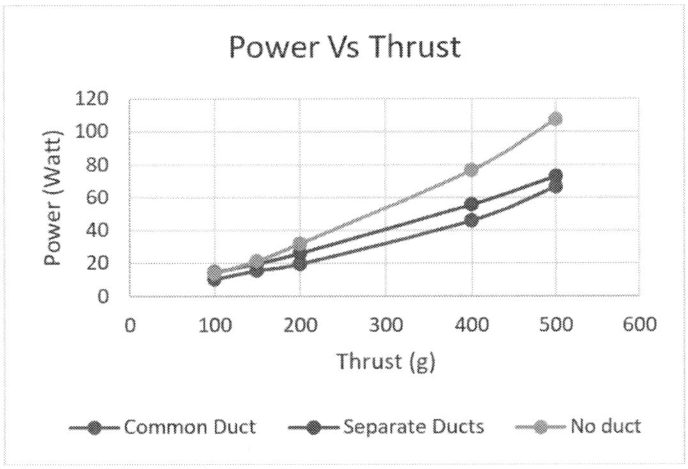

Fig 16. Power vs Thrust comparison graph

From No duct experiment forward flight configuration, obtained results are tabulated in Table 1.

Table 1. No duct forward flight results

Thrust (g)	Voltage (V)	Current (A)	M1 RPM	M2 RPM	Wind speed (Mt/sec)	Power (Watt)
63	16.2	0.55	4130	410	4.5	8.91
100	16.1	0.86	5040	5050	5.4	13.8
150	15.9	1.34	6210	6220	7	21.3
200	15.3	2.03	7262	7289	8.2	31.8
400	15.3	5	9940	9950	10.2	76.5
500	14.9	7.2	10967	10936	12.8	107.2

From Common duct experiment forward flight configuration, obtained results are tabulated in Table 2.

Table 2. Common duct forward flight results

Thrust (g)	Voltage (V)	Current (A)	M1 RPM	M2 RPM	Wind speed (Mt/sec)	Power (Watt)
65	15.2	0.5	4158	4146	4.2	7.6
100	15.1	0.68	4760	4735	5	10.2
150	15.1	1.03	5822	5797	6.3	15.5
200	15	1.3	6496	6480	7	19.5
400	14.8	3.1	9020	9000	9.3	45.8
500	14.6	4.55	10260	10264	12	66.43

From Separate duct experiment forward flight configuration, obtained results are tabulated in Table 3.

Table 3. Separate ducts Forward Flight

Thrust (g)	Voltage (V)	Current (A)	M1 RPM	M2 RPM	Wind speed (Mt/sec)	Power (Watt)
55	16.6	0.59	4452	4450	4.6	9.79
100	16.5	0.89	5441	5451	5.9	14.6
150	16.4	1.2	6347	6362	6.7	19.68
200	16.3	1.6	7035	7043	7.8	26
400	15.9	3.5	9550	9570	8	55.6
500	15.5	4.7	11350	11390	10.1	72.85

Where, g= grams; V=Voltage; A= Ampere; Mt/sec= Meter/Second; RPM= Rotations per minute.

6. Conclusions:

Ducted coaxial propellers increases thrust and decreases power consumption. Common duct for two motors is very effective compared to the motors with no duct and separate ducts. Motors with no ducts generated 500g of thrust at 107.2watt power input and separate ducts generated 500g of thrust at 72.85watt, which means 32% of power consumption in common duct experiment is reduced compared to the no duct experiment. Whereas in the case of common duct, two motors generated 500g of thrust at 66.43watt of power input, which means 38% of power consumption was reduced compared to the motors with no duct. Using ducts effectively increases thrust and reduces power consumption but adds weight to the UAV. If the extra thrust generated at same power consumption by using ducts is more than the weight of the ducts, then UAV efficiency will be increased. Coaxial ducted propellers make Vertical take-off easy which is Hover flight.

Future works based on this project can be, Reduction in the weight of duct can be done by using other lightweight materials like balsa wood. H/D value is 0.7, whereas it can be varied and tested by doing iterations. Number of blades of the propellers can be changed along with pitch angle, length of the propellers can be varied and tested. Clearance value between propeller and duct inner wall can be reduced and tested. Separately ducted coaxial propellers might enable UAV to move during hover flight without using any rudders. This might be possible by making lower duct tiltable and can be used as an alternate method where rudders cannot be used. But the power consumption will be 4% less than common duct ones. Other configurations like downward and backward flight configurations can be tested by making suitable changes.

References

[1] Jonathan Elliot Holzsager. The Effects of Coaxial Propellers for the Propulsion of Multirotor Systems. A thesis submitted to the Graduate School of New Brunswick Rutgers, The State University of New Jersey. October 2017.

[2] Adam Bondyra, Stanis law Gardecki, Przemys law G¸asior, and Wojciech Giernacki. Performance of coaxial propulsion in design of multi-rotor UAVs. Springer International publishing Switzerland.DOI: 10.1007/978-3-319-29357-8_46.

[3] B. Theys∗ , G. Dimitriadis, P. Hendrick, J. De Schutter KU Leuven, Leuven, Belgium. Influence of propeller configuration on propulsion system efficiency of multi-rotor Unmanned Aerial Vehicles. June 2016, DOI: 10.1109/ICUAS.2016.7502520.

[4] P. Beaumier. A low order method for co-axial propeller and rotor performance prediction. HAL Submitted on 3 Nov 2014. Id: hal-01079555. March 2016.

[5] Endrowednes Kuantama , Radu Tarca. Quadcopter thrust optimization with ducted-propeller.

MATEC Web of Conferences 126, 01002 (2017). ORADEA 2017. DOI: 10.1051/matecconf/201712601002.

[6] Andria, G., Nisio, A. D., Lanzolla, A., Spadevecchia, M., Pascazio, G., Antonacci, F., & Sorrentino, G. M. (2018). Design and Performance Evaluation of Drone Propellers. 2018 5th IEEE International Workshop on Metrology for AeroSpace (MetroAeroSpace).

[7] Pereira, and Jason Louie. "Hover and Wind-Tunnel Testing of Shrouded Rotors for Improved Micro Air Vehicle Design." Digital Repository at the University of Maryland, 29 Aug. 2008, drum.lib.umd.edu/handle/1903/8752.

[8] Hugo F. Bento, Reynard de Vries and Leo L. Veldhuis. Aerodynamic performance and Interaction Effects of Circular and Square ducted Propellers. Published online on 5th January 2020. https://doi.org/10.2514/6.2020-1029.

[9] Jing Ze Han, Bing Yan Jiang, Chao Wei Zhuang, Xiong chun du. Experimental Investigation on Aerodynamic Characteristics of Propeller in a Ducted Fan type VTOL UAV. December 2014. https://doi.org/10.4028/www.scientific.net/AMM.711.12

[10] Tadeusz Koronowicz, Prof. Zbigniew Krzemianowski, Ph. D. Teresa Tuszkowska, Ph. D. Jan A. Szantyr, Prof. A complete design of ducted propellers using the new computer system. POLISH MARITIME RESEARCH 2(60) 2009 Vol 16; pp. 34-39 10.2478/v10012-008-0019-3.

[11] Characterization of ducted contra-rotating propeller propulsions 11:175682931983766; January 2019. DOI: 10.1177/1756829319837661

[12] Leishman J. G., Ananthan S., "Aerodynamic Optimization of a Coaxial Proprotor", EUA, Maryland, University of Maryland, Department of Aerospace Engineering, May 2006.

[13] Reviewing and Investigating the Use of Co-Axial Rotor Systems in Small UAVs Stephen D. Prior Autonomous Systems Laboratory, Department of Product Design and Engineering, School of Engineering and Information Sciences, Middlesex University, Trent Park Campus, London N14 4YZ, UK. s.prior@mdx.ac.uk

[14] Optimizing a Coaxial Propulsion System to a Quadcopter Cédric Martins Simões Dept. Engenharia Mecânica, Instituto Superior Técnico, Av. Rovisco Pais, 1049-001 Lisboa, Portugal

[15] Mahadevan P., "The Military Utility of Drones", Center for Security Studies (CSS), ETH Zurich, July 2010.

[16] Coleman C. P., "A Survey of Theoretical and Experimental Coaxial Rotor Aerodynamic Research", NASA Technical Paper 3675, California, Moffett Field, Ames Research Center, March 1997.

Hexacopter design for carrying payload for warehouse applications

M. Aswath[1*], S. Jeevak Raj[1]

School of Electronics Engineering, Vellore Institute of Technology, Chennai

E-mail: m.aswath08@gmail.com

Abstract. Airborne vehicles are gaining traction in research fields and in bringing it for commercial use. Multi-rotors for payload transport is one of the highly researched topics. In this paper, a hexacopter was prototyped based on commercially available components in the market and dynamic analysis was performed with constraints derived from ideal conditions, such as neglecting the effect of air resistance. The experiment shows the flight of the hexacopter to be stable with and without payload attached to it. Also, various computer vision methods were explored to detect and identify waypoints, in the form of gates, for the hexacopter to follow during its flight mission. The detection mechanism employed is HSV mask based filtering on input image and performing moment calculation on the extracted contour to determine the center of the gates. Localization of the drone with respect to the environment is performed with a monocular visual SLAM method known as ORBSLAM2.

Keywords: Hexacopter, Multi-rotor design, Computer vision, ROS, Modelling and Simulation

1. Introduction

Unmanned Ariel Vehicles(UAVs) are aiding us in a plethora of fields such as agriculture, swarm-based collaborative robots for construction sites, warehouses management, package delivery, and many more. A particular type of UAV called VTOLs(Vertical Take-Off and Landing) has made it possible to get airborne without a runway. It is high time that progress and improvement in these fields are facilitated. For agriculture[1], drones are used to monitor crop health and estimate harvest size. It is also used to spray pesticides over the entire crop region. For this, the drone must carry the heavyweight of the pesticide tank with the accompanying mechanism. At the same time, it must also be robust to turbulence caused by the wind. In construction and warehouse management [2], the drones must be agile and have a fast control response because it might have to avoid any debris, falling objects in the air. By having good agility, it makes a safe working environment. An active are of research is in the domain of swarming behaviors. Robot swarming refers to agents that forms a network and are able to reorganize dynamically as the agents leave or enter the network. They provide greater flexibility in terms of cooperation among multi-agents. Swarming may open the path to the fusion of data, such as mapping of the environment with multiple bots simultaneously. Some bio-inspired flight swarms utilize steering and flocking behaviors. For the control of the UAVs, first, an overview of existing controllers [3] and their effectiveness should be studied. This would help us decide which type of controllers would be most helpful for this implementation. The most common type of control mechanism used is the PID controller which can be used to achieve basic hovering and attitude control. The UAVs has autonomy features such as the ability to maintain a flat position, maintaining altitude, position hold, headless mode, care-free, automatic takeoff and landing, return to home, follow-me mode, GPS way-point navigation, Orbit around target object and pre-programmed aerobatics such as flips and sharp maneuvers. ROS(Robot Operating System) is acts as a middle-ware between heterogeneous systems and processes. ROS processes are in the form of nodes, connected by edges called topics. ROS nodes can pass messages between each other through topics. The messages do not pass through the master, rather the master enables peer-to-peer communication for topics. This decentralized architecture and flexibility of ROS makes it well suitable for robotics applications. The main advantage introduced is that the robot may also easily use off-board computers for heavy computations. Visualization tools are also provided in ROS such as rivz and rqtgraph which are used extensively by robotics developers. A simulation software called Gazebo is also a main package in ROS. The novelty of this study lies in the ability to integrate open-sourced research algorithm such as computer vision, SLAM and other image processing algorithms(CLAHE, HSV mask tuning), which can be both user-friendly and helps in accelerating research in commercial products as they are ubiquitously available in the market and easily procurable by individuals.

2. Literature review

The study on "Hexacopter for heavy payload transport" by Bhakti Yudho Suprapto et al. [4] provided us with useful insights and information to approach this problem and tailor a variant of the drone for our objective.

Yu Miao et al. [5] introduced the faster and efficient algorithm of CLAHE by introducing the parameter T in bilinear interpolation, which speeds up the interpolation and improves the speed of the entire algorithm.

S. Li et al. [6] developed a faster pose estimation algorithm and has tolerance to noise using the n-point method. This algorithm detects the gates' four corner points even in blurred vision and uses a predictive feed-forward strategy to find the gate that is not in the visibility.

Adrian Rosebrock [7] at pyimagesearch proposed an algorithm for estimating the distance between two objects, which uses the triangular similarity to find the distance.

Zhaleh Sadreddini et al. [8] proposed a method to estimate the distance to an object from the camera using feature lines extracted from the environment, the tiled floor, in their case. The feature line is then used to calculate the distance to the object that lies on the horizontal horizon line of the image.

Raul Mur-Artal, Juan D. Tardos came up with SLAM based on ORB features [9]. ORB-SLAM2 supports monocular, stereo, and RGB-D cameras. Because stereo and RGB-D cameras are expensive, bulky, and more power-consuming, a more economical approach with a ubiquitously available monocular camera was chosen for this purpose.

Ebrahim Karami, Siva Prasad and Mohamed Shehata [10] performed a comprehensive study on various feature detectors like SIFT, SURF, BRIEF and ORB. The study illustrates the performances and computation time involved executing each of these algorithms. Further, feature matching capability of the algorithms was also tested by identifying the keypoints from the given two images at when subjected to scaling, rotation, distortions and noise. This helps in deciding the type of feature detector that can work best in the visual SLAMpipeline.

Pawel Burdziakowski [11] presented his implementation of ORBSLAM2 running on a Jetson TX2 onboard computer mounted on the UAV. He uses a stereo rig setup and uses the stereo cam package offered by the ORBSLAM2 API.His study evaluates errors that are caused by inaccurate camera calibration. He obtains the intrinsic parameters from each calibration session and then compares the path mapped by the UAV under these different camera parameters. The path comparison was also done with GPS guided mode and with the ground truth path. From this study, the performance of ORBSLAM2 can be understood and the correct method of camera calibration is verified and supported by the obtained result.

Victor M. Becerra compiled the study on uses and applications of UAVs [12], which suggests that they are increasing rapidly, and the market is about to hit $14 billion by 2027. Growth in the production of UAVs is expected to reach $13.1 billion by 2027. Seeing the demand and the growth of autonomous features in UAVs alongside the UAV market, the applications are widespread from military to civilian, including surveillance, remote sensing, delivery services, agricultural sector, target acquisition, aerial imaging, and infrastructure monitoring, industrial inspection, and emergency medical aid.

Ram Prasad Padhy et al. proposed a solution [13] for using the Convolutional Neural Network model, which is trained over dataset images and responsible for controlling the movement in the left and right direction. Since the dataset wouldn't be available every time, our method uses image processing algorithms to control it's movement.

3. Design of Hexacopter

3.1. Hexacopter configuration

Hexacopter has six motors attached to the frame with arms, these motors rotate the propeller at very high speeds to generate thrust. To achieve stability, three motors are made to rotate in clockwise(CW) and counterclockwise(CCW) directions. With this setup, it is able to cancel the net moment about the drone's yaw axis. The type of motors used in multi-rotors is BLDC motors, the rotations speed is controlled by using 3 different phases of current. The set speed of rotation of the motors is maintained with the help of ESC(Electronic Speed Controllers). ESCs constantly monitor the feedback current from the motor and make minute changes to the input current for the motors. Inertial Measurement Unit(IMU) which comprises of accelerometer and gyroscope, is used to measure short displacements and determine the orientation of the drone. This is done based on accelerations experienced by the body. Pixhawk is open-sourced flight controller, with on-board sensors that has many inbuilt algorithms to filter the raw sensor data and provides a precise data estimate. It also supports high level protocols such as MavLink which is used to communicate with a computer that may send commands to move the drone to a certain location in world's GPS coordinates. Batteries are used to power the on-board computer, sensors and motors. Optical flow sensor is used to measure very short displacements at very high frequency, this can be thought of as an odometry present in the wheeled robots. The model of optical flow sensor used is the PX4Flow, this sensor is chosen because it is widely used in UAV applications and has robust filtering algorithms like the EKF(Extended Kalman Filter), which has gained traction in recent years and finding many applications in real world implementations. Along providing the displacement along the ground plane, it also measures the altitude of the drone with the help on inbuilt sonar sensor. The ultrasonic sensor measures the distance from ground to the UAV's frame to determine the altitude of the flight. Moreover, the sensor is also capable of providing correct readings in indoor as well as outdoor environments, even without the need of illumination of LEDs. The optical flow methods attempt to calculate the motion at each voxel position between two image frames, which are taken at t and $t+\Delta t$. These methods are referred to as differential because they are based on local image signal approximations of the Taylor series; that is, partial derivatives are used with regard to spatial and temporal coordinates. Jetson nano is a computer that has powerful computation capabilities given its small footprint and weight profile. It is capable of running Neural Networks for image classification, object detection, semantic segmentation. The power consumption of this computer is as low as mere 5 watts. Lithium Polymer batteries are most commonly used due to their ability to provide high discharge rates and relatively procurable battery type. The specification of the battery chosen is 11.4 volts provided by three cells placed in series, with a discharge rate of 30C. The main frame chosen for the hexacopter design is the DJI F550. This frame is intended towards hobbyist which enables to attach custom components such as GPS, telemetry, LIDARs(Light Detection and Ranging), on-board computers and attach custom motors. Based on these mentioned components, a physical model of the hexacopter was designed in Solidworks and the final 3D CAD(Computer Aided Design) model is exported as URDF(Unified Robotic Description Format) format to load it into the ROS-Gazebo simulation world. ROS, Robot Operating System, allows heterogeneous computers to communicate with each other and provide low level device controls and message passes between processes.

Table 1. Weight distribution of Hexacopter

S.no	Item	Weight(gm)
1	F550 Hexa-Copter Frame	620
2	LiPo 8000mAh 3S 30C/60C	615
3	Jetson Nano	249
4	BLDC Motors EMAX RS2306	204
5	ESC Emax BLHeli 30A	168
6	Pixhawk	40
7	Optical Flow Sensor	30
8	Camera 8MP IMX219-170	22
9	Tri Blade Propellers	15
	Total	1900

3.2. Hardware composition

It is necessary to evaluate the total weight distribution of the hexacopter; this allows for analysis of the components and highlights the importance for each part and track its performance with previous and future iterations of the Multi-rotor designs. The heaviest component is all multirotor designs is the battery. For achieving a longer flight time, a battery with larger capacity may be chosen. However, under the additional weight of the larger battery pack, it in-turn directly affects the flight time. However, since the implementation is done without LIDAR, significant amount of weight that would otherwise occur is eliminated. LIDARs also consumes an expensive amount of power from the battery. The weight distribution of all the components of the hexacopter is as shown in Table 1.

3.3. Structural and Stability Analysis

The physical dynamics of the system must be studied to understand the expected motion when different forces are exerted to the hexacopter during its flight. Two significant studies can be conducted to understand the structural integrity and the stability of the hexacopter during its flight. The structural analysis is used to determine the integrity of the frame, arms, and landing gear when subjected to forces that may be caused due to gravity, thrust and collisions that the hexacopter would be able to withstand. For the structural analysis in Figure 1, the arm of the hexacopter is isolated and the base plate was fix and force of 10 Newtons is applied at the end of the arm where the motor exerts force to simulate a situation in flight when it generates thrust to carry its weight along with the payload. The displacement of the plastic material after applying the force is found to be 8mm.

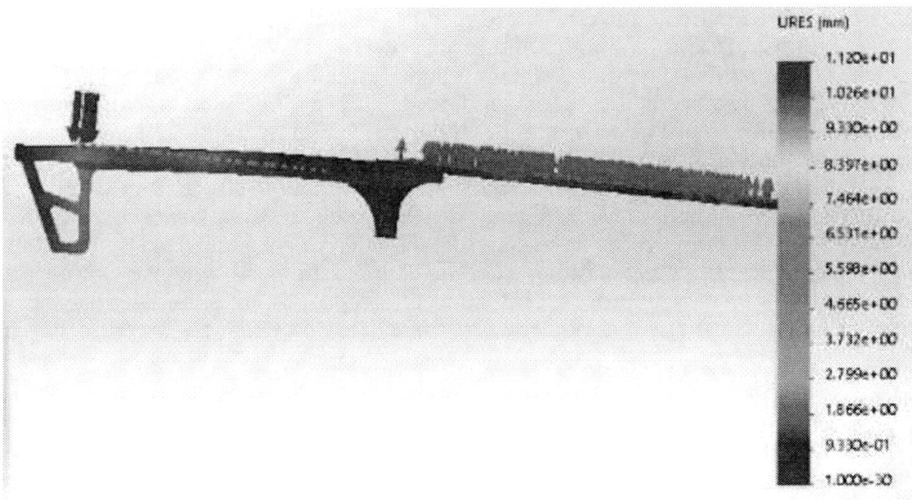

Figure 1. Stress test

The stability of the physical body is done to check the placement of COM(Center Of Mass) of the structure. By identifying the location of COM, the direction of motion of the hexacopter can be predicted when the frame is subjected to forces around it's COM. By having the COM in the middle, the thrust by the propellers acts evenly around the COM, thus cancelling any net torque that may cause undesired rotation and drift. Thus, for stability analysis, the center of mass of the hexacopter with and without the payload is evaluated. The payload is attached well below the arms and the body frame. The mass of the payload is given as 2 Kgs. The resulting location of the COM with payload is shown in Figure 2 and COM location without the payload is shown in Figure 3. Adding the payload towards one side of the hexacopter will cause the hexacopter to drift or put more strain on the motors present directly above the payload. This is because, the motors above the payload need to provide more thrust to balance the frame and prevent it from toppling over during the flight.

IOP Conf. Series: Materials Science and Engineering 1012 (2021) 012025 doi:10.1088/1757-899X/1012/1/012025

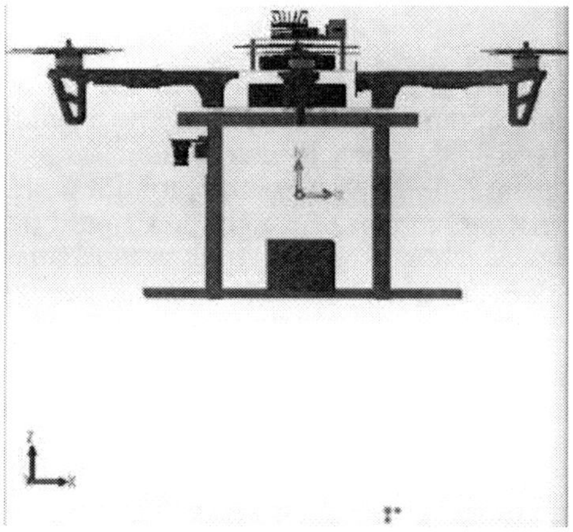

Figure 2. COM without payload

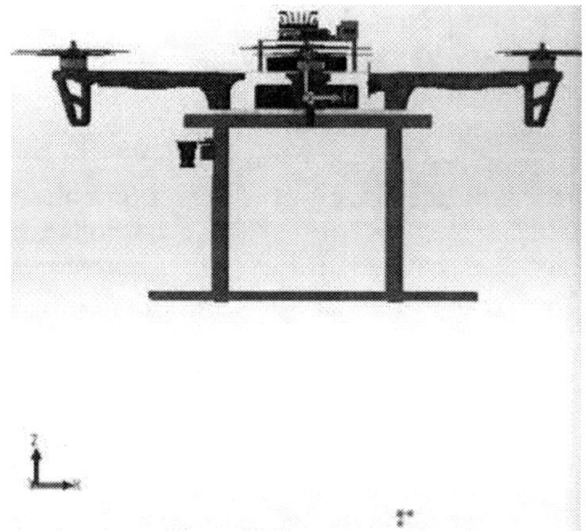

Figure 3. COM with payload

3.4. Control system modelling

The CAD model is loaded into a simulation environment called Gazebo where the dynamics of bodies can be simulated. Since the CAD file contains the mass properties like the center of mass, the moment of inertia it is possible to estimate the real-world responses safely in a simulation. The initial objective must be to test and take-off and landing of the hexacopter; this means that it must reach a set-point altitude. To achieve this, a PID(Positional, Integral and Derivate) feedback control was used. The PID controller receives the data from the output and feeds it back into the input to generate actuator command for the next time step. The controller has three main parameters Kp, Ki, Kd through which the error of the output data is multiplied and added together. The parameter Kp denotes the proportional error constant, the higher the error, a larger thrust command is sent to the motors. The integral gain Ki is used to nullify any errors that are accumulated overtime such as, a constant wind that may force the hexacopter away form the target position or attitude. The derivative gain Kd is the amount by which sensitivity to changes in physical quantities such as changes in wind velocity can be accounted for changing the thrust commands appropriately. The concerned output parameter in this case is the current altitude of flight in the world, and the error is calculated with respect to the desired set-point height of the hexacopter to reach as it's final state.

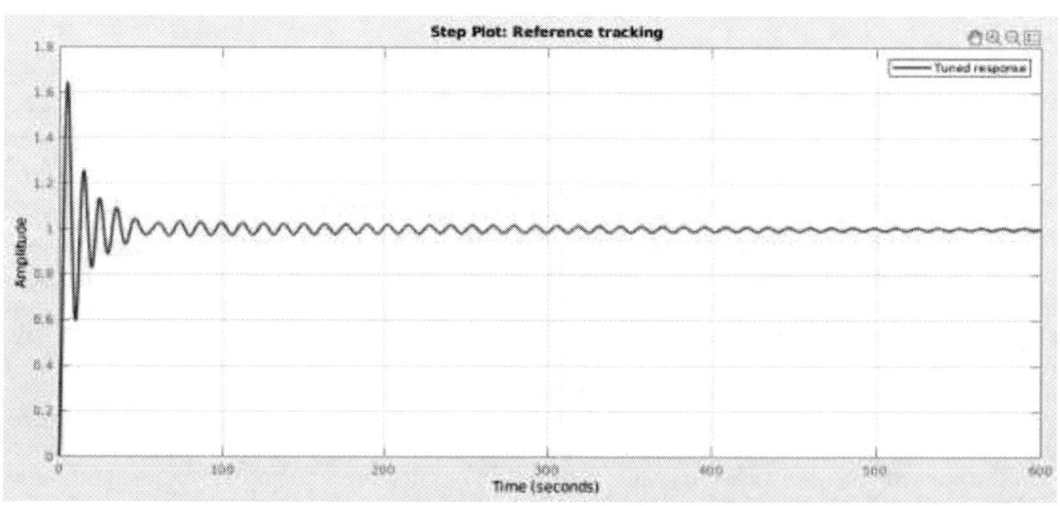

Figure 4. Initial response

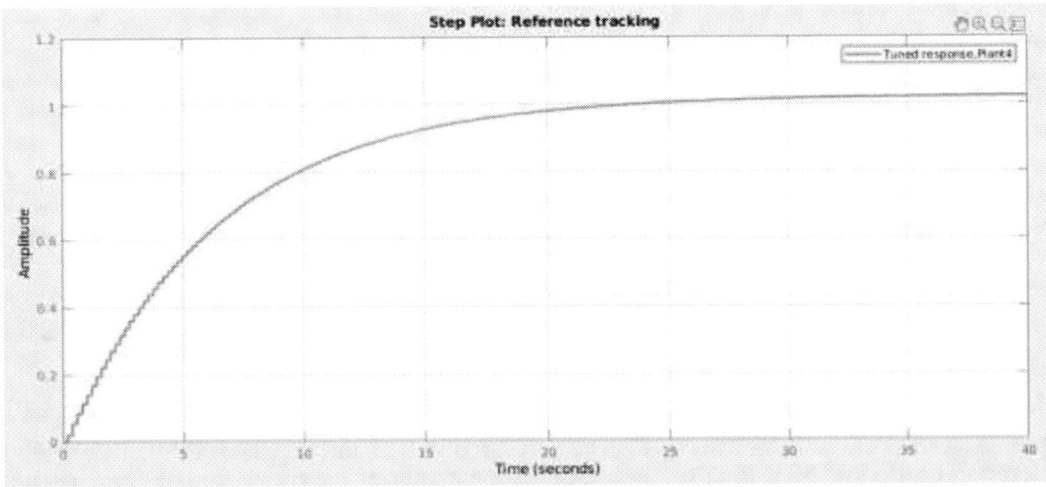

Figure 5. Tuned response

The Gazebo simulation publishes the current altitude of the hexacopter at small and discrete timesteps of around 1 milliseconds. The simulation was performed to obtain the initial response, as shown in Figure 4 is obtained. The hexacopter model was made to reach a desired setpoint height of 1 meter with random $K p$, Ki, Kd values as the PID constants. Then, the system was made to linearize around the desired state by using the error response. The MATLAB's PID tuner, proportional, integral, and derivative gains are adjusted until a smooth transient response, as shown in Figure 5 is obtained.

4. Navigation with Computer Vision

Computer vision plays an important role in detection and alignment to the gate. The block diagram as shown in Figure 6, summarizes and provides an overview of the process described below. So, the flow of computer vision is as follows. First, the image is captured from the camera. Then, CLAHE (Contrast Limited Adaptive histogram equalization) is applied to the image to improve the visibility and ensure that there is no glare or disturbances in the image. CLAHE is applied to tiles(smaller regions of the image) and combined using Bilinear Interpolation. Since CLAHE needs to be applied to the color image, the image was separated into three channels and for each channel in HSV, CLAHE was applied and then the channels were recombined. Next, the HSV mask is obtained by the following method. First, to detect the gate by it's yellow color frame, the HSV values were tuned with trackbars with a python a script to create a mask. After obtaining the values, the HSV mask was applied such that now, only the yellow color gets detected. Then, the contours are detected from this image with canny edge detection, each of the contours is looped by checking the solidity, dimension, and aspect ratio. If the value matches the requirement then the distance to the gate from the camera is estimated using triangular similarity. The rectangle with the maximum area and the minimum distance is chosen and the trajectory is generated.

To estimate the distance to the gate, the triangle similarity method has been used. This method works as follows, first the perceived focal length of the camera is calibrated and the width of the gate is known. So, by applying the variable the length in pixels of the gate, obtained from the processed image, to the Formula 1, the actual distance d to the gate is obtained. Where W is known width of the gate, f is the focal length of the camera lens, and p is the width in pixel, by using these values, the distance can effectively predict.

$$d = \frac{W f}{p} \quad (1)$$

IOP Conf. Series: Materials Science and Engineering 1012 (2021) 012025 doi:10.1088/1757-899X/1012/1/012025

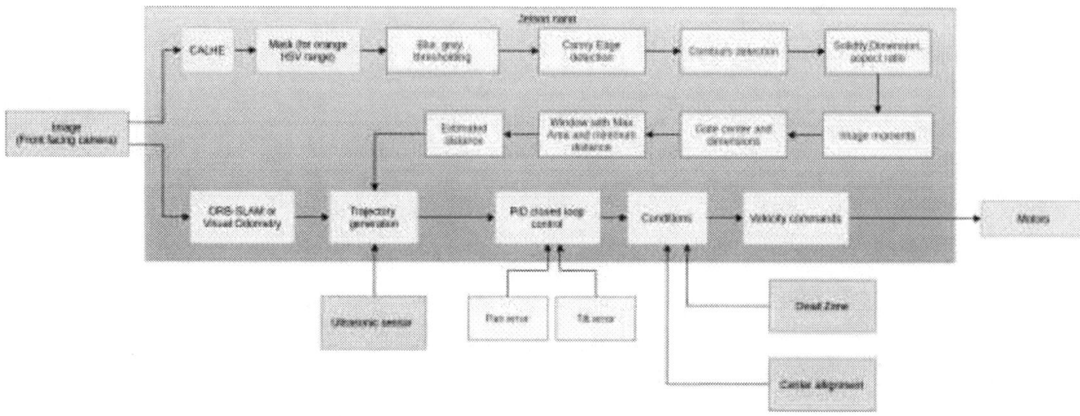

Figure 6. Block diagram showing the vision system pipeline

The hexacopter aligns itself using the PID control, it also uses the previous error values to find both pan error and tilt error. As it is carrying a payload, the center of the frame for the drone should be a little high so this is done by setting appropriate values to the dead zone if the drone finally gets aligned to the center then it passes the gate.

4.1. Localization of the camera frame

As the implementation requires high-level understanding from digital images, videos for ensuring that the robot traverses in the right direction. So, practically, the robot has to capture the entire essence of its surroundings in order to choose a path that seems less likely to have any obstacles and there are a lot more factors that have to be analyzed. This puts forward a necessity to implement novel computer vision algorithms. So, in order to have these tasks performed, ORB feature detector and descriptor is one of the tools that marks the basis for the rest of the perception stack.

For localization of the drone with respect to the environment, ORBSLAM2 was used. This method is able to localize the camera by identifying ORB features in the image, as shown in Figure 7. Oriented FAST and Rotated BRIEF(ORB) was developed by Ethan Rublee, Vincent Rabaud, Kurt Konolige, and Gary R. Bradski at OpenCV labs, as an initiative to use open-source algorithm for feature extraction unlike SIFT and SURF which were patented. Incorporating FAST keypoint detector and BRIEF descriptor makes the ORB feature detector attractive, both of these together represent an object.

4.1.1 FAST Keypoint Detector The implementation of the drone requires real-time video processing for executing corner detectors. FAST stands for Features from Accelerated Segment Test, and it is basically a method for corner detection. Moreover, features aren't something that can be given a definition as it is something that varies with applications. They are typically an interesting part of an image and to put it simply, they are necessary fragments of information (image) that are relevant for a particular application. Features in images can assume any form, be it edges, corners or regions of interests. FAST features, however, do not have an orientation component or multiscale characteristics. Given a pixel P in an image, the brightness of P is quickly compared to the neighboring 16 pixels that are around P in a small circle. The circle's pixels are then sorted into three classes (lighter than p, darker than P, or P-like). If more than 8 pixels are darker or brighter than p, the keypoint will be selected. So keypoints found quickly give us information on the location of an image's edges. A multiscale image pyramid is used in the ORB algorithm. An image pyramid is a multi-scale representation of a single image consisting of arrangements of images at different resolutions, all of which are versions of the image. The downsampled version of the image is included in each pyramid level as compared to the previous level. Once a pyramid has been created by the orb, the FAST algorithm is used to identify keypoints in the picture. By detecting keypoints at different resolutions of the image, ORB is partly scale invariant in this manner. After locating keypoints, ORB assigns orientation to each keypoint like left or right, depending on the direction the pixel intensity level changes around that keypoint. Intensity centroid is used by ORB which calculates the changes in intensity levels. The intensity centroid adopts that a corner's intensity as offset from its center, so orientation is assigned with this vector. So, for our application, features would be something that would help in object recognition, motion detection, video tracking, and other purposes. So, FAST is used this to extract such feature points which would be later used to track and map objects.

4.1.2. BRIEF Descriptor BRIEF, which stands for Binary Robust Independent Elementary Features, is a descriptor that gives out feature descriptions or feature vectors. Ideally, it encodes the feature points in such a way that they are invariant under any image transformation, such as shear, rotation and scaling. It is also robust to different lighting condition to an extent. As the drone moves in its surrounding with varying lighting conditions, the camera may have a slightly different exposure levels at each frame, it has to be ensured that the feature points are kept tracked. This adds the necessity of including image descriptors that suits well for such situations. BRIEF takes all the keypoints found by the FAST algorithm and transforms them into a vector of binary features so that they can represent an object together. Binary vector features are also known as a binary feature descriptor, a feature vector containing only 1 and 0. In short, a function vector that is a string of 128 to 512 bits describes each keypoint. To make the descriptor from being invariant to high frequency noise, BRIEF starts by smoothing the image using a Gaussian kernel. In a defined neighborhood around that keypoint, then BRIEF selects a random pair of pixels. A patch is known as the defined neighborhood around pixels, which is a square with a certain width and height of pixels. The first pixel from the random pair is drawn with a standard deviation or spread of sigma from a Gaussian distribution centered around the keypoint. The second pixel from the random pair is drawn with a standard deviation or spread of sigma by from a Gaussian distribution centered around the first pixel. Now, if the first pixel is brighter than the second pixel, the value 1 is assigned, else 0 is assigned to the corresponding bit. For a 128-bit vector, this method is repeated for a keypoint 128 times. For each keypoint in an image, BRIEF develops a vector like this. However, BRIEF is not invariant to rotation, so rBRIEF(Rotation-aware BRIEF) is used by the ORB algorithm. Without losing out on the speed aspect of BRIEF, ORB attempts to add this feature. Similarly, BRIEF requires far less storage capacity with a speed that is well above what the conventional feature descriptors can offer. ORB does is a sort of fusing FAST detector and BRIEF descriptor. Though both of them have several advantages, there are few drawbacks, and ORB fuses them in a way that would overcome the major drawbacks. For example, BRIEF has been observed to perform poorly with rotation, and ORB overcomes the drawback by steering BRIEF accordingly with the orientation of feature points. Hence the name rotated BRIEF. This is how ORB is able to provide faster and more efficient performance than other algorithms such as SIFT (Scale-Invariant Feature Transform), SURF (Speeded-Up Robust Features). Hence, in effect, ORB provides: a faster and accurate oriented component to FAST and a computation that is more efficient by orienting BRIEF features. Thus the ORB feature extraction has proven to be advantageous for our implementation.

4.1.3. SLAM SLAM, also known as Simultaneous Localization and Mapping, is a computational problem of constructing a map of an environment that usually is not known while simultaneously keeping track of the location of an agent within that environment. Now a visual SLAM based on the feature points is implemented. There is a need of a constructing a map of the environment that is initially unknown to the robot, wherever the robot is supposed to navigate, and that's how the robot will be guided along a path without having to run into any obstructions. So, SLAM is something that totally fits into the requirements.

Even though a camera many contain many number of lenses placed before the CMOS sensor, these cameras can be reduced to a simple pin hole model. However, before the output image from the camera can be used, the distortions introduced by the lens must first be corrected. This can be done by calibrating the camera with a checkered board pattern. The result of the calibration with checkered board pattern gives the intrinsic camera parameters. The intrinsic parameters comprises of the focal length in x and y directions of the image frame and the principal point(x,y), which is the center of the image denoted in pixel space. There is also a shear coefficient s, which denotes the amount of shear present in the image pushed out by the camera. The extrinsic matrix of the camera is used to denote the transformation of the camera with respect to the world coordinate frame. The extrinsic matrix, comprising of rotation and translation, is computed by matching ORB features that is extracted in each frame at subsequent time instances.

The features of bundle adjust algorithm can briefly explained as, refining both the camera poses and the point cloud with the help of each other. Two types of Bundle Adjust is performed in ORBSLAM2: full bundle adjust and local bundle adjust. When a set of images is rendered, describing a set of 3D points from several viewpoints, bundle adjust is used for simultaneous refinement of 3D coordinates depicting the scene geometry, the thresholds of relative motion, including the optical parameters, intrinsic and extrinsic, of the cameras installed to procure the images. Bundle adjust has a very essential role to be played in feature-based monocular SLAM. In several new era SLAM pipelines, bundle adjustment is known to perform to have a proper estimate of the 6-DOF camera trajectory as well as 3D map (sparse 3D point cloud) directing from the input feature tracks. In the ORBSLAM2 pipeline, The local bundle adjust is performed after it receives a new keyframe. After loop closing is detected, full bundle adjust is performed to correct drift in translation of the camera's trajectory and correct the

Figure 7. ORBLSAM2 feature detection

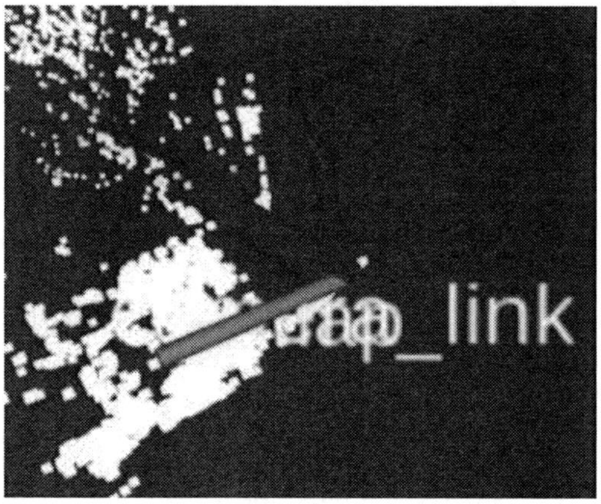

Figure 8. Camera frame localization

location of world points in 3D which were reconstructed from the acquired images during traveling in the loop. Considering all the aspects mentioned, two basic weaknesses circumvents SLAM pipeline based on bundle adjustment. Firstly, the requirement of carefully initializing bundle adjustment depicts that all variables, basically the map, must have an estimation as precise as possible, as well as maintained over time, that makes the overall algorithm very complex. Secondly, as estimating the 3D structure (which needs sufficient baseline) is considered to be inherent in bundle adjustment. The visual SLAM algorithm will face challenges during the time of slow motion or pure rotational motion.

Bundle adjustment gets down to the fact of minimizing the projection error between the image positions of analyzed and predicted image points. The error is computed as the sum of squares of a greater number of nonlinear, real-valued functions. Hence, the minimization is attained involving nonlinear least-squares algorithms and approaches. For error calculation and correction in these three types of bundle adjust, Levenberg-Marquadt has shown to be one of the most efficient due to its ease of incorporation and its utilization of an effective damping strategy that renders the capability to converge to the solution in very short period of time, from a wide range of initial prompts. Through iterative linearization of the function to be reduced in the neighborhood of the current estimate, the Levenberg–Marquardt algorithm uses the concepts of linear systems, known as the normal equations. These all can be exploited to produce high computational benefits by installing a sparse variant of the Levenberg–Marquardt algorithm which takes an edge of the normal equations' zeros arrangement, while avoiding storing and operating on the zero-elements.

Hence, with minor translations of the camera, the features are compared with the successive frames to estimate the rotation and translations of the camera in pixel space. The localization of the camera frame with respect to the world is shown in Figure 8. To convert from pixel space to real-world coordinates, a scale factor is multiplied with the translation vector. This scale factor is obtained during the initialization of the ORBSLAM2 algorithm.

5. Results

The hexacopter is tested for takeoff and landing with the payload attached to it, in the simulation world as shown in Figure 9. Initially, the hexacopter is placed at origin and touching the ground. After the simulation starts, the thrust commands are sent form Simulink to the hexacopter's motors in the Gazebo simulation. The response plot of the hexacopter's altitude at each time step attained by the is obtained. The response is plotted along with the desired setpoint altitude is shown in Figure 10. The yellow coloured lines represent the setpoint given by us, and the blue coloured plot is the response of the hexacopter. The hexacopter was able to rise from ground to an altitude of 1 meter in about 1.2 seconds in a smooth transient behavior. The hovering was smooth without any oscillatory behavior. The landing also exhibited the same characteristics as the takeoff.

The image processing part was tested for a single gate detection as shown in Figure 11, the model was tested using single gate, adjusting it's position by moving left and right, which worked fine. And once it is aligned, it moved forward after calculating the distance. Since the single gate detection

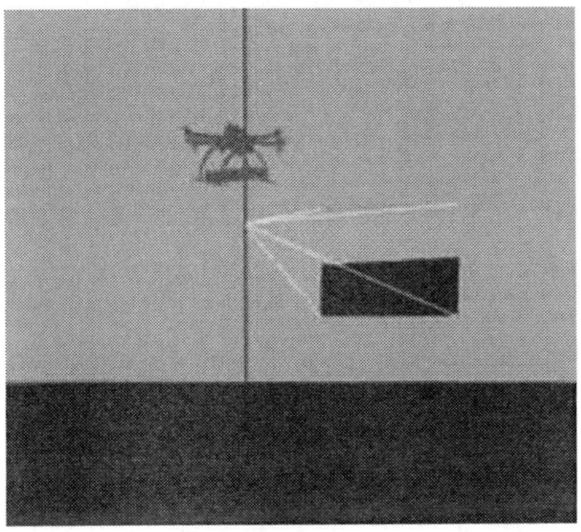

Figure 9. Gazebo simulation world

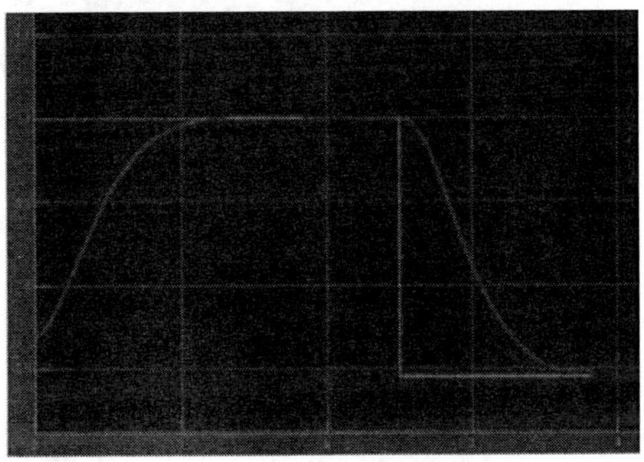

Figure 10. Takeoff and Landing response plot

approach was working fine with single gate, testing was done with 15 gates to obtain the desired outcome. Now, the constraint is to detect multiple gates in a single camera frame. First, the camera calculates the area of each of the gates in its field of view, it then chooses the gate closest to it, the gate with the maximum area, and aligns itself to the center of that gate and passes through it. The image in the first quadrant is the original input image, the second quadrant depicts the HSV mask for detecting the yellow color of the gate, the third quadrant shows the edge detection and the fourth quadrant shows the alignment direction towards the center. Thus as shown in Figure 12, it is evident that the multi-rotor tries to align itself towards the center(by moving left as shown in the arrow mark) of the gate and passes through it. The multi-rotor passes through 15 gates covering a total distance of 15 meters set on a straight aisle. The average speed was found to be 0.7 m/s.

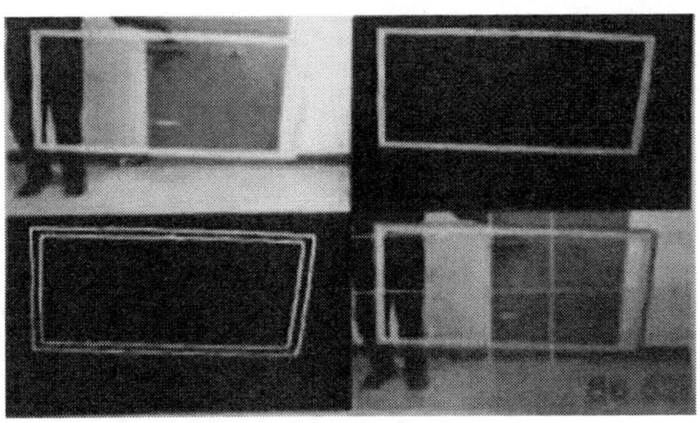

Figure 11. Single gate

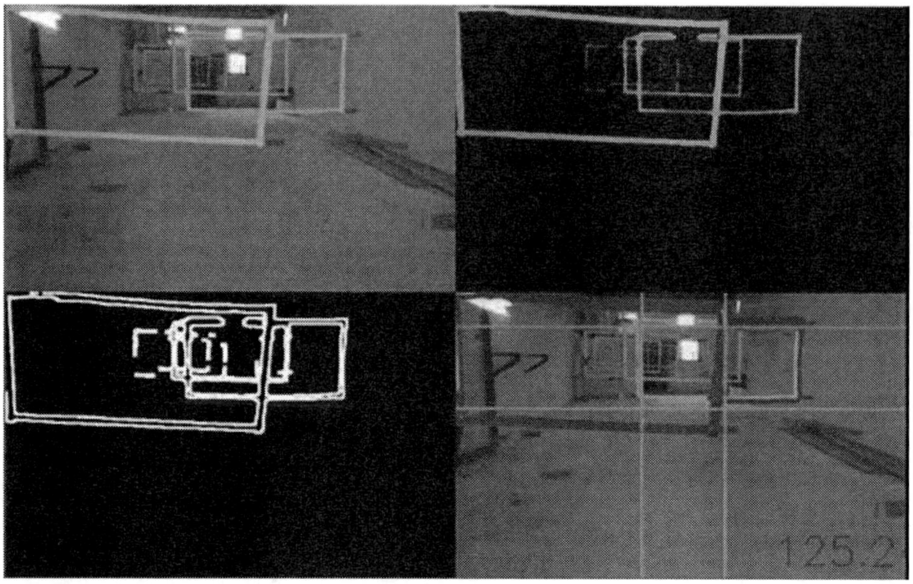

Figure 12. i)Top left: raw input image. ii)Top right: HSV mask. iii)Bottom left: border extraction. iv)Bottom right: arrow indicates the direction of motion to align the multi-rotor to the center of the gate.

6. Conclusion

The physical design process of a hexacopter with a CAD software was explored and structural and stability analysis was performed. The 3D model was then used to perform simulation to validate the design. Moreover, the implementation and verification of computer vision algorithms to locate the waypoints in the form of gates was done and the monocular visual SLAM to localize drone with respect to the environment was also performed.

References

[1] M. Reinecke and T. Prinsloo. "The influence of drone monitoring on crop health and harvest size". In: *2017 1st International Conference on Next Generation Computing Applications (NextComp)*. 2017, pp. 5–10.

[2] J. H. Ong, A. Sanchez, and J. Williams. "Multi-UAV System for Inventory Automation". In: *2007 1st Annual RFID Eurasia*. 2007, pp. 1–6.

[3] B. Han et al. "A Review of Control Algorithms for Quadrotor". In: *2018 IEEE International Conference on Information and Automation (ICIA)*. 2018, pp. 951–956.

[4] B. Y. Suprapto et al. "Design and development of heavy-lift hexacopter for heavy payload". In: *2017 International Seminar on Application for Technology of Information and Communication (iSemantic)*. 2017, pp. 242–247.

[5] Yu Miao et al. "Application of the CLAHE Algorithm Based on Optimized Bilinear Interpolation in Near Infrared Vein Image Enhancement". In: *Proceedings of the 2nd International Conference on Computer Science and Application Engineering*. CSAE '18. Hohhot, China: Association for Computing Machinery, 2018. ISBN: 9781450365123. DOI:10.1145/3207677.3277957. URL:https://doi.org/10.1145/3207677.3277957.

[6] Shuo Li et al. "Autonomous drone race: A computationally efficient vision-based navigation and control strategy". In: *CoRR* abs/1809.05958 (2018). arXiv:1809.05958. URL:http://arxiv.org/abs/1809.05958.

[7] Adrian Rosebrock. *Find distance from camera to object using Python and OpenCV*. URL:https://www. pyimagesearch.com/2015/01/19/find- distance- camera- objectmarker- using- python- opencv/(visited on 09/12/2020).

[8] Z. Sadreddini, T. Çavdar, and H. B. Jond. "A distance measurement method using single camera for indoor environments". In: *2016 39th International Conference on Telecommunications and*

Signal Processing (TSP). 2016, pp. 462–465.
[9] Raul Mur-Artal and Juan D Tardós. "ORB-SLAM2: An open-source slam system for monocular, stereo, and RGB-D cameras". In: *IEEE Transactions on Robotics* 33.5 (2017), pp. 1255–1262.
[10] Ebrahim Karami, Siva Prasad, and Mohamed Shehata. *Image Matching Using SIFT, SURF, BRIEF and ORB: Performance Comparison for Distorted Images*. 2017. arXiv:1710.02726 [cs.CV].
[11] P. Burdziakowski. "Towards Precise Visual Navigation and Direct Georeferencing for MAV Using ORB- SLAM2". In: *2017 Baltic Geodetic Congress (BGC Geomatics)*. 2017, pp. 394–398.
[12] Victor Becerra. "Autonomous Control of Unmanned Aerial Vehicles". In: *Electronics* 8 (Apr. 2019), p. 452. DOI:10.3390/electronics8040452.
[13] Ram Padhy et al. "Deep Neural Network for Autonomous UAV Navigation in Indoor Corridor Environments". In: *Procedia Computer Science* 133 (Jan. 2018), pp. 643–650. DOI:10.1016/j.procs.2018.07.099.

Acknowledgment
The authors would like to express their gratitude to their faculties for guidance on this paper and parents, family, and friends for their moral support throughout this project.

Detection of an Explosive Material In Landmine, Aqueous And Air Medium Through Sensor Operated Unmanned Guided Vehicle

Jayashri V.Chopade[1*], Kanchan D.Ganvir[2], Dr. Harish U. Tiwari[3]
[1]Mechanical Engineering Department, Pimpri Chinchwad College of Engineering and Research, Ravet, Pune, India.
[2]Mechanical Engineering Department, Priyadarshini Bhagwati College of Engineering, Nagpur, India.
[3]Pimpri Chinchwad College of Engineering and Research, Ravet, Pune, India.

E-mail: chopadejv91@gmail.com

Abstract: Various explosives and harmful materials are most dangerous for human beings; detection of such materials is growing concern with curiosity as well. This paper identifies the broad array of possible mediums of explosions. As it is very difficult to find explosive material which is hidden, saving the life of an individual or whole from chemical explosion is need for the safety of our nation. The study focuses on detection of explosives which works on landmine, water and air. Searching for an explosive material manually is difficult task for any person; hence an attempt has been made to fulfil the necessity of need to design such equipment which will help researcher and engineers to find the explosive material to save lives. This work intends to design the unmanned vehicle i.e. drone which will work under ground, water and in an air with remote control. The sensors are used to find the hidden explosion in various medium. To detect non-contact type metal object, proximity sensor is used. A clarification of the crucial investigation for each approach is joined by its overall appropriateness to the pertinent the relevant scientific findings.

1. Introduction

Terrorism is of the major concern in the world, because of the terrorist one cannot live their life with full of freedom and also in the communal zones or places there is fear to have any explosive in that areas and the bomb squad cannot reach to the explosive on time [2]. Inductive proximity sensors issued to find out metallic objects which are not contacted physically. Active principle is based on a coil and oscillator that generates an electromagnetic field in the nearby areas of the detecting surface. The existence of a metallic thing (actuator) in the functioning area causes a stifling of fluctuation generosity. The increase or tumble of such oscillation is predictable by a threshold circuit which vagaries the output of the sensor. The modules are selected from the variety of brushless motors, battery technologies and cell arrangements, and static area propellers appropriate to use in a four rotor suspended vehicle.
A quad copter is a floating vehicle where four rotors are used for lift, steering, and steadiness. Like other in-flight vehicles, a quad copter can attain upright flight in more steady form. There is no effect

by the torque concerns that a plane practices on focal rotor. Likewise because of quad copter's persistent design, it is calmer to build and sustain.
In many areas like army, military engineers, scientists, used emerging Quad copters to apprehend various practical areas. They are used for scouting and gathering information [6]. It is having kind of finding for persistence fatalities in a catastrophe zone to check the condition of power lines. Quad copters in making today can hold light cargoes, such as foodstuff and medicinal supplies, and deliver where ordinary planes cannot reach easily.

2. Engineering Findings

2.1 Failure in handling Handheld Metal Detector:

There is a failure of handheld detector as it is having insufficient power supply, loose of connection of complicated circuit may happen and dicey mission to detect bomb by disposal person.

Scope of the Finding

- Modifications in design can be done so as to eradicate the intricate nature of the system.
- Use of higher range sensor and component for the sake of having inclusive range of detection capacity.
- Use of Camera and GPS module to find the location of bomb decorously.

2.2 Detection of Hidden explosive material

Figure 1 Land Mines detection

Finding of explosive material under land mine is possible due to AR drone [6]. Unknowingly human lose their life. AR drone is useful to capture the images from particular height and detect availability of explosive material.

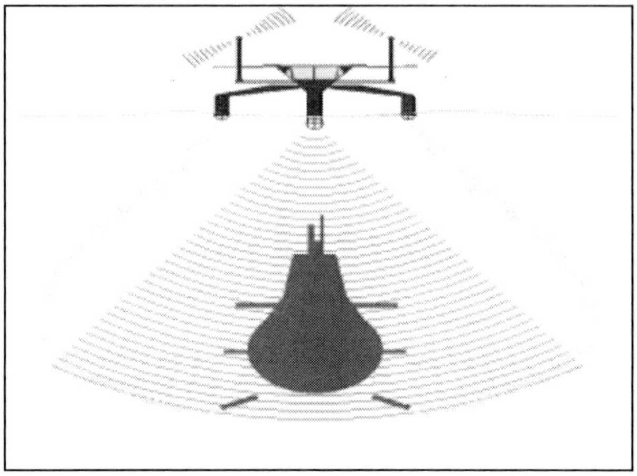

Figure 2 Under Water Mines detection

3. Objectives of UAV:

i) The most important objective is to detect explosives with the help of explosive detect sensors and to achieve versatility of drone to work in different mediums such as air, under water and on land.
ii) It will be helpful to the police force, Navy and Defence and to raise the power of the nation to avoid terrorist presence thereby giving full liberty to people to live their life without fear.
iii) Cease to human corporeal from terrorist attacks.
iv) Find the exact location of explosive which are unseen and not detected certainly

4. Literature Review

To detect hidden explosive nanosensors are used with nanochemical sensors for achieving sensitivity and selectivity. These sensors are versatile in nature and they are based on receptor [2].

Finding of explosive expedients (IEDs) on the side of road is skirmish against terrorism. Localization and magnetic moment valuation of a static ferromagnetic IED by a PMAS. The PMPS combines more than 21 arrays of 3 axis fluxgate array. Nonlinear equation set is employed for framing the difficult by assuming magnetic dipole model of IED. The data is detected by sensor and these data that is localization and detection of IED is calculated by appropriate process magnetic dipole model and nonlinear optimization algorithm. The result obtained by numerical simulation displays the long baseline is useful to enlarge recognition area and the more revealing range is of 12m for group of 6 fluxgate sensors [3].

Applied analysis is done by AR drone which is used as an integrated tool. This tool is placid by AR drone quad rotor and base station for spotting the operation. After collection of data from onboard camera, algorithm is done for landmine detection. The system is collected the data averagely from landmine for outdoor state in three different constraints: flight speed, visibility of the object and flight altitude. Base on result it is decided that whether the explosive material are present or not on land [6].

The research is done on design of Quad copter. The modification is done on simple Quad copter. The frame is essential part because all other parts are mounted on that. To check whether the frame can sustain the load of vehicle or not, static analysis is done on that. A small deformation was occurred on plate but that is up to tolerable limit and design is safe for use. [7]

The rotor is used in searching hidden explosive and rescue operation. The rotor is used in unmanned aerial vehicle (UAV). Here Aerodynamic analysis is done on rotor. In rotor vehicle, fixed pitch propeller is used and to model the aerodynamics blade element and vortex theory was investigated combined. It is clear that the propeller can be suitably utilized for quad rotor UAV[11]

Study is focuses on robot working under water for explosion detection. Here mainly three types of superior wireless control systems are used for robot to operate under water. ZigBee communication is the first system, second system is light approach system and last one is magnetic field. ZigBee communication system is used to transmit and received the data. Data is provided to the ARM controller and depending upon nature of command the robot get operated [10].

This paper discusses experimental and numerical vibration analysis of a multirotor frame and application of the modal analysis technique on a multicopter chassis. Analysis of vibration is done on six arm multirotor. It focuses on experimental model analysis of structural components of hex copter frame. For analysis hit impact and shaker excitation is used. Very accurate and reliable result shows the balance between experimental and numerical work [19].

5. Architecture of the proposed work

The research is based on drone, working in three medium in air, under water and on land. The drones are available which are working in all three medium separately but combination is not available. The intention of work is to design single drone which can work in three medium. For this work initially literature review is done and what methodology they have adopted is understood. From study it is decided to work on finding hidden explosive from single drone in all medium. Experimentation is done by using different components and sensors like PID controller, wireless communication, Single controller etc. As a result of this the components are properly mounted on frame and after analysis it is found that the design of drone is safe and it successfully worked in all three medium to find explosive material.

6. Blades:

6.1 Directions of Rotation:

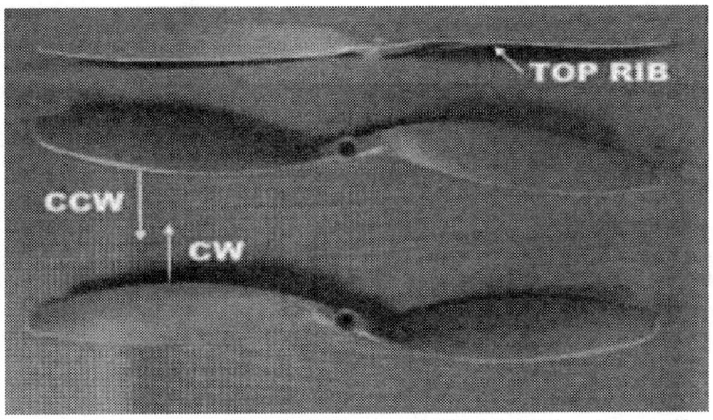

Figure 3 Counter Clock Wise and Clock Wise Propellers.

Number of Blades:

- As mention in figure 3, for mini quad copter, 3-blade (or tri-blade) propellers are normally used in racing as well as for free-style flying. The tri blades have more grips in the air.
- If more blades are used, the surface area added more and hence it creates larger thrust in expense of and more drag better current draw.

Table 1 Selection of components

Frame Size	Prop Size	Motor Size	KV
150mm or smaller	3″ or smaller	1105 -1306 or smaller	3000KV or higher
180mm	4″	1806	2600KV– 3000KV
210mm	5″	2204-2208, 2306	2300KV-2600KV
250mm	6″	2204-2208, 2306	2000KV-2300KV
350mm	7″	2208	1600KV
450mm	8″, 9″, 10″ or larger	2212 or larger	1000KV or lower

6.2 H – Type Frame:

For durability the aluminum frame is used with carbon fiber plate. To support motors, the frame is designed with many different configurations so that components can be mounted on that. The arms are prepared of consistent 5/8 inch aluminum bars which are hallow and square in size and customs mutual nuts and bolts to clamp the frame together.

- Length of Rod – 18cm
- Frame Type – H- Type Quadcopter
- Weight – 10 gm

6.3 Metal Touch Sensor:

There are three important components on circuit board of sensor. One module deals with the surrounding area and analog signals are sending to the second unit and amplifier. The amplifier magnifies the signals as per resistant value of potentiometer; signals are sending to analog output of component. Third module of sensor which acts as a comparator is used to switch on the digital output and LED.

7. Experimental Validation

7.1 UAV Quad copter

UAV Quad copter is having four rotating rotors which are used for movement and lift purpose. For stabilization it uses electronic control system and electronic sensors. [11]

7.1.1 Flight Control

Quad copter contains of four motors uniformly dispersed on frame of quad copter as shown in figure 4. The circular shape signifies the spinning rotors of quad copter and the arrows denote the direction of rotation [19]. First and third number motor rotates in a clockwise route by consuming pusher rotors. Second and fourth rotor rotates in an anti-clockwise direction with the help of puller rotors. All motor generates a thrust and torque around the centre of quad copter. Because of the reverse spinning orders of the motors, the clear torque about the centre of quad copter is preferably nil, creating nil angular acceleration. It removes the need of yaw stabilization.

IOP Conf. Series: Materials Science and Engineering 1012 (2021) 012026 doi:10.1088/1757-899X/1012/1/012026

An upright force is form by growing the speed of all motors by similar quantity of throttle. As the upright forces overwhelmed the gravitational forces of earth, the quad copter creates to upsurge in altitude. Figure 4 shows the vertical movement of the quad copter. As above, the rounds epitomize the spinning rotors, loftier arrows symbolize the path the rotors are spinning, and the black arrows epitomize the forces caused by the spinning rotors.

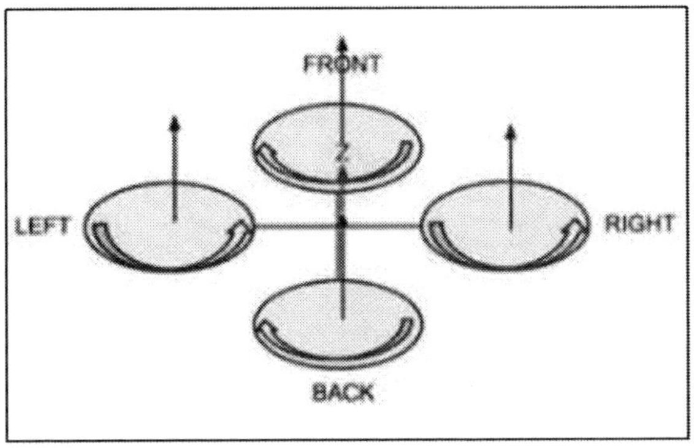

Figure 4 Quad copter: Vertical thrust movement.

The front or rare motor speed is increased or decreased which is given by pitch, result of this the Quad copter is turn along x-axis. Due to left and right motors the overall thrust is same as hovering, so only pitch angle acceleration is changed. In figure 4 example of pitch movement of quad copter is given. If the speed of front motor goes down the forces produced by respective rotor are minimum than the forces produced by back rotor. These forces are indicated by blue arrows. The forces indicated by blue arrows are responsible to tip quad copter in forward direction and this movement indicated by red arrow.

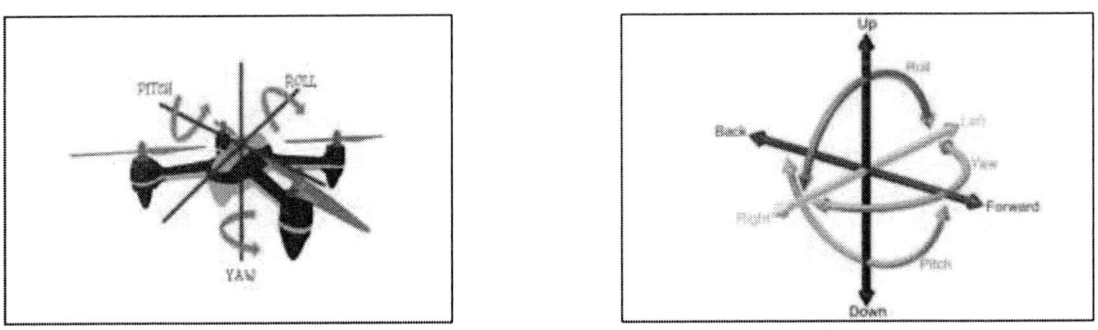

Figure 5 Gyroscopic motions in quad copter

Due to high or low speed of left rotor and right motor roll is provided, this cause the quad copter turns along y axis. The back and front motors balanced the overall upright thrust as a result of this only roll angle get changed. In Figure 5 it is seen that if the speed of right motor decreases, the corresponding rotor creates less force as compared to left rotor. The blue arrows denoted the force created by right rotor. Because of this force, the tip of quad copter moves in right side which is denoted by red colour.

IOP Conf. Series: Materials Science and Engineering 1012 (2021) 012026 doi:10.1088/1757-899X/1012/1/012026

8. Design

- Length of rod = 18cm,
- Length of landing gears = 5.4cm

8.1 Frame

- Specifications:-

Control board Size = 3.4*3.4cm
Thickness = 0.5cm

- Arm:-
- Size = 8.4*0.5cm
- Weight = 0.25 gm each
- Thickness = 0.2
- Outside corner hole = 1.5cm
- Inside hole = 0.4cm

8.2 Electronic Speed Controller

According to battery Specification =2200 mAh
The C rating is 30C
30C = 30 * 4 = 120 A
Safe current flow is 120 A.

8.3 frame size

Propeller size = n×m
Where, n = Diameter of propeller
m = Pitch of propeller
Propeller size = 9×4.5
Radius (R) = 4.5
X = 2.R
X = 2×4.5 = 9 cm

9. EXPERIMENTATION

9.1 Fundamentals of Aerodynamics:

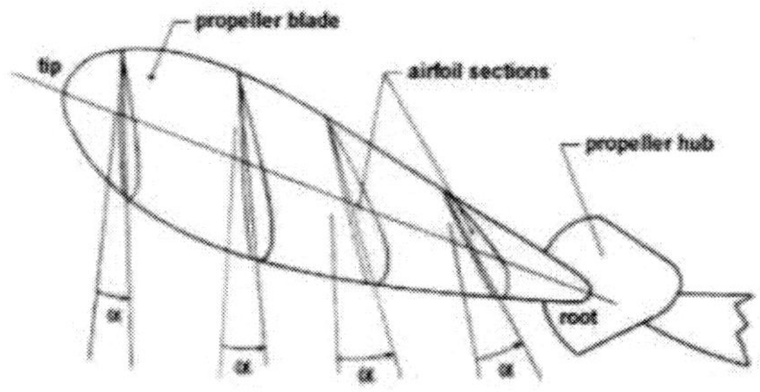

Figure 6 Fixed pitch propeller geometry

If there are variation of β and c throughout the length of the radius, the angle of attack α is also varied. [11]

- Pitch angle β=blade angle
- Fixed pitch
- Propeller as 'air screw
- $P = 2\pi r \tan\beta$

- P/ D= πx tanβ
- X = r/ R

9.2 Efficiency of the propeller and an Advance Ratio

The efficiency is represented as,

$$\eta = \frac{V}{nD}\frac{C_T}{C_P}$$

Where, V= free stream fluid velocity=8m/s,

n=rotation per seconds =150rpm,

D=Diameter of Propeller=0.09m,

C_T=Coefficient of thrust=0.06,

C_P=Coefficient of power =0.05.

$$\eta = 71.11\%$$

The velocity ratio is also known as the advance ratio, J

$$J = \frac{V}{n * D} = 0.59$$

9.3 Results and Discussion

Table 2 Result table

Parameter	Frame	Propeller
Analysis	Structural	Structural
Load / Speed Applied	10 N	1480 rpm
$S_{ut}$for ABS	44 MPa	44 MPa
Max equivalent stress	1.85 MPa	1.375 MPa

From above result, as the load increases the Stress is also increases on both frame and propeller.

10. Analysis of Frame:

10.1 Loading Condition FBD:

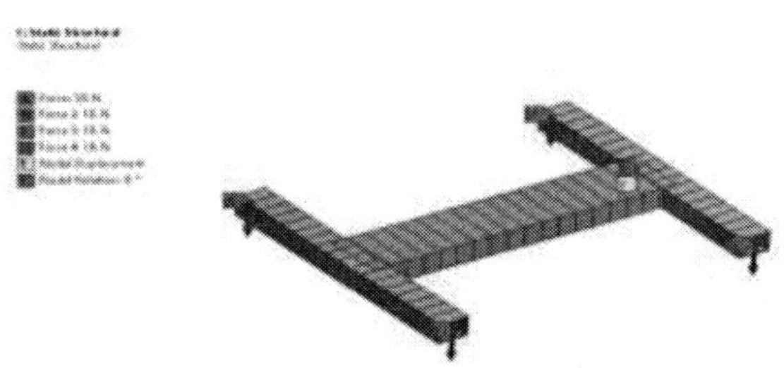

Figure 7 Free Body Diagram of Frame.

10.2 Total Deformation:

Maximum Deformation = 0.26753 mm

Figure 8 Deformation of Frame.

10.3 Maximum Combined Stress:
Maximum Combined Stress = 1.8505 MPa

Figure 9 Maximum combined Stress on Frame

10.4 Minimum Combined Stress:

Minimum Combined Stress = 1.8505 MPa

Figure 10 Minimum combined Stress on Frame.

10.5 Total Bending Moment:
Maximum Bending Moment = 950N-mm

Minimum bending Moment = 0 N-mm

Figure 11 Total Bending Moment on Frame.

10.6 Modal Analysis:

Figure 12 Modal Analysis of Frame.

Frequency due to inertia forces

$$\omega = \sqrt[2]{\frac{K}{M}}$$

Table 3 Modal Analysis.

Sr No.	MODE	FREQUENCY{HZ}
1	1	360.52
2	2	360.53
3	3	445.07
4	4	445.07
5	5	510.29
6	6	510.36

Excitation Frequencies Due to propeller

$$\omega = \frac{2 * \pi * N}{60}$$

$$\omega = \frac{2 * \pi * 1480}{60}$$

$$\omega = 154.98\, rad/sec$$

FREQUENCY

$$fn = \frac{\omega}{2\pi}$$

$$fn = \frac{154.98}{2\pi}$$

$$fn = 24.66\ Hz$$

Since excitation frequency is less than frequency so resonance will not occur hence the system is safe.

10.7 Stress Analysis of Propeller:

Maximum Equivalent Stress: 1.375 N/mm^2

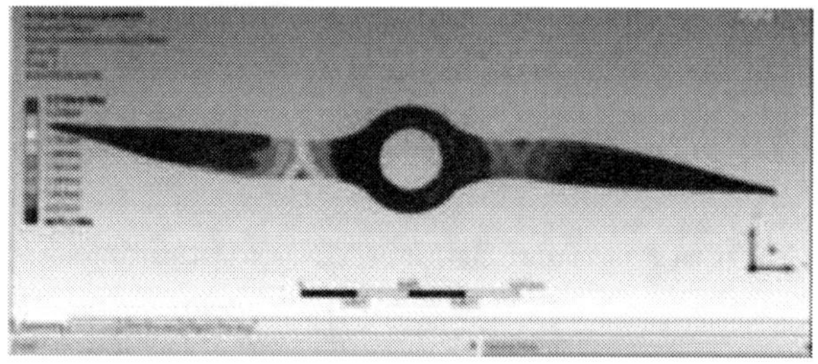

Figure 13 Stress Analysis of Propeller.

11. Conclusion

In the conclusion, the present study demonstrates the design and analysis of the single unmanned vehicle i.e. drone which can work in water, underground and in air with remote control system. Various components are selected in such a way so that it can work in all three medium. All the components are tested, interface and verified. The sensors are used to find non-contact hidden explosion in various medium. Pitch, roll and yaw are carefully tuned with PID controller for proper stabilization. Wireless communication system is used to receive and transfer the data. Single controller is used to fly in air and drive on land in the drone so as to achieve the compactness of system. The frame is made with light material and analysis of frame is done for checking deformation, combined stress and bending. After experimentation it is observed that the system is safely design. The device showcases the potentialities of the drone to detect an explosive material in aqueous, air and water medium.

References

[1] Salimzhan A. Gafurov, Evgeniy V. Klochkov," Autonomous unmanned underwater vehicles development tendencies" ,Samara State Aerospace University, 34, Moskovskoye Shosse, Samara, 443086, Russian Federation.

[2] Larry Senesac and Thomas G. Thundat," Nanosensors for trace explosive detection" , Oak Ridge National Laboratory, Oak Ridge, TN 37831-6123, USA.

[3] Renhao Liu, Hua Wang," Detection and localization of improvised explosive devices based on 3-axis magnetic sensor array system" School of Astronautics, Beijing University of Aeronautics and Astronautics, Beijing 100191, PR China,2010.

[4] Damien Rembelskia, Christelle Bartheta, Céline Frénoisa, Geoffrey Gregisa," Improvement of Explosive Detection with a Fluorescent Sensor Using a Heating Device", aCEA, DAM, Le Ripault,F-37260, Monts, France, EUROSENSORS 2014, the XXVIII edition of the conference series .

[5] Austin Choi-Fitzpatrick, Tautvydas Juskauskas," Up in the Air: Applying the Jacobs Crowd Formula to Drone Imagery", aAssistant Professor, School of Public Policy, Central European University, Budapest

1054, Hungary b MPA student, School of Public Policy, Central European University, Budapest 1054", Hungary,2015.

[6] C. Castiblanco, J. Rodriguez,I. Mondragon,C. Parra, and J. Colorado, "Air Drones for Explosive Landmines Detection", School of Engineering, Pontificia Universidad Javeriana Cr 7 No 40-62, Bogota, Colombia.

[7] Anudeep M, G Diwakar, Ravi Katukam ," Design of A Quad Copter and Fabrication", Prasad V Potluri Siddhartha Institute of Technology, Vijayawada, IJIET.

[8] ParagParihar, PriyanshuBhawsar, PiyushHargod," Design and Development Analysis of Quad copter", COMPUSOFT, An international journal of advanced computer technology, 5 (6), June-2016 (Volume-V, Issue-VI) .

[9] Endrowednes KUANTAMA, Dan CRACIUN, Radu TARCA," QUAD COPTER BODY FRAME MODEL AND ANALYSIS", ANNALS OF THE UNIVERSITY OF ORADEA Fascicle of Management and Technological Engineering ISSUE #1, MAY 2016.

[10] Suraj Namdeo Rathod, Dagade Megha T, Nakhate Mayur V, Lokare Supriya S," Wireless Underwater Metal Detector Robot", International Journal of Innovative Research in Computer and Communication Engineering Vol. 4, Issue 4, April 2016.

[11] Yogianandh Naidoo, RiaanStopforth, Glen Bright," Rotor Aerodynamic Analysis of a Quadrotor for Thrust Critical Applications" 4th Robotics and Mechatronics Conference of South Africa (ROBMECH 2011) 23-25 November 2011, CSIR Pretoria South Africa.

[12] Ali Jameel AL-Mousawi ,Haider. K. AL-Hassani," A survey in wireless sensor network for explosives detection", Computer Science Department, College of Education, University of Al-Mustansirya, Baghdad, Iraq. November 30, 2017.

[13] Moses Bangura, Marco Melega, Roberto Naldi and Robert Mahony," Aerodynamics of Rotor Blades for Quadrotors", Tuesday 5th January, 2016.

[14] NanyapornIntaratep, W. Nathan Alexander, William J. Devenport, Sheryl M. Grace, Amanda Dropkin," Experimental Study of Quad copter Acoustics and Performance at Static Thrust Conditions", 30 May - 1 June, 2016, Lyon, France Aeroacoustics Conferences.

[15] Hairuniza Ahmed Kutty and Parvathy Rajendran," 3D CFD Simulation and Experimental Validation of Small APC Slow Flyer Propeller Blade", School of Aerospace Engineering, UniversitiSains Malaysia, Penang 14300, Malaysia 25 February 2017.

[16] John Steven Monk," a propeller design and analysis capability Evaluation for high altitude application".

[17] UrškaTomšič," Detection of Explosives: Dogs vs. CMOS Capacitive Sensors"

[18] M. Hassanalian, A. Abdelkefi," Classifications, applications, and design challenges of drones: A review", Department of Mechanical and Aerospace Engineering, New Mexico State University, Las Cruces, NM 88003, USA.

[19] J. Verbeke, S. Debruyne," Vibration analysis of a UAV multirotor frame", KU Leuven, Faculty of Engineering Technology Zeedijk 101, B-8400, Oostende, Belgium.

Autonomous Snake Robot with Serpentine type Navigation

M G Bangar[1], H S Nirgude[1*], S P Ghodake[1] and S S Ohol[1]

Department of Mechanical Engineering, College of Engineering Pune, Pune, India

Email: nirgudehs16.mech@coep.ac.in

Abstract. Biological snakes can easily maneuver through complex and undulating terrains which gives the motivation to create a bio mimic robotic snake which can easily slither through narrow spaces and uneven surfaces unlike the wheeled or legged mechanisms that have terrain limitations for its maneuverability. Such robots can be used in rescue operations in earthquake areas where it can easily crawl, surveillance and maintenance of complex or dangerous structures like pipelines in nuclear power plants. In this paper we discuss the development of autonomous snake robot with serpentine type navigation. The paper discusses in detail about the design of the robot, mathematical modelling, and motion study in virtual environment using MATLAB. The paper also discusses the controlling of the robot using Arduino Nano. The aim of the project is to develop a bio mimic snake which can maneuver over any terrains.

1. Introduction

Biological snakes can uniquely perform variety of tasks. Their sleek body and limbless motion allow them to adapt to various types of environments, terrains and climate. The motion of snake is an inspiration to define a new term of locomotion i.e. wheel less or limbless locomotion. A typical snake robot can slither into narrow spaces or uneven surfaces to perform complex tasks like pipe cleaning in nuclear plants or surveillance in earthquake affected areas. Whereas wheeled or legged devices have certain terrain limitations also these devices may not be able to access to certain complex locations. Unlike in a legged mechanism where legs help to push and in wheeled mechanism where wheels aid in motion but in snake robot body motions itself promote locomotion. With stability, terraianbility, good traction and high redundancy snake robots can serve various critical operations like exploration, medicine and inspection.Development of such kind of autonomous snake robot requires study of various mechanisms that can mimic the gait of a biological snake. Design of conceptual model and motion study for the same is required. This paper focuses on the development of a four module autonomous snake robot which can achieve serpentine motion of snake. In serpentine motion snake thrusts its body from side to side, moving forward in a waved motion.

IOP Conf. Series: Materials Science and Engineering 1012 (2021) 012027 doi:10.1088/1757-899X/1012/1/012027

2. Design

The snake robot consists of four modules. Each module is powered using two motors. For the first prototype Dynamixel servo motors are used as they provide a torque of 15.3 kg-cm and provides position feedback so that it makes the control of the overall robot easier. Each module consists of two Dynamixel motors coupled together by 3D printed clamps as shown in Figure 1. The overall snake robot body is connecting these modules to form a chain. The dynamixel servo motors are controlled using Arduino Nano R3 and L298 motor driver. Motion control of the robot is achieved using PS2 remote controller. Due to the simple mechanism and sleek design the weight of the robot body is 1.5kg and dimensionally, the robot is 1m in length and about 50mm in diameter. In this robot we are using total 8 dynamixel servo motors (Figure 2). The purpose of using this many number of actuators is to provide compliance to the robot to achieve various gaits that can mimic snake locomotion.

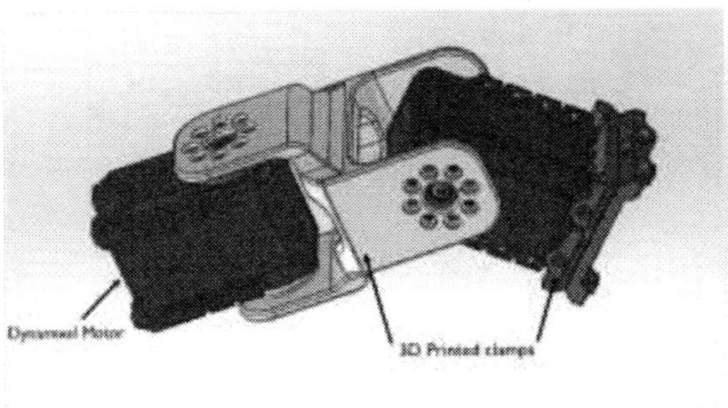

Figure 1: Module of Snake Robot

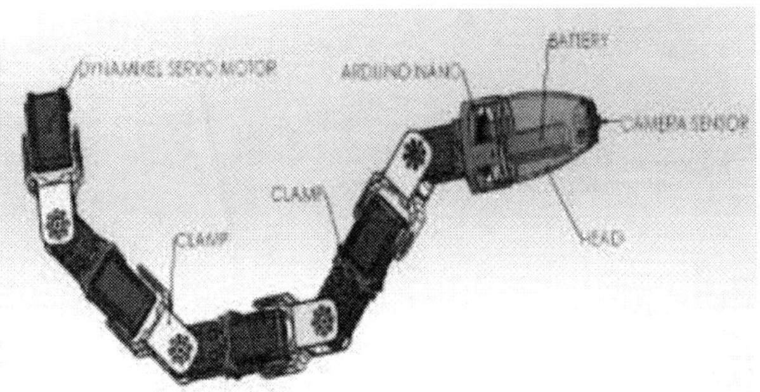

Figure 2: Assembly

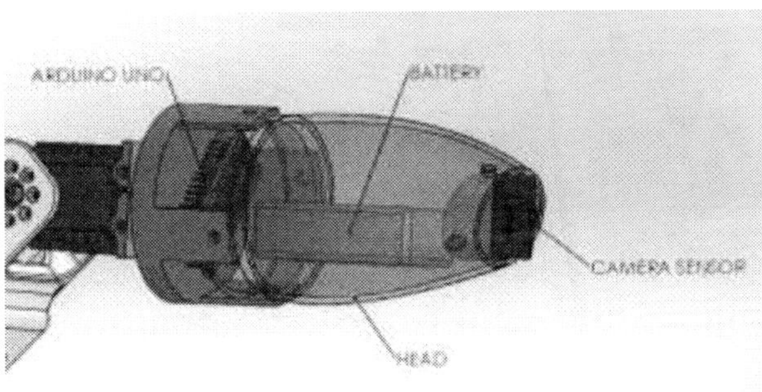

Figure 3: Head

3. Mathematical Modelling [7]

$\O_y = \theta_{y+1} - \theta_y = \alpha \sin(i\beta + \beta/2) + \Upsilon$ ---------(1)

Here-

α(Alpha)= $a^* |\sin(\beta/2)|$ ------------(2)

β (beta)=b/n -------------(3)

Υ(gamma)=-c/n -------------(4)

From (1), the serpentine motion shape is related sine curve equation . In this amplitude is alpha.Phase difference is β(beta). Values of a, b and c, affects the α, β and γ values respectively .

We can find position of each joint with respect to X-Y coordinate system at any instant with the help of of (5), differentiating we get velocities which will help us getting Feedback for Simultaneous Localization and mapping(SLAM)

$x_t = x_0 + \Sigma_{t=1 \text{ to } t=n} 2*l*\cos\theta_t$------(5)

$y_t = y_0 + \Sigma_{t=1 \text{ to } t=n} 2*l*\sin\theta_t$ -------(6)

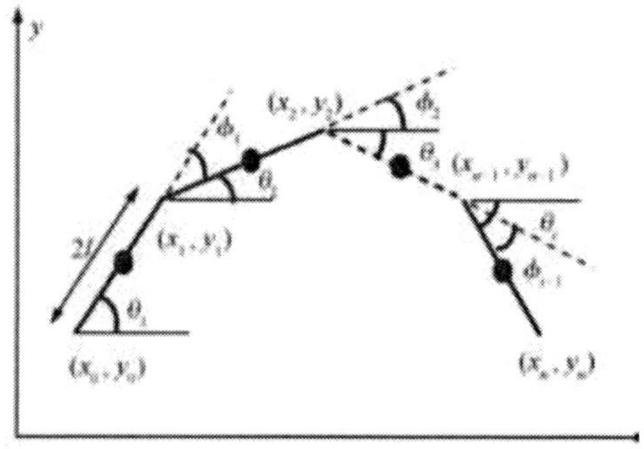

Figure4: Link Angles representation

θt, absolute angle of ith link and (x, y) is the coordinates of tail. With the help of above equation we can find snake joints position in x-y plane and the x and y components of velocities are seen by Differentiating (5) and (6). (Figure 4)

$y_i = y + \Sigma_{t=1 to t=n} 2*l*\theta t*\cos\theta t$ --------------(7)
$x_i = x + \Sigma_{t=1 to t=} n2*l*\theta t*\sin\theta t$ --------------(8)

3.1. Actuator Torque Calculations

$\varnothing(t) = b_o + b_1 t + b_2 t^2 + b_3 t^3$

Where, $\varnothing = \theta_{n+1} - \theta_1$

Solving Differential Equation with Initial conditions

$\varnothing(t_i) = \varnothing_i$=Initial absolute angle Difference

$\varnothing(t_f) = \varnothing_f$= Final absolute angle Difference

$d\varnothing(t_i)/dt = 0$

$d\varnothing(t_i)/dt = 0$

Considering these initial conditions and solving the equations

We get

$b_0 = \varnothing(t_i)$

$b_1 = 0$

$b_2 = 0.1875[\varnothing(t_f) - \varnothing(t_i)]$

$b_3 = 0.03125[\varnothing(t_i) - \varnothing(t_f)]$

IOP Conf. Series: Materials Science and Engineering 1012 (2021) 012027 doi:10.1088/1757-899X/1012/1/012027

Hence, we get Trajectory Planning Equation as

$Ø(t)= Ø(t_i)+0t+0.1875[Ø(t_f)- Ø(t_i)]t^2+0.03125[Ø(t_i)- Ø(t_f)]t^3$ I

Here, $Ø(t)=\theta_{n+1} - \theta_1$

Angular Velocity(W)=d Ø(t)/dt

$=0.375[Ø(t_f)- Ø(t_i)]t+0.09375[Ø(t_i)- Ø(t_f)]t^2$......II

And

Angular Acceleration= d W/dt

$= 0.375[Ø(t_f)- Ø(t_i)]+0.1875[Ø(t_i)- Ø(t_f)]t$..........III

From Mathematical Modelling Equations We Calculated each Increment in θ at each instant/step and at each joint

Table 1: Theta Values

Step	Θ1	Θ2	Θ3	Θ4	Θ5	Θ6	Θ7	Θ8
1	-35	-39	-25	0	25	39	35	13
2	-39	-25	0	25	39	35	13	-35
3	-25	0	25	39	35	13	-39	-25
4	0	25	39	35	13	-25	0	25
5	25	39	35	13	0	25	39	35
6	39	35	13	25	39	35	13	0
7	35	13	39	35	13	25	39	35
8	13	35	13	39	35	13	25	39

Considering 7 steps happening in 1 second every step happens in 1/7 second (Table 1)
Hence respective angular velocity= absolute difference*7
Respective Angular acceleration = change in angular velocity*7 (Table 2)

Table 2: Calculated values

Step	Θ1	Θ2	Θ3	Θ4	Θ5	Θ6	Θ7	Θ8
1	-35	-39	-25	0	25	39	35	13
2	-39	-25	0	25	39	35	13	-35

Here angular acceleration are obtained of each joint ,(I)moment of inertia of structure about the axis of rotation is obtained in our design as $I=3.2*10^{-5}kgm^2$.Hence Torque at each joints are evaluated=I*angular acceleration.Hence from market survey we selected Dynamixel servo having resolution 0.29-degree, Stall Torque=1.5Nm & with Position, Temperature & Load Voltage feedback.

Table 3: Torque Values

Joint Number	Maximum Calculated Torque(Nm)
1	$2.8*10^{-4}$
2	$3.2*10^{-4}$
3	0
4	$3.2*10^{-4}$
5	$2.735*10^{-4}$
6	$4.92*10^{-4}$
7	$7.66*10^{-4}$
8	$1.259*10^{-3}$

The Torque obtained are less as our structure manufactured is of 3D Printed part of PLA material which has high strength to density ratio.Strength=59MPa
density=0.00105kg/mm^2(Table 3 and 4)

Table 4: Calculated values

Absolute Difference In °	4	14	25	25	14	4	22	48
Respective Speed in °/sec	28	98	182	182	98	28	154	350
Speed in rad/sec	0.48	1.71	3.17	3.17	1.71	0.48	2.68	6.11
Angular acceleration rad/sec2	8.55	10.27	0	10.27	8.55	15.4	23.95	39.35

4. Motion Study

Before manufacturing the actual robot, it is important to study the motion of the designed robot in a virtual environment. By doing so we can study the gait cycle that can be achieved in the robot and also verify the design. The Simulation results could be further used for controlling the motion of the actual robot.

4.1. Motion Study using MATLAB SIMULINK

Prior to manufacturing of the prototype, it is necessary to analyze the motion of the robot virtually in order to get the required gait cycle. Motion study will also help us in getting the required equations to feed in the code to run the motor accordingly.For this purpose, we are using MATLAB Simulink to study the motion of the robot.The Solidworks cad model is imported in MATLAB using *sm_import* feature from MATLAB and corresponding block diagram is generated. Figure5 shows the block diagram generated when single module was imported in MATLAB.

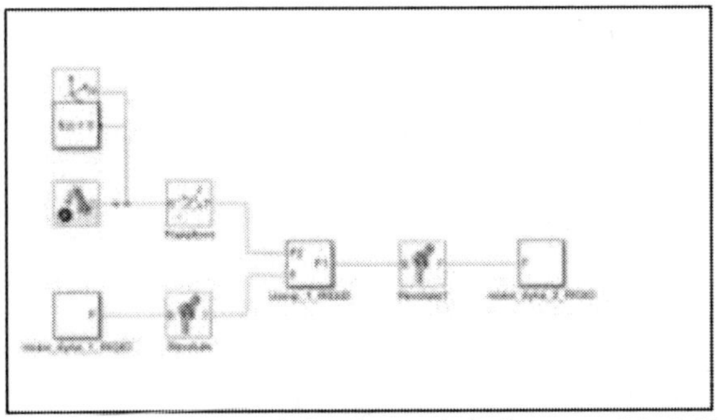

Figure 5:Block diagram of single module imported in MATLAB

In this robot, in each module, we require that the motor should trace a limited angle and not complete rotation. Hence, we modified the block diagram to give a time dependent function as input to the motor i.e. revolute joint in block diagram.
The modified block diagram is as shown in Figure 6.

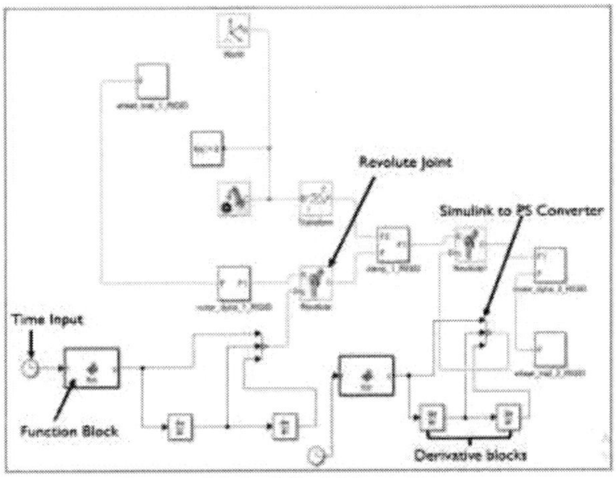

Figure 6:Modified block Diagram

Since we are using time dependent equation as input we need to convert it to acceleration to feed it to the motors hence we have used two derivative blocks on each module coupled to revolute joint using Simulink-PS converter.
In the functional block diagram we have used time dependent equations. The MATLAB code in the functional block is given below.

```
function y = fcn(t)
y= 0.5;
if(t>0)&&(t<4)
    y= -0.5*t;
else
    y= -2.5 + 0.2*t;
end
end
```

Using this code, we can simulate a 4sec long cycle in which motor traces angle of 120 degrees.

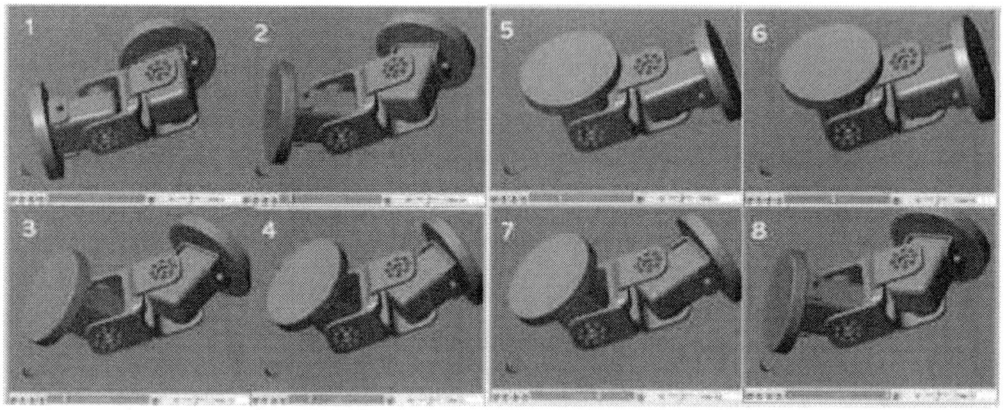

(a) (b)

Figure 7:Simulation output window

Here, we can see that the motor traces an angle of 60 deg and returns to its original position which is our desired type of motion. Shown in Figure 7.
Similarly, all the modules altogether were simulated using the same method to study serpentine type navigation.

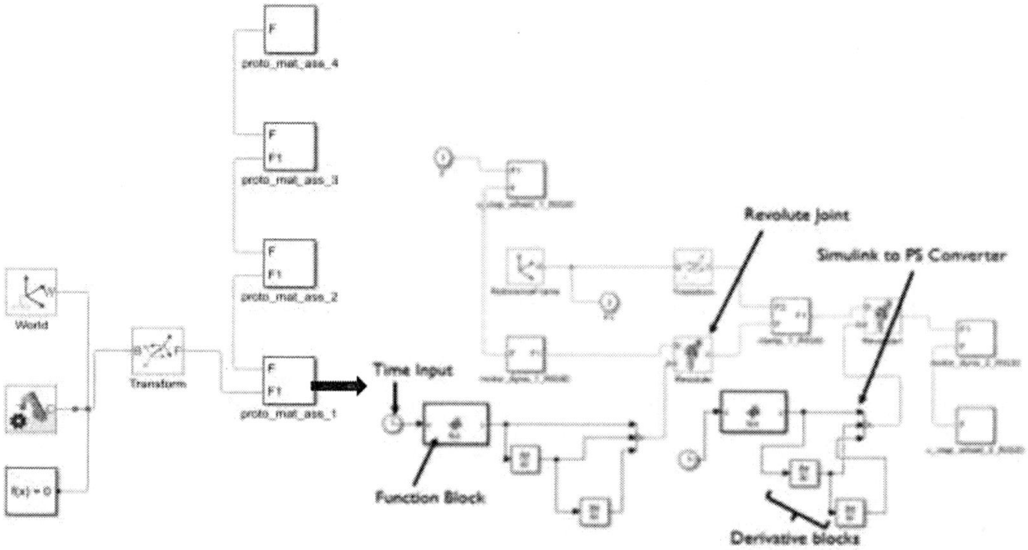

Figure 8:Block diagram for four module design of snake robot

Each port block represents 1 module and internally consists of block diagram as shown in Figure8.
The output for the block diagram can be seen in MALAB output window as shown in Figure 8.

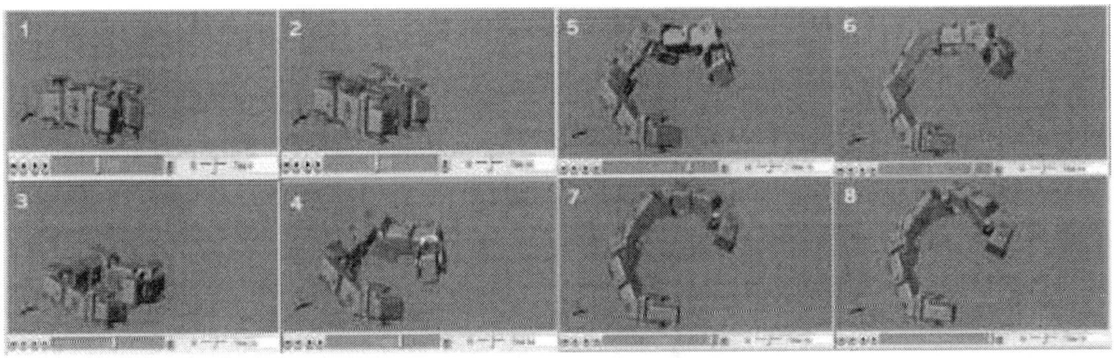

(a) (b)

Figure 9: Simulation output window

Here we can see that the snake robot opens up and closes and returns back to original position by changing the sign if the equation fed to the function block we can achieve reverse orientation of the above Figure 9.

*4.1.1. Motion Studyusing Denavit-Hartenberg Parameters*Denavit – Hartenberg parameters are used to determine position of end effector of the robot when joint variables are numerically specified. For any complex link mechanism we need to define only four parameters i.e. α, a, d, θ. Our snake robot mechanism is way much similar to industrial robot mechanism. But, with 8 actuators calculating position using D-H matrix becomes complex.MATLAB Robotics system toolbox helps to study the various orientation of robot for various gaits. MATLAB code for the same is given below

```
>> r= SerialLin
k(dh)
r =
noname:: 8 axis, RRRRRRRR, stdDH, slowRNE
+---+-----------+-----------+-----------+-----------+-----------+
| j |     theta |         d |         a |     alpha |    offset |
+---+-----------+-----------+-----------+-----------+-----------+
|  1|         q1|          0|          1|       1.57|          0|
|  2|         q2|          0|          1|       1.57|          0|
|  3|         q3|          0|          1|       1.57|          0|
|  4|         q4|          0|          1|       1.57|          0|
|  5|         q5|          0|          1|       1.57|          0|
|  6|         q6|          0|          1|       1.57|          0|
|  7|         q7|          0|          1|       1.57|          0|
|  8|         q8|          0|          1|       1.57|          0|
+---+-----------+-----------+-----------+-----------+-----------+
>> r.plot([0 0 0 0 0 0 0 0])
>> r.teach
```

The command r.teach gives us a teach pendent to control the joint variables. Thus, by changing joint variables we can study various types of orientations of the robot.

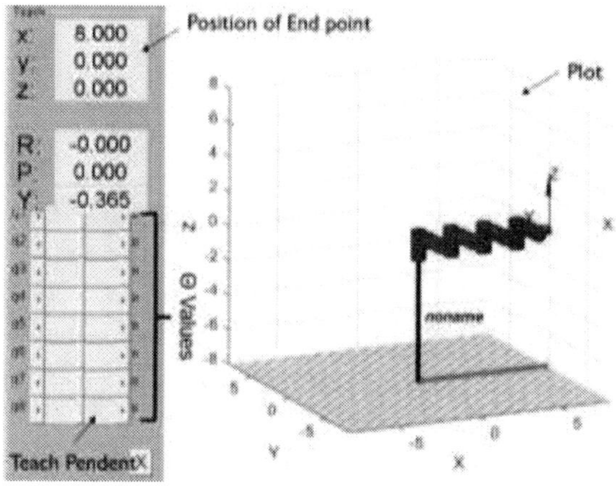

Figure 10: Plot for D-H matrix table

On the Teach pendent we can control the values of θ and thus for various angles traced by motor we can get different orientations of the robot (Figure 10). Thus, we can study virtually the motion of the robot. We tried various orientations of the robot keeping Z axis co – ordinates constant and Y axis co - ordinates constant, respectively. The result is as shown in above figure. But we cannot precisely determine the exact values of θ that are required for serpentine motion just by this trial and error method. The θ values required for desired serpentine motion can be calculated using method mentioned in pervious sections. Assuming link length of 1 unit we found values of θ as given in table.

Table 5: θ Values

Step	Θ1	Θ2	Θ3	Θ4	Θ5	Θ6	Θ7	Θ8
1	-35	-39	-25	0	25	39	35	13
2	-39	-25	0	25	39	35	13	-35
3	-25	0	25	39	35	13	-39	-25
4	0	25	39	35	13	-25	0	25
5	25	39	35	13	0	25	39	35
6	39	35	13	25	39	35	13	0
7	35	13	39	35	13	25	39	35
8	13	35	13	39	35	13	25	39

These steps are repeated to continue motion of the robot. The plot corresponding to these values of θ is as shown in Figure 11 and 12.

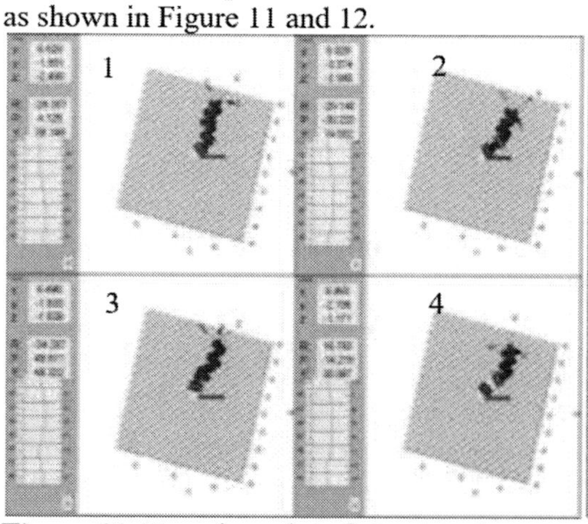

Figure 11: Top view plot of step 1 to 4

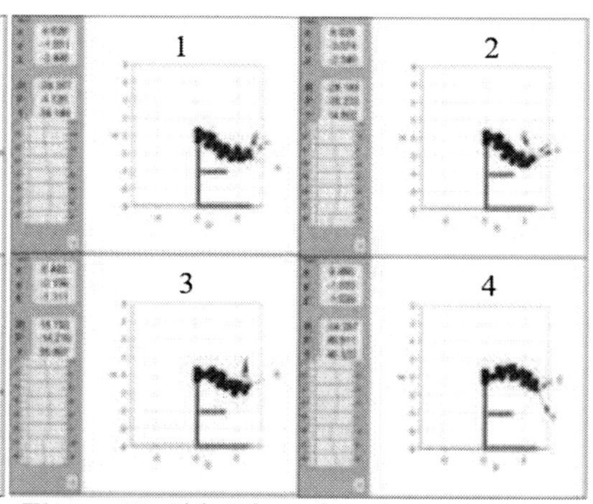

Figure 12: Side View plot of step 1 to 4

The plot of steps 5 to 8 are mirror images of steps 1 to 4 in reverse order which is similar to snake motion. Above results clearly shows that the robot performs sidewinding gait when we are using the calculated values of angular positions of the motor shaft. Thus, by varying θ values on the teach pendant we can study the motion of the snake robot in virtual environment. According to the above study we can infer that the snake robot can perform serpentine gait from the calculated values. Hence, the calculated values are correct and can be used to feed to the motors through code for actual run of the robot.

5. Control System

The Dynamixel servo motors are controlled through Arduino nano and is powered by 12 Volt Lithium Polymer (LiPo) battery.The Arduino nano is small, complete and breadboard friendly board based on ATmega328 (Arduino Nano 3.x.), it has the same functionality as that of Arduino UNO but in a small size. This is helpful in building a compact circuit for our snake robot considering its number of actuators and compact structure. Arduino nano has an operating voltage of 5V however, the input voltage may vary from 7V to 12V.

Arduino nano can be easily programmed using Arduino IDE which is Integrated Development Environment.Arduino software IDE has library functions through which we can control the positions of Dynamixel servo motors according to the type of motion required for the snake robot.

6. Conclusions

The Dynamixel Servo AX-12A used gives a torque of 15.3 kg-cm and repeatability up to 90% and 100% accurate positional feedback for tracing required angular position of motor shaft. The number of modules used are 4 and hence 8 actuators are used providing a compliant motion to the robot as we get eight Degrees of freedom. High Degree of freedom ensures that the serpentine gait is achieved, and the robot can manuever like a snake. Glossy PLA material is used for 3D printed parts it has a coefficient of friction of 0.42.PLA is less prone to warping during print and is much more 'stickier' than ABS. As it holds the layers firmly hence there is very less probability of breakage thus it gives required strength and material properties as compared to ABS and normal PLA material. The behaviour of the snake robot in a virtual environment was studied using MATLAB Simulation and Robotics Toolkit which generated results corresponding Denavit Hartenberg Parameters of the robot these results were used to feed angular positions to the motors and soft code was generated using the same angular positions. Arduino Nano is used as controller and code is generated in Arduino software which has functions to control motion of Dynamixel Servo motors.

References

[1] C. Ye, S. Ma, B. Li, Y. Wang and J. Tao, "*Dynamics analysis on serpentine locomotion of a new snake-like robot*", Robot, vol. 27, no. 6, (2005).

[2] C.Holden, O.Stavdahl, and J. T. Gravdahl. "*Optimal dynamic force mapping for obstacle-aided locomotion in 2D snake robots*". In: Proc. of the IEEE/RSJ International Conference on Intelligent Robots and Systems (IROS), Chicago, Illinois, United States. 2013, pp. 321–328.

[3] F. Sanfilippo et al. "*A review on perception-driven obstacle-aided locomotion for snake robots*". In: Proc. of the 13th International Conference on Control, Automation, Robotics and Vision (ICARCV), Phuket, Thailand. 2016, pp. 1–7

[4] J. Burdick, J. Radford, and G. Chirikjian, "*A 'sidewinding' locomotion gait for hyper-redundant*

robots" in Proc. IEEE Int. Conf. Robot. Autom., May, 1993, pp. 101–106.

[5] A.A. Transeth and K. Y. Pettersen, "*Developments in snake robot modeling and locomotion,*" in Proc. IEEE Int. Conf. Control, Autom., Robot. Vis., Singapore, Dec. 2006, pp. 1393–1390.

[6] O. Egeland and J.T. Gravdahl, *Modeling and Simulation for Automatic Control.* Trondheim, Norway: Marine Cybernetics, 2002, pp.306- 311.

[7] M.H.A. Majid, M.R. Khan, S.N. Sidek "*Development of Wheel- Less Snake Robot with Two Distinct Gaits and Gait Transition Capability*'. In International Journal of Automation and Computing. December 2013, pp. 255-259.

[8] X. Zhao, L. Dou, Zhong Su and N. Liu," *Study of the Navigation Method for a Snake Robot Based on the Kinematics Model* ", 2018, pp.1255-1252.

[9] L. Pfotzer, S. Klemm, A. Roennau, J. M. Zöllner, and R. Dillmann, "*Autonomous navigation for reconfigurable snake-like robots in challenging, unknown environments,*" Robotics and Autonomous Systems, 2017,pp. 610-613.

[10] Manzoor S, Cho YG, Choi Y "*Neural oscillator based CPG for various rhythmic motions of modular snake robot with active joints*", 2019, pp. 1124-1132.

Acknowledgements

We express our sincere gratitude towards Dr S.S. Ohol for guiding us throughout the project. We also thank Dr M.R. Nandgaonkar and Department of Mechanical Engineering for providing the facilities. We are immensely grateful towards Hon. Director Dr B.B. Ahuja for being a constant source of inspiration. We thank the Robot Study Circle of the Production Engineering for providing state of the art facilities which have proved to be invaluable for the completion of this report.

Health monitoring of automotive clutch system by using Bayes algorithms

Ganjikunta Chakrapani[1], **V Sugumaran**[1*]
[1]School of Mechanical Engineering, Vellore Institute of Technology, Chennai

E-mail: v_sugu@yahoo.com

Abstract. Clutch is an essential component in an automotive transmission system that helps in power transmission from the engine to the gearbox. Continuous operation of the clutch leads to degradation, damage, and reduced lifetime of internal components. Such factors can provoke the occurrence of various faults, which left unmonitored will result in clutch damage and seizure. Hence, continuous monitoring of the clutch is necessary to minimize unwanted breakdowns. This paper presents a condition monitoring technique based on vibration analysis to monitor the fault occurrence of different components. A machine learning approach is carried out to classify various faults. Such as fingers worn, pressure plate broken, pressure plate worn, friction material loss, and tangential strip bent. Feature extraction is performed on the acquired vibration signals using statistical learning process. The most important and significant features are selected from the extracted features by J48 decision tree algorithms. Further, feature classification is done with the help of Bayes based classifiers and the obtained results are compared to predict the best in class classifier for real-time.

Keywords: Clutch system, vibration analysis, machine learning, fault diagnosis, Bayes classifier

1. Introduction

A clutch is a mechanical system that precisely engages and disengages power transfer from the engine to the gearbox. The major function of clutch is witnessed during the time of gear shift and off-road driving conditions. Varying loads on the pressure plate in the course of gear shift tends to create a large amount of heat that can reduce the reliability, lifetime and performance of transmission system. Such factors can cause damage to internal components resulting in fault occurrences thereby affecting the performance of clutch. Unexpected transmission system failures lead to unanticipated vehicle failures which possess human safety risks, prolonged downtime, and high cost of maintenance. To minimize the ill effects of aforementioned failures, condition monitoring of clutches is necessary [1]. The method of measuring a state parameter in machinery is condition monitoring. A loss would thereof, implies a major shift in the physical framework. The use of condition monitoring will be used to schedule maintenance and take steps to minimize the effects of losses. State control devices only can assess the degradation of the situation as fault occurrence. Several techniques were used to determine the most possible fault occurrences in a system, including vibration examination, the examination of the oil crystals, thermal imaging, etc. Among these techniques, the study of vibration is most commonly used [2].

Many researchers have attempted to perform condition monitoring on various machinery using vibration analysis which is presented as follows. Using vibration characteristics that signify the state of the brake system, it is possible to track the state of the brake system. To obtain the statistical characteristics of vibration signals, the C4.5 decision tree algorithm was used. [2], Vibrational analysis was performed on a mono-block centrifugal pump to identify the fault occurrence [3]. Misfires build a unique pattern of vibration associated with a single cylinder. To detect misfires, functions can be extracted and analyzed. Using the J48 algorithm, the useful features have been recognized and the features selected were used in different functional trees[4]. The authors in [5] performed condition monitoring on bearing parts. Visual representation of the vibration signal frequency-domain characteristics may be necessary to identify the faults; however comprehensive domain awareness is needed. Automatic diagnostic techniques allow critical decisions to be taken by relatively unskilled operators. In this context, machine learning algorithms combined with vibration signal analysis have been successfully used in the condition monitoring of various components like wind turbine blades[6], planetary gearbox[7], milling tool[8] and flywheel [9]. Hence the vibration analysis is suitable to classify faults in the clutch.

The objective of this paper is to classify the faults of the clutch using vibration analysis and machine learning algorithms. Vibration signals are collected through a piezoelectric sensor and data acquisition device from the test setup. The features are extracted from the acquired vibration signals using the statistical learning process and the relevant useful features are selected from the extracted features by the use of J48 algorithm. The feature classification process is established by the use of Bayes classifier. For real-time execution, the average classification performance of the classifiers is contrasted and the better classifier is recommended.

2. Experimental setup

The experimental setup used in the present study is depicted in Figure 1. Single plate friction clutch, AC motor, flywheel, load cell, and bearings are assembled together to form the setup. The clutch used is a commercially available Maruti Zen dry friction clutch attached to a constant speed AC motor of 1400 rpm. The flywheel is coupled to the motor with the help of a 200 mm long shaft. Load variations are simulated in the setup by using a dynamometer connected directly to the clutch by a 250 mm long shaft. The shafts are prone to external stresses and moments that can lead to shaft breakage. Journal bearings are used on both the ends of the shaft to resist shaft breakage. Using direct adhesive method, an uniaxial piezo-electric sensor (model-4517, measuring - 500g, sensitivity - 10 mV/g is attached on the flat surface of the test setup. The vibration signal from the accelerometer is obtained with the help of a wireless National Instruments Data Acquisition system (DAQ-Model NI 9234 with CDAQ chassis NI9191, 4 channel, 51.2 k Samples / sec) equipped with signal conditioning unit. The signals are passed in to the load transformer and an analog to digital converter which is directly stored in to the storage device. To eliminate various irrelevant features, the signal was further processed from the storage device.

Figure 1. Experimental setup of a clutch system

2.1 Experimental procedure:

The vibration signals are measured at a steady rotating speed of 1400 rpm from the clutch system running in normal state. The efficiency characteristics of the clutch operating under normal conditions were analyzed. Initially, a healthy good condition clutch was attached to the setup and the vibration signals are established using a transducer and NI DAQ. The above process was repeated for all the test conditions for different loads. Fingers worn, pressure plate broken, frictional material loss, pressure plate worn and tangential strip bent are the test conditions considered, For all clutch test scenarios, the frequency of sampling is 24 kHz and the sampling size is 8192. More the number of observations greater were its value of statistical measures. However, the number of samples should be optimal such that computational time is reduced. For each of the clutch conditions, 100 measurements were taken and vibration signals were stored in the data files. Figure 2. represents the overall procedure of clutch fault monitoring. Figure 3. and Figure 4. shows the time domain signals for various clutch conditions.

IOP Conf. Series: Materials Science and Engineering 1012 (2021) 012028 doi:10.1088/1757-899X/1012/1/012028

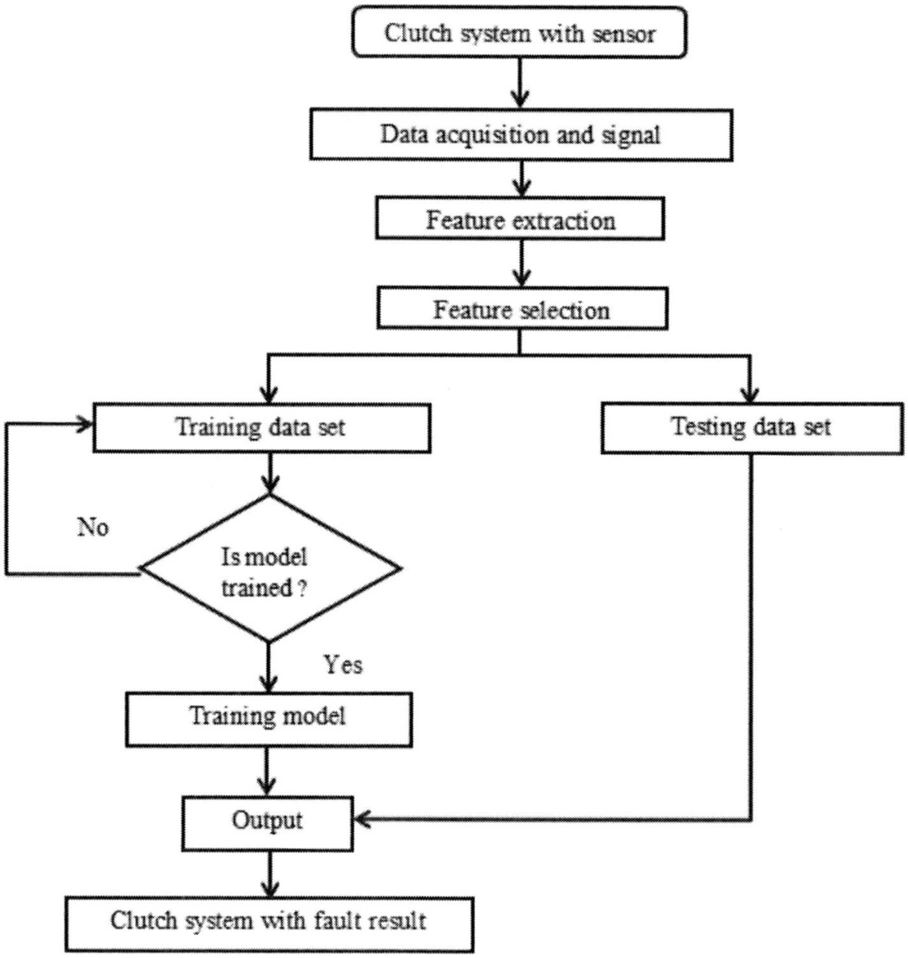

Figure 2. Methodology

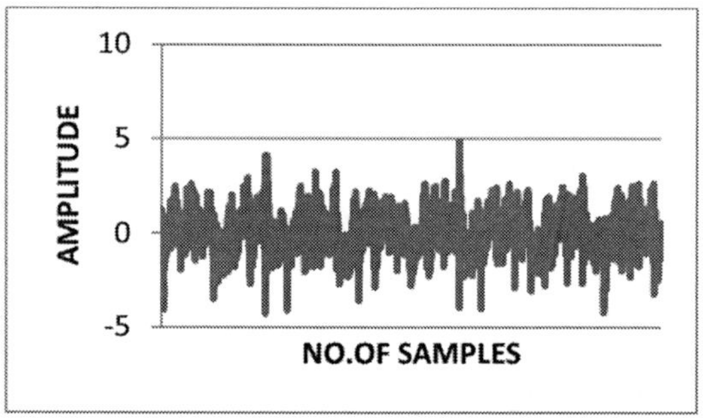

Figure 3. Vibration signal- good

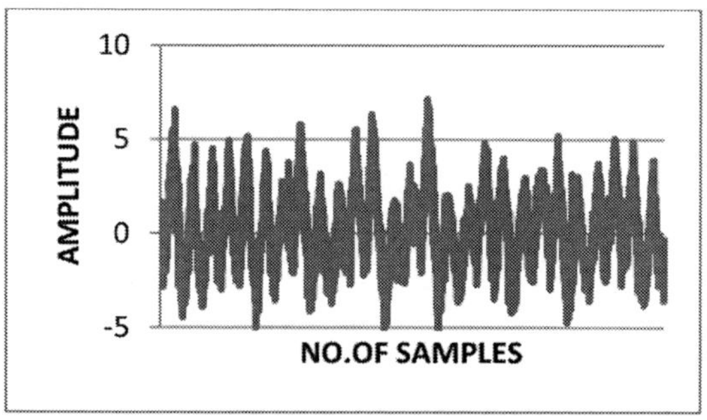

Figure 4. Vibration signal- pressure plate worn

3. Feature extraction and selection

The process of extracting features from a large number of vibrational signals is feature extraction. A total of 8192 samples from the experimental setup were collected. The whole of vibration signals cannot be fit into building the model as it is time consuming. Hence, these large numbers of signals must be converted to 12 statistical features by using extraction process. The statistical features are mean, mode, standard deviation, kurtosis, skewness, standard error, maximum, minimum, sample variance, sum and range.

Out of all the extracted features only certain features contribute in classification faults. Such significant features are selected from the extracted features with the help of J48 decision tree algorithm. A sample depiction of decision tree is illustrated in Figure 5. The number of selected features varies with every load conditions applied. In the present study, the numbers of features selected are 5, 7 and 5 for no load, 5 kg load and 10 kg load respectively.

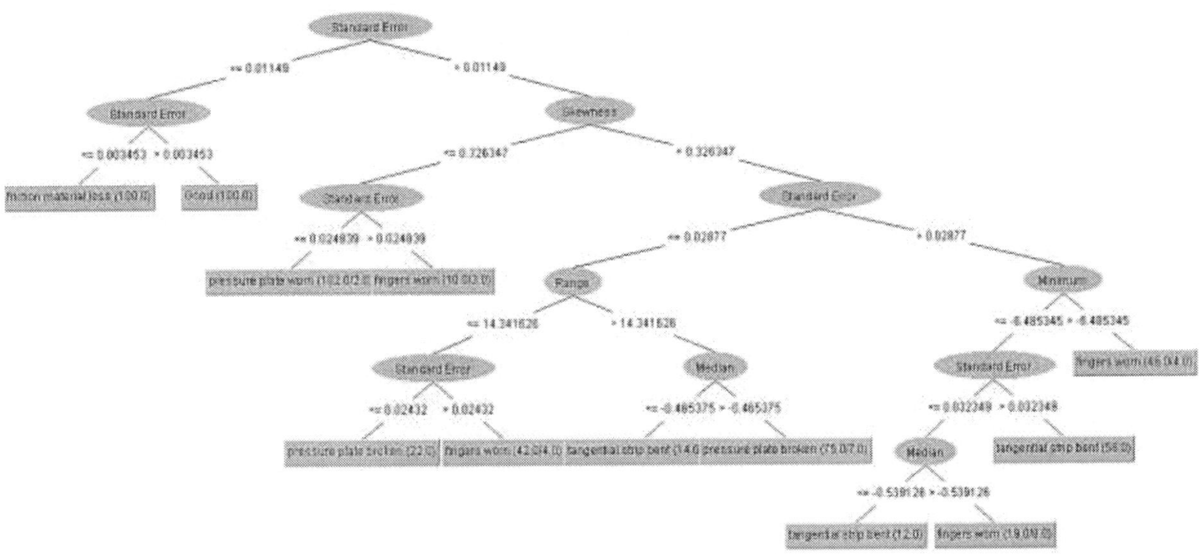

Figure 5. Decision tree for statistical features

3.1 Feature classification:

Bayes classifier:

A Bayes rule-based classification algorithm is the Naïve Bayes algorithm, which assumes that the attributes $X_1 ... X_n$, given Y, all are constitutionally distinct from one another. The advantage of such a statement has been that the interpretation of $P(X|Y)$are significantly simplified and the issue of measuring it from the effects of testing. For eg, consider the condition if $X = X_1, X_2$

$$P(X|Y) = P(X_1, X_2 |Y) = P(X_1 |X_2, Y)P(X_2|Y) = P(X_1 |Y)P(X_2|Y) \cdots\cdots (1)$$

Most commonly, when *X* comprises n attributes which are *Y*-independent constitutionally,

$$P(X_1 \dots X_n|Y) = \prod_{i=1}^{n} P(X_1|Y) \cdots\cdots (2)$$

Note that since *Y* as well as X_i is Boolean variables, the description of $P(X_i = x_{jk}|Y = y_j)$ required *i, j, k* or needs *2n* parameters. If we do not allow a conditional assumption of independence, this is a drastic decrease relative to the $2(2^n - 1)$ the requisite parameters for characterizing $P(X|Y)$.

Now let's extract the Naïve Bayes classifier usually assumes that *Y* is any single-value vector and that $X_1 \dots X_n$ attributes are any particular attributes or attributes of real value. We aim to train the classifier to emit the distribution of probability over potential values of *Y*, which we ask to classify for each new *X* case. The representation for probability which the *Y* will take over the prospective k^{th} significance, as per Bayes law, is

$$P(Y = y_k|X_1 \dots X_n) = \frac{P(Y = y_k)P(X_1 \dots X_n|Y = y_k)}{\sum_j P(Y = y_j)P(X_1 \dots X_n|Y = y_j)} \cdots\cdots (3)$$

Where, the sum of all potential y_j of *Y* values are taken over. Now, provided that the X_i is conditionally independent of *Y*, we can rewrite this as

$$P(Y = y_k|X_1 \dots X_n) = \frac{P(Y = y_k)\prod_i P(X_i|Y = y_k)}{\sum_j P(Y = y_j)\prod_i P(X_i|Y = y_j)} \cdots\cdots (4)$$

Above Eq (4) is the basic equation for the classification of the Naïve Bayes. This equation illustrates well how determine the likelihood the *Y* would have on any specified value based on known X^{new} attributes and provided the estimated distributions of $P(Y)$ and $P(X_i|Y)$from training results, given a new instance of $X^{new} = (X_1 \dots X_n)$. The classification law of Naïve Bayes for obtaining the critical value of *Y* based on one's interest is given by

$$Y \leftarrow \underset{y_k}{\arg max} \frac{P(Y = y_k)\prod_i P(X_i|Y = y_k)}{\sum_j P(Y = y_j)\prod_i P(X_i|Y = y_j)} \cdots\cdots (5)$$

Simplifying to follows (since y_k doesn't really rely on a denominator)

$$Y \leftarrow \underset{y_k}{\arg max}\, P(Y = y_k)\prod_i P(X_i|Y = y_k) \cdots\cdots (6)$$

The Bayes Net consists of a collection of variables among variables forming a directed acyclic graph *G*= *(V, E)*, $V = (A_1, A_2 \dots A_n)$ and a collection of guided edges, *E*, in which a cumulative allocation of possible factor is described by a conditional distribution to every parent variable. A node $A_i \in V$ is a random variable and the conditional dependency between A_i and A_j is a directional edge from A_i toA_j, (A_i, A_j). Each vector in a Bayesian network is independent from its non-descendants, provided the value of its parents will be in *G*. In order to correctly infer subsequent distribution, such *G*-encoded independence decreases the amount of parameters needed to describe a joint distribution. The joint distribution *P (V)* in the Bayesian network over for each vector a conditional distribution has a parametric form which can be learned by calculating the maximum probability. All the conditional distributions defined in the Bayes Net are the product of *P(V)* is

$$P(A_1, A_2 \dots A_N) = \prod_{i=1}^{N} P(A_i/Pa_i) \cdots\cdots\cdots\cdots\cdots\cdots\cdots\cdots\cdots\cdots\cdots\cdots \quad (7)$$

When the conditional distribution of A_i is $P(A_i/Pa_i)$, provided that Pa_i represents that parent collection of A_i. For each vector, a conditional distribution does have a parametric structure that must be obtained by calculating the possible probability.

4. Result and discussion:

4.1 Varying load conditions:

A total of 3 load conditions considered namely no load, 5kg load and 10kg load. No external load is applied to the clutch in no load condition, at 5kg load condition and 10kg load condition the external load is applied on clutch with help of dynamometer. Classification process is performed using Bayes Net, Naïve Bayes classifiers. Table.1 represents the classification accuracy of Bayes classifier at different load conditions. From Table 1, it can be inferred that the classification accuracy of Bayes Net classifier for no load, 5kg load and 10kg load are 91%, 90.5% and 94.5% respectively, similarly the corresponding classification accuracy of naïve Bayes classifier is found to be 90.1667%, 85.1667% and 93.5% respectively. Considering the obtained results, the overall classification accuracy average for Bayes Net classifier is higher than naïve Bayes classifier. The Figure 6 shows the comparison of various load conditions with accuracy of Bayes Net and Naïve Bayes classifier.

Table 1. Classification accuracy of Bayes classifier at different loads

Load condition	No-load		5kg load		10kg load	
Classifier type	Bayes Net	Naïve Bayes	Bayes Net	Naïve Bayes	Bayes Net	Naïve Bayes
Time taken to build the model	0.06	0.01	0.08	0.01	0.17	0.01
Classification accuracy	91.00%	90.16%	90.50%	85.16%	94.50 %	93.50 %

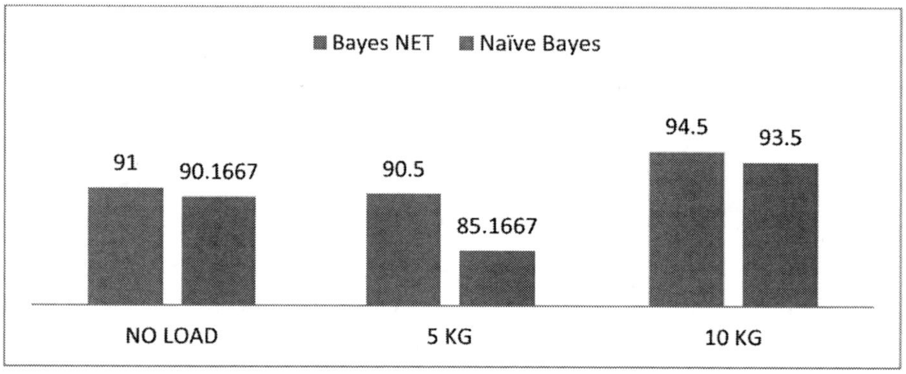

Figure 6. Classification accuracy of Bayes classifier

4.2 Performance evaluation of classifiers:

Case 1: No load condition

Feature classification is done by using Bayes classifiers. 600 instances for each condition were considered and 100 instances for each class for classification. At no-load condition, Bayes Net classifier found maximum classification accuracy of 91.00%. Out of 600 instances, 546 instances are classified correctly and the remaining 54 instances are misclassified with other classes. In confusion matrix (Table.3). Correctly categorized instances represented by diagonal elements, and misclassified instances are represented by non-diagonal elements.

Table 2 . Classification summary of no load condition

Total Number of Instances	600
Correctly Classified Instances	546 (91.00 %)
Incorrectly Classified Instances	54 (9.00 %)
Root relative squared error	37.09 %
Relative absolute error	12.30 %
Mean absolute error	0.0342
Root mean squared error	0.1382
Kappa statistic	0.892

1. For fingers worn category 84 instances are classified correctly and the remaining 16 misclassified as 6 pressure plate broken and 10 tangential strip bent.

2. Friction material loss category 100 instances classified correctly.

3. At good category 100 instances are classified correctly.

4. Pressure plate broken condition 72 instances classified and the remaining 28 instances misclassified as 16 fingers worn, 1 pressure plate worn and 12 tangential strip bent.

5. Pressure plate worn category 99 instances classified correctly and the remaining 1misclassified as fingers worn.

6. Tangential strip bent condition 91 instances classified correctly and the remaining 9 misclassified as 6 fingers worn and 3 pressure plate broken.

Table 3. Confusion matrix for NO load

CLASS	FW	FML	GOOD	PPB	PPW	TSP
FW	**84**	0	0	6	0	10
FML	0	**100**	0	0	0	0
GOOD	0	0	**100**	0	0	0
PPB	16	0	0	**72**	0	12
PPW	1	0	0	0	**99**	0
TSB	6	0	0	3	0	**91**

Case 2: 5 kg load condition

Bayes Net classifier found 90.50% of classification accuracy at 5 kg load condition, 543 instances classified correctly out of 600 instances, and the remaining 57 instances are misclassified with other classes.

Table 4. Classification summary of 5kg load condition

Total Number of Instances	**600**
Correctly Classified Instances	**543 (90.50 %)**
Incorrectly Classified Instances	57 (9.50 %)
Root relative squared error	40.48 %
Relative absolute error	12.59 %
Mean absolute error	0.035
Root mean squared error	0.1509
Kappa statistic	0.886

Table 5. Confusion matrix for 5kg load

CLASS	FW	FML	GOOD	PPB	PPW	TSB
FW	**76**	0	0	0	16	8
FML	0	**100**	0	0	0	0
GOOD	0	0	**100**	0	0	0
PPB	0	0	0	**93**	7	0
PPW	3	0	0	9	**80**	8
TSB	3	0	0	0	3	**94**

Case 3: 10 kg load condition

Bayes Net classifier found 94.50 % of classification accuracy at 10 kg load condition, 567 instances classified correctly out of 600 instances, and the remaining 33 instances are misclassified with other classes.

Table 6 . Classification summary of 10 kg load condition

Total Number of Instances	**600**
Correctly Classified Instances	**567 (94.50 %)**
Incorrectly Classified Instances	33 (5.50 %)
Root relative squared error	30.34%
Relative absolute error	9.92 %
Mean absolute error	0.0276
Root mean squared error	0.1131
Kappa statistic	0.934

Table 7. Confusion matrix for 10kg load

ClASS	FW	FML	GOOD	PPB	PPW	TSB
FW	**96**	0	0	0	0	4
FML	0	**100**	0	0	0	0
GOOD	0	0	**100**	0	0	0
PPB	7	0	0	**89**	9	2
PPW	2	5	0	0	**86**	7
TSB	1	0	0	0	3	**96**

5. Conclusion

State control of a clutch using vibration signals was performed and the statistical characteristics were derived. Using Naïve Bayes and Bayes Net classifiers, each of them was categorized. The Decision Tree was used to minimize dimensionality. The statistical function classification precision using the Bayes Net classifier was found to be 91.00 %, 90.50 % and 94.50 % respectively at no load, 5 kg load and 10 kg load settings. It can be established that the Bayes Net classifier provides more reliable results compared to the Naïve Bayes classifier by analyzing the results. It can further be inferred that, under all three load scenarios, the Bayes Net classifier can be implemented to estimate the condition of the clutch. The findings of the classification are promising for related research to be carried out.

References

[1] A. P. Ompusunggu, J. M. Papy, S. Vandenplas, P. Sas, and H. Van Brussel, 2013. "A novel monitoring method of wet friction clutches based on the post-lockup torsional vibration signal," vol. 35, no. 1–2, pp. 345–368.

[2] R. Jegadeeshwaran and V. Sugumaran, 2013, "Comparative study of decision tree classifier and best first tree classifier for fault diagnosis of automobile hydraulic brake system using statistical features," vol. 46, no. 9, pp. 3247–3260.

[3] V. Sugumaran and K. I. Ramachandran, 2011, "Effect of number of features on classification of roller bearing faults using SVM and PSVM," vol. 38, no. 4, pp. 4088–4096.

[4] A. Sharma, V. Sugumaran, and S. Babu Devasenapati, 2014, "Misfire detection in an IC engine using vibration signal and decision tree algorithms," vol. 50, no. 1, pp. 370–380.

[5] V. Sugumaran, G. R. Sabareesh, and K. I. Ramachandran, 2008, "Fault diagnostics of roller bearing using kernel based neighborhood score multi-class support vector machine," vol. 34, no. 4, pp. 3090–3098.

[6] A. Joshuva and V. Sugumaran, 2019, "A Lazy Learning Approach for Condition Monitoring of Wind," Measurement, p. 107295.

[7] Y. Lei, D. Han, J. Lin, and Z. He, 2013, "Planetary gearbox fault diagnosis using an adaptive stochastic resonance method," vol. 38, no. 1, pp. 113–124.

[8] C. Zhou et al., 2019, "Vibration singularity analysis for milling tool condition monitoring," Int. J. Mech. Sci., vol. 166, p. 105254, 2020.

[9] T. Hasegawa, M. Saeki, T. Ogawa, and T. Nakano, 2019, "Vibration-based fault detection for flywheel condition monitoring," vol. 17, pp. 487–494.

A Bayes learning approach for monitoring the condition of suspension system using vibration signals

P Arun Balaji[1], **V Sugumaran**[1*]
[1]School of Mechanical Engineering, Vellore Institute of Technology, Chennai

E-mail: v_sugu@yahoo.com

Abstract- The suspension system is an important component in an automobile, aimed to provide better comfort and good road handling characteristics during vehicle motion on uneven road conditions. Varying load conditions, prolonged operations, continuous road shocks absorption, and time-based degradation of internal components can create faults in a suspension system. Fault occurrences in the suspension system can damage the internal components that lead to malfunctioning of the system causes failure and endanger vehicle safety. Thus, condition monitoring has become an essential part of identifying fault occurrences in a suspension system. This paper is focussed on condition monitoring of suspension system using vibration signal analysis and faults are classified with the help of machine learning-based Bayesian classifiers. A test setup is fabricated to simulate the working of the suspension system under different load conditions with one good and six faulty conditions. Vibration signals are recorded and used to extract statistical features. Further J48 decision tree algorithm was utilized to identify the most significant features during feature selection process. Bayes classifiers such as NaiveBayes and BayesNet classifiers were used to find the type of fault occurrence from the selected features. The overall results of the aforementioned classifiers are compared and the best in the Bayesian classifier is suggested for real-time application.

1. INTRODUCTION

Damping of unwanted vibration remains a prime concern in automobiles. The suspension system is used to control the vibrations transmitted from road wheels to the vehicle body due to the road irregularities during vehicle motion and to maintain the contact between wheel and the road. The purpose of the sound suspension system is to provide better comfort and improve road handling characteristics. The efficient working of the suspension system is degraded by various reasons such as fatigue, wear, external damage, improper fitting, and uneven wheel pressure resulting in causes for damaging other components. Since, the suspension system is operated under heavy loads and prone to continuous vibration transmission, it is subjected to various fault occurrences. To achieve the proper functioning of the suspension system and reduce the occurrence of faults, frequent monitoring is necessary. Several condition monitoring techniques carried out in recent times vary based on the type of suspension and components used. Various literatures suggest that condition monitoring based on vibration signal analysis is predominantly used to monitor the condition of the suspension system. The suspension system consists of many parts and monitoring each component with a specific fault diagnosis method makes condition monitoring a time-consuming process [1]. McPherson suspension system is widely used in most modern front-wheel-drive cars due to its simplicity and lightweight construction [2]. The suspension system

IOP Conf. Series: Materials Science and Engineering 1012 (2021) 012029 doi:10.1088/1757-899X/1012/1/012029

consists of struts, lower arm, knuckles, ball joints, and tie rod. The performance of the suspension system degrades due to various reasons such as fatigue, wear, external damage, improper fitting, etc. Such factors can result in fault occurrences leading to damage to internal components and raise safety concerns. Early detection of fault can improve system reliability and safety also will minimize down time[3][4]. Condition monitoring of each component with a specific fault diagnosis technique will be a time-consuming process [5]. Hence, an efficient fault diagnosis technique is the need of the hour. The condition monitoring of the suspension system can be categorized into two types namely, the conventional method and machine learning method. In the conventional method, Using Fast Fourier Transform (FFT) the vibration signals in time domain signals are converted to the frequency domain signals and the faults can be classified by monitoring the peak frequency in a spectrum. The frequency components in the traditional FFT approach tend to change continuously due to the frequent speed variations and wear and tear of components. Such variations have made the traditional approach complex and time-consuming[6]. While the machine learning approach uses a continuous learning process and adapts to varying test conditions. The vibration signals acquired from all the test conditions are classified with the help of machine learning algorithms such as Bayesian classifiers, Decision trees, lazy classifiers, etc.

Several researches have been carried out in damper,ball joint and spring in three category based on the conventional approach which is discussed as follows. Ventura *et al* discussed two methods for gaining access to the condition of the shock absorber using Micro-electro-mechanical systems (MEMS) sensor. In the first method, sprung and unsprung mass acceleration measurement is used to determine the transmissibility as a function of the damping coefficient to estimate the condition of the damper. In the second method, the pressure inside the compression chamber is measured along with vehicle acceleration using sensors to monitor the condition of the damper being assessed [7][8]. Luis Carvalho *et al* proposed a method to determine the condition of the ball joint used in automobiles by calculating the ratio of signals acquired from both good and faulty conditions in the frequency domain [9]. Burdzik *et al* proposed a method to monitor the condition of suspension by means vibration signal analyzer where the signal from the exciter, wheel, and car body for the given time domain [10]. Wang *et al* proposed a cluster based method to identify the faults in the spring of suspension system. By acquiring accelerometer signal from 4 corner of the vehicle. The proposed method online condition monitoring using data driven approach by means probability c mean clustering to identify the fault the spring and used Fisher discrimination analysis to isolate and label the fault [11].

From the literature survey, the following observations were made.

- Most of the research findings based on data driven method talks about the detection of a single fault such as faults in damper, spring and ball joints.
- Model based method of fault identification is not suitable for suspension system due to the dynamic nature of the suspension system.
- The fault detection in semi and active suspension systems are used to design fault tolerance system to maintain uninterrupted working of the suspension system
- Machine learning approach is widely used in active suspension system

Based on the observations made a novel technique is proposed for the condition monitoring of the suspension system based on the Machine learning approach. The following technical contributions are presented as follow:

1. Vibration-based condition monitoring is performed with help of a machine learning algorithm.
2. Multiples fault occurrences in a component can be identified.
3. The proposed method will pay way to online condition monitoring of suspension system.
4. Classification is carried out with the help of BayesNet and Naive Bayes algorithms.

This article is constructed with the following sections: Experimental setup and procedure along with a brief introduction of faults considered are explained in Section 2. Feature extraction and selection processes are described in Section 3, a brief description of the Bayesian classifier is discussed in section 4. Section 5 projects the derived results and the necessary discussion. The conclusion of the present study is given in Section 6.

2. EXPERIMENTAL STUDIES

The experimental setup is designed to simulate the working of suspension system in the uniform road profile at a constant speed. The fault condition component is replaced for each fault study and their signals are acquired from accelerometer placed in the lower arm. The measured signals are sent to analog to digital converter to collect data for recording. The features were extracted from the data acquired for fault classification. The basic work flow of the study is given figure 1.

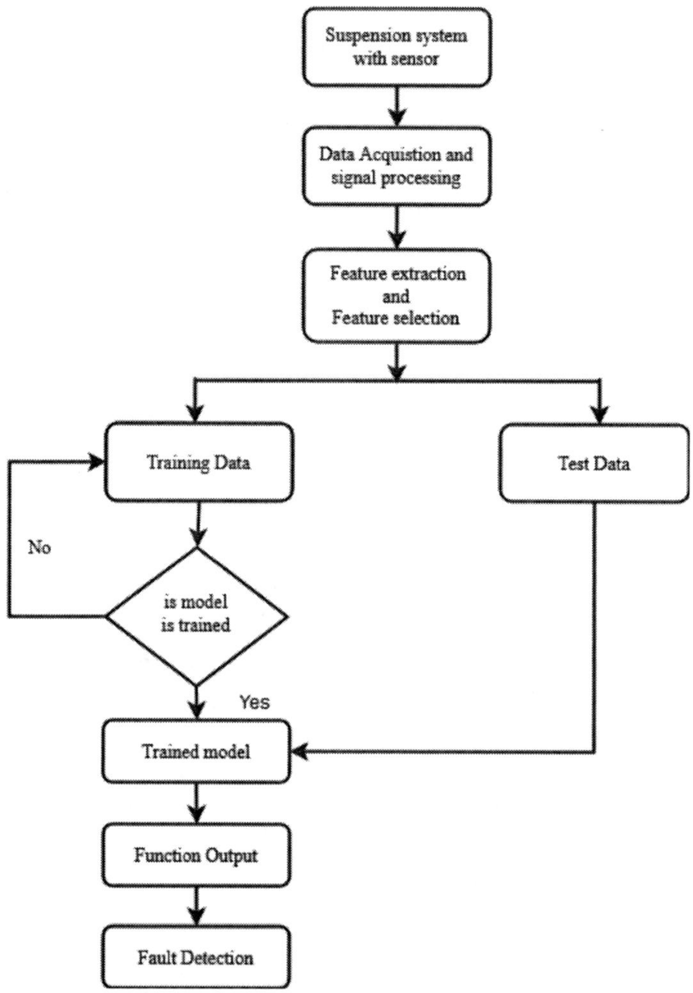

Figure 1 Flow chart of suspension system fault diagnosis system.

2.1MCpherson suspension system

The McPherson suspension system is used in the present study due to its lightweight and simplicity. The strut, lower arm, knuckle, driveshaft and tie rod constitute the McPherson suspension system. The present study equips a commercially available Hyundai i10 suspension system. Other components like idle roller, pulley, and frame are fabricated such that the drive from constant speed AC motor of 1400 rpm is transmitted to the drive shaft through belt and pulley arrangement. The driveshaft is connected to the wheel by knuckle bearing through which the power from the motor is transmitted to the road wheel. For increasing the sprung mass, the idle roller is raised with the help of a hydraulic jack and supporting column. A lock nut is attached to the hydraulic jack to prevent any further movement. The increase in height induces load variations in the suspension system which has been measured using a pressure gauge attached to the hydraulic jack. The overall setup is illustrated in Figure 2.

Figure 2 Experimental setup

2.2 Data aquiction system

A piezoelectric type accelerometer is attached to the lower arm as shown in Figure 2 with help of general purpose adhesive. The drive from the motor is transmitted to the wheel via belt drive and the drive shaft so that final speed of 1200 rpm is maintained.The vibration signals from the accelerometer are acquired by the Data Acquisition system (DAQ) where the signal conditioning is carried out. The signals are amplified and filtered from unwanted noises during signal conditioning. These signals are acquired at a sample length of 10,000 at the rate of 20 kHz from a sensor of 10.26 mV/g sensitivity are recorded. The first 100 trials from the acquired signals are taken into consideration for feature extraction. The above process is repeated for different faults under 3 load conditions namely, no load, 200 psi, and 400 psi pressure.

The faults considered in the study for condition monitoring of the suspension system are tie rod Ball joint worn out, loweram ball joint, lower arm bush worn out, strut mount failure, strut external damage,strut worn out and wheel low pressure.The faults discussed in the study and their brief introductions of all the faults considered with the reason for failure and symptoms with typical image are discussed in Table 1

IOP Conf. Series: Materials Science and Engineering 1012 (2021) 012029 doi:10.1088/1757-899X/1012/1/012029

Table 1 Description of various faults in a suspension system

S.No	Fault Condition	Components affected	Reason for failure	Symptoms	Figure
1	Strut external damage	Strut	Physical damage	Dent	
2	Strut mount failure	Strut mount	Fatigue, sudden road shocks	Visible crack	
3	Lower arm bush worn-out	Lower arm	Varying loads, harsh driving conditions, degradation over time	Cracking noise, loss in directional control	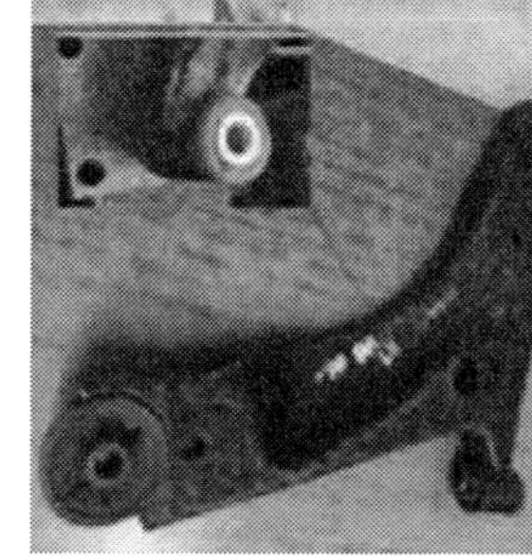

IOP Conf. Series: Materials Science and Engineering 1012 (2021) 012029 doi:10.1088/1757-899X/1012/1/012029

4	Strut worn-out	strut	Prolonged usage, lubrication loss,	Difference height,bouncy ride, once side steering pull	
5	Ball joint worn-out	Lower arm	Lubrication loss, heavy loads	Wheel wobbling, steering vibration,uneven tyre wear,	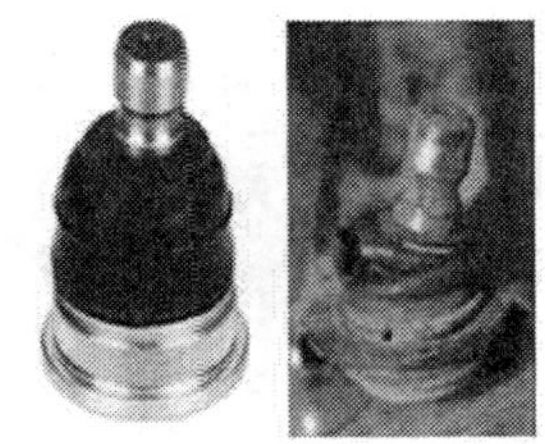
5	Ball joint worn-out	Tie rod	Driving in uneven road, Heavy loads	Wheel wobbling, steering vibration,uneven tyre wear,	
6	Wheel low pressure	Tyre	Check valve failure, puncture	The difference in vehicle height, hard steering	

3. EXTRACTION OF FEATURE AND FEATURE SELECTION

Feature extraction is the process of extract useful parameter from a raw signal which can establish the information required to characterize the signal. The fault in the system has its own vibration pattern which can be detected using statistical parameter of vibration signals this has been proved by many researchers reported in the literature [12][13][14][15]. Thus a statistical feature will be best method of features extraction with less complexity and less processing time.Statistical features extract various parameters through descriptive statistics [16]. Table 2 presents the parameter and their brief definitions which are extracted using descriptive statistics

Table 2 Description of statistical features

Parameters	Description
Mean	The ratio between sum of the data and the total number of values in a dataset
Standard error	Error during the distribution of statistic sampling
Median	Median is the intermediate value in a dataset that differentiates the lower and higher values.
Range	Difference between the maximum value to the minimum value
Sample variance	The distribution measure of a given dataset
Skewness	Asymmetry in the distribution of statistical data
Kurtosis	The spread of data around the mean.
Standard deviation	The variation of data set point with the mean value
Mode	The value which occurs repeatedly in a data set
Maximum	The maximum value possessed in the data set
Sum	The total sum of all the data set=(x1+x2+x3……xn)
Minimum	The least value possessed in the data set

Out of these descriptive features, only some features will contribute positively to classification. Adding more number of features will increase computation time and complexity. To reduce computation time and increase the classification accuracy, unwanted features are removed and the most significant features are determined. The J48 decision tree algorithm a clone of C4.5 algorithm is used in the process for feature selection process for its superior classification performance was utilized to select important features in the feature selection process [17][18] [19]. In the decision tree, features are ranked from top to bottom with increasing significance. Figure 3 shows the decision tree of selected features for the no-load condition using j48 algorithm.

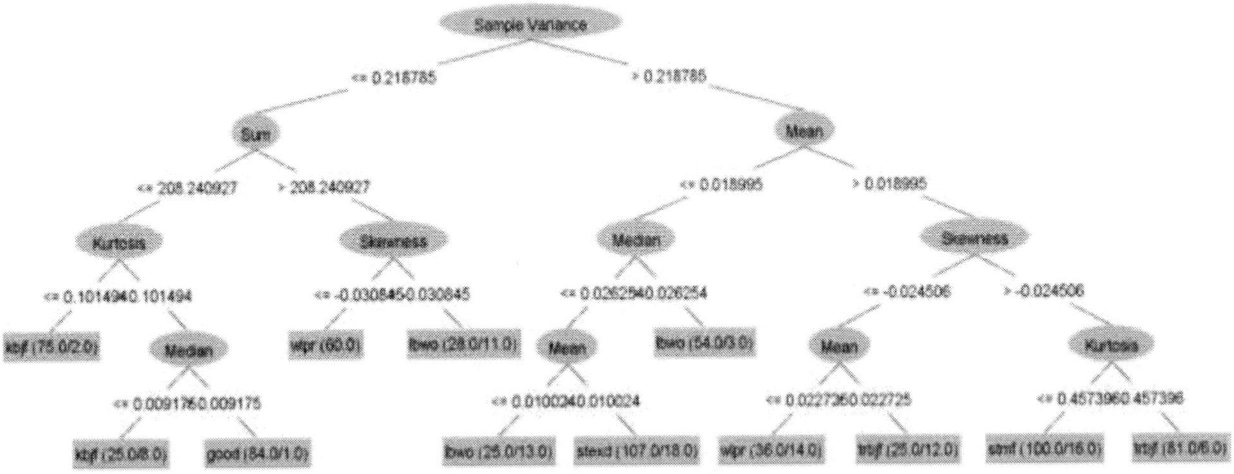

Figure 3 Decision tree representing the most significant feature from top to bottom

4. CLASSIFIERS DESCRIPTIONS

4.1. Bayes classifier:

A Bayes rule-based classification algorithm is the Naïve Bayes algorithm, which assumes that the attributes $X_1 \dots X_n$, given Y, all are constitutionally distinct from one another. The advantage of such a statement has been that the interpretation of $P(X|Y)$are significantly simplified and the issue of measuring it from the effects of testing. For eg, consider the condition if $X = X_1 , X_2$

$$P(X|Y) = P(X_1 , X_2 \, |Y) = P(X_1 \, |X_2, Y)P(X_2|Y) = P(X_1 \, |Y)P(X_2|Y) \cdots\cdots (1)$$

Most commonly, when X comprises n attributes which are Y-independent constitutionally,

$$P(X_1 \dots X_n|Y) = \prod_{i=1}^{n} P\,(X_1 \, |Y) \cdots\cdots\cdots\cdots\cdots\cdots\cdots\cdots\cdots\cdots\cdots\cdots\cdots\cdots\cdots\cdots\cdots\cdots (2)$$

Note that since Y as well as X_i is Boolean variables, the description of $P(X_i = x_{jk}|Y = y_j)$ required i, j, k or needs $2n$ parameters. If we do not allow a conditional assumption of independence, this is a drastic decrease relative to the $2(2^n - 1)$ the requisite parameters for characterizing $P(X|Y)$.

Now let's extract the Naïve Bayes classifier usually assumes that Y is any single-value vector and that $X_1 \dots X_n$ attributes are any particular attributes or attributes of real value. We aim to train the classifier to emit the distribution of probability over potential values of Y, which we ask to classify for each new X case. The representation for probability that Y will take on its prospective k^{th} value, according to Bayes law, is

$$P(Y = y_k|X_1 \dots X_n) = \frac{P(Y = y_k)P(X_1 \dots X_n|Y = y_k)}{\sum_j P(Y = y_j)P(X_1 \dots X_n|Y = y_j)} \cdots\cdots\cdots\cdots\cdots\cdots\cdots\cdots (3)$$

Where, the sum of all potential y_j of Y values are taken over. Now, provided that the X_i is conditionally independent of Y, we can rewrite Eq (3) as

$$P(Y = y_k|X_1 \dots X_n) = \frac{P(Y = y_k)\prod_i P(X_i|Y = y_k)}{\sum_j P(Y = y_j)\prod_i P(X_i|Y = y_j)} \cdots\cdots\cdots (4)$$

Above Eq (4) is the basic equation for the classification of the Naïve Bayes. This equation illustrates well how determine the likelihood the Y will have on any specified value based on observed X^{new} attribute values and provided the estimated distributions of $P(Y)$ and $P(X_i|Y)$from training results, given a new instance of $X^{new} = (X_1 \dots X_n)$. The classification law of Naïve Bayes for obtaining the critical value of Y based on one's interest is given in Eq(5)

$$Y \leftarrow \underset{y_k}{\arg max} \frac{P(Y = y_k)\prod_i P(X_i|Y = y_k)}{\sum_j P(Y = y_j)\prod_i P(X_i|Y = y_j)} \cdots\cdots\cdots (5)$$

Simplifying to follows (since y_k doesn't rely on a denominator)

$$Y \leftarrow \underset{y_k}{\arg max}\ P(Y = y_k)\prod_i P(X_i|Y = y_k) \cdots\cdots\cdots (6)$$

The Bayes Net consists of a collection of variables among variables forming a directed acyclic graph *G*= *(V, E)*, $V = (A_1, A_2 \dots A_n)$ and a collection of guided edges, *E*, where a cumulative distribution of variables is described by a conditional distribution to every parent variable. A node $A_i \in V$ is a random variable and the conditional dependency between A_i and A_j is a directional edge from A_i toA_j, (A_i, A_j). Each vector in a Bayesian network is independent from its non-descendants, provided the value of its parents will be in *G*. To correctly infer subsequent distribution, such *G*-encoded independence decreases the amount of parameters needed to describe a joint distribution. The joint distribution *P (V)* in the Bayesian network over for each vector a conditional distribution has a parametric form which can be learned by calculating the maximum probability. All the conditional distributions defined in the Bayes Net are the product of *P(V)* is

$$P(A_1, A_2 \dots A_N) = \prod_{i=1}^{N} P(A_i/Pa_i) \cdots\cdots\cdots (7)$$

When the conditional distribution of A_i is $P(A_i/Pa_i)$, provided that Pa_i represents that parent collection of A_i. For each vector, a conditional distribution does have a parametric structure that must be obtained by calculating the possible probability.

5. RESULT AND DISCUSSION

5.1. Effect of features in classification

The extracted features were used to distinguish between good and faulty conditions from the vibration signals. During the process, a few of the statistical features may be less significant in the classification of faults which can be removed. The significance of the extracted features were observed in the tree by depth-first technique which is use to find the rank of features in classifying the class efficiently [10]. It was visualised that the significance of features in the process of classification is in the sequence of decreasing order i.e. Sum, Mean, Sample Variance, Skewness, Kurtosis, and Median. Further addition of features reduces the accuracy and increases the computation time. The features selected were utilized to establish a model that provides the best possible classification accuracy for a practical application which required less computational power. The Figure 4 represent the effect of number of features in J48 classifier performance in no load condition and in Table 3 compared the variation in classification accuracy before and after feature selection process.

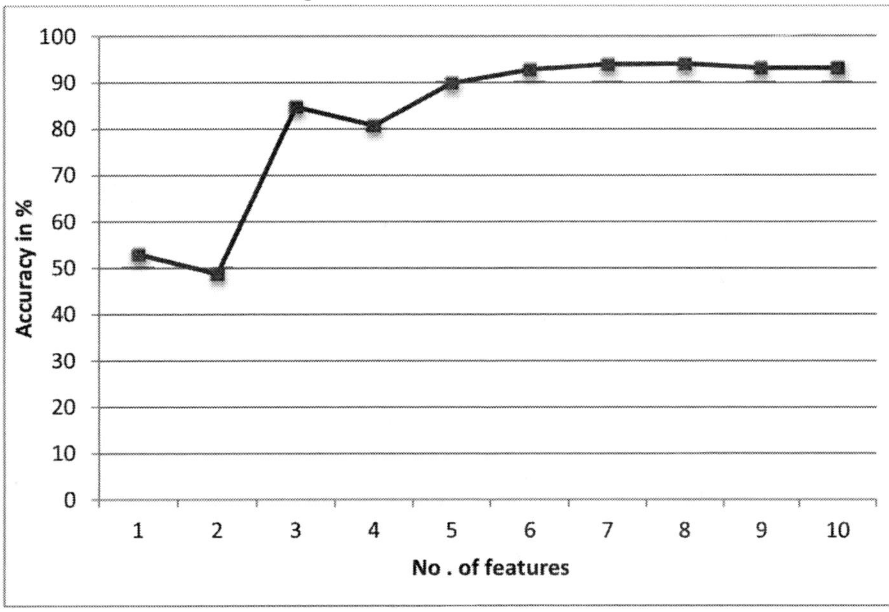

Figure 4 Effect of features in classification

Table 3 Effect of features in classification accuracy using J48 algorithm

J48 Classifier- with minimum number of obj 2	Accuracy in %	
	Using all the statistical features	Using selected statistical features
No-load	93.125	94
Load at 200psi	92.25	92.375
Load at 400 psi	91	91.375

5.2. Performance evaluation of classifiers to varying load

In the present study, Bayesian algorithms namely, Naive Bayes and BayesNet were used to classify faults from the selected features. During this process, the default parameters were kept uniform for both BayesNet and Naive Bayes classifiers. The resulting classification accuracy shows that the BayesNet classifier outperformed the Naive Bayes classifier by classifying the maximum number of instances correctly. Table 4 represents the comparison of Bayesian classifiers for different load conditions. The comparative plot of the classifier is shown in Figure 5.

Table 4 Comparison of classifier performance to varying load

Condition	No Load		200 Psi		400Psi	
Classifiers	BayesNet	Naive Bayes	BayesNet	Naïve Bayes	BayesNet	Naive Bayes
Time- sec	0.06	0.01	0.02	0	0.01	0
Accuracy %	88.5	86.25	88.5	86.37	88.25	87.5

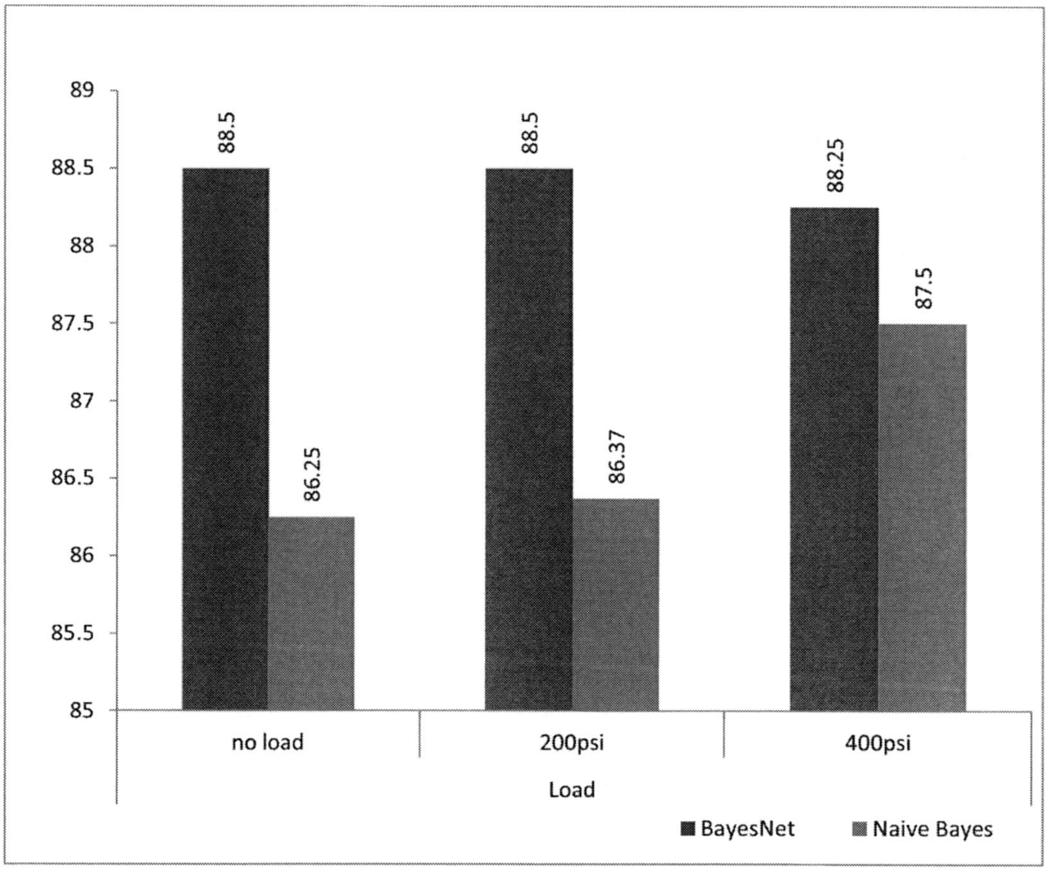

Figure 5 Performance of the classifier to varying load condition

Table 5 overall performance of Bayesian classifiers

Classifier	Overall Accuracy %	Time to build model (sec)
BayesNet	88.41	0.03
Naive Bayes	86.70	0.003

5.3. Classification using BayesNet

The performance of the BayesNet classifier at no load condition is represented in the form of the confusion matrix in Table 5. In the confusion matrix 'a' denotes Good condition and alphabet 'b' to 'h' represent Lower arm ball joint fault (labjf), lower arm bush worn-out (labwo), strut external damage (sted), strut mount fault (stmf), strut worn-out (stwo), tie rod ball joint worn-out(trbjf), and wheel low pressure(wlp) respectively. During classification process, the parameter is set to the default value, and verified with 10 fold cross-validation, and found that BayesNet classifier classify fault conditions with 88.5% accuracy i.e. out of 800 instances the 708 instances were correctly classified rest 92 instances were misclassified., in the case of Naive Bayes classifier with no load condition classifier has accuracy of 86.25%.

The accuracy of 88.5% was attained by following classification by a BayesNet classifier in no load condition. The table 6 showns the confusion matrix bayes net classifier in no load condition,the diagonal element in the confusion matrix below denotes the correctly classified instance, other element in the matrix denotes misclassified fault instances.

Table 6 Confusion matrix of BayesNet classifier at no-load condition

a	b	c	d	e	f	g	h	classified as
99	0	0	0	0	0	0	1	a = good
0	**77**	19	0	0	1	0	3	b = labjf
0	7	**83**	6	4	0	0	0	c = labwo
0	0	11	**72**	3	0	14	0	d = sted
0	1	8	1	**90**	0	0	0	e = stmf
0	3	1	0	0	**96**	0	0	f = stwo
0	0	0	4	1	0	**95**	0	g = trbjf
1	0	3	0	0	0	0	**96**	h = wlp

The trend of superior performances of BayesNet classifiers is followed in other loading conditions. Bayes net classifier has a classification accuracy of 88.5% and 88.25% ,Naive Bayes has 86.37% and 87.5% classification accuracy for 200 psi and 400 psi load conditions respectively, the classifier classifies the fault conditions with 88.25% accuracy. The confusion matrix of BayesNet classifier at 200 psi and 400 psi load is given in Table 6 and Table 7 respectively from the confusion matrix we can understand that the fault signal low arm ball joint fault (b = labjf) and strut external damage(d = sted) has maximum misclassification comparing all the load condition, the signal pattern for the particular faults is been interfere by the external noise causing misclassification since signal taken in dynamic situations, also there is chance for misclassification is due to the algorithm due to limitation in number of data set taken for the particular studies, is not sufficient to establish the relationship between independent variables.From this one can understand that there is a need for study to reduce misclassification further.

Table 7 Confusion matrix of BayesNet classifier at 200 psi load

a	b	c	d	e	f	g	h	classified as
91	3	2	2	1	0	1	0	a = good
12	**75**	0	1	9	0	0	3	b = labjf
0	0	**93**	0	1	0	0	6	c = labwo
4	0	0	**76**	7	12	1	0	d = sted
0	6	0	3	**90**	0	1	0	e = stmf
0	0	0	4	0	**96**	0	0	f = stwo
1	1	0	0	2	0	**96**	0	g = trbjf
5	1	1	1	1	0	0	**91**	h = wlp

Table 8 Confusion matrix of BayesNet classifier at 400 psi load

a	b	c	d	e	f	g	h	classified as
83	0	0	3	0	10	4	0	a = good
0	**92**	2	0	0	0	0	6	b = labjf
0	2	**97**	0	0	0	0	1	c = labwo
0	3	1	**78**	9	0	5	4	d = sted
2	0	0	2	**94**	0	2	0	e = stmf
12	0	0	0	0	**88**	0	0	f = stwo
0	0	0	1	5	0	**94**	0	g = trbjf
0	13	0	7	0	0	0	**80**	h = wlp

6. CONCLUSION

This article describes an approach based on Bayes classifier that identifies the variations in vibration signals pattern for different suspension fault conditions. The statistical features from vibration signals were extracted and further feature selection was carried out using J48 algorithm. Classification was performed on selected features with the help of two Bayesian based machine learning classifiers. A 10-fold cross-validation was adopted to verify the BayesNet and Naive Bayes classifiers. The performances between the classifiers are evaluated and it is observed that the BayesNet classifier displayed maximum over all classification accuracy of 88.4%, under all loading conditions when compared with Naive Bayes classifier having 86.70%. The faults lower arm ball joint fault and external strut damage are the reason for the low accuracy so there is further need of study on misclassific ation. We can conclude that the BayesNet classifier will be more suitable for identifying other five faults in the suspension system effectively with given number of data set. Thereby increase the reliability, comfort of the vehicle and reduce the downtime of the suspension system.

References

[1] Solanki V V and Patel P N S 2014 Static Analysis and Geometric Optimization of Independent Suspension Link Int. J. Thesis Proj. Diss. **2** 47–57
[2] Colombo D, Gobbi M, Mastinu G and Pennati M 2009 Analysis of an unusual McPherson suspension failure Eng. Fail. Anal. **16** 1000–10
[3] Borner M, Isermann R and Schmitt M 2002 A sensor and process fault detection system for vehicle suspension systems SAE Technical Papers
[4] Khode S S 2017 A Review on Independent Suspension System of Light Commercial Vehicle IOSR J. Mech. Civ. Eng. **17** 14–9
[5] Konieczny L and Burdzik R 2017 Modern suspension systems for automotive vehicles and their test methods Vibroengineering Procedia **14** 233–7
[6] Joshuva A and Sugumaran V 2020 A lazy learning approach for condition monitoring of wind turbine blade using vibration signals and histogram features Meas. J. Int. Meas. Confed. **152**
[7] Ventura P J C, Ferreira C D H, Neves C F C S, Morais R M P, Valente A L G and Reis M J C S 2008 An embedded system to assess the automotive shock absorber condition under vehicle operation Proc. IEEE Sensors 1210–3
[8] Ferreira C, Ventura P, Morais R, Valente A L G, Neves C and Reis M C 2009 Sensing methodologies to determine automotive damper condition under vehicle normal operation Sensors Actuators, A Phys. **156** 237–44
[9] Luís Carvalho, Sérgio Santos C F 2019 A new methodology for detection of a loose or worn ball joints used in vehicles suspension system Biosyst. Eng. **1** 1–7
[10] Burdzik R and Konieczny Ł 2014 Application of vibroacoustic methods for monitoring and control of comfort and safety of passenger cars Solid State Phenomena vol 210 pp 20–5
[11] Wang G and Yin S 2014 Data-driven fault diagnosis for an automobile suspension system by using a clustering based method J. Franklin Inst. **351** 3231–44
[12] Alamelu Manghai T M and Jegadeeshwaran R 2019 Vibration based brake health monitoring using wavelet features: A machine learning approach JVC/Journal Vib. Control
[13] Alamelu Manghai T M and Jegadeeshwaran R 2019 Vibration based real time brake health monitoring system – A machine learning approach IOP Conf. Ser. Mater. Sci. Eng. **624** 012027
[14] Sharma A, Sugumaran V and Babu Devasenapati S 2014 Misfire detection in an IC engine using vibration signal and decision tree algorithms Meas. J. Int. Meas. Confed. **50** 370–80
[15] Anoop P S and Sugumaran V 2017 Classifying machine learning features extracted from vibration signal with logistic model tree to monitor automobile tyre pressure SDHM Struct. Durab. Heal. Monit. **11** 191–208
[16] Karabulut E M, Özel S A and İbrikçi T 2012 A comparative study on the effect of feature selection on classification accuracy Procedia Technol. **1** 323–7
[17] Aravanis T C I, Sakellariou J S and Fassois S D 2020 A stochastic Functional Model based

method for random vibration based robust fault detection under variable non–measurable operating conditions with application to railway vehicle suspensions J. Sound Vib. **466** 115006

[18] Praveenkumar T, Jasti A, Saimurugan M and Ramachandran K I 2014 Vibration based fault diagnosis of automobile gearbox using soft computing techniques ACM Int. Conf. Proceeding Ser. **10-11-Oct**

[19] Sugumaran V and Ramachandran K I 2011 Effect of number of features on classification of roller bearing faults using SVM and PSVM Expert Syst. Appl. **38** 4088–96

Fault Detection in aerial images of photovoltaic modules based on Deep learning

S.Naveen Venkatesh[1], **V.Sugumaran**[1*]

[1]School of Mechanical Engineering, Vellore Institute of Technology, Chennai, India

E-mail: v_sugu@yahoo.com

Abstract. Operation and maintenance of photovoltaic (PV) modules are currently the prime concerns of the expanding photovoltaic industry. Unmanned aerial vehicles (UAVs) are applied in the field of inspection and monitoring of faults that occur in a photovoltaic module (PVM). Such inspections can significantly reduce time and human interference to provide accurate classification results. Technological advancements and innovative techniques in the fast moving world expect instantaneous results. Fault diagnosis is one such technique that provides instantaneous results and assures enhanced lifetime of various critical components. This paper presents the fault detection in PVM based on deep learning with the help of aerial images acquired from UAVs. Convolutional neural networks (CNN) are adopted to extract high level features from the images which are classified using the softmax activation function. The feature extraction and fault classification is carried out by using a pre-trained VGG16 network. A total of six test conditions are considered in the study. Burn marks, delamination, discoloration, glass breakage, good panel and snail trail are the several test conditions considered. The classification result for the pre-trained CNN model is exhibited and performance of the model is evaluated.

1. Introduction

Futuristic power generation depends essentially on renewable energy sources. Apart from water and wind energy, solar energy is considered as one of the crucial technologies that supply about two percent of global power demand. Solar energy is harnessed from large solar power plants spread across various geographical regions with an estimated power output of 1GW. Solar farms are constructed with a large number of photovoltaic modules (PVM). The overall power output in solar PVM tends to degrade during their life span due to faults. Ever-changing environmental conditions with an increase in natural hazards make assessing PVM conditions and identifying faults necessary (1). Fault diagnosis is an appropriate technique to determine the faults that occur in a PVM. Several conventional fault diagnosis techniques are available; however, such techniques are time consuming and are limited to single fault detection. Thus, an advanced fault diagnosis technique is the need of the hour. Various authors have suggested the using an image based fault diagnosis technique with the application of unmanned aerial vehicles (UAVs) (2,3). Visual inspection of the acquired aerial images consumes a lot of manpower. Also, the accuracy of fault detection in manual inspection depends highly on the examiners experience and varies with every examiner.

Considering the constraints mentioned above, a reliable, fast and low cost evaluation of PVM is proposed in the present study. An automatic fault diagnosis technique utilizing deep convolutional neural networks (CNN) is employed to classify the PVM faults from the acquired aerial images. Rapid rise in innovative technologies and fault detection techniques have minimized human interventions in various fields of inspection. Accurate and precise outcomes delivered by the innovative techniques have paved way for the replacement of conventional techniques with automatic approaches. Deep learning technique automatically delivers the classification results in identifying the type of faults. Several literatures have described the use of deep learning in fault diagnosis of PVM which are discussed as follows. A deep learning based approach was performed by Li et.al for automatic detection of faults in large scale PV plants (4). The authors considered six test conditions and overall classification accuracy was found to be 97.9%. The major advantages of deep learning are utilizing pre-trained models or build models from scratch. Several authors in (5–8) have utilized deep learning in the field of fault diagnosis by building models from scratch. Also, various pre-trained models like AlexNet (9), ResNet 50 (10), VGG16 (11) etc have been applied in various image classification applications. Artificial neural networks have been put into use effectively for diagnosing loss in the output power of PVM with the help of power estimation using the sun's position. Similarly, the condition of PVM is diagnosed using low resolution satellite images with the help of CNN producing higher classification rate exhibits the ability of deep learning based techniques to assess images of PVM (12). In general, CNN require large amount of training data such that the model is trained well to perform classification. Certain challenges persist while using CNN based classifications which are listed as follows.

- Acquisition of good quality sound data is the most challenging task in CNN.
- Multiclass classification requires uniform dataset. Bias in datasets can result in misclassification.
- Nominal hardware is required for performing training and classification.

The above mentioned challenges can be addressed by implementing data preprocessing and data augmentation techniques. Preprocessing image data removes unwanted noise from the image data, thereby creating quality dataset. Data augmentation artificially expands the image dataset such that the model can be trained effectively. Based on the literature survey, the following observations were made.

- The conventional techniques available are confined for identification of a single fault at an instance. Detection of multiple faults with advanced techniques can be considered as an up and coming research interest in the field of fault detection and diagnosis.
- Most of the literatures focus on diagnosis of electrical faults but only a few have reported on visual faults.
- Internet of things and artificial intelligence are the emerging techniques that provides a wide scope in the development of advanced fault detection and diagnosis techniques.
- A combination of several conventional fault detection and diagnosis techniques generally termed as hybrid techniques can be time saving and cost effective with more impact in fault detection and diagnosis.

In the present study, the feasibility of CNN is examined to detect faults in a PVM using aerial images. A pre-trained VGG16 (13) model is used to perform the classification task on the aerial images. VGG16 in comparison with AlexNet contains a number of convolutional layers with smaller filters stacked together to learn and extract complex features from the images provided. The pre-trained VGG16 exhibits better performance compared to other pre-trained models with fewer convolution layer (14). The major advantage of CNN is its ability to work with low resolution and low quality images. A total of six test conditions namely, burn marks, delamination, discoloration, glass breakage, good panel and snail trails are considered. The list of technical contributions made is listed as follows.

- A pre-trained VGG16 model is used to classify various faults from aerial images of PVM.
- Multiple faults are considered.

- Data augmentation technique is used to expand the image dataset.
- Uniform dataset is considered such that misclassification can be reduced.
- The superior classification performance of the model is evaluated with the help of confusion matrix. Softmax is used as classification function.

2. Experimental procedure

2.1. UAV based Data Acquisition and Monitoring System

Unmanned aerial vehicles (UAVs) have been employed in a wide range of fields such as surveillance, security, infrastructure inspection and emergency conditions. In recent years, UAV have been employed in the field of fault diagnosis due to their ability to acquire precise data over large geographical regions in limited time. Conventional visual inspection techniques were performed with the help of skilled inspectors for long period of time. Such methods had certain drawbacks like: (1) large areas require huge man power (2) availability of trained examiners (3) human error rate (4) safety concerns of examiners (5) delayed results (15,16). The above mentioned drawbacks have paved way for innovative methods to perform monitoring on PVM. The world is moving at a fast pace and expects instantaneous results in almost every real time applications. Thus, several researchers and capitalists have suggested the use of UAV as an advanced solution. The overview of UAV based data acquisition and monitoring system is provided in Figure 1. UAV are widely used across the globe for inspection purposes due to the following reasons.

- Minimized danger and health risks for inspectors
- High and in depth data collection
- Faster launch or deployment
- Versatile usage
- Quick and easy accessibility to data
- Inspection of operating machineries and reduces downtime
- Cost effective and Time efficient

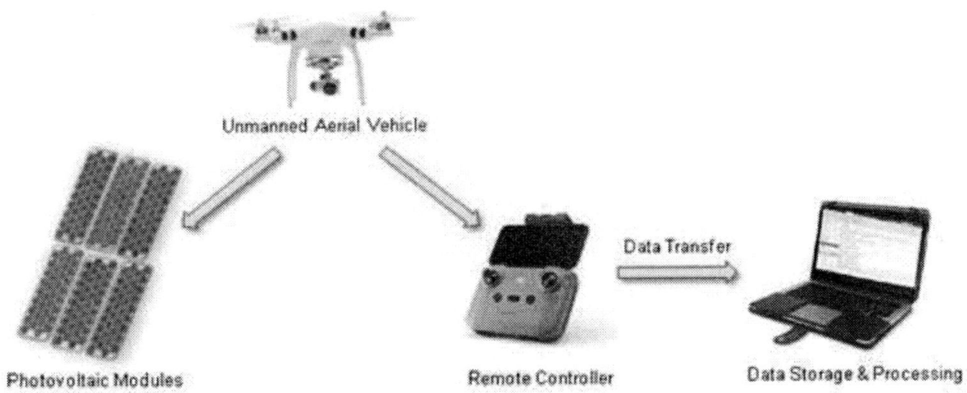

Figure 1. Overview of UAV based Data Acquisition and Monitoring System

2.2. Visible faults in a PVM

Extended outdoor operation under varying environmental conditions is the major reason for fault occurrences in a PVM. Faults in a PVM will have a direct effect on the reliability, lifetime and performance. Detection and diagnosis of faults at an early stage will resist the performance degradation of PVM by the application of necessary preventive measures. Figure 2 depicts the most common visible faults that occur in a PVM. Burn marks, delamination, discoloration, glass breakage and snail trail (17–21)are the visible faults considered in the study. A brief description of the visible faults with their cause and effects are presented in Table 1.

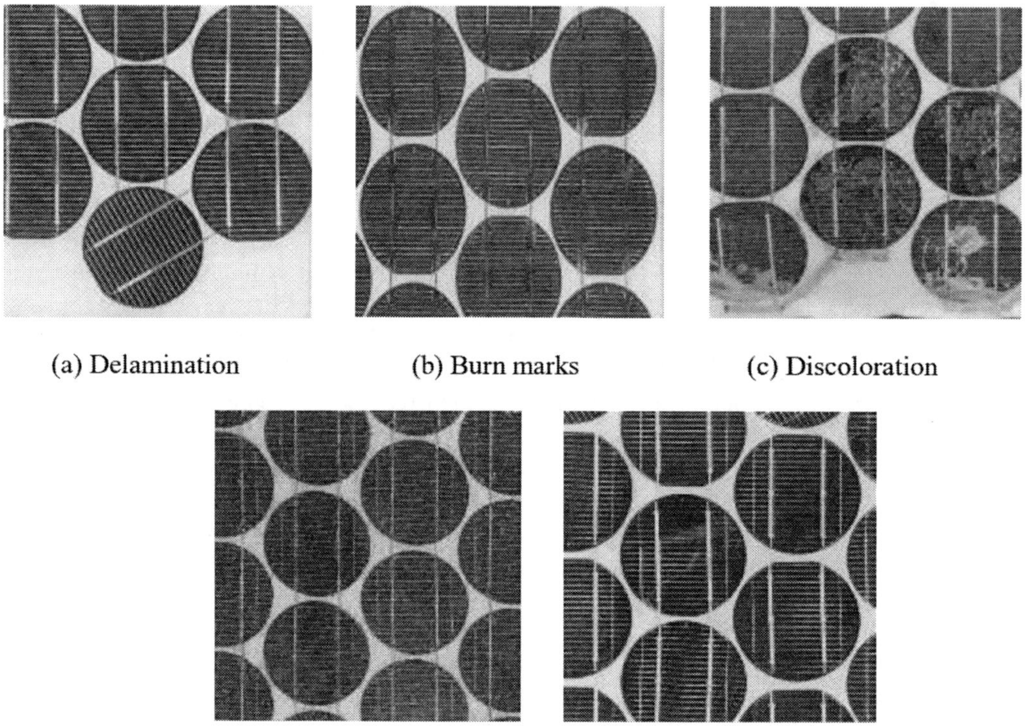

(a) Delamination (b) Burn marks (c) Discoloration

(d) Glass breakage (e) Snail trail

Figure 2. PVM visible faults (22)

Table 1. Brief description of visible faults in a PVM

S.No	Visible Faults	Reason for fault occurrence	Effects
1	Burn marks	Failure in solder bonds, local heating and ribbon breakage	Output power loss and Fire hazards
2	Delamination	Loss in adhesion between glass, encapsulant and back cover	Corrosion
3	Discoloration	High exposure to ultraviolet radiation, humidity and heat	Yellowing and browning of PVM
4	Glass breakage	Thermal stresses, impacts during transportation and installation	Moisture penetration, corrosion and lower irradiance
5	Snail trails	Stress induced on micro cracks along edges	Accelerated degradation

2.3. Preprocessing and Data augmentation of aerial images
The aerial images collected with the help of UAVs will be of different sizes and resolutions. Further, factors such as external noises, panel reflections, vibrations due to flight motion and other external

factors hinder the quality of images obtained. To improve the resolution of images and provide effective classification of various faults, such disturbances must be removed from the images. Also, larger images cannot be fit into CNN models as it increases the complexity and computational time. Image pre-processing is one technique which helps in solving the aforementioned uncertainties. Noise removal, reshaping and resizing are some of the methods involved in image preprocessing. The pre-trained VGG16 architecture accepts images of size 224x224 pixels. Hence, all the acquired images must be resized to a uniform size that is acceptable by VGG16 architecture. The UAV acquired 100 images for each test condition accounting for a total of 600 images. However, such small custom datasets might be insufficient for training CNN networks and can result in overfitting. Data augmentation is adopted in this study to artificially expand the acquired aerial images dataset. Using this technique, several image transformation functions like rotation, blur, horizontal shift, vertical shift and crop were applied on the acquired images. After the application of data augmentation, the image dataset expanded from a total of 600 images to 3150 images (525 images for each test condition). Table 2 represents the data augmentation functions utilized.

Table 2. Data Augmentation functions

Transformation functions	Value
Rotation (Clockwise, Anticlockwise)	0° - 180°
Flip (Horizontal, Vertical)	90°
Noise	Random
Blur	Gaussian
Warp	40

3. Pre-trained VGG16 CNN Architecture

The VGG16 network architecture was proposed by Karen Simonyan and Andrew Zisserman in the annual ImageNet Large Scale Visual Recognition Challenge (ILSVRC) (13). The architecture contains a stack of 13 convolutional layers, 5 max pooling layers, 3 fully connected layers and 1 output/classification layer. The overall architecture of VGG16 network is depicted in Figure 3.

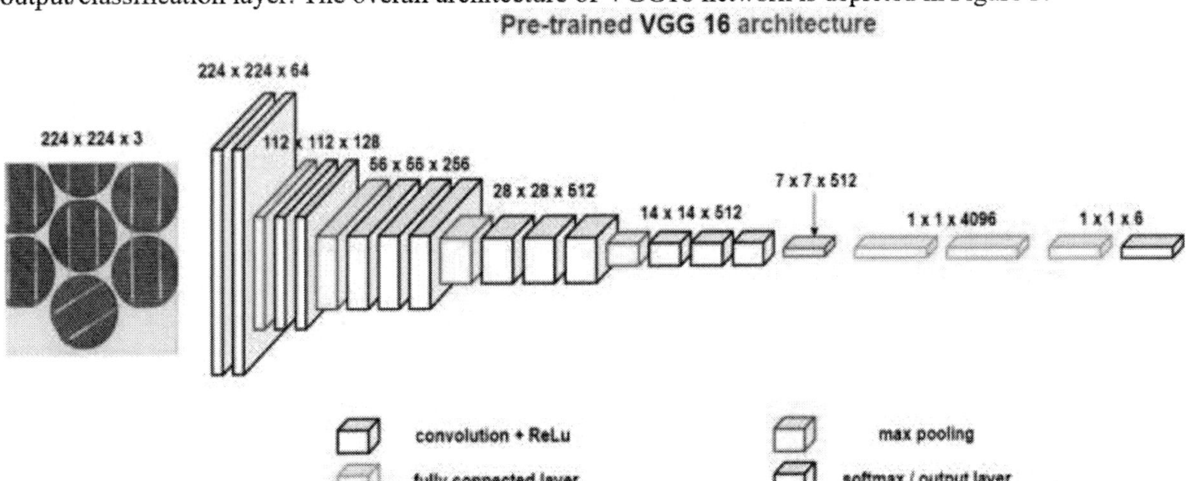

Figure 3. Overview of pre-trained VGG16 model

VGG16 architecture contains a total of 13 convolution layers arranged in a particular designed pattern for classification of images. A kernel of dimension 3×3 containing learnable parameters such as A and c are transferred over an image of y pixels to perform the convolution operation that delivers an output x. The movement of the kernel is determined by the stride that commands the kernel to move either pixel wise or skipping of several pixels. The following equation represents the simple working of the convolutional operation.

$$x = f(Ay + c)$$

Convolutional neural networks have the capability to extract features and patterns automatically from the aerial images for discriminating every PVM test condition. Simple features like edges are learnt by the initial convolutional layers that combine with the features extracted by the following convolutional layers resulting in complex features extraction. For every convolution layer passed the image gets resized and is broken down into simpler volume such that the most significant features are extracted. The size of the images depends on the filter size passed through and the stride considered. Each convolution layer contains a non-linear activation function called Rectified Linear Unit (ReLU) that introduces uncertainty (23). Each convolution layer is supported by a max pooling layer to down sample and reduce the size of activation map. The stacked arrangement of convolution layers ends with a classification layer. In the present condition, the convolution layers in the VGG16 architecture are connected to two fully connected layers consisting of 4096 neurons. Additionally, the two fully connected layers are followed by another fully connected layer with 6 neurons representing the number of considered PVM test condition. A final output layer consisting of softmax function performs the classification on the images. A padding of 1 pixel is performed after every convolutional layer to prevent spatial feature of images.

. A pre-trained VGG16 model present in deep learning toolbox of Matlab 2020a version is used to perform the classification process. The hyper parameters settings used for fine tuning the VGG16 architecture is displayed in Table 3.

Table 3. Hyper parameters used for fine tuning VGG16 architecture

Hyper parameters	Values
Number of epochs	5
Mini batch size	10
Initial learning rate	0.0001
Weight learning rate	0.003
Momentum	0.9

4. Results and Discussion

The PVM dataset created was split into two categories namely, training and validation dataset. The split ratio used for the process is 0.8 which means the dataset is split into 80% training data and 20% validation data. The reason for selecting such split is that more training data can make the network learn more features. The validation dataset images are selected in a random order in every PVM condition. Further, the model is compiled using adaptive moment estimation (ADAM) optimizer with a validation frequency of 100 iterations. Figure 4 represents the overall performance of the pre-trained VGG16 model. The plots depict the variation of accuracy and loss for both training and validations sets with respect to the number of iterations. The proposed VGG16 model is initially assessed with the validation dataset prior being adopted for fault classification. The plots show that the training of the pre trained VGG16 model converges quickly for the validation dataset. The model also exhibits minimum error thereby approving the feasibility of network for fault classification problems. From Figure 4 one can observe that the increase in number of epochs and iterations tends to show a gradual rise in both training and validations accuracy. The trend of training accuracy seems to be quite uniform after 400 iterations and the overall average training accuracy was found to be 98.47%. Faster convergence of network depicts the accelerated learning ability of the network resulting in minimized loss function. The overall validation accuracy of the trained VGG16 network was observed to be 95.40% which is also evident from the confusion matrix depicted in Figure 5.

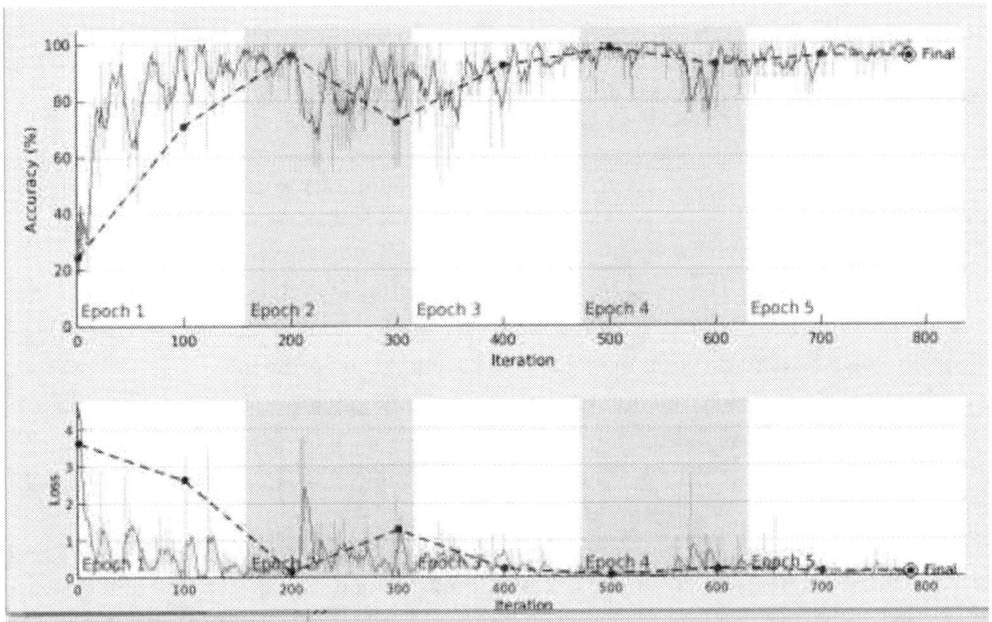

Figure 4. Overall performance of pre trained VGG16 network

From Figure 5, one can infer that only one test conditions of PVM namely, good panel has misclassified to a greater extent. The rate of misclassification indicates that the features extracted are not sufficiently trained and requires more amount of training such that the model learns the features well. Out of 630 instances taken into validation, 601 instances are correctly classified and 29 instances are misclassified.

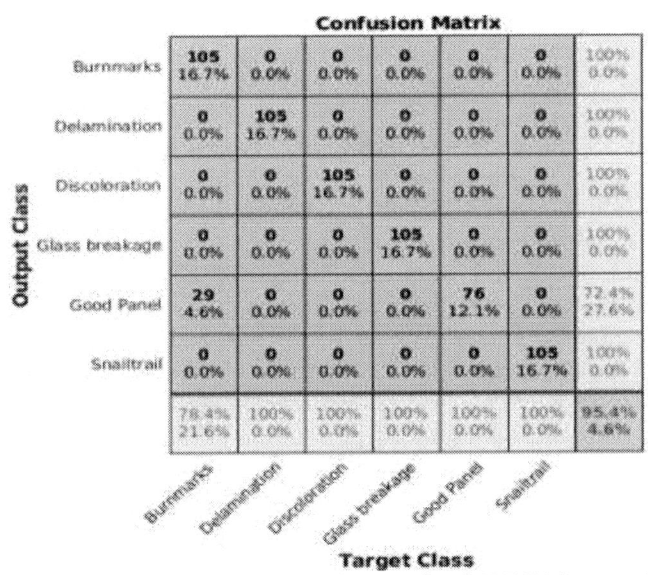

Figure 5. Confusion matrix of pre trained VGG16 network

From the confusion matrix it can be observed that good panels are misclassified as faulty panels. The model can be regarded as a better working model, since the good panels are considered as faulty. Such situation gives rise for immediate inspection of the panels. On the other hand, if the model misclassifies faulty panels as good panels then it will lead to a situation of safety concern. Faulty panels being misclassified as good panels are considered dangerous rather than good panels being misclassified as faulty panels. In the former case, the need for inspection is eliminated and can result in safety hazards. Based on the obtained results, one can suggest the use of pre trained VGG16 for fault classification problems to be used in real time applications.

5. Conclusion

This paper presented a fault classification model of PVM conditions in aerial images with the help of a pre trained VGG16 model. The outcomes of the present work are listed as follows.

- An effective image preprocessing and data augmentation technique applied in the work is exhibited. The model is trained for a limited number of iterations.
- A multiclass classification is performed for all the conditions and the numerical results confirm that the classification model is more accurate and efficient.
- Typical PVM conditions including burn marks, delamination, discoloration, glass breakage, good conditions and snail trail are considered.
- The results explained that the model produced high classification accuracy of 95.40% in classifying all the PVM conditions.

The exhibited solution can be put into use on a real time basis after enhanced training and various folds of validation. On further development, such kind of models can even be integrated on UAV platform for performing inspection on large scale PVM farms. Automation of inspection techniques with UAVs can minimize human interference, reduce time consumption and eliminate manual errors. Several other pre trained models like ResNet50, GoogLeNet, VGG19 etc can be used for performance comparison in future works.

References

[1] Madeti S R and Singh S N 2018 Modeling of PV system based on experimental data for fault detection using kNN method Sol. Energy 173 139–51

[2] Grimaccia F, Leva S, Dolara A and Aghaei M 2017 Survey on PV Modules' Common Faults after an O&M Flight Extensive Campaign over Different Plants in Italy IEEE J. Photovoltaics 7 810–6

[3] Grimaccia F, Leva S, Niccolai A and Cantoro G 2018 Assessment of PV Plant Monitoring System by Means of Unmanned Aerial Vehicles Proc. - 2018 IEEE Int. Conf. Environ. Electr. Eng. 2018 IEEE Ind. Commer. Power Syst. Eur. EEEIC/I CPS Eur. 2018 1–6

[4] Li X, Yang Q, Lou Z and Yan W 2019 Deep Learning Based Module Defect Analysis for Large-Scale Photovoltaic Farms IEEE Trans. Energy Convers. 34 520–9

[5] Pierdicca R, Malinverni E S, Piccinini F, Paolanti M, Felicetti A and Zingaretti P 2018 Deep convolutional neural network for automatic detection of damaged photovoltaic cells Int. Arch. Photogramm. Remote Sens. Spat. Inf. Sci. - ISPRS Arch. 42 893–900

[6] Deitsch S, Christlein V, Berger S, Buerhop-Lutz C, Maier A, Gallwitz F and Riess C 2019 Automatic classification of defective photovoltaic module cells in electroluminescence images Sol. Energy 185 455–68

[7] Tang W, Yang Q, Xiong K and Yan W 2020 Deep learning based automatic defect identification of photovoltaic module using electroluminescence images Sol. Energy 201 453–60

[8] Correa-Jullian C, Cardemil J M, López Droguett E and Behzad M 2020 Assessment of Deep Learning techniques for Prognosis of solar thermal systems Renew. Energy 145 2178–91

[9] Krizhevsky A, Sutskever I and Hinton G E 2017 ImageNet classification with deep convolutional neural networks Commun. ACM 60 84–90

[10] Wen L, Li X and Gao L 2019 A transfer convolutional neural network for fault diagnosis based on ResNet-50 Neural Comput. Appl. 0123456789

[11] Kim K W, Hong H G, Nam G P and Park K R 2017 A study of deep CNN-based classification of open and closed eyes using a visible light camera sensor Sensors (Switzerland) 17

[12] Lu X, Lin P, Cheng S, Lin Y, Chen Z, Wu L and Zheng Q 2019 Fault diagnosis for photovoltaic array based on convolutional neural network and electrical time series graph Energy Convers. Manag. 196 950–65
[13] Simonyan K and Zisserman A 2015 Very deep convolutional networks for large-scale image recognition 3rd International Conference on Learning Representations, ICLR 2015 - Conference Track Proceedings pp 1–14
[14] Krishnaswamy Rangarajan A and Purushothaman R 2020 Disease Classification in Eggplant Using Pre-trained VGG16 and MSVM Sci. Rep. 10 1–11
[15] Guerriero P, Cuozzo G and Daliento S 2016 Health diagnostics of PV panels by means of single cell analysis of thermographic images EEEIC 2016 - Int. Conf. Environ. Electr. Eng.
[16] Zhang D, Wu F, Li X, Luo X, Wang J, Yan W, Chen Z and Yang Q 2017 Aerial image analysis based on improved adaptive clustering for photovoltaic module inspection 2017 Int. Smart Cities Conf. ISC2 2017
[17] Rajput P, Tiwari G N, Sastry O S, Bora B and Sharma V 2016 Degradation of mono-crystalline photovoltaic modules after 22 years of outdoor exposure in the composite climate of India Sol. Energy 135 786–95
[18] Bouraiou A, Hamouda M, Chaker A, Lachtar S, Neçaibia A, Boutasseta N and Mostefaoui M 2017 Experimental evaluation of the performance and degradation of single crystalline silicon photovoltaic modules in the Saharan environment Energy 132 22–30
[19] Han H, Dong X, Li B, Yan H, Verlinden P J, Liu J, Huang J, Liang Z and Shen H 2018 Degradation analysis of crystalline silicon photovoltaic modules exposed over 30 years in hot-humid climate in China Sol. Energy 170 510–9
[20] Bouraiou A, Hamouda M, Chaker A, Neçaibia A, Mostefaoui M, Boutasseta N, Ziane A, Dabou R, Sahouane N and Lachtar S 2018 Experimental investigation of observed defects in crystalline silicon PV modules under outdoor hot dry climatic conditions in Algeria Sol. Energy 159 475–87
[21] Dolara A, Lazaroiu G C, Leva S, Manzolini G and Votta L 2016 Snail Trails and Cell Microcrack Impact on PV Module Maximum Power and Energy Production IEEE J. Photovoltaics 6 1269–77
[22] S N V and Sugumaran V 2020 Fault diagnosis of visual faults in photovoltaic modules : A Review Int. J. Green Energy 00 1–14
[23] Jiang X, Pang Y, Li X, Pan J and Xie Y 2018 Deep neural networks with Elastic Rectified Linear Units for object recognition Neurocomputing 275 1132–9

Comparative study on different balancing conditions of an air filled tyre using statistical features and classification via regression algorithm

P S Anoop[1] V Sugumaran[1*]

[1] School of Mechanical Engineering, Vellore Institute of Technology, Chennai campus, Chennai, India.

E-mail: v_sugu@yahoo.com

Abstract: Tyre condition monitoring systems (TCMS) are the safety systems used in a vehicle for measuring the condition of tyre like tyre pressure, temperature, balancing etc. In this era of increasing vehicle accidents, these systems are having paramount importance in terms of safety. The current technology TCMS uses direct sensors like pressure sensors or wheel speed sensors etc, which are highly expensive. This paper puts forward an innovative indirect TCMS system using condition monitoring techniques and machine learning. For different five balancing conditions vertical vibrations from fixed wheel hub were extracted from a moving air filled tyre with the help of an accelerometer. The tyres filled with air are considered with different pressure values to represent puncture, normal, idle and high pressure conditions. In feature extraction process, the statistical features were extracted from the acquired signals and the prominent features were selected using J48 algorithm. Selected features were classified with the help of classification via regression algorithm and reasonable high accuracy was obtained. This paper attempts to study the effect of unbalance of the wheel on the classification accuracy of an indirect TPMS system. The results are compared and presented

1. Introduction

Tyres are circular parts attached to the rims for enhancing the axial load of the automobile to the ground. Pneumatic and non-pneumatic tyres are the two basic types of it. Most of the tyres, used in passenger automobiles and bicycles are air filled structures. These tyres absorb vertical vibrations as it roll and give the passenger a flexible cushioning effect. Non pneumatic tyres are airtight, combining the wheel and tyre into a single component. These are used in special type of vehicles such as tractors, riding lawn mowers etc. In terms of advantages, pneumatic tyres are valued more than non-pneumatic tyres because it finds noise-reducing ride, durability and ability to absorb shock. Depending on the method of production air-filled tyres are categorized as belted-ply, bias-ply and radial-ply.

Radial tyre is a type of tyre where the fabric piles (nylon layer) are arranged at right angles to the direction of travel. Main advantages of radial tyres are that

- The side walls are flexible.
- The contact area of radial tyres with road is high, so the vehicle stability will be higher.
- Fuel consumption will be less as the rolling distance is less.
- Heat production and vibration are low, so tyre life will be high.

The specification of the air-filled tyre used for the test is coded 165/80 R14 85 S. The various parameters in the tyre code are clearly illustrated in Figure. 1. The tyre selected for the current test is shown in Figure.2.

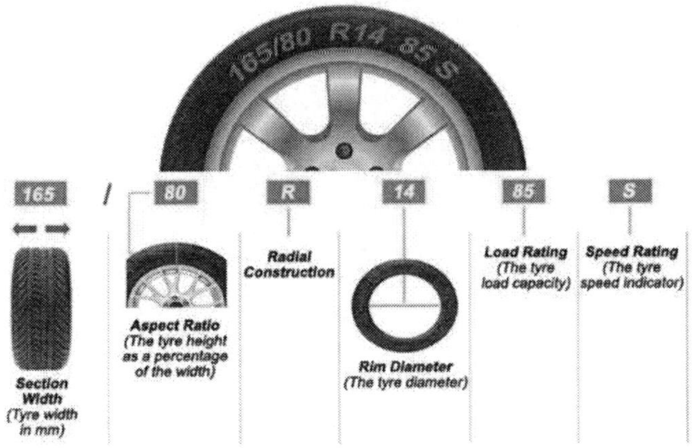

Figure 1 Tyre specification diagram

Figure 2 Tyre selected for current experiment

1.1 Tyre Condition Monitoring System (TCMS)

In this fast-growing automobile world, safety has always been a priority. Here TCMS finds its place. A study led by the UK tire industry gathering found that between 2003 and 2005, around 50% of light vehicle crashes were caused by tyres[1]. TCMS alarms tyre-allied problems and thus ensuring a safer mode of transportation. TCMS are small electronic subsystems that observe the tyre pressure and tyre balancing on individual wheels of a vehicle. A TCMS-mounted automobile helps the driver to check the condition of the tyre at low pressure or if something bad has happened to the tyre. An effective TCMS can easily predict tyre failures. This reduces the risk of tyre-related road accidents. Depending on the data acquisition method, TCMS can be classified into-Direct TCMS (dTCMS) and Indirect TCMS (iTCMS). As the name Implies, a direct TCMS pressure sensor measures the pressure and tyre condition and transfers the data to a controlling unit attached to the dashboard. In the Indirect TCMS, pressure and tyre condition are measured indirectly from the wheel speed sensors and then the data transferred to the control unit. The advantages of iTCMS are affordable price, easy maintenance and reliability. This study depends on indirect TCMS so as to prove that iTCMS have more advantages than that of dTCMS.

The direct TPMS is expensive than its indirect counterpart. Indirect TPMS requires less programming/maintenance over the years than a direct TPMS in which resynchronization may require costly tools. In this case of indirect TPMS, overall installation hassles are less than its direct counterpart. But in the case of direct TPMS, sensors should change periodically once battery is worn out and there is a high chance of damage to the sensor during uninstalling or installing. So, we have developed an Indirect TPMS by using an accelerometer module. Vibrations are used to manage the system.

While driving a car the tyres rotate uniformly due to the equal distribution of the balancing mass around the wheel. This method is mentioned as wheel balancing. Severity of unbalance in a tyre results in vibrations of varying degrees which reverberates through the vehicle chassis. If the severity of unbalance is more, it becomes very dangerous to drive because the steering wheel experiences severe vibrations. Tyre imbalance leads to higher fuel consumption at higher speeds. An increase in wheel imbalance causes a peak vibration at the wheels, which will result in severe damage to the vehicle.

In recent years, iTCMS research opportunities in current trending technologies such as machine learning and artificial intelligence have gained wide acceptance[2]. iTCMS initially operates on the principle that low-pressure tyres have a reduced diameter compared to properly inflated tyres. The main drawback of this system is that all four tyres fail to detect inflation at the same time. Subsequent studies, however, have addressed this shortcoming through spectrum analysis and advanced signal processing techniques [3]. The rolling resistance of a tyre is the amount of energy required to keep the tyre moving. Schuring and Futamura said that when the rolling resistance coefficient decreases, the fuel efficiency increases due to the hardness of the tyre.[4].

Experiments using accelerometer are having wide acceptance these days. Robinson et al., developed an experimental setup in which accelerometer was attached to the tyre and the vibrations were artificially applied to the tyre with a hammer[5].In another experiment Wei et al used an accelerometer and a piezoresistive pressure sensor for detecting the tyre pressure. When the vehicle start moving, accelerometer detects the motion and activates the pressure sensor[6]. An inductive powered transducer method and a moving piston method are the two systems suggested by Hill et al for measuring indirect tyre pressure. The main drawback of this system is that pressure can be detected only while the vehicle moves[7]. Kanwar et al proposed a piezoelectric vibration harvesting system, which powers the tyre pressure monitoring sensors[8].

In another study Craighead developed a technique in which he made use of vibration measurements and frequency spectrum analysis of normal tyre to detect the tyre pressure and wheel balance. The accuracy obtained was relatively less than 12% and the measure of unbalance was determined [9]. Hasan et al suggested a technique that uses air pressure in pneumatic tyres. If tyre pressure varies, an alert warning will be registered. The user can modify the minimum and maximum tyre pressure level[10].

Tyre life span is closely related to tyre pressure. One minute deflection in tyre pressure (17%) reduces tyre life by 25% and increases the overall fuel consumption by 2%[11]. Wu et al. proposed different energy harvesting methods using TPMS and determined the importance of piezoelectric sensors in TPMS[12]. Some commercial vehicle-only systems that determine tyre pressure and tyre radius using a Vehicle Stability Sensor and GPS[13]. Garcia-pozuelo et al. (2017) developed a strain-based technique to monitor the rolling speed, vertical load, inflation pressure by adding fuzzy logic[14].

Studies based on indirect tyre pressure monitoring system need to be accelerated. With the help of supervised machine learning, Svensson et al suggested an iTPMS. In this work the proposed method permits to detect thread depth and incorrect tyre pressure. The results show that the proposed system has an accuracy of 90.54%[15]. In general, several important experiments were carried out in the areas of both dTCMS and iTCMS[16]. Only a few studies have been conducted to detect both tyre pressure and tyre balancing simultaneously using iTCMS. Moreover proper emphasis on experiments based on machine learning was also very limited. But these are the needs of the hour. In this context, this study proposes a new approach which claims that both tyre pressure and tyre balancing can be measured using indirect TCMS with the help of machine learning techniques. Detailed methodology of the proposed system as shown in Figure. 3.

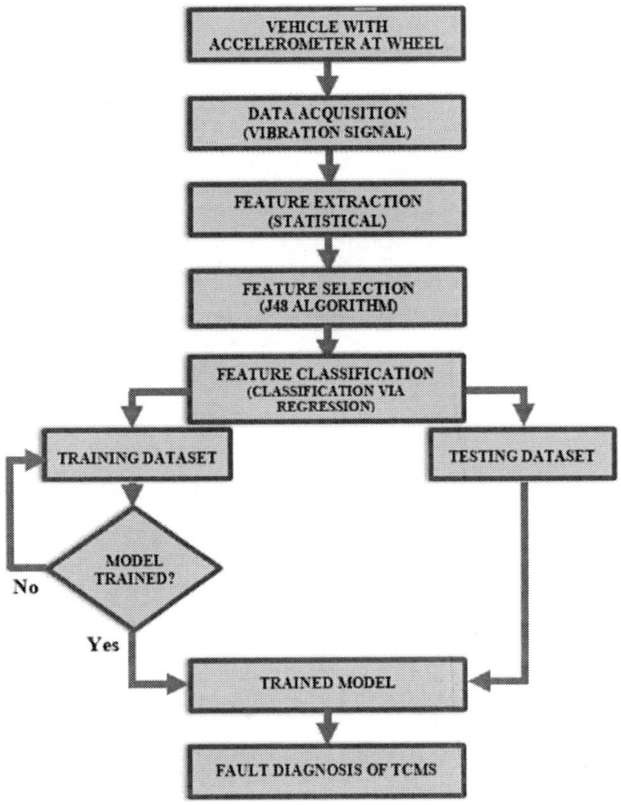

Figure 3 Methodology of proposed iTCMS

2. Experimental Setup

An accelerometer is mainly used to acquire vibration signals. In this experiment, a tri-axial accelerometer was fixed to the back left wheel hub of the selected automobile to avoid unwanted engine vibration signals (Refer Figure 4). 'NI USB-6001' was used as data acquisition (DAQ) device. To avoid the external electronic noise, a shielded wire was connected between accelerometer and DAQ. The specifications of the MEMS accelerometer and NI USB-6001 are given in [17]

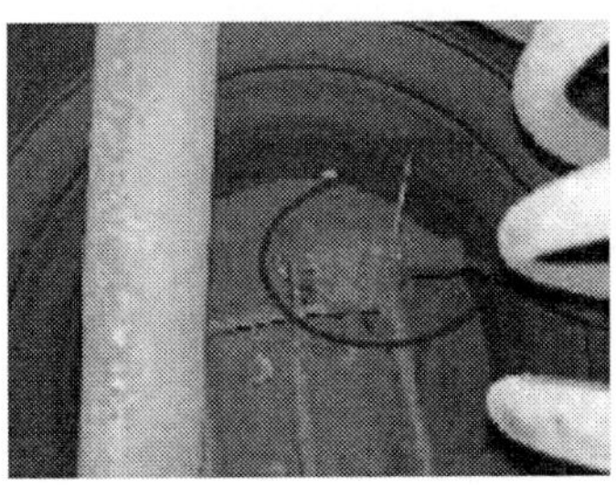

Figure 4 MEMS Accelerometer fitted on the fixed wheel hub of the rear left wheel

The tyre selected for this study was set to balanced condition, which means that there was no additional weight rather than the balancing weight fitted with the tyre. The total balancing weight fitted was 40g and the tyre was inflated with air. The data acquired here was termed as 0 balance condition. Similarly different balancing conditions were obtained by adding or removing the balancing weights. The signals were acquired for different pressure conditions such as High, Normal, Puncture and Idle. 19psi, 31psi and 40psi were taken as the tyre pressure of puncture, normal and high conditions respectively. Riding a speed below 10 km/hr, without sufficient amplitude and vibration, was taken was Idle. As shown in Table 3, the conditions -40 balance, -20 balance, 0 balance, +20 balance and +40 balance where chosen for data collection of High, Normal, Puncture and Idle cases (refer Table 1). The experiments were conducted within the speed limit of 10 km/hr and 100 km/hr.

Statistical features were extracted from different class of signals. The feature selection was done using j48 algorithm and features were classified using Classification via Regression Algorithm. A balanced study was conducted by collecting 60 samples in all cases mainly High, Normal, Puncture and idle. On the whole, a total of 1200 samples were obtained. Each of these signals was having 5000 data points and 1 kHz sample rate[18].

Table 1 Balance conditions mass level

Conditions	Weight Added(g)	Wheel Position
-40 Balance	0g	Rear Left
-20 Balance	20g	Rear Left
0 Balance	40g	Rear Left
+20 Balance	60g	Rear Left
+40 Balance	80g	Rear Left

2.1 Sampling rate calculation

The tyre used for this particular study was 165/80 radial. The tyre radius(R) was rated as 30.1 cm. The maximum and minimum speed of the study was 100km/hr and 10km/hr respectively. Based on the study[18], the minimum and maximum frequency is 1.47 Hz and 14.73 Hz respectively. Considering Nyquist Shannon sampling theorem, the minimum sampling rate will have to be greater than or equal to 29.46 Hz [19] . So this study prefers to take 1 kHz as sampling rate. Vibration signals for zero Balance condition is plotted in Figure 5.

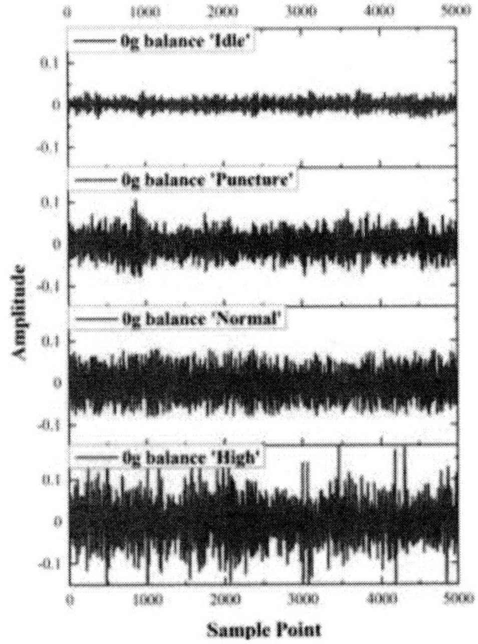

Figure 5 Vibration signals for zero Balance condition

3. Feature Extraction

Machine learning consists of 3 steps-feature extraction, feature selection and feature classification. Feature extraction is a feature reduction process in which it aligns the available features according to their importance in classification. Any fault in the mechanical components will induce a considerable change in vibration signals and that change can be detected from the statistical parameters of vibration signals. This has been proved by many researchers reported in the literature. Thus statistical features can be an effective for extracting meaning information from the vibration signals with less complexity and less computational time.

Targeted vibration signals for the conditions -40 Balance, -20 Balance, 0 Balance, +20 Balance and +40 Balance were chosen for data collection of High, Normal, Puncture and Idle cases. This study used statistical feature extraction, which is one of the best and common feature extraction techniques. Statistical features extracted were mean, kurtosis, mode, median, sample variance, range, sum, standard error, maximum, standard deviation, minimum and skewness.

4. Feature Selection

The features extracted during feature extraction technique were tested using j48decision tree algorithm. In J48 algorithm, the features are selected through the calculated entropy and information gain values. These parameters will decide upon the selection of most contributing features out of the given input features and thereby the numbers of input variables are reduced while developing a predictive model[20]. The involvement of each feature tested and the features influencing less for the classification were filtered for decreasing computational load of the system. In [21] the detailed procedure of feature selection using J48 decision tree algorithm was clearly explained. Figure 6 describes the features selected for different conditions. Figure 7 shows the decision tree generated from j48 decision tree algorithm.

Balance Condition	Features Selected
40 Balance	Mean, Standard Error, Median, Mode, Minimum, Kurtosis, Standard Deviation, Sample Variance and Skewness,
20 Balance	Mean, Mode, Standard Deviation, Kurtosis, Minimum, Maximum
0 Balance	Mean, Standard Error, Median, Mode, Minimum and Sample Variance
20 Balance	Standard Deviation , Standard Error, Skewness, Median, Mode and Kurtosis,
40 Balance	Mean, Standard Error, Median, Kurtosis, Range, Maximum

Figure 6 Features selected for different Balance conditions

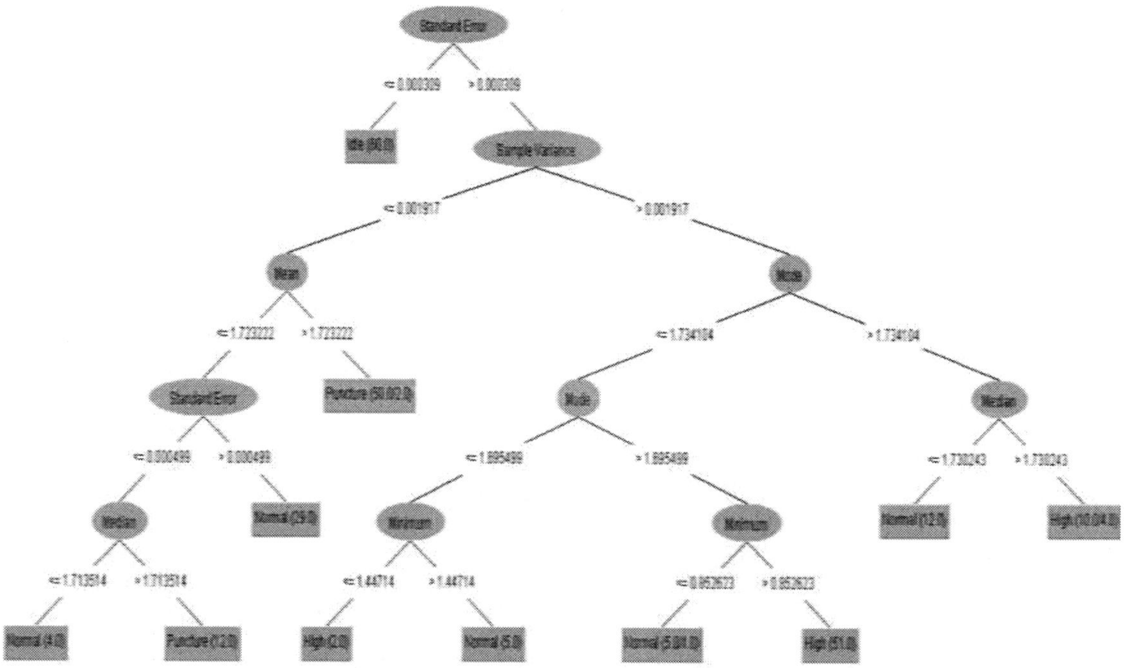

Figure 7 Decision tree generated from j48 decision tree algorithm.

5. Feature Classification – Classification via Regression Algorithm (CVR)

The classifier used for feature classification in this specific study is Classification via Regression Algorithm. In this algorithm classification problem is converted to regression functions. It binaries and builds a regression model for each class value. The principles of both linear regression and decision tree algorithm are combined in this classifier[22]. The two stages of this method are

1. Create a normal decision tree, by increasing the difference and variations of attributes in accordance with the output values. The deviation reduction standard must be evaluated during the decision tree building.
2. Pruning the main decision tree in to possible sub trees and merging it with regression function[23].

During the training time of Classification via regression algorithm, enormous numbers of decision trees are generated. Every single tree provides its classification, which is taken as a vote for that class; the object is assigned to the class that has the maximum number of votes. This algorithm used in many mechanical vibrational studies. It is a powerful tool for predicting and it does not over fit. This has been proved by many researchers reported in the literature[24].

5.1 Performance Parameters
Classification accuracy is the main performance parameter for evaluating the performance of an algorithm at different conditions. It is ratio of number of correct prediction and total number of predictions, which is indicated as the percentage. In every experiment, the average of the difference between expected outcome and real outcome is termed as Mean absolute error (MAE). It is an important parameter while considering the performance of an algorithm. The average magnitude of the error calculated using a quadratic scoring rule is known as Root mean squared error (RMSE)[25]. Kappa statistics is a commonly accepted statistics to measure the settlement between two or more observers. In this measurement, the observers may sometimes agree or disagree simply by chance. A kappa statistics of 0.8 to 1.0 indicates perfect agreement, 0.61 to 0.8 indicates substantial agreement whereas a kappa of less than 0 indicates agreement equivalent to chance[26][27].

6. Results and Discussions
6.1 Zero balance condition using Classification via Regression Algorithm

From Figure 7, it is evident that Classification via Regression Algorithm provides the maximum classification accuracy of 91.25% for zero balanced air condition. The confusion matrix for zero Balance air condition generated by trained Classification via Regression Algorithm is shown in Figure 8. Figure 9 shows class-wise detailed accuracy for the trained Classification via Regression Algorithm for zero Balance air condition.

From Figure 8, the correctly classified instances (out of 60) were represented as diagonal elements in the confusion matrix. From the matrix, one can notice that 59/60 of 'idle' instances were correctly classified followed by 'puncture' instances where 57/60 are correctly classified. 56/60 instances are correctly classified for 'high' state and 47/60 instances for 'normal' state[28].

Predicted Class

Actual Class — Classified as	0AirHig	0AirNor	0AirPun	0AirIdl
0AirHig	56	4	0	0
0AirNor	10	47	3	0
0AirPun	0	3	57	0
0AirIdl	0	0	1	59

Figure 8 Confusion matrix for zero Balance air condition generated by trained Classification via Regression Algorithm

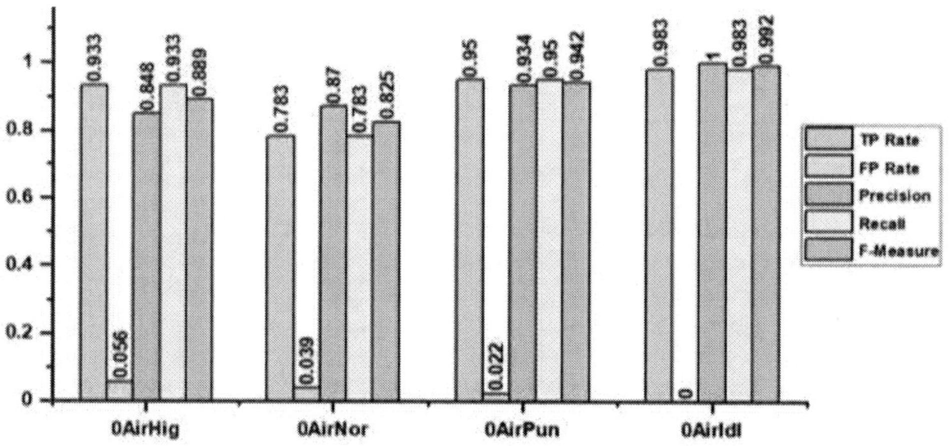

Figure 9 Class-wise detailed accuracy for the trained Classification via Regression Algorithm for zero Balance air condition

6.2 -20 Balance condition using Classification via Regression Algorithm classifier

In this condition the total balancing weight added to the selected tyre was 20g instead of 40g (Refer section 6.1). The confusion matrix for -20 Balance air condition generated by trained Classification via Regression Algorithm is shown in Figure 10. Figure 11 shows class-wise detailed accuracy for the trained Classification via Regression Algorithm for -20 Balance air condition.

		Predicted Class			
	Classified as	-20AirHig	-20AirNor	-20AirPun	-20AirIdl
Actual Class	-20AirHig	56	4	0	0
	-20AirNor	8	48	4	0
	-20AirPun	0	8	52	0
	-20AirIdl	0	0	0	60

Figure 10 Confusion matrix for -20 Balance air condition generated by trained Classification via Regression Algorithm

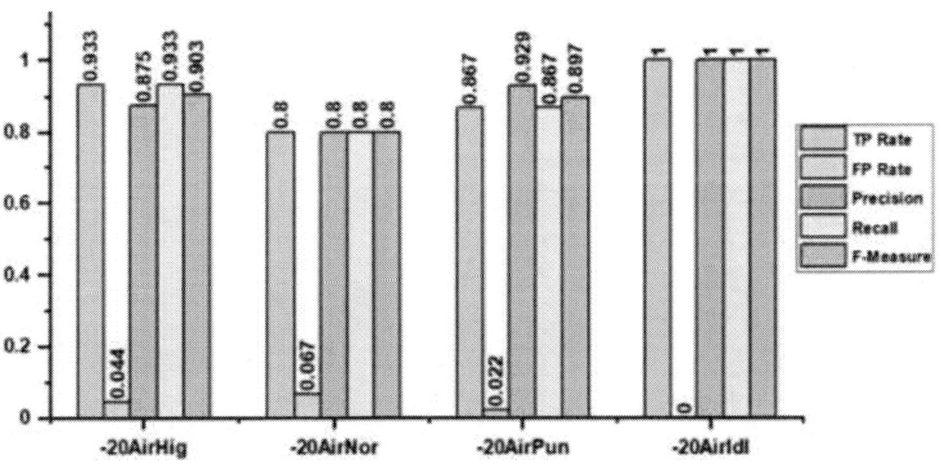

Figure 11 Class-wise detailed accuracy for the trained Classification via Regression Algorithm for -20 Balance air condition

6.3 -40 Balance condition using Classification via Regression Algorithm classifier

In this condition the total balancing weight added to the selected tyre was 0g instead of 40g (Refer section 6.1). The confusion matrix for -40 balance air condition generated by trained Classification via Regression Algorithm is shown in Figure 12. Figure 13 shows class-wise detailed accuracy for the trained Classification via Regression Algorithm for -40 Balance air conditions.

Predicted Class

Actual Class / Classified as	-40AirHig	-40AirNor	-40AirPun	-40AirIdl
-40AirHig	46	6	8	0
-40AirNor	19	38	3	0
-40AirPun	13	6	39	2
-40AirIdl	0	0	1	59

Figure 12 Confusion matrix for -40 Balance air condition generated by trained Classification via Regression Algorithm

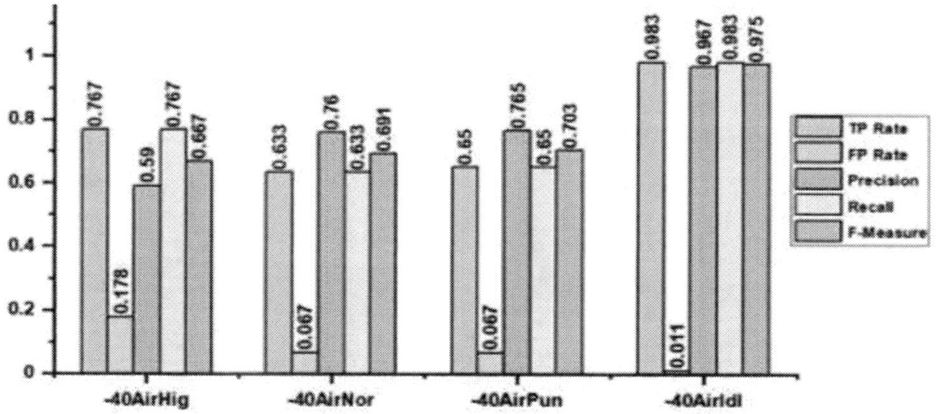

Figure 13 Class-wise detailed accuracy for the trained Classification via Regression Algorithm for -40 Balance air condition

6.4 +20 Balance condition using Classification via Regression Algorithm

In this condition the total balancing weight added to the selected tyre was 60g instead of 40g (Refer section 6.1). The confusion matrix for +20 Balance air condition generated by trained Classification via Regression Algorithm is shown in Figure 14. Figure 15 shows class-wise detailed accuracy for the trained Classification via Regression Algorithm for +20 Balance air condition.

Predicted Class

Actual Class — Classified as	+20AirHig	+20AirNor	+20AirPun	+20AirIdl
+20AirHig	49	4	7	0
+20AirNor	8	48	4	0
+20AirPun	15	2	43	0
+20AirIdl	0	0	0	60

Figure 14 Confusion matrix for +20 Balance air condition generated by trained Classification via Regression Algorithm

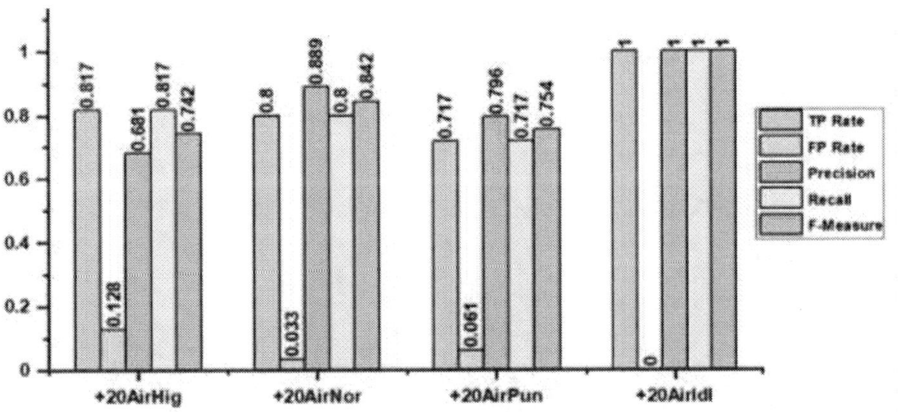

Figure 15 Class-wise detailed accuracy for the trained Classification via Regression Algorithm for +20 Balance air condition

6.5 +40 balanced condition using Classification via Regression Algorithm classifier

In this condition the total balancing weight added to the selected tyre was 80g instead of 40g (Refer section 6.1). The confusion matrix for +40 Balance air condition generated by trained Classification via Regression Algorithm is shown in Figure 16. Figure 17 shows class-wise detailed accuracy for the trained Classification via Regression Algorithm for +40 Balance air condition.

Predicted Class

Actual Class: Classified as	+20AirHig	+20AirNor	+20AirPun	+20AirIdl
+20AirHig	52	6	2	0
+20AirNor	3	46	11	0
+20AirPun	7	15	38	0
+20AirIdl	0	0	0	60

Figure 16 Confusion matrix for +40 Balance air condition generated by trained Classification via Regression Algorithm

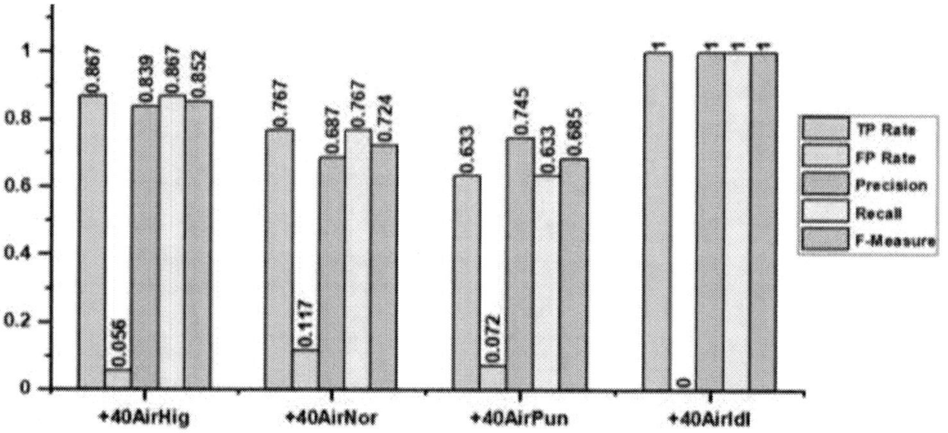

Figure 17 Class-wise detailed accuracy for the trained Classification via Regression Algorithm for +40 Balance air condition

6.6 Effect of Classification via Regression algorithm on different balance conditions

For the zero balance condition, kappa statistics was measured to be 0.8833. Kappa statistics measured the arrangement of likelihood with the true class. Closeness of prediction to the ultimate result was measured by mean absolute error. In this particular classification, the mean absolute error was 0.1234.The root mean square error value was measured to be 0.2156. Similarly one can easily understand the different errors for different balance conditions from Figure 18 and Figure 19.

From Figure 19, one can identify the trends of change in classification accuracy as the unbalance in wheels is changed by adding mass. For this setup, the lowest classification accuracy is obtained at -40 Balance condition which is 75.83%. The highest classification accuracy is obtained at the zero balance condition, which is 91.25%.

As the balancing mass is increased above or decreased below the zero balanced condition, there is a significant reduction in classification accuracy.

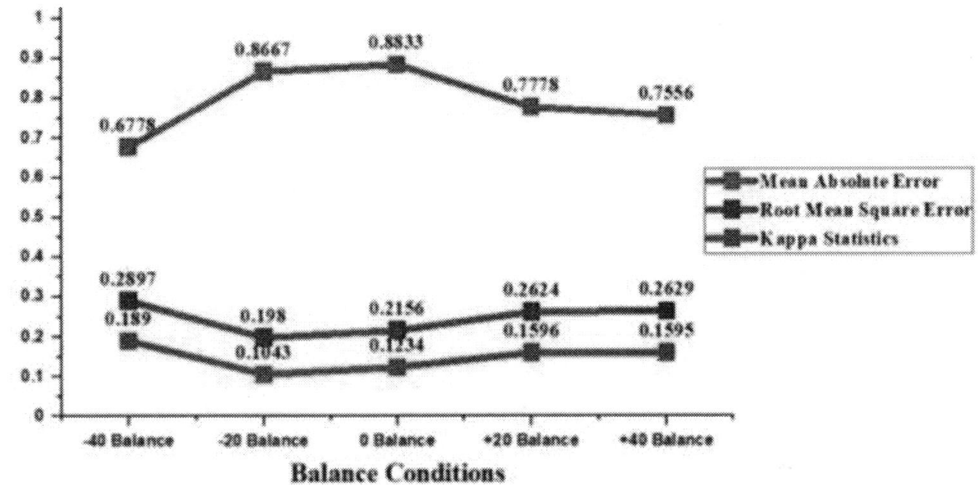

Figure 18 Performance parameters for different Balance conditions

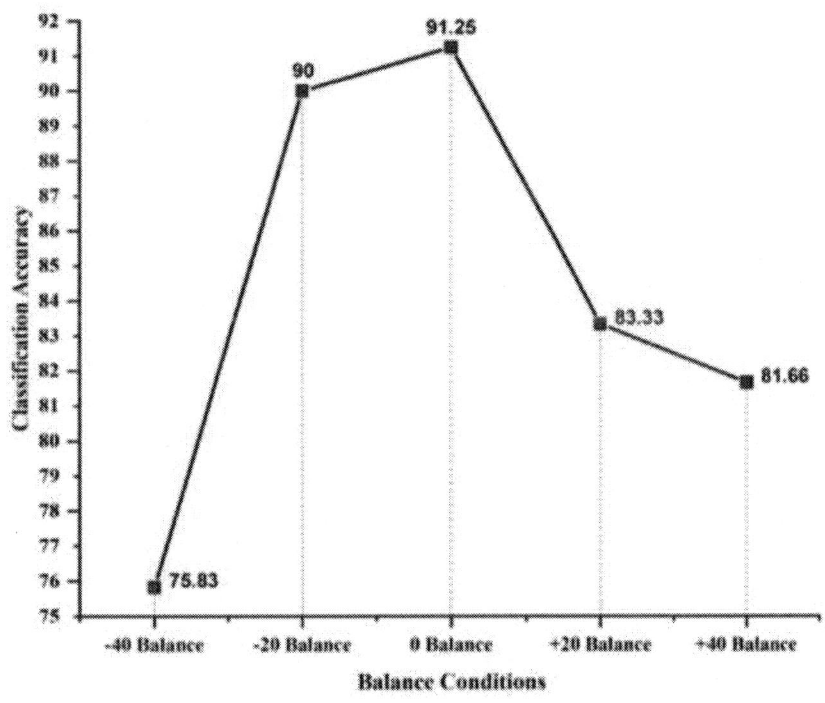

Figure 19 Effect of Balancing in terms of variation in the classification accuracy

7. Conclusion

Nowadays safety is an important aspect in the growing automobile industry. So devices like tyre condition monitoring systems have its own importance. This work showed an algorithm based comparison of different vibration signals obtained from air filled tyres under five different balancing conditions. Total five data models were created using Classification via Regression Algorithm and compared using data modeling technique. From the Zero Balance condition at 40g, other balance conditions were created by adding or removing the balancing mass. 10-fold validation method was used to validate the models and classification accuracy of 91.25% was obtained for Classification via Regression Algorithm using zero balance air condition. This methodology may be adopted for tyre condition monitoring because the error rate was considerably less. In future studies, one can modify the methodology by increasing or decreasing the mass by 5g to improve the accuracy.

References

[1] Paine M, Griffiths M and Magedara N 2007 *The Role Of Tyre Pressure In Vehicle Safety, Injury And Environment*

[2] Anoop P S. V Sugumaran and Hemanth Mithun Praveen 2016 Tyre Pressure Monitoring System Using Machine Learning Approaches – A Review *International Journal of Control Theory and Applications* **9** 371–82

[3] Premarsha B 2016 Survey on Tire Pressure Monitoring System *International Journal of Engineering and Technical Research (IJETR)* **0869** 120–1

[4] Schuring D J and Futamura S 1990 Rolling Loss of Pneumatic Highway Tires in the Eighties *Rubber Chemistry and Technology* **63** 315–67

[5] Howard, H. R., NcGinnis, T. A., Daugherty R H 1993 Remote Tire Pressure Sensing Technique

[6] Wei C, Zhou W, Wang Q, Xia X and Li X 2012 TPMS (Tire-Pressure Monitoring System) Sensors: Monolithic Integration of Surface-Micromachined Piezoresistive Pressure Sensor and Self-Testable Accelerometer *Microelectron. Eng.* **91** 167–173

[7] Hill M, Malson P R W C and Turner J D 1990 The development of a low cost system for monitoring tyre pressures *IEE Colloquium on Chassis Electronics* pp 2/1-2/3

[8] Singh K B, Bedekar V, Taheri S and Priya S 2012 Piezoelectric vibration energy harvesting system with an adaptive frequency tuning mechanism for intelligent tires *Mechatronics* **22** 970–88

[9] Craighead I A 1997 Sensing tyre pressure, damper condition and wheel balance from vibration measurements *Proceedings of the Institution of Mechanical Engineers, Part D: Journal of Automobile Engineering* **211** 257–65

[10] Hasan N N, Arif A and Pervez U 2011 Tire pressure monitoring system with wireless communication *2011 24th Canadian Conference on Electrical and Computer Engineering(CCECE)* 99–101

[11] Marton Z, Fodor D, Enisz K and Nagy K 2015 Frequency analysis based Tire Pressure Monitoring *2014 IEEE International Electric Vehicle Conference, IEVC 2014* 0–4

[12] Wu B, Fang Y and Deng L 2019 Summary of Energy Collection Application in Vehicle Tire Pressure Monitoring System *Proceedings of the 2019 4th International Conference on Automation, Control and Robotics Engineering* CACRE2019 (New York, NY, USA: Association for Computing Machinery)

[13] Ryan J and Bevly D 2012 *Tire radius determination and pressure loss detection using GPS and vehicle stability control sensors* vol 8 (IFAC)

[14] Garcia-Pozuelo D, Yunta J, Olatunbosun O, Yang X and Diaz V 2017 A strain-based method to estimate slip angle and tire working conditions for intelligent tires using fuzzy logic *Sensors (Switzerland)* **17**

[15] Svensson O, Thelin S, Byttner S and Fan Y 2017 Indirect Tire Monitoring System - Machine Learning Approach *IOP Conference Series: Materials Science and Engineering* **252**

[16] Anoop P.S. and V. Sugumaran 2018 Tyre Pressure Monitoring System Using Statistical Analysis And Rotation Forest Algorithm *Pakistan Journal of Biotechnology* **15** 36–9

[17] Anoop P S and Sugumaran V 2020 The Influence of Tyre Balancing In Nitrogen Filled Tyres Using Statistical Features And Random Forest Algorithm *International Journal of Mechanical and Production Engineering Research and Development (IJMPERD)* **10** 6679–90

[18] Anoop P S, Sugumaran V and Mithun Praveen H 2016 Implementing K-Star Algorithm to Monitor Tyre Pressure using Extracted Statistical Features from Vertical Wheel Hub Vibrations *Indian Journal of Science and Technology* **9** 1–7

[19] McLean R F, Alsop S H and Fleming J S 2005 Nyquist-overcoming the limitations *Journal of Sound and Vibration* **280** 1–20

[20] Joshuva A, Kumar K R, Gangadhar G S S, Dhanush S S and Arjun M 2020 Rough Set Theory Based

Blade Condition Classification on Wind Turbine through Statistical Features *{IOP} Conference Series: Materials Science and Engineering* **923** 12010
[21] Anoop P S and Sugumaran V 2017 Classifying machine learning features extracted from vibration signal with logistic model tree to monitor automobile tyre pressure *SDHM Structural Durability and Health Monitoring* **11**
[22] Frank E, Wang Y, Inglis S, Holmes G and Witten I H 1998 Using model trees for classification *Machine Learning* **32** 63–76
[23] Yulita I N, Fanany M I and Arymurthy A M 2019 Comparing Classification via Regression and Random Committee for Automatic Sleep Stage Classification in Autism Patients *Journal of Physics: Conference Series* **1230** 12010
[24] Joshuva A and Sugumaran V 2019 Selection of a meta classifier-data model for classifying wind turbine blade fault conditions using histogram features and vibration signals: A data-mining study *Progress in Industrial Ecology* **13** 232–51
[25] Joshuva A and Sugumaran V 2019 A Lazy Learning Approach for Condition Monitoring of Wind *Measurement* 107295
[26] Viera A J and Garrett J M 2005 Anthony J. Viera, MD; Joanne M. Garrett, PhD (2005). Understanding interobserver agreement: the kappa statistic. Fam Med 2005;37(5):360-63. *Family Medicine* **37** 360–3
[27] Joshuva A, Kumar R S, Sivakumar S, Deenadayalan G and Vishnuvardhan R 2020 An insight on VMD for diagnosing wind turbine blade faults using C4 . 5 as feature selection and discriminating through multilayer perceptron *Alexandria Engineering Journal* **59** 3863–79
[28] Anoop P S. V Sugumaran and Hemanth Mithun Praveen 2018 Analyzing Vertical Vibrations Of Automobile Wheel Hub To Monitor Tyre Pressure Using Statistical Features And Support Vector Machine Algorithm *Pakistan Journal of Biotechnology* **15** 10–3

Vibration based Data Analysis of Single Acting Compressor through Condition Monitoring and Multilayer Perceptron – A Machine Learning Classifier

Aravinth Sivakumar[1], **Sugumaran Vaithiyanathan**[2*]

[1]Department of Mechanical Engineering, Rajalakshmi Institute of Technology, Kuthambakkam, Chennai - 600124
[2]School of Mechanical Engineering, Vellore Institute of Technology, Vandalur – Kelambakkam road, Chennai – 600127

E-mail: v_sugu@yahoo.com

Abstract- The air compressor is one of the desired mechanical equipment used for producing compressed air, which is utilized for performing various industrial and domestic functions. Its operation involves several rotating and fluctuating members which fail due to several miscellaneous reasons as the members prone to dynamic working environment quite frequently. The deficiencies create huge impact over the overall performance and thus leads to economic losses associated with system seizure. It is now essential to predict the occurrence of faults at earlier stages in order to avoid major shutdowns. Hence, in this article, a data modelling study using a machine learning algorithm is proposed. Initially, the vibration signals are measured as physical parameters from the compressor test rig as it contains critical information regarding the system working conditions instantly. The statistical features were extracted from the acquired signals and by using the J48 algorithm the most prominent features were selected. These selected features were classified using Multilayer Perceptron and its performance in fault classification was presented

Keywords: Fault diagnosis, vibration signals, machine learning algorithms, condition monitoring, statistical feature extraction, Multilayer Perceptron.

1. Introduction

Air compressor is one of the important mechanical equipment which are widely used in critical industrial and domestic applications. The faults associated with such crucial systems indulge in process malfunctions which may also lead to severe critical causalities. Hence, monitoring of the system conditions continuously has gained importance and which is done through condition monitoring that helps to detect, analyze and diagnose the roots of the failures in advance. The system conditions are monitored through the real time physical parameters associated with the systems, where a change in signal pattern insists upon a considerable development of system faults. Over the years, many researchers have conducted various fault diagnosis study using different techniques and methodologies to predict the occurrence of faults in advance. Some of the literatures pertaining to the compressor fault condition monitoring are discussed through the section below.

Data capturing is the heart of fault diagnosis where the physical parameters are measured to effectively diagnose the condition of the machine [1]. The compressor system was divided into four phases namely, piston head, non-return valve (NVH), opposite of non-return valve and opposite of flywheel side. On each phases, six sensor positions are considered. A statistical approach was carried out on those 24 positions and by the values of peak amplitude, standard deviation, variance and root mean squared (RMS). The results suggested that, the collection of data is found effective when the sensor is positioned over the piston head [2].

The valves are considered as the frangible part of reciprocating compressor where the periodic failure happens. Therefore, the valve fault diagnosis is crucial to avoid major causalities and downtime losses [3]. The valve cracks and valve breaks are the most common valve fault conditions, Kurt Pitchler et al.., presented a diagnostic approach to detect the faults through vibration signals. At first, the obtained information was changed into multi-dimensional vector space and thereby metric data was defined. They have utilized data to calculate the variation in gap between the testing compressor and an actual compressor. Higher the variation indicates an abnormality in the working of compressor. [4]. Yuefei Wang conducted an experiment to diagnose the typical compressor valve faults. Leakage, valve flutter, delayed closing and improper fit was the conditions investigated through acoustic emission and simulated valve motion [5]. S Meenakshi Sundaram conducted a fault diagnosis study to diagnose rotating machine faults. A total of 24 fault classes was investigated through acoustic and vibratory parameters. Here, the decision tree of C4.5 algorithm was utilized to narrow down the most contributing features for classification and those features was fed into Ada-boost algorithm for classification. The results indicate that, 90% automation was achieved in determining rotating machine fault diagnosis [6]. Milad Golmoradi presented a fault diagnosis study on air compressor. Daubechies wavelet transform was used for the analysis of the compressor conditions and through J48 algorithm, the 93.33% accuracy was obtained in classifying compressor faults [7]. Kotha prashanth et al.., conducted a study on vibration based fault monitoring in compressor system. A total of 4 test conditions was investigated under all tree based classifiers and it was found that Random Forest Tree classifier showed the highest fault classification result of 86% accuracy in classifying compressor faults [8]. Sumit Kumar Sar et al.., conducted a fault diagnosis study on a rotating machinery through vibration signals. This study has considered room mean square (RMS), kurtosis and crest factor as an input features and those features effectiveness in fault classification was compared with the machine learning classifiers like; Probabilistic Neural Network (PNN), the Decision tree, K-nearest neighbour classifier and Radial Basis Network (RBN) classifier. They concludes that, the decision tree performs better as compared with the other classifiers and they have suggested it as an important tool for the problems that dealt with non-linear data classification [9]. Abdenour Soualhi conducted a study on bearing health monitoring. Hilbert – Huang Transform technique was implemented along with Support Vector Machines (SVM) and Regression classifiers. The experimental results gives the clear indication that, the degradation state of bearings was detected efficiently through the developed technique [10] W S Yang et al.., conducted a fault diagnosis study on air compressor using Probabilistic Neural Network (PNN) as classifier. At first, the features were extracted using the Wavelet packets (WPD) and the Continuous wavelet transform (CWT) as the feature extraction tool and the results indicates that, the lifting wavelet transform produces a comprehensive result in reflecting the fault conditions and this method is suggested as a tool for online fault diagnosis of air compressor [11]. Joshuva et al.., conducted a study on wind turbine blade condition monitoring through statistical features. Here, total of six different blade faults was investigated through vibration signals. The statistical feature extraction technique was used and the J48 decision tree algorithm is used for the selection of features and Rough Set Theory (RST) feature classification tool is used as a classifier. The classification accuracy was found to 75.5% and since it produces a minimal mean absolute error, they suggested that RST can be employed to identify the blade conditions [12]. The step by step process of the data modeling study are displayed through the Figure 1.

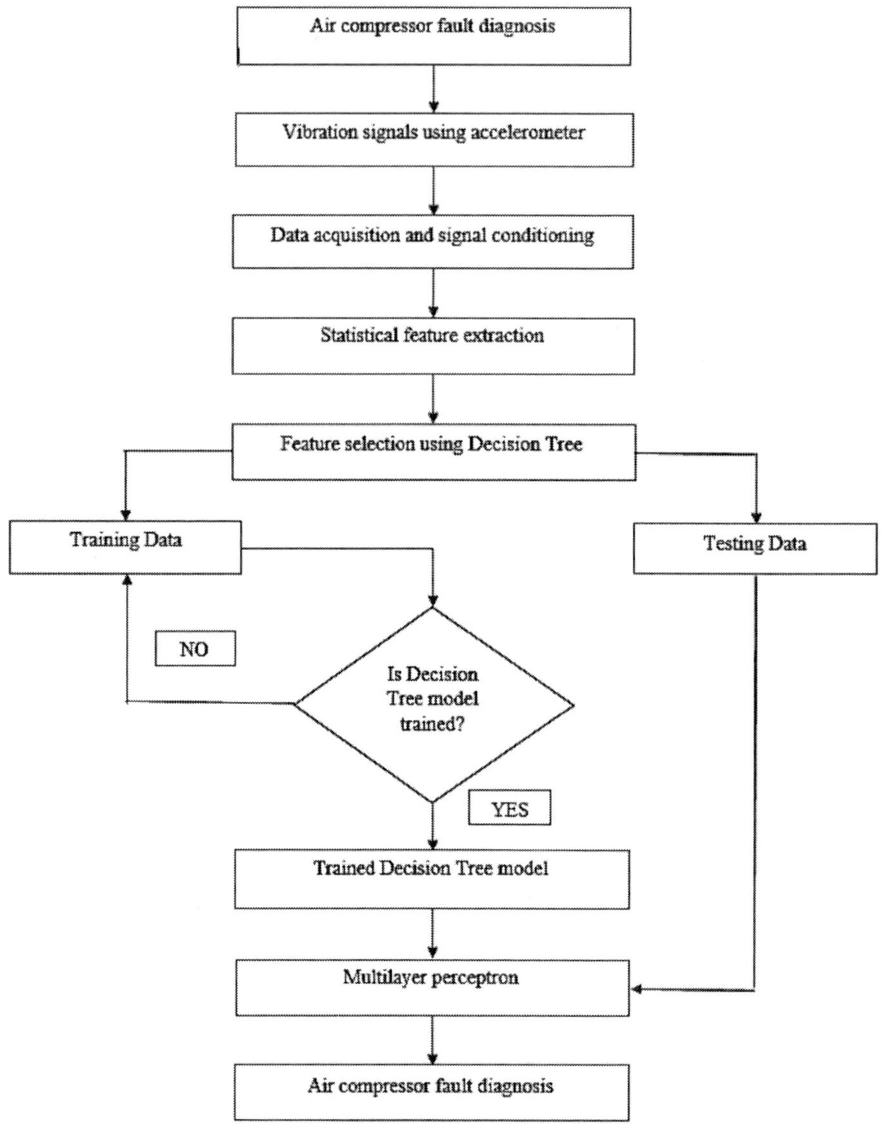

Figure 1. Methodology

2. Experimental studies

A data modeling study on the compressor system is developed to monitor the dynamic characteristics while operating under five different fault conditions. This section explains in detail about the experimental setup and the procedure adopted for the effective conduct of the experiment.

2.1. Experimental setup

A single-stage reciprocating air compressor illustrated in Figure 2 is taken as an experimental setup. The overall experimental arrangement includes; a piezoelectric transducer, a data acquisition (DAQ) module, signal processing setup with NI Lab View. The vibration signals are acquired for six different test conditions of the compressor using an accelerometer sensor at 500g range, 100mV/g sensitivity and with a resonant frequency of 40 Hz.

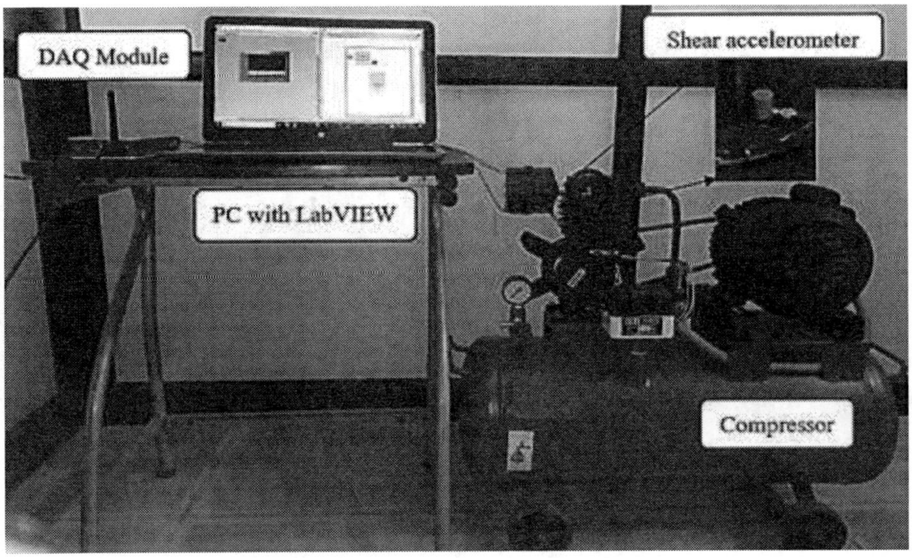

Figure 1. Experimental setup

2.2. Experimental procedure

Initially, the vibration patterns are recorded for the good working compressor arrangement and once it is done, the remaining conditions (faults) are incorporated into the system one by one and the vibration signals were recorded of 100 each. Except for the healthy (good condition) state of the compressor, all the other test conditions like inlet & outlet valve fluttering (IVF, OVF & IOVF), valve plate leakage (VPL), and check valve faults (PRV) are created artificially [13]. Figure 3 shows the pictorial representations of the faults created.

3a Inlet & outlet valve plate fluttering (Inverting valve plates upside down)

3b. Removal of bonding seal

3c. Check valve fault

Figure 3. Simulated fault conditions.

3. Statistical feature extraction process

Once the vibration signals were collected from the experimental test rig, these signals are processed to extract meaningful information through the number of features present internally. Here, in this article, the statistical features are extracted using descriptive statistics [14]. This extraction process is effective enough to detect the changes in vibration signals for any kind of mechanical failures and thus it has been used to detect the fault occurrence of an air compressor system. [15]. Standard deviation, skewness, standard error, kurtosis, sum, mean, range, sample variance, mode, medium, minimum and maximum were the features computed out of this extraction process. Out of this extracted features, the most contributing ones are identified through the feature selection process and those features will be served as an input to the classifier in order to determine the fault classification performance.

4. Decision tree based feature selection

Several data mining process are handled nowadays to retrieve collectible information from the available data set. One such technique is the decision tree. Its structure comprises of the root, a number of nodes, and leaves and it follows the "Top-Down Induction on Decision Tree" system where top nodal feature contribution towards fault classification is high when compared with the other subsequent nodal positions. The existence of an attribute in a decision tree structure gives information about the importance of an associated attributes during classification. The C 4.5 algorithm follows two important phases during feature selection namely; a building phase and a pruning phase. The following subsection provides a detailed explanation of these phases.

4.1. Building phase

This phase explains in detail about the construction of the decision tree [16]. Here, the top nodal position showcases the feature with a maximum contribution towards fault classification. Now, the root node is partitioned into two internal nodes or decision nodes (binary decision tree) based on the trail on an attribute, and those internal nodes are connected through the branches. These attributes are selected through the estimated entropy-based information gain. For each and every partitioning, an additional node is attached to the existing structure and this process continues until an identical class appear over the partitioning.

4.2. Pruning phase

The features that present in the structure of the decision tree don't exhibit equal contribution in classification. Therefore the features which exhibit the low or negligible contribution are to be pruned for an effective fault classification outcome. Here, the J48 algorithm follows an error-based pruning process where the error rate is calculated at each nodal position of a decision tree based on the overall aggregate of misclassification. From the calculated error rate, the features with the lowest error rate values were selected for further classification, and the features that exhibit the maximum error rate are pruned out of the decision tree. Figure 4 shows the post pruned decision tree structure where only 7 out of 12 input features are present.

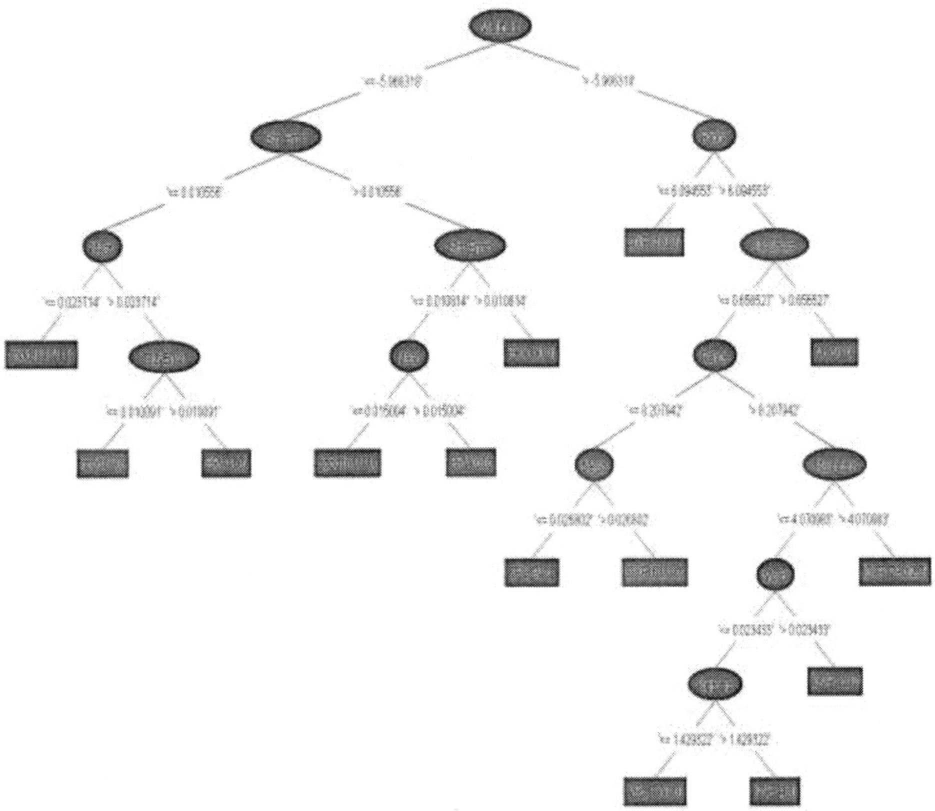

Figure 4. Decision tree for feature selection

5. Feature classification process

It is a class of feed-forward artificial neural networks (ANN) and it is often called as a neural network or a multilayer perceptron (MLP). This classifier can be trained to approximate virtually any smooth and measurable functions. And moreover it doesn't make prior assumptions as in case of other classifiers concerning to the data distribution. This classifier is suitable for processing a non-linear functions which can also be trained to accurately generalize when given with a new and unseen data for a non-linear applications [17]. The structure of a MLP possess more than one hidden layer of the perceptron. Each perceptron is a single neuron model (Figure 5a) which accepts the weighted input for further activation. The multilayer perceptron neural network model possesses three important layers namely; input, hidden and output layers (Figure 5b).

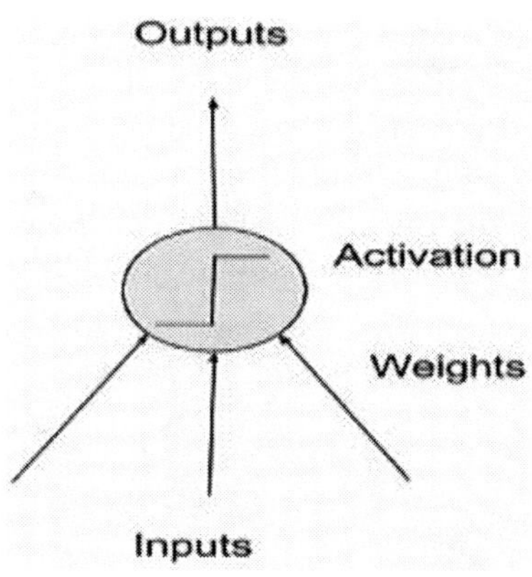

5a. Single neuron model

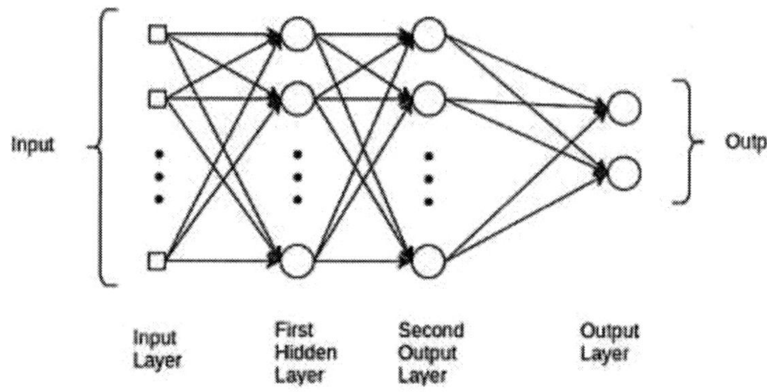

5b. Model of simple network

Figure 5. Multilayer Perceptron model

The multilayer perceptron neural network model possesses three important layers namely; input, hidden and output layers (Figure 5b). The hidden and output layer uses an activation function for its operation. Here the data sets are trained through back-propagation; a supervised learning technique. Where a weighted biased sum is added to the given inputs and the overall activation function used as a transfer function for producing the output [18]. Initially, the neural networks are trained on the datasets and the procedure for the preparation of data for the training is as follows;

- Input data should be numerical
- The network process the input and upon activating the activation function, the neurons finally produces the desired output value
- The obtained results are compared with the expected outcome and the deviation or error in calculation were determined
- The network model is trained again and again to have an effective outcome

6. Result and discussion

The vibration data were acquired for six different test conditions of air compressor and the statistical features associated with the signals were extracted using descriptive statistics. And the decision tree (J48 algorithm) is used for the selection of the most contributing features for fault classification from the computed features. And it is clear that, minimum, standard error, range, mean, skewness, maximum and kurtosis were the features present out of the given input features (Figure 3).

6.1. Effect of the number of input features

The standard rule of decision tree is that, the feature which contributes to the maximum in fault classification present at the top nodal position and the subsequent positions contribution reduces as compared to the previous nodal position of a decision tree. Therefore, it is essential to study the effect on the number of input features of decision tree in fault classification. Hence the features are combined from top nodal position towards the bottom nodal position of the decision tree (top to bottom hierarchy) and the classification accuracy for each and every combination were determined through J48 algorithm. Table 1 shows the suitable combinations and their corresponding classification accuracy

Table 1. Number of statistical features and corresponding classification accuracy

Combinations	Statistical feature combination	Accuracy (%)
1	Minimum	66.83
2	Minimum and standard error	89.83
3	Minimum, standard error and range	92.67
4	Minimum, standard error, range and mean	94
5	Minimum, standard error, range, mean and skewness	95.67
6	Minimum, standard error, range, mean, skewness and maximum	95.67
7	Minimum, standard error, range, mean, skewness, maximum and kurtosis	95.67

Referring to Table 1, it is clear that, the feature combinations minimum, standard error, range, mean and skewness exhibit the highest fault classification rate of 95.67% and those features were selected as input features for the classifier

6.2. Feature classification

The features; minimum, standard error, range, mean and skewness (Referring to Table 1) were served as an input to the classifier Multilayer Perceptron and an accuracy of 94.67% obtained for the fault classification at a computation time of 0.64 seconds. The internal parameters associated with Multilayer Perceptron like learning rate and momentum are varied from 0.1 to 1 and the corresponding change in classification accuracy were noted. Figure 6 gives the graphical plot generated for the variation in learning rate at momentum value as 0.2.

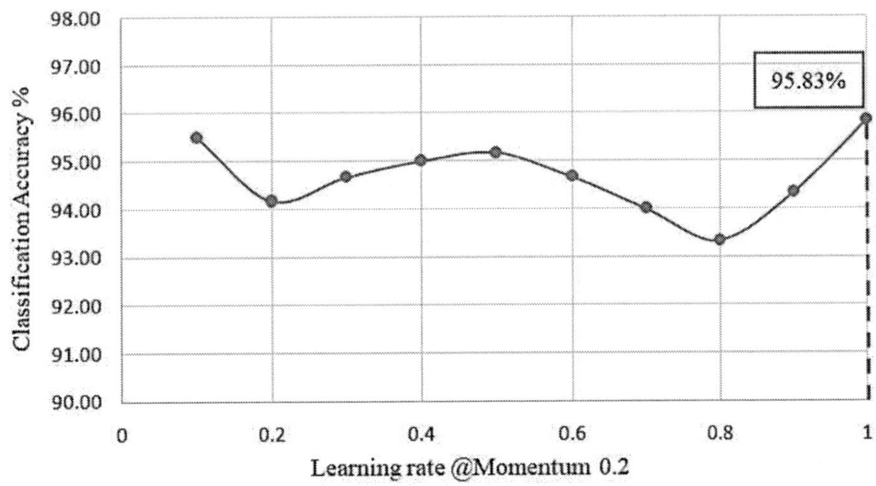

Figure 6. Learning rate versus classification accuracy

Referring to Figure 6, it is clear that, the rate of classification accuracy was obtained at a rate of 95.83%, which shows an improved accuracy for the learning rate value at 1 with momentum value as 0.2. The overall performance of the data model can be analyzed through the correctly classified and misclassified instances of confusion matrix shown in Table 2. The Table 3 shows the detailed class wise accuracy for each classes which indicates the classifiers performance with different rate of global measures [18].

Table 2. Confusion matrix of Multilayer perceptron

	GOOD	IVF	OVF	PRV	VPL	IOVF
GOOD	99	0	0	1	0	0
IVF	0	96	0	1	1	2
OVF	0	0	100	0	0	0
PRV	5	0	0	95	0	0
VPL	0	0	1	0	95	4
IOVF	1	0	0	0	9	90

From the confusion matrix (Table 2), it is clear that, 575/600 samples were classified correctly to their respective classes and 25/600 samples were misclassified. The diagonal elements shows the instances that are correctly classified out of the given input instances and the non-diagonal elements shows the instances that are misclassified. This misclassification in confusion matrix indicates that during the machine learning process, 25 samples were misclassified or identified as other faulty condition of the study. In condition outer valve fluttering (OVF), the total input samples were correctly classified to their respective classes, but in condition inlet & outlet valve fluttering (IOVF), 10/100 samples were misclassified into valve plate fluttering (VPL - 9) and the remaining one as GOOD condition. In condition valve plate leakage (VPL) and pressure relief valve (PRV), 5/100 samples were misclassified. In PRV, 5 samples are misclassified as the condition GOOD and in VPL, 4 samples are misclassified as the condition IOVF and the remaining one sample is misclassified as the condition OVF.

Table 3. Accuracy by class of Multilayer Perceptron

TP Rate	FP Rate	Precision	Recall	F-Measure	ROC Area	Class
0.99	0.012	0.943	0.99	0.966	0.999	GOOD
0.96	0	1	0.96	0.98	0.975	IVF
1	0.02	0.99	1	0.995	1	OVF
0.95	0.004	0.979	0.95	0.964	0.999	PRV
0.95	0.02	0.905	0.95	0.927	0.99	VPL
0.9	0.012	0.938	0.9	0.918	0.99	IOVF

7. Conclusion

A vibration based fault diagnosis on air compressor system have been performed in this article. A total of six different compressor test conditions were investigated using 100 samples of vibration data. From the acquired signals, the statistical features have been extracted and the decision tree of J48 algorithm have been utilized as a feature selection tool and the best contributing features were selected and served as an input to the classifier Multilayer perceptron to measure its effectiveness towards fault classification outcomes. The important highlights of the obtained results are listed below,

- Out of 12 statistical features; minimum, standard error, mean, range and skewness alone are sufficient enough to produce a better classification outcomes

- The algorithm produces 95.83% as the classification accuracy with a computational time of 0.56 seconds

Hence, the results clearly indicates that, the compressor fault classification under statistical features exhibits the substantial result and performance. And it can be implemented in real time applications for an effective multi fault classification outcomes.

References

[1] Nishchal K Verma, Kadambari J, Abhijit B, Tanu S and Al Salour 2011 Finding Sensitive Sensor positions under faulty condition of Reciprocating Air Compressors *IEEE International Conference on System Engineering and Technology (ICSET).* 242 – 246

[2] Nishchal K Verma, Kadambari J, Abhijit B, Tanu S and Al Salour 2011 Finding Sensitive Sensor positions under faulty condition of Reciprocating Air Compressors *IEEE International Conference on System Engineering and Technology (ICSET).* 242 – 246

[3] Houxi Cui, Laibin Zhang, Rongyu Kang and Xinyang Lan 2009 Research on fault diagnosis for reciprocating compressor valve using information entropy and SVM method *Journal of Loss Prevention in the Process Industries. 22,* 864 – 867

[4] Kurt Pichler, Andrea Schrems, Thomas Buchegger, Matthias Huschenbett and Markus Pichler 2012 Fault detection in reciprocating compressor valves for steady state load conditions *IEEE International Symposium on Signal Processing and Information Technology, IEEE Publisher.* 224 – 229.

[5] Yuefei Wang, Ang Gao, Sulu Zheng and Xueyuan Peng 2015 Experimental Investigation of the fault diagnosis of typical faults in reciprocating compressor valves *Journal of Mechanical Engineering Science. 0(0)* 1 – 15

[6] Meenakshi Sundaram S and Saimurugan M 2019 Performance improvement of classifier in fault diagnosis of rotating machines using sensor fusion techniques *International Journal of Engineering and Advanced Technology*. Vol. 8, Issue. 6, 1136 – 1141.

[7] Milad Golmoradi, Ebrahim Ebrahimi and Mohamad Javidan 2017 Fault diagnosis of compressor based on decision tree and fuzzy inference system *Journal of Vibroengineeering.* Vol. 12, 54 – 6

[8] Kotha Prashanth and Elangovan M 2019 Vibration based fault monitoring of a compressor using Tree based algorithms *IOP Conference Series: Material Science and Engineering*. 577, 1 – 7.

[9] Sumit Kumar Sar and Ramesh Kumar 2020 Performance analysis of classification algorithms for fault diagnosis in rotating machines *International Journal of Innovative Technology and Exploring Engineering*. Vol. 9, Issue 6, 452 – 455

[10] Abdenour Soualhi, Kamal Medjaher and Noureddine Zerhouni 2015 Bearing health monitoring based on Hilbert – Huang Transform, Support Vector Machine and Regression. *IEEE Transactions on Instrumentation and Measurements*. Vol. 64, Issue 1, 52 – 62

[11] Wang W S, Su Y X and Chen Y P 2019 Air compressor fault diagnosis based on lifting wavelet transform and probabilistic neural network *IOP Conference Series: Materials Science and Engineering*. 657, 1 – 7

[12] Joshuva A, Rajesh Kumar K, Sriram Gangadhar G S, Dhanush S S and Arjun M 2020 Rough set theory based blade condition classification on wind turbine through statistical features *IOP Conference Series: Material Science and Engineering*. 923, 1 – 7

[13] Aravinth S and Sugumaran V 2018 Air compressor fault diagnosis through statistical feature extraction and Random Forest classifier *Progress in Industrial Ecology – An International Journal*. Vol. 12, Nos. 1/2, 192 – 205

[14] Joshuva A, Sugumaran V 2019 Crack detection and localization on wind turbine blade using machine learning algorithms: A data mining approach *Structural Durability & Health*

Monitoring. Vol. 13, no. 2, 181 - 203

[15] Joshuva A, Sugumaran V 2017 A comparative study on Bayes classifier for blade fault diagnosis in wind turbines through vibration signals. *Structural Durability & Health Monitoring*. Vol. 12, No. 1, 69 – 90

[16] Joshuva A and Sugumaran V 2017 A data driven approach for condition monitoring of wind turbine blade using vibration signals through best-first tree algorithm and functional trees algorithm: A comparative study *ISA Transactions*. 67 160 – 172

[17] Joshuva A, Sathish Kumar R, Sivakumar S, Deenadayalan D and Vishnuvardhan R 2020 An insight on VMD for diagnosing wind turbine blade faults using C4.5 as feature selection and discriminating through multilayer perceptron *Alexandria Engineering Journal*. 1 – 17

[18] Gaurang Panchal, Amit Ganatra, Kosta Y P and Devyani Panchal 2011 Behavior analysis of Multilayer Perceptron with multiple hidden neurons and hidden layers *International Journal of Computer Theory and Engineering*. Vol. 3, No. 2, 332 – 337

IOP Conf. Series: Materials Science and Engineering 1012 (2021) 012033 doi:10.1088/1757-899X/1012/1/012033

Adapted wavelets by parameterization of wavelet bases for estimates of feature extraction of signals

Manju B R[1*], Athul Rajendran[2]

[1]Department of Mathematics, Amrita Vishwa Vidyapeetham, Amritapuri, India

[2]Department of Electrcal and Electronis Engineering, Amrita Vishwa Vidyapeetham, Amritapuri, India

E-mail: manjubr@am.amrita.edu

Abstract- The work demonstrates the parameterization of filters coefficients of compactly supported wavelet to implement in the design of the best wavelet selection. The technique determines the best wavelet by means of a mathematical description and gives a representation in a single parameter for wavelets of finite length. The approach takes an input wavelet that is approximated to the best match. Through a choice of the parameter to adapt to the wavelet coefficients the perfect adaptation of the wavelet is achieved. For wavelet selection, the adaptive approximation carried out through parametrization addresses the challenge of visualization by setting up a Matlab programme that relies on the best selection with high potential to feature extraction of arbitrary signals.

1. Introduction

In this time, the primary reason of death around the globe is malignant growth or carcinoma. The central problem in approximation theory is to represent a functionby convenient family of functions. Wavelets are well suited to approximate one-dimensional piece-wise signals when non-linear approximation is allowed. As the original signal can be represented in terms of a wavelet expansion by using wavelet coefficients as a linear combination of wavelet functions, data operations can be performed using just the corresponding wavelet coefficients. Moreover if a best wavelet is chosen so as to adapt to the data, then it gives the best results. The fundamental idea of decomposition of $\mathbf{L}^2(\mathbf{R})$ into approximation or detail spaces is that a function can be decomposed in to an approximation of the original and the details needed to recover the original from the approximate. Wavelets are compactly supported in a diadic space and so it is possible to generate a set of parametric coefficients. The study of parameterization of trigonometric functions were conducted by many authors [7][3][9][10]. These formulae are a convenient way to generate a collection of wavelets with varying approximation properties. The explicit formulae for parameterization can be found in that were constructed for orthogonal wavelets for various lengths. Constrained solution for filter length eight and ten was developed in [5]. Pollen parametrization of wavelet systems is found in [2]. Parameterization of four, six, eight and ten are presented in [4].

As wavelet analysis deals with the expansion of functions in terms of basic functions of a fixed function called mother wavelet, they permit a closer connection between the function being represented and their coefficients. This powerful wavelet basis function created a new approach for mathematicians to answer some of the difficult problems. It has further helped for efficient methods of computation in many applications. Here we study the fault detection of a very important component of all machines- the roller bearings. To know the status of a machine, it is important to monitor the component which has great influence on the accuracy, reliability and life dependency of the machine. Studies [6] have shown that 90 percentage of the faults in roller bearings are related to either an inner or outer race fault. Such defects generate a series of impact vibrations every time a running roller passes over the surfaces of the defects. Hence we focus on the above two faults and their combinations. The construction of optimal wavelets using parameterization and its application to biomedical signal compression, pattern recognition can be found in [8][1][4].

Vibration signals are seen to be a preferable source in discovering the patterns of defects. In other words, any defect away from the normal or perfect condition will result in a change in waveform. Our interest lies in this changed waveform. Therefore we look forward to mathematically describe or model the problem. As long as the pattern is identified, the machine fault associated with the pattern can also be identified. Hence wavelet stands best that can fit a close connection to any arbitrary signal. The class of wavelets functionsψ^{α} : $\alpha \in [0,2\pi]$ of _filter length four depending on a parameter α. We determine an approximation of a given function in terms of the wavelet basis functions{ $\psi^{\alpha}_{j,k}$, $j, k \in N$ }. The approximation is carried out by fixing the number of terms. This turns to be an excellent method to help select the best wavelet basis for an approximation. The description is given a better understanding through a Matlab programme

Theorem 1.1 [1] Suppose $h = \{h_k\}_{k \in Z}$ is a sequence satisfying
1.$\sum_{k \in Z} |k|^{\epsilon} \, |h(k)| < \infty$ for some $\epsilon > 0$.
2.$\sum_{k \in Z} h_k = \sqrt{2}$
3.$\{R_{2k}h\}_{k \in Z}$is orthonormal in $l^2(Z)$
Then $\prod_{j=1}^{\infty} H\left(\frac{\omega}{2^j}\right)$ converges uniformly on bounded subsets of R to a function
$\widehat{\emptyset} \in L^2(R)$,where $H(\omega) = \frac{1}{\sqrt{2}} \sum_{k \in Z} h_k \, e^{-ik\omega}$. (1)
Let $\emptyset = (\widehat{\emptyset}_n)^{\vee}$ .Then for each $j \in Z$, V_j defined by
$V_j = \{\sum_{k \in Z} z(k) \emptyset_{j,k} : z = z(k)_{k \in Z} \in l^2(Z)\}$, is a
Multi resolution analysis with scaling function $\emptyset$ and scaling sequence h. (2)

2. Algorithm for the construction of Wavelets from Finite length Filters

The finite length sequence is$h = \{h_k\}_{k \in Z}$,$h = (h(0), h(1) \ldots.. h(N-1))$
and(h_k) satisfies the following (i) $\sum_{k \in Z} h_k = \sqrt{2}$
(ii)$\{R_{2k}h\}_{k=0}^{N/2}$ is orthonormal in $l^2(Z)$.
Suppose $H(\omega) = \frac{1}{\sqrt{2}} \sum_{k \in Z} h_k \, e^{-ik\omega} = \frac{1}{\sqrt{2}} \sum_{k=0}^{N-1} h_k \, e^{-ik\omega}$ (3)
Then $\prod_{j=1}^{\infty} H\left(\frac{\omega}{2^j}\right)$ converges uniformly on R to a function $\widehat{\emptyset} \in L^2(R)$.
The steps listed out in the setting up of a wavelet system in L^2 is as follows:

Step 1: Fix a finite sequence (h_k) satisfying the conditions
1.$\sum_{K=0}^{N-1} h_k = \sqrt{2}$
2.$\{R_{2k}h\}_{k=0 \text{ to } N-1}$is orthonormal in $l^2(Z)$

Step 2: Determine $\widehat{\emptyset}(\omega) = \frac{1}{\sqrt{2}}\prod_{j=1}^{\infty} H\left(\frac{\omega}{2^j}\right)$ where $H(\omega) = \frac{1}{\sqrt{2}}\sum_{K=0}^{N-1} h_k\, e^{-ik\omega}$
Step 3: Determine $\emptyset = (\widehat{\emptyset})^{\vee}$
Step 4: Determine $\psi(t) = \sqrt{2}\sum_{K=0}^{N-1} g_k\, \emptyset(2t-k)$ where
$g_k = (-1)^k h_{1-k}, k = 0 \text{ to } N-1$

This gives the wavelet function ψ and thus the wavelet basis $\psi_{j,k}$ can be determined.

3. Construction of Pattern Adapted Wavelet

We consider the function

$f(x) = \sin x$, $f(x) = \cos x$, and the triangular wave $f(x) = \begin{cases} x \text{ for } 0 \leq x < 1/2 \\ 1-x, \text{ for } 1/2 \leq x < 1 \end{cases}$

As examples and determine the best approximation wavelets for various choices of the number of terms in the series for approximation.
The program is presented for the case of filter length4. The following are the steps of the computation. The codes generated are named asCode1, NwCode 1, Code2, Code3 and Code 4.
Code1: Generation of filter sequence
Fix a value for the parameter $\alpha \in [0,2\pi]$. Then the filter coefficients are given by

$$h_0 = \frac{1}{2\sqrt{2}} + \frac{1}{2}\cos\alpha,\ h_1 = \frac{1}{2\sqrt{2}} + \frac{1}{2}\sin\alpha,$$
$$h_2 = \frac{1}{2\sqrt{2}} - \frac{1}{2}\cos\alpha,\ h_3 = \frac{1}{2\sqrt{2}} - \frac{1}{2}\sin\alpha \qquad (4)$$

The Code contains the formulae for the corresponding filter coefficients of length 4.The file contains $H = [h_0, h_1, h_2, h_3]$. (5)

If an arbitrary value for the argument $\alpha \in [0,2\pi]$ is given in Code1, it returns four values. For the present calculation we consider the argument α to vary in $[0,\pi]$.

NwCode1:Construction of Scaling Function and Wavelet Function

The sequence $\{h_0, h_1, h_2, h_3\}$so generated is then used for the construction of the trigonometric polynomial $H(\omega)$given by the algorithm

$$H(\omega) = \frac{1}{\sqrt{2}}\sum_{k=0}^{3} h_k\, e^{-ik\omega} \qquad (6)$$
$$= \frac{1}{\sqrt{2}}(h_0 + h_1 e^{-i\omega} + h_2 e^{-2i\omega} + h_3 e^{-3i\omega})$$

Our construction of the scaling function $\emptyset$ proceeds from the Fourier transform $\widehat{\emptyset}$ given by Theorem1.1

$$\widehat{\emptyset}(\omega) = \frac{1}{\sqrt{2}}\prod_{j=1}^{\infty} H\left(\frac{\omega}{2^j}\right) \qquad (7)$$

$\widehat{\emptyset}(\omega)$is constructed using the infinite product in j . We can restrict the range of j to a chosen value p so that

$$\widehat{\emptyset}(\omega) = \frac{1}{\sqrt{2}}\prod_{j=1}^{p} H\left(\frac{\omega}{2^j}\right) \qquad (8)$$

The programme is carried out using NwCode1.The value of $\widehat{\emptyset}$ is obtained by the function HH4. For the present programme $p = 100$.
Further we find $\emptyset$ by finding the inverse Fourier Transform. It is given by the function ifftHH4. Thus NwCode1 gives the scaling function$\emptyset(\omega)$.

Verification for Scaling function: We check if $\emptyset(x) = \sqrt{2}\sum_{k=0}^{3} h_k \emptyset(2x-k)$ to confirm that $\emptyset(x)$ is the solution of the scaling equation.
If not, increase the value of p in (6) and get a new value for $\emptyset$. This is continued till the verification is satisfactory. In the present programme this value is obtained as $p = 100$.
In continuation to this, the next programme generates ψ.
The construction of the wavelet function $\psi$is carried out by

$$\psi(x) = \sqrt{2}\sum_{k=0}^{3} g_k \emptyset(2x-k) \quad (9)$$

where $g_k = (-1)^k h_{1-k}$. (10)

$$= (h_1, -h_0, h_3, -h_2)$$

The values of g and ψ are displayed.

This completes the construction of the wavelet function of filter length 4 for an arbitrary angular parameterα.
The following step gives the approximation of an input function using the wavelet produced.
Code2: Fixing input functionf
Three input functions are considered. They are the sine wave named as input 1, cosine wave named as input 2, and a triangular wave named as input 3. For convenience of computation all input functions are considered as defined in [0,1]. When $f(x) = \sin x$ is considered in [$[0,2\pi]$, we regard this as the function
$f(x) = \sin 2\pi x$in [0,1].
Similarly cosine function is considered as
$f(x) = \cos 2\pi x$ in [0,1].
Here we have Code2(inp_number, n) whereinp_number represent the number given for the input and n represents the number of values of the input function.
For example Code2(inp_number 1,10) represents the function
$f(x) = \sin 2\pi x$ represented by 1024values spread at equal intervals in the domain.
Code3: ErrorEstimation
When a function f is approximated by wavelet basis$\psi_{j,k}$, the error is

$$\left\|f - \sum_{j=0}^{n}\sum_{k=0}^{2^j} c_{j,k}\psi_{j,k}(t)\right\| \quad (11)$$

For convenience of computation we consider the square of the norm
$\left\|f - \sum_{j=0}^{n}\sum_{k=0}^{2^j} c_{j,k}\psi_{j,k}(t)\right\|^2$ to represent the error.
Now the norm in (9) is approximated by the mean square error (mse) which is given by
$mse = \frac{1}{N}\sum_{i=1}^{N}[f(t_i) - \tilde{f}(t_i)]^2$where

$$\tilde{f} = \sum_{j=0}^{n}\sum_{k=0}^{2^j} c_{j,k}\psi_{j,k} \quad (12)$$

We determine the value of α for which the mse is minimum.

Code3 is used for approximating the signal.
From the fact that wavelet basis $\psi_{j,k}$is
$\psi_{j,k}(t) = 2^{j/2}\psi(2^j t - k)$for integers j, k and $k = 0,1,2,\ldots.2^j - 1$
and$c_{j,k} = \int_0^1 f(t)\psi_{j,k}(t)dt$
and $f = \sum_{j=0}^{\infty}\sum_{k=0}^{2^j-1} c_{j,k}\psi_{j,k}$ (13)

In order to carry out the procedure, we need to fix the number of terms ineqn(10) by fixing the maximum of j to be s.
Now the signal f has the approximation
$f(t) \approx \sum_{j=0}^{s} \sum_{k=0}^{2^j-1} c_{j,k}\psi_{j,k}(t) = \tilde{f}(t)$ (11)
In this sum for $j = 0, k = 0$
when$j = 1, k = 0$ to 1
when$j = 2,\ k = 0,1,2,3$
and when $j = s, k = 0,1,2, \dots 2^s - 1$
Then $N = 2^{s+1}$ is the number of terms in the sum

(14)

For convenience of presentation, we consider the number of terms in (11) as
$N = 2^s - 1$ by assuming that an initial term of value0 is added in the sum.

Here in Code3 we may fix the pattern, and select the number N of coefficients $c_{j,k}$for approximation for any argument $\propto$.
The mean square error is taken as a measure in approximating the signal.
Let $n = n_f$ be the number of points in $[0,1]$ at which the values of the input signal f is given. Let $f_j = f(t_j), j = 1,2, \dots\dots n$
then $mse = \frac{1}{n}\sum_{j=1}^{n}(\tilde{f}_j - f_j)^2$ [19]
(15)
where$\tilde{f}_j = \sum_{j=0}^{s} \sum_{k=0}^{2^j-1} c_{j,k}\psi_{j,k}(t_j)$and $N = 2^s - 1$
ThusCode3outputs the mean square error when approximating the pattern with N number of coefficients and n number of input values for the input function.

Code4.Optimization

The value for the parameter $\propto \in [0,1]$ that gives the best approximation is intended in,Code4.For a fixed number of coefficients N, in the wavelet approximation, the best value of the parameter say α_{best} is extracted, the value of α that gives the minimum error when approximating with N terms.The code optimizationn.m can be used for this purpose. An initial argument for α and a desired step size are included in the code. In this code it is fixed as $arg = 0.001$ and step size $= 0.001$. The code namely Code4 is called to execute the program. In Code4 we select the input functionand the number of sample points of the input function, that is n. In addition, the number of coefficients chosen for approximation, that is N, is also provided.

The objective function is minimize (mse) of the selected input function given by Eqn (12). As the initial value for the argumentα is 0.001, the mse is calculated for the above fed data to generate the error.In the next step, to the initial argument, the step size is added to get the new argument $\alpha$to calculate the new mse. The newly obtained mse is compared with the earlier mse and the lower of the two is now stored as the error minimum. The code runs from the initial value$arg = 0.001$till $arg = 3.14$, since the interval considered is [0,1].The mse values so obtained are compared every time and the final value of the mse gives the lowest of all the error values. This is the minimised mse and the argument corresponding to this is given asα_{best}.

4. Illustration and Results:

Consider the input function sin(x), $f(t) = \sin 2\pi t, t \in [0,1]$. The values of f(t) are stored in $n = 1024$ values at points $t_j, j = 1,2 \dots 1024$ spread at equal intervals in [0,1].The number of terms in the approximation sum in eqn (13) is taken as N. Various values of N = 2,4,8 ....1024 are considered. Note that when $n = 1024$, the indexing variable j varies from 0 to 9.

The parameter $\alpha \in [0,1]$ is considered with initial value $\alpha = 0$ and incremented by .001. In Code4, input 1 is chosen as the sine wave. It fetches a mean square error for some arbitrary value of the argument$\propto \in [0,1]$. Here the input function f and the approximated function $\tilde{f}$ are shown at certain values of t. The wavelet is considered with argument $\alpha = 0.8$. The mean square error is calculated for the points of the input function taken as$n = 1024$. In Table 1, $N = 3$ and in Table 2, $N = 5$.

Table 1

t	f(t)	$\tilde{f}$ (t)	error
0	0	-0.3960	0.3960
0.1429	0.7818	0.3800	0.4018
0.2857	0.9749	0.7051	0.1698
0.4286	0.4339	0.7156	-0.2817
0.5714	-0.4339	-0.6869	0.1530
0.7143	-0.9749	-0.7075	-0.2674
0.8571	-0.7818	-0.3966	-0.3852
1	0	0.3908	-0.3908

The mse = 9.8134×10^{-5}

Table 2

t	f(t)	$\tilde{f}$ (t)	error
0	0	0	0
0.9680	0.5713	0.8109	-0.2396
0.2581	0.9987	0.8232	-0.1755
0.4194	0.4853	0.7180	-0.2327
0.5806	-0.4853	-0.8121	0.3368
0.7419	-0.9987	-0.8417	-0.1570
0.9032	-0.5713	-0.3286	-0.2427
0.9355	-0.3944	-0.3285	-0.0759
0.9677	-0.2013	-0.3250	0.1237
1	0	-0.3249	0.3249

The mse = 1.6213×10^{-5}

Table 3 gives the result of the calculation of the best wavelet with parameter α_{best}. This is given for various choices of the number of terms in the wavelet approximation series which is denoted as N.In all these cases the input function f is specified by values of 1024 points in [0,1].

Table 3

N	∝	MSE
2	.8970	.0658
4	.8750	.0350
8	.8140	.0126
16	.8740	.0045
32	.7800	.0012
64	.7940	$3.3707\text{x}10^{-4}$
128	.7880	$8.416\text{x}10^{-5}$
256	.7860	$2.0031\text{x}10^{-5}$
512	.7850	$3.8900\text{x}10^{-6}$
1024	.7850	$1.0507\text{x}10^{-8}$

Table 4 gives the values of the filter coefficients and mean square errors for different values of parameter ∝

Table 4

N	α	$H = (h_0, h_1, h_2, h_3)$	Mse
32	.5	(0.7923,0.5933,-0.0852,0.1138)	0.0072
64	.5	(0.7923,0.5933,-0.0852,0.1138)	0.0041
32	.75	(0.7194,0.6944, -0.0123,0.0127)	0.0016
64	.75	(0.7194,0.6944, -0.0123,0.0127)	4.9813x10-4

Table 5, Table 6, Table 7 and Table 8 gives the corresponding results for the cosine wave$f(t) = \cos 2\pi t$ for $t \in [0,1]$.

Table 5

t	f(t)	$\tilde{f}(t)$	error
0	0	0.3974	0.3974
0.1429	0.6235	0.4091	0.2144
0.2857	-0.2225	-0.4805	0.2580
0.4286	-0.9010	-0.4935	-0.4175
0.5714	-0.9010	-0.4942	-0.4068
0.7143	-0.2225	-0.4943	0.2718
0.8571	0.6235	0.8799	0.2564
1	1	0.9001	0.9001

Table 6

t	f(t)	$\tilde{f}$ (t)	error
0	1	-0.0031	0.0031
0.9680	0.8208	0.4201	0.4007
0.2581	-0.0506	0.4264	0.3758
0.4194	-0.8743	-0.7927	0.2339
0.5806	-0.8743	-0.6404	-0.2239
0.7419	0.0506	0.0838	0.0332
0.9032	0.8208	0.6469	0.1739
0.9355	0.9190	0.6470	0.2720
0.9677	0.9795	0.6562	0.3233
1	1	0.6564	0.3436

Table 7

N	∝	MSE
2	0.8910	0.0863
4	1.000	0.0432
8	1.010	0.0146
16	0.8880	0.0060
32	0.7820	0.0018
64	0.7880	$4.3822\text{x}10^{-4}$
128	0.7810	$9.9225\text{x}10^{-5}$
256	0.7830	$2.2395\text{x}10^{-5}$
512	0.7810	$4.2584\text{x}10^{-6}$
1024	0.7800	$8.3015\text{x}10^{-9}$

Table 8

N	α	$H = (h_0, h_1, h_2, h_3)$	Mse
32	0.5	(0.7923,0.5933, -0.0852,0.1138)	0.0034
64	0.5	(0.7923,0.5933, -0.0852,0.1138)	0.0052
32	0.75	(0.7194,0.6944,-0.0123, 0.0127)	0.0023
64	0.75	(0.7194,0.6944,-0.0123, 0.0127)	$5.812\text{x}10^{-4}$

For triangular wave , $f(x) = \begin{cases} x \text{ for } 0 \le x < 1/2 \\ 1 - x, \text{ for } 1/2 \le x < 1 \end{cases}$ the table 9 gives the validation of the mean square errors for various approximations that gives the best value of the parameter α.

Table 9

N	∝	MSE
2	0.9000	0.0038
4	0.8750	0.0020
8	1.0010	$7.3481\text{x}10^{-4}$
16	0.8710	$2.4706\text{x}10^{-4}$
32	0.7670	$6.7606\text{x}\ 10^{-5}$
64	0.7830	$1.8873\text{x}10^{-5}$
128	0.7890	$4.9013\text{x}10^{-6}$
256	0.7860	$1.1788\text{x}10^{-6}$
512	0.7850	$1.9863\text{x}10^{-7}$
1024	0.7850	$7.4405\text{x}\ 10^{-10}$

The table 10 gives the filter coefficients and the corresponding mean square errors for a few approximations.

Table 10

N	α	$H = (h_0, h_1, h_2, h_3)$	Mse
32	0.5	(0.7923,0.5933,-0.0852,0.1138)	$4.8678\text{x}10^{-4}$
64	0.5	(0.7923,0.5933,-0.0852,0.1138)	$2.2792\text{x}10^{-4}$
32	.75	(0.7194,0.6944,-0.0123, 0.0127)	$2.3678\text{x}\ 10^{-5}$
64	.75	(0.7194,0.6944,-0.0123, 0.0127)	$1.2174\text{x}\ 10^{-5}$

5. Application of approximation using parameters

Effective utilization of the vibration signals helps in identifying and estimating the performance in working of the machine. In comparison to graph plot study, wavelets provide a flexible tool, while graphs do not take in to consideration the particular class of signals to be processed. Moreover wavelets are robust to moderate changes, small shifts and deformations that can reflect a great variation in graphs.

The main advantage of time frequency analysis is discovering the patterns of frequency changes, which usually represent the nature of the signal. Hence any defect away from normal or perfect condition will result in a change of waveform, be it the ECG Signal, or EEG signal or the bearing signals [18][21]. As long as this pattern is identified, the machine fault associated with this pattern can also be identified. Misinterpretation of results may lead to false alarms or the failure to detect anomalous signals. Such negligence may cause fatal breakdown of machines, which could interrupt production and services. An experiment has been designed for use in vibration-based machine fault diagnosis [11][12][13].

5.1 Experimental Setup

In the application considered [11], the vibration signal from the piezoelectric pickup mounted on the test bearing was taken, after allowing for initial running of the bearing. The sampling frequency was 12000 Hz and sample length was 8192 for all speeds and all conditions. The sample length is chosen long enough to ensure data consistency. Sample length of around 10000 was chosen in each case. In techniques like wavelet based feature extraction, the number of samples is preferably 2n. The nearest 2n to 10000 is 8192 and hence, it was taken as sample length. The vibration signal was stored in the data file.

Four cases were considered –
i) Normal bearing (without any fault),
ii) Bearing with inner race defect,
iii) Bearing with outer race defect
iv) Bearing with both inner race and outer race defects.

The data file containing the sample signals of each of the four classes are collected. One sample of each is taken to fit as the pattern for our approximation study.

5.2. Implementation:

Signal transmission is based on transmission of a series of numbers. The series representation of a function is important in all types of signals. The data samples are collected for the four classes of faulty bearings. Consider one single sample in each for our study. The first sample is the good ones (GO) coded as input 4, the inner race (IR) is input 5, the outer race (OR) is input 6, the inner outer race (IOR) is named as input 7 in Code2.

In Signal processing, sampling is the reduction of a continuous signal to discrete signal [20]. A sample refers to a value or set of values at a point in time and/or space. Sampling frequency/rate defines the number of samples per second taken from a continuous signal to a discrete signal. For functions varying with time, if $f(t)$ is a continuous function(or signal) to be sampled and if sampling be performed by measuring the values of $f(t)$ every T seconds, then the sampled function is given by the sequence $f(nT)$, for every integer value of n.

In our experiment the sampling frequency is 12,000 Hz and so the sampling time is $T = {}^{1}/_{12000}$ seconds. In Code2 , the sampling sequence nT is taken along the $x-$axis and the data values are plotted along the $y-$axis by taking the corresponding values from the data set to obtain the patterns in each classes of data.

Consider one sample signal each from the four classes mentioned above. By selecting inputs 4,5,6,7 respectively inCode 4, the corresponding wavelet can be generated that outputs the mean square error between the sample and the wavelet [14][15][16]. The mean square error between the wavelet and each pattern needs to be optimized. The optimizationn.m file could be used to further obtain the least error for a wavelet of filter length 4 thus giving the parameter α.Let α_1, α_2 , α_3 , α_4 be the corresponding parameters obtained in that order. They are assumed to be the standard values to discriminate the faults for any later study [17].
On applying the code optimizationn.m we obtained
$\alpha_1 = 0.6010, \alpha_2 = 0.4650, \alpha_3 = 0.5000, \alpha_4 = 0.001$

6. Conclusion

The samples GO, IR, OR, IOR, is now representable in terms of the parameters α_i , $i = 1,2,3,4$which is the feature extracted for the detection of faulty bearings. The above obtained values of α_i can be taken as a measure. For any arbitrary pattern to be verified for faultiness the parametric feature has to be extracted. Subsequently a comparative study with the above values identifies the class in which this pattern falls. A pre assigned limit of variation may be fixed to discriminate the amount of fault in the bearing. The procedure presented here bares the simplicity in its representation in terms of a single parameter. This has met with the challenge of the apparent lack of quantitative results in wavelet analysis. Thus the paper could give an easy to use wavelet analysis tool kit for feature extraction.

IOP Conf. Series: Materials Science and Engineering 1012 (2021) 012033 doi:10.1088/1757-899X/1012/1/012033

References

[1] Dario Farina, Omar Feix do Nascimento,Marie-Francoise Lucas, Christian Doncaarli. 2007. Optimization of wavelets for classi_cation of movement-related cortical potentials generated by vibration of forcw-related parameters, Elseiver, Journel of Neuroscience Methods 162, 357-363.

[2] David Pollen. 1989. Parameterization of compactly supported wavelets, Company Report, Aware, Inc., AD890503,1-4.

[3] David Pollen, SUI. 1990. (2, F[z, 1/z]) for F a subfield of C, J. Amer. Math. Soc.3, 611-624.

[4] Haihui Wang, Lizhong Peng. 2006. Parameterization of Univariate Wavelet tight frames with short support, Elseiver,Communications in Nonlinear Science and Numerical Simulation 11, 663-677.

[5] Lai, Ming-Jun, Roach, David W. 2002. Parameterizations of univariate orthogonal wavelets with short support", Approximation Theory X, 369{384,Innov. Appl. Math., Vanderbilt Univ. Press.

[6] Laurent Brechet, Marie-Francoise Lucas, Christian Doncaarli, Dario Farina. 2007. Compression of Biomedical Signals with mother wavelet optimization and Best-basis wavelet Packet selection, IEEE Transactions on Biomedical Engineering, 54.

[7] Lina, Jean-Marc. 1993. Mayrand, Michel, Parametrizations for Daubechieswavelets. Phys. Rev. E (3) 48, 6, 4160-4163.

[8] Mogens Nielsen, Ernest NlanduKamavuako,MichaelMidtgaardAnderson,Marie-Francoise Lucas,Dario Farina. 2006. Optimal Wavelets for Biomedical signal compression, Medical Biological Engineering Computation 44, 561-568.

[9] S. H. Wang, A. H. Tewk, and H. Zou. 1994. Correction to `parametrization ofcompactly supported orthonormal wavelets, IEEE Trans. Signal Process. 42, 208-209.

[10] Wells,R.O.,Jr. 1993. Parameterizing smooth compactly supported waveletsTrans.Amer.Math.Soc.338, 919-931.

[11] V. Sugumaran, V.Muralidharan, K.I. Ramachandran. 2007. Feature selection using Decision tree and classification through Proximal Support Vector Machine for fault diagnostics of roller bearings, Mechanical Systems and Signal Processing 21, 930-942.

[12] Manju B.R, A. R Rajan, V.Sugumaran. 2011. 'New Wavelet feature for Fault Diagnosis of Roller Bearings using Decision tree' , Journal of Mechanical Engineering and Technology IJMET, 2 , 70-84.

[13] Manju B.R, A. R Rajan, V.Sugumaran. 2010. 'Wavelet Design for Fault Diagnosis of Roller Bearings using Continuous Wavelet Transform' International Journal for Mechanical Engineering and Technology IJMET, 1, 38-48.

[14] Manju B.R, A. R Rajan, V.Sugumaran. 2012. Optimizing the parameters of wavelets for pattern matching using GA' , Journal of Advanced Research in Engineering and Technology (IJRET) 3, 75-85.

[15] Amos Gilat. MATLAB- An introduction with applications, 5th Edition, John Wiley and Sons INC ,USA.

[16] Vinay K. Ingle, John G. Proakis. Digital Signal Processing using MATLAB, third Edition, CENGAGE Learning, USA.

[17] R K Bansal, A K Goel, M K Sharma. 2009. MATLAB and its applications to Engineering, Pearson India Publications.

[18] Manju, B. R, and Anju R. Nair. 2019. Classification of Cardiac Arrhythmia of 12 Lead ECG Using Combination of SMOTEENN, XGBoost and Machine Learning Algorithms, *2019* 9th International Symposium on Embedded Computing and System Design (ISED*)*.

[19] Manju, B. R, and M. R. Sneha. 2020. ECG Denoising Using Wiener Filter and Kalman Filter. Procedia Computer Science 171, 273-281.

[20] Manju, B. R, and Parvathy Rema. 2020. A Performance based comparative study on the Modified version of Empirical Mode Decomposition with traditional Empirical Mode Decomposition, Procedia Computer Science 171, 2469-2475.

[21] Manju, B. R, and B. Akshaya. 2020. Simulation Of Pathological ECG Signal Using Transform Method, Procedia Computer Science, 171, 2121-2127.

Efficient multi-level lung cancer prediction model using support vector machine classifier

Manju B R[1*], Athira V[2], Athul Rajendran[3]

[1]Department of Mathematics, Amrita Vishwa Vidyapeetham, Amritapuri, India

[2]Centre for Wireless Networking and Applications, Amrita Vishwa Vidyapeetham, Amritapuri, India

[3]Department of Electrcal and Electronis Engineering, Amrita Vishwa Vidyapeetham, Amritapuri, India

E-mail: manjubr@am.amrita.edu

Abstract- This paper aims at the requirement for an interactive learning framework which empowers the successful checking of disorder in a patient. Principal component analysis stands out as an outstanding algorithm to significantly classify the target classes. PCA blends associated characteristics and makes a dissipated showcase of its components well. Scree plot examination gives solidarity of how many principal components are to be retained. Support Vector Machines (SVM) is a fast and dependable classification algorithm that outperforms other techniques with a limited amount of data. The obtained components will be served to Support Vector Machine for further classification. The pre-dangerous stage will remind the clinical experts to give additional consideration to those patients. The expectation ability is estimated in terms of the confusion matrix. The model developed gives a high and uncompromising accuracy in early detection of different levels of malignancy

Keywords: Premalignant, PCA, SVM, confusion matrix

1. Introduction

In this time, the primary reason of death around the globe is malignant growth or carcinoma. Among the different types of existing cancers like breast, liver, stomach and colorectal, majority of deaths are caused by lung carcinoma (1.6 million) [1]. According to the overall information in 2012, lung cancer contributes 13% all cancers considered. Lung Carcinoma seems to be a propagation of morphological changes [8]. The two varieties of lung cancer are Small Cell and Non-Small Cell Cancers (SCLC and NSCLC) [2]. The reasons for lung disease incorporate hereditary and non-hereditary reasons [3]. Tobacco smoking is the far leading cause of lung cancer. It can severely affect the health of chain smokers as well as people in contact with smokers [4]. The lungs of the non-smokers can be affected by this malignancy due to air pollution, working atmosphere with radon, asbestos or any other chemical exposure [4]. Hereditary causes include the gene mutations in lungs. Individuals who have had taken radiation therapy for other diseases can also be the victims of this malignancy [4].

Mostly this condition is recognized at the advanced stage. This can be limit the cure of therapeutic intervention [5]. The different stages of lung cancer [27] are tabulated below with explanation in Table.1:

Table 1. Lung Cancer stages explained

Stages	Explanation
Stage 1	The cancer is confined only in the lungs.
Stage 2	The cancer is in the lungs and can have metastasized to the lymph nodes adjacent to the lungs.
Stage 3	Cancer is localized in the lungs and has spread to the lymph nodes in the center of the chest.
Stage 3A	The cancer has become extensive, however, is localized to one side of the lung.
Stage 3B	The cancer has metastasized to the lymph nodes on both the sides of the lungs.
Stage 4	The cancer has metastasized to both the lungs or has spread to other parts of the body. Stage 4 is the most advanced stage of lung cancer.

The primary care of medical practitioners is needed for the fast survival. The identification of lesion type will gives the signals for the further treatment. The malignant stage is red signal, benign stage counts for green stage and an intermediate stage can be introduced as pre-malignant stage which is indicated as yellow signal. If a yellow signal is appeared, the doctors should continuously monitor those patients and it will be helpful for the fast cure. The popular examples declared by World Health Organization (WHO) are typical adenomatous hyperplasia; diffuse idiopathic pulmonary neuroendocrine cell hyperplasia, squamous metaplasia with dysplasia and carcinoma in-situ [6]. These conditions can be advance to harmful stages like carcinoid tumors, squamous cell carcinoma and adenocarcinoma [7]. Manual evaluation of lung features for the qualitative result is mentioned in [10]. Ground Glass Opacity [GGO] is considered to be pre- malignant, but this shadow is not differentiable by radiologist [9]. People reached a stage to be aware of the pre-malignant stage.

Detection using screening procedures includes Computed Tomography (CT) and other bronchoscopic techniques [6]. CT visualizes the spongy nodules of the windpipe within single breath threshold [12]. These techniques can capture 2-3 mm of lesions and it can create confusions on determining whether it is malignant, benign or the intermediate stage. Until this point in time, screening preliminaries have had no critical effect on survival [7]. Machine learning and feature selection has a golden chair in Lung cancer Detection, nowadays. A numerical measurement fed to a most accurate machine learning system is more preferable than the distorted image obtained by radiologists. In [9], GGO is extracted for the study using Deep Learning, which is as intelligent as human brain. Automated system to view histopathological part of respiratory tract is explained in [10]. The trending state of art to do lung cancer diagnosis with the aid of computer knowledge is explained in [11], touching the technical concerns and validation. CAD procedure has a golden standard in terms of speed and accuracy, when compared to other methods [12].

In the world, main part of datasets are available via web based networking and from this tremendous piece of information, features useful for the tasks like classification and prediction can be fetched [13]. These types of feature vectors are classified into target classes using Random Forest Classifier in [14]. Along with the existing blocks of classifiers, weighted models of SVM, ANN and NB created sequential diagnostic model for lung cancer patients in [15].

Some of the reported studies regarding machine learning applications on lung cancer is tabulated in TABLE.2. This paper introduces an automated platform for the lung cancer victims belongs to premalignant stage. The application is overlying on a PCA – SVM combined model and all the statistical parameters are well drawn for the further analysis.

Table 2. Reported studies on Lung Cancer Detection

Sl.No.	Name	Year	Overview
1.	Comparison of Lung Cancer Detection Algorithms [19].	2019	Applied Principal Component Analysis, K-Nearest Neighbors, Support Vector Machines, Naïve Bayes, Decision Trees and Artificial Neural Networks machine learning methods before and after preprocessing to detect anomaly.
2.	A Deep Learning Based Approach to Lung Cancer Identification [16].	2018	A 3D neural network based diagnosis with tensorflow library which effectively works even with less number of samples.
3.	Lung Cancer Detection Using Image Processing and Machine Learning HealthCare [17].	2018	SVM classifier and image processing is collaborated on CT scan lung images to enable early detection.
4.	Small-Cell Lung Cancer Detection Using a Supervised Machine Learning Algorithm [18].	2017	A novel neural-network based algorithm with entropy degradation method (EDM), to detect small cell lung cancer (SCLC) from CT images.
5.	A Survey on Lung Cancer Detection using Image Data Analysis and Machine Learning [25].	2017	This study aims to highlight the significance of data analytics and machine learning.
6.	Lung cancer classification using deep learned features on low population dataset [20].	2017	A deep Autoencoder classification mechanism which works on low population an unstructured data.
7.	Deep learning application trial to lung cancer diagnosis for medical sensor systems [22].	2016	Personal diagnosis is enabled using sensing data and deep learning techniques for the feature selection
8.	Convolutional neural networks for lung cancer screening in computed tomography (CT) scans [23].	2016	Special properties of CNN networks like spatial invariance which extract features easily is applied for lung cancer detection.
9.	Computer aided diagnosis system based on machine learning techniques for lung cancer [21].	2012	Feature extraction using wavelet transform is applied on lung cancer dataset to achieve high accuracy.
10.	Predictive Analysis of Lung Cancer Recurrence [24].	2011	The predictive analysis of lung cancer recurrence based on non-small cell lung cancer carcinoma gene expression data using data mining and machine learning techniques.

2. Methodology

The strategy followed in this work is drawn in Fig.1

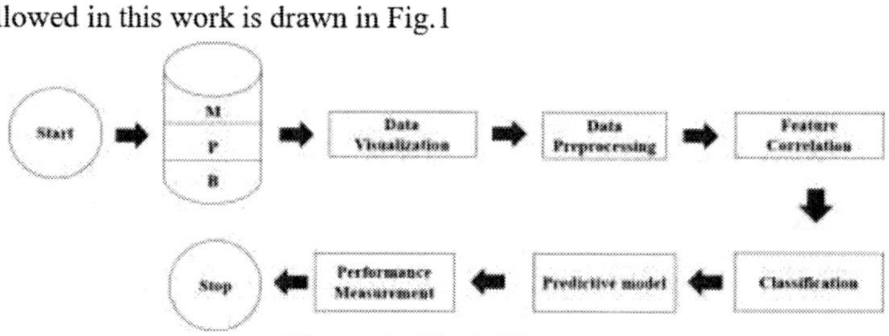

Figure 1. Block Diagram

2.1 Data Collection

Database used in this research work is obtained from Lung Cancer Dataset UCI repository [26]. This multivariate data consists of 600 instances with three target classes like Malignant (M, count-215), Pre-malignant (P, count-194) and Benign (B, count-190) indicates three conditions of lung cancer tumors. The features listed in this csv dataset are Air Pollution, Alcohol use, Dust Allergy, Occupational Hazards, Genetic Risk, Chronic Lung Disease, Obesity, Smoking, Chest Pain, Coughing of blood, Fatigue, Weight loss, Shortness, Wheezing, Swallowing difficulty, Clubbing of fingers, Frequent cold, Dry cough and Snoring.

2.2 Data Visualization

Visualizing the data is the method of converting the data into abstract images which follows certain patterns or trends. It encourages the analysts to think of huge choices in application level. Python offers various incredible diagramming libraries like pandas, Matplotlib etc., which creates various highlighted histogram, scatter plot and density curve [28]. The features that are skewed in TABLE.2 are stretched as combination of histograms and density curves in Fig. 2. The followed pattern of each feature corresponding the target class through histograms. The normal or exponential distribution of features helps in parametric analysis in terms of maximum value, standard deviation, minimum value, mean etc.

2.3 Data Skewness.

The skewness of the data represents the asymmetry in statistics. The curve will seems to be skewed to either left or right [29]. The result can be quantified to analyse the difference between the obtained distribution and normal distribution.

- If mean > median, then "Positively skewed".
- If mean < median, then "Negatively skewed".

Table 3. Skewness of features

Features	**Skew Values**
Air Pollution	0.195541
Alcohol use	0.023802
Dust Allergy	-0.563203
Occupational Hazards	-0.170883
Genetic Risk	-0.085225
Chronic Lung Disease	-0.109587
Balanced Diet	-0.053694
Obesity	0.061515
Smoking	0.407958
Passive Smoker	0.372199
Chest Pain	0.189506
Coughing of Blood	0.162642
Fatigue	0.857201
Weight Loss	0.370156
Shortness of Breath	0.380384
Wheezing	0.222926
Swallowing Difficulty	0.480487
Clubbing ofFinger Nails	0.787004
Frequent Cold	0.456050
Dry Cough	0.231364
Snoring	0.546970

IOP Conf. Series: Materials Science and Engineering 1012 (2021) 012034 doi:10.1088/1757-899X/1012/1/012034

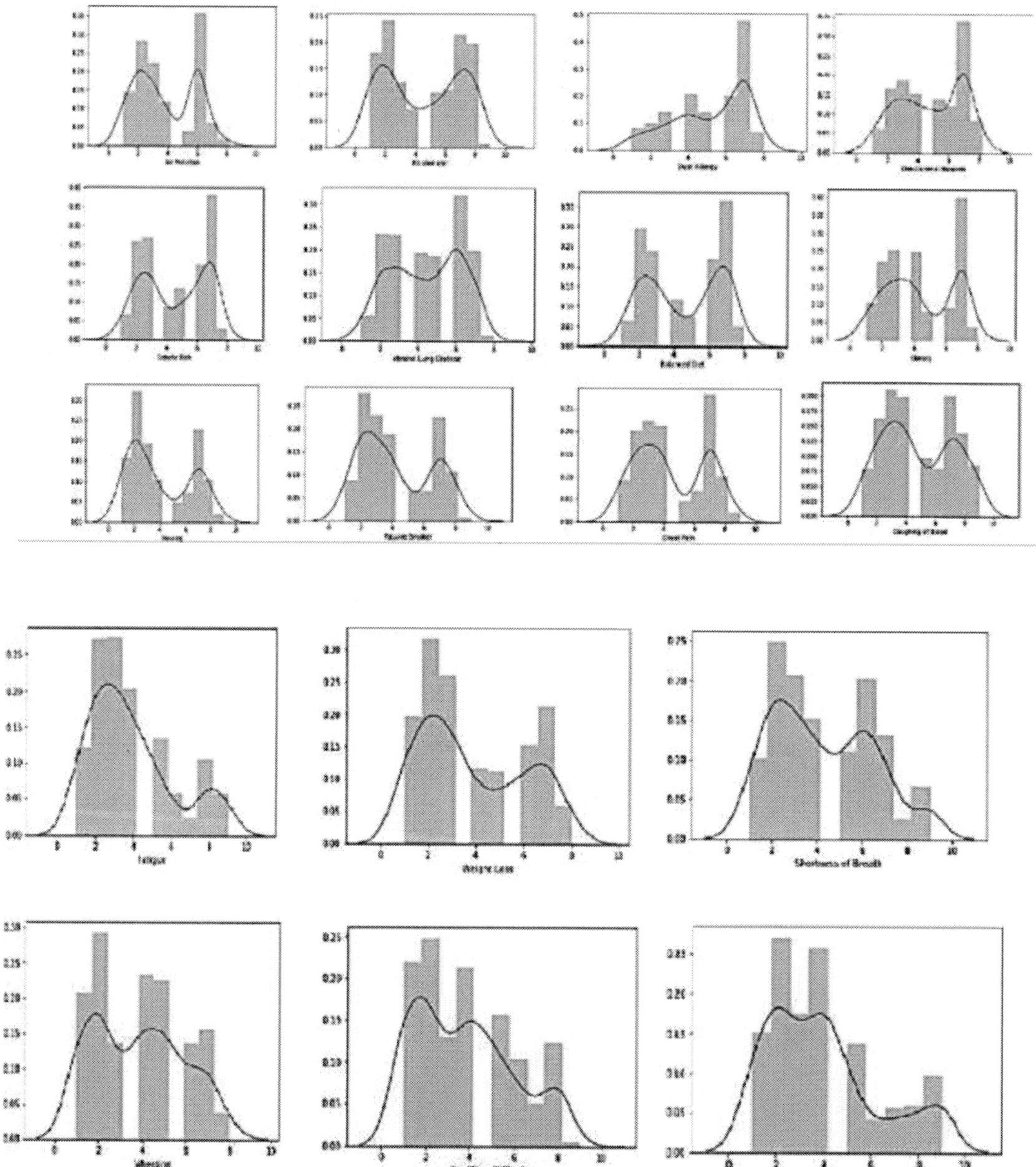

Figure 2. Visualizing the feature

2.4 Data Preprocessing

Preprocessing is an important pre-requirement for any data examination. It is generally an excellent plan to set up the information in such a manner to uncover the structure of the data to the machine learning calculations that needs to use. Data preprocessing techniques are well known in enhancing the capability power of classification systems [31]. This includes various exercises [30] like:

- Allotting numerical qualities to target.
- Dealing with missing numerical.
- Normalizing the highlights

Some unimportant highlights like patient-id, Age and Gender are cleared out to expand the effectiveness of the model we consider. As of now the information contains 3 targets, 21 characteristics and 599 cases. The cleaned csv dataset completes the preprocessing, after reducing the dimensionality.

Dimensionality Reduction – Principal Component Analysis (PCA). PCA handles the feature selection strategies through the view of reduced dimensionality. The unfocussed features can bring tremendous decline in the performance rate of prediction models. Diminishing the dimensionality of a data [38] by picking the significant features present in the underlying dataset are named as feature selection. These informational indexes are simpler to investigate and break down information in simpler way for calculations without superfluous factors to process. PCA is figured as a powerful data representation tool in [37]. PCA works on the mathematical basis of linear algebra which analyses the correlation of features [39]. In Fig. 3, PCA feature space transformation is done with two target classes namely Malignant (M) and Benign (B) and samples are spread along the axes. This type of transformation is also applicable for the Pre-malignant class, which is displayed along with the conventional classes in Figure.5. Some of the relevant markings of PCA in medical domain are:

For distinguishing normal and abnormal prostate cells, PCA is used along with signal processing in [32]. The irrelevant points are cleared out from the original feature space using PCA in breast cancer detection [33]. PCA pre-forms the data and concentrates on relevant features for training the model [34]. The PCA based algorithm achieved improved accuracy with the absence of overfitting and outliers in [35]. In [36], a PCA-FNN system is proposed to tune parameters for categorizing liver cancer.

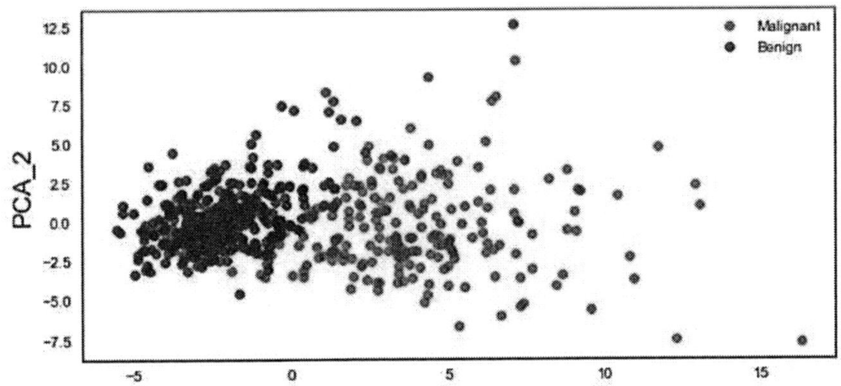

Figure 3. PCA feature space

Scree Plot: The numbers of PCA components displayed in the Fig. 3. are PCA_1 and PCA_2, which indicates the two components, used to transform the feature space. The significance of choosing principal component number in PCA can be vividly seen in the scree plot in Fig. 4. A scree plot is an analytic device to check whether PCA functions are good indicators or not [40]. The components are arranged by the measure of variety they spread.

PCA_1 catches the most variety, PCA_2 the second most, etc. Fig.4 shows an elbow curve at the point 2, which points to the generation of PCA_1 and PCA_2 for the fed input.

IOP Conf. Series: Materials Science and Engineering 1012 (2021) 012034 doi:10.1088/1757-899X/1012/1/012034

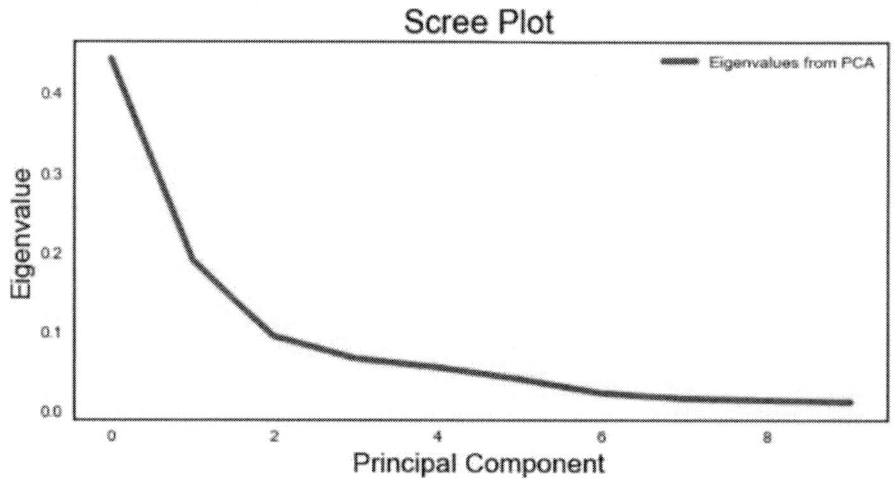

Figure. 4. Scree Plot

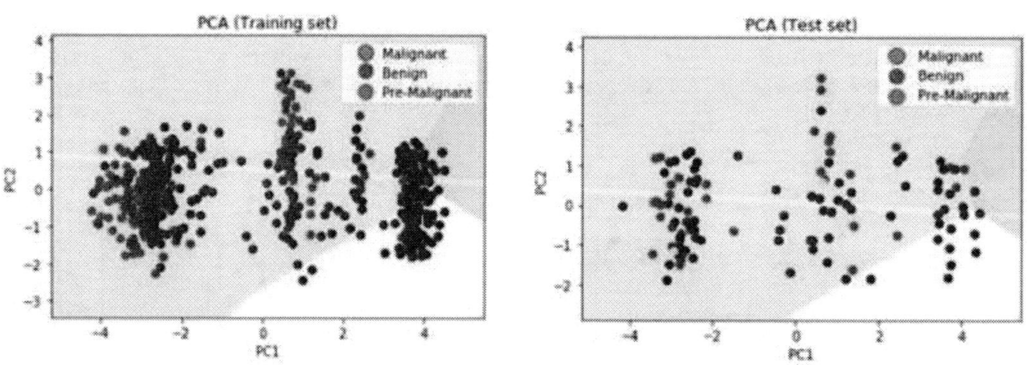

Figure 5. Three target classes with PCA

2.5 Selection of features through Feature correlation

Machine learning algorithms are always acknowledged with framework of features that are quantifiable. Data Scientists cleanse data, by eliminating the unwanted features or features having less contribution to the target class. Feature correlation is the inevitable step which indicates the contribution of each feature to the target class. Correlation can be positive or negative. Redundancy and relevancy of features are pestered in [42]. Positive correlation values indicate the direct proportionality of features with the class, while negative correlation tells about the inverse proportionality. Correlation value 1 means "perfectly correlated". If two features x and y are directly proportional to malignant class, then one of the feature can be removed by considering their correlation value. To visualize the correlation values together on a plane, Heatmap is generally used by the analysts. The Heatmap for the lung cancer prediction model is displayed in Fig. 6.

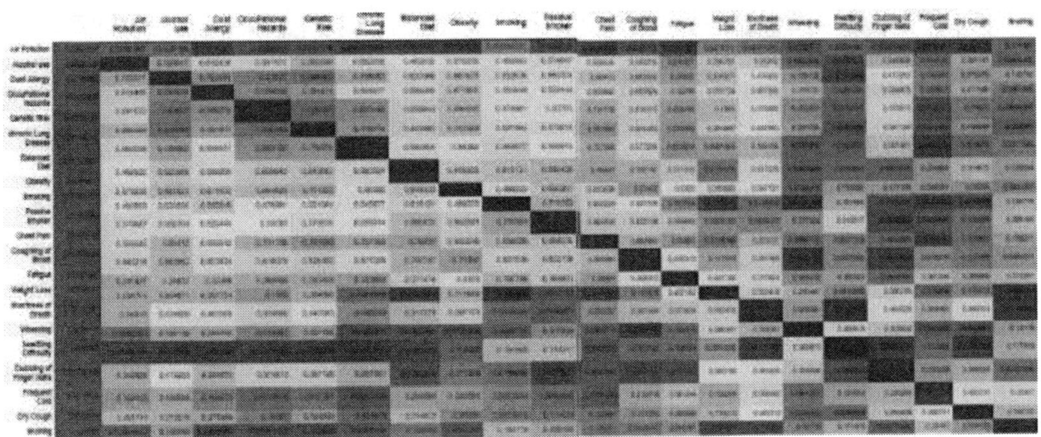

Figure 6. Heatmap representation

2.6 Classification

Classification is the logical grouping of data. It has the capability to make decisions on real and unstructured data. It plays a major role in health care and data security. CVDs and diabetes are classified evidently using ANN in [43]. Neural network synced with fuzzy system to detect asthma in [44]. Diagnosis of epilepsy using signaling techniques and machine learning can be seen in [45]. The possibilities of Alzheimer's disease are explored using classification task in [46]. In the field of data security, variable indexes are used to classify in [47] and mobility of big data is enhanced to improve data security in [48]. In [49], data mining is explored to eradicate malware and DoS attacks. Texts are pruned and classified to ensure security in an automated way [50]. Conventionally, the cancer dataset was classified as Malignant (M) and Benign (B), but this paper throws light into an intermediate condition called "Pre-malignant", which provides an extra care for the patients. In data security, there will be a continuous monitoring of categories like copied, transmitted and retrieved data. Classification involves labeling information to make it effectively accessible and identifiable. It can eliminate duplications which reduces storage and backup costs. It can incredibly decrease the processing time.

Support Vector Machine (SVM): SVM is one of the most popular classification algorithms which have an elegant way of transforming nonlinear data. Classification strategy of SVM is well explained in [52]. Hyper plane is the important tool of SVM that separates the data points in such a manner that the margin between two classes will be wide and the data points will be as far as possible. In this way, hyper plane will be creating a decision boundary with support vector points nearer to the left and right hyperplane. Linear SVM model is used for this lung cancer prediction study. A sample of SVM classification is shown in Fig. 7. Class 1 belongs to Malignant (M) and class -1 belongs to Benign (B).

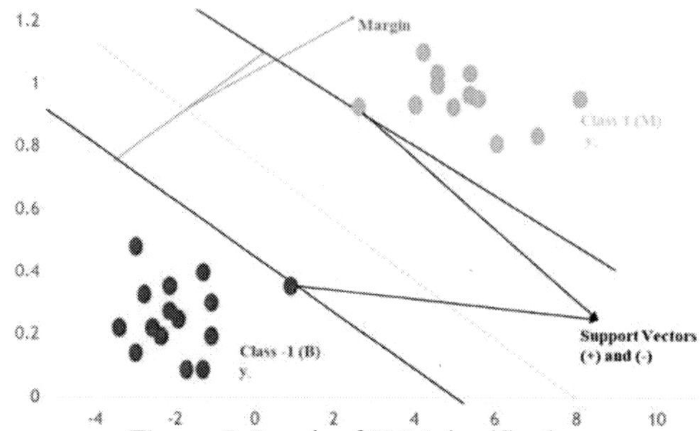

Figure. 7. Sample of SVM classification

Advantages of SVM [51] are:

- Clear separation of data points into classes.
- Expertise in high dimensional analysis.
- Support vectors needs minimum memory storage.
- Parameters are tunable and user friendly.

SVM classifiers are prominent in health domain. 2-level DWT images using a GLCM (Gray level Co-occurrence Matrix) effected by means of an SVM (Support Vector Machine) in [53]. Chip based system with SVM flavor is designed in [54]. The various strategies of SVM to detect breast cancer in an accurate way is explained in [55] [56] [57].

3. Results – Performance measurement

Results are obtained through confusion matrix and classification report. Confusion matrix is generated from the binary classification outcomes [58]. The highlighted parameters like accuracy, error rate, truth positive rate (TPR), false positive rate (FPR), truth negative rate (TNR) and false negative rate (FNR) can be calculated on the basis of this matrix. Accurate model with high score is the uncompromising factor for detection systems. GPS based map matching systems [59], and processors with frequency clocks [60] question the score of accuracy and ask for improvement. So, data scientists will try to minimize the error rate to a great extent. Accuracy score and error rate of this model are equated below as eqn [1] and eqn [2]. Commonly, the cancer prediction is deployed by 2 x 2 matrices. Here, it is innovated as 3 x 3 matrix and drawn in Fig. 8 along with the classification report. Classification Report will gives the precision, recall, f1-score and support of this classification system. Parametric definitions are in Table 4.

Table 4. Parametric Definitions

Sl.No.	Parameters	Deftnitions
1	True Positive (TP)	Correct positive predictions
2	False Positive (FP)	Incorrect positive predictions
3	True Negative (TN)	Correct negative predictions
4	False Negative (FN)	Incorrect negative predictions
5	Error Rate/Misclassification rate	Total number of incorrect predictions
6	Accuracy	Total number of correct predictions
7	Sensitivity	The number of correct positive predictions (TP) divided by the total number of positives (P).
8	Specificity	The number of correct negative predictions divided by the total number of negatives.
9	Precision	The number of correct positive predictions divided by the total number of positive predictions.
10	F1-score	Harmonic mean of precision and recall.

$$\text{Accuracy} = (\text{TPR} + \text{TN})/\text{Total} = 0.87 \quad (1)$$

$$\text{Error Rate} = (\text{FP} + \text{FN})/\text{Total} = 0.3 \quad (2)$$

```
confusion matrix
[[160   4   6]
 [  0 178  14]
 [  6  36 136]]
```

Confusion matrix plot of SVC

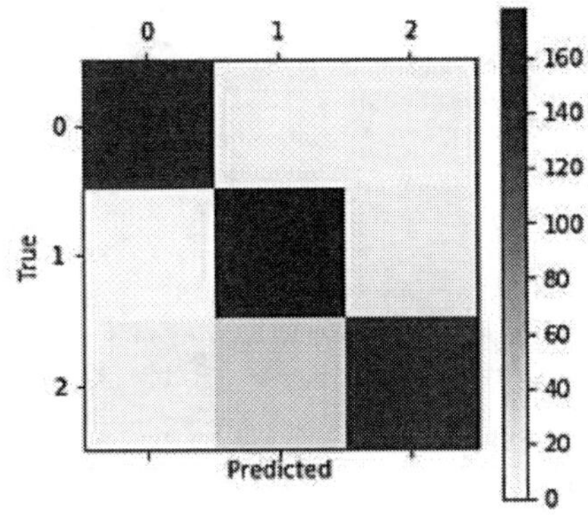


```
Classification report
             precision    recall  f1-score   support

          B       0.96      0.94      0.95       170
          M       0.82      0.93      0.87       192
          P       0.87      0.76      0.81       178

avg / total       0.88      0.88      0.88       540
```

Figure 8. Performance of SVM classification model

4. Conclusion and Future Work

This investigation draws thoughtfulness to the significance of pre-malignant stage in early detection of lung carcinoma. The real and inconsistent data is processed and cleansed. The precision of the SVM classifier could clearly signature with high accuracy of the prediction with high distinction. The confusion matrix obtained could effectively label and identify the information hidden in the dataset. Here the performance is coined in terms of classification report toasted on confusion matrix and its calculations. Accuracy score and error rate of this model is presented as 3 x 3 matrix in contrast to the conventional 2x2 matrix. Classification Report gives the precision, recall, f1-score and support of this classification system. The work gives a scope for extension to optimization level that can ensure leading accuracy in machine learning diagnostics.

References

[1] Ferlay, J.; Soerjomataram, I.; Dikshit, R.; Eser, S.; Mathers, C.; Rebelo, M. 2015. Cancer incidence and mortality worldwide: sources, methods and major patterns in GLOBOCAN 2012. Int J Cancer, 136(5), 359 –386.

[2] Gadgeel, S.M.; Ramalingam, S.S; Kalekerian, G.P. 2012. Treatment of lung Cancer, adiol Clin North Am, 961-74.

[3] Sruthi Ignatious, Robin Joseph . 2015. Computer Aided Lung Cancer Detection System, Proceedings of 2015 Global Conference on Communication Technologies (GCCT 2015), 7342723.

[4] American Cancer Society, URL: https://www.cancer.org/cancer/non-small-cell-lung-cancer/causes-risks-prevention/what-causes.html.

[5] W. Wang and S. Wu. 2016. A Study on Lung Cancer Detection by Image Processing, International Conference on Communications, Circuits and Systems, Guilin, 371-374.

[6] A. Sarkar, A. Sadhu, S. B. Thakur, S. Sengupta, A. Mukherjee and J. Chatterjee. 2016. Multimodal characterization of radiologically detectable lung lesions, 38th Annual International Conference of the IEEE Engineering in Medicine and Biology Society (EMBC), Orlando, FL, 1422-1425.

[7] Greenberg, A. K., Yee, H., & Rom, W. N. 2002. Preneoplastic lesions of the lung. Respiratory Research, 3(1).

[8] Klebe, S., & Henderson, D. W. 2013. Facts and fiction: premalignant lesions of lung tissues. Pathology, 45(3), 305–315.

[9] K. Hirayama et al. 2017. Extraction of GGO regions from chest CT images using deep learning , 17th International Conference on Control, Automation and Systems (ICCAS), Jeju, 2017, 351-355.

[10] Yu, K.-H., Zhang, C., Berry, G. J., Altman, R. B., Ré, C., Rubin, D. L., & Snyder, M. 2016. Predicting non-small cell lung cancer prognosis by fully automated microscopic pathology image features. Nature Communications, 7, 12474.

[11] El-Baz, A., Beache, G. M., Gimel'farb, G., Suzuki, K., Okada, K., Elnakib, A., Abdollahi, B. 2013. Computer-Aided Diagnosis Systems for Lung Cancer: Challenges and Methodologies. International Journal of Biomedical Imaging, 1–46.

[12] Bhavanishankar .K1 and Dr. M.V.Sudhamani2. 2015. TECHNIQUES FOR DETECTION OF SOLITARY PULMONARY NODULES IN HUMAN LUNG AND THEIR CLASSIFICATIONS -A SURVEY, International Journal on Cybernetics & Informatics (IJCI), 4.

[13] S. Shandilya and C. Chandankhede. 2017. Survey on recent cancer classification systems for cancer diagnosis, International Conference on Wireless Communications, Signal Processing and Networking (WiSPNET), Chennai, 2590-2594.

[14] M. Sharma, R. Singh and M. Bhattacharya. 2017. Classification of breast tumors as benign and malignant using textural feature descriptor," IEEE International Conference on Bioinformatics and Biomedicine (BIBM), Kansas City, MO, 1110-1113.

[15] Hosseinzadeh, F., KayvanJoo, A., Ebrahimi, M., & Goliaei, B. 2013. Prediction of lung tumor types based on protein attributes by machine learning algorithms. SpringerPlus, 2(1), 238.

[16] CENGIL, E., & CINAR, A. 2018. A Deep Learning Based Approach to Lung Cancer Identification. International Conference on Artificial Intelligence and Data Processing (IDAP).

[17] W. Rahane, H. Dalvi, Y. Magar, A. Kalane and S. Jondhale. 2018. Lung Cancer Detection Using Image Processing and Machine Learning HealthCare, International Conference on Current Trends towards Converging Technologies (ICCTCT), Coimbatore, 1-5.

[18] Q. Wu and W. Zhao. 2017. Small-Cell Lung Cancer Detection Using a Supervised Machine Learning Algorithm, International Symposium on Computer Science and Intelligent Controls(ISCSIC),Budapest, 88-91.

[19] Ö. Günaydin, M. Günay and Ö. Şengel. 2019. Comparison of Lung Cancer Detection Algorithms, Scientific Meeting on Electrical-Electronics & Biomedical Engineering and Computer Science (EBBT), Istanbul, Turkey,1-4.

[20] S. M. Salaken, A. Khosravi, A. Khatami, S. Nahavandi and M. A. Hosen. 2017. Lung cancer classification using deep learned features on low population dataset, IEEE 30th Canadian Conference on Electrical and Computer Engineering (CCECE), Windsor, ON, 1-5.

[21] H. R. H. Al-Absi, B. B. Samir, K. B. Shaban and S. Sulaiman. 2012. Computer aided diagnosis system based on machine learning techniques for lung cancer, International Conference on Computer & Information Science (ICCIS), Kuala Lumpeu, 295-300.

[22] R. Shimizu et al. 2016. Deep learning application trial to lung cancer diagnosis for medical sensor systems, International SoC Design Conference (ISOCC), Jeju, 191-192.

[23] P. Rao, N. A. Pereira and R. Srinivasan. 2016. Convolutional neural networks for lung cancer screening in computed tomography (CT) scans, 2nd International Conference on Contemporary Computing and Informatics (IC3I), Noida, 489-493.

[24] Shweta Srivastava, Manisha Rathi, and J.P. Gupta. 2011. ACC 2011, Part I, CCIS 190, 260– 269.

[25] Apoorva Mahale, Chetan Rawool et.al. A Survey on Lung Cancer Detection using Image Data Analysis and Machine Learning, International Journal of Innovative Research in Computer and Communication Engineering.

[26] UCI Lung Cancer Dataset, URL: https://archive.ics.uci.edu/ml/datasets/Lung+Cancer

[27] Stages and Survival Rate of Lung Cancer, URL: https://www.epainassist.com/chest-pain/lungs/stages-and-survival-rate-of-lung-cancer.

[28] Introduction to Data Visualization in Python, URL: https://towardsdatascience.com/introduction-to-data-visualization-in-python-89a54c97fbed

[29] Machine Learning Basics: Data Skewness and handling, URL: https://medium.com/mlrecipies/machine-learning-basics-data-skewness-and-handling-778ec7adb18c.

[30] What Steps should one take while doing Data Preprocessing? URL: https://hackernoon.com/what-steps-should-one-take-while-doing-data-preprocessing-502c993e1caa.

[31] D. H. Deshmukh, T. Ghorpade and P. Padiya. 2015. Improving classification using preprocessing and machine learning algorithms on NSL-KDD dataset, International Conference on Communication, Information & Computing Technology (ICCICT), Mumbai, pp. 1-6.

[32] Ghosh, A., & Barman, S. 2013. Prediction of Prostate Cancer Cells based on Principal Component Analysis Technique. Procedia Technology, 10, 37–44.

[33] S. Ghosh, S. Mondal and B. Ghosh,. 2014. A comparative study of breast cancer detection based on SVM and MLP BPN classifier," 2014 First International Conference on Automation, Control, Energy and Systems (ACES), Hooghy, 1-4.

[34] S. Jhajharia, H. K. Varshney, S. Verma and R. Kumar. 2016. A neural network based breast cancer prognosis model with PCA processed features, International Conference on Advances in Computing, Communications and Informatics (ICACCI), Jaipur, 1896-1901.

[35] Akshya Yadav, Imlikumla Jamir, Raj Rajeshwari Jain, Mayank Sohani. Breast Cancer Prediction using SVM with PCA Feature Selection Method, International Journal of Scientific Research in Computer Science, Engineering and Information Technology.

[36] Ms. G.Ilakkiya and Mrs.B.Jayanthi. 2013. Liver Cancer Classification Using Principal Component Analysis and Fuzzy Neural Network, International Journal of Engineering Research & Technology (IJERT), Vol. 2 Issue 10.

[37] F. Song, Z. Guo and D. Mei. 2010. Feature Selection Using Principal Component Analysis, International Conference on System Science, Engineering Design and Manufacturing Informatization, Yichang, 27-30.

[38] M. Partridge and M. Jabri. 2000. Robust principal component analysis, Neural Networks for Signal Processing X. Proceedings of the IEEE Signal Processing Society Workshop (Cat. No.00TH8501), Sydney, NSW, Australia, 289-298.

[39] S. Sehgal, H. Singh, M. Agarwal, V. Bhasker and Shantanu. 2014. Data analysis using principal component analysis, International Conference on Medical Imaging, m-Health and Emerging Communication Systems (MedCom), Greater Noida, 45-48.

[40] How to read PCA biplots and scree plots, URL: https://blog.bioturing.com/2018/06/18/how- to-read-pca-biplots-and-scree-plots/.

[41] Why Feature Correlation Matters .... A Lot! URL: https://towardsdatascience.com/why- feature-correlation-matters-a-lot-847e8ba439c4.

[42] G. Qu, S. Hariri and M. Yousif. 2005. A new dependency and correlation analysis for features, IEEE Transactions on Knowledge and Data Engineering, 17,1199-1207.
[43] B. Alić, L. Gurbeta and A. Badnjević. 2017. Machine learning techniques for classification of diabetes and cardiovascular diseases, 6th Mediterranean Conference on Embedded Computing (MECO), Bar, 1-4.
[44] A Badnjevic, M Cifrek, D Koruga, D. Osmankovic. 2005. Neuro-fuzzy classification of asthma and chronic obstructive pulmonary disease, BMC Medical Informatics and Decision Making Journal.
[45] S. Avdakovic, I. Omerhodzic, A. Badnjevic, D. Boskovic. 2014. Diagnosis of Epilepsy from EEG Signals using Global Wavelet Power Spectrum, MBEC, Dubrovnik, Croatia.
[46] A Aljovic, A Badnjevic, L Gurbeta. 2016. Artificial Neural Networks in the Discrimination of Alzheimer's disease Using Biomarkers Data IEEE MECO, Bar, Montenegro.
[47] F. Fatemi Moghaddam, M. Yezdanpanah, T. Khodadadi, M. Ahmadi and M. Eslami. 2014. VDCI: Variable data classification index to ensure data protection in cloud computing environments, IEEE Conference on Systems, Process and Control (ICSPC 2014), Kuala Lumpur, 53-57.
[48] I. Hababeh, A. Gharaibeh, S. Nofal and I. Khalil. 2019. An Integrated Methodology for Big Data Classification and Security for Improving Cloud Systems Data Mobility IEEE Access, 7,9153-9163.
[49] M. Monshizadeh and Z. Yan. 2014. Security Related Data Mining, IEEE International Conference on Computer and Information Technology, Xi'an, 775-782.
[50] K. Alzhrani, E. M. Rudd, C. E. Chow and T. E. Boult. 2016. Automated big security text pruning and classification," IEEE International Conference on Big Data (Big Data), Washington, DC, 3629-3637.
[51] What do support vector machines trump other classification methods, URL: http://www.simafore.com/blog/bid/112816/When-do-support-vector-machines-trump-other-classification-methods.
[52] Chen Junli and Jiao Licheng. 2000. Classification mechanism of support vector machines, WCC 2000 - ICSP 2000. 5th International Conference on Signal Processing Proceedings. 16th World Computer Congress 2000, Beijing, China, 3, 1556-1559.
[53] D. P. Kaucha, P. W. C. Prasad, A. Alsadoon, A. Elchouemi and S. Sreedharan. 2017. Early detection of lung cancer using SVM classifier in biomedical image processing, IEEE International Conference on Power, Control, Signals and Instrumentation Engineering (ICPCSI), Chennai, 3143-3148.
[54] S. Afifi, H. GholamHosseini and R. Sinha. 2017. SVM classifier on chip for melanoma detection, 39th Annual International Conference of the IEEE Engineering in Medicine and Biology Society (EMBC), Seogwipo, 270-274.
[55] Z. Luo, X. Wu, S. Guo and B. Ye. 2008. Diagnosis of Breast Cancer Tumor Based on PCA and Fuzzy Support Vector Machine Classifier, Fourth International Conference on Natural Computation, Jinan, 363-367.
[56] M. Sewak, P. Vaidya, C. Chan and Zhong-Hui Duan. 2007. SVM Approach to Breast Cancer Classification, Second International Multi-Symposiums on Computer and Computational Sciences (IMSCCS 2007), Iowa City, IA, 32-37.
[57] M. Hussain, S. K. Wajid, A. Elzaart and M. Berbar. 2011. A Comparison of SVM Kernel Functions for Breast Cancer Detection, Eighth International Conference Computer Graphics, Imaging and Visualization, Singapore, 145-150.
[58] Basic evaluation measures from the confusion matrix, URL: https://classeval.wordpress.com/introduction/basic-evaluation-measures/.
[59] M. Hashemi and H. A. Karimi. 2016. A Machine Learning Approach to Improve the Accuracy of GPS-Based Map-Matching Algorithms (Invited Paper),IEEE 17th International Conference on Information Reuse and Integration (IRI), Pittsburgh, PA, 77-86.
[60] Y. Tanaka, K. Oka, T. Ono and K. Inoue. 2016. Accuracy analysis of machine learning-based performance modeling for microprocessors, 2016 Fourth International Japan-Egypt Conference on Electronics, Communications and Computers (JEC-ECC), Cairo, 83-86.

[61] Manju, B. R., and Anju R. Nair. 2019. Classification of Cardiac Arrhythmia of 12 Lead ECG Using Combination of SMOTEENN, XGBoost and Machine Learning Algorithms." 9th International Symposium on Embedded Computing and System Design (ISED). IEEE.
[62] Manju, B. R., and M. R. Sneha. 2020. ECG Denoising Using Wiener Filter and Kalman Filter, Procedia Computer Science 171, 273-281.
[63] Manju, B. R., and Parvathy Rema. 2020. A Performance based comparative study on the Modified version of Empirical Mode Decomposition with traditional Empirical Mode Decomposition, Procedia Computer Science 171, 2469-2475.
[64] Manju, B. R., and B. Akshaya. 2020. Simulation Of Pathological ECG Signal Using Transform Method, Procedia Computer Science 171, 2121-2127.

Effect of Vibration on Surface Roughness of Drilling on Glass Fiber Reinforced Plastic

M S Sureshkumar[1], S Krishna prasath[1*], C Lokeshkumar[1], S Loheshwaranath[1]

[1]Department of Mechanical Engineering, Sri Ramakrishna Engineering College, Coimbatore, India

E-mail: krishnaprasath.1604107@srec.ac.in

Abstract. In present scenario, Glass Fiber Reinforced Plastics (GFRP) has found to be highly influential in the field of manufacture of automotive and aerospace parts due to its high strength and lightweight characteristics. Processed GFRP components are assembled to the main structure by means of mainly rivets and fasteners. So, drilling in a GFRP material becomes inevitable in manufacturing for its wide applications. The purpose of this experiment is to conduct a study of the effect of vibration on surface roughness while drilling a GFRP plate using a 10mm HSS drill. The experiment has been implemented by varying the drill speed and feed accordingly in a 10mm thick GFRP plate and the respective vibration and surface roughness results were obtained. Furthermore, the results obtained have been processed in Minitab to provide a comprehensive understanding. ANOVA has been performed to establish a relation between the vibrations developed and surface roughness of the hole.

1. Introduction

GFRP are widely employed in automotive, aerospace and sporting equipment industries owing to its low weight, high strength and specific stiffness [1, 2]. A number of research papers have focused on the study of delamination, cutting force, thrust force and damage developed due to drilling in GFRP which have provided a broad understanding in machining GFRP. The surface roughness over a drilled hole in a GFRP is indispensable for assessing the hole quality as it affects the functionality of a component such as wear resistance, fatigue resistance and friction [3, 4]. Cutting speed, feed rate, thrust force, thickness of plate and drill point angle are some factors which have direct influence on surface roughness of the material [5, 6]. Among the various drilling parameters, feed and spindle speed are the major parameters affecting the surface roughness of the hole [7, 8]. Experiments determined that drilling methods and drill bits used also contribute significantly to the delamination of GFRP [9-11]. Analysis of Variance(ANOVA) has been utilized to study the effect of drilling parameters on surface roughness of the GFRP [12-14].

Sureshkumar M S [15] investigated surface roughness of drilled holes by varying the drill speed, feed, thickness of GFRP plate using Taguchi's L27 full factorial design. Further, ANOVA was performed to test the results and a mathematical model was designed based in Box-Behkn design method. The results indicated that low speed and low feed rate has less effect on surface roughness.

Isik Birhan [16] carried out experiments to investigate damage factor in drilling on hole entrance and exit using 8mm cemented carbide drill. The results revealed that the damage occurred at both hole entrance and exit were reduced when cutting speed is increased. On the other hand, increasing feed decreases hole quality at exit and affects inversely at hole entrance.

Ulas Hasan Basri [17] observed that the vibration developed in drilling process using HSS and carbide drill bit are directly proportional to the cutting speed and feed rate. It has been concluded that vibration effects should be minimized in order to maintain both precise dimensions and quality of the surface.

The vibration developed while drilling a GFRP cannot be eliminated but can be minimized to a considerable extent by selecting appropriate drilling parameters. When GFRP workpiece is subjected to external vibration, it has been found that the workpiece vibration affects thrust force and tool wear, which further influences the surface roughness of drilled holes [18, 19]. So, it is noticeable that excluding other factors, workpiece vibration also contributes towards surface roughness of a drilled hole. Experiments have been conducted by varying the speed and feed of the drill, and the respective vibrations occurred are noted. This study is aimed to find the direct effect of vibration on surface roughness developed while drilling a GFRP plate. Obtained results are interpreted via different forms of graph processed using Minitab and ANOVA has been performed to study the relationship between surface roughness and vibration occurred. Considering the quality of drilled holes, the results obtained from this study can be made use for dampening the workpiece vibration along a particular axis amplifying the surface quality of the drilled holes while maintaining other effective parameters.

2. Experimental work

2.1. Specimen preparation

The GFRP specimen used in this study is manufactured by hand lay-up technique. The glass fibers are reinforced with isophthalic resin with 30% of reinforcement and fiber length between 20mm to 30mm. The GFRP specimen size of 80mm × 120mm has been cut for conducting the experiment. Further, two 4mm holes were drilled alongside and tapped to fix the accelerometer rigidly with the help of screws.

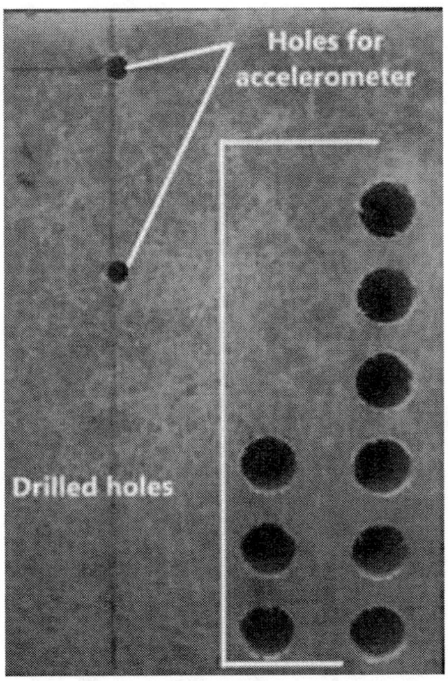

Figure 1. GFRP specimen plate.

2.2. Methodology

The drilling experiments were carried out according to full factorial design consisting of 9 drilled holes on the GFRP plate. In this study, the drilling parameters, cutting speed and feed rate are used as control parameters and each parameter has three distinct levels as shown in Table 1. The workpiece has been drilled using a 10mm HSS drill bit in a vertical machining center(Siemens 802D) and the vibrations developed during drilling along x, y and z axes have been measured with the help of an accelerometer(Dytran 7543A). The vibrations were measured by placing the accelerometer as close as possible to the drilled holes to acquire precise measurements. Positioning the accelerometer at a different location does not directly affect the results but provides the results in terms of multiples and shows minimized values, so it is effective when placed close to the drilled holed. Finally, surface roughness tester(Mitutoyo SJ210) has been utilised to measure the surface roughness of each hole at the inner surface to ensure that the delamination at the rim does not affect the measurements. Three trials were taken for each hole and average of the three reading were considered for analysis.

Table 1. Control factors and three levels of variation.

Factors	Level 1	Level 2	Level 3
Speed (rpm)	1000	2000	3000
Feed (mm/min)	40	80	120

Figure 2. Experimental setup.

3. Results and discussion

The results obtained from the experimentation are represented in Table 2. The results are fed into Minitab for statistical analysis and to establish regression models. With the help of these analyses and mathematical models, the effect of control parameters was studied and optimum values are found.

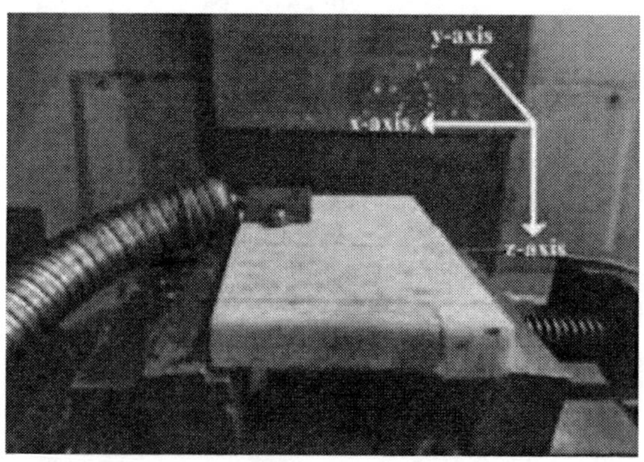

Figure 3. Axes representation.

IOP Conf. Series: Materials Science and Engineering 1012 (2021) 012035 doi:10.1088/1757-899X/1012/1/012035

Table 2. Experimental results.

Hole no.	Speed (rpm)	Feed (mm/min)	Vibration (mm)			Surface roughness (μm)
			x-axis	y-axis	z-axis	
1	1000	40	-0.8069	1.561	9.42	6.648
2	1000	80	-0.7571	1.627	10.21	7.635
3	1000	120	-0.797	1.619	10.16	8.098
4	2000	40	-0.792	1.627	10.50	7.961
5	2000	80	-0.788	1.654	10.47	8.367
6	2000	120	-0.807	1.770	10.69	5.496
7	3000	40	-0.748	1.662	10.70	9.30
8	3000	80	-0.826	1.639	15.17	7.981
9	3000	120	-0.739	1.670	9.34	6.848

3.1. Analysis Of Variance (ANOVA)

ANOVA is widely used to determine significant relationship between two groups of variables. Table 3 represent the results of ANOVA conducted between the vibrations developed along the three axes and surface roughness of the drilled holes. When considering vibration as a factor, number of external interferences such as the bed of drilling machine, acoustic noise and temperature may have some effect on measuring vibration. The probability values exhibit the reliability of the derived results. The coefficients obtained are used to derive the regression equation which is used to derive predicted results for the required inputs.

Table 3. Analysis of Variance table.

(a)

	Coefficients	Standard Error	t Stat	P-value
Intercept	37.45652	15.64965	2.393441	0.062123
x-axis	23.29576	14.44509	1.612711	0.167726
y-axis	-9.54506	6.519095	-1.46417	0.203028
z-axis	0.385479	0.248452	1.551523	0.181476

(b)

	DOF	SS	MS	F	Significance F
Regression	3	4.751306	1.583769	1.537689	0.313926
Residual	5	5.149834	1.029967		
Total	8	9.90114			

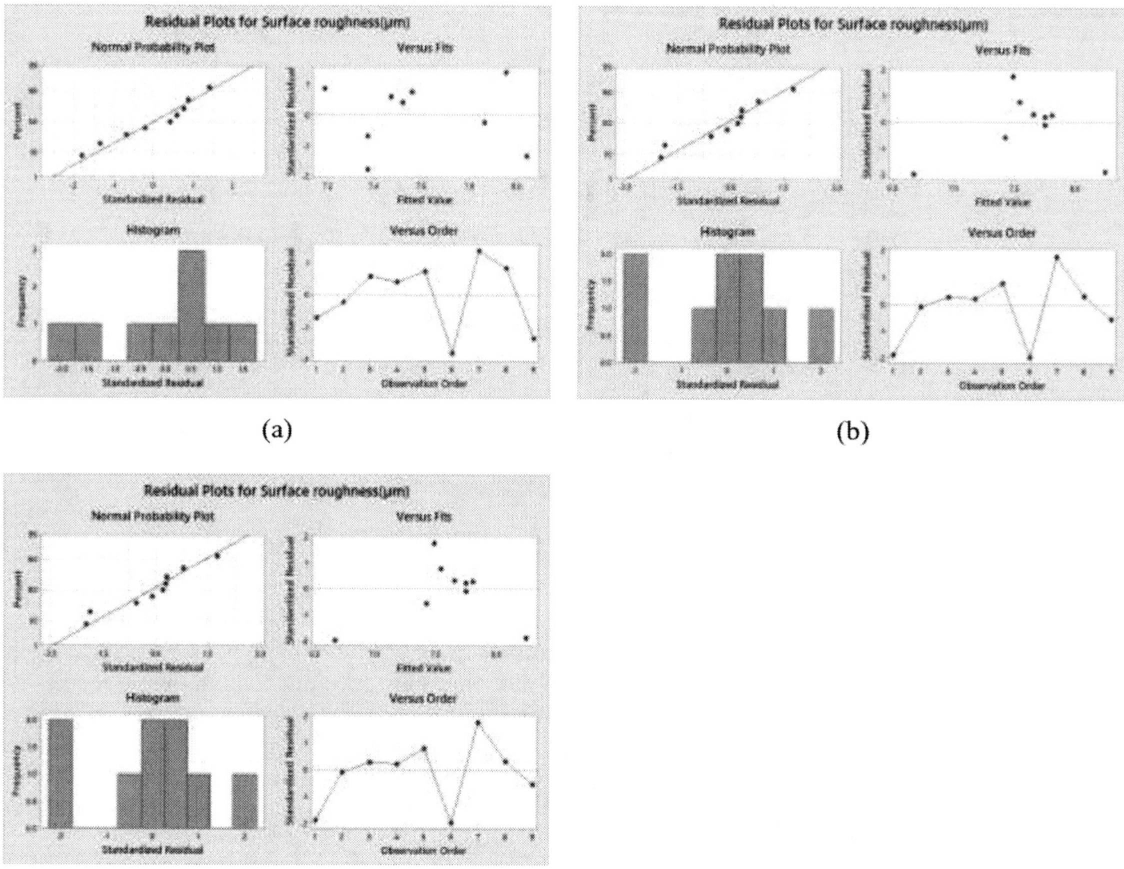

(a) (b) (c)

Figure 4. Residual plots for surface roughness considering (a)x-axis, (b)y-axis and (c)z-axis vibrations

Figure 4 represents the normal probability plots in which the points lie closer to the normal line. The versus fit has a symmetrical section between the upper and lower section of the versus line. The histogram figures follow a curve similar to the normal curve. The residuals in the versus order follows a non-uniform pattern and shows no trend. Thus, the residual plots show that there is a significant relation between the vibration developed along each axis and surface roughness.

3.2. Mathematical model

$$R_a = 37.46 + 23.29(x\text{-}axis) - 9.54(y\text{-}axis) + 0.38(z\text{-}axis) \quad (1)$$

The coefficients from ANOVA table is utilized to derive a regression equation to predict the surface roughness. Equation (1) is the derived regression equation for surface roughness. The equation also indicates that vibration along x-axis has more influence over R_a. Since significance F value is comparatively lower than F value, it is clear that null hypothesis can be rejected.

The regression equation has been utilized to predict the surface roughness of the hole, provided that the vibrations along the three axes are given which in this case, are the measured results. Under a similar case, when the position of the accelerometer is adjusted, it provides results which are of multiples to that of the original results. The differences between the measured and predicted results indicate the residuals. The residual of each hole and the variance percentage between the measured and predicted are shown in Table 4. The maximum and minimum variance percentages are 15% and 5% respectively, and the variance percentage changes averaged to 9.2%. Figure 5 represents the measured vs. predicted graph indicating that the predicted line follows a similar path to that of the measured line.

Table 4. Residuals.

Measured	Predicted	Residuals	Variance percentage
6.648	7.390542	-0.74254	11%
7.635	8.225226	-0.59023	8%
8.098	7.352811	0.745189	-9%
7.961	7.523992	0.437008	-5%
8.367	7.347894	1.019106	-12%
5.496	5.882854	-0.38685	7%
9.3	8.292024	1.007976	-11%
7.981	8.417581	-0.43658	5%
6.848	7.901075	-1.05307	15%

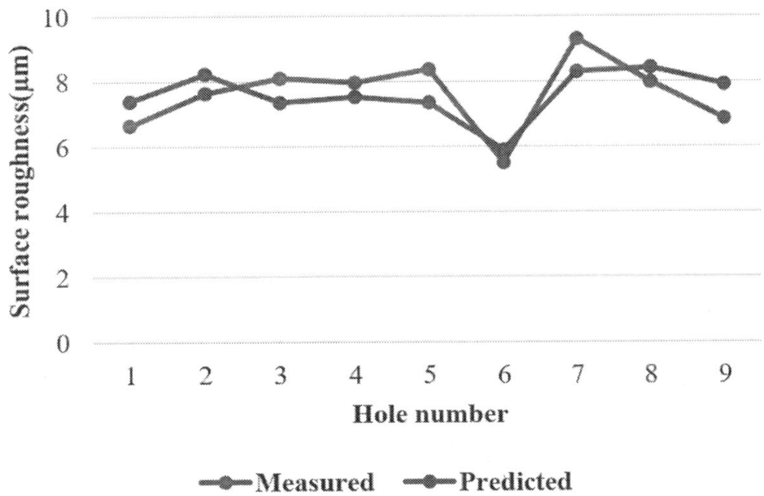

Figure 5. Measured results vs. predicted results.

4. Conclusion

The relationship between the surface roughness of drilled holes and vibration developed in GFRP composite has been studied. Mathematical models were generated in order to provide analytical results and are found to be reliable. Regression equation decisively indicates that the vibration along the x-axis influences surface roughness to a greater degree. So, it is evident that reduced vibration along x-axis while drilling can provide better surface roughness of the hole. Investigations in vibration dampening of work piece would further emphasize the reduction of surface roughness leading to an improved quality of drilled holes. In addition to that, GFRP being highly abrasive in nature, tool wear also plays a crucial role in producing surface quality. Further research can be done taking tool wear into consideration to attain high surface quality of drilled holes.

References

[1] Sureshkumar M S, Lakshmanan D and Murugarajan A 2014 Experimental investigation and mathematical modelling of drilling on GFRP composites *Materials Research Innovations* **18**(sup1) S1-94-S1-97

[2] Shivanagere Abhay, Sharma S and Goyal P 2018 Modelling of glass fibre reinforced polymer (Gfrp) for aerospace applications *Journal of Engineering Science and Technology* **13**(11) 3710-28

[3] Wang M Y and Chang H Y 2004 Experimental study of surface roughness in slot end milling AL2014-T6 *International Journal of Machine Tools and Manufacture* **44**(1) 51-57

[4] Latha B and Senthilkumar V S 2010 Modeling and analysis of surface roughness parameters in drilling GFRP composites using fuzzy logic *Materials and Manufacturing Processes* **25**(8) 817-27

[5] Mohan N S, Shettar M and Hiremath P 2015 Influence of drilling conditions of Glass Fiber-Reinforced Plastic [GFRP] composite materials *4th World Conf. on Applied Sciences, Engineering & Technology* 24-26

[6] Kilickap E 2010 Optimization of cutting parameters on delamination based on Taguchi method during drilling of GFRP composite *Expert systems with applications* **37**(8) 6116-22

[7] Shunmugesh K, Panneerselvam K and Jospaul Thomas 2014 Optimising drilling parameters of GFRP by using grey relational analysis *International Journal of Research in Engineering and Technology* **3**(6) 302-05

[8] Palanikumar K 2011 Experimental investigation and optimisation in drilling of GFRP composites *Measurement* **44**(10) 2138-48

[9] Kavad B V, Pandey A B, Tadavi M V and Jakharia H C 2014 A review paper on effects of drilling on glass fiber reinforced plastic *Procedia Technology* **14** 457-64

[10] Hocheng H and Tsao C C 2003 Comprehensive analysis of delamination in drilling of composite materials with various drill bits *Journal of materials processing technology* **140**(1-3) 335-39

[11] Hussain S A, Pandurangadu V and Kumar K P 2011 Machinability of glass fiber reinforced plastic (GFRP) composite materials *International Journal of Engineering, Science and Technology* **3**(4) 103-18

[12] Vankanti V K and Ganta V 2014 Optimization of process parameters in drilling of GFRP composite using Taguchi method *Journal of Materials Research and Technology* **3**(1) 35-41

[13] Davim J P, Reis P and António C C 2004 Drilling fiber reinforced plastics (FRPs) manufactured by hand lay-up: influence of matrix (Viapal VUP 9731 and ATLAC 382-05) *Journal of Materials Processing Technology* **155** 1828-33

[14] Davim J P, Reis P and Antonio C C 2004 Experimental study of drilling glass fiber reinforced plastics (GFRP) manufactured by hand lay-up *Composites Science and Technology* **64**(2) 289-97

[15] Sureshkumar Manickam Shanmugasundram, Lakshmanan Damodhiran, Arunkumar Veluchamy, Vignesh Kathirvel, Vinothkumar Thirugnanam, Damodhiran L, Veluchamy A, Kathirvel, V and Thirugnanam V 2013 Experimental investigation and Mathematical modelling for Surface Roughness of drilling on GFRP composites *Applied Mechanics and Materials* **391** 46-50

[16] Işık B and Ekici E 2010 Experimental investigations of damage analysis in drilling of woven glass fiber-reinforced plastic composites *The International Journal of Advanced Manufacturing Technology* **49**(9-12) 861-69

[17] Ulas H B, Ozkan M T and Malkoc Y 2019 Vibration prediction in drilling processes with HSS and carbide drill bit by means of artificial neural networks *Neural Computing and Applications*

31(9) 5547-62
[18] Ogawa K, Aoyama E, Inoue H, Hirogaki T, Nobe H, Kitahara Y, Katayama T and Gunjima M 1997 Investigation on cutting mechanism in small diameter drilling for GFRP (thrust force and surface roughness at drilled hole wall) *Composite Structures* **38**(1-4) 343-50
[19] Ramkumar J, Malhotra S K and Krishnamurthy R 2004 Effect of workpiece vibration on drilling of GFRP laminates *Journal of Materials Processing Technology* **152**(3) 329-32

Generative Design and Topology Optimization of Analysis and Repair Work of Industrial Robot Arm Manufactured Using Additive Manufacturing Technology

M. Kumaran [1*], **V. Senthilkumar** [1]
[1] Department of Production Engineering, National Institute of Technology Tiruchirappalli, Tamil Nadu, India – 620015

E-mail: kumumech@gmail.com

Abstract. Today's industrial robots work in a wide range of industries, the industrial robots are developed to perform operations such as pick and place, manufacturing assembly line, welding work, metal forging, and painting operations. The constraints of available robots are in terms of structural complexity, more material in the body, over weightage, and space. To overcome the difficulties of the current scenario, the generative design and topology optimization based industrial robot arm is manufactured by Additive Manufacturing (AM) technology using the Powder Bed Fusion (PBF) process. The scope of the research is when the generatively designed robot arm getting crack or broken, the Direct Energy Deposition (DED) processes use to help the repair of the PBF parts. In the present research, studies were conducted on the topology optimized industrial robot arm using SolidWorks software, and to verify the significance of DED to extend requirements needed for PBF parts, sample parts were extended and their metallurgical characteristics, tensile properties, and hardness observed.

Keywords: Additive Manufacturing, Generative Design, Topology Optimization, Robot Arm, Industrial Robot

1. Introduction

An industrial robot [1] structure is the main drawback, it has an over material quantity and more body weight of the robot. Modern industrial robot production with AM technology [2, 3] is used to resolve the major drawback. The technology of AM directly produces the component using laser-powder technology with the help of a generative design and topology optimization method. Generating design and topology optimization of additive components can address unwanted materials and overcome the 30% - 70% weight reduction range. The scope of the research is to analyze the case study of when getting the external crack or broken part in additive manufactured robot arm to resolve using repair work by the DED process.

Unless it is topologically optimized, the weight of each part in the assembly may exceed the required weight. The extra weight means too much material is used, the load on the moving parts exceeds the necessary level, energy efficiency is compromised, and the cost of transporting parts is higher. Now, using topology optimization techniques, it is necessary to lightweight components, design durable for any application. To ensure that manufacturing specifications are met, set minimum material thicknesses, and identify exclusion areas, you may easily define targets and apply controls.

2. Design Methodology

1. Generative design and topology optimization

Through generative design and topology optimization [4], these simulation techniques allow the use of simulation-driven design methods to customize lightweight designs and high-performance parts. Advances in manufacturing technology have also allowed the use of processes such as metal additive manufacturing to build these sometimes complex designs [5].

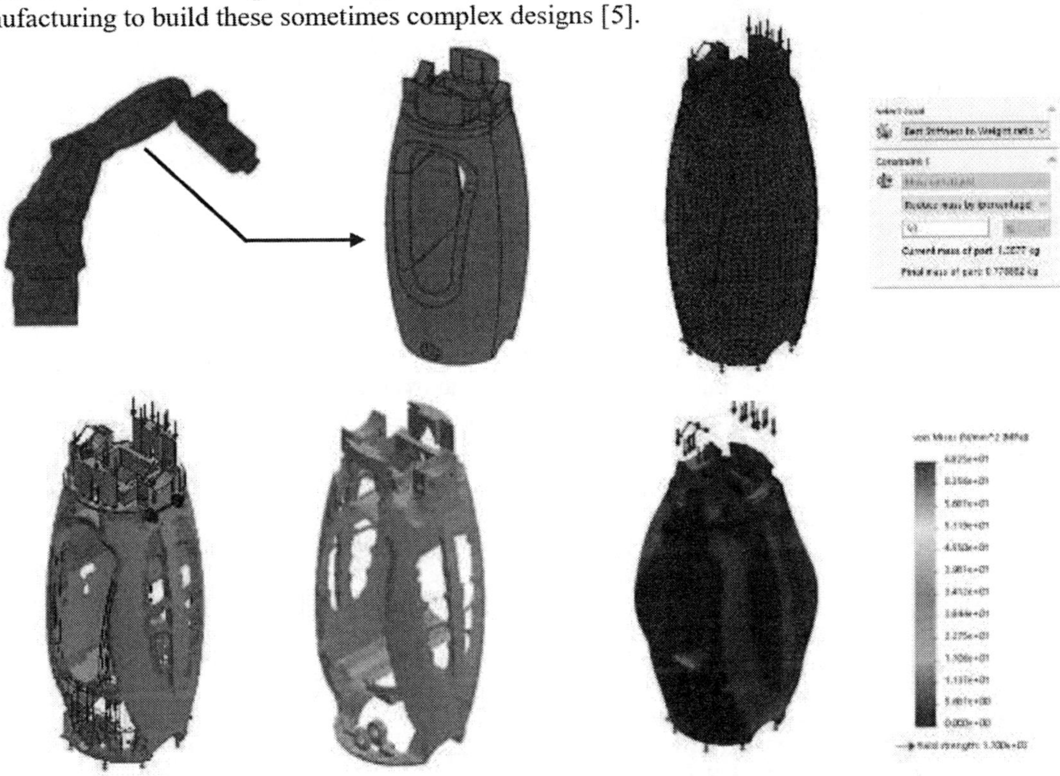

Figure 1. Topology optimization designed robot arm and stress analysis.

As shown in **Figure 1.** the topology optimization design used by SolidWorks software help to reduce the material consumption at 50% from the original design and robot arm weight is minimized from 1.55kg to 0.77kg. The latest SolidWorks software utilizes the topology optimization function to eliminate the design weight once, thereby reducing material consumption by 50%. In the design analysis, Stainless Steel 316L (SS 316L) material was used. Von-Mises stress is the value used to determine whether a given material yields or breaks. After topology optimization, the design part is analyzed under the action of external force, and the maximum Von-Mises stress is 68.25 MPa. Use the finite element model of the robot arm for stress analysis, which can control both stress and strain loading conditions. The stress distribution is affected by the material and physical properties.

2. Simulation of metal-based additive manufacturing processes

Figure 2. shows the manufacture based simulation process in AM using altair inspire print3D software [6]. It is used to optimize the process parameters of the manufacturing conditions. While varying primary process parameters such as laser power (W) 190, 200 and 210 respectively; scanning speed (mm/s) 800, 900 and 1000 respectively; layer thickness (μm) 200, 300 and 400 respectively.

Based on the results of the minimum residual stress (50 Mpa), the laser power is 195 W, the scanning speed is 900 mm/s and the layer thickness is 400 μm.

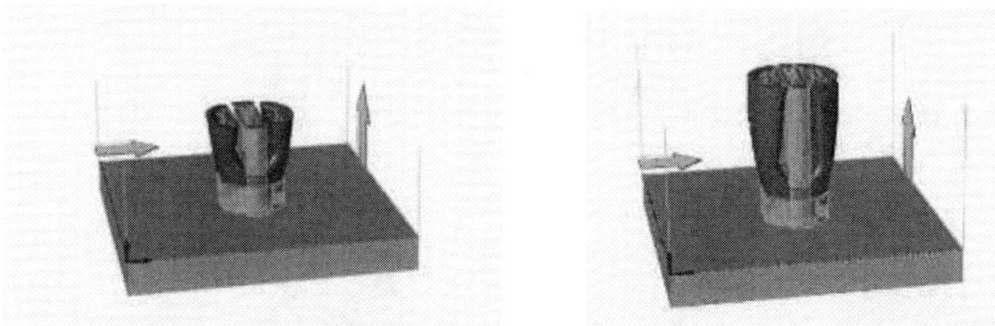

Figure 2. Simulation of building the processes of additive manufacturing.

3. Experimental Results

1. Making the component using additive manufacturing (PBF Process)
In this research, spherical granules of SS 316L [7] in the range of 5 - 45 μm diameter were used for PBF process, with the chemical composition as follows Fe 68.34; Cr 16.62; Ni 10.57; Mo 2.11; Si 0.60; Mn 1.58; C 0.030; N 0.10; P 0.035; S 0.020. As shown in **Figure 3(a)**, the morphology of SS 316L powder was observed by a Field Emission Scanning Electron Microscope (FE-SEM), which specified approximately spherical shaped powder particles. **Figure 3(b)**, are the histograms with an average particle size of 22 μm diameter by intensity size distribution. The study results shown in **Figure 3(c)**, An Energy-Dispersive X-ray spectrometer (EDAX) (Elemental analysis) measured the powder for element ratios, indicating that the chemical composition meets the standard value of SS 316L.

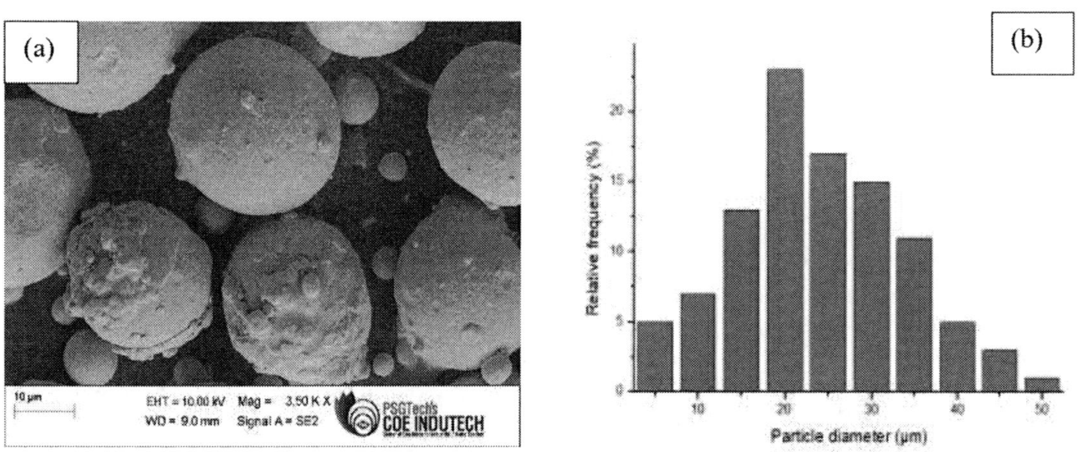

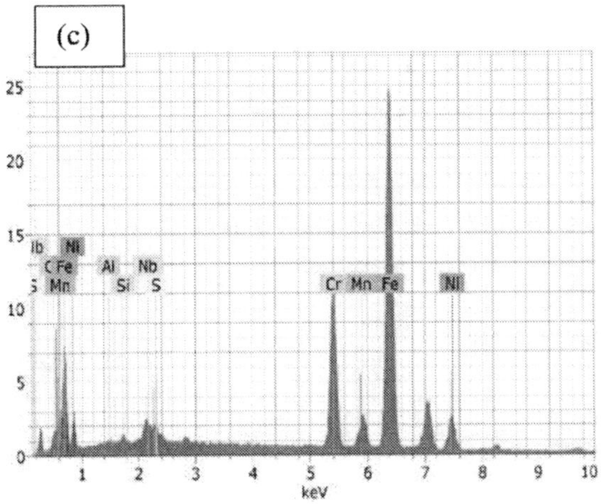

Figure 3. SS 316L powder: (a) FE-SEM morphology for PBF; (b) Histogram of particle size distribution; and (c) EDAX results of SS 316L powder for PBF.

It was used to make a robot arm of PBF with a scanning speed 900 mm/s, laser power of 195 W, hatching pitch of 100 μm, layer thickness of 400 μm and laser diameter of 0.2 mm. These values are the same as those used for the simulation of metal-based additive manufacturing processes using the altair inspire print3D software based on the least residual stress and the best mechanical properties. Specimens were manufactured in a highly clean argon gas chamber. **Figure 4(a)** in this study, tensile samples were prepared vertically using the ASTM (E8) standard [8]. **Figure 4(b)** shows the tensile test result for manufactured PBF specimen [9]. **Figure 4(c)** magnifies the PBF of the fractured Sample. Dimples are observed in this area. **Figure 4(d)** shows the result of the stress-strain value of the tensile specimen sample of the yield stress of the PBF sample is 564 MPa, at a total elongation of 45.5%. When the ultimate force used was 7440 N, it withstood the ultimate stress at 675 MPa.

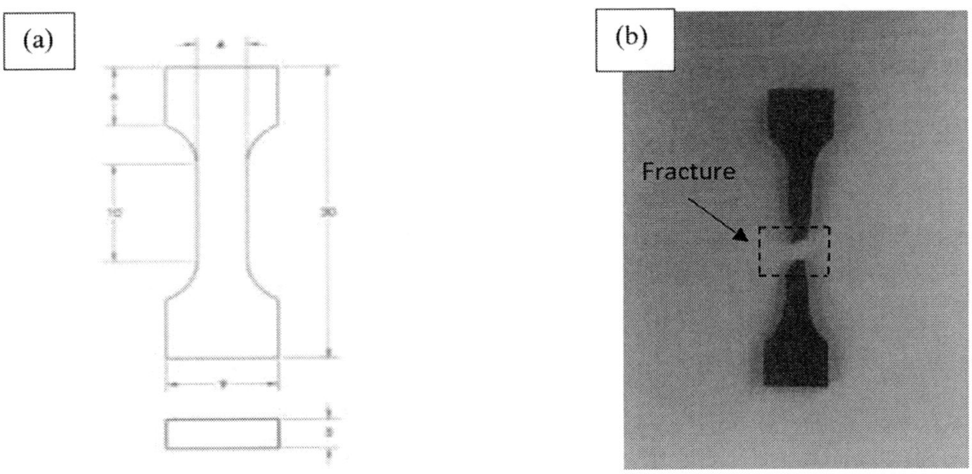

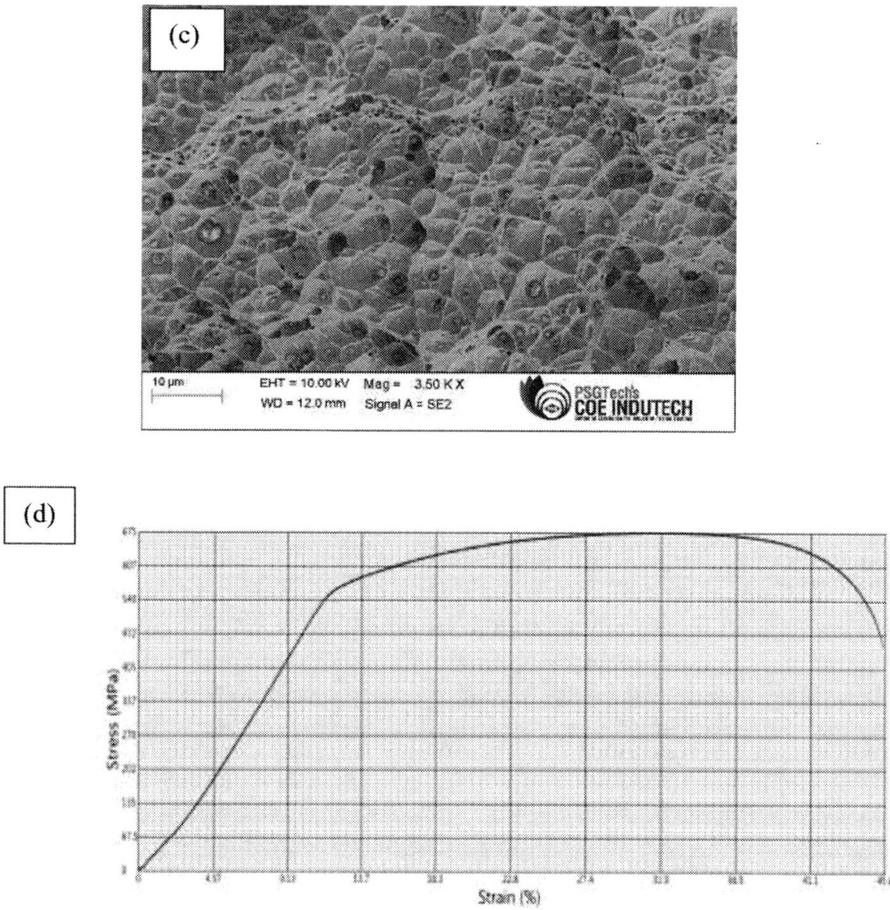

Figure 4. (a) Configuration of tensile specimen (b) Macro image of PBF specimens, after tensile testing (c) Micro-fractography of fractured specimen, and (d) Engineering stress-strain curve of PBF SS 316L.

2. Repairing the component using additive manufacturing (DED Process)

Repairing work experiments [10, 11] were conducted to establish the best condition for extension while varying primary process parameters such as powder feed rate (PR) 2, 3 and 4 g/min respectively; scanning speed (SS) 300, 500 and 700 mm/min respectively; and laser power (P) 400, 600 and 800 W respectively. The objective was to evaluate the deposition characteristics according to the requirements of the L9 Taguchi-method. Cross-section of the DED product under different process conditions. Based on the harness analysis (225.4 HV) and microstructural results, processing conditions of powder feed rate of 4 g/min, 500 mm/min scanning speed and 600 W laser power were selected as optimal conditions. Spherical granules of SS 316L in the range of 50 - 150 µm diameter were used for the DED process, the histograms with an average particle size of 80 µm diameter. **Figure 5(a)** To extend the DED additive manufacturing process, the PBF substrate used various parameters. The PBF substrate prepared were of 15 mm height (Z-axis, upright position), and using the DED process extended the remaining 15 mm height above the substrate. Finally, DED extended above the PBF surface area. The PBF specimen was extended up to a height of 15 mm using the DED process as shown in **Figure 5(b)** and **Figure 5 (c)**. Portions A and B are a cross-section of PBF specimen extended by DED, based on the extended height. It shows the Sandwich Structure (SWS) portion C [12]. The extended DED showed no cracking, and revealed excellent metallurgical bonding between the substrate and the extended region.

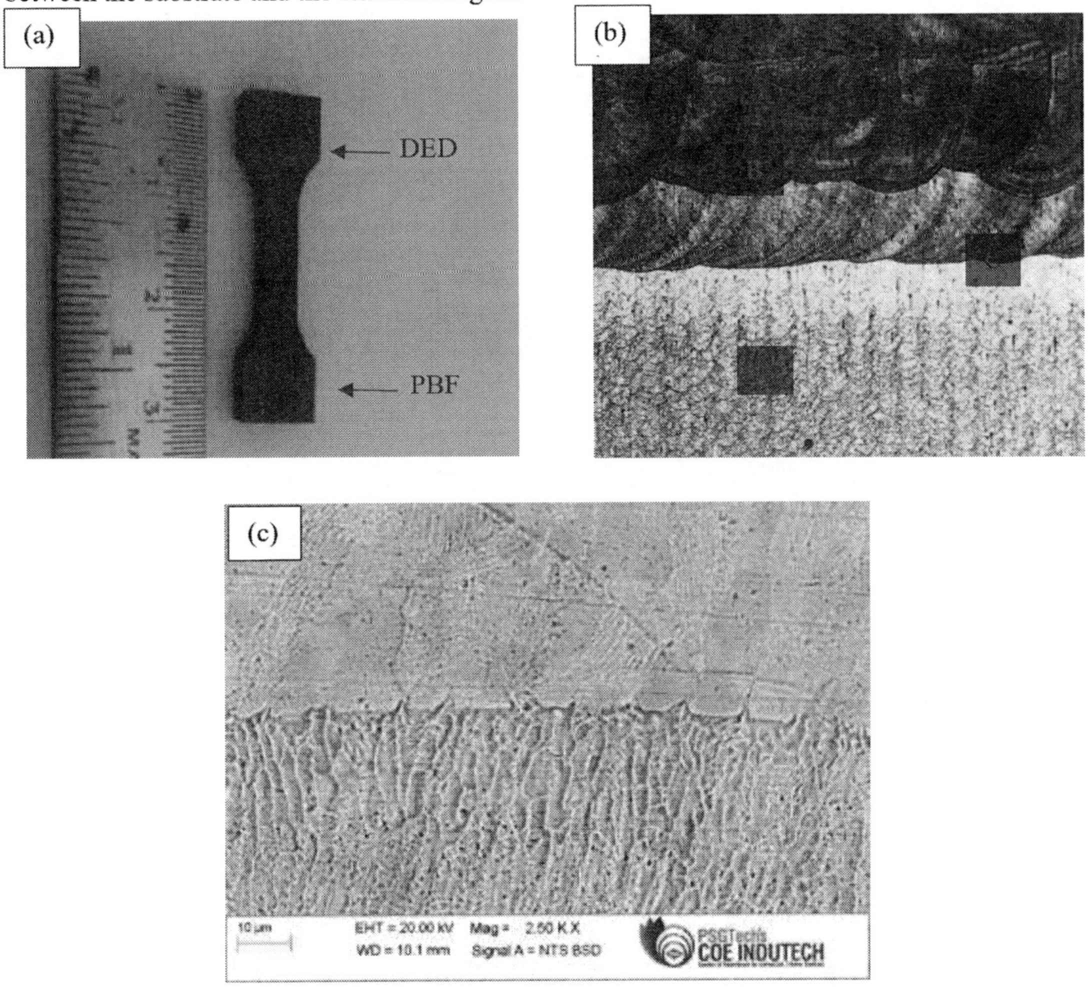

Figure 5(a) DED extended specimen (SWS) (b) and (c) Microstructural images of extended SWS specimens.

Figure 6(a) shows the tensile test result for manufactured the extended process (SWS) specimen. **Figure 6(b)** magnifies the extended area of the fractured specimen. Dimples are observed in this area. **Figure 6(c)**The results of the extended process (SWS) tensile test specimens, the sandwich structure specimen's yield stress was 380 MPa, at a total elongation of 23.4%. When the ultimate force used was 6170 N, it withstood the ultimate stress at 536 MPa.

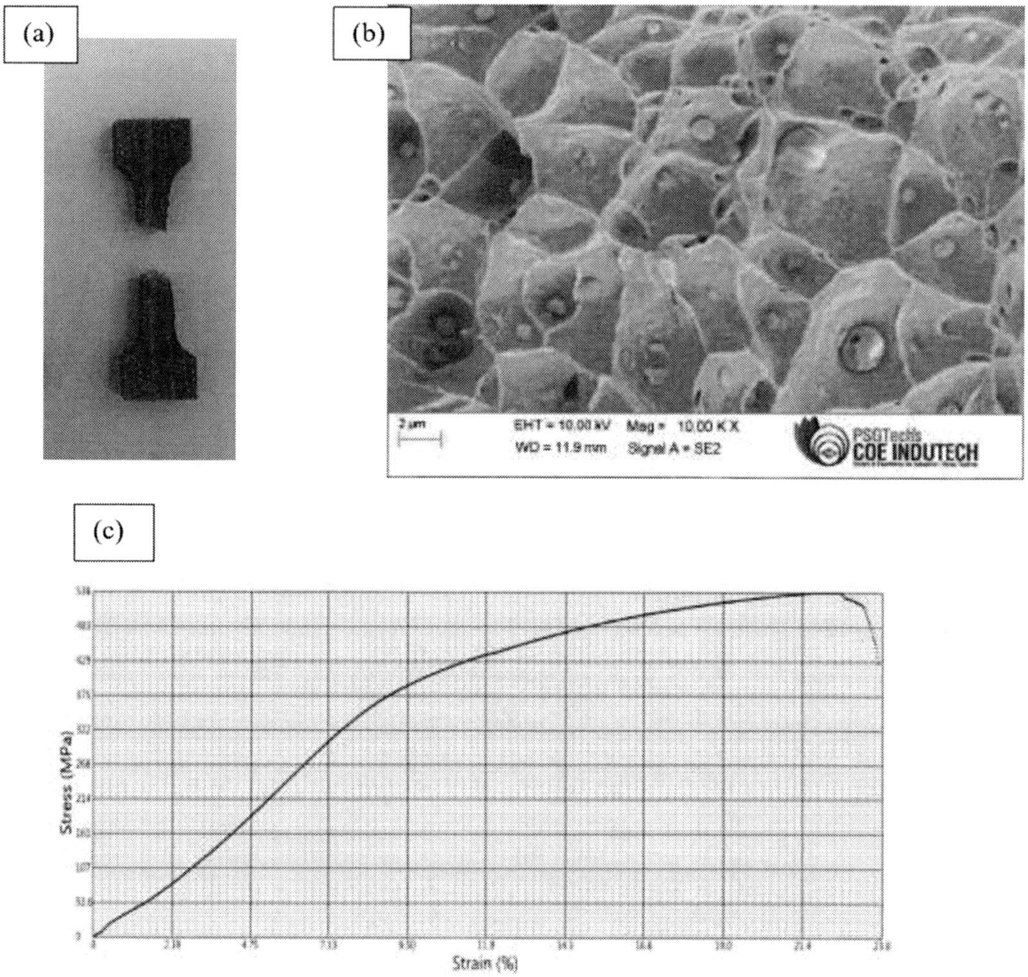

Figure 6(a) Macro image of PBF specimens, after tensile testing (b) Micro-fractography of fractured specimen, and (c) Engineering stress-strain curve of SWS SS 316L.

4. Conclusion

This paper has studied the generative design and topology optimization of robot arm fabricated by metal additive manufacturing using SS 316L powder. To evaluate the design, mechanical properties and microstructures were analyzed. The robot arm optimized with low residual stress also fabricated with 50% of weight reduction using metal additive manufacturing. Finally, we conclude in this paper that the metal additive manufacturing is suitable for robot arm also sandwich structure specimen of SS 316L is suitable for extended or repairing applications.

References

[1] Francesco De Pace, Federico Manuri, Andrea Sanna and Claudio Fornaro 2020 A systematic review of Augmented Reality interfaces for collaborative industrial robots Computers & Industrial Engineering **149** 106806

[2] Amit Bandyopadhyay and Kellen D Traxel 2018 Invited review article: Metal-additive manufacturing-Modeling strategies for application-optimized designs Additive Manufacturing **22** 758-774

[3] V. Senthilkumar, C. Velmurugan, KR. Balasubramanian and M. Kumaran 2020 Additive Manufacturing of Multi-Material and Composite Parts Additive Manufacturing Applications for Metals and Composites, IGI Global 127-146
[4] Mustafa Bugday and Mehmet Karali 2019 Design optimization of industrial robot arm to minimize redundant weight Engineering Science and Technology an International Journal **22** 346-352
[5] Fei Weng, Shiming Gao, Jingchao Jiang, JianJian Wang and Ping Guo 2019 A novel strategy to fabricate thin 316L stainless steel rods by continuous directed energy deposition in Z direction Additive Manufacturing **27** 474-481
[6] Noha Peter, Zachary Pitts, Spencer Thompson and Ankit Saharan 2020 Benchmarking build simulation software for laser powder bed fusion of metals Additive Manufacturing **20** 30903-9
[7] Yong-Deok Im, Kyung-Hoon Kim, Kyung-Hwan Jung, Young-Kook Lee and Kuk-Hyun Song 2019 Anisotropic Mechanical Behavior of Additive Manufactured AISI 316L Steel Metallurgical and Materials Transactions A **50** 2014-2021
[8] Zhaopeng Tong, Xudong Ren, Jiafei Jiao, Wangfan Zhou, Yunpeng Ren, Yunxia Ye, Enoch Asuako Larson and Jiayang Gu 2019 Laser additive manufacturing of FeCrCoMnNi high-entropy alloy: Effect of heat treatment on microstructure, residual stress and mechanical property Journal of Alloys and Compounds **785** 1144-1159
[9] Zan Li, Thomas Voisin, Joseph T. McKeown, Jianchao Ye, Tom Braun, Chandrika Kamath, Wayne E. King and Y. Morris Wang 2019 Tensile properties, strain rate sensitivity, and activation volume of additively manufactured 316L stainless steels International Journal of Plasticity **120** 395-410
[10] Wook Jin Oh, Wook Jin Lee, Min Seob Kim, Jong Bae Jeon and Do Sik Shim 2019 Repairing additive-manufactured 316L stainless steel using direct energy deposition Optics & Laser Technology **117** 6-17
[11] G.F. Sun, X.T. Shen, Z.D. Wang, M.J. Zhan, S. Yao, R. Zhou and Z.H. Ni 2019 Laser metal deposition as repair technology for 316L stainless steel: Influence of feeding powder compositions on microstructure and mechanical properties Optics & Laser Technology **109** 71-83
[12] Xinchang Zhang, Wei Li, Xueyang Chen, Wenyuan Cui and Frank Liou 2018 Evaluation of component repair using direct metal deposition from scanned data The International Journal of Advanced Manufacturing Technology **95** 3335-3348

Acknowledgment
Thanks are largely due to Mr. A.R.Vinod, Scientist-C and Mr. B.N. Manjunath, Scientist-B at the Additive Manufacturing Technology Division, Central Manufacturing Technology Institute, Bangalore, India. for conducting the characterization and experiments of the PBF and DED processes. Also, I gratefully acknowledge the support of staff members at PSG Tech's COE INDUTECH Laboratory, Coimbatore, India. for conducting the FESEM and EDAX studies.

IOP Conf. Series: Materials Science and Engineering 1012 (2021) 012037 doi:10.1088/1757-899X/1012/1/012037

Python implementation of fuzzy logic for artificial intelligence modelling and analysis of important parameters in drilling of hybrid fiber composite (HFC)

Vimal Samsingh[1], Achyuth Ramachandran[1], Anirudh Selvam[1] and Karthick Subramanian[1]

[1] Department of Mechanical Engineering, Sri Sivasubramaniya Nadar College of Engineering, Chennai, Tamilnadu, India

E-mail: selvamanirudh@gmail.com

Abstract. Composite materials present the advantage of being able to be specially designed for a particular application by combining appropriate reinforcement materials with a matrix material suited to withstand the operant conditions. The use of Hybrid-Fiber Composites (HFCs) addresses the need for greener manufacturing processes while also meeting product specifications in a wide range of applications, all for nominal prices. In order to improve our understanding of the machining processes compatible with HFCs, this paper presents findings from a study in which the effects of drilling on glass-flax-hemp fibre hybrid composite samples are observed and modeled. Pivotal parameters in drilling, namely drill bit diameter, spindle speed and feed rate are studied, and a fuzzy-logic inference system (FIS) coded in Python is used to model the thrust force and torque acting on the composite sample. A comparison between experimentally obtained and model-generated values of the same indicate very good correlation, thus verifying the effectiveness of the FIS.

1. Introduction

Composites have the ability to be custom-made according to the application necessity, are lightweight and easy to manufacture. This adaptable nature is achieved by changing the material matrix or reinforcement or the mixing ratio of reinforcement to matrix material [1]. This has diverse industrial applications in fields such as aerospace, civil engineering, electrical and electronics, automotive and many more. End-of-life action for conventional composite materials such as GFRP, CFRP etc. consist mainly of disposing the samples in landfills and dumps; this is problematic and potentially dangerous to local flora and fauna, owing to the non-biodegradable nature of such materials and their tendency to linger within the biomass present and contaminate it. The use of hybrid fiber composites containing naturally occurring reinforcement materials, such as flax, jute, hemp fibre, sisal etc. pose an elegant and easy-to-implement solution to this problem.

Ever-advancing research is being done in the area of hybrid fiber composites, both to understand their properties as well as to analyse their feasibility to replace conventional materials. HFCs containing cellulosic fibres are found to improve mechanical properties while driving down costs [2]. Hybrid Fiber Composites have been identified as one of the best replacements for single-fiber reinforced composites. HFCs have been found to possess enhanced mechanical, thermal and damping properties when compared to single-fiber reinforced composites, thereby making them good replacements for wood, wood fiber composites and conventional materials [3].

Conventional machining processes like turning, facing, grinding, milling, drilling, welding, etc. are cornerstones in the field of component production and manufacturing. Machining of composite materials requires extreme care over critical process parameters such as material removal rate, machining forces acting on the billet, torque produced due to twisting motion in processes like drilling, milling etc. in order to avoid production of defects in the form of delamination of layers, internal and external cracks, surface burrs and irregularities among many others [4]. Consequently, the effects of various process parameters on composite materials are extensively studied and recorded.

Being an intensive process in terms of contact between tool and billet, and machining forces produced, drilling of fiber polymer composites HFCs has been widely studied for the ways in which it affects the integrity of the composite sample [5]. Of the methods used to join polymer composites, drilling is the most common method used, making it a vital process which needs extensive research on what process parameters must be used to machine the composites safely [6].

The work done by Davim et al. on drilling of GFRP indicates that feed rate is one of the significant parameters that dictates the thrust force generated [7, 8]. Nagarajan et al. analysed the drilling parameters of silk-glass hybrid composite and found that thrust force varies inversely with spindle speed, but directly with drill bit diameter [9]. Similar effects are observed in GFRP [11].

In this work, a composite sample with hemp fibre - flax - E-glass hybrid reinforcement is fabricated using the hand lay-up process and subjected to drilling. Three process parameters, namely spindle speed (rpm), feed rate (mm/min) and drill bit diameter (mm) are chosen for analysis. Different values for each parameter are decided and experiments are conducted with all possible combinations of the values. The thrust force and torque acting on the sample are experimentally recorded, taking the average of three trials.

Based on the observed data, a Fuzzy-Logic Inference System (FIS) is created, in which the observed input and output parameter ranges are defined and divided into "fuzzy" non-sharp sets. A set of qualitative "fuzzy" rules is created by experimentation and analysis of the experimental data, to map the correlation between different ranges of input and output parameters.

The newly developed system is fed the values of the set input process parameters, and the output values predicted by the model are recorded. A comparison is made between the experimentally obtained data and the predicted values, and the effectiveness of the model in predicting the thrust force and torque for a given set of inputs is quantified. The excellent correlation observed between the two sets suggests that the model is a very good fit to the data and that an accurate capture of the process parameters is possible.

2. Experimentation

The raw materials used in the fabrication of the hybrid fiber composite were:

(i) Epoxy Resin (Araldehyde LY 556)
(ii) Hardener (Araldite) - HY 951
(iii) Fiber layers (Glass fiber- 2, Flax fiber- 3, hemp fiber- 2)
(iv) Gel coat- liquidity epoxy resin
(v) Releasing agent

Hand layup process was used to fabricate the composite due to minimum investment in the mould. The individual thicknesses of the reinforcement mats were 0.2 mm (e-glass fiber), 0.35 mm (flax fiber) and 0.6 mm (hemp fiber). The final sample was 12 mm thick, and the fibers were stacked in the order as shown in figure 1. Taking the example of work done by Palanikumar et al. [8], Nagarajan et al. [9] and Samsingh et al. [13], and also owing to financial constraints, the volume fraction of the resultant composite was neither determined nor considered for analysis.

A hand roller was used to impregnate the reinforcement layer. The roller is used each time to obtain the desired thickness and prevent the formation of bubbles. After the layering process, the sample was cured at room temperature for 24 hours. After all these processes, drilling operations were performed on the fabricated workpiece, using the drilling machine as in figure 2.

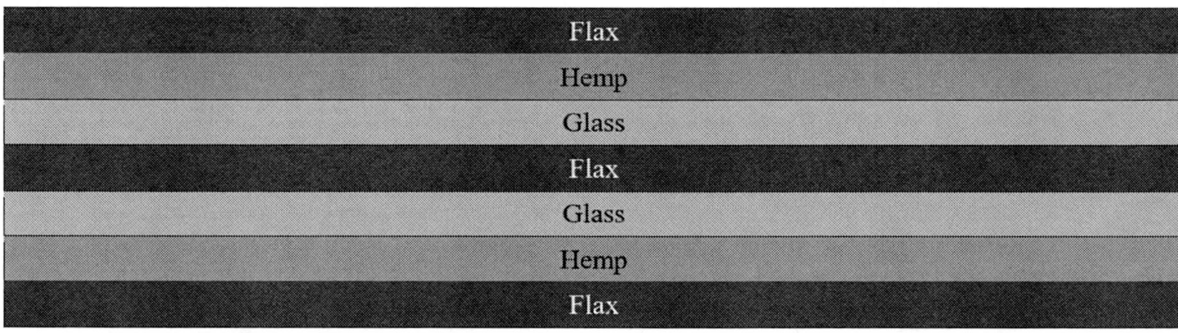

Figure 1: Fiber stacking sequence used in the composite sample

Figure 2: Hand-layup process & Experimental set-up

3 operant parameters, namely spindle speed (rpm), drill bit diameter (mm) and feed rate (mm/min) were chosen and 3 possible values were chosen for each, leading to a total of 27 unique combinations of process parameters. The operant parameters were chosen based on the results obtained by Vantaki et al. [4], Davim et al. [7] and Palanikumar et al. [8]. Considering the choice of 3 control parameters each with 3 levels of values, a Taguchi L_{27} Orthogonal array was chosen for parameter specification, analysis and experiment design. This is an orthogonal array capable of considering 13 variables at three levels and three interactions. The degrees of freedom of the experiment is $27 - 1 = 26$. The degrees of freedom of the scenario considered for the three parameters at three levels is $3 \times (3 - 1) = 6$. The chosen values and assigned levels are specified in table 1, and the L_{27} array is shown in table 2.

Table 1: Selected control parameters for experimentation

S.No	Parameter	Levels		
		1	**2**	**3**
1	Spindle Speed (rpm)	500	1000	1500
2	Feed Rate (mm / min)	60	120	180
3	Drill Bit Diameter (mm)	4	6	8

A Kistler Quartz 3-Component piezo-electric dynamometer of type 9257B was used for measurement of thrust force and torque acting on the sample. The drill bit used in each case was an 8-facet solid carbide drill bit with 4 faces having 2 cutting edges each, having 2 point angles of 90° and 180°. The drilling procedure was carried out using a CNC vertical machining center, details of which are given in table 3.

For each combination of the control parameters, the experimental procedure was conducted three times, and the value of the thrust force and torque measured in each trial was recorded on local storage. The average of the three trials for each experiment was computed and considered for analysis. Considering the results of the work done by Fernández-Pérez et al. [10], lubrication was not provided in order to aid in the thermal softening and easy drilling of the HFC sample.

Table 2: Taguchi L27 orthogonal array for Design of Experiments

Expt.	P1	P2	P3	P4	P5	P6	P7	P8	P9	P10	P11	P12	P13
1	1	1	1	1	1	1	1	1	1	1	1	1	1
2	1	1	1	1	2	2	2	2	2	2	2	2	2
3	1	1	1	1	3	3	3	3	3	3	3	3	3
4	1	2	2	2	1	1	1	2	2	2	3	3	3
5	1	2	2	2	2	2	2	3	3	3	1	1	1
6	1	2	2	2	3	3	3	1	1	1	2	2	2
7	1	3	3	3	1	1	1	3	3	3	2	2	2
8	1	3	3	3	2	2	2	1	1	1	3	3	3
9	1	3	3	3	3	3	3	2	2	2	1	1	1
10	2	1	2	3	1	2	3	1	2	3	1	2	3
11	2	1	2	3	2	3	1	2	3	1	2	3	1
12	2	1	2	3	3	1	2	3	1	2	3	1	2
13	2	2	3	1	1	2	3	2	3	1	3	1	2
14	2	2	3	1	2	3	1	3	1	2	1	2	3
15	2	2	3	1	3	1	2	1	2	3	2	3	1
16	2	3	1	2	1	2	3	3	1	2	2	3	1
17	2	3	1	2	2	3	1	1	2	3	3	1	2
18	2	3	1	2	3	1	2	2	3	1	1	2	3
19	3	1	3	2	1	3	2	1	3	3	1	3	2
20	3	1	3	2	2	1	3	2	1	1	2	1	3
21	3	1	3	2	3	2	1	3	2	2	3	2	1
22	3	2	1	3	1	3	2	2	1	1	3	1	1
23	3	2	1	3	2	1	3	3	2	2	1	2	2
24	3	2	1	3	3	2	1	1	3	3	2	3	3
25	3	3	2	1	1	3	2	3	2	2	2	2	3
26	3	3	2	1	2	1	3	1	3	3	3	3	1
27	3	3	2	1	3	2	1	2	1	1	1	1	2

IOP Conf. Series: Materials Science and Engineering 1012 (2021) 012037 doi:10.1088/1757-899X/1012/1/012037

Table 3: Details of machining center

CNC Vertical Machining Center	
Manufacturer / Model	Bharat Freitz Werner, Bangalore / BMV 40 T20
CNC Control System	SIEMENS 802D
Table size	710 mm × 400 mm
Traverse (x, y, z)	510 mm × 410 mm × 460 mm
Magazine capacity	20 tools
Spindle speed range	60 – 6000 rpm
Spindle motor	AC 10 kW
Total connected load	25 kVA

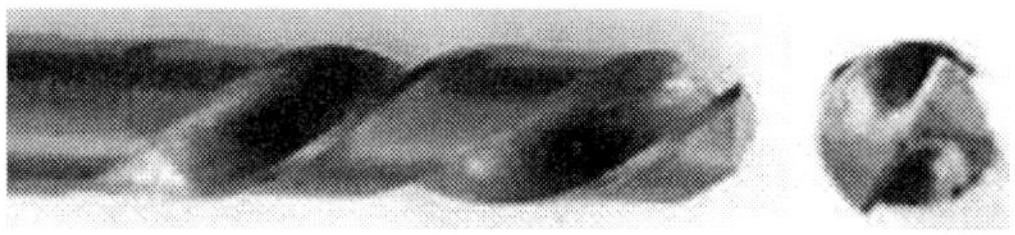

Figure 3: 8-facet solid carbide drill bit

3. Construction of Predictive Computer Model in Python using Fuzzy Logic

Usage of fuzzy logic modelling to predict process parameters has been widely studied, with algorithms largely being implemented in MATLAB, using the Fuzzy Logic toolbox [12, 13]. In this work, a novel approach to the same has been taken, where Python code was developed open-source, which made use of the scikit-fuzzy (skfuzzy) module in tandem with other popular Python data-analysis packages such as numpy, pandas and matplotlib.

Unlike regular classification models which unilaterally determine the category to which the result of a set of inputs belongs, fuzzy modelling does not make sharp distinctions between classes. Instead, fuzzy logic works based on a set of rules, which map a certain range of input parameters to one or more such ranges in the output space, which is then disambiguated by evaluating the "membership" (0 being a non-member, 1 being a definite member) of the obtained output value. A general statement of the rules may be as shown below:

Rule 1: If x_1 is A_1, x_2 is B_1. . . x_n is Z_1, then y_1 is K_1
Rule 2: If x_1 is A_2, x_2 is B_2. . . x_n is Z_2, then y_2 is K_2
. . .
Rule n: If x_1 is A_n, x_2 is B_n. . . x_n is Z_n, then y_n is K_n

Here, $x_1, x_2 \ldots x_n$ are the inputs given to the model and $y_1, y_2 \ldots y_n$ are the outputs corresponding to each of the n cases, $A_1, A_2 \ldots A_n$, $B_1, B_2 \ldots B_n$ and $Z_1, Z_2 \ldots Z_n$ are the fuzzy membership functions for the inputs, and $K_1, K_2 \ldots K_n$ are the fuzzy membership functions for the outputs.

In the MATLAB Fuzzy Logic Toolbox, the inputs and outputs are defined using a visual GUI, while code snippets with appropriate ranges for each fuzzy set are written for defining each membership function of the input and output parameters when using the Python skfuzzy module.

The skfuzzy module is an open-source Python module with a framework to define input and output parameters and "fuzzify" them, and establish a set of rules to map a given combination of inputs to a fuzzy membership function in the output. It allows the definition of the required input parameters by taking the range of variation of values, and subdividing them into fuzzy membership functions which overlap with each other over the range of inputs. As a result, each input value does not have a definite category, but rather has varying degrees of membership (from 0 to 1, being a non-member to being a member of only that category) in multiple categories. The same goes for possible output values.

The program written to perform the fuzzy-logic modelling follows this algorithm:

(i) Define the Antecedents (input parameters) by specifying their labels (eg. 'Spindle Speed') and their ranges of variation (eg. 500 to 1500 rpm).

(ii) Create 3 triangular membership functions for each Antecedent and assign qualitative labels to each function / category such as 'Low', 'Medium', and 'High'.

(iii) Define the Consequents (output parameters) by specifying their labels (eg. 'Torque') and their ranges of variation observed experimentally.

(iv) Create 9 triangular membership functions (for improved accuracy) for each Consequent and assign qualitative labels to each function / category as needed.

(v) Define the rules of operation as a Rule set which maps each combination of Antecedent categories to a Consequent category (In this work, rules were defined by considering the experimentally observed data and using trial and error to determine the rule set which gave the best fit mapping).

(vi) Assign Antecedents, Consequents and Rule set to a Fuzzy logic Inference System (FIS) which can take input values, map them to the Antecedents, find the suitable Consequent categories and 'defuzzify' the Consequent values using the degree of membership of the Antecedents and Consequents.

(vii) Perform this mapping for all input parameters given in table 1 to get predicted output values.

(viii) Plot relevant images using separate data visualization modules.

In the model, the ranges of variation of the control parameters (drill bit diameter, spindle speed, feed rate) were fed into the code and split into 3 membership functions each, with qualitative labels namely 'Low', 'Medium' and 'High'.

In order to improve accuracy and correlation between observed and predicted values, the range of experimentally obtained output values was imported and split into 9 fuzzy membership functions with qualitative labels namely 'Ultra Low', 'Very Low', 'Low', 'Low-Medium', 'Medium', 'High-Medium', 'High', 'Very High' and 'Ultra High' categories.

The fuzzy sets for the control parameters and output parameters are respectively shown in figures 4 and 5.

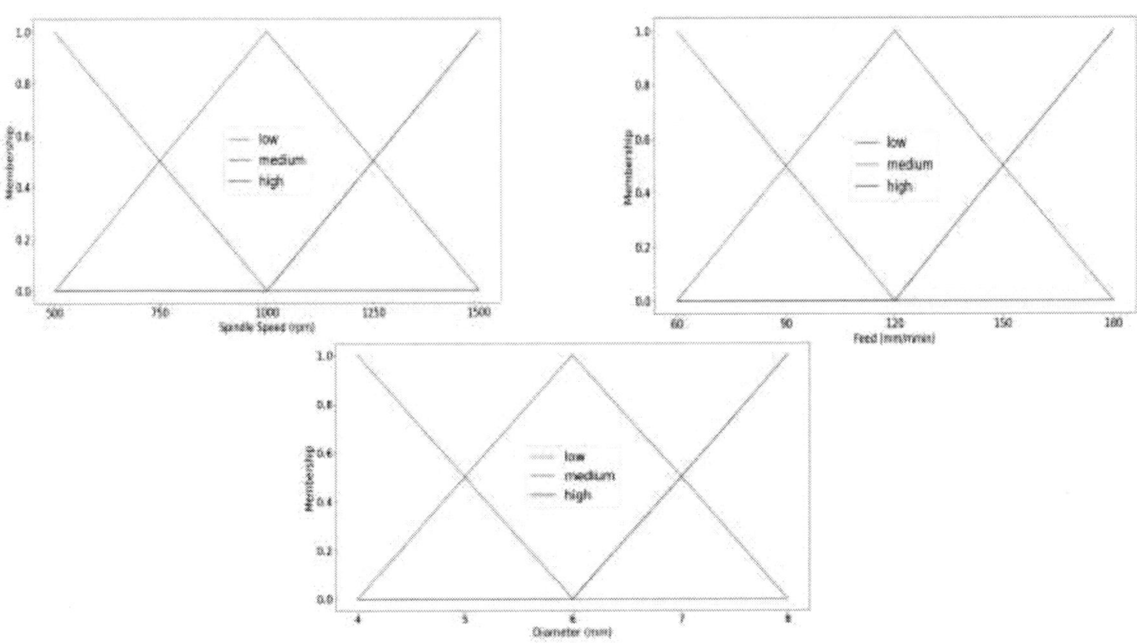

Figure 4: Fuzzy membership functions for control parameters

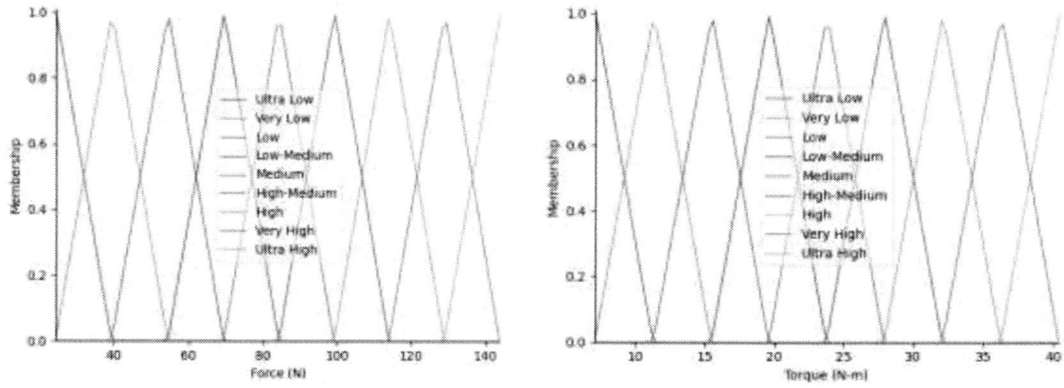

Figure 5: Fuzzy membership functions for output parameters

4. Results and Discussion

The experimental values of thrust force and torque acting on the sample are given in table 4.

Table 4: Experimental Values for Thrust Force and Torque

S.No	Spindle Speed (rpm)	Feed Rate (mm/min)	Drill Diameter (mm)	Thrust Force (N)	Torque (N-m)
1	500	60	4	39.98	7.17
2	500	120	4	61.42	7.12
3	500	180	4	92.32	9.54
4	1000	60	4	26.09	11.28
5	1000	120	4	38.07	11.12
6	1000	180	4	48.45	11.29
7	1500	60	4	24.64	9
8	1500	120	4	29.07	10.27
9	1500	180	4	38.45	10.07
10	500	60	6	53.94	16.17
11	500	120	6	86.67	20.45
12	500	180	6	108.8	18.77
13	1000	60	6	38.68	18.77
14	1000	120	6	51.65	21.31
15	1000	180	6	65.77	19.99
16	1500	60	6	31.59	14.57
17	1500	120	6	45.47	19.3
18	1500	180	6	53.18	18.69
19	500	60	8	72.33	28.15
20	500	120	8	106	32.07
21	500	180	8	143.9	39.75
22	1000	60	8	55.39	23.5
23	1000	120	8	71.18	40.4
24	1000	180	8	83.62	31.59
25	1500	60	8	40.44	19.76
26	1500	120	8	58.14	37.78
27	1500	180	8	69.12	33.19

The values obtained were input into the FIS devised in Python, and the fuzzy mapping rules were defined based on the values observed. The general syntax of the rules was as follows

If Spindle Speed in [$SetS_1$] and Feed Rate in [$SetS_2$] and Drill Diameter in [$SetS_3$], then Thrust Force in [$SetO_1$]

If Spindle Speed in [$SetS_1$] and Feed Rate in [$SetS_2$] and Drill Diameter in [$SetS_3$], then Torque in [$SetO_1$]

After defining the rules to map every possible combination of input fuzzy sets to appropriate output fuzzy sets, the input values were fed to the model and the thrust force and torque were predicted. The predicted values and deviation from the observed values are given in table 5.

Table 5: Predicted Values and % error

S.No	Spindle Speed (rpm)	Feed Rate (mm/min)	Drill Diameter (mm)	Thrust Force (N)		Torque (N-m)	
				Prediction	%deviation	Prediction	%deviation
1	500	60	4	39.56	1.05	8.51	-18.66
2	500	120	4	54.45	11.35	8.51	-19.46
3	500	180	4	99.17	-7.42	8.51	-10.77
4	1000	60	4	29.62	-13.53	11.29	-0.09
5	1000	120	4	39.56	-3.91	11.29	-1.53
6	1000	180	4	54.45	-12.38	11.29	0.00
7	1500	60	4	29.62	-20.21	8.51	5.48
8	1500	120	4	29.62	-1.89	11.29	-9.94
9	1500	180	4	39.56	-2.89	11.29	-12.12
10	500	60	6	54.45	-0.95	15.44	4.51
11	500	120	6	84.27	2.77	19.60	4.16
12	500	180	6	114.09	-4.86	19.60	-4.42
13	1000	60	6	39.56	-2.28	19.60	-4.42
14	1000	120	6	54.45	-5.42	19.60	8.02
15	1000	180	6	69.37	-5.47	19.60	1.95
16	1500	60	6	39.56	-25.23	15.44	-5.97
17	1500	120	6	39.56	13.00	19.60	-1.55
18	1500	180	6	54.45	-2.39	19.60	-4.87
19	500	60	8	69.37	4.09	27.92	0.82
20	500	120	8	99.17	6.44	32.08	-0.03
21	500	180	8	138.92	3.46	39.01	1.86
22	1000	60	8	54.45	1.70	23.76	-1.11
23	1000	120	8	69.37	2.54	39.01	3.44
24	1000	180	8	84.27	-0.78	27.92	11.62
25	1500	60	8	39.56	2.18	11.29	42.86
26	1500	120	8	54.45	6.35	36.24	4.08
27	1500	180	8	69.37	-0.36	32.08	3.34
				Mean error	-2.03	**Mean error**	0.69

Line graphs were drawn by overlaying the experimental values with the predicted ones for each experiment trial. The plots are shown in figure 6.

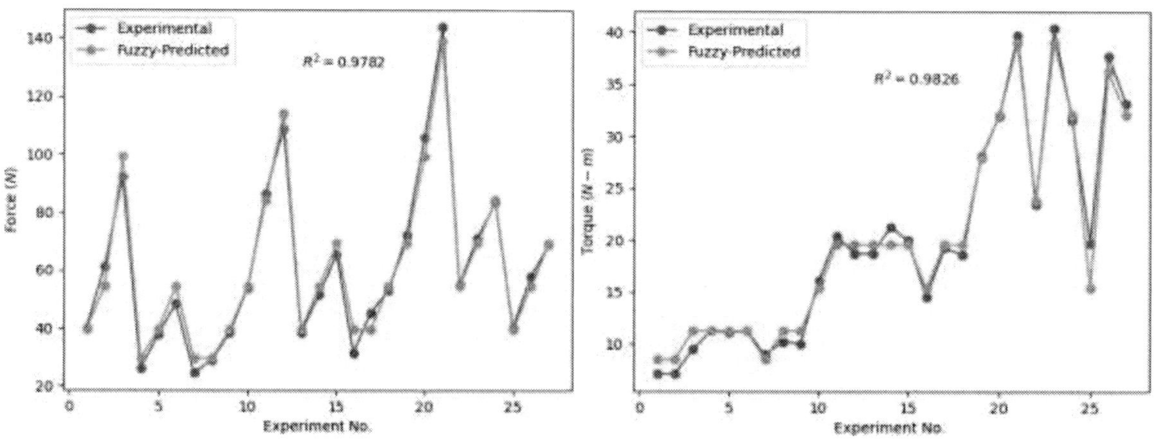

Figure 6: Comparison of experimental values and values predicted by the fuzzy model

The figures indicate a relatively predictable variation in thrust force and torque with the control parameters, namely drill bit diameter (mm), spindle speed (rpm) and feed rate (mm/min). The thrust force appears to vary directly with feed rate and drill bit diameter, and inversely with spindle speed, while the torque appears to generally increase with spindle speed and feed rate. This observation is in agreement with the results of the work done by Nagarajan et al [9].

A more thorough comparison is made by producing scatter plots between the experimental and predicted values, and fitting a linear trend curve. A bar chart comparison of 5 randomly selected data points (to eliminate the possibility of spurious correlations) is also done for better corroboration. The goodness of fit of the model to the data is apparent in the plots, which are shown in figures 7 and 8.

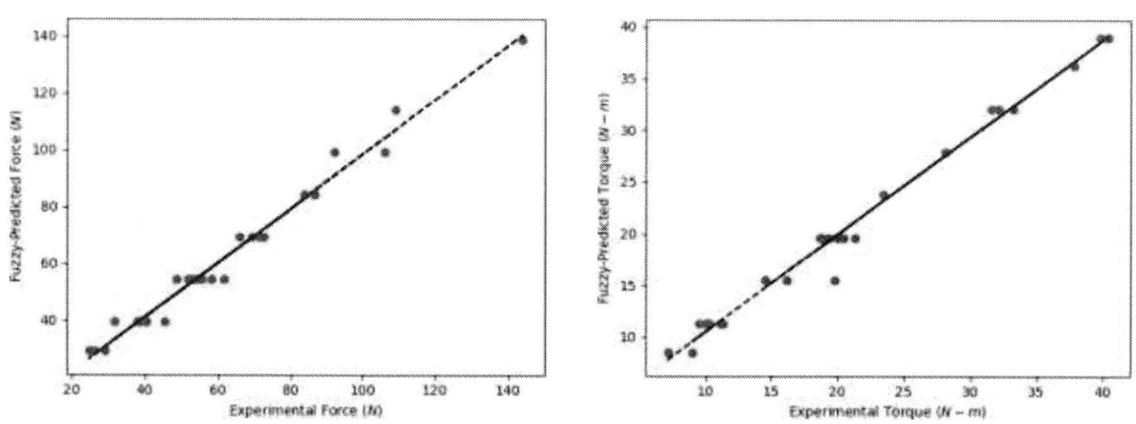

Figure 7: Trend line between experimental and predicted values

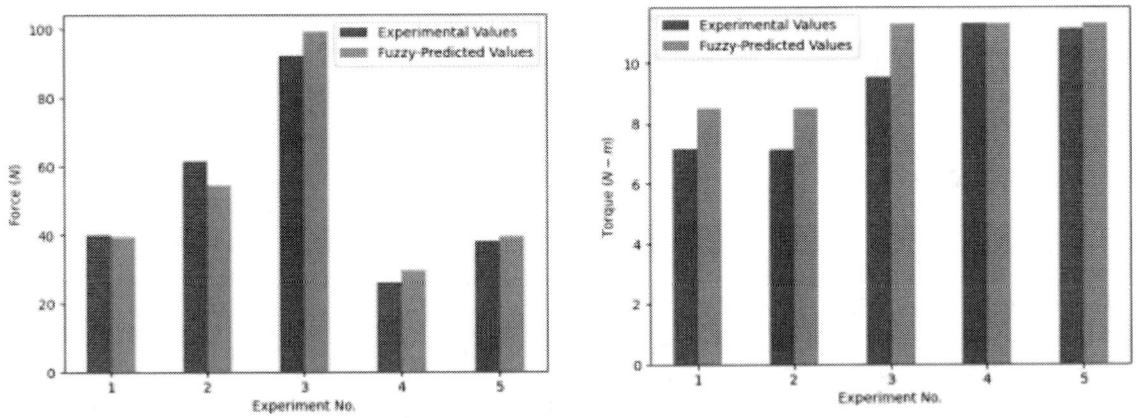

Figure 8: Examination of randomly chosen experimental trials

In order to compare the capabilities of the skfuzzy module in Python with the more popular Fuzzy Logic Toolbox in MATLAB, the same model was created using that toolbox in MATLAB and the experimental values were used to train it. A comparison between the predicted results obtained in each case and the experimental results was made. The plots obtained in the comparison are shown below in figure 9.

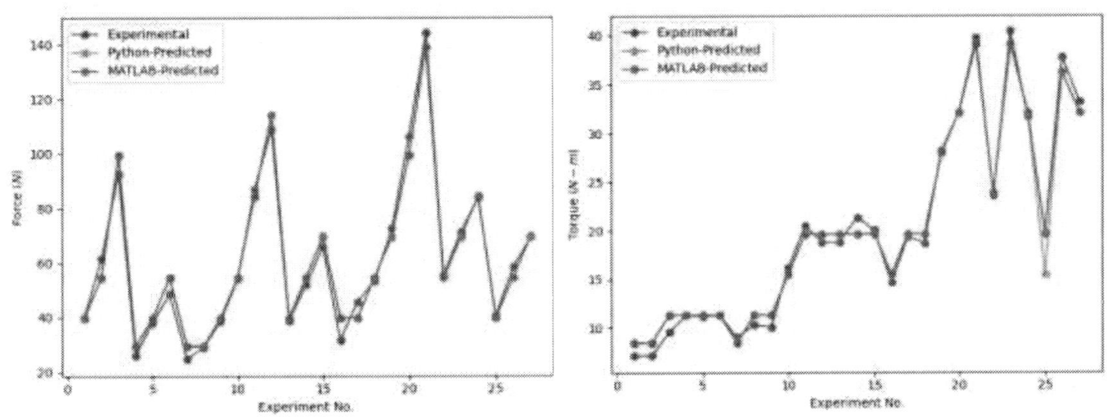

Figure 9: Comparison of Python skfuzzy module and MATLAB Fuzzy Logic Toolbox

The plots make it clear that although both tools are very good predictors of the output values, the Fuzzy Logic Toolbox has a slight edge over the skfuzzy module, and it seems to be able to better handle the random variations of the output, especially in the case of the torque prediction, which may be attributed to the higher computational suitability and capacity of MATLAB in general. This is further quantitatively demonstrated by calculating the R^2 correlation coefficient in each case. The obtained values are given in table 6.

Table 6: Comparison of Python model with MATLAB model

S. No	Predicted Quantity	Correlation R^2 with experimental values	
		Python FIS	**MATLAB FIS**
1	Thrust Force (N)	0.9782	0.9801
2	Torque (N-m)	0.9826	0.9915

The obtained values are very close to 1 in using both the Python and MATLAB models, indicating that the model can account for virtually all unpredictability occurring in the system.

5. Conclusions and Future Work

In summary, drilling tests were conducted on glass-flax-hemp fiber hybrid composite fabricated with the hand-layup process. From previous work, the significant control parameters to be analyzed were chosen to be the spindle speed, feed rate and drill bit diameter. Experiment design and assignment of control parameters was done using a Taguchi L_{27} orthogonal array. Experimentally measured values of thrust force and torque acting on the composite sample were recorded, and a fuzzy-logic inference system (FIS) composed of 3 inputs, 2 outputs and 54 rule sets was designed in Python to make predictions for the same.

The FIS model was found to have near-perfect correlation with experimentally obtained values of thrust force and torque, thus establishing its suitability for use in modelling of drilling parameters for HFCs. The modelling process can be used to reduce the chance of damage of the composite sample during drilling in the way of delamination and internal cracks caused due to excessive thrust force and torque.

The high degree of correlation observed between the values with minimal computational effort in the model suggests that this technique can be used for other composite materials than the one under study. Obviously, the specific interactions between the input and output parameters and the fuzzy rules for mapping will vary from one material to another, but the general principles of the fuzzy logic modelling can still be applied to any composite by collecting enough data for training.

For better performance of the model, but at the risk of over-fitting, more process parameters such as coolant use, heat generated etc. can be integrated into the model; also, the membership functions for inputs and outputs can be further granularized for better results at the added expense of computational complexity. Further, the nature of the membership functions used can be varied and the effect on the computational efficiency of the predictive model could be studied. For instance, instead of using triangular membership functions, trapezoidal or Gaussian bell-curve membership functions could be considered.

In this work, only three levels of variation were considered for each input parameter taken. In forthcoming work, the range of input parameters considered could be widened (e.g. spindle speeds of 2000 and 2500 rpm may be included). It is also possible to conduct tests and perform modelling for different composite samples and engage in a differential study, analyzing the input-output interactions for each sample and noting the influence of other possible significant control parameters.

References

[1] Shin PS et al. 2015 Optimum mixing ratio of epoxy for glass fiber reinforced composites with high thermal stability *Compos. B Eng.* **79(4)** 132-37

[2] Jawaid M and Abdul Khalil 2011 Cellulosic/synthetic fibre reinforced polymer hybrid composites: a review *Carbohydr. Polym.* **86(1)** 1-18

[3] Sathishkumar T, Naveen J, and Satheeshkumar S 2014 Hybrid fiber reinforced polymer composites – a review *J. Reinf. Plast. Compos.* **33(5)** 454-71

[4] Vantaki VK and Ganta V 2014 Optimization of process parameters in drilling of GFRP composite using Taguchi method *J. Mater. Res. Technol.* **3(1)** 35-41

[5] Kavad BV, Pandey AB, Tadavi MV and Jakharia HC 2014 A review paper on effects of drilling on glass fiber reinforced plastic *Proc. Technol.* **14** 457-64

[6] Panchagnula KK and Palaniyandi K 2018 Drilling on fiber reinforced polymer/nanopolymer composite laminates: a review *J. Mater. Res. Technol.* **7(2)** 180-89

[7] Davim JP, Reis P and António CC 2004 Experimental study of drilling glass fiber reinforced plastics (GFRP) manufactured by hand lay-up *Compos. Sci. Technol.* **64(2)** 289-97

[8] Palanikumar K, Prakash S and Shanmugam K 2008 Evaluation of delamination in drilling GFRP composites *Mater. Manuf. Process.* **23(8)** 858-64

[9] Nagarajan V, Kumar V and Samsingh RV 2017 Artificial intelligence modelling and analysis of pivotal parameters

in drilling hybrid fiber composite (HFC) *FME Trans.* **45(4)** 641-46
[10] Fernández-Pérez J, Cantero JL, Díaz-Álvarez J and Migueléz MH 2019 Hybrid composite-metal stack drilling with different minimum quantity lubrication levels *Materials* **12(3)** 448-60
[11] Latha B and Senthilkumar VS 2008 Fuzzy rule based modeling of drilling parameters for delamination in drilling GFRP composites *J. Reinf. Plast. Compos.* **28(8)** 951-64
[12] Aamir M, Tu S, Tolouei-Rad M, Giasin K and Vafadar A 2020 Optimization and modeling of process parameters in multi-hole simultaneous drilling using Taguchi method and fuzzy logic approach *Materials* **13(3)** 680-95
[13] Samsingh RV, Latha B and Senthilkumar VS 2009 Modeling and analysis of thrust force and torque in drilling GFRP composites by multi-facet drill using fuzzy logic *International Journal of Recent Trends in Engineering* **1(5)** 66-70

A Focus-Measurement Based 3D Surface Reconstruction System for Dimensional Metrology

S.M.Vignesh[1*], R. Senthilnathan[1]

[1]Department of Mechatronics Engineering, SRM Institute of Science and Technology SRM Nagar, Kattankulathur, 603203, Kanchipuram, Chennai TN, India.

E-mail: vigneshs1@srmist.edu.in

Abstract. Image focus is an important parameter which depends on various other parameters of an imaging system. One among them is the distance of the object plane from the lens. This parameter is referred as the depth in computer vision literatures. Since depth affects the focus level of pixels in an image, focus measurement can be used to interpret the depth of points of an object. The paper presents the details of the novel indigenously developed automated measurement system which can perform high-precision movements based on feedback control. The measurement system is inspired by classical systems used in techniques such as shape from focus. This system fundamentally being a mechatronics system integrates a high-precision multi-stage translation mechanism which is actuated with a servo motor with sensors such as camera, linear encoder, optical limit switches. The hardware interface is developed for automated measurement through a USB data acquisition system. The paper presents the methodology of surface reconstruction and detailed description of the various challenges encountered in the integration. The surface reconstruction is performed for various millimetres and sub millimetres level engineered specimens. The paper presents the future direction of research and also various potential application possibilities for such measurement setup.

Keywords: metrology, surface reconstruction, shape from focus, depth from defocus.

1. Introduction

In modern manufacturing industries, non-contact dimensional measurement techniques plays a vital role. The use of non-contact measurement techniques preserves the surface quality of the part and stability of the measurements which in turn improves the final value of the products.

In many industrial applications, machine vision systems are becoming increasingly relevant, such as automatic inspection, robotic manipulations etc. All these activities are carried out in a 3-dimensional environment, which in turn requires the reliable information of the 3D depth of the scene. For some cases sparse 3D depth information is all require to understand the scene, whereas for other cases it requires dense 3D depth map is required. Sometimes the accurate shape of the surface of an object in the scene may be required instead of estimating the surface's mere range value. As a result, the computer vision and robotics community has developed a great deal of research interest in the acquisition of depth information. Currently available vision based range finder can be classified into two broad categories, active and passive techniques. In active technique special lighting devices illuminates the scene, whereas in passive technique the ambient light provides lighting for the scene.

1.1. *Shape from focus*
Shape from focus (SFF) technique is one of the passive method for depth estimation, which requires stack of images from an object captured at different focal setting of the imaging acquisition device. As SFF is the passive approach, there is no physical contact between the surface of the target and the imaging system. In addition to this fact, SFF uses single camera, this over-comes the correspondence problem usually occurs in the stereo vision imaging device. The local variation of the focus in the image is the cue for depth.

According to the thin lens camera model the focus in the camera is adjusted by changing the distance, v, between the lens plane and the sensor plane. During image formation only points a particular distance will be in focus this depends on the focal setting of the lens. The relationship between focal length, distance from the image and distance from the target is given by the thin lens equation:

$$\frac{1}{f} = \frac{1}{u} + \frac{1}{v}$$

Since each point in the image plane has the corresponding point in the object plane, the maximum focus will only happen at a particular distance.

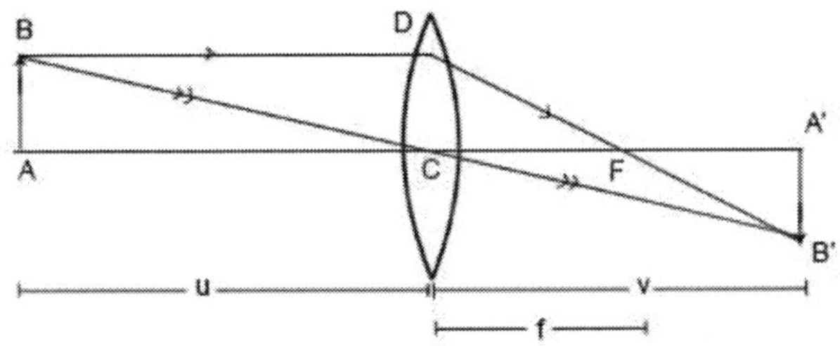

Figure 1. Formation of image.

The degree of focus is the function of lens setting and the depth of the scene. Hence the focus of an image is cue to estimate the depth with the known lens setting. Shape from focus (SFF) technique is used to interpret the depth of a 3-dimensional object. For the SFF technique a sequences or stack of images are required which are taken at different focal distance from a reference plane. These stack of images are captured by fixing the optical elements fixed and by translating the object at different stages. And the translation of different stages are recorded as the focal distances for the stack of images. Then the focus measure operator is applied for all the pixels along the stack of images. The focus measure values obtained at each pixel along the focal stack are interpolated by the Gaussian curve to fit the peak value of the focus measure. After obtaining the peak values of each pixel the depth of each pixel is interpolated by Gaussian 3 point approximation.

1.2. *Shape from defocus*
When an object point is not in focus, a blurred circle would be formed in the image whose radius is defined by a blur parameter 'σ' is the image of the point, given by.

$$\sigma = \rho r v \left(\frac{1}{f} - \frac{1}{v} - \frac{1}{D}\right)$$

Where f is focal length, D is the distance between the point of the object and the lens, v is distance between the lens and image detector, r is the aperture radius of the lens, and ρ is a camera constant [6].

Shape from defocus (SFD) is another technique to interpret the depth of a 3-dimensional object. SFD requires two defocused images of an object captured at different lens settings. The relation between two defocused images is given by:

$$\sigma_1 = \alpha\sigma_2 + \beta$$

Where α and β are known parameters depending in camera settings. The relative blur between two defocus pictures is evaluated by assuming a suitable optical transfer function model. Over the local region depth is assumed to be constant [6]. It is possible to calculate the depth of the object corresponding to the local area by understanding the lens setting and the blur parameters.

In [1], Peter Kinnel et al. used a 3D metrology vision system to automatically detect the location of the target object. Victor Alonsoa et al. in [2], expressed the application of machine vision system for metrology in industry 4.0.

This paper focuses on development of an automated image capturing system with a translation mechanism required to capture stack of image from an object at various focal distances. The linear translation mechanism is coupled with a linear potentiometer for collecting the feed-back of the distance travelled by the translation stage. The hardware system is connected to the computing system through the USB data acquisition system.

2. Experimental Setup

The experimental set-up consists of linear translation table attached with high-precision multi-stage translation mechanism which is actuated with a servo motor and optical limit switches. And to measure the translation distance a linear potentiometer is attached with the linear translation table. For image acquisition USB camera attached with tele- centric lens is used. The hardware setup is interface with the computer through USB data acquisition system.

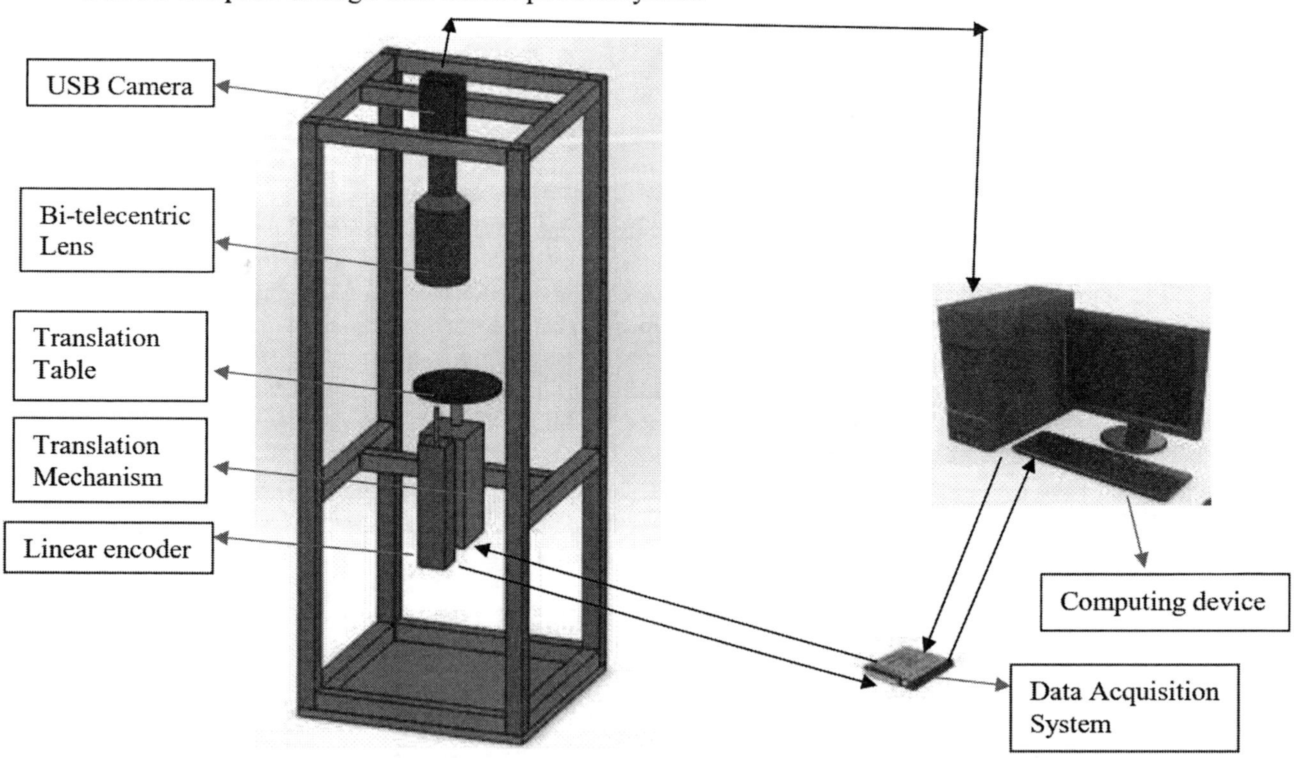

Figure 2. Image Acquisition System.

2.1. *Specifications*
All images are captured by Contras tech VT-EX500CS USB 2.0 camera with the 2592 X 1944 image resolution and 5fps frame rate. The object is focused through a TC23036 bi-tele centric lens 0.243x magnification, working distance of 102.5 mm, and field depth of 11 mm. The distance travelled by the translation table is measured by the linear potentiometer with useful electrical stroke 100 mm, repeatability 0.01 mm, and displacement speed 3 m/s. Reversing the direction of the table movement is achieved by the optical limit switches placed at the extreme ends of the translation mechanism. The various inputs and outputs of motor for the translation table, linear potentiometer, and optical limit switch are connected to the system with the help of NI USB-6009, 8 inputs, 14-bit data acquisition device.

3. Motion Control and Image Acquisition

3.1. *Characteristics of the translation*
The rectilinear motion of the translation table is achieved by the worm wheel and belt drive connected with the servo motor. The characteristic of this rectilinear motion is analyzed for the better understanding of the system. In order to analyze, the translation stage is allowed to translate multiple times and the corresponding voltage change in the linear potentiometer which is coupled with the stage is recorded through the data acquisition device. The graphical representation of the voltage change is shown in the following figure. 3 (a) and (b). From the graph it is observed that during the start of the table motion both from bottom as well as top there is no significant change in voltage in the potentiometer. This happens because of the initial friction force which resisted the motion of linear potentiometer coupled with the translation table.

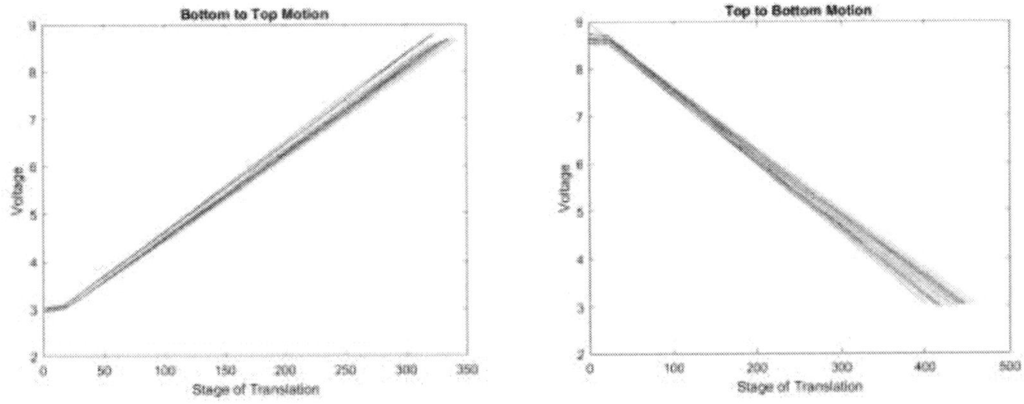

Figure 3. (a) Bottom to Top Motion (b) Top to Bottom Motion.

3.2. *Correction strategy*
If the selection of object height is lesser than the maximum sweep of the translation table. Then along the region where the motion of translation table is restricted due to friction force may not be considered for image acquisition. Moreover all the points of image acquired in this region will not be in focus.

3.3. *Feedback control*
The remaining linear region of the graph shows the repeatability of the linear potentiometer. From this it is observed that voltage corresponding to the distance is not consistent for the repeated cycles. So the various distance of the translation table has to be controlled by the different voltages for each trail.

3.4. *Image Acquisition*

Image acquisition of the object takes place at various translational stages. The object is placed over the translation stage and allowed to move along z-axis towards the lens. The distance d, between the fixed reference plane and the translation stage is noted. Depending on the optical parameter, working distance, the focused plane will be fixed at a specific distance from the lens. The distance between the focused plane and the reference plane is also fixed. As the object reaches the focused plane, all points lying on it will be perfectly focused while other points will be blurred. Images of the object will be captured at regular interval of distances.

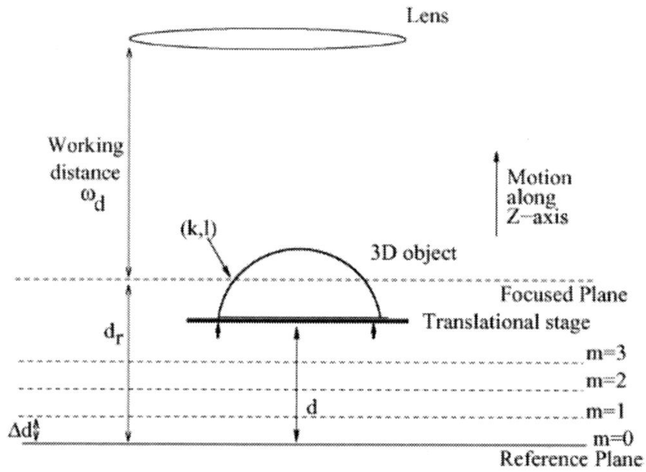

Figure 4. Schematic of Image Acquisition in SFF.

3.5. *Image Stack Preparation*

As the object moves towards the focused plane, images are captured at different translation stage. For each captured image corresponding distance from the reference plane is noted. This distance value helps to stack the image in sequence. Each point in the image will be perfectly focused on any one of the image along image stack. The point which is at the best focus along the image stack has to be estimated. The estimated value is known as the focus measure of the point in an image.

3.6. *Focus Measurement*

In order to estimate the focus measure, a focus operator is applied for each image along the stack of images. The focus measure operator o (x, y) is applied to the defocused image Id (x, y), the result will be a new image r (x, y) that may be expressed as:

$$I_d\ (x,y) = h\ (x,y) * I_f\ (x,y)$$

$$r\ (x,y) = o\ (x,y)\ * (h\ (x,y) * I_f\ (x,y))$$

This can be rewritten as:

$$r\ (x,y) = h\ (x,y)\ * (o\ (x,y) * I_f\ (x,y)$$

Since convolution is linear and shift-invariant. From the above expression it is obvious, applying a focus measure operator to a defocused image is equivalent to defocusing a new image obtained by convolving the focused image with the operator. The focus operator should be a high pass filter which highlights the high frequency variation along the image. Several operators are proposed for this focus measurement. Among which the most widely used operators are the Modified Laplacian, Gray- Level variance, Modified Gray Level Variance, etc. In this work Modified Gray Level Variance is used for focus measurement.

3.7. *Gaussian Interpolation and Depth Estimation*

Once the focus measure of all the pixels using any one of the focus operator is determined, next is to estimate the depth of each point of the object. Gaussian interpolation is the widely used method for is task. For each pixel the image with highest focus measure is evaluated and depth corresponding to the pixel is interpolated using Gaussian function around the position. In the figure. 5, different focus measure of the pixel at a coordinate (i, j) are shown along the image stack. For the Gaussian 3 point interpolation maximum focus measure value F_m, previous value Fm-1 and the next value Fm+1 are used. The depth of the pixel is assumed to be the peak value of the Gaussian.

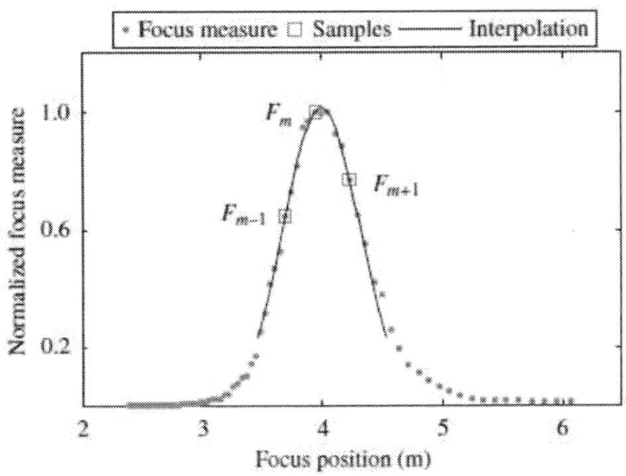

Figure 5. Gaussian 3 Point Interpolation.

3.8. *Filtering*

After obtaining the depth map using shape from focus it would be noticed that some regions will be inaccurate and corrupted by noise. To order to filter this several smoothening technique are available like median filter, bilateral filter, etc.

3.9. *Confidence level estimate*

Depending on the optical configuration and image content the quality of the depth map obtained will be affected. The response of focus measure operation on the image stack is also based in the two effects. These factors determines the accuracy of the depth map obtained. So confidence level estimate has to detect the variation due to optical configuration and imaging condition. In addition to that it should be able to detect the focal variation when the focal operator fails to detect.

4. Results and Discussion

The image stack of the 3D object is acquired from the image acquisition device as mentioned in section 2. And the stack of images are sequenced according to the focal configuration of the optics in the image acquisition system as in figure. 6. Then the focal measure operator is applied to the stack of images. In this work Modified Gray Level Variance focus operator is used for the focus measurement. Then by using the Gaussian 3 point interpolation the peak values of the focus corresponding are estimated. Figure. 7. Shows the depth plot of the 3 object using shape from focus technique.

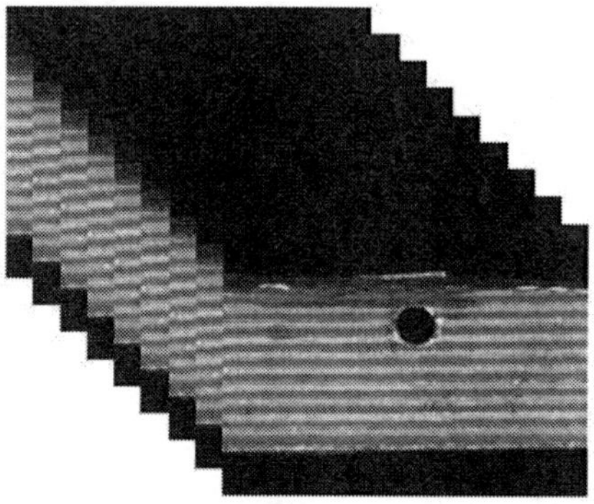

Figure 6. Image Stack of 3D Object.

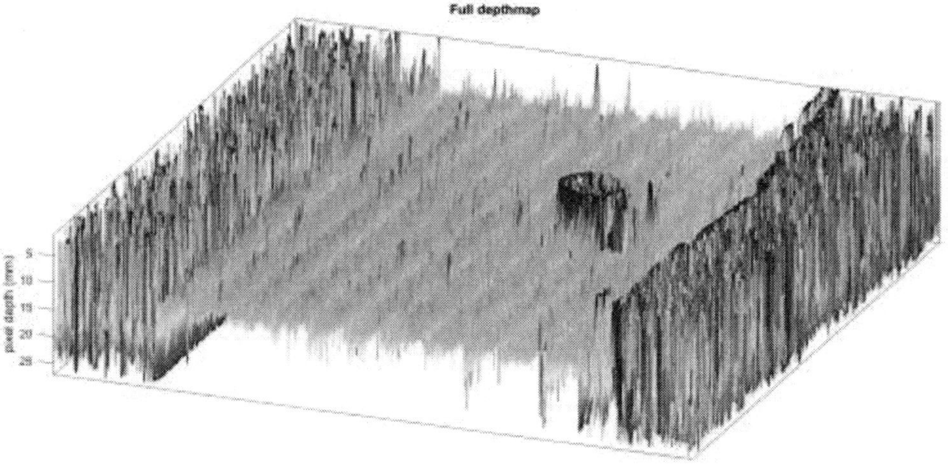

Figure 7. Depth map of Image Stack.

5. Conclusions

The depth map plot can be used in various industrial applications surface measurement, surface reconstruction, reverse engineering etc. By improving the calibration of the image acquisition system the accuracy of the depth map with close tolerance can be achieved. In this work the 3D object with good surface texture was chosen for the experiment to show the better visualisation of surface variation. Whereas the experiment can be extended for the smooth surfaces. In case of smooth surface the focus measure operator need to be modified accordingly to capture the focal variation of the image stack. The same technique can be extended for various environment by designing an image acquisition according to the requirement. Accordingly the suitable focus measure operator need to be developed for capturing the focal difference. In case for industrial application the environment for the measuring device will be fixed, this makes the measurement system for robust application for different surface characteristics.

References

[1] PeterKinnell TomRymer JohnHodgson LauraJustham MikeJackson, Autonomous metrology for robot mounted 3D vision systems.

[2] VictorAlonsoa AngelDacal-Nietoa LuísBarretobc AntónioAmara ldEduardoRiveroa, Industry 4.0 implications in machine vision metrology: an overview.

[3] S.K. Nayar and Y. Nakagawa. Shape from Focus. *IEEE Transactions on Pattern Analysis and Machine Intellegence* 16(8):824-831, 1994.

[4] S. Pertuz, D. Puig, M. A. Garcia. Reliability measure for shape-from focus. *Image and Vision Computing* 31:725–734, 2013.

[5] S. Pertuz, D. Puig, M. A. Garcia. Analysis of focus measure operators for shape-from-focus. *Pattern Recognition*. 46(5): 1415-1432, 2013.

[6] Subhasis Chaudhuri, A. N. Rajagopalan. "Depth From Defocus: A Real Aperture Imaging Approach", *Springer Nature*, 1999.

[7] RR Sahay, AN Rajagopalan. Dealing with parallax in shape-from-focus. *IEEE Transactions on Image Processing* 20 (2), 558-569.

[8] Aamir Saeed Malik, Tae-Sun Choi. Anovel algorithm for estimation of depth map using image focus for 3D shape recovery in the presence of noise. *Pattern Recognition* 41 (2008) 2200 – 2225.

[9] Murali Subbarao, Tae Choi, Arman Nikzad. Focusing techniques. *Optical Engineering* 32(11), 2824—2836.

[10] Usman Ali, Muhammad Tariq Mahmood. 3D Shape Recovery by Aggregating 3DWavelet Transform-Based Image Focus Volumes Through 3DWeighted Least Squares. *Journal of Mathematical Imaging and Vision* 10851-019-00918-8.

[11] Zhiqiang Ma, Dongjoon Kim, Yeong-Gil Shin. Shape-from-focus reconstruction using nonlocal matting Laplacian prior followed by MRF-based refinement. *Pattern Recognition* 103 (2020) 107302.

[12] Usman Ali, Ik Hyun Lee, Muhammad Tariq Mahmood. Guided image filtering in shape-from-focus: A comparative analysis. *Pattern Recognition* 111 (2021) 107670.

[13] Sung-An Lee and Byung-Geun Lee. Accurate 3D Surface Reconstruction for Smart Farming Application with an Inexpensive Shape from Focus System. *Journal of Sensors* Volume 2020, Article ID 1859512, 7 pages.

[14] Murali Subarao, Gopal Surya. Depth from Defocus: A Spatial Domain Approach. *International Journal of Computer Vision*, 13, 3, 271-294.

Manufacturing of 3 D Shrouded Impeller of a Centrifugal Compressor on 3D-Printing machine using FDM Technology

K Aruna Prabha [1*], P Sai Rohit [1], Sai Chand Nitturi[1], Bobba Nithin[1]

[1]Department of Mechanical Engineering, VNR Vignana Jyothi Institute of Engineering and Technology, Telangana, India

E-mail: arunaprabhakolluri@gmail.com

Abstract. Blades of impellers are intricate and are having complex profiles which are difficult to be manufactured by convectional milling processes. To produce such multifaceted precise dimensional functional components, expertise combined with new technology machining such as 5-Axis CNC machining is required which is costly and tedious task. The impellers components have maximum diameters of 200mm and thickness of blades are less than 1.5mm. A rapid manufacturing technique aids in manufacture of blade profiles which are robust in design and prototypes can be easily made for testing and analysis this aid in design validation. In particular, the investigation was focused on shrouded and unshrouded 3D impeller of centrifugal compressors. The present paper describes about the basic functioning of FDM based 3D printer while printing open and shrouded impeller and various steps involved in it and finally advantages and disadvantages associated with 3D printing technology.

Keywords: Shrouded 3D impeller, FDM, Rapid manufacturing, Centrifugal Compressors

1. Introduction

Additive manufacturing (AM) technology has been considered a breakthrough for future product development and production. There has been tremendous effort in recent years to introduce technology advancements to reorganize and establish businesses and its allied fields globally. One of the studies indicates that usage of AM technology is restricted to only few sectors but soon will become a future, integrating all sectors of manufacturing. Prototyping components from AM based rapid technology can create and build models based on CAD files and computer data. This technology detects and constructs layers which befall one by one thus representing and building the replica or idea of original product. This process doesn't require tooling or fixture set ups as in convectional machining. This delivers successful commodities in shorter times with add on replacement facilities unlike regular machining process which involves cutting or removal of material example milling. Now days, large number of materials like plastics, metals, ceramics, etc, can be used for 3D printing technologies. Rapid prototyping covers different fields in its application point of view such as health care, research investigations, armed forces, structural engineering, fashion and jewellery industry, design and prototype making, CAE etc.

Figure 1. 3-D printer

1.1. Manufacturing of impeller

Shang Liangchen et al[1] applied latest and advanced CAD/CAM methods to manufacture blades of centrifugal compressor impeller. Reverse engineering, Unigraphics NX (CAD/CAM) software, coordinate measuring machines (CMM) technologies are discussed in this paper. Finally they concluded that cavity milling in conjunction with five axis milling technology adapted could increase efficiency and accuracy in manufacturing impeller blades. Aruna prabha et al[13] used 5-Axis machines in conjunction with UNIGRAPHICS software to optimize tool paths for manufacturing steam turbine blades. Klaus bodden berg [2] discussed about the manufacturing methods of impeller and how the casting and forging manufacturing processes are useful for impeller blank. Moreover, they discussed about what are the methods for joining of blades and hub in industries. Those methods are welding, brazing and riveted methods. Selection of materials and heat treatment operations are discussed. Finally, in this paper they have explained on entire view of manufacturing system of impeller for turbo compressors. Toru kaino[3] discussed and dealt about machining large size impellers used in centrifugal compressors which require 5-axis machining set ups. This improved efficiency and accuracy of products by retaining its shape. By introducing the contour machining in line with cutting resistance- levelling arrangement the machining time was reduced and can also chatter was easily detected. This paper depicts the use of new technology methods applied to efficiently machine large impellers with ease. B.Sajjan, et al[4] designed and analysed impeller type centrifugal pump on ANSYS work bench. The initial objective is to test for deflections and stresses in impeller as well as centrifugal pump housing by applying load onto the belt. Various materials are selected for pump assembly and their corresponding induced stresses are individually evaluated ensuring that it does not exceed its elastic limit. This gave various factors as results to determine how different materials vary and which material is best for pump building. These include natural frequency, stress, strain, deformation factors and modal values. Modelling is done in design modeller and analysis in work bench both available in ANSYS. This FEM analytical tool can be used to solve many structural, static and dynamic problems and also can find deformations, stresses in impellers throughout its profile. Syam Prasad el al[5] analyzed static and dynamic factors in centrifugal pump impeller. They used CATIA software for modeling of impeller. They did analysis on impeller using ANSYS software to find out the stress, strain, deformation, deflection and mode shapes of impeller. They used three types of alloy materials to estimate the impeller performance. Those alloy materials are Inconel 740, Inconel 830, and Warpaloy. Finally, they concluded that comparing the results of three alloys, Inconel alloy 740 is best for impeller making. Dinesh Tare al[6] investigated static and dynamic analysis of impeller of centrifugal blower. They used CATIA Software for modeling of blower. They did analysis on blower by using ANSYS software to find out the stress, strain, deformation, deflection and mode shapes of impeller. In this paper, meshing of blower is done by using hyper mesh software. They used three types of alloy materials to estimate the impeller performance. Those alloy materials are MS, SS, and SS316L Steel. Finally they concluded that comparing the results of three alloys, SS316L STEEL is best for blower design. Tao Wu, Yan long Cao, Jiangxi Yang[7] investigated about the core part in centrifugal compressor impeller and found that quality of blade effects its function as in structural potential and its aerodynamic characteristics in compressor. Numerical analysis is performed on different impeller profiles and dimensions and simulation is done on NUMECA software where centrifugal impeller 3-dimesional viscous force is analysed. This checks profile inaccuracies on various size and design impellers. A comparative study of designed impeller determines its aerodynamic behaviour and analysis result will validate the design process. This showed that machining errors effects impeller's functional performance, in terms of pressure ratio and its effectiveness. Determining the S-type profile error on blade surface is the important factor which affects the performance aspects hence the design can be made accordingly. Min-Ho Cho-al[8] discussed about the case based impeller machining strategies (CBIMS). CBIMS support system replaces efficient machining strategies with the existing machine strategies. It creates impeller strategies in machining following a stepwise reasoning procedure. Finally, this paper concluded that through CBIMS, latest machining strategies are promptly developed making use of both cooperation base and E-manufacturing work of CBIMS. Rech al[9] analyzed the sound levels during the operation of milling. In his paper identified that during milling operation, sound pressure levels cross more than 100 DBA compared to drilling, tapping and reaming. They arranged anechoic chamber and microphones to analyze the sounds in different levels. After the experimental work the evaluation of sound levels, noise signals in milling, 92 DBA level of sound waves are obtained. Finally concluded that air leakages, damped noises and depth of cut has little effect on noise emission. Mill wills having less diameter provide good stability and reduce turbulence which are main contributing factor for producing undue noise. Francis quail et al[10] presented a rapid manufacturing technique for prototype building of a regenerative pump impeller. They found that intricate, multifaceted blade profiles of impellers built by RP technology can be used for analysis, testing and in developing optimized design. Vinod G. Gokhare et al[11] in their literature review mentioned and presented different 3-D printing techniques, characteristics and its significance and selection of the best materials suitable for 3 D printing machine. Ana Pavlovic et al[12] in their paper adopted polymers based additive manufacturing technique in prototyping water pump impeller. This assessed the functional requirement and facilitated in manufacturing of original parts. The polymer-based technology is very helpful in manufacturing and prototyping water pump impellers.

2. Methodology

TECHB Weaver 3D Printer using Fused Filament Fabrication or Fused deposition modelling method for manufacturing a light weight and high strength 3D Shrouded Impeller for centrifugal Compressors with Industrial Grade thermoplastics are used. This equipment has a wire made of plastic substance coiled in form of a spool and the material ejects from nozzle whenever needed and deposited onto layers during printing. The heat is supplied which melts the substance used for printing and with the help of machine it can be moved in horizontal and vertical direction via controller-controlled mechanism. The entity is formed by extracting melted material and layered until it solidifies and hardens due to cooling action taking after it is being released from nozzle. This technology is mainly used to print materials having low melting temperatures, plastics like Polylactic Acid (PLA) and Acrylonitrile Butadiene Styrene (ABS).

3. Processes

Fused Deposition Modeling (FDM) fabrication process works in this fashion as shown in figure 2. Thermoplastic material in thin film form called spool which is fed or inserted into the 3D-printer. As the required temperature is attained by the nozzle the filament it is supplied onto top for extrusion called head and in the nozzle it liquefies. The head of extrusion can travel in three directions where they follow in x, y, z axis. The liquefied substance extrudes out in the form of thin filament layers which are built on to predefined positions where it hardens due to cooling. Additional cooling facility is provided by fixing fans onto the extrusion head. For the region to be covered numerous passes are needed as in colouring a rectangle with colour pencils. When a layer is finished, the build platform descends and a fresh layer is deposited but in some cases extrusion head lifts up to deposit the layer. This procedure continues till the component printing is finished. Acrylonitrile Butadiene Styrene [ABS] the material used in 3-D printing is very resilient, elastic, stretchy and lightweight making the extrusion process very simple and easier for making small parts. To print ABS materials the temperature should reach about 210 - 250°C and during this there are hazardous gases produced which cause respiratory problems in humans and hence this material usage is limited. The alternative to it is Poly lactic acid (PLA) material, a biodegradable thermoplastic that is derived from corn. PLA is biocompatible, tougher when solidified and melts at lower temperature compared to ABS material.

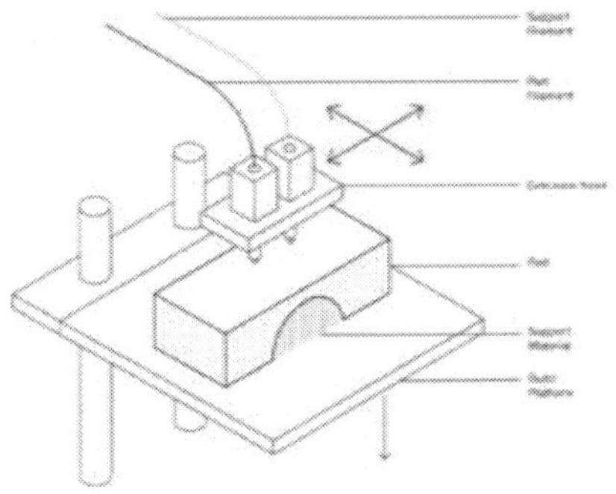

Figure 2. FDM Printer Configuration

4. Experimentation/Design & Modeling

Modelling for part generation are done in CAD software's such as CATIA, UNIGRAPHICS etc or by using 3D scanners. The geometric data collected for 3D modelling in computer graphics is like mould preparation in manufacturing. In 3D modelling, analysis and shape data are collected as in the look of the object. Scanners can also provide data relevant to 3D models. Prior to model printing in 3D .STL file, it is pre-processed by "slicer" software (figure 6) in which the 3D model is sliced into layers of millimetre thickness aided by a G-code file then .STL file sends commands to the printer. The open source softwares available for slicing are Slic3r, KISSlicer, and Cura. A set of G-code commands as shown in figure 4 are given to printer to place layer after layer either for sintering, liquid material deposition or in form of sheets which represent the cross sections of original model. When they are placed accordingly and combining all layered fused is the exact and finished model.

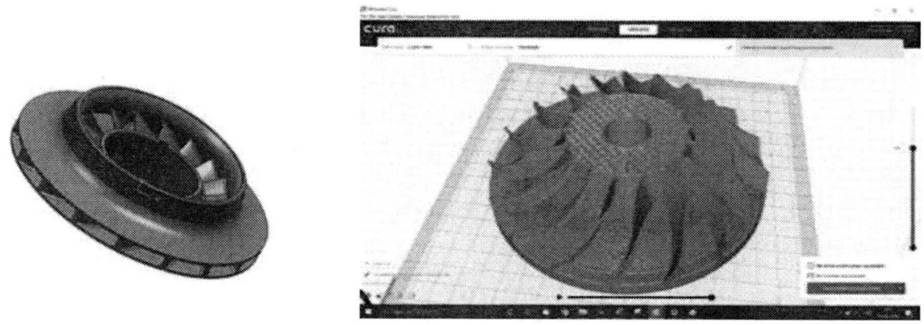

Figure 3. 3D-Models of Open and Shrouded Impeller

Most FDM systems are flexible and allow the adjustment of various parameters, such as temperature of nozzle, build platform, raster speed, the rate of cooling, layer thickness which can differ around 50 to 400 μm [3]. The less thick layers can be used to produce curved, smooth, finer and dimensionally accurate components whereas thicker layers can rapidly print parts which are economical. Generally FDM uses layers upto 200 μm and generally not used for complete solid printing. Alternatively it used concepts of outer and inner filling concepts where in outer area the shell is printed and inner layers are infilled as seen in figure 5 with low density type structures. This reduces long hours of printing and lessen material use. A twenty-five percentage in fill density is set for a shell thickness of 1 millimeter which ensures tough structure and rapid print. The fill percentage is 20% and printing time is about 18 hours.

Figure 4. Sample of G-Code

Figure 5. Open Impeller during printing process

Figure 6. Slicing Software CURA with Parameters

Figure 7. Open and Shrouded impeller after printing

Support structure in Fused Deposition Modelling aids in developing geometries which require overhangs. Structures cannot be built in air and hence in 3D printing also liquid layer can't be deposited without a support structure. Hence supports play a major role in 3D printing as these can be easily removed and object is printed. In impeller printing by FDM each and every layer is deposited using heated filaments which hold and remain onto other layers either below or surrounding it. Every layer is printed with some offset to subsequent layer. This facilitates 45°angle building in models making prints flexible so that it could spread afar from before layers width. As the component is printed having an overhang further than 45°, it hangs down and necessitates support material underneath to clutch up which is used in impellers building. By using the above mentioned FDM based 3D Printer an open Impeller of 150 mm diameter is printed and closed or shrouded impeller of 150 mm diameter and 45 mm as shown in figure 7.

5. Conclusions and Future Scope

Complex 3-D impeller blade profiles are easily printed by FDM technique unlike 5- Axis machining which are expensive and requires large set ups. By using the above mentioned FDM based 3D Printer an open Impeller of 150 mm and closed or shrouded impeller of 150 mm diameter and 45 mm height is printed. This prototype built allows in verifying, validation and test the impeller helping in design improvisation whenever required. This ensures time to market idea projection, reduces expensive investments but with limitations such as unable to produce large components, post processing damage evaluation etc. But overall, the complex and intricate designs of impeller blades of centrifugal compressors are easily modelled by CAD software and printed using Fused Deposition Modelling method.

Present method of 3D printing plastic patterns can be used for making shells for investment casting of actual impeller by choosing suitable plastic material such as PMMA, which can be heated up and removed from the ceramic shells. In future direct metal 3D printing will be the process to be adopted for printing of actual impeller using 17-4PH material powder using powder bet LASER sintering process.

References

[1] Shang Liangchen and Wes-Tsai Wang (2001), "Computer aided manufacturing technologies for centrifugal compressor impellers", Journal Of Materials Processing Technology, Vol. 115, 284-293.

[2] Klaus bodden berg, Manager of Quality and Materials, Compressor Division, On the manufacture of Impellers for Turbo Compressors, , Mannesmann – DEMAG, Duisburg, Germany, 9-23.

[3] Toru kaino, Takasago Machining & Instrumentation Assembly Department, Machinery center machinery Business, Machining Technology for Large Impellers of Centrifugal Compressors, Kobelco technology 2015, Review no.33.

[4] B.Sajjan, A.Santhosh, M. Jaya ram, K. Anusha, Linear and static analysis of impeller type centrifugal compressor with different material, IJESC 2016, Vol. 6 No.11.

[5] A Sham Prasad, BVVV Lakshmipathi Rao, A Babji, Dr. P. Kumar Babu "Static and Dynamic analysis of a Centrifugal Pump Impeller, International Journal of Scientific & Engineering Research, Volume 4, Issue 10, October-2013, ISSN 2229-5518.

[6] Dinesh Tareet, Vaibhav Bhagat & Basavaraj Talikotti, Static and dynamic analysis of impeller of centrifugal blower, IJIRSET 2016 June, volume 5.

[7] Tao Wu, Yan long Cao, Jiangxi Yang, Analysis of the influence of 29 blade's machining error on Aerodynamic performance of impeller based on NUMECA procedia, CIRP 27(2015), pp 155-162

[8] Min-Ho Cho, Dong-Won Kim, Chan-GieLee, Eun-Young Heo CBIMS: Case based impeller machining strategy support system, Robotics and Computer-Integrated Manufacturing, Volume 25, Issue 6, December 2009, Pp 980-988.

[9] J Rech, F. Dumont , A. Le Bot, P.J. Arrazola, Reduction of noise during milling Operation, CIRP Journal of Manufacturing Science and Technology, Volume 18, August 2017, Pages 39-44.

[10] Francis quail, Thomas Scanlon, Mathew Stickland, Development of a Regenerative Pump Impeller using Rapid Manufacturing Techniques, Rapid Prototyping Journal ,Volume 26 Issue 9.

[11] Vinod G. Gokhare, Dr. D.N. Raut and Dr. D.K. Shinde, A review paper on 3 D –Printing aspects and Various Processes used in 3D, International Journal of Engineering Research & Technology (IJERT), ISSN: 2278-0181, Volume 6 Issue 06, June – 2017.

[12] Ana Pavlovic, Milan Sljivic, Milija Kraisnik, Jovica Ilic and Jelica Anic, Polymers in Additive Manufacturing : The case of a Water Pump Impeller January 2017, FME Transactions 45(3), pp 354-359

[13] K Aruna Prabha, Balla Srinivasa Prasad, Machining of Steam Turbine Blade on 5-Axis CNC Machine, Materials Today: Proceedings, volume 8, pp 3001-3007.

Garbage Waste Segregation Using Deep Learning Techniques

Sai Susanth G[1], Jenila Livingston L M[1*], Agnel Livingston L G X [2]

[1]School of Computer Science and Engineering, Vellore Institute of Technology, Chennai, Tamil Nadu, India -600127
[2]Assistant Professor, CSE Department, St. Xavier's Catholic College of Engineering, Chunkankadai, Nagercoil, Kanyakumari District, Tamil Nadu, India -629003

Email: jenila.lm@vit.ac.in

Abstract. Waste segregation is one of the primary challenges to recycling systems in major cities in our country. In India, 62 million tons of garbage is generated annually. Of this 5.6 million tons of wastes consist of plastic materials. About 60 percent of this is recycled every year. In addition, 11.9 million tons are recycled from 43 million tons of solid waste produced. Though the numbers sound good, a serious problem in the recycling industry is the segregation of waste before recycling or any other waste treatment processes. In India, at present situation waste is not segregated when collected from households. So a lot of workforce and effort are needed to separate this waste. In addition to this people working in this industry are prone to various infections caused due to toxic materials present in the waste. So the idea is to decrease the human intervention and make this waste segregation process more productive. The proposed work is aimed to build an image classifier that identifies the object and detects the type of waste material using Convolutional Neural Network. In this work, four different models of the CNN, such as ResNet50, DenseNet169, VGG16, and AlexNet, trained on ImageNet, are used to extract features from images and feed them into a classifier to make predictions and distinguish a type of waste from its corresponding category. The experimental results showed that the performance of DenseNet169 was significantly greater than all four models and the performance of ResNet50 is closer to DenseNet169.
Keywords: Convolutional Neural Network, Deep learning, Waste segregation, Pre-trained models

1. Introduction

Recycling is a very important phenomenon in a healthy and green environment. With raising awareness among the citizens of India regarding the importance of using recycling items to decrease the consumption of natural resources and garbage disposal, the recycling industry is booming. People are willing to use more recycled products and also contribute their part to the society by disposing of their waste to recycle. As a result, there is a need to divert more waste to recycling industries which can only be done after proper segregation [1].

Even the non-recyclable waste is treated with methods like burying, combustion, plasma gasification only after waste segregation which removes objects that may destroy the quality of nearby water bodies or soil or air when burnt or burying. Hand-picking is not possible once the amount of waste increased with increasing population

To counter this problem we can automate the entire process by building an image classifier using a convolutional neural network (CNN) and thereby decrease the time for the waste segregation and make it cost-effective. CNN is an artificial neural network which is very popularly used in analyzing images. CNN relatively minimizes the pre-processing for an image dataset [3]. The key to CNN's success is due to its ability to learn high-level abstractions from raw input image data [5]. There are some automated methods for garbage segregations existing already [7, 12], but they have mainly focused on hardware implementations with less accurate waste classification.

The four pre-trained CNN models used here are ResNet50, DenseNet169, VGG16, and AlexNet. These pre-trained models are previously trained on ImageNet dataset that consists of almost 22,000 object categories for the purpose of computer vision research [4]. The custom CNN architecture requires a large dataset and also takes a lot of time to train. Pre-trained models overcome these issues as its weights are already optimized when trained on the ImageNet dataset. Though the models can't achieve 100% accuracy, it is built to achieve the best possible classifier and to minimize the errors to the maximum possible extent.

The organization of the paper is as follows. Section 2 deals with the Survey of Literature. Dataset collection and modifications are shown in Section 3, and Methodology is presented in Section 4. Findings are presented in Section 5, followed by a conclusion in Section 6.

2. Related Work

Of the existing pre-trained models trained on ImageNet database, AlexNet, which won ImageNet Large Scale Visual Recognition Challenge (ILSVRC) in 2012 is a highly competent architecture and proven to work efficiently on most image datasets. It is commonly used for the segregation of waste but its efficiency is not comparable to other pre-trained models.

Mindy Yang and Gary Thung [6], collected images for the dataset used Machine Learning and neural networks to segregate waste. Using Support Vector Machine (SVM), they classified waste into six different categories and an 11-layer CNN. SVM performed well compared to CNN with 63 percent accuracy, while CNN is only 23 percent accurate.

Yash Desai et al. [7] proposed a method that uses CNN to classify degradable and non-degradable waste. A flap can be used to insulate the grouped waste into different classifications. Afterward, the computerized order helps throughout the time spent on sanitation. The ordered waste can also be arranged and handled by the companies in a concomitant manner.

Micro-controller based automatic waste garbage segregation is proposed by Pushkar Sathe et al. [12], using custom neural networks. This model classifies waste into four classes namely paper, glass, plastic, and metal. There are four dust bins connected to a servo motor. The type of waste is recognized by the system and the respective dust bin is opened. The drawback here is if there any non-recyclable waste exists, it will be classified into the existing four classes.

Zabir et al.[13] introduced three methods, Convolutional Neural Network (CNN), AlexNet and Bag of Features (BoF). They used the MCIndoor20000 dataset, which consists of doors, signage, and stairs images of Marshfield Clinic. BoF generates almost the same precision as CNN, which shows that machine learning is almost as good as deep learning. Experimental results showed that AlexNet achieves higher precision than CNN and BoF.

Hoque et al. [14] implemented a CNN based system in order to classify the type of waste and open the dustbin when the waste is to be poured into the waste bin. Using this method the separated waste at the basic level becomes easier. They have segregated four categories of waste, namely glass, paper, plastic, and metal.

Osiany et al. [15], designed and made an Automatic Garbage Collector (AGATOR), a rotor robot model to counter the efficiently and effectively flow-free garbage accumulation in the river. The full load pushes up to 5 kg at the garbage receptacle. The Robot average speed when collecting the garbage is 0.26 m/s.

Another project by G.Mittal et al. [16], designed a smartphone application which detects a pile of waste in an image. The authors collected the images in Bing Image search and used the AlexNet model and achieved an accuracy of 87.69%.

Sachin Hulyalkar et al. [17] proposed a system that automatically segregates the waste at the source itself, thereby reducing physical effort. The program is based on Machine Learning principles, image recognition and the Internet of Things (IoT). The aim of this project is to capture images of single waste material and to identify and segregate it effectively into four classes, i.e. metal, glass, paper, and plastic

Olugboja Adedeji and Zenghui Wang have implemented SVM and CNN using ResNet architecture on the same dataset. Clearly, ResNet architecture gave better results compared to SVM and also non-pre-trained CNN models used by Gary Thung and Mindy Yang. So it is proved that pre-trained models are performing better than conventional CNN [18].

Pavithra [19] has developed a smart trash system that aims to use IR sensors and gas sensors where the gas sensors detect if there are any poisonous gases in the waste and IR sensor detects the level of waste-filled. RFID is placed at a municipal corporation office interface with PC. If the trash is filled, the information is sent to the corporate office to empty the bin.

Anjali et al. [20], developed a method of detecting solid waste that uses the motion of an Unmanned Aerial Vehicle (UAV). Here drone is used to detect waste using image processing techniques that detect waste contaminated places and send notification of location coordinates. They have interfaced Arduino that collects images and GPS, GSM modules for communication and location tracking. The scope of this project is only a mass waste detection and segregation is out of scope.

Chandaluru Priyanka & Sri Ramya [2] used Machine Learning models like SVM, KNN, Decision Tree and compared them to CNN. CNN performed well over the mentioned Machine Learning algorithms. Their dataset is better compared to other researchers. Previous researchers used less than 1000 images whereas they have used more than 3000 images to train the model.

3. Dataset and Data Collection

The images used in the proposed work are taken from Gary Thung and Mindy Yang's dataset while working on waste segregation [6]. Initially, this dataset has 2527 images. Since the dataset is very small, we added more images to this dataset from Google Images. Later on in this project, some images collected from Google Images are irrelevant to that particular class as mentioned in Table 1. They are contributing to a less accurate model. So these misclassified images are removed. The dataset contains six classes namely cardboard, glass, metal, paper, plastic, trash with about 4,163 total images.

Table 1. Details of Dataset

Class	Number of images in existing dataset	Number of images of modified dataset
Cardboard	403	651
Glass	501	769
Metal	410	819
Paper	594	909
Plastic	482	878
Trash	137	137
Total images	2527	4163

4. Methodology

Classification is a method by which features are extracted from the dataset. This is achieved by splitting the knowledge into various clusters, based on the characteristics. A novel model is built which performs predictions and classifies it by training it on known data.

The proposed system involves three basic modules such as pre-processing, image augmentation, and feature extraction. Image Augmentation is to increase more pictures by re-sizing, zooming, turning pictures, and so on to build new pictures. With this methodology, the model will be ready to catch a larger number of 'features' than previously and will ready to predict images much better. During feature extraction, the system characterizes the unlabeled data with maximum possible exactness.

4.1 Architecture of proposed system

Figure 1 displays pre-trained models that work on a given dataset. The pre-trained models trained on the ImageNet database are used for the extraction and classification of features after processing and data-augmentation.

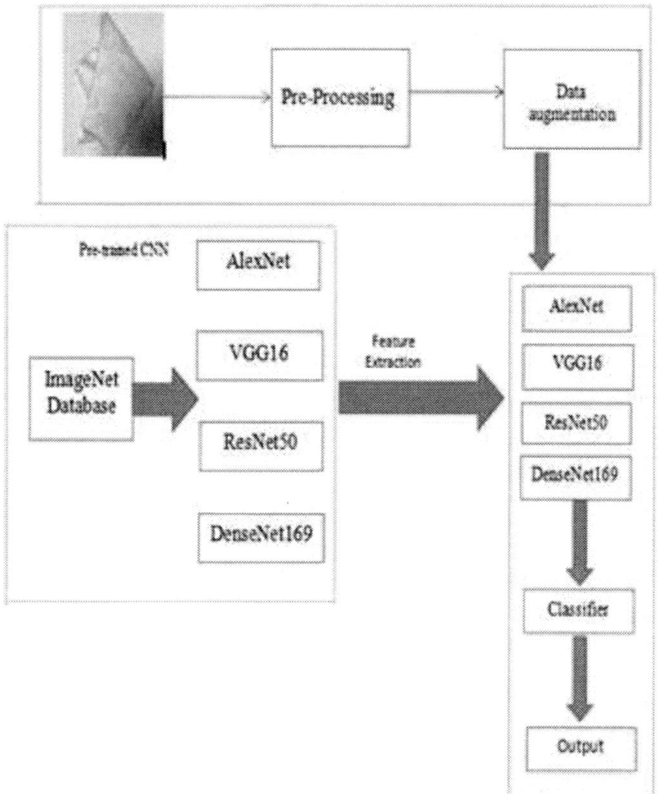

Figure 1. Methodology of proposed work

4.2 Pre-processing and data augmentation

The dataset is relatively small for pre-trained models to handle. Overfitting can be a concern. Therefore some steps must be taken before the training model. Dataset size is doubled by adding images from the Google images. Additionally, certain techniques of augmentation such as Random Re-sized Crop, Random Horizontal Flip, are used.

4.3 Convolutional Neural Network
Convolutional Neural Network (CNN) is prominently used for image analysis. Its hidden layers called convolutional layers to make it more special. Each convolutional layer contains a set of filters. These filters detect patterns or features in the images. A simplest CNN has the following layers:
4.3.1 Convolutional layer: A convolutional layer extracts features of images by using filters. Filters are small matrices of dimensions of our wish filled with random values. These filters detect patterns by striding over the input images after which we get the resultant feature map which is passed to *the* next layer.
4.3.2 Pooling layer: In this layer, a window normally a size of 2x2 is placed over the feature map and the maximum value is selected in the window neglecting all other values. It results in a decreased picture scale.
4.3.3 Fully connected layer: The actual image recognition and classification happens in this sheet. The shrink images are taken and inserted into a single vector. This vector is compared and the image is classified with the vectors obtained from the trained images. Below are the CNN architectures that are used in this research,

AlexNet. It is eight layers deep, the first five layers being convolutional layers and last three layers are fully connected layers. There is a max-pooling layer after each convolutional layer. It uses Rectified Linear Unit (ReLU) activation function. ReLU is preferred over Tanh to add non linearity. It accelerates speed by eight times without accuracy change. It accepts only images of size of 256x256. If the images are of any other size then they are converted to 256x256 and then used [9].

VGG16. VGG networks are developed to increase the depth of existing CNN architectures though ResNet is introduced later.3x3 sized filters are used in convolutional layers with a stride of size one, which reduces the number of parameters in the networks.VGG networks are more difficult to train, compared to AlexNet. Here 16 layered VGG network is used for the segregation of waste [11].

ResNet50. Residual Network (ResNet) architecture is considered as a starting point to transfer learning. There are five stages in ResNet50 architecture. Each stage has a convolutional block and an identity block. Each convolutional block has three convolutional layers. It is introduced to simplify training deeper networks. It is eight times deeper than VGG architecture, but still has lesser complexity [8].

DenseNet169. The speciality of DenseNet architecture is each layer is connected to every other layer in a feed-forward way. Each layer will get feature maps from all other previous layers. Due to this, the features are not lost even after passing through lots of layers. In DenseNet architecture, there is a reduction in parameters due to reuse of functions [10].

4.4 Network Training
The dataset is split into 80% training set and 20% testing set. Google Colab is used for this project as it provides GPU free of cost. Google Colab has 1000s of cores running simultaneously. It provides with 12.72 GB RAM and 358.27 hard disk spaces in one run time. We took advantage of Fastai library to train CNN models. The Learning rate is determined after plotting learning rate vs loss graph (Figure 2). From the graph, suitable learning rate is selected and each CNN model is trained for 50 epochs using data augmentation.

4.5 Problems Addressed
The two key concerns that were found most are:
1. The dataset is too small to train. So there can be a problem of insufficient training.

2. There is a problem of overfitting because of less data.

Image Scraping and removal of misclassified images is performed to combat these problems.

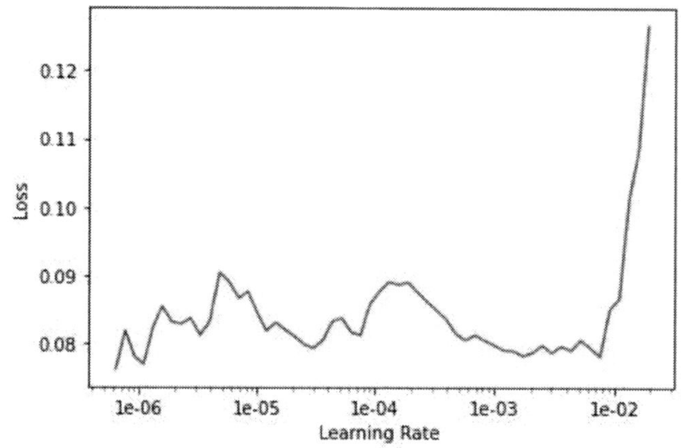

Figure 2. Loss vs Learning rate for DenseNet169

4.6 Image Scraping

The models performed considerably well. Still, there is a scope to increase accuracy. To improve the model's accuracy, image scraping is done to increase the dataset set, and some of the most misclassified images are removed.

Image scraping is nothing but collecting images from the web and preparing a dataset from those images. In previous work, authors have followed this procedure but later, they neglected due to less accuracy. Therefore, using their expertise, we wanted to use the dataset alongside it by scraping photos from Google images, we doubled the number of images.

We downloaded about 500-600 images for each class and inserted 300-350 manually by removing images that aren't connected at all. We need to be more precise when looking to get more images that are linked and match best.

4.7 Removing the most misclassified images

Scraping the images from the web and adding them to the dataset, sometimes give adverse effects to the dataset. So we need to take necessary precautions while doing web scraping. One is the elimination of the most misclassified images. Therefore, the top 20 most misclassified images are removed from the dataset and instead of them 16 images are newly added during web-scraping.

5. Results and Discussion

Once model is built, the next step is to assess the performance of the developed model using some evaluation metrics. The following metrics were used for performance assessment

Accuracy: Accuracy is the most commonly used and best output metric to compare various models. It is a simple ratio of correctly predicted samples to total number of samples.

$$\textbf{Accuracy} = \frac{\textbf{Correctly predicted samples}}{\textbf{Total number of samples}}$$

(Or)

$$\textbf{Accuracy} = \frac{\textbf{TP + TN}}{\textbf{TP + TN + FP + FN}}$$

True Positives (TP): These are correctly predicted positive values which mean the sample which belongs to a particular class is predicted that it belongs to original class. For example, if an object belongs to the paper class, it is correctly predicted by the algorithm / model.

True Negatives (TN): These are correctly predicted negative values. When a sample is not really part of a class, the model also predicts that the sample is not part of that class. For instance, if an object does not belong to the paper class, the predicted class correctly classifies the same thing.

False Positives (FP): When a sample belongs to a negative class and the prediction model incorrectly classified as positive. For example, when an object is not a plastic material but prediction says it is a plastic material.

False Negatives (FN): When a sample belongs to a positive class and the prediction model incorrectly classified as negative.

Table 2. Confusion Matrix

		Predicted Class	
		Class=Yes	Class=No
Actual Class	Class=Yes	True Positive(TP)	False Negative(FN)
	Class=No	False Positive(FP)	True Negative(TN)

True positive and true negatives are correctly predicted samples whereas false positives and false negatives are incorrectly predicted values and the confusion matrix is shown in Table 2. So we need to minimize false positives and false negatives to build a better model.

Precision: It measures how good the model is at assigning positive events to the positive class. Precision is the ratio of true positive values to total predicted positive values.

$$\text{Precision} = \frac{\mathbf{TP}}{\mathbf{TP + FP}}$$

Recall: Recall measures how good the model is in detecting positive events. Therefore, the formula for recall is the same as sensitivity. Recall is the ratio of true positive values to total actual positive values.

$$\text{Recall} = \frac{\mathbf{TP}}{\mathbf{TP + FN}}$$

Initially AlexNet, VGG16, ResNet50, DesnseNet169 accuracy and performance are summarized by Table 3. The dataset is too small and there is a problem of over fitting.

Table 3. Initial test set accuracy

Model	**Recall (%)**	**Precision (%)**	**Accuracy (%)**
AlexNet	79.8	82.7	83.7
VGG16	85.1	88.2	86.9
ResNet50	84.2	89.6	89.7
DesneNet169	92.1	92.1	92.6

After bulking up images by adding images and removing misclassified images, the models performed considerably well, the results are summarized in Table 4.

Table 4. Accuracy after Scraping

Model	Recall (%)	Precision (%)	Accuracy (%)
AlexNet	84.4	88.8	89.3
VGG16	86.6	90.1	91.7
ResNet50	88.2	90.6	93.4
DesneNet169	92.5	94.5	94.9

Though there is no considerable change of accuracy in DenseNet169, it is nearly 6% high in AlexNet, 5% in VGG16 and 4% in ResNet50. Thus the web scraping and elimination of misclassified images performed well to develop models with low accuracy.

Table 5. Evaluation metrics for AlexNet

Material	Precision	Recall
Cardboard	0.98	0.95
Glass	0.82	0.77
Metal	0.80	0.88
Paper	0.90	0.85
Plastic	0.83	0.83
Trash	0.69	0.88

Looking at the precision and recall Table 5, cardboard, paper is predicted well by the model. Trash is not predicted well as the number of images is not sufficient to train the model. Glass, Plastic and Metal's performances are satisfactory. Below graph (Figure 3) shows how accuracy increases with the number of batches processed in AlexNet Model.

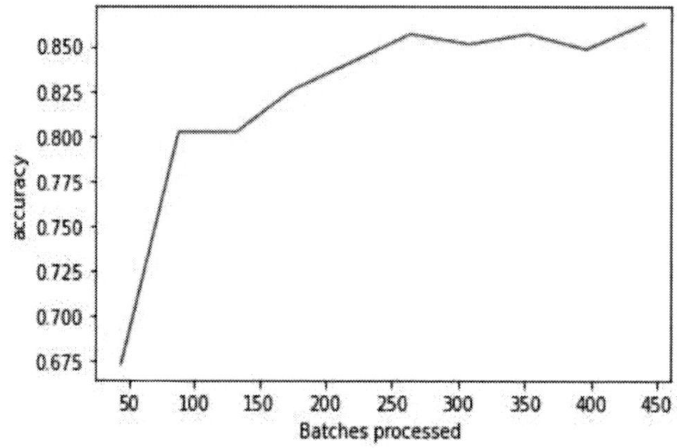

Figure 3. Batches processed vs Accuracy of AlexNet model

A confusion matrix is a table often used to define a classification model's output on a collection of test data for which the true values are known. It enables the output of an algorithm to be visualized.

Confusion matrix of AlexNet model on the **test dataset** is given in Figure 4. It is the least performing model of all four models.

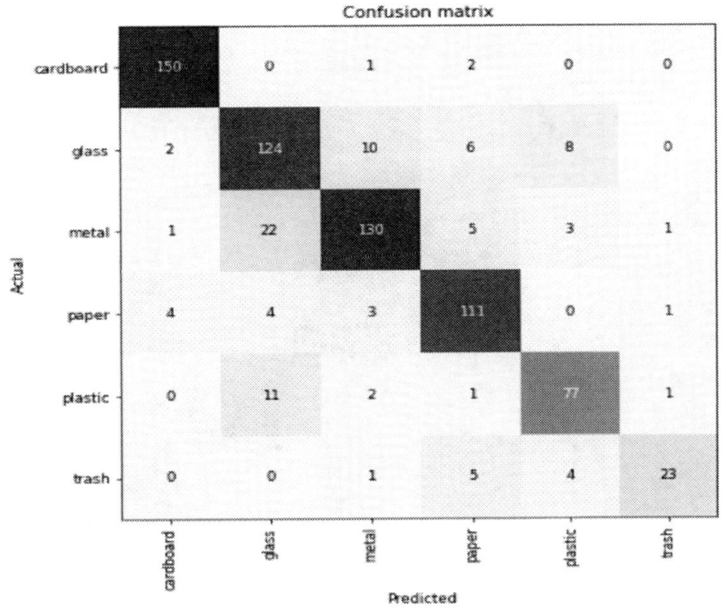

Figure 4. Confusion matrix of AlexNet model

Table 6. Evaluation metrics for VGG16

Material	Precision	Recall
Cardboard	0.98	0.93
Glass	0.91	0.91
Metal	0.91	0.89
Paper	0.94	0.95
Plastic	0.92	0.92
Trash	0.61	0.84

From the above precision and recall Table 6, we can conclude that the model with VGG16 architecture performed well for almost all the class except trash class. Below graph (Figure 5) shows how accuracy increases with the number of batches processed with respect to VCG16.

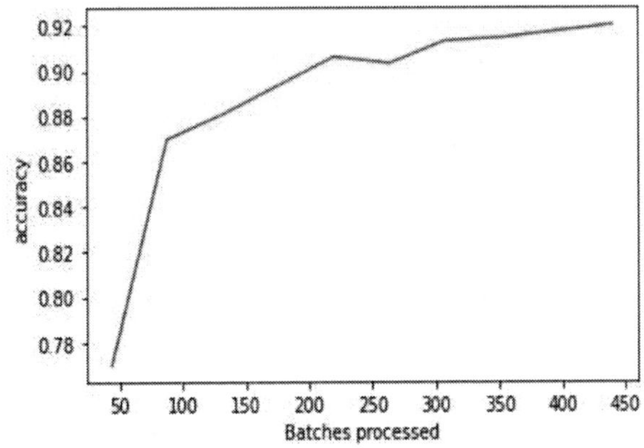

Figure 5. Batches processed vs Accuracy of VGG16 model

Figure 6 displays Confusion matrix of model VGG16 on **test dataset** and it really performed well for cardboard and Paper.

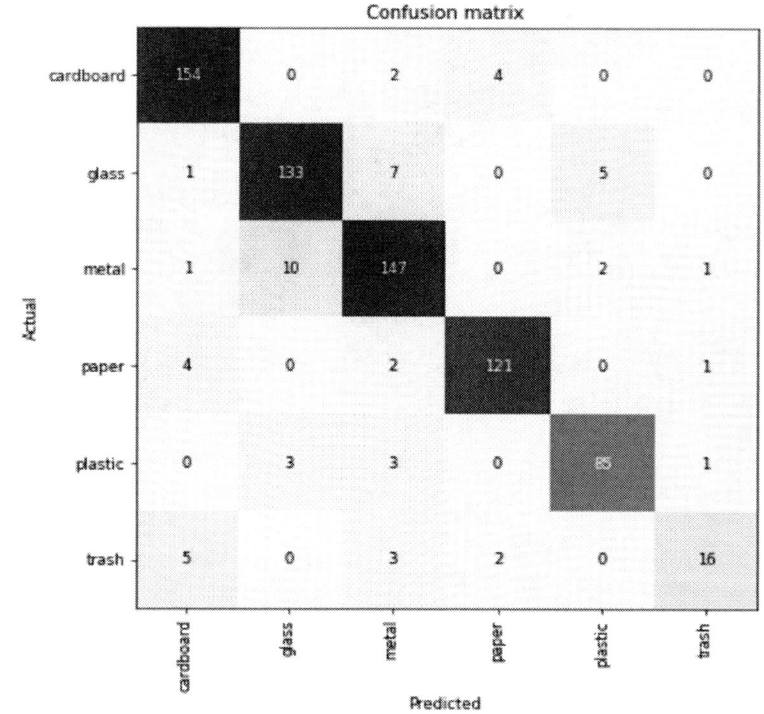

Figure 6. Confusion matrix of VGG16 model

Table 7. Evaluation metrics for ResNet50

Material	Precision	Recall
Cardboard	0.96	0.97
Glass	0.83	0.86
Metal	0.88	0.85
Paper	0.90	0.86
Plastic	0.85	0.86
Trash	0.44	0.57

Looking at the precision and recalling Table 7, we can conclude that ResNet50 model worked well for cardboard and Paper. Its performance on Glass, Metal and Plastic is satisfactory. Like the other models, the model couldn't detect Trash as well. Graph (Figure 7) shows how accuracy increases with the number of batches processed in ResNet50 model.

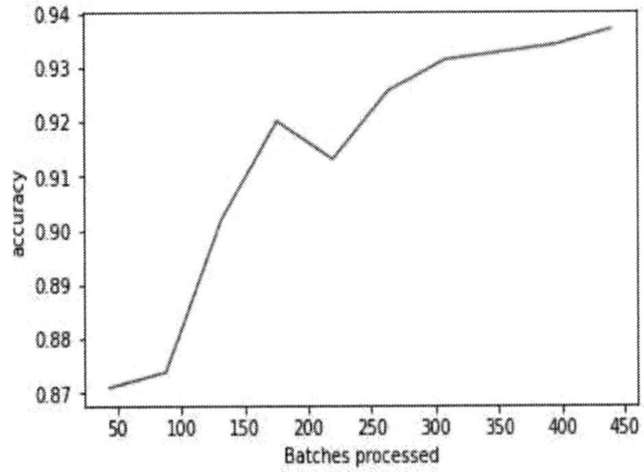

Figure 7. Batches processed vs Accuracy of ResNet50

ResNet50's performance is very nearer to that of DenseNet169. Confusion matrix ResNet50 model on **test dataset** is given in Figure 8.

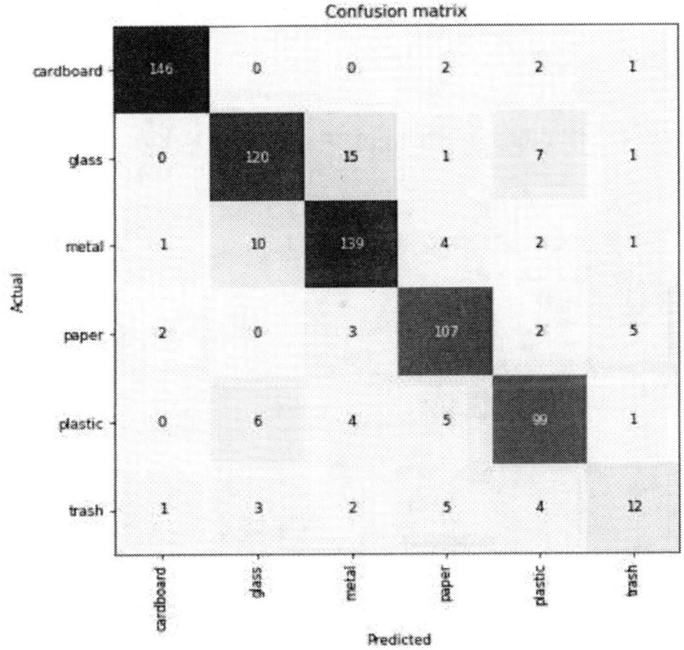

Figure 8. Confusion matrix of ResNet50 model

Table 8. Evaluation metrics for DENSENET169

Material	Precision	Recall
Cardboard	1.00	0.99
Glass	0.93	0.87
Metal	0.95	0.90
Paper	0.93	0.95
Plastic	0.83	0.88
Trash	0.80	0.80

Assessing the above table 8, DenseNet169 predicts all the classes reasonably well. Trash and Plastic are the most mispredicted classes. Graph (Figure 9) shows how accuracy increases with the number of batches processed with respect to DenseNet169 model.

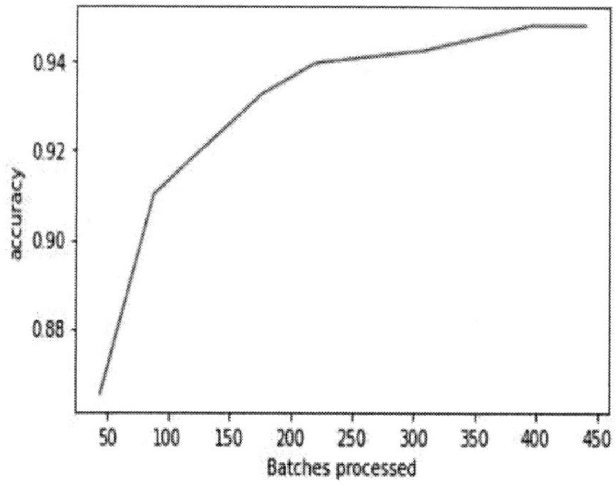

Figure 9. Batches processed vs Accuracy of DenseNet169

We can assess more by understanding confusion matrix of best performed model DenseNet169 using **test dataset**.

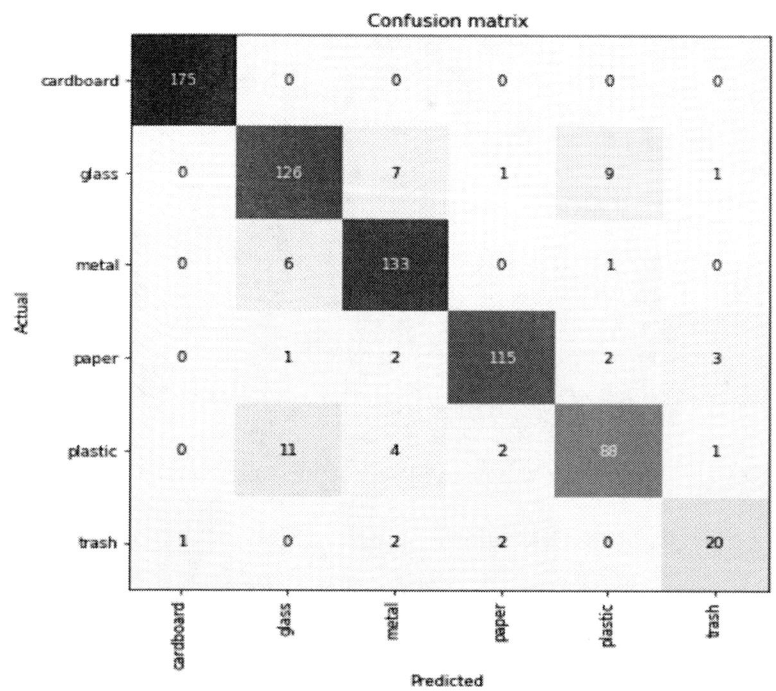

Figure 10. Confusion matrix of DensNet169

It is clear from confusion matrix (Figure 10) that glass is frequently confused. So glass is the area of concern and improvisation to be made in classification of glass.

IOP Conf. Series: Materials Science and Engineering 1012 (2021) 012040 doi:10.1088/1757-899X/1012/1/012040

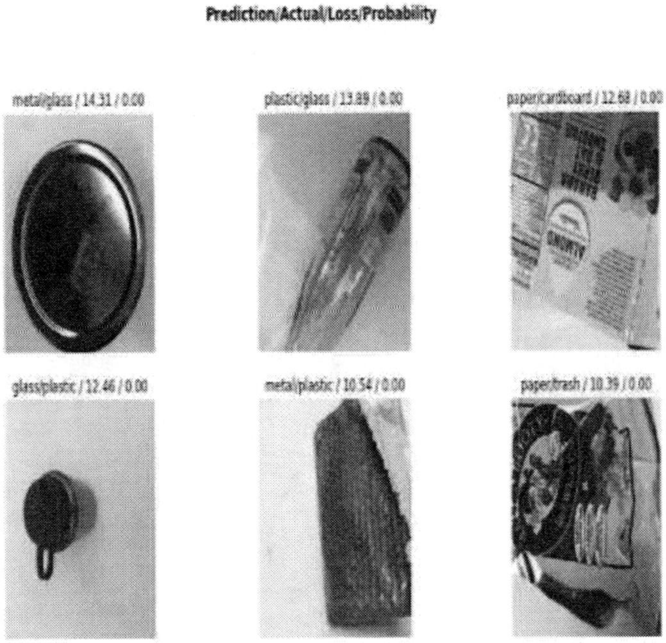

Figure 11. Some of the mis-classified images

Sometimes glass is classified as plastic or metal and in some cases metal or plastic is classified as glass (Figure 11). Also trash is often misunderstood because of insufficient learning.

6. Conclusion

In this research, our goal is to segregate the waste materials using machine learning. We made use of the pre-trained architectures and six waste classification categories to accomplish this goal. Web scraping is done for removal of misclassified images and it has been working well in developing low precision models. Compared with other versions DenseNet169 performed better. The output of ResNet50 is likewise closer to DenseNet169. The most misclassified or misunderstood of the six categories is 'glass.' We need to add more and clear 'glass' images to enhance the model.

References

[1] Miranda, Ruben, Blanco and Angeles. 2010. "Environmental awareness and paper recycling". *Cellulose Chemistry and Technology*.Vol. **44**. pp. 431-449.

[2] Chandaluru Priyanka, Sri Ramya P. 2020. Image Based Classification of Waste Material Using Convolution Neural Network, *International Journal of Advanced Science and Technology* Vol. **29**, 5, pp. 2967 - 2975

[3] Albawi S, Mohammed T A and Al-Zawi S. 2017. "Understanding of a convolutional neural network,", *International Conference on Engineering and Technology (ICET), Antalya*, pp. 1-6

[4] Deng J, Dong W, Socher R, L.-J. Li, Li K, and Fei-Fei L. 2009. "ImageNet: A Large-Scale Hierarchical Image Database", *IEEE Conference on Computer Vision and Pattern Recognition*, pp. 248-255.

[5] Bakator M, Radosav D. 2018. "Deep Learning and Medical Diagnosis: A Review of Literature", *Multimodal Technologies and Interaction*.Vol. **2**, 47, pp. 1-12.

[6] Thung, Gary and Ming Xiang Yang. 2016. "Classification of Trash for Recyclability Status.", *Corpus* Id-27517432.

[7] Yesha Desai, Asmita Dalvi, Pruthviraj Jadhav, Abhilasha Baphna.2018. "Waste Segregation Using Machine Learning", *International Journal for Research in Applied Science and*

Engineering Technology (IJRASET), ISSN : 2321-9653, Vol. **6**, pp. 537-541.

[8] He K, Zhang X, Ren S and Sun J, 2016. "Deep Residual Learning for Image Recognition", *IEEE Conference on Computer Vision and Pattern Recognition (CVPR)*, Las Vegas, NV, pp. 770-778.

[9] Alex Krizhevsky,Ilya Sutskever and Geoffrey E Hinton, 2010. "ImageNet Classification with Deep Convolutional Neural Networks", *ImageNet Large Scale Visual Recognition Challenge.*

[10] Gao Huang, Zhuang Liu, Laurens van der Maaten and Kilian Q Weinberger. 2017. "Densely Connected Convolutional Networks", *IEEE Conference on Computer Vision and Pattern Recognition.* pp. 4700-4708.

[11] Karen Simonyan and Andrew Zisserman. 2015."Very Deep Convolutional Networks for Large-Scale Image Recognition", *International Conference on Learning Representations (ICLR).* pp. 1-14.

[12] Pushkar Sathe, Omkar Tawade, Tanmay Kale, Syed Samar Abbas and Diksha Thakur. 2019. "Waste Segregation using Convolutional Neural Network", *International Journal for Research in Applied Science and Engineering Technology (IJRASET),* Vol. **7**, pp. 932-937.

[13] Zabir M, Fazira N, Ibrahim, Zaidah, Sabri and Nurbaity. 2018. "Evaluation of Pre-Trained Convolutional Neural Network Models for Object Recognition". *International Journal of Engineering and Technology(UAE).*Vol. **7**. pp. 95-98.

[14] Shanjenbam Brojendro Singh, Abu Salah Muslaha Uddin Laskar, Biltu Roy, Aminul Hoque Choudhury, Zahidul Islam, Jakir Hussain Mollah, Shadeed Masood Hoque, Mohsin Ali and Pranav Kalita. 2016. "Design of Municipal Dry Waste Segregating Machine Using Conveyor", *International Journal of Innovative Research in Science, Engineering and Technology (IJIRSET),* Vol. **5**, pp. 7156-7162.

[15] Nurlansa, Osiany, Dewi Anisa Istiqomah and Mahendra Astu Sanggha Pawitra. 2014. "AGATOR (Automatic Garbage Collector) as Automatic Garbage Collector Robot Model". *International Journal of Future Computer and Communication*, Vol. **3**, pp. 367-371.

[16] Mittal Gaurav, Kaushal B Yagnik, Mohit Garg and Narayanan C Krishnan. 2016. "SpotGarbage: smartphone app to detect garbage using deep learning". *Proceedings of the ACM International Joint Conference on Pervasive and Ubiquitous Computing.* pp. 940-945.

[17] Hulyalkar Sachin, Rajas Deshpande, Karan Makode and Siddhant Kajale. 2018. "Implementation of Smartbin using Convolutional Neural Networks". *International Research Journal of Engineering and Technology (IRJET)* ,Vol. **5**, pp. 3352-3358.

[18] Adedeji, Olugboja, and Zenghui Wang. 2019. "Intelligent Waste Classification System Using Deep Learning Convolutional Neural Network." *Procedia Manufacturing*. Elsevier B.V., Vol. **35**. pp. 607–612.

[19] Pavithra. 2014. "Smart Trash System:An Application using ZigBee", *International Journal of Innovative Science,Engineering & Technology*,Vol. **1**, pp. 319-323.

[20] Anjali Pradipbhai Anadkat, Monisha B V, Manasa Puthineedi, Ankit Kumar Patnaik, Shekhar R and Riyaz Syed. 2019. "Drone based Solid Waste Detection using Deep Learning & Image Processing", *Alliance International Conference on Artificial Intelligence and Machine Learning (AICAAM),* pp. 357- 364.

Evolution in E-Commerce with Augmented Reality

Navneet Garg[1], Ankita Pareek[1*], [2]Ajinkya Lale[2], S K Jagannath Charya[1]

[1]University Department, Rajasthan Technical University, KotaRajasthan, India, Pin 324010
[2]Ramrao Adik Institute of Technology, Navi Mumbai, Maharashtra, India Pin 400706

E-mail: ankitapareek26@gmail.com

Abstract- New shopping methods with various attributes compared with traditional internet shopping can improve the administrations of business websites or E-commerce system. This paper is a dense abstract of the broadest parts of elective innovation dependent on AR collaborations that avoids the utilization of physical interaction with the product. The utilization of internet shopping has been developing significantly over some time. Augmented reality (AR) may assume the most significant job in improving the method of web-based shopping. The advancement of new AR innovation can be made valuable if it can imitate the highlights that have made web-based shopping the most favorable shopping source in present times. AR is equipped for working in a gadget utilizing Google's AR Core or Apple's AR Kit SDKs. It is a straightforward technique to create the 3D model before an individual that can be effectively equivalent which results in a better impression of the item by recreating in reality. 3D models are being delivered utilizing a game motor with AR SDKs in the client gadget which is controlled through a lot of projects written in an IDE. The execution of AR shopping is changing internet business by helping the web retailers to decrease the developing expense of profits and furnishing their clients with all the more captivating and advantageous approach to shop through their gadgets. Imaginative AR arrangements permit clients to envision their item continuously, from the comfort of their place. And it may play a major role in the industrial revolution with a specialization area.

Keywords: Augmented Reality, AR kit SDKs, E-commerce, Industrial revolution.

1. Introduction

AR is a helpful perception strategy to overlay computer graphic illustrations in the real world. AR can consolidate the representation strategy to apply to numerous applications. A vision-based AR framework was introduced for representation connection in a device; AR additionally empowers perception of imperceptible ideas or occasions by superimposing virtual items or data onto physical articles or situations.

Despite the positive attitude for Augmented Reality technology and the abundant shopper experimentation that the innovation offers, augmented reality techniques selection and utilization to improve the client involvement with E-commerce is somewhat restricted. Investigation of authoritative appropriation of augmented reality for an internet business will bring out significant variables associations should zero in on while adopting the usage of augmented reality techniques to upgrade the shopping experience of their buyers. The interesting challenge for web-based/online business companies is to give their online clients the "try before you buy" clients with different needs who are progressively inquisitive and requesting visual and material reenactment. Thus, rich media content, for example, high-resolution item pictures, recordings, and 3D designs have been incorporated into sites to upgrade the client experience. Online business organizations are investigating the capability of the abundant mass communication (broadcasting, publishing, and the internet) and especially vivid substance proposed by Augmented Reality to give an improved natural interface and an enhancing experience to clients, subsequently making another worldview in the space of E-Commerce. Rich media content in the online business setting by and large characterizes any type of substance that is an improvement from typical content and static pictures and is intended to draw in the client visually and emotionally. [1]

2. Definition of Augmented Reality

Augmented Reality (AR) exhibits the mix of innovations that empower to join the PC created content with a live video display in real-time.[2] Augmented Reality is the combination of virtual and real worlds[3], enclosing the real world[4] with computer-generated[5] virtual objects in real-time.[6] As per one of the most ordinarily acknowledged definitions, AR is supposed to be an innovative technique having three key prerequisites: a combination of the real and virtual article/objects in a real environment[7], adjusting real and virtual article/objects[8] with each other, and real-time interaction.[9] On account of versatile AR, innovation includes the expansion of advanced components to this present reality through a cell phone camera. [10] AR was perceived as a rising innovative technology of 2007, and with today's smartphone and AR programs, we are starting to get a handle on this astoundingly new and invigorating kind of human-PC collaboration.[11] An AR framework is officially characterized as an application that satisfies the accompanying three properties, to be specific, (a) ready to mix genuine and virtual substance in a genuine climate, (b) is continuous and intuitive, and (c) can enlist virtual substance in a 3D climate. An AR framework is officially characterized as an application that satisfies the accompanying three properties, to be specific, (a) combination of virtual and real substance in a real world, (b) is realistic & intuitive, and (c) can enlist virtual substance in 3D environment. [12]

3. Literature Review & AR with E-commerce

The term Augmented reality was first used for advertising in the automotive industry. A major evolution possible within E-commerce by using the application of augmented reality with it. 77%of customers prefer to use AR capability to the pre-view product, variation such as color, size, style, and difference. AR marketing & advertising is a significant idea that incorporates computerized (digital) data or items into the subject's view of the real world, regularly in union with other media, to uncover, articulate, or show shopper advantages to accomplish hierarchical objectives. [13] The market value for Augmented Reality was 640.4 Million out of 2015 and is required to create $120 Billion in income by 2020. In that capacity, AR is encountering a tremendous prevalence among organizations and customers. [14]

The uses of AR are broad across businesses, for example, producing, correspondence, wellbeing, retail, route, military, instruction, gaming, and internet business. This assessment bases on the utilization of AR in web business considering the tremendous ability of this striking development in providing an in-store shopping experience. The device is able to modulate 3D objects in different spaces, permitting users to collaborate with advanced delivery to their own place with consolation. Organizations like IKEA and converse are utilizing augmented reality to assist clients with imagining household items in their homes progressively utilizing cell phone applications. This expansion in the development of AR applications can be ascribed to the buyer's apparent advantage and positive experience. E-commerce is developing at a fast movement, which is apparent in the measurements expressing that over a billion Web clients bought products through internet business sites in 2013. Truth be told, retail online business deals added up to $1.85 trillion every 2016, and e-retail incomes are extended to develop to $4.50 trillion out of 2021. This emotional change in the advanced retail scene with a quickly developing client base has evoked a solid impulse in online business companies to separate themselves among their rivals and receive inventive strategies to take into account their buyers. With regards to AR innovation in web-based business firms, if the E-commerce firms wish to incorporate AR into their internet business portable application, they have to upgrade their mechanical competency by creating 3D item models utilizing 3D modeling software and programming tool and have a solid innovative team. [1]

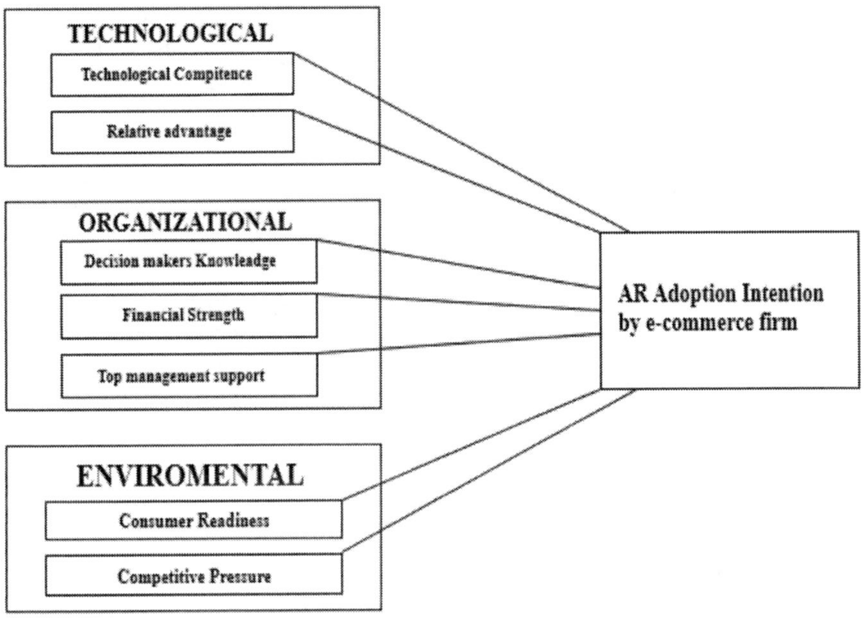

Figure 1: Framework for the adoption of Augmented Reality by E-Commerce Industry[1]

4. Mobile Augmented Reality

As computer increment in force and diminishing in size, a new portable, wearable, and unavoidable figuring applications are quickly getting practical, giving individuals admittance to online assets consistently and everywhere. This new adaptability makes a conceivable new class of uses that abuse the individual's encompassing setting. Augmented Reality as of now presents an especially incredible (UI) to setting mindful figuring conditions. AR frameworks coordinate virtual data into an individual's physical condition so the person will see that data as existing in their environmental factors. Portable AR frameworks offer this assistance without compelling the person's whereabouts to an exceptionally prepared territory. [2] AR in the Smartphone has become an extremely appealing stage for Augmented Reality technology in recent years. As before AR was being utilized in Smartphones, some exploration had been done in the field of portable AR to supplant the first rucksack in addition to head-mounted showcase arrangements. These current tools are known as super portable PCs. This advancement was then preceded by the substitution of UMPCs to PDAs lastly the current day PDA. As indicated by research done by, advanced mobile phones are focusing on an alternate market in AR when contrasted with the underlying long stretches of UMPCs. The smart mobile phones are intended for a huge and wide scope of customers with shockingly powerful and silly-check AR execution. Most PDAs have an inherent camera which permits the full usefulness of AR to be utilized by a cutting edge customer. Notwithstanding, the quality of computer tracking is still exceptionally impacted by its camera and the fundamental picture sensor attributes like casing size, update rate, or focal point mutilation. [15]

5. Tools and Methodology used

The methodology can be implemented by using various tools described below:

Blender-3D: It is an open-source 3D creation suite which is available free of cost and supporting in the integrity of the 3D pipeline—demonstrating (modeling), fixing, animation, simulation, delivering compositing, and movement tracking video altering and game creation. Blender is another platform and runs similarly well on Linux, Windows, and Macintosh PCs. In this assessment, Blender bolsters the 3D work measure, for instance, demonstrating, fixing, liveliness, reenactment, video altering, and game turn of events. This designing, can be done in a window, known as viewport. The designing is convenient because, mainly the Blender application has characteristics that make it simpler for architects to complete the planning cycle, which is quite easy for designers/developers to finish the designing. [16] Blender was picked, that is programming chipping away at the standards of Open Source reasoning, which make it more adaptable, more extensive serviceable, and moderate. It utilizes GNU General Public Licensing for totally free software arrangements and it doesn't rely upon the motivation behind using. Blender, based on the programming language Python, is general programming apparatus for a depiction of 3D PC illustrations with a combination of highlights for animations, acknowledgment of movement frameworks in understanding material science, and so forth for program creation, it utilizes visual scripting. [17]

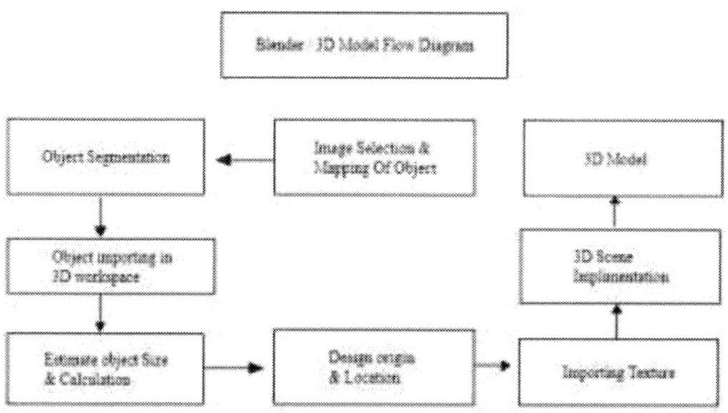

Figure 2: Blenders 3D working Framework

Unity-3D: It is a cross-platform to incorporated 3D game-engine created by Unity technology Co.Lt. It can superpose the virtual onto reality and acknowledges human-computerized collaboration with some Augmented Reality tools. It permits Vuforia SDK augmentation modules to distinguish and follow under the relating ports and makes AR applications and games. It gives plentiful advancement box systems to make games and other intuitive 3D content. Solidarity 3D can add daylight, mist, wind, skybox, water, and other physical materials, surrounding sound, and enlivened video to the virtual scene. Then, you can peruse, test, and alter 3D application situations. Additionally, it is accessible to delivery to the necessary stages, for example, Windows, iOS, Android, etc. [18] Unity is an illustration and material science motor which is utilized to fabricate scale-capable applications that can be worked for numerous stages with the equivalent codebase. Solidarity likewise permits the client to choose a designed API of their decision. [19]

Vuforia AR SDK: It is a complete, versatile enterprise Augmented Reality platform. Its broad-extent solution is suitable for ensuring that it can provide correct augmented reality techniques to each & every client reliant on their business requirements. Vuforia is an SDK that makes AR applications simpler by dealing with the identification and following picture targets or fiducial markers. Vuforia is an SDK that gives the location and following picture focus by utilizing highlight recognition. An element is any point in a picture that is on the edge of various hued areas. [19] Vuforia with AR techniques having various enterprising applications—where to start relies upon understanding where you'll get the fastest, most meaningful ROI. These utilization cases stand apart as offering simple reception, clear returns, unmistakable advantages, and a guide to scale up. Vuforia endows the quickest, easiest, and most exceptional AR content development resolutions, helpful in industrial enterprise customers address work-force opportunities & challenges, and meet business objectives.

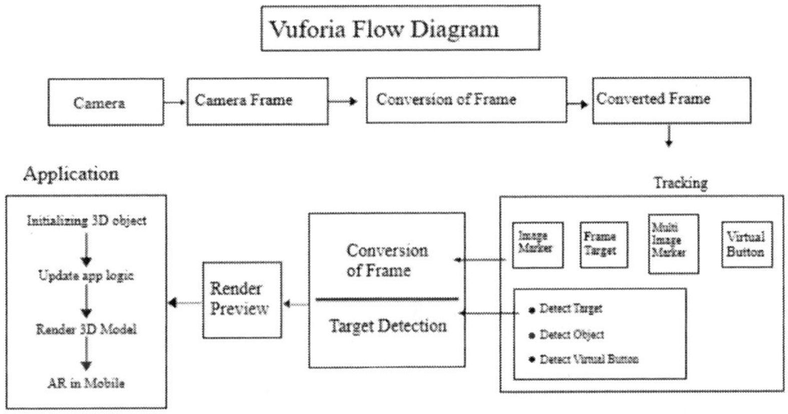

Figure 3: Vuforia Working Framework

Visual Studio Code: This is a cross-stage code supervisor for composing pioneer cloud & web applications that will work on Linux, OS X, and Windows. It offers developer in-built support for various dialects and as Microsoft noted in the present Build featured discussion, the composer will include rich code help and route for these dialects. JavaScript, Node.js, and ASP.NET 5 designers will likewise to get a lot of extra instruments.

6. Working Procedure

The formation of a 3D model takes place in 3D model software Blender, which is available free of cost. Coloring and texture also take place in Blender3D, which can be further exported in the desired extension or file formatting. The desired 3D model can be exported in the desired file format like.obj. fbx. blend. Now the exported assets can be managed in-game engine unity 3D. Unity 3D usage combination of Vuforia plus visual studio plus game assets to perform the augmented reality task. The composting process results in Augmented reality product and experience:

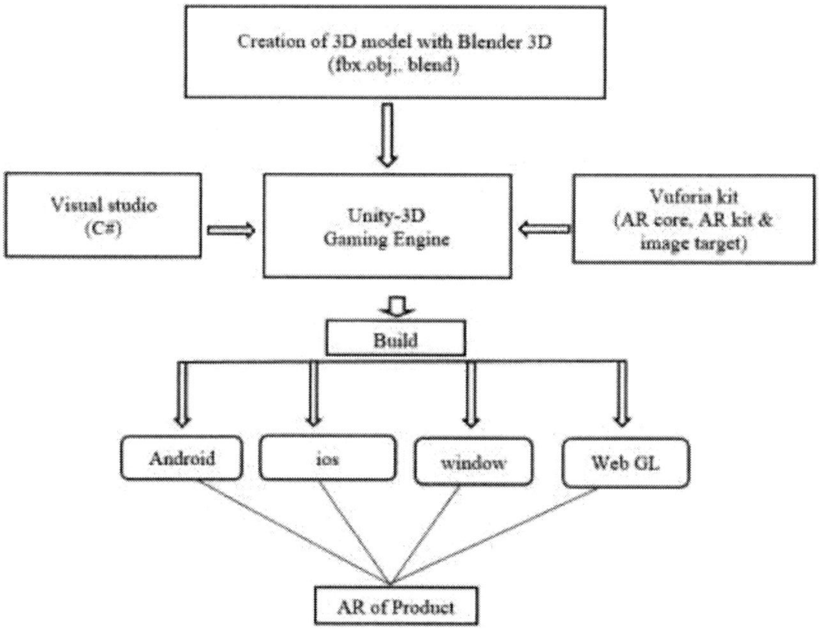

Figure 4: Methodology Framework

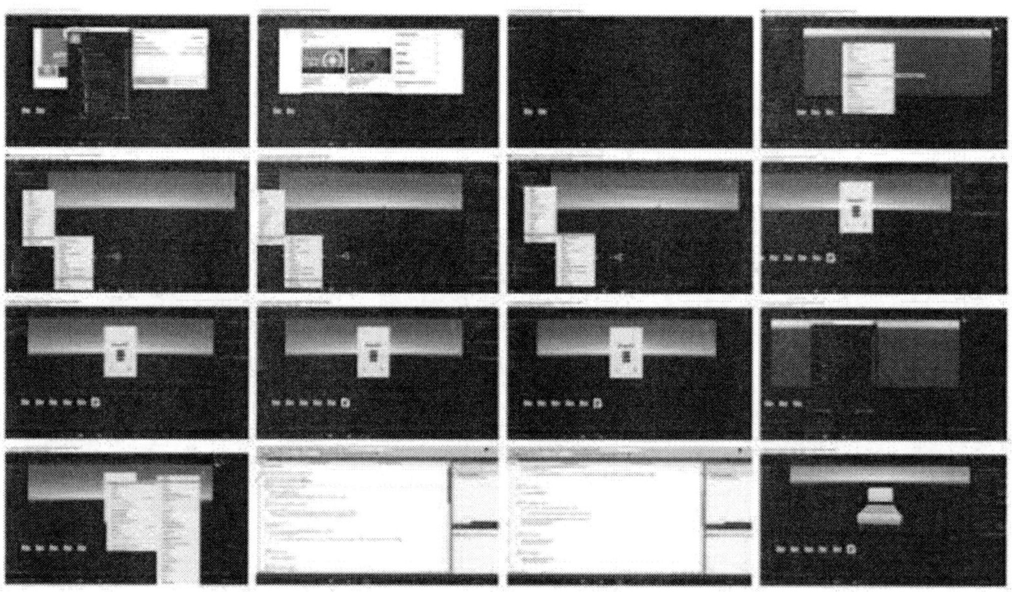

Figure 5: Implementation of 3D modeling of Products

7. Relative advantage & overhead cost Analysis

Relative bit of space is how much advancement is viewed more helpful than rivalry. The apparent advantages of AR will give an impetus to innovation appropriation. Past examination shows a positive connection between the apparent bit of leeway of utilizing the innovation over rivalry (relative preferred position) and appropriation of IS developments. AR reality offers a few advantages to clients and improves the experience of shopping. AR spares time taken to peruse the items, giving them a shot whenever expected, and makes shopping helpful for E-customers. Data innovation is known to be a force device to increase the feasible upper hand. In the present exceptionally. [20]

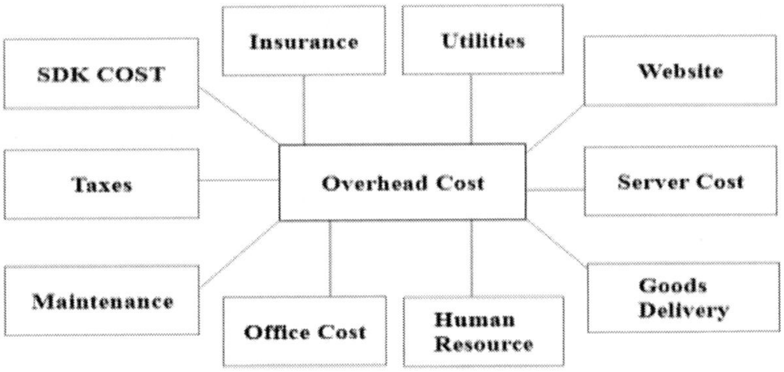

Figure 6: Overhead Cost Analysis

8. Result

The execution of the item/ Product with Augmented reality in e-commerce permits a client to review items or experience administrations by the reasonable appearance of the item in their daily life before purchasing. Implementation of this paper can assemble an innovative promoting effort to catch the client s consideration and impact, attitude, and behavior. This can assemble attention to mark credits and offer clients the opportunity to essentially encounter the advantage of those characteristics for themselves. Clients cannot just shop for an item/product online through genuine interfaces but also attempt the item. Retailers can conquer physical restrictions and offer admittance to each item includes, eventually bringing more clients into the selling funnel for higher transformation.

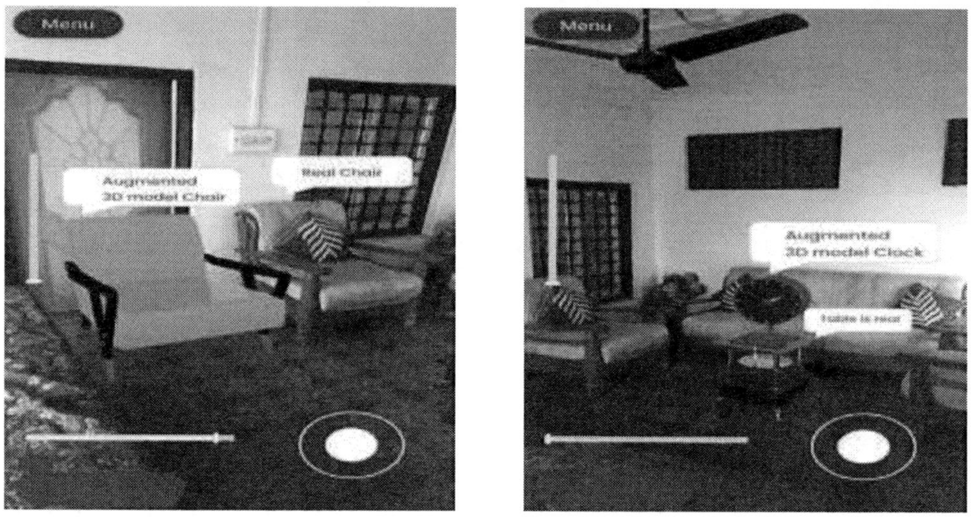

(a) (b)

Figure 7: (a) & (b) Representation of 3D object in Augmented Reality

9. Conclusion & Future scope

By utilizing augmented reality precisely, clients can be propelled to select the correct choice for buying items/products. This is advantageous to the retailer to persuade their intended interest group besides; clients will have the option to get extensive data like surveys, and related items. All the more critically, AR specifically, can provide clients to in-store shopping experience, paying little heed to their zone service can superimpose 3D objects in various spaces, permitting customers to interfacing with advanced delivery to their own place with consolation. In this domain of the present scenario of more products, shorter runs immensely accelerate the variation in products, and enhance the business rivalry. Data generated from the application of AR methods in marketing can give useful feedback even to define top-down manufacturing policy as heuristic about future customer needs.[21] Resilience in the application of above mentioned AR-based marketing method lies in its ability to impart an intelligent fast and effective decision-making thought process in the mind of application user who may not be even technically sound about the usage of the product and its relations with other elements in its vicinity after being procured. In the future, a dynamic framework consisting of various stakeholders such as a user with an E-commerce interface who may be or even can be a potential customer, various professionals, and concerned entrepreneurs related with various domains ranging from logistic and supply chain to manufacturing, R&D, customer support can be proposed and can be simulated to hypothetical conditions and responses given by them can be recorded and studied for various desired objectives.

Appendix:

Coding used in visual Studio:

```
using System.Collections;
using System.Collections.Generic;
using UnityEngine;

public class ObjectScaler: MonoBehaviour
{
    public void Scaler(float value)
    {
        float z = value * 3;
        if (z < .5) { z = .5f; }
        this.gameObject.transform.localScale = new Vector3(z, z, z);
    }
    public void Rotator(float value)
    {
        value -= 0.5f;
        float z = this.gameObject.transform.localRotation.y + (value * 360);
        this.gameObject.transform.localRotation = Quaternion.Euler(0,z,0);
    }

    void Update()
    {
        Joystick joystick = FindObjectOfType<Joystick>();
        this.transform.localPosition += new Vector3(joystick.Horizontal/400, 0,
joystick.Vertical/400);
    }

    public void Exit()
    {
        Application.Quit();
    }
}
```

References

[1] S. Chandra and K. N. Kumar, "Exploring factors influencing organizational adoption of augmented reality in e-commerce: Empirical analysis using technology-organization-environment model," *J. Electron. Commer. Res.*, vol. 19, no. 3, pp. 237–265, 2018.

[2] M. Mekni, "Augmented Reality : Applications , Challenges and Future Trends," pp. 205–214.

[3] J. Garzón, J. Pavón, and S. Baldiris, "Augmented reality applications for education: Five directions for future research," *Lect. Notes Comput. Sci. (including Subser. Lect. Notes Artif. Intell. Lect. Notes Bioinformatics)*, vol. 10324 LNCS, no. October, pp. 402–414, 2017, doi: 10.1007/978-3-319-60922-5_31.

[4] A. Estapa and L. Nadolny, "The Effect of an Augmented Reality Enhanced Mathematics Lesson on Student Achievement and Motivation," *J. STEM Educ.*, vol. 16, no. 3, pp. 40–49, 2015.

[5] D. Nandigam, S. S. Tirumala, and N. Baghaei, "Personalized learning: Current status and potential," *IC3e 2014 - 2014 IEEE Conf. e-Learning, e-Management e-Services*, no. December, pp. 111–116, 2015, doi: 10.1109/IC3e.2014.7081251.

[6] M. Akçayır and G. Akçayır, "Advantages and challenges associated with augmented reality for education: A systematic review of the literature," *Educ. Res. Rev.*, vol. 20, pp. 1–11, 2017, doi: 10.1016/j.edurev.2016.11.002.

[7] E. Solak and R. Cakir, "Istraživanje uloge tehnologije proširene stvarnosti u nastavi jezika," *Croat. J. Educ.*, vol. 18, no. 4, pp. 1067–1085, 2016, doi: 10.15516/cje.v18i4.1729.

[8] H. Y. Wang, H. B. L. Duh, N. Li, T. J. Lin, and C. C. Tsai, "An investigation of university students' collaborative inquiry learning behaviors in an augmented reality simulation and a traditional simulation," *J. Sci. Educ. Technol.*, vol. 23, no. 5, pp. 682–691, 2014, doi: 10.1007/s10956-014-9494-8.

[9] X. Zhou, L. Tang, D. Lin, and W. Han, "Virtual & augmented reality for biological microscope in experiment education," *Virtual Real. Intell. Hardw.*, vol. 2, no. 4, pp. 316–329, 2020, doi: 10.1016/j.vrih.2020.07.004.

[10] T. Khan, K. Johnston, and J. Ophoff, "The Impact of an Augmented Reality Application on Learning Motivation of Students," *Adv. Human-Computer Interact.*, vol. 2019, 2019, doi: 10.1155/2019/7208494.

[11] D. W. F. Van Krevelen and R. Poelman, "A Survey of Augmented Reality Technologies, Applications and Limitations," *Int. J. Virtual Real.*, vol. 9, no. 2, pp. 1–20, 2010, doi: 10.20870/ijvr.2010.9.2.2767.

[12] M. Billinghurst, A. Clark, and G. Lee, "A survey of augmented reality," *Found. Trends Human-Computer Interact.*, vol. 8, no. 2–3, pp. 73–272, 2014, doi: 10.1561/1100000049.

[13] P. A. Rauschnabel, R. Felix, and C. Hinsch, "Augmented reality marketing: How mobile AR-apps can improve brands through inspiration," *J. Retail. Consum. Serv.*, vol. 49, no. March, pp. 43–53, 2019, doi: 10.1016/j.jretconser.2019.03.004.

[14] M. Y. C. Yim, S. C. Chu, and P. L. Sauer, "Is Augmented Reality Technology an Effective Tool for E-commerce? An Interactivity and Vividness Perspective," *J. Interact. Mark.*, vol. 39, pp. 89–103, 2017, doi: 10.1016/j.intmar.2017.04.001.

[15] L. W. Shang, M. H. Zakaria, and I. Ahmad, "Mobile phone augmented reality postcard," *J. Telecommun. Electron. Comput. Eng.*, vol. 8, no. 2, pp. 135–139, 2016.

[16] D. R. Anamisa, M. Yusuf, F. A. Mufarroha, and N. Rohmah, "Design of Virtual Reality Application for Taharah Using 3D Blender," *J. Phys. Conf. Ser.*, vol. 1569, no. 2, 2020, doi: 10.1088/1742-6596/1569/2/022071.

[17] J. N. Marcinčin, J. Barna, M. Janák, L. N. Marcinčinová, and V. Fečová, "Utilization of Open Source tools in assembling process with application of elements of augmented reality," *Proc. VRCAI 2011 ACM SIGGRAPH Conf. Virtual-Reality Contin. its Appl. to Ind.*, pp. 427–430, 2011, doi: 10.1145/2087756.2087836.

[18] X. Liu, Y.-H. Sohn, and D.-W. Park, "Application development with vuforia and unity 3D," *Int. J. Appl. Eng. Res.*, vol. 13, no. 21, p. 43, 2018.

[19] S. S. Deshmukh, C. M. Joshi, R. S. Patel, and Y. B. Gurav, "3D Object Tracking And Manipulation In Augmented Reality .," *Int. Res. J. Eng. Technol.*, vol. 5, no. 1, pp. 287–289, 2018.

[20] ; Supreeth and S. Manava, "FACTORS INFLUENCING ADOPTION OF AUGMENTED REALITY TECHNOLOGY FOR E-COMMERCE Recommended Citation," *Factors Influ. Adopt. Augment. Real. Technol. E-Commerce*, vol. 342, 2016.

[21] W. Skinner, "Manufacturing: Missing Link in Corporate Strategy," Harvard Business Review, Vol. 47, No. 3, pp. 136-145 May 1969.

Vehicle Accident Emergency Alert System

B Sumathy[1*], L Sundari[2], S Janani Priyadharshini[2], G Jayavarshini[2]

[1]Department of Instrumentation and Control Engineering, Sri Sairam Engineering College, Chennai – 600044.
[2]Department of Electronics and Communication Engineering, Sri Sairam Engineering College, Chennai – 600044.

E-mail: bsumathy.ice@sairam.edu.in

Abstract. Today's world is moving in such a speed that no person can spare a minute to stop and look what is happening. The rapid increase in population has inevitably increased the number of vehicles on road and more road accidents with life-loss. The report from World Health Organization (WHO) revealed that approximately 1.35 million people are losing their life each year, because of road accidents. The major reason for accidents is due to drunken driving, distracted driving, over speeding etc. Also, nearly 55% of the life-loss in the highways occurs due to the negligence of accident report of the passer-by vehicles especially during the night or in deserted places remain unnoticed till dawn. A person surviving from the accident, tends to lose their life due to lack of immediate medical response. A Vehicle Accident Emergency Alert system is proposed for immediate attention which could save their life. As soon an accident occurs the vibration sensor or the accelerometer present in the system, transmits the signals to Arduino controller. Latitudes and longitudes data are collected from the GPS system, passed using the GSM module to the emergency centre, and sends a text message to all the people listed in the emergency List. Getting the exact location would help the ambulance to reach the spot with shortest route and time. The proposed alert system could be implemented with less cost and incorporated in all vehicles in near future so that the rate of life-loss could be minimized.

Keywords: Emergency alert, accelerometer, GSM module, Arduino, Embedded C

1. Introduction

Accidents are quite common on Indian roads. The major reasons behind these accidents are poor roads, rash driving, drunken drive, distracted driving, and lack of implementation of traffic rules. As the population in cities increases and urbanization expands, the number of people using automobile are increasing at a rapid rate, this is directly connected to the increasing incidence of road accidents [1]. In 2008, it was found that road accidents were the fourth major reason causing deaths in the world. About 1-1.3 million people die in road accidents and about 20 to million people suffer from injuries that are not fatal, with many people sustaining life with abnormalities which is consequence of accident. Road accidents are the major causes of death among youth aged 15-29 years and there is 1-3% drop in gross domestic products (GDP) [6].During the years from 1970 to 2011, the road accidents increased about 4.4 times. Among which the death toll has increased by 9.8 times and injury cases has increased by 7.3 times , even after the backdrop of greater than 100 times increase in the number of registered automobile and to a four-fold increase in the road way.

India has second largest road network in the world which is about 3.3 million km. Since 65% of freight and 87% of passengers traffic as India is supported by the road network and may grow over 14% every year, supporting the automotive telematics. About two-thirds of road traffic injury (RTI) deaths are reported in the age-group 15-44 years. In 2017, [14] officially reported road accidents were 464,910, claiming 147,913 deaths and 470,975 injured persons that is 405 life losses and 1,290 injuries per day from 1,274 road accidents. If no action is taken, death caused due to road crashes is around 19 million annually by 2020. A survey taken by Alcohol and Drug Information Centre (AIDC)India, revealed that around 40% of the road accidents is caused by drunk and drive. Accidents are a major sight in big cities, about 52% of accidents occur in the highways which are about 5% of the total road network [13]. Major causes of accidents on highways are due to high vehicle speeds or higher volume of traffic on the roads. The fatality can be reduced to a greater extent by giving immediate medical assistance to the victims in the accident spot. This can be done by the automatic alert system.

The main objective of this is to reduce the accidents by sending a message to the registered mobile using wireless communication techniques. GPS system helps in finding the location of the accident spot. Once the intensity of accident is known, the system will check for the nearest health centre and notify them about the accident. The medical team can rush to the spot immediately without any delay as the location is shared. This system is considered useful as it immediately intimates to the person related to the victim, the nearest health centre, police station and ambulance. As soon as accident occurs, the vibration sensor or the accelerator present in the system transmits signals to the Arduino. The latitudes and longitudes are collected from the GPS system, passed using the GSM module to the emergency centre [4] [7]. This sends the text message to all the people listed in the emergency list. The microcontroller AT89S52 is code in embedded C using Arduino support. A person surviving an accident may lose his/her life due to vulnerability of immediate primary health care; this can be avoided by this system.

1.1 APPS Developed to Prevent Accident [15]:

[1] Kruzr: The purpose of this app is to make emergency calls by declining other calls and initializing an alert message.

[2] Safely Home: This app is used for location tracking, monitor speed and to detect road crashes to help the victims.

[3] Party hard drivers: This prevents drunk and drive.

[4] One touch response: This app is used in an accident spot, to dispatch their own team and connect with the police and emergency services.

[5] Ham Safar road safety: The app collects the details of hospitals, police stations and other emergency services and sends out precise map locations.

2. Materials and Methods

Few citizens are misusing the technology development which causes more adverse impact in the society. For example, using the mobile phones while driving, rash driving, drunk and drinking etc. have increased. Most of the accident happens because of the carelessness of people. Major cause is either drunken drive or due to over speed. To detect the accidents in highways, isolated areas and to give quick information, the Accident alert system is introduced. The sensors are placed in the vehicles to detect the abnormal changes in vehicles [1]. Sensor outputs are collected, processed, and transferred as control signal to different devices like to generate the alert message through tracking system by GSM module. The controller used is the Arduino controller and vibration sensor i used for accident detection at the spot area. In [2], the authors proposed a system that helps to alert the nearest medical centre, where accident happened and status of the victim, to enable for give proper first aid/

IOP Conf. Series: Materials Science and Engineering 1012 (2021) 012042 doi:10.1088/1757-899X/1012/1/012042

emergency treatments. The system is implemented by using accelerometer to get details about accident nature and the heartbeat sensor output that gives the condition of the victim. If the heartbeat is abnormal the sensor immediately sends the message containing the pulse rate details to the victim's mobile. An App developed, will alert the nearest hospital and people of close contact, and shares the exact location for further action. Also, they claimed that it is a low-cost device.

The proposed system [3] is designed for an autonomous detection, effective report system and gives emergency help for the road accident victims. Also, the system uses a low-cost SMS/message alerting system to safeguard the life of victim for immediate medical assistance. The components used are heartbeat sensor, accelerometer for motion detection and Raspberry Pi board as controller. The system [4], aimed to give emergency alert and services to automotive systems. The unit comprises of GPS system, GSM Module, CAN transceiver (Controller Area Network), Alcohol sensor. In case of emergency, the people of close contact would get the alert call for the occurrence of accident and condition of victims. The CAN transceiver is a device which can stop the vehicles through the Bus pins. Also, Arduino uses the GSM modem and checks whether the vehicle has crossed the geo fence, and the system is a self-programming and a low voltage detector. The work [5], explored the Automatic Doppler Shift Adaptation (ADSA) and refined the same for Vehicle ad-hoc networks. The authors claimed that the system proposed is robust with minimum time. Thus, the authors achieved a standard for vehicular communication. In [6], the authors devised a system whcih detects the nature of the driver, namely drunken by using an alcohol sensor. Accident occurrence is monitored by the accelerometer and depends on the signal received, air bag opens to avoid head injury and controlling part done by the AVR controller.

In paper [7], the authors proposed a system to determine the nature of the accident occurred and decision can be taken upon it. For this vibration sensor are integrated with ARM processor to send the alert information through GSM module. Victor in [8], developed a system called as "Vehicle Accident Alert and Locator", shortly called as VAAL is helpful in emergency rescue during accidents of auto crash. The system is installed with machine to machine (M2M) device namely GSM modem, mobile operator along with emergency operating devices to find the exact location. Supriya and Ashutosh [9] has discussed about the Automated Accident Intimation and Rescue System (AAIRS) which uses the vibration sensor and GSM modem to rescue the victim. When the sensor senses the vibration, it checks whether the sensor value is above the threshold or below it and sends the alert message accordingly. In [10], the authors explained a Geographical Information System-Transport system instead of traditional Geographic Information Systems. Also, they included the traffic models based on the time to time variation and not on the fixed slots. They made a study with environmental compatibilities to enable to avoid weather related accidents also. In [11], the authors achieved the Smart Summoning of Ambulance during a vehicle Accident, by the integrated unit which comprises of Arduino, communication module along with crash sensor. The same is combined in a single box structure to increase the reliability and exact location is send to the people of close contact and emergency nodes.

3. Proposed Methodology

The block diagram of Vehicle Accident Emergency Alert System proposed is shown in **Figure 1** and the corresponding circuit diagram is shown in **Figure 2.** The block diagram gives the overall working of the system. The system consists of an Arduino UNO where the commands to be passed are coded using embedded C using Arduino IDE. The Arduino is connected to an LCD in which each process undergoing is displayed on it. The accelerometer ADXL335 MEMS gives the XYZ coordinates of the vehicle and is converted to digital using ADC via SPI protocol and it is sent to the Arduino. GPRS system is used to enable the internet connection and the latitudes and longitudes are obtained from the nearest mobile tower. This data obtained is sent to the members listed while registering in the Arduino. The Arduino is also connected to the driver's IC which is used to handle the current through the engine and the alarm. When the car meets with an accident, the Arduino processes the commands and the GPRS system immediately obtains the latitude and the longitude from the cell-phone towers as well as the GPS and sends the text message instantaneously. LCD is used to view the coordinates of

the vehicle and when the vehicle crosses the threshold value of the co-ordinates the entire process is activat
ed.

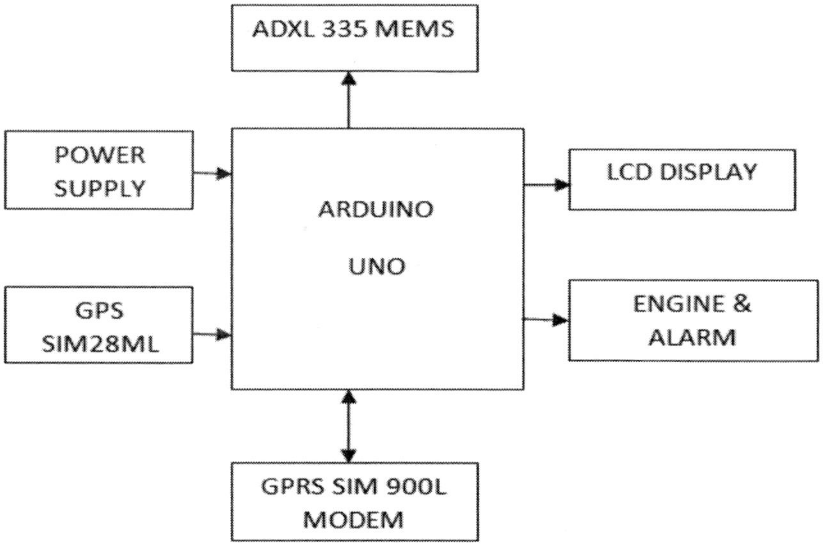

Figure 1. Block Diagram of the Accident Emergency Alert System.

The circuit diagram is shown below.

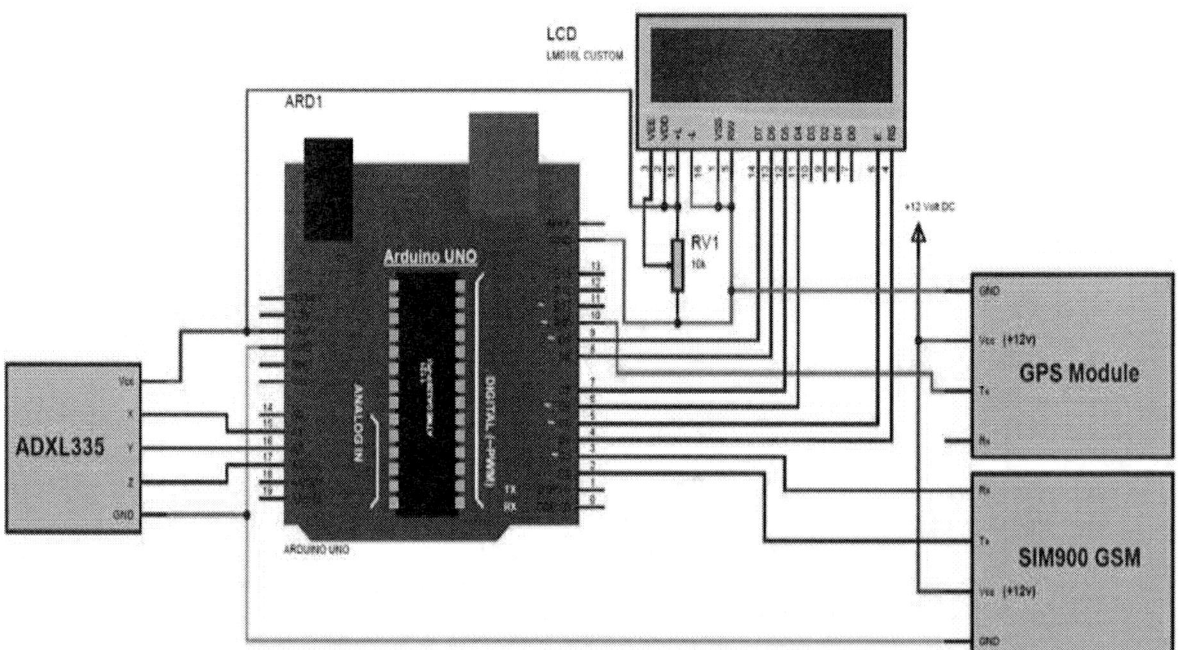

Figure 2. Circuit Diagram of the Proposed System

The ADXL 335 MEMS is a triple co-ordinate accelerometer. The accelerometer obtains the values of the plane in which the car is travelling and feeds it to the MCP3208 12-bit Analog-to-Digital Converter (ADC) where the signals obtained are converted digital format (**Figure 3**). If the plane values obtained crosses the value beyond which the safety of the car is on the danger line (i.e. threshold value), and the emergency protocol is summoned immediately.

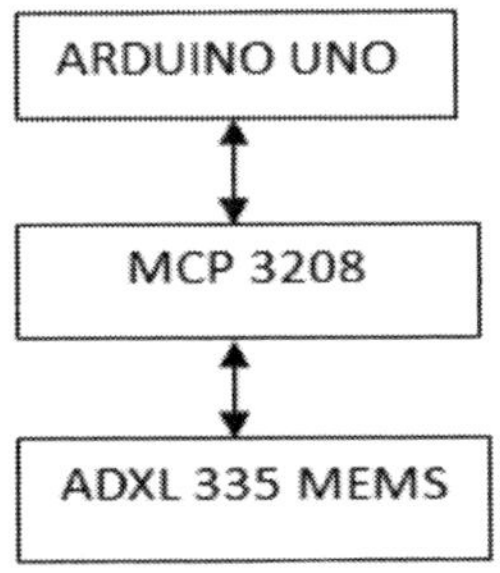

Figure 3. Accelerometer Signal Flow Diagram

The GPRS modem is used to gather information on the location of the car obtained from the GPS and push it as a text message. It makes use of the local mobile network towers (**Figure 4**). The coordinates obtained from the accelerometer fed to the micro-controller have collaborated with the location obtained from the GPRS modem. This action takes place within a few seconds after a crash has been detected. The MAX232 IC acts as an intermediate link between the Arduino and the GPRS modem by converting and maintaining voltage signals. MAX232 IC makes sure that the data is transmitted properly without any fluctuations as it is the main voltage stabilizer of the system.

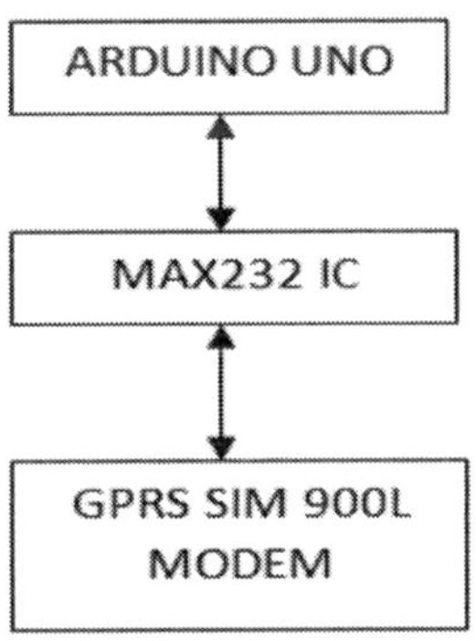

Figure 4. Location information to Arduino

The data obtained and combined is next sent to the driver's IC, an NPN transistor (BC547). This transistor is mainly used for switching and amplification (**Figure 5**). Here it amplifies the signal obtained from the Arduino and sends it to the alarm so that it is ringing sound makes sure that an accident has been confirmed. If no serious accident has occurred and the person is stable enough to help themselves, then, the device can be turned off manually.

IOP Conf. Series: Materials Science and Engineering 1012 (2021) 012042 doi:10.1088/1757-899X/1012/1/012042

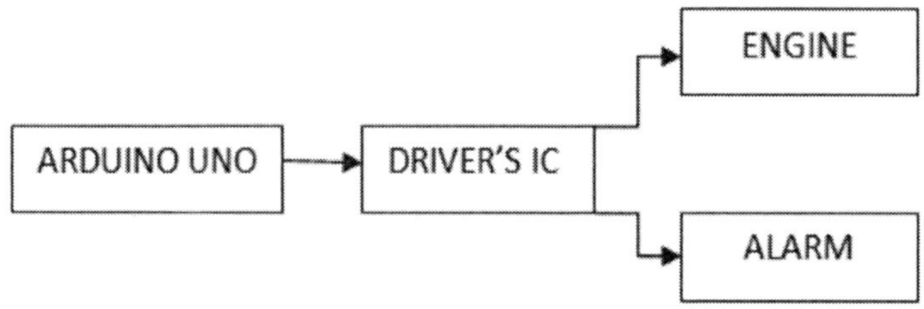

Figure 5. Arduino information to Alert system

4. Hardware and Software Details

The following part explains about the hardware and software details. The major components are Arduino, accelerometer, voltage level controller, GSM module etc.

4.1. ARDUINO UNO

Arduino is a hardware platform or a development board, which has a microcontroller based on it. It can be used to communicate between Arduino boards or between a computer and an Arduino or between microcontrollers and Arduino etc. It consists of ATmega328p microcontroller along with components like crystal oscillators, voltage regulators, and serial communicators to support the microcontroller. The voltage range is 3.3V to 5V with 6 analog pins and 14 input/output pins.

4.2. ADXL 335 MEMS

Figure 6. shows the ADXL 335 MEMS, which is a high range, low frequency accelerometer that is used in shock and blast detection applications. It is used to measure the static acceleration of gravity in the tilt-sensing applications as well as dynamical acceleration from motion shock or vibration. It has triple axes coordinating system i.e. all the three planes (XYZ) are considered. The device measures the acceleration due to gravity and the range is ±3 g. The device has a small ball enclosed within and the walls enclosing the ball are made up of piezo-electric elements. So, whenever the ball touches the wall due to acceleration, the co-ordinates along all the 3 axes are recorded and sent to the Arduino via MCP3208 (**Figure** 7).

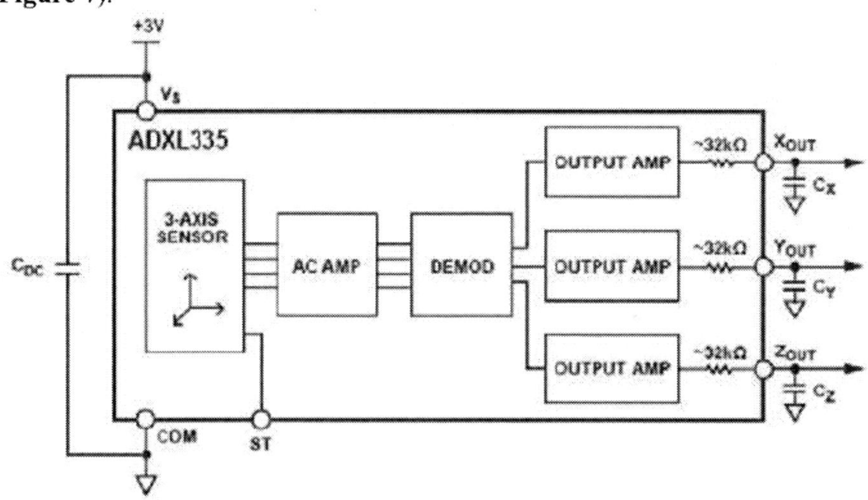

Figure 6. MEMS Accelerometer Circuit Diagram

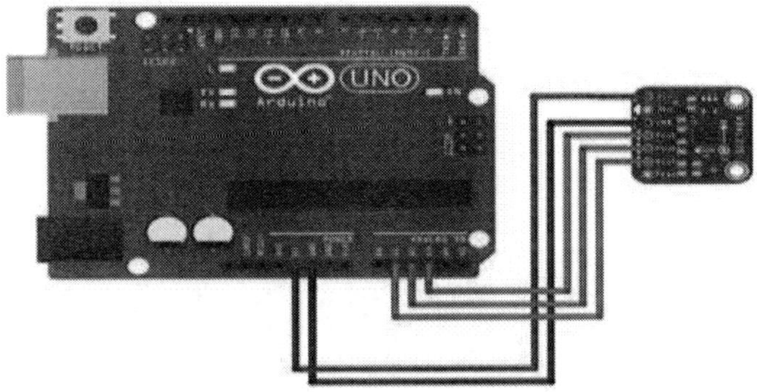

Figure 7. Accelerometer with Arduino

4.3. MPC3208 Circuit

Figure 8. shows the circuit diagram of MCP3208 circuit. It is an electronic Multi-Chip Package IC which is predominately used to convert analog signals to digital signals which consists of 8I/P channels and has 12-bit input. It is used in sensor interfaces, to control the sensor and to acquire data from various sensors and co-ordinate them accordingly.

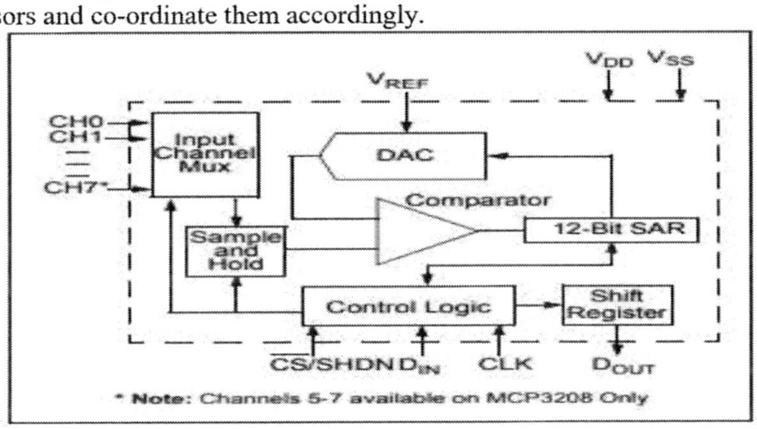

Figure 8. Circuit Diagram of MCP 3208

The data received is in analogous form which must be converted into digital format so that the Arduino understands the process and makes, only the right commands are passed. The operating voltage in a region between 2.7V and 5.5V and has an industrial temperature of -40°C to +85°C. The analogous inputs are programmable either to give 8 single ended inputs or 4 pseudo-differential input pairs. The maximum rate of sampling is at VDD=5V.

4.4. MAX 232 IC Circuit

Figure 9. shows the circuit diagram of MAX 232 IC, which is an Integrated circuit acting as a voltage level converter. This IC is used in RS232 communication system which is the standard protocols for series communication. This communication system is used to connect the signals between data terminal equipment's such as modem, computers. This entire setup is utilized for voltage conversion of TTL devices such that it is made compatible with PC serial port and vice-versa. This has dual drivers and dual receivers to support two conversions in parallel. Average speed is 120kBit/s and has an operating current of 8mA. The device is a 16-pin configuration with 4 external capacitors that has a range between 8μF to 10μF and withstands a voltage of 16V.

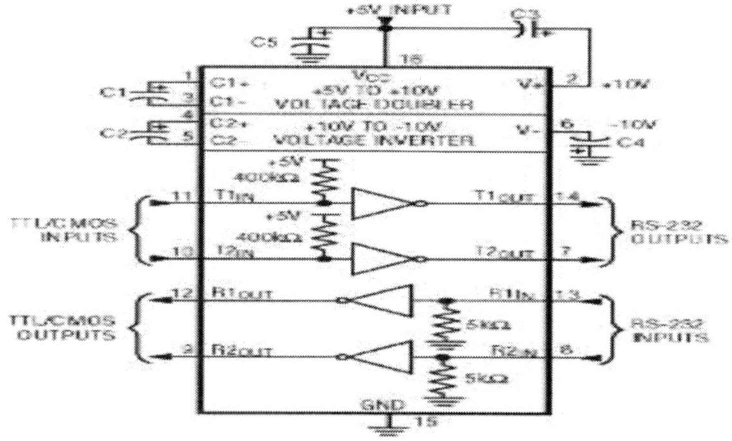

Figure 9. Circuit Diagram of MAX 232

4.5. GPRS SIM 900L

The most effective miniature form of the GSM modem is the SIM900L GPRS module. This can be used to do a lot of activities such as send text SMS, receive, or make calls and even connect to the internet. The operating voltage of this model is 3.4V to 4.5V, consumes less power and it is constructed with the help of LiPo battery. The dimensions of the device is 23mm*35mm*5.6mm. It can connect to any GSM even with the basic 2G network connection. Thus, it makes the device handy and persistent even in low bandwidth. The GPRS connects to the internet using "http:/". They communicate with the Arduino using a UART port. It also supports an internal GPS Module which makes use of the mobile towers to get the latitudes and longitudes of the device and hence can be used as a tracker. The device with the location data can be sent as link by accessing mobile data present in a text message.

4.6. GPS SIM28ML

GPS SIM28ML is a high-performance stand-alone receiver with a built in LNA (**Figure 10**). It does not require an antenna for transmission and can track signals as low as -165dBm without network assistance. The device consumes very less power and it supports various navigation applications. The mapping application used in all devices makes use of this GPS predominantly as it can track the location without the support of internet connection i.e. via A-GPS. This model is cost efficient and is reliable. The accuracy of this device is very much precise and can give the data of latitude longitude and altitude. This model is preferred as it gives the data with a more accurate timestamp also.

Figure 10. GPS SIM28ML Diagram

4.7. Embedded C

The most popular language used to program Arduinos is the embedded C. This code is widely used by the software industries to code many electronic gadgets as the code is easy to understand, reliable, portable, and scalable. The system proposed makes use of embedded C in Arduino IDE. The flow of the program is as follows: the maximum and the minimum limits are defined after which all the header files and the various declaration and preliminary calculations are done. Several void functions are created for example, a void function for GSM module is created to initialize the GSM module and a separate void function is created to establish communication between the hardware and software system. Using the loops, the accelerometer calibration is done along with reference values of the threshold axis. A separate void function is created for the GPS to get the exact location, extract, and convert them to decimal values and to send it to the LCD. By this way, the system is skilfully so that the entire setup runs flawlessly.

4.8. Flow chart

Figure 11. shows the flow chart of the vehicle accident alert system.

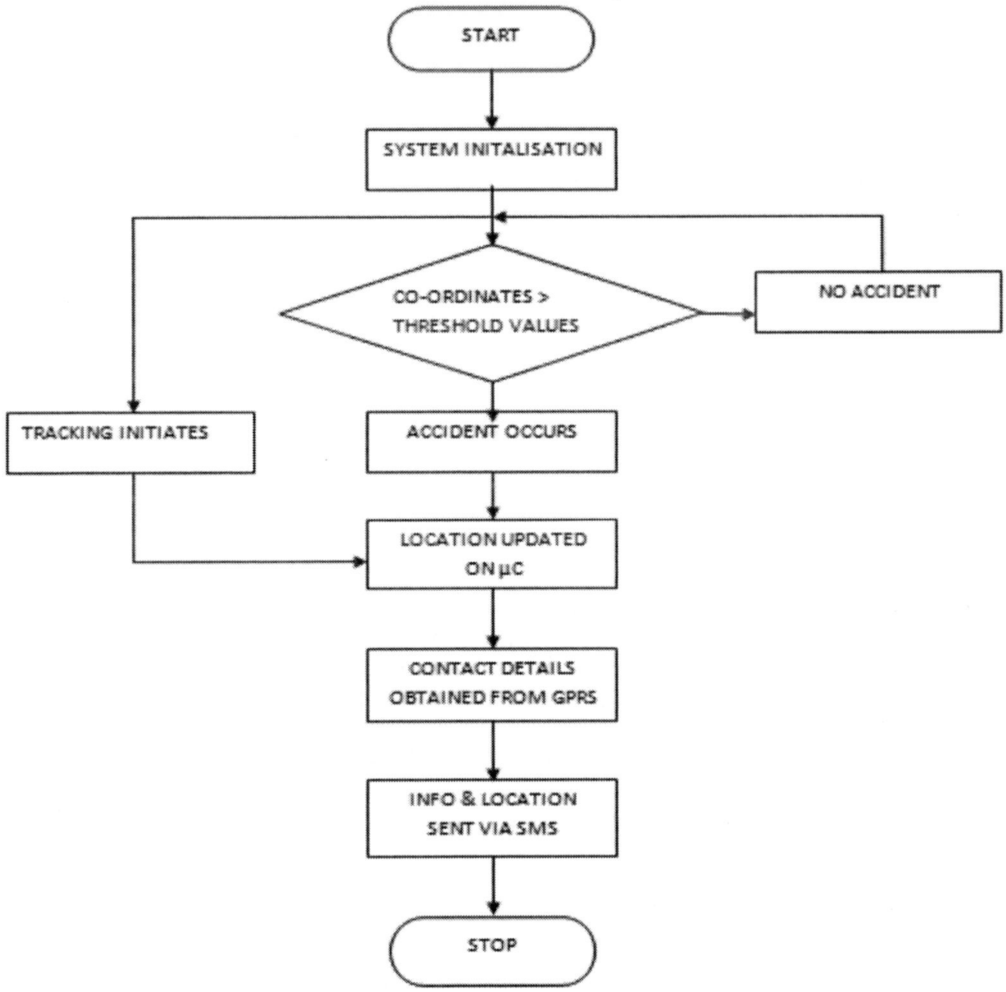

Figure 11. Flow chart of the Proposed System

5. Results and Discussion

The Table I, gives the range values of accelerometer, along all the possible axes at which the accident occurs. In all the three axes the voltage recorded is 3.24 volts.

Table 1. Accelerometer Output

Axis	Range	Voltage Recorded
X	0-10 or 170-180 (ROLL)	3.24V
Y	0-10 or 170-180 (PITCH)	3.24V
Z	0-10 or 170-180 (YAW)	3.24V

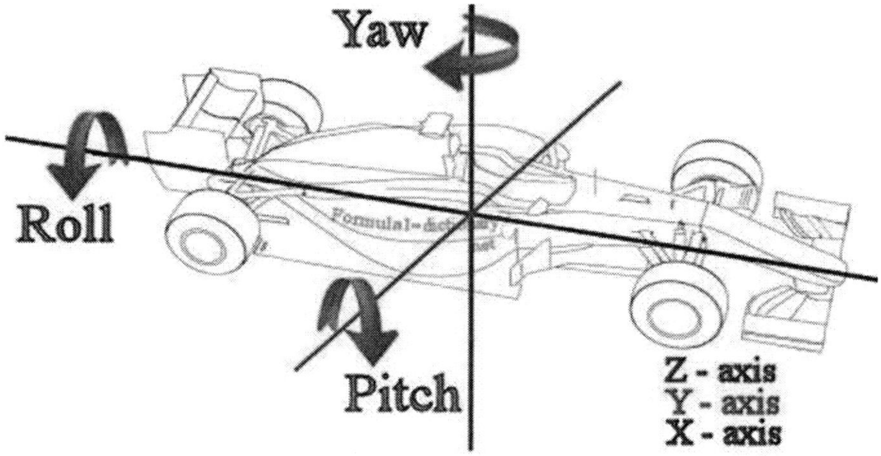

Figure 12. Car Longitudinal Movements

The longitudinal movement on three axes is shown in the **Figure 12.** Consider the longitude possibilities, if the vehicle is travelling along all the axes at 90 degrees, the vehicle is safe, and no accident message is initiated. Now, if the vehicle is tilted between the range 0-10 and 170-180, if it is along the X axis, the vehicle rolls either on the right or left and accident occurs. Else if it is along the Y axis the vehicle pitches back or forth causing a front/back crash. Else the vehicle yaws along the Z axis without a proper control over the drift occurred and hence an accident takes place. The values of these co-ordinates are updated every second to the Arduino. By making use of the Piezo-electric property present in the MEMS accelerometer, any abnormalities in the system is detected easily by comparing these values with the threshold value measured by the accelerometer and fed into the board.

Figure 13. shows the flow chart for GPS working. Due to the presence of the GPS module, an abrupt change in the latitude and longitude can be used to detect an accident. For instance, if the vehicle is hit by another vehicle on any of the four sides, and if the car does not roll, yaw or pitch, but instead, moves along the direction of hit at a high speed, the sudden change is sensed and an error is immediately sent to the Arduino and it immediately starts the emergency protocol as soon as it crosses verifies the values obtained from the GPS with the pre-coded threshold value. The threshold values of the latitude and longitude are calculated with the help of mathematical formulae such as the Harsevine's formula and Vincity's algorithm. The geo-fencing limit of vehicles can be approximately set to 2.5 m. If the vehicle crosses this limit instantaneously, the entire system starts it action.

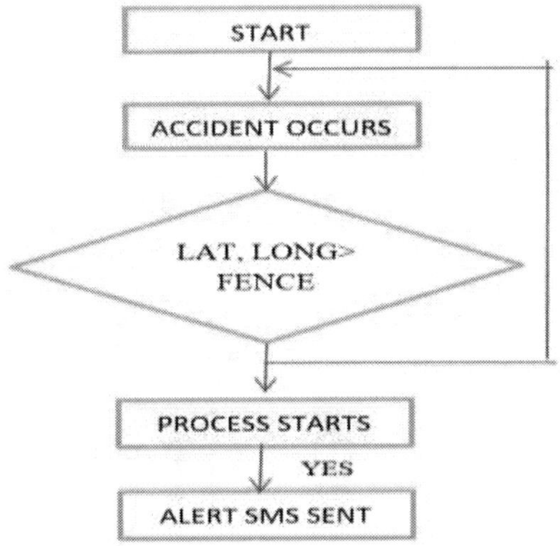

Figure 13. Flowchart for GPS module working

The following table (Table No. II) gives the distance output moved for change in latitude and longitude.

Table 2. Distance Output for Latitude change

Latitude Change	Distance (m)	Longitude Change	Distance (m)
0.001'	1.84	0.001'	2
1'	1840	1'	1805
1^0	110160	1^0	108299
10^0	1101600	10^0	1082990

From the above table, it is inferred that a small change in the latitude makes a greater impact on the distance. GPS is used as a secondary system to recheck if the accident has occurred or not. It proves to be an excellent backup also. The same process of sending an alert SMS is followed here as soon as the information from the GPS module is passed to the Arduino. The program in the Arduino is suitably coded in embedded C such that, there is exact readings of the data. As soon as the accident alert is obtained from the Arduino, the GSM module sends the location of the incident spot obtained from the GPS as a link in a text message [**Figure 14**], to the registered numbers as well as the nearest EMC.

IOP Conf. Series: Materials Science and Engineering 1012 (2021) 012042 doi:10.1088/1757-899X/1012/1/012042

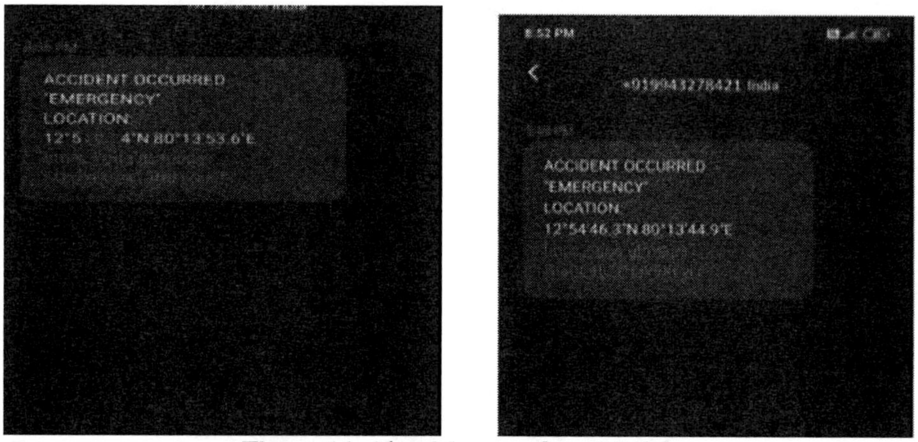

Figure 14. Alert Message from Module

If an accident occurs the entire process of the system is initiated, and the text message is sent as follows. Using the link provided in the text message, one can obtain the exact location of the incident spot and reach there quickly by taking the shortest route. The text message is also forwarded to the nearest EM centre, so that an ambulance is sent to the location as quickly as possible and death rate can be reduced. This system can be installed in all types of vehicles such as two-wheelers, three-wheelers as well as four wheelers.

6. Conclusion and Future works

The Vehicle Accident Alerting and Detecting System could be a safer system and about two third of the lives from dangerous road accidents could be saved, especially in remote areas where the human activity is less. The GPS tracker attached in the system gives the information of the exact geographical location that could specify the latitude and longitude. The SMS contains the details about the information of occurrence of accident and condition of patient by measuring the vital bio signals and the registration number of the vehicle. SMS alert is sent immediately to the nearby hospitals, ambulance, police station, also to the family members of the victim. The ambulance could be arrived at the accident spot immediately by using the location details and quick medical help could be provided to the victim. If the victim is not injured severely, then that person can switch off the alert system by pressing a button which is placed in the side of the unit. Thus, a simple way is achieved to reduce the frequency of accidents and immediate alerting system, a low-cost way to safe high cost lives. As a future works, the method could be devised as a low-cost product and will be installed in all type of vehicles.

References

[1] Kalyani T, Monika S, Naresh B and Mahendra Vucha. 2019. Accident Detection and Alert System. *International Journal of Innovative technology and Exploring Engineering* **8**(4S2).

[2] Nicky Kattukkaran, Arun George and Mithun Haridas T P. 2017. Intelligent Accident Detection and Alert System for Emergency Medical Assistance *International conference on computer communication and Informatics.*

[3] Vani A, Yahdidya N, Yaswanth I and Kareema Pasha M D. 2020. Implementation of Accident Detection and Alert System for Emergency Medical Assistance using Mobile Application with MIT app Inventor *International Journal for Research in Accident Science and Technology* **8**(4).

[4] Tanushree Dalai. 2013. Emergency Alert and Service for Automotive for India *International Journal of advanced Trends in computer science and emergency* **2** 08-12.

[5] Feukeu E A, Djouani K and A Kurien. 2014. Performance and comparative Analysis of ADSA in a vehicular Network: MAC Approach in IEEE 802.11 *International Conference on ambient Systems, networks, and Technologies.*

IOP Conf. Series: Materials Science and Engineering 1012 (2021) 012042 doi:10.1088/1757-899X/1012/1/012042

[6] Aboli Ravindra Wakure, Apurva Rajendra Patkar, Manisha Vitthal Dagale and Priyanka Pradeep Kumar Solanki. 2014. Vehicle Accident Detection and Reporting System using GPS and GSM *International Journal of Engineering Research and Development* **10** (4) 25-28.

[7] Varsha Goud and Padmaja V. 2012. Vehicle Accident Automatic Detection and Remote Alarm Device *International Journal of Reconfigurable and Embedded Systems* **1** 49-54.

[8] Victor Olugbemiga Matthews and Emmanuel Adetiba. 2011. Vehicle Accidents Alert and Locator *International Journal of Electric and Computer Science*.

[9] Supriya Sawwashere and Ashutosh Lanjeswar. 2015. Automated Accident Intimation and Rescue System for Vehicles *International Journal of Application or Innovation in Engineering and Management 4* (5).

[10] Joana Maria Segui Pons and Maurici Ruiz Perez. 2014. Geographic Information Systems and Intelligent Transport Systems: Technologies used to form new Communication networks *Networks and Communication Studies* **17** 53-70.

[11] Sivaraman Karthikeyan, Shreyas Ramachandran Srinivasan, Syed ali J, Veeraraghavan A K and Uvais karni. 2018. Smart Summoning of Ambulance during a vehicle Accident *International Conference on Green Computing and Internet of Things*.

[12] Megha Nirbhavane and Shashi Prabha. 2014. Accident Monitoring System using Wireless Application *International Journal of Advanced Research in Computer Engineering and Technology* **3**.

[13] Manisha Madhukar Ruikar. 2013. National Statistics of Road Traffic Accidents in India *Journal of Orthopedics Traumatology and Rehabilitation* **6** (1).

[14] Sharma N, Kumar A. 2018. Road accidents in India: Dimensions and issues online.

[15] https://sites.ndtv.com/roadsafety/5-indian-apps-and-services-that-are-making-roads-a-little-safer-1676/

IOP Conf. Series: Materials Science and Engineering 1012 (2021) 012043 doi:10.1088/1757-899X/1012/1/012043

Design of Portable Air Purge Level Transmitter with Built-in Calibration Feature

Dr. A. K. Joshi, Dr. V. M. Upadhye and Dr. S. P. Madhe

Instrumentation and Control Department, Cummins College of Engineering, Karvenagr, Maharashtra, Pune-411052

E-mail: atul.joshi@cumminscollege.in, vaishali.upadhye@cumminscollege.in, swati.madhe@cumminscollege.in

Abstract. In almost all the process industries level is the prominent process parameter which is being monitored and controlled. For level measurement a wide variety of level sensors are available. For level measurement of chemically corrosive liquids or liquids containing suspended solids, specific sensors are used. In such applications, the direct contact of sensor with the fluid is avoided. Air purge or bubbler pipe is the most suitable level measurement method in such cases. In this method, compressed air is forced through the bubbler pipe which is placed in the level measurement tank. This compressed air/gas emerges from the other end of the tube in the form of bubbles. The pressure in the pipe is equal to the head pressure created by the height of liquid column in the tank. The conventional air purge level measurement method has limitations like lack of portability and non-electrical output. This also consumes more power, lacks in scheduled measurement and requires recalibration for the changes in the range. To overcome these limitations a novel portable programmable air purge level transmitter is designed in this paper. The designed system consists of an on board compressor to improve the portability. The electrical output is obtained using MEMS pressure sensor which is mounted on the top of the bubbler pipe. The electrical signal given by the MEMS pressure sensor is converted to standard output using Microcontroller and Signal Conditioning Unit. The system is made programmable using keyboard and display interface to microcontroller. Due to this recalibration and scheduled measurement is possible and this results in reduced power consumption.

1. Introduction

Liquid level measurement is an integral part of process industries like chemical industry, petroleum industry, fertilizer industry, food processing industry, etc. In process industries, there are many challenges in liquid level measurement due to hazardous environment [1]. Liquid level measurement techniques are classified as contact type and non-contact type [2]. The most commonly used contact type liquid level sensors are float type, air purge, differential pressure cell, capacitance probe. The conventional level sensing techniques are reliable and less expensive [3]. However, these sensors require for maintenance and the life span of these sensors is limited. Capacitive-level sensors are reliable, sensitive and simple in construction. So they are most commonly used [4]. However, frequent re-calibration is required for capacitance probe due to changes in temperature, density and dielectric media. Ultrasonic type is the most commonly used non-contact liquid level sensor owing to it's simple construction [5]. But the readings are affected due to interference. Ultrasonic level measurement in narrow vessels like bore well gives false reflections leading to incorrect readings. Due to complicated electronics this method becomes expensive.

For level measurement of chemically corrosive liquids or liquids containing suspended solids, direct contact of sensor with the fluid is avoided. Air purge or bubbler pipe is the most suitable level measurement method in such cases since in this method compressed air/gas is forced through the bubbler pipe which is placed in the level measurement tank. The back pressure in the bubbler pipe which is directly proportional to the level in the tank is measured by a pressure sensor in the developed system. Thus the pressure sensor thus does not come in direct contact with the fluid. Also due to purging of air, contamination of the probe due to the liquid having suspended solids is avoided. In such applications, if capacitance probe is used, then due to contamination of the probe the readings are affected. However in conventional air purge level measurement method there are some limitations like lack of portability, non-electrical output. This also consumes more power, lacks in scheduled measurement and requires re-calibration for change in range. To overcome these limitations a novel portable programmable air purge level transmitter is designed in this paper. The designed system consists of an on board compressor to improve the portability. The electrical output is obtained using MEMS pressure sensor mounted on the top of the bubbler pipe. The electrical signal given by the MEMS pressure sensor is converted to standard output using Micro-controller and Signal Conditioning Unit. The system is made programmable using keyboard and display interface to micro-controller. Due to this re-calibration and scheduled measurement is possible and this results in reduced power consumption.

2. Literature Review

The sensors used for liquid level measurement can be classified based on various parameters [5], [6], [7], [2]. Based on the measurement mode it can be classified as consistent or point esteem value. In consistent measurement mode the level is measured continuously using continuous level sensors. In case of point esteem value, the level is detected if it is below or above a specified point by the point level sensor [4]. The level sensors can also be classified as contact or non-contact type. The level sensors classified based on the principle of application are ultrasound , capacitive, optical, Vibrating or Tuning Fork, Conductivity or Resistance, Float Switch. Most of these sensors are invasive and are affected by various elements in the environment. A remote grounded capacitive sensor for the liquid level measurement was designed by Reverter et al. [8]. This system was based on contact type. Khan et al. [9] have designed non-contact capacitance type liquid level transducer with low cost. To measure oil level Lu et al. [10] have developed a sensor based on inductive and capacitive principle. K. V. Santhosh et al has designed Liquid Level Measurement instrument Using Multi sensor Data Fusion [11]. Level transmitter using force resistive sensor was presented by Anamika Lata et al. [3].

As seen from the literature survey, the existing methods are based on capacitive sensors, ultrasonic sensors, potentiometric sensors. The level transmitters are generally calibrated for a fixed SPAN. The accuracy of the output current from these transmitters is specified for % of Full scale (F.S.). Most of the industrial applications demand accuracy of around 1% of F.S. If the full span of level is not used for measurement, then the span of output reduces and hence the overall accuracy degrades. Capacitive sensors are susceptible to changes in dielectric constant because of ambient temperature variation, chemical compositions and radio frequency interference (RFI). These sensors are affected by clogging problem. Ultrasonic sensors cannot be used in the tanks having narrow cross sectional area like tube-well. The side wall of the well reflects the ultrasonic signal back to the receivers. It is also required to fit the sensor head inside the tank. The potentiometric sensors need mechanical arrangements for tracking the level in tank. It is generally used with float and counter balancing weight. This sensor has sliding contact hence it has mechanical ware and tare problem. The sensor electrically represents

Thevenin's voltage source whose output resistance changes according to the level. This affects the accuracy of measurement.

Out of all these sensors the bubbler tube is the simplest and a low cost solution to the level measurement. But these sensors are not portable and require frequent calibration. To address these issues we have designed Portable Programmable Air Purge Level Transmitter in this paper.

3. System Details

The detailed block diagram of the developed system is shown in Figure 3. The developed system consists of measurement unit, signal processing unit with user interface and standard transmission unit. The MEMS pressure sensor senses the back pressure in the bubbler pipe. The output signal of this sensor is conditioned in the signal conditioning block and is given to the Micro-controller for further processing. The signal is then converted it to standard industrial current range of $4 - 20mA$.

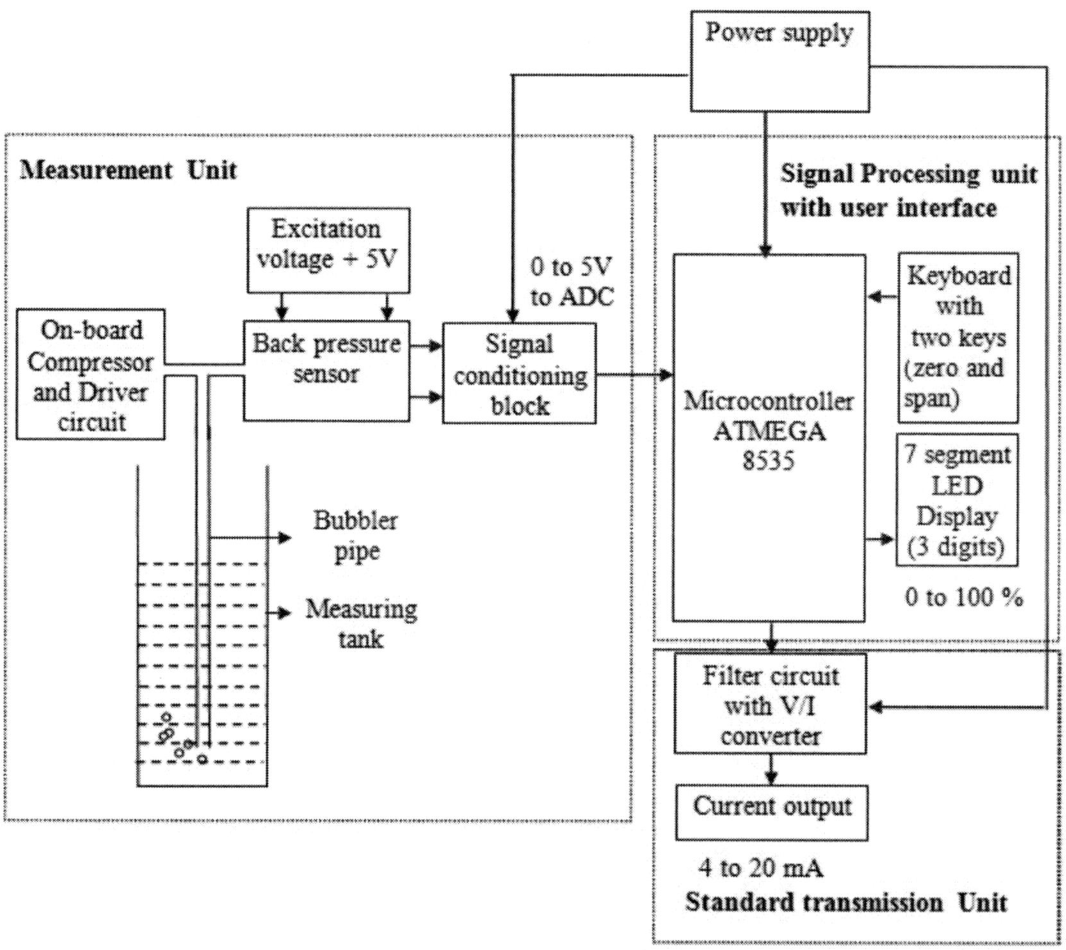

Figure 1. Block Diagram of Developed System

3.1. Static pressure measurement unit

Measurement unit consists measuring tank, bubbler pipe, on-board compressor with driver circuit and MEMS pressure sensor MPXV10. Measuring tank is used for storing liquid of which the level is measured. A PVC pipe is used as a bubbler pipe on which the back pressure sensor and the compressor are mounted. The advantage of using PVC pipe is that it is corrosion resistant. A purge gas is normally an air supply used specifically in areas which are susceptible to contamination or an oxidative reaction with the medium. Compressed air is used as the purge gas in our system. Nitrogen gas can also be used in case of flammable liquids. In the developed system a diaphragm compressor is used which is as shown in Figure 2. It is commonly used for domestic aquariums as an aerator. It is a positive displacement type diaphragm pump. It is also deployed in wrist type blood pressure monitor. The specifications of the same are given in the Table 1. The compressor is mounted on top of the bubbler pipe in order to pump air into it. Compressor is driven by a regulated power supply built using IC TIP 122.

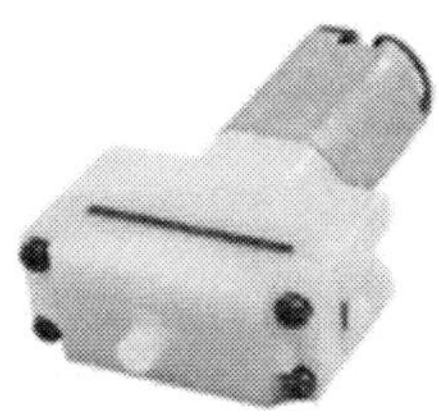

Figure 2. Diaphragm compressor

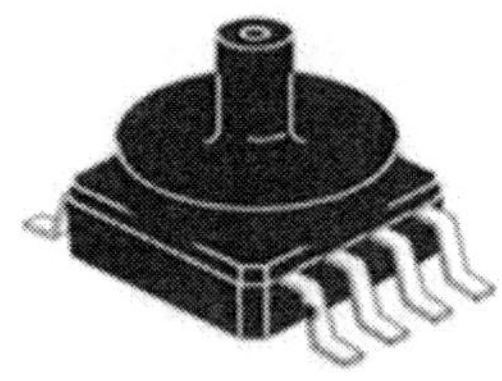

Figure 3. MPXV10 series pressure sensor

Table 1. Specifications of Compressor

Sr. No.	Parameter	Specification
1.	Rated voltage	3V
2.	Voltage range	2.0 to 3.2 V
3.	Current consumption	below 450 mA
4.	Highest pressure	Above 47kPa

MPXV10 series back pressure sensor is shown in Figure 3. MPXV10 is silicon piezo-resistive pressure sensor. It's output voltage changes with changes in the back pressure which is due to changes in the level. It's differential pressure range is 0 to 10kPa. The sensor is powered with a regulated 5V DC source. This sensor is mounted on the bubbler pipe which is connected to the compressor using T-connector. The back-pressure in the bubbler-pipe increases proportionally with the level. The back pressure is equal to the height of a liquid column. This static back pressure (P) is related to the level (h) in the tank by the equation 1.

$$P = h\rho g \tag{1}$$

where ρ is the density of the liquid and g is the gravitational acceleration. The output voltage signal obtained from pressure sensor is in milli volt range. The signal is amplified to match the further circuit. Instrumentation amplifier AD620 is used to amplify the signal. Gain (G) of AD620 is set using RG resistor connected between pin 1 and 8 as shown in Figure 4. The output of AD620 is connected to SPAN and ZERO circuit as shown in Figure 5 to provide output as 0-5V.

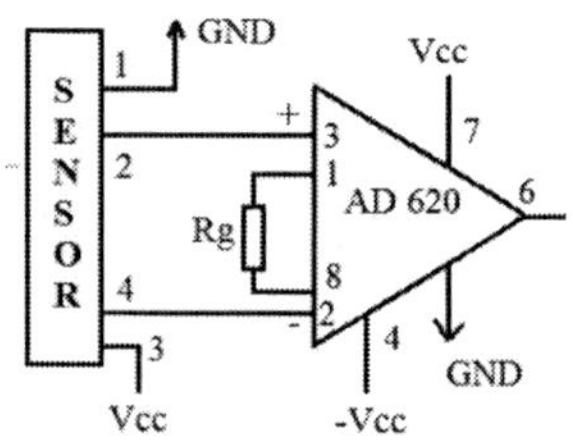

Figure 4. Connection diagram for IA-AD620

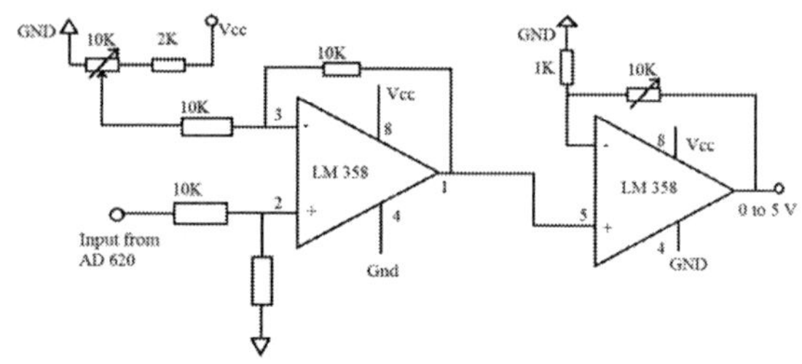

Figure 5. Circuit for SPAN, ZERO adjustment

3.2. Signal Processing unit with user interface

The signal processing unit consists of ATmega8535 micro-controller, LED Display and Keyboard. Microcontroller ATmega8535 is the heart of the system which acts as the signal processor. The system has two parts namely indicator section and current transmitter section. These sections are controlled by the microcontroller. The microcontroller performs mapping of input level range and the output current of $4 - 20mA$. The sensor-signal conditioner drives the ADC of controller. The analog signal from the measuring unit is converted into digital form using ADC and is given to micro-controller. We have used inbuilt 8 channel, 10 bit analog to digital converter of ATmega8535 micro-controller. The micro-controller processes the ADC Data. It is also responsible for synchronization of operation of various devices like LED display, keyboard, V to I converter connected to the system. 3-digit seven segment LED displays are used for indicating the liquid level in range of 0 to 100%. The system is provided with user-friendly pair of keys on-board. The keys are used to calibrate and set the user defined range of measurement. Thus, the display device in conjunction with the keyboard enables the operator to set the span and zero values for calibration.

The beauty of this system is that, it can be calibrated for any span of input level just by clicking the ZERO and SPAN keys. The calibration counts are stored in to EEPROM of the controller hence, the system holds the calibration even in case of power failure. The analog signal from the measuring unit is converted into digital form using ADC and is given to micro-controller. We have used inbuilt 8 channel, 10 bit analog to digital converter of ATmega8535 micro-controller. The micro-controller processes the ADC Data. It is also responsible for synchronization of operation of various devices like LED display, keyboard, V to I converter connected to the system. 3-digit seven segment LED displays are used for indicating the liquid level in range of 0 to 100%. The system is provided with user-friendly pair of keys on-board. The keys are used to calibrate and set the user defined range of measurement. Thus, the display device in conjunction with the keyboard enables the operator to set the span and zero values for calibration.

3.3. Standard Current Transmission Unit

The developed system is not only a level indicator but also a level transmitter. The standard transmitter output of $4 - 20mA$ is generated using filtering and V to I converter circuit. The digital values corresponding to 0 and 100% of level in the tank are used to drive the internal pulse width modulator (PWM) of ATmega8535. The PWM output signal from ATmega8535 is averaged by using a passive low-pass filter. Then output is then buffered and scaled to drive the current booster. The boosted $4 - 20mA$ output is provided as a process variable to various process controllers. Figure 6 shows the detailed circuit diagram of standard current transmission unit. The current booster section of the system is implemented with non-inverting amplifier with unity gain. The transmitter section uses 100Ω resistance as a loop-current sensing resistance. It makes voltage drop proportional to the output current. The same voltage drop is fed back to the booster amplifier. The PWM signal varies from 0 to 5V in 256 steps. But the required voltage drive for current booster varies from $0.4V$ to $2V$ (for 4 -20 mA with 100Ω R-sense resistance). The scaling is done by using pair of voltage buffers, which then drive the current booster. The boosted current output is sourced by transistor TIP 122 which is darlington transistor with current gain of 1000. The PCB design and fabrication of the developed system is carried out

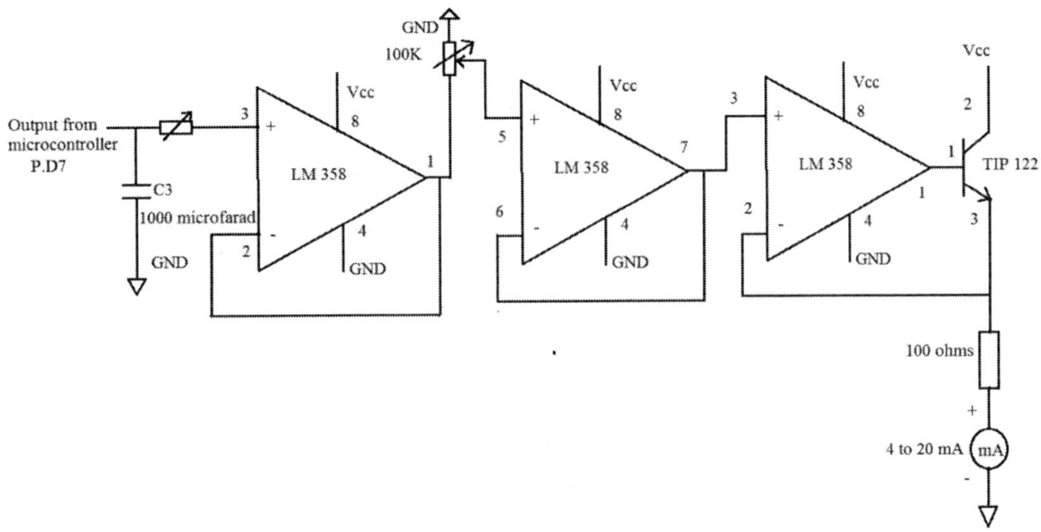

Figure 6. Circuit diagram of Standard Transmission Unit

in-house in our laboratory. The photograph of the same is shown in Figure 7.

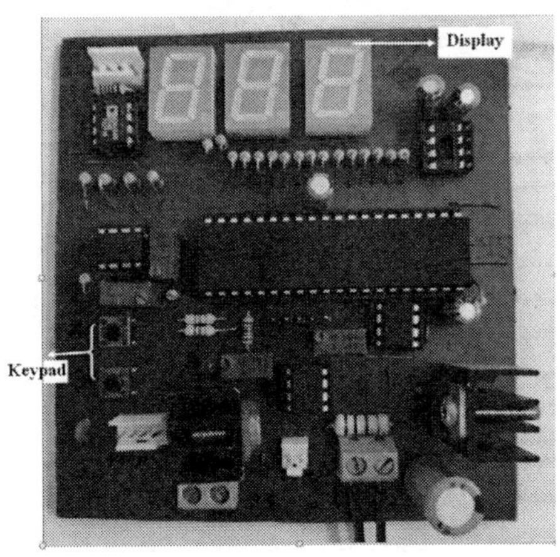

Figure 7. PCB of Developed System

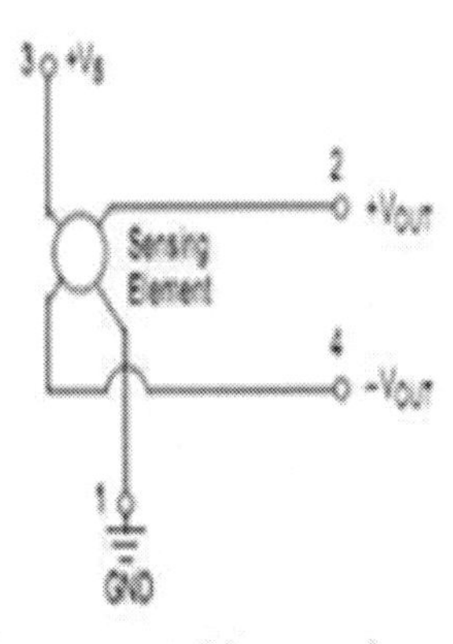

Figure 8. Pin Diagram of MEMS Pressure Sensor

3.4. System Design Calculations

The heart of the system is the signal processing unit. It is based on ATmega 8535 micro-controller. It has 8-channel 10-bit Successive Approximation Register (SAR) type ADC. The

DAC is implemented by using 8-bit Fast-PWM module of the controller. The design calculations for 0-100 cm range are given below. The ADC has Vref of 5 V. Hence the least count is calculated as

$$5V/1023 = 0.004887V = 4.88mV/step \quad (2)$$

The output of pressure sensor is amplified to match with the Vref = 5V. The ADC provides 1023 samples for full scale output of the sensor. The Pulse Width Modulator (PWM) module is driven from 0-100 count to produce proportional PWM signal. The basic PWM module is of 8-bit hence it provides 0-255 steps to produce 0-100% PWM signal. The averaging circuit connected at the output pin of PWM based DAC provides 0-5 V linear output. The basic accuracy of the DAC is calculated as

$$1/255 \times 100 = 0.392\%(F.S.) \quad (3)$$

The current transmitter unit generates 0-100 % signal by producing 0.4 V to 2 V. It means that 0.4 V = 0 % which is converted to 4 mA (ZERO) and 2 V = 100 % converted to 20 mA (SPAN). The desired accuracy of the output current is 1% of F.S. The V to I converter is driven with an increment of 16 mV. The $100k\Omega$ potentiometer in Figure 6 is adjusted to set 2 V.

$$2V - 0.4V = 1.6V \quad (4)$$

$$1.6/100 = 0.016V = 16mV/step \quad (5)$$

The ZERO of measurement = 4 mA hence the required DAC output is 400 mV. The SPAN of measurement = 20 mA hence the required DAC output is 2000 mV.

$$400mV/16mV = 25 \quad (6)$$

Hence the initial count of 25 in PWM unit produces 4 mA current.

$$2000mV/16mV = 125 \quad (7)$$

Hence the final count of 125 in PWM unit produces 20 mA current. The above calculation shows that 2 V signal is required from DAC for 125 count in PWM but the actual output can be calculated as

$$125/255 * 5V = 2.45V \quad (8)$$

Hence the $100k\Omega$ potentiometer shown in Figure 6 is set to drop the voltage from 2.45 V to 2 V which then drives the input of V to I converter. The performance of the current transmitter unit is shown in Figure 11. The signal conditioner of pressure sensor provides 0-5 V output for span of 0-100 cm level in the tank. The following experimentation shows the performance for different calibration setting of the system.

For 0-80 cm range the calculations are as follows. Level of 0-80 cm produces 0-4V which corresponds to 0-819 count. The calibration count for PWM is calculated as 819/100 = 8.19. This shows that the PWM count changes for every 8.19 count of ADC.

The existing systems are permanently set to a fixed range thereby shortening the range which results in degradation in accuracy of output current.The results are shown in Table 2.

From comparison of graphs shown in Figure 9, it is clear that the output accuracy of current is maintained to 1% of F. S irrespective of the range. The calibration procedure for the transmitter is simple, screw-driver less and user friendly. It is done by clicking the keys namely ZERO and SPAN. The deigned system can be programmed for min range of 10 cm as per following calculations with the same accuracy of 1% of F.S.:
ADC output = 4.88 mV/step Current output is in 100 steps $Accuracy = 1\% of F.S.$ 100 steps x 4.88 mV/step = 488 mV The Max. Output of sensor = 5000 mV at 100 cm 488/5000*100 = 9.76 cm can be rounded off to 10 cm.

Table 2. Observation of ADC count and signal conditioning output for 100 am and 80 cm level

Level in cm	ADC count	Signal conditioner output (mV)	PWM count for 0-100cm	PWM count for 0-80cm
0	2	0	24	25
5	51	248	30	31
10	102	497	36	38
15	152	741	41	44
20	202	985	45	50
25	256	1249	50	57
30	306	1493	54	63
35	357	1742	61	67
40	410	2000	66	75
45	462	2254	69	81
50	509	2483	75	88
55	560	2732	79	93
60	615	3001	85	100
65	666	3250	89	107
70	715	3489	96	112
75	765	3733	102	119
80	820	4001	105	125
85	870	4245	110	-
90	923	4504	116	-
95	974	4753	120	-
100	1022	4987	125	-

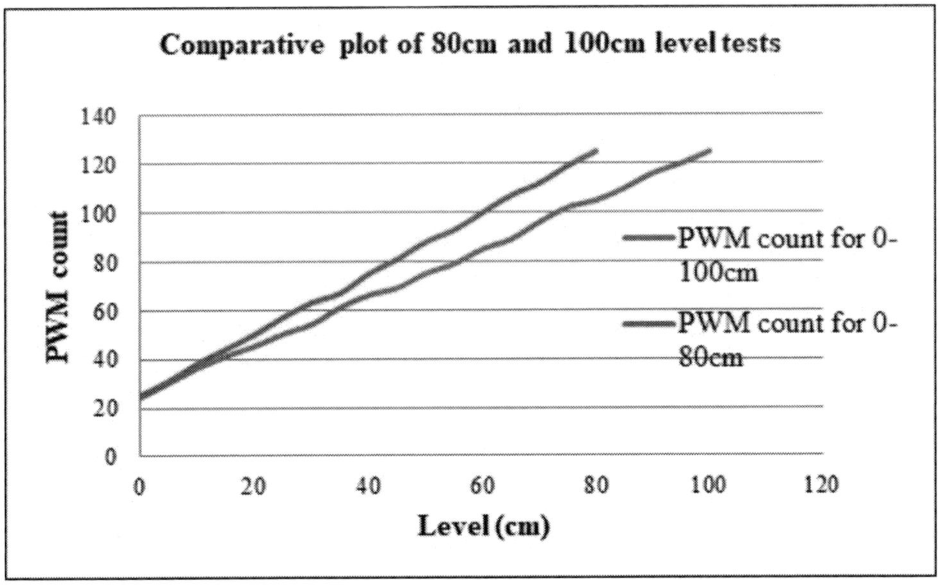

Figure 9. Graph of 100cm and 80cm PWM output

4. Experimentation and Observations

Various experiments were performed to calibrate and test the developed system. The system is designed to measure level in the range of 0 cm to 100 cm.

4.1. Compressor Testing

Here we tested the maximum voltage at which the compressor can be driven without much fluctuations in output. For this the level was maintained constant at 70 cm in the measuring tank. It is observed that at 3V, the output voltage of compressor is stable. Also according to this maximum voltage we calculated the maximum level which the transmitter can measure. The maximum drivable range of the compressor is tested by following procedure:

(i) Compressor attached to a pipe sealed at one end.
(ii) A pressure gauge attached to other end of pipe
(iii) Compressor supply voltage is adjusted to 3V to get stable output voltage.

The pressure observed on the pressure gauge was $0.7kg/cm^2$ i.e. $6999.332mm$ of water column. This pressure corresponds to a level of approximately $7m$ of water column. Thus the Maximum range of drivable water column by the compressor is $7m$.

4.2. Characterization of pressure sensor

Figure 8 shows the pin diagram of MEMS pressure sensor MPXV10. The output voltage of the pressure sensor is observed for minimum and maximum level in the measuring tank. The output voltage obtained for the given range was 25mV to 60mV.

4.3. Calibration of Signal Conditioning Circuit

The output of MEMS pressure sensor MPXV10 is signal conditioned to generate micro-controller compatible signal of 0V to 5V. Zero adjustment is done to get 0V for 25mV and span adjustment is done to get 5V for 60mV using the circuits shown in Figure 4 and Figure 5. The readings were taken for changing the level from 0 mm to 100 mm. Figure 10 shows the graph of the output voltage versus level. This 0V to 5V output voltage is given as an input to the inbuilt ADC of micro-controller 8535. The internal reference voltage for the ADC was 2.5V. Using the keyboard and display the system can be calibrated for different ranges of level.

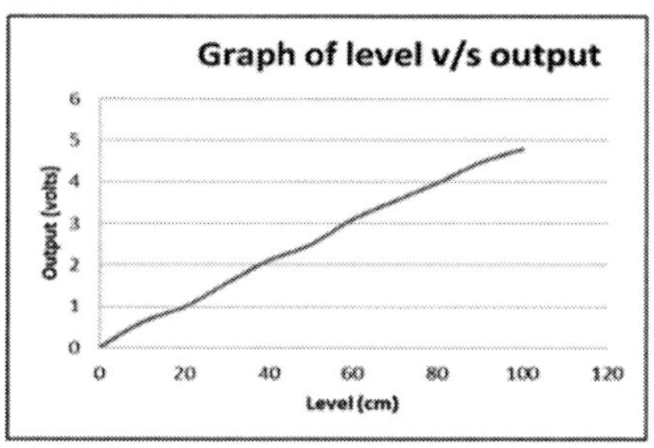

Figure 10. Gragh of Signal Conditioning Circuit Output Versus Level

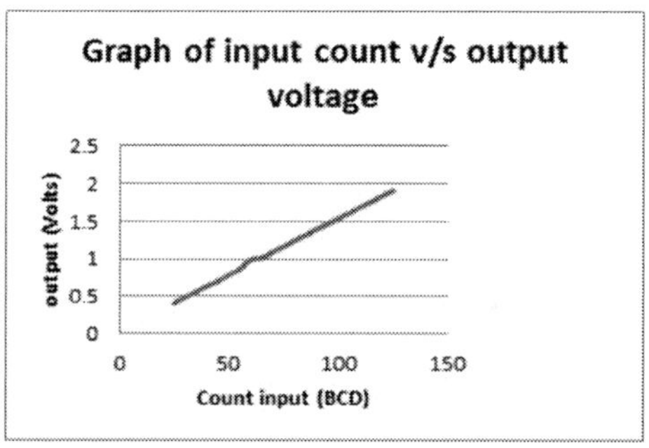

Figure 11. Characteristic of Transmitter Section

4.4. PWM and V to I Circuit

The PWM section is used as a digital to analog converter (DAC) which is used to drive the current booster for 4 to 20 mA output. The basic accuracy of output current is designed for 1 % of full scale. It means that the 4-20mA output has 100 discrete steps. It is a live zero system hence at 4mA, the PWM is proportional to 25 decimal count. And hence for 20 mA the PWM count is 125. The mapping of output current is done by referring to the ADC count from the input process variable. In the proposed system the input process variable is level. The level signal is transformed into voltage signal by using MEMS pressure sensor.

Following steps are followed for testing of standard transmission unit:

(i) The pulse width modulator is loaded with 25 decimal count in its output control register (OCR).

(ii) The input voltage to current booster is measured. It is found that it is 0.4V and the output current is 4mA.

(iii) The pulse width modulator is loaded with 125 decimal count in its OCR.

(iv) The input voltage to current booster is measured. It is found that it is 2.0V and the output current is 20mA.

Figure 11 shows the Characteristic trend of the transmitter section. It is linear with slope of 0.16 mA/count.

4.5. Calibration and Testing Procedure of the Developed System

Testing of the whole set-up is implemented for two settings (100cm and 80cm) and in two phases namely **calibration** and **performance testing**.

4.5.1. Calibration for 100cm level

(i) The dip-tube is fixed in to the empty tank to set the 0 % level by clicking the ZERO key on the circuit board shown in Figure 7. The display showed "0".

(ii) The tank is filled with tap-water up to 100cm level. The 100% level is set by clicking the SPAN key shown in Figure 7. The display showed "100"

4.5.2. Performance testing for 100cm level

(i) The tap is opened to drain the water up to 50cm and the output current is measured which was found to be 12mA and the display showed "50"

(ii) The level in the tank is then varied by pouring and draining the water and the output current is measured and the display is observed.

4.6. Calibration for 80cm level

(i) The dip-tube is fixed in to the empty tank to set the 0 % level by clicking the ZERO key on the circuit board shown in Figure 7. The display showed "0".

(ii) The tank is filled with tap-water up to 80cm level. The 100 % level is set by clicking the SPAN key shown in Figure 7. The display showed "100"

4.7. Performance testing for 80cm level

(i) The tap is opened to drain the water up to 40cm and the output current is measured which was found to be 12mA and the display showed "50".

(ii) The level in the tank is then varied by pouring and draining the water and the output current is measured and the display is observed.

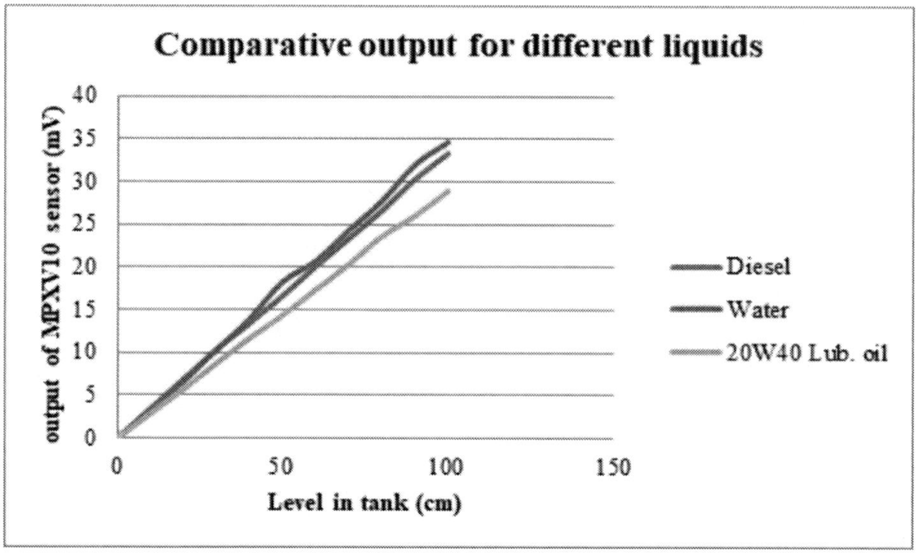

Figure 12. Graph of sensor output vs level for different liquids.

4.8. System setup testing for different liquids

The proposed system is tested for measurement of level in a fabricated calibration test set-up. There are three liquids used for testing the performance of the system. The results are as shown in Figure 12.

The proposed sensor can be used for measurement of level in a tank having narrow opening on the top. It can be used for acid, oil, water, lubricating oils etc. The output of the sensor dependent on the density of the subject liquid. Hence it is required to calibrate the system before use. The performance of the system is tested for Lubricating oil 20W40, tap-water and Diesel.

5. Conclusion

The performance of the system is tested on an experimental setup. It is found that the overall characteristic of the system is linear within ±1% accuracy. The system can be calibrated for range of 0-10cm to 0-100cm without degrading output current accuracy of 1% of F.S. The total current consumed by the system is 120mA at 12Volts. Hence, the overall consumption of the system is 1.4W which is substantially less. Due to low power consumption the system is operated on 12Volt batteries. The design and construction of the developed system is simple and is cost effective. The developed transmitter can be used for measuring level upto 7m. The system can be deployed for vessels having narrow openings. It will need opening just to insert the dip-tube. The system is made programmable using keyboard and display interface to micro-controller. It can be used for corrosive liquid by using SS316 grade tube. The system has an on-board indicator. The system can be safely used in hazardous area. The air is continuously flooded through the system hence the pressure sensor does not expose to the corrosive fumes from the liquid. The system is portable due to on board compressor. Due to this re-calibration and scheduled measurement is possible and this results in reduced power consumption. It is sensitive to specific gravity of the subject liquid. The system cannot be used for powder material. It cannot be used for viscous and sticky liquids. There is a scope for future development. The system has a micro-controller hence can be adapted to IoT platform for remote monitoring. The proposed system can be used for measurement of Specific gravity of the subject liquid by inserting the dip-tube at a fixed level. SD-card can be provided for data logging.

IOP Conf. Series: Materials Science and Engineering 1012 (2021) 012043 doi:10.1088/1757-899X/1012/1/012043

References

[1] Ali S F and Mandal N 2019 Design and development of an electronic level transmitter using inter digital capacitor vol 19 pp 5179–5185

[2] Doeblin E O 1966 *Measurement systems application and design* (New York: McGraw-Hill.)

[3] Lata A, Kumar B and Mandal N 2018 Design and development of a level transmitter using force resistive sensor as a primary sensing element vol 12 pp 118–125

[4] Singh Y, Raghuwanshi S and Kumar S 2018 Review on liquid-level measurement and level transmitter using conventional and optical techniques pp 1–12

[5] Patranabis *Principles of Industrial Instrumentation,2e*

[6] Webster J 1999 *The Measurement, Instrumentation, and Sensors: Handbook* (CRC Press) Electrical engineering handbook series ISBN 9783540648307 URL `https://books.google.co.in/books?id=b7UuZzf9ivIC`

[7] Liptak B 2003 *Instrument Engineers' Handbook, Volume One: Process Measurement and Analysis* (CRC Press) Instrument Engineers' Handbook ISBN 9781420064025 URL `https://books.google.co.in/books?id=a05NAvaqfcUC`

[8] Reverter F, Li X and Meijer G 2007 Liquid-level measurement system based on a remote grounded capacitive sensor vol 138 pp 1–8

[9] Khan S, Khalifa, Htike K K, Zahirul Alam A H M, Islam M R, Khalifa O O and Khan S S 2008 A non-contact capacitance type level transducer for liquid characterization *2008 International Conference on Computer and Communication Engineering* pp 1264–1269

[10] Lu G, Hu H, He B and Chen S 2009 A new-type sensor for monitoring oil-water interface level and oil level pp 2–981

[11] K V Santhosh Blessy Joy a S R 2020 Design of an instrument for liquid level measurement and concentration analysis using multisensor data fusion

A Low-Cost Portable Handheld System for Monitoring and Verification of Vehicle Documents

Vibin Mammen Vinod [1*], Govindasamy Murugesan[1], Mekala V[1], Deepa D[1], Abinaya S[1], Bairavi K[1]

[1]Electronics and Communication Engineering, Kongu Engineering College, Perundurai, Erode – 638060.

E-mail: vibin.ece@kongu.edu

Abstract. The 2013 World Traffic Safety Status Report from WHO reports that more than 231000 people in India are killed annually in road accidents. Our research seeks to develop a system/portal from which we can track a person's driving characteristics, such as traffic violation records, license details, insurance data, etc. Whenever residents refuse to comply with laws, a finger print scanner can help the police quickly access their records. Payment of penalty is initiated from his/her account immediately through a secured gateway. A two-level authentication and security scheme are employed such that the citizen would not have to compromise with their credentials. Whenever a citizen is caught breaking a law, the person's biometric scan is carried out to open his e-License account with OTP which is available as an application. Account provides name, age, location, photo, information of license valid or not and amount has to be debited from bank account. This application can be easily used by police officers to avoid the citizens from breaking the rules. The proposed design is compact and portable and has been tested with a minimal dataset of 79 persons and found to provide satisfying results.

Keywords: E-License, Android Application, Biometric Scan.

1. Introduction

The Global status report on road safety 2013 of WHO estimates that more than 231000 people are killed in road traffic crashes in India every year. Approximately half of all deaths on the country's roads are among vulnerable road users - motorcyclists, pedestrians and cyclists. The major reasons for road accidents can be attributed to multiple factors with reckless driving and disobedience to traffic rules being the primary causes. Our study aims to create a system/portal wherein we would be able to keep a tab on the driving traits of a person like his/ her history in traffic offences, license particulars, insurance particulars, etc [1, 6, 8]. The proposed system creates an easy and portable facility for the law enforcement agencies to identify a law-breaking citizen. The main objective of the proposed work is to prevent non-license citizens from driving and causing accidents. An important and very reliable human identification method is fingerprint identification. Fingerprint identification is one of the most popular and reliable personal biometric identification methods [6,8]. In this work we have developed the citizen's account attached with their license for payment of penalty charges of traffic rule breaking.

Whenever citizens break rules, the applicable charges of rule will be sent on citizens registered mobile numbers using SMS. OTP will be sent to citizens to conform further payment processes. The information about citizens is stored online on the server and citizens can have access to their account on the website of RTO. Citizens will get a message on their registered mobile number about insufficient balance on their account.

Indrajeet Nagesh Amble, et. al., [2], proposed IOT based e-license verification system using fingerprint sensor. Whenever a resident is found in breach of the statute, a biometric scan of the individual is carried out to access his e-License account from the database. The account includes name, DoB, License No., address and blood community and offers background records on infringements of fine settlement regulations. If the contract has been signed, the individual will be told of the breach of the law and the fine to be imposed for it. The resident would then collect OTP on the registered mobile number for payment of the fine. Then the citizen tells the received OTP to the police, the OTP is firstly verified by the system and then the settlement of the fine is done from the citizens' bank account or RTO wallet. Citizens are provided with their login ID and passwords so that they can also log in to their account on the database website to obtain information about their violated laws, all fine settlements, bank account deductions and RTO wallets. The resident is entitled to add, withdraw or change his deposit.

Komal Chorghade, et. al., [3] proposed RTO automation using QR code. The framework consists of three major components, i.e. admin, traffic police and customer. The customer signs with the service manager and the manager verifies the customer and documents the details of the user in the RTO database. The traffic police are also licensed by the controller. After authentication, the user signs in to the mobile interface and creates the QR code for the certificate. The QR code is downloaded from the RTO folder. The traffic police search the QR code and enter the vehicle number to verify the license and the car records. The traffic police will assess the challan and add it to the driver, and the challan data will be added to the RTO database.

Raghavendra, Sheddi, et. al., [7] proposed e-verification of driving license through Aadhaar database. In this article, the procedure is that a new applicator must send a hard copy of his text to the RTO administrator. The hard copy includes documents such as driving license, bank account numbers, and Aadhaar number. RTO officer's ought to take a thumb picture of the driver using a fingerprint scanner. Both of these documents are stored in the registry database via online registration. Since the police log in to the machine, they have to take the thumb print of the driver to check the thumb print in the RTO database. If it suits the records of the driver, the driving license will be checked on the basis of this detail. If no action has been taken as a result of the authentication, otherwise depending on the status of the driving license, a fine amount is produced and deducted from the online bank account of the driver.

Amrutha G. Bakale, et. al., [10] proposed on cross verification of vehicle and driver for RTO. The concept behind this paper is inspired by the E-RTO management framework and there is no requirement for the vehicle driver to hold the papers. Traffic police are gathering information directly from the Smartphone programme. Initially, the new applicator will supply the RTO administrator with the hard copy version of his paper. This information is maintained in a cloud database by online registration. And there's a server-side limit in JSP. An android programme will be issued to the police on the client side. The RTO database will collect car and license-related details after police logins into the system. If authentication fails, the police will be given the identification to find out more details about the person.

2. Methodology

The system proposed by us incorporates a biometric sensor(fingerprint) to access the data. The system is a rugged and sturdy handheld, easily portable such that the law enforcement authorities can carry around with ease. Connectivity with the server is brought about in two ways: i) Using GSM and ii)

Using a WiFi, if available. Provision of two sources of connectivity enables the user to have a choice of network depending on the data traffic.

Whenever a resident is found in violation of a statute, a biometric scan of the individual is done to access his e-License account from the database. The account also includes name, DoB, licence no, address and blood group with the past knowledge about the fine payments breach of laws. When the contract is signed, the party becomes aware of the broken law and the fine payable for it. The resident then collects OTP for paying the fine on the registered mobile no. Then the resident says the issued OTP to the authorities, the machine first verifies the OTP and then the fee is paid from the bank account or RTO wallet of the people. The person is supplied with his / her username identification number and passwords so that he / she can also report on the network page and collect updates about his / her violations, all fine transfers, bank account deductions and RTO wallet descriptions. The person has provision for inserting or deleting from the database or modifying their bank account. Block diagram below describes this device well.

Figure 1. below shows the block diagram of the portable verification device. This figure depicts the integration of the various input devices and sensors with the microcontroller. It also depicts the inter-relation between the microcontroller and the cloud server.

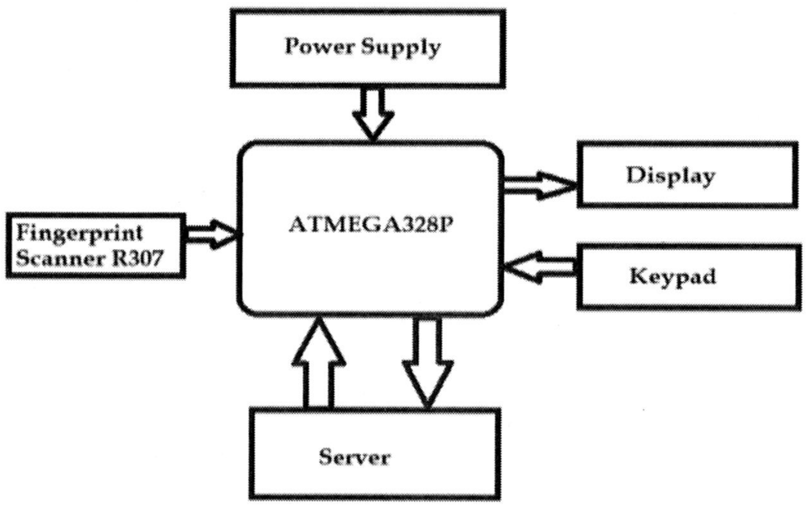

Figure 1. Block Diagram of the Proposed Verification Device

Figure 2. below shows the flow diagram of the proposed workflow. This flowchart depicts the flow of the process sequentially.

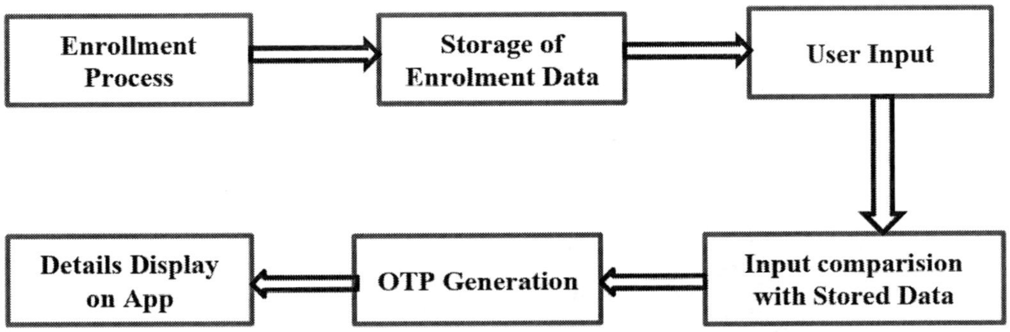

Figure 2. Software flowchart of the proposed Verification Device

The below diagram is the complete circuit diagram of the proposed system, a low cost portable handheld system for monitoring and verification of vehicle documents. This circuit diagram consists of ESP8266, R307 Fingerprint sensor, Arduino and Liquid Crystal Display.

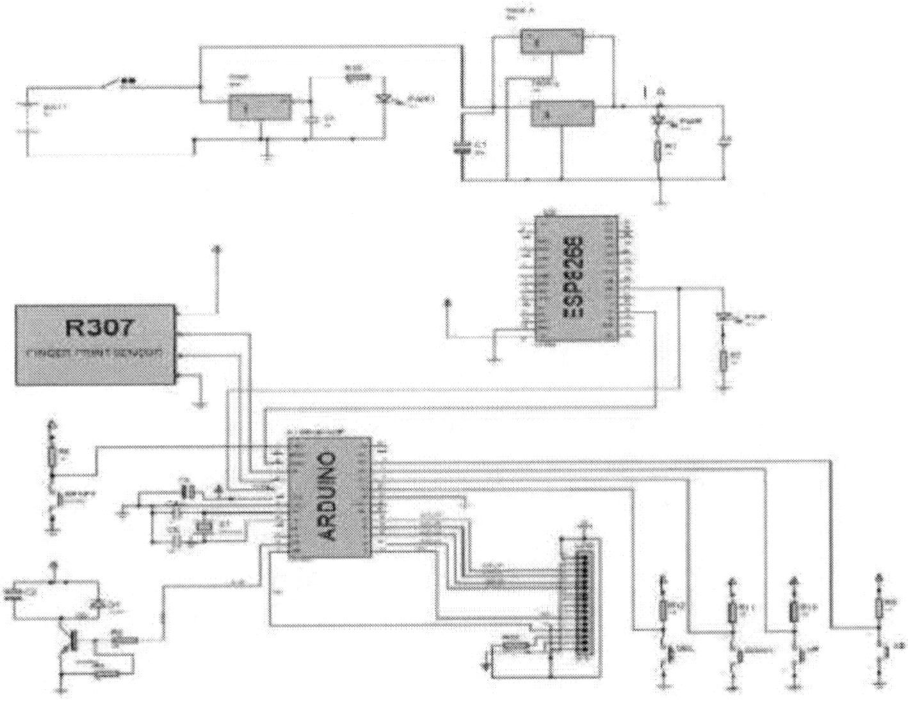

Figure 3. Circuit Schematic of the Design on Proteus Simulator

3. Results

A hardware setup of the design is created as shown in **Figure 4.** ESP8266 is the heart of the system ensuring proper access and communication with the webpage. Software development is created with an android app builder that acts as a drag and drop method based on the MIT App inventor. It is also known as the Makeroid app. It acts as a top-level component in the android version and suitable for mobile app development. It also hosted as a google cloud platform in which the result is safely stored in the server. Without any code language, the result can be directly monitored by using Wi-Fi module IP Address After installing a modular companion, QR code scanning process will be performed for seeing the result in offline mode.

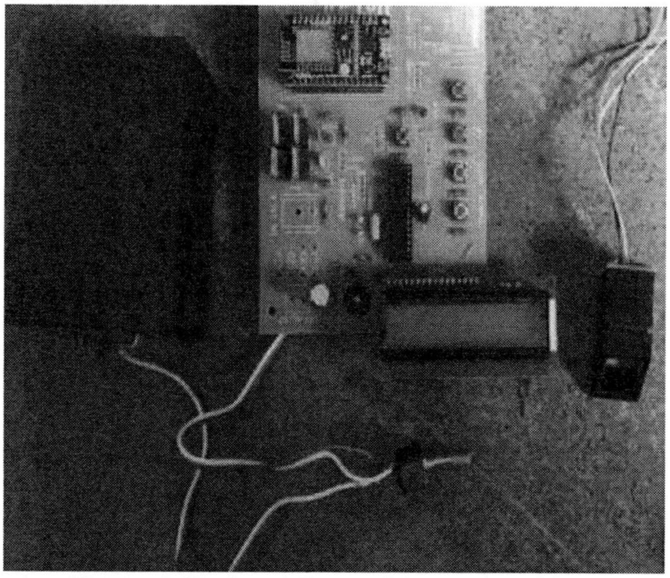

Figure 4. Hardware Setup of the proposed system.

It is installed for no internet access and QR code is configured. After scanning QR code, automatically Network ID password will be displayed. By using this password, we can directly connect to the android phone for monitoring the result. As the result is displayed in an online builder called Arduino IDE. Without exporting, a modular companion is enabled to display the monitoring value by IP address by connecting it with the builder. It instantly has more features for accessing the result. It is also used as an extension IDE for advanced uses. Here, monitoring result using Arduino ide and modular companion has similar result for accessing purpose.

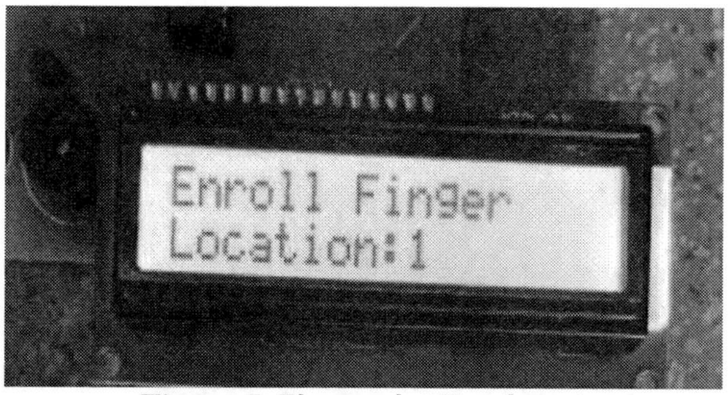

Figure 5. Fingerprint Enrolment

Figure 6. Fingerprint Match

Provision to enrol and delete the biometric feature from the system is provided. Users can opt for two different modes of authentication either through biometric scan or an OTP based method.

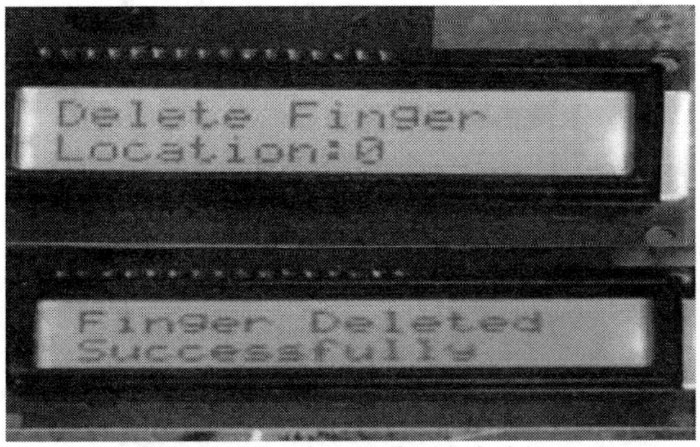

Figure 7. Fingerprint Deletion

Figure 8. Screenshot of the proposed App

The verification and authentication procedures are made easy and convenient with a custom-built android app. Separate consoles are available both for the law enforcement agencies and the citizen. Based on the level of access, varying features are provided.

4. Conclusions

The designed system has drastically reduced the manual work of the RTO system. The scheme incorporates a traffic control service to review and take measures to verify the driving license. It also lets RTO authorities efficiently keep documents and removes a lot of red tape and manual labour. It stops RTO abuse and also gives an advantage that citizen need not carry documents always.

With the various technologies, including face recognition and number plate recognition, the functionality can be improved. It can be further improved to give drivers reminders of expiry dates of records, search car details including the RC book, pollution checks, insurance and others. It can be tested in an environment in real time.

References

[1] Sailaja D, Sri MN, Samuel P. An Biometric-Based Embedded System for E-Verification of Vehicle Users for RTO. In. *International Conference on Intelligent Computing and Communication* 2019 Jun 7 (pp. 711-718). Springer, Singapore.

[2] Kellow Pardini, Joel J. P. C. Rodrigues, Sergei A. Kozlov, Neeraj Kumar & Vasco Furtado. 2019. IoT-Based Solid Waste Management Solutions: A Survey *Journal of Sensor and Actuator Networks* **8**(1), 5.

[3] Manju Mohan, RM. Kuppan Chetty, Vijayram Sriram, Azeem, P. Vishal, and G. Pranav. 2019. IoT Enabled Smart Waste Bin with Real-Time Monitoring for efficient waste management in Metropolitan Cities In. *International Journal of Advanced Science and Convergence* **1**(3) 13-19

[4] Sharma GA, Sarade AB, Gupta SO, Janbhare SA, Mukhopadhyay NI. E-driving license and RC book verification system using QR code. *InProceedings of 65th IRF International Conference*, ISBN 2016 Nov (Vol. 978, p. 93).

[5] Ratakonda B, Therala A, Hanumanthu CK. Driving license detection using QR code. *InE3S Web of Conferences 2020* (Vol. 184, p. 01010). EDP Sciences.

[6] Vinod VM, Thokaiandal S, Sindhuja CS, Mekala V, Manimegalai M, Prabhuram N. A Comprehensive Study On Academic And Industry Authentication And Attendance Systems. *International Journal of Scientific & Technology Research*, Vol. 9, Issue 3, 2020.

[7] Orazio Miglino, Henrik Hautop, Stefano Nolfi. Evolving Mobile Robots in Simulated and Real Environments *Artificial life* **2**(4) 417-434.

[8] [8]Mekala V, Vinod VM, Manimegalai M, Nandhini K. Face Recognition Based Attendance System. International Journal of Innovative Technology and Exploring Engineering(IJITEE), Volume-8 Issue-12, October 2019, pp. 520 - 525.

[9] Michel, O. 2004. Cyberbotics Ltd. Webots™: Professional Mobile Robot Simulation *International Journal of Advanced Robotic Systems* **1**(1) 5.

[10] Marigowda CK, Thriveni J, Karangi JK. A secured transport system by authenticating vehicles and drivers using RFID. In *Advances in Computer Science, Engineering & Applications* 2012 (pp. 313-321). Springer, Berlin, Heidelberg.

[11] Kumar MV, Kumar SR. Fingerprint based licensing system for driving. *International Journal of Advanced Research in Computer and Communication Engineering*. 2014 Sep;3(9).

[12] Divya D, Padmasarath S. Finger vein based licensing and authentication scheme using GSM. *IOSR Journal of Computer Engineering*. 2013 Nov;15(3):30-5.

[13] Rubella JA, Suganya M, Senathipathi K, Kumar BS, Gowdham KR, Ranjithkumar M. Fingerprint based license checking for auto-mobiles. In *Fourth International Conference on Advanced Computing (ICoAC)* 2012 Dec 13 (pp. 1-8). IEEE.

[14] Mudholkar SS, Shende PM, Sarode MV. Biometrics authentication technique for intrusion detection systems using fingerprint recognition. *International Journal of Computer Science, Engineering and Information Technology (IJCSEIT)*. 2012 Feb;2(1):57-65.

[15] Lourde M, Khosla D. Fingerprint Identification in Biometric SecuritySystems. *International Journal of Computer and Electrical Engineering*. 2010 Oct 1;2(5):852.

Smart Guidance System for Blind with Wireless Voice Playback

B Sumathy[1*], K M Pavithran[1], N Nizam[1], V A Surya[1]

[1]Instrumentation and Control Engineering Department, Sri Sairam Engineering College, Chennai – 600044.

E-mail: bsumathy.ice@sairam.edu.in

Abstract. The visually impaired/blind people face many challenges and seeks help from the other people even for their daily routine life due to various complications. Also, they could not go independently outside, even for their basic needs and mentally depressed due to their current inability. A wearable navigation system/device is proposed and developed to help the blind people as well as the partially blind people or challenged people. The devised unit/ prototype consists of the displacement measuring sensor for the detection of the obstacles and depth. The data received from the IR and Ultrasonic sensor is feed into the Arduino UNO Microcontroller and it is processed. The output from the controller is given to the impaired people through the wireless voice playback modem, to alert them when they found obstacle or collision. The receiver end receives the corresponding current position values through the navigation systems and the static, dynamic condition of the impaired people is measured using mems accelerometer and the same information (latitude, longitude) is shared as SMS to their family members or guardian to monitor them, during any emergency. The proposed guidance cane is a cheap, simple, user friendly and smart device.

Keywords: Visually impaired - Arduino - accelerometer- voice playback – latitude-longitude.

1. Introduction

All people are equal before nature and world but due to various reasons like birth disabilities, accidents, poor nutrition some people lost their eyesight or partially blind. The percentage of vision blindness is more in developing countries and under development countries when compared to developed countries. Also, childhood blindness is crucial, and those children should be given their basic rights and claims to live in the world. So, considering the overall difficulties and social movement of these people, the researchers tried out many research ideas and model to help them with the most priority of findings. Visually impaired people are those who cannot be able to identify the things which is near to them. In this case they seek help from the family members, and they need special care. Family members need to spend more time with them to do their daily activities and for family members to do the personal work, is an additional burden. The blind people hesitate to come out, but they have lot of talents when compared to others. As part of our society, responsibilities should be given to normal people, to help them to explore their knowledge. In olden days, the dogs were used to guide them while travelling. Now the technology has grown, and lot of instruments have been invented for guidance purpose. It is a low-cost instrument, and it is user friendly for both visually challenged people and for family members.

Visually impaired people need some aiding devices related to the blindness disorders. Out of 100 people, 10 people are having poor eyesight and eye related complications. They are also finding hard to walk on their own safely, to do their own needy works, without others help [9]. To solve such kind of issues, the smart devices are designed. The visually impaired people are facing many difficulties while comparing to others and the family members also facing difficulties to monitor them. Now-a-days the visually impaired people are proving their talents in various fields. The impaired people are also a part of our society so, we have some responsibility and contribution to make their life happy in this world. By the contribution of many peoples, the following devices are found namely Sound based or Sonar instruments[10], Walking cane which smart decisions and reporting [11], Intelligent observation/vision systems [12], walking/ governing stick [13], These devices use the displacement findings technique for detecting the hurdles and barriers around their surroundings by considering all the distance parameters. Most of the situations, the depth of the position of particular location on their pathway is used by sending and receiving the ultrasonic waves Thus location could be precisely measured and the same can be transferred to their attendees when they are in needy circumstances. Systems like VOICE [14], Sound View [15], SVETA [16] and CASBLIP [10] are also developed due to research advancement in this area. The basic operation of above units is explained below. The images are captured by the single camera and it is processed and converted into audio message and it is transferred to impaired people. This audio messages will really help these to find out their whereabouts.

There are many challenges and difficulties faced by the impaired people and all cannot be listed here. But the research findings and models are trying to fulfil the needy people at least for their basic support to do their work independently without others help. There are some difficulties in detecting the obstacles such as downward stairs, holes are hidden but they are dangerous to the blind people. Usually vibration or sound signals are coming out as the feedback information. Many devices are invented using the principle of sound, frequency, and vibration and those are gaining importance in recent years. Any accidents or hit can be sensed by the particular sensors attached to their smart devices or cane with the help of the change in frequency or disturbances/vibrations. So, the Smart Guidance System is proposed to overcome the disadvantages. The format of the paper is as follows with the following contents. First division of the contents is introduction and second one is literature survey which explained the various contributions related to Smart Guidance System for Blind, third division dealt with the presented methodology which consists of all basic to important components. In fourth division outputs and explanation are narrated and finally future works and conclusion are explained.

2. Literature Survey

In [1], the authors proposed a blind stick for visually impaired people for navigation. They designed the system and implemented the same in the smart pole for proper hurdles identification and path finding purposes. The ultrasonic sensor is used for detection of obstacle and the buzzer is used for alert. The system is used for limited distance and the visually challenged person can locate only with that limited area. Also, the stick provides the personal independence to the impaired people. In [2], the developed a cheap, user friendly smart blind guidance system and used for the mobility of visually challenged people in the limited distance. Infrared sensing element is utilized to measure the reflected range by the barrier. The reflected range and overall system are controlled with the microcontroller. Effective smart stick for Blind people [3], is developed using infrared sensor to detect the up and down movement in the stairs. Sensing element using ultrasonics is meant for obstacle detection and the overall system is controlled by microcontroller. The entire integrated system is wired, and the future scope may be extended for wireless.

Ultrasonic Blind walking Stick [4], is developed using Ultrasonic sensor. The main aim of this instrument is to overcome the disadvantages of traditional stick. The ultrasonic sensing element is utilized to sense the obstacle present in sight of visually impaired individual. This device can sense accurately the slopes, up and down places. This blind stick can detect the hindrances and hurdles especially when the impaired people walking alone in the higher sloppy regions by their own. Mishra and Koley [5] developed a very low-cost equipment to find the location and whereabout of the sight affected or blind person. The aim is to completely minimize the use of the guide for path finding of the location for the vision weaken person in remote area and to develop the low-cost equipment. GPS tracking system is used for location information and the ultrasonic sensor for the obstacle detection. The system also incorporated a voice-based system communication protocols for transmission and reception of data.

In [6], the authors developed an electronic traveller aid for visually blind people .The above system consist of single board processing system .In this kit the author capture the image which is in front of blind and it is processed by the Fuzzy clustering algorithms as real time images and it is transferred to blind people with help of stereo earphones the image processed and result is given as stereo sounds. The system is based on the various image processing steps which automatically locate the position and location which is based on the intelligent decision-making techniques namely fuzzy logic systems. Smart Stick for Blind People [7], is proposed and the same is used to change the lifestyle of sight affected population. In this system the ultrasonic sensing element is utilized to identify the barrier and Arduino uno is used as microcontroller. The location tracking is implemented in the system to track the visually impaired people by the family members it is particularly useful to monitor them.

Many methods and devices are proposed and implemented by the researchers in helping the visually challenged people, still it is open for more findings and developments due to the latest technological developments. A day today survey is really required from those people to find their difficulties, pain and struggles to march further in this findings and developments. One of the recent and notable development in this smart guidance system is Indriya meaning an organ, which is a handheld small device or companion for the blind people. AI technologies are also incorporated in recent development of products for visually impaired. One important product may be considered is the AI glasses which incorporates intelligent techniques for location finding and hindrance identification and it is a real aid or friend to the blind people. A small attempt is made to fulfil the basic needs and requirements of the visually challenged people is reported in the presented method.

3. Proposed System

In the proposed methodology, the Smart guidance system is designed using the Arduino uno microcontroller to process the input data. The MEMS accelerometer gives its output in all three coordinates for any roll off or of normal motion of the people under investigation. This accelerometer collects the data in three axes and combined with obstacle detection parameter values, decision is taken based on the optimum values. A predefined roll off or normal motion of the people is fed as the reference values. When these optimum values are exceeded both in the case of accelerometer or vibration sensor, or due to the combined effect of both, the controller will transmit the necessary instruction and information. The ultrasonic sensor detects the obstacle and the depth of the route or pathway and it is given as an input to the Arduino uno microcontroller and the given data are processed and the output is given to the impaired people through the wireless voice playback modem. The wireless voice playback is used for the internal communication between the device and the impaired people. The GPRS identify the position and the location of the people by longitude and latitudinal basis. A proper internal power supply unit is also built upon in the device to energise the same. Safety precautions and limit of power supply is optimized for biomedical applications. The overall working flow of the presented methodology is manifested in **Figure 1.**

The proposed methodology is followed by hardware and software descriptions. The following section explains about the hardware components and their specifications involved in the proposed system. The most important components are MEMs accelerometer, Ultrasonic sensor, GSM module, Arduino Controller etc.

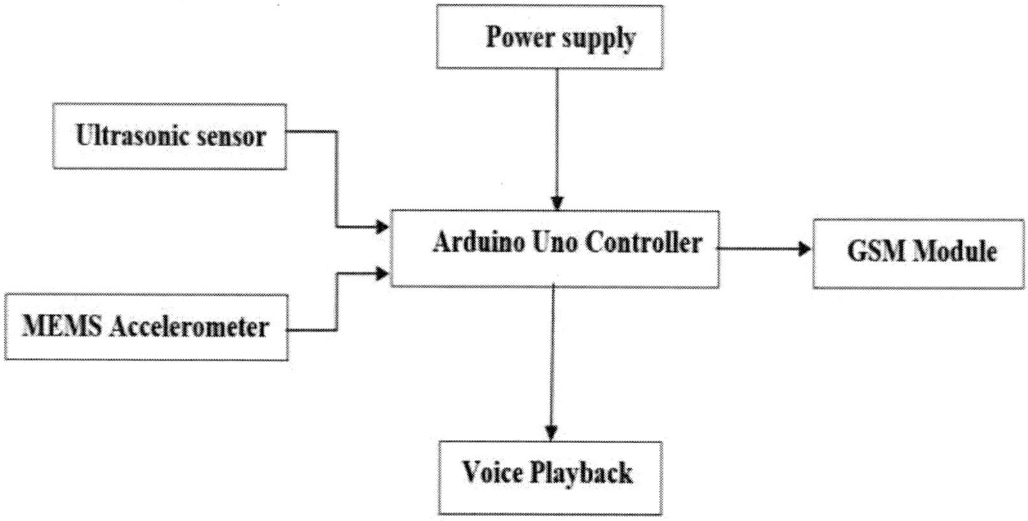

Figure 1. Overall Working Flow Model of the Presented Methodology

3.1. MEMS Accelerometer

The structural representation namely the functional blocks of MEMS accelerometer is manifested in **Figure 2**. In the proposed system the MEMS accelerometer is used to find the static and dynamic position of the impaired people.

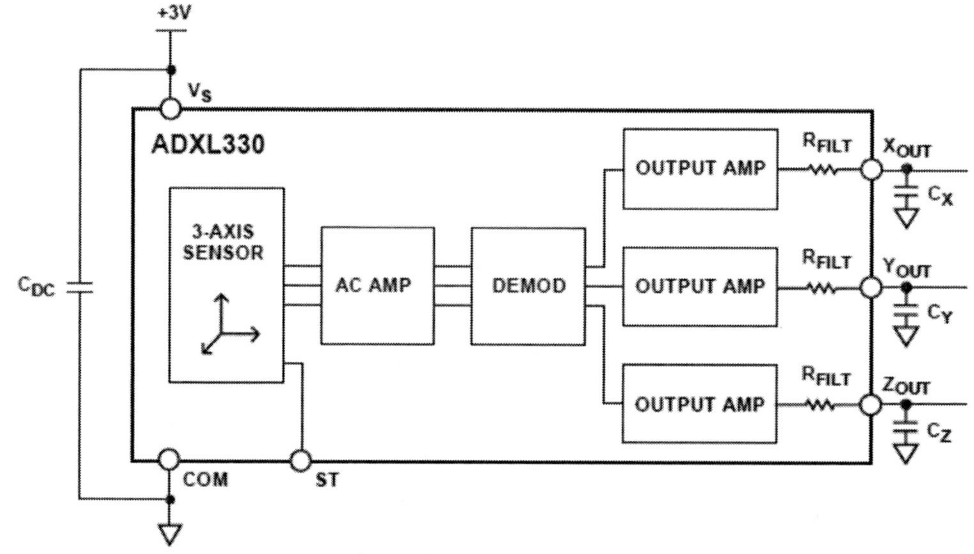

Figure 2. Block diagram of MEMS accelerometer

The MEMS accelerometer is electromechanical component, which is used to find the motion. The input of the MEMS accelerometer is mechanical moving, and the corresponding output is in the form of 3D graphical representation. The output of the MEMS accelerometer is sent to their family members via the GSM modem whether the impaired people in the static or dynamic position. MEMS accelerometer used is the 3- axis sensor because it measures the motion of the object in all the three co-ordinates and shows the output in only one axis. MEMS accelerometer is electromechanical component it is also used as vibration sensor. The MEMS accelerometer is used to measure both static and the dynamic motion of the person. If the person is moving, the corresponding movement is shown in the X-axis as variation. The accuracy rate of accelerometer is remarkably high.

3.2. Ultrasonic Sensing Element
The operation of ultrasonic sensing element is shown in the **Figure 3**. The ultrasonic sensing element is utilized to recognize the obstacle, disturbances, holes, and cavities on the road while making the movements. The sensor has better accuracy with respects to obstacle detection.

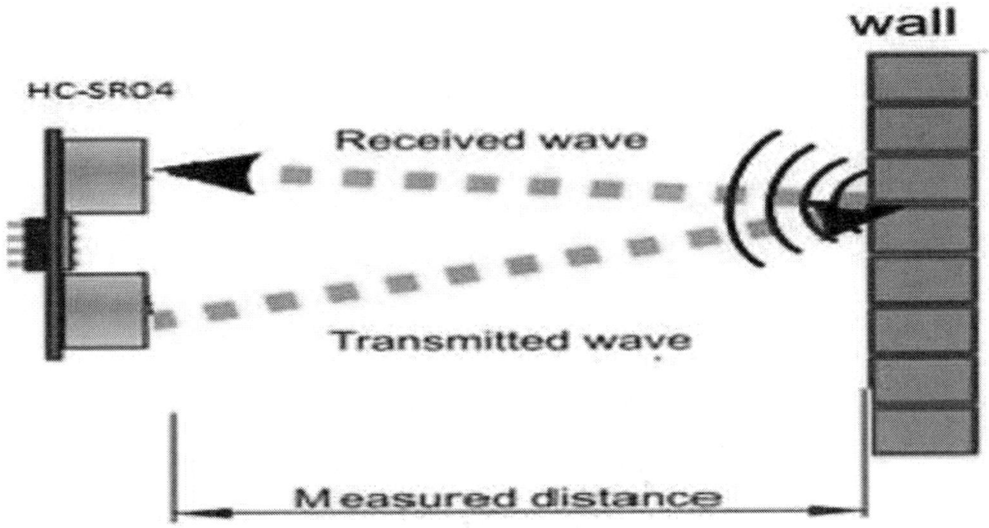

Figure 3. Ultrasonic Sensor

The short and high frequency sound waves is emitted by the ultrasonic sensor and it fall on the obstacle and it reflect as the echo signals to the receiver end sensor. The distance is calculated based on the time duration between the emitting and receiving of the echo signal. The ultrasonic sensor detects the cavity in the road and the distance range of the obstacle, and it is transmitted to the microcontroller. The output is processed, and the information is sent to the visually impaired person through wireless communication. The distance covered is between 2 cm to 4 m for finding the obstacles. This ultrasonic sensor basic principle is utilized in many smart guidance systems because of its simplicity in operation and able to detect the obstacles and hurdles well in advance to avoid the collision or fall off during movements.

3.3. Standard Mobile communications System part
Standard global mobile communication system shortly called as GSM part projects and works with a standard range of frequency extending from 850 MHz to 1900 MHz.Via GPS module the GSM module send the location information as message to the family members or to the guardian. This module part is very small and able to accommodate with in a small space in the user application devices like smart mobile sets, tablets, smart watches etc.

3.4 ZIGBEE
Zigbee is used to transfer the data from MP3 module to the visually impaired people. The high-level protocols communication used by ZIGBEE is formulated from IEEE802.1 which is the IEEE standard for WPAN. It is the abbreviation for Wireless Personal Area Networks. For the device which requires the wireless network, ZIGBEE is easily designed and used for short personal distance. ZIGBEE supports the transfer of data coming from the sensor at the rate of 250 kbps. ZIGBEE can operate at various frequencies such as 868MHz,902-968MHz and 2.4MHz. The battery life is significantly

improved due to it consume extremely low power. It is used to transmit the instruction via wireless voice playback to the visually impaired people. It is used to share the instruction to the visually impaired people for short distance communication.

3.5 ARDUINO UNO

Arduino uno **(Figure 4)** is used in the proposed system as a micro controller to control the process. It is an open source microcontroller based on the microchip AT-mega 328P microcontroller and developed by Arduino cc. It is an 8-bit microcontroller and has both analog and digital inputs and outputs. Various expansion boards and circuits can interface with this board. This board contains 14 input and output lines which is in digital form and 6 lines of it are pulse width modulated lines, six analog input lines are also available and has programmable IDE. The Arduino uno collects the data from the sensor and GPRS, GSM and it is processed, and the processed data is transmitted through the transmitting device to alert the impaired people.

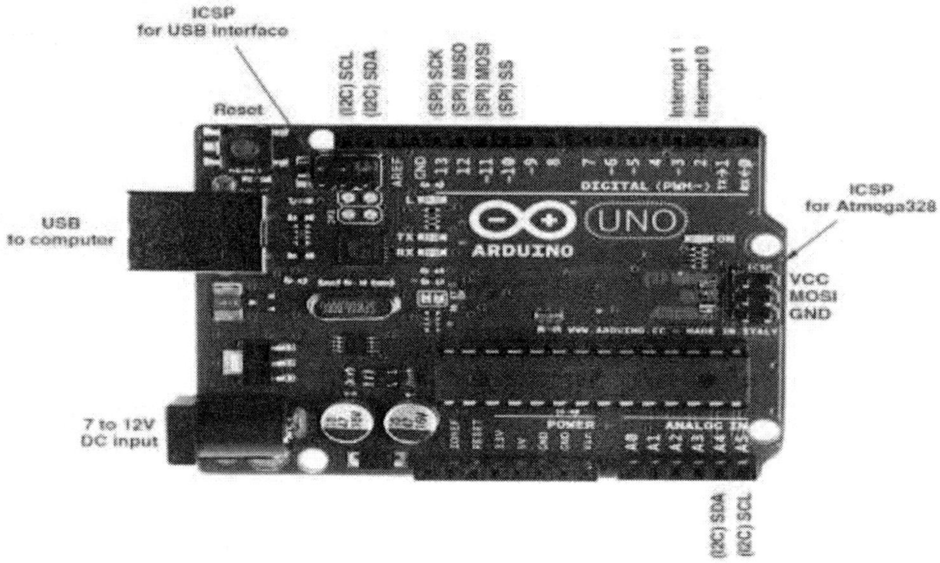

Figure 4. ARDUINO UNO

3.6 Programming in Embedded C

Programming in Embedded C is one of the important software module part and it is widely applied in many real time applications. It is beneficial to utilize the high-level programming and optimization of this embedded language. The signals are generated in the trigger pin of the pulse sensor and the trigger pulse falls on the object and it will return through the air medium. The pulse which return is known as the Echo pulse. The echo will reach the receiver the distance of the echo pulse is calculated by the received signal. The distance of the obstacle is conveyed via voice playback. The loop is continued till the person reach the destination.

3.7 Flow Chart

In **Figure 5**, the flowchart of the proposed system is shown. The trigger pulse is generated from ultrasonic sensor which is giving a high frequency sound waves and is focussed to fall on the object and received back as echo signal. There are two methods available one is based on time transit method and other is based on the frequency method. In time transit method, the difference between the transmission and reception of data is used. The frequency-based signals are transmitted and received if any obstacle is found. Based on the distance between transmitted and received signal, the output is generated. The output can be generated as a voltage in terms of milli volts. Depends on the output the voice play back message is sent. The output values along with three axes accelerometer values a decision is taken and sent as an alert message for any deviations from optimum values.

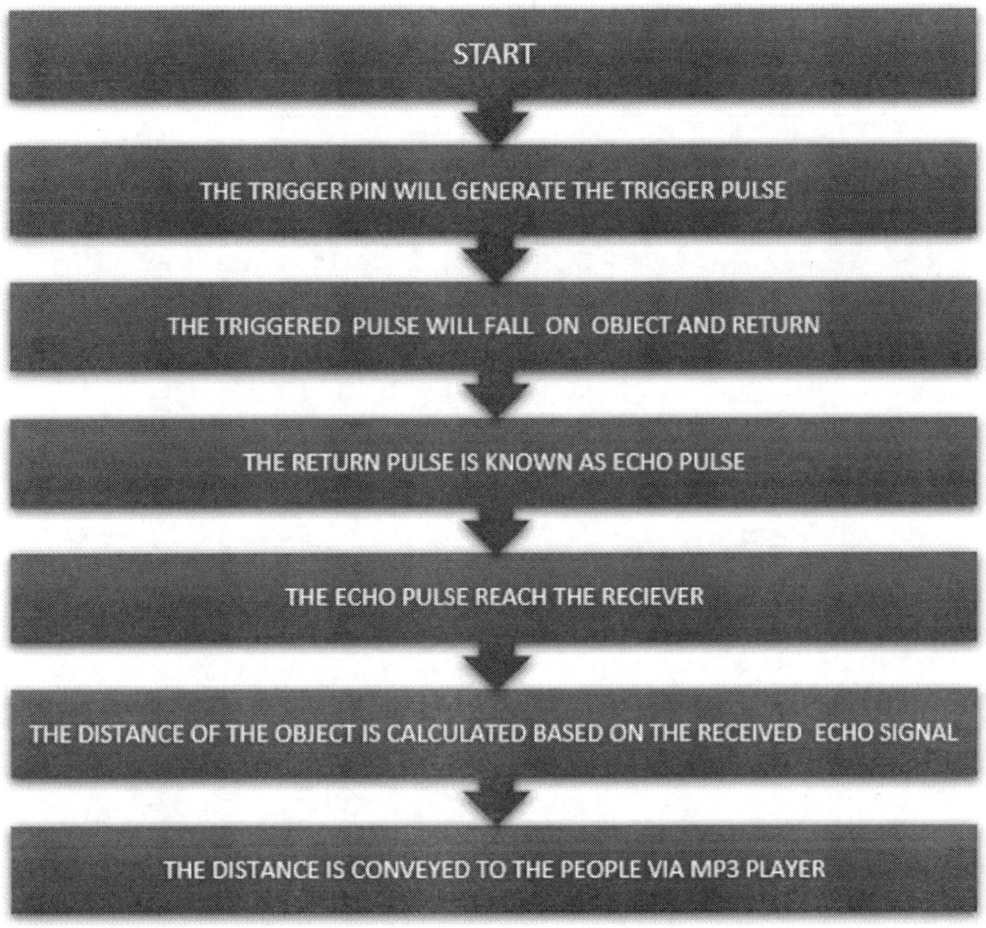

Figure 5. Flow chart of the Proposed System

4 Outputs and Explanation

The output from all the sensors, controller is verified separately, and cumulative value is tested and verified. The ultrasonic sensor is working based on the echo/reflection principle, so more consideration is given to study about the back scattering and receiving of waves for different objects. The measurement is starts with the microcontroller and the microcontroller send the triggering pulse to the ultrasonic sensor, the sensor will send out the signal with 40KHz and it is received back as reflected wave after falling upon the obstacle if there is no obstacle the emitted wave does not reflect back. The simulation test is done, and the analog value is calculated for different distance and the theoretical value is also calculated. The visually impaired people can use this smart guidance system to the maximum coverage area of 4 m.

The output from the ultrasonic sensor called as displacement measurement sensor is tabulated as below. (**Table No I**). The output is the variation of displacement as a functional value of voltage. The error value is noted between the output voltage between measurement of calculated and measure value. The output from the ultrasonic sensor gave the correct when compared with the calculated value. As a result, the output accuracy is strengthened. Also, the output is a linear value.

Table 1. Results of Displacement Measurement Sensor

Displacement (cm)	Output voltage Theoretical(mV)	Output voltage Actual (mV)	Difference
5	24	23	1Mv
10	48	46	2mV
20	98	95.3	2.7mV
30	147	143.8	3.2mV
40	195	190.5	4.5mV
50	247	241	6mV
75	355	348	7mV
100	500	489	11mV
150	745	730	15mV
200	950	926	24mV
250	1250	1220.7	29.3mV
300	1450	1410.6	39.4mV
350	1735	1691	44mV
400	2000	1953.2	46.8mV

The below graph shown in the **Figure 6**. shows the variation difference between the actual and calculated value for various distances, for samples (X-axis). Y-axis indicates the measured and calculated value as an integrated output.

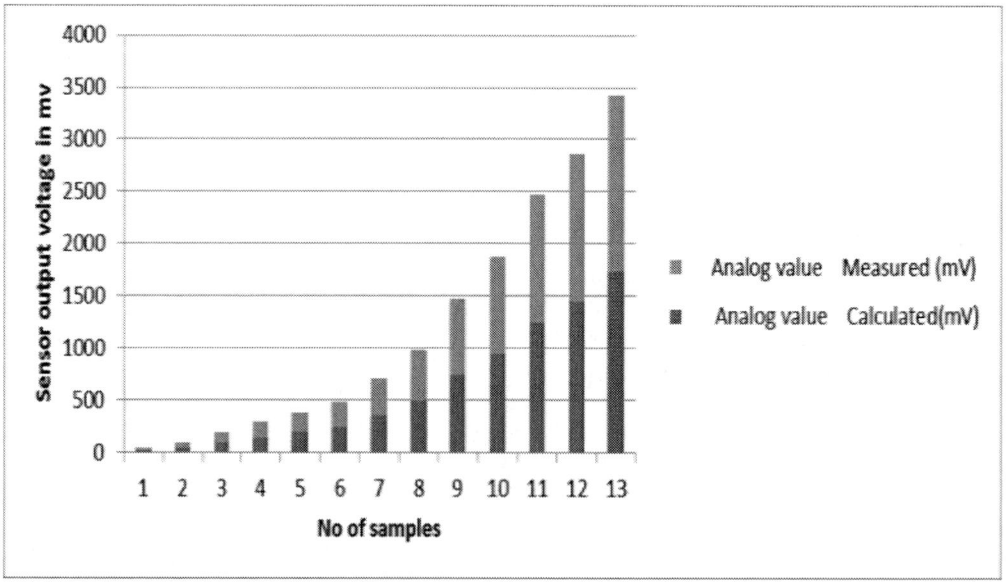

Figure 6. Ultrasonic sensor output (calculated and measured value)

The graph below shown in the **Figure 7,** shows the increase in the analog value of output from the sensor. A standard linear function is maintained throughout the measurement for respective distance.

IOP Conf. Series: Materials Science and Engineering 1012 (2021) 012045 doi:10.1088/1757-899X/1012/1/012045

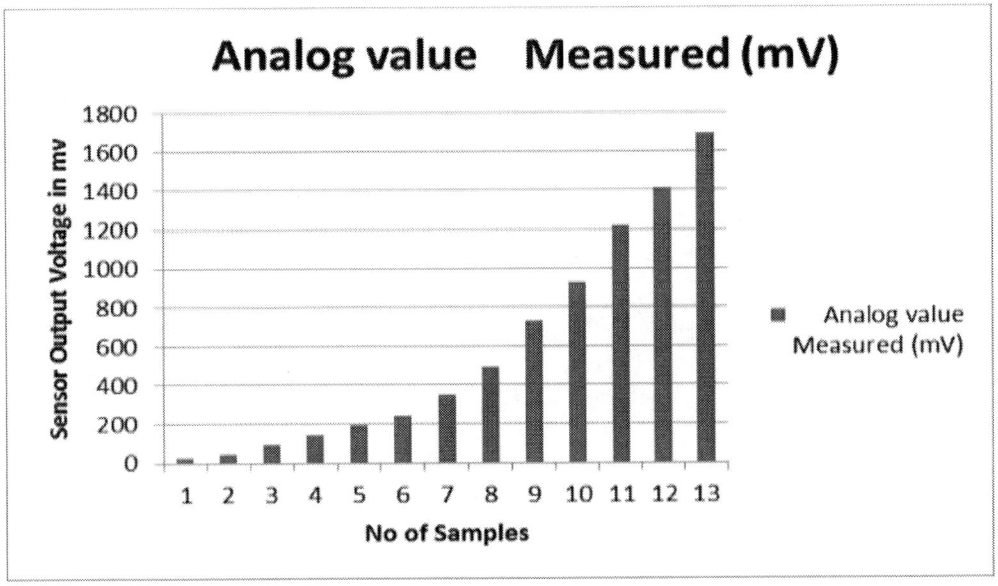

Figur
e 7. Analog Measured output voltage

MEMS accelerometer is used to find the static or dynamic position of impaired people with its 3-axis sensor. The accelerometer senses the position values in all the three quadrants X, Y, Z. If the accelerometer output is a constant value/ static position of the impaired people for prolonged time, automatically it will send the message to the guardian and family members for their respective registered mobile number via GSM modem. Location data will be conveyed through their relative and people of close contact as lengthwise and breadthwise values. (Latitude, Longitude values). The family members will be receiving an alert message and depends on the distance and traffic, the person could be safeguarded immediately., with continuous monitoring. Also, the device is a user-friendly device for both educated and uneducated peoples and for various age group peoples. The output as location data and alert message to emergency contact 1 and contact 2 is shown in the **Figure 8.** and **Figure 9.** Also, the alert message as a guidance is sent as a audio playback to the impaired people also.

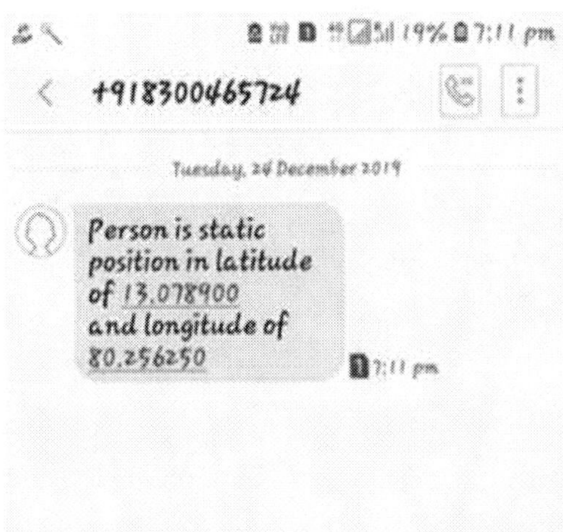

Figure 8. MEMS Accelerometer sample output- Contact-1

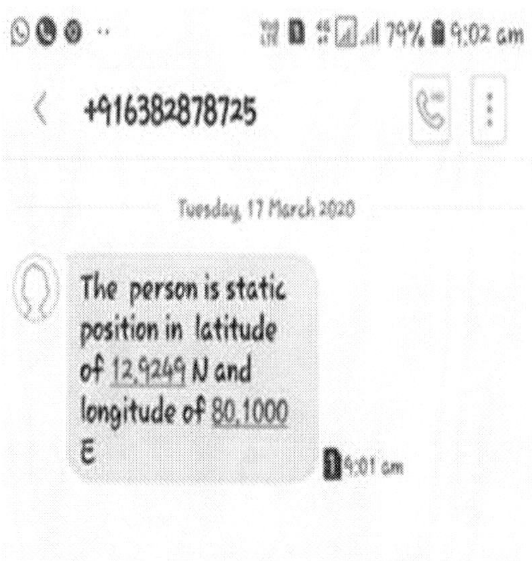

Figure 9. MEMS Accelerometer sample output- Contact-2

Finally, If the visually impaired people is in static position, for a predetermined period of time and also any collision/ any emergency situations, the current latitude and longitudinal data of the current whereabouts of the blind people is conveyed through the registered mobile number for their guardian. Latitude and longitude have specific value, and it does not merge with any other location values. Also, it is quite easy to access by local authorities.

5 Conclusion and Future works

The main objective of devising the blind and visually impaired people safety navigation system is proposed and the same is implemented and explained in this paper. The small and light weighted device is quite easy to carry with them while they travel alone and it will completely guide them during the entire travel or independently do their own work, without others help. Also, their family members feel free while they travel alone because it gives information/alert to their family in case of emergency. The MEMS accelerometer will respond even for a small shock, so that the difference output or error could not be easily predicted. As, a future work, the same system could be extended for avoidance of small external disturbances as a function of output. As an extendable work, in case of any accidents/emergency, the same information could be integrated to the nearest hospital/ health centres to immediately act for life saving of the people concerned. Also, along with the latitude and longitude values, the correct address will also be included in the future works.

References

[1] Manikanta K S, Phani T S S and Pravin A. 2018. Implementation and Design of Smart Blind Stick for Obstacle Detection and Navigation System *International Journal of Engineering Science and Computing* **8** (8).

[2] Amjed S. Al-Fahoum, Heba B. Al-Hmound and Ausaila A. Al-Fraihat. 2013. A Smart Infrared Microcontroller-Based Blind Guidance System *Hindawi Publishing Corporation Active and Passive Components* **2**.

[3] Ayat Nada, Samia Mashelley, Mahmoud A Fakhr and Ahmed F Seddik. 2015. Effective Fast Response Smart for Blind People *Second International conference in Bioinformatics and environmental engineering*.

[4] Vijayalakshmi Badre, Romba Chhabria, Tanmay Kandam and Kritika Karamchandani. 2016. Ultrasonic Blind Walking Stick with Voice Playback *International Research Journal of Engineering and Technology* **03** (04).

[5] Somnath Koley and Ravi Mishra. 2012. Voice Operated Outdoor Navigation System for Visually Impaired Persons *International Journal of Engineering Trends and Technology* **03** (02).
[6] Sainarayanan G, Nagarajan R and Sazali Yaacob. 2007. Fuzzy Image Processing Scheme for Autonomous Navigation of Human Blind *Applied Soft Computing* **07** (01) 257-264.
[7] Manikandan Shanmugam, John Victor, Mayank Gupta and Saravana Kumar K. 2017. Smart Stick for Blind People *International Journal of Trend in Research and Development.*
[8] Mohamedarif Regade, Bibi Ayesha khazi S, Sushmita Sunkad, Mrityunjay C K and Sharda K S. 2019. Smart Stick for Blind using Machine Learning *International Journal of Innovative Science and Research Technology* **04** (05).
[9] National Disability Policy: A Progress Report. 2014. *National Council on Disability.*
[10] Terlau T and Penrod W M. 2008. Sonar Curriculum Handbook
[11] Whitney L. 2009. Smart cane to help blind navigate.
[12] Hans du Buf J M, Barroso J, Jojo M F Rodrigues, Paredes H, Farrajota M, Fernandes H, Jos J, Teixeira V and Saleiro M. 2011. The Smart Vision Navigation Prototype for Blind *Users International Journal of Digital Content Technology and its Applications*, **05** (05) 351 – 361.
[13] Ulrich I and Borenstein J. 2001. The guide cane-Applying mobile robot technologies to assist the visually impaired *IEEE Transaction on Systems, Man, and Cybernetics-Part A: Systems and Humans* **31** (02) 131-136.
[14] Meijer P. 1992. An Experimental System for Auditory Image Representations *IEEE Transactions on Biomedical Engineering*, **39** (02) 112-121.
[15] Nie M, Ren J and Li Z. 2009. Sound View: an auditory guidance system based on environment understanding for the visually impaired people *Proceedings of the 31st Annual International Conference of the IEEE Engineering in Medicine and Biology Society: Engineering the Future of Biomedicine* 7240–7243.
[16] Balakrishnan G, Sainarayanan G, Nagarajan R and Yaacob S. 2007. Wearable Real-Time Stereo Vision for the Visually Impaired *Engineering Letters* **14** (02).
[17] Fajarnes G P, Dunai L, Praderas V S and Dunai I. 2010. CASBLiP- a new cognitive object detection and orientation system for impaired people *Proceedings of the 4th International Conference on Cognitive Systems, ETH Zurich, Switzerland.*

A Low-Cost Portable Smart Card Based Attendance System

Vibin Mammen Vinod [1*], Govindasamy Murugesan[1], Mekala V[1], Thokaiandal S[1], Vishnudevi M[1], Siddharth S M[1]

[1]Electronics and Communication Engineering, Kongu Engineering College, Perundurai, Erode – 638060.

E-mail: vibin.ece@kongu.edu

Abstract. Managing attendance may be a vital record-keeping activity in any institution. There are different methods of maintaining the attendance from manual system where attendance is marked in sheets to automated attendance on the likes of biometrics. The combination of frequency Identification (RFID) with fingerprint biometric technology was to enhance the safety level and integrity of the records. The designed system not only makes the system design simpler but also enhances the productivity of the institution both in terms of man power and time. The system doesn't only simplify the method of taking attendance but reduces error and allows for faster verification of student attendance, all with minimal human interaction. This technique will help the authorities manage the attendance system in a more organized, efficient and time saving manner. The planning of the system is straightforward and portable making it an honest candidate for commercial and academic purpose. A GUI based version and app-based version has been designed to make the system both PC and mobile friendly. The system is built with various levels of access permissions and security. The system was tested for a small group of 100 students and was found to provide accurate marking of attendance for all users.

Keywords: Biometric, RFID Tags, Fingerprint, Smart card, Autonomous system.

1. Introduction

Attendance is a very significant criterion in many organizations, and it is used for different reasons. The reasons are varied as ranging from maintaining of records, assessing grades to promoting adequate and reliable school attendance. In the majority of schools, a minimum percentage of attendance for a class is required before students can sit for an examination. But due to the lacunae in the existing schemes of manual attendance monitoring, this requirement is many a times flown in air. The conventional way of monitoring attendance envisages to mark signature manually each and every time a test or a class is conducted. System like the one discussed is lagging in its functionality and hence leads to multiple causes of failures. Excessive time needs to be spent by each student in locating their names in the attendance sheet and then signing. The system also leads to increased rate of impersonation. A foolproof system capable of eliminating all these above-mentioned issues is the need of the hour.

RFID ensures a wireless medium of communication between a reader and a tag or multiple tags making use of EM waves. Multiple systems [21 - 28] are available in literature that makes use of RFID based card systems for attendance entry to keep the process simple and hassle free but all fails on a single front wherein impersonation cannot be eliminated.

The problem of impersonation has created a new dimensionality to the process of automated attendance management. This has led to the conception of a hybridization of techniques. An additional level of access scheme employing biometrics will play out an enhanced security measure. The process of hybridization ensures that the advantages of both schemes are leveraged effectively towards the better design of the system. The combination of frequency Identification (RFID) with fingerprint biometric technology was to enhance the safety level and integrity of the records.

The proposed system is aimed at designing and implementing a new hybrid approach in automating the attendance management process. The design is built around RFID and fingerprint-based access schemes. RFIS ensures easy hassle-free access whereas fingerprint ensures that impersonation is kept at minimum. The proposed design had been utilized for automating attendance management in an examination hall. The attendance details are collected through the portable device and then transferred through the server, thereby updating those attendance details to the website that we had created.

1.1. Fingerprint based technology

A portable attendance system was introduced using the GSM Network with Fingerprint Technique to control and monitor student attendance and to send a list to student parents [1]. Talaviya et al., have proposed hybrid structure involving Zigbee and fingerprint to monitor and record student attendance at a university [2]. Yadav et al., presented and introduced a compact attendance system for fingerprint detection [3]. Kamaraju et al., also introduced time sensitive and cost-effective solution for workers' attendance in an industrial plant using Zigbee and biometrics [4]. Zainal et al., showed that design and development of the attendance monitoring and management framework using fingerprint technology and Arduino microcontrollers provided a safe and compact solution method [5]. Potadar et al., have proposed to register, control and report student attendance using a fingerprint technique to minimize the likelihood of error-related attendance tracking for academic sectors [6].

1.2 QR Code based Technology

Rakhi Joshi et al., have proposed a smart learning and attendance system by implementing a QR code-based access to a central server built with .Net and visual studio technologies. An android based app is also created which allowed students to access contents from anywhere and anytime. The system also provided for attendance report generation both to parents and individuals as messages [8]. Nath & Mukhopadhyay have proposed a mobile-enabled on-line attendance system [9]. Noor et al., have proposed Android-based application to take, handle student attendance records by using camera system as sensor [10]. Buddhiwant, have presented Android based and GPS application for attendance monitoring of students [11]. Dinesh Kumar et al., have proposed smart phone-based attendance system for employees in which the QR Code is used as an identification proof. In Future this method may implement in academic for proper maintaining of the attendance recording management [12]. Emmanuel C et al., have presented an offline Identity Card authentication System using QR code, Smartphone is presented as solution to authenticate the bonafide student of tertiary institutions which results in faster authentication [13]. Jacob et al., using NFC mobile attendance system, launched an attendance control system for university students [14].

1.3. Barcode Based Technology

Y. K. Saheed et al have proposed an attendance management system using barcode which is to be printed on the student's identity card. By scanning their card in the device, the attendance of the student is updated, which results in reducing effort on paper work and it saves the valuable time [7]. K. Lakshmi Sudha et al., had developed an attendance management for the students while entering lab using barcode scanner which helps the teacher to maintain a registry book [5]. P Rahmah Al Sheikh et al., have proposed barcode-based attendance management using a unified modeling language which then helps in maintaining the details of the students [15].

1.4 Bluetooth based Technology

Vishal Bhalla et. al., have presented a Bluetooth based attendance system which is a method of taking attendance using instructor's mobile telephone. It is paperless, quick, and the student entries can be confirmed [16]. Kumar & Kumar have developed a system to track, track and locate students on campus using a novel Bluetooth Face Recognition feature [17]. Tiwari et al., have a student attendance system based on GPRS which the lecturers can easily access through the internet to verify and track attendance tracking for their students [18]. Idachaba F.E et al., have designed a system which identifies the user using his fingerprint and the data is sent to a database using the Bluetooth for large halls and auditorium [19]. Riya Lodhaa et al., have designed a system utilizing radio frequency and microchip-based techniques to create a secure environment for inventory management. The advantages incorporated in the system are low power consumption and high data transfer rate [20].

1.5 RFID based Technology

Patel et. al., proposed an intelligent system to record students' attendance in real time which results in low time consumption [21]. Singhal et al., proposed a design using RFID tag identification along with a SMS capability [22]. Arulogun O. T et al., proposed a solution to attendance monitoring of lecture hours by a coordinated approach involving handshake between RFID tag and reader with a central PC [23]. Chiagozie, & Nwaji, proposed a door system based on RFID technologies to handle time-attendance [24]. Arbiain et al., proposed an attendance system for automatically tracking/ managing pupil attendance in laboratory settings using RFDUINO method [25]. Kurniali, built a student attendance management system using RFID to control student engagement in an Indonesian university with a web-based approach [26]. Yuru et al., created a participant inspection system based on ARM and RFID technologies has been developed [27]. Yadav et al., presented to students and teachers an automated attendance management system that sends alerts to parents via GSM. [28].

1.6 Face Recognition based Technology

Mekala et al. implemented a face recognition-based class attendance monitoring system that relies on Principal Component Analysis technique of classification [29]. A detailed comprehensive survey on the attendance systems in literature classified based on the techniques is presented by Vinod V.M. et al.,[30]. A face-recognition based attendance system is introduced by using DIP and a Euclidean distances filtering system by A. Cathrine Joe Silvia et al., [31]. Wang et al., developed an Automatic attendance method with two Faster Face Detection and SeetaFace face recognition algorithms built into the deep learning algorithm. [32]. Khatun et al., have an educator implemented and introduce an Iris recognition method for tracking and handling participants and have submitted a report in the programming environment to the predefined e-mail address [33]. Eigen face and Principal Component Analysis is found to improve face recognition rate and accuracy as is evident form the work proposed by Wagh et al., [32]. Sayanekar et al., have an NFC technology and face recognition technique developed for the biometric attendance control system introduced for student monitoring and recording at different campus locations [34].

2. Methodology

The proposed system uses the RFID Tag and Fingerprint sensor for marking the attendance of the students. In the proposed system, a database of all the students admitted is collected from the admission details and another biometric database is also appended to it. The students' attendance in the examination hall is taken by either scanning the tag or by using fingerprint of the student using the portable device designed for attendance management. The components of hardware, such as RFID

Tags, RFID Scanners, fingerprint scanner, Raspberry pi and LCD Displays, used to simplify the assistence device for students to monitor their attendance and for exam marking. If the student's RFID Tag or fingerprint is read by the system, automated attendance marking will be performed on the local host. As and when the entire attendance of the hall/ class is noted, the report is uploaded to the server from the host by the concerned administrator. Bult upload of report is advisable as it reduces the data transmission as well as improves security.At the server end, the report can be accessed for storage and retrieval by the competent authorities.

Figure 1. below represents the block diagram of the attendance system in which the raspberry pi is interfaced with RFID Reader and Fingerprint sensor and LCD. After scanning their tag or fingerprint in the device, the attendance of the students will be updated in the website by the authenticated person.

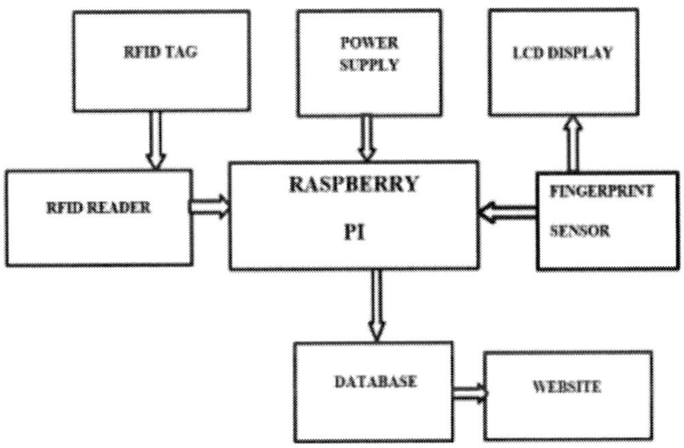

Figure 1. Block Diagram of the Proposed Attendance Device

Figure 2. below represents the flow diagram of the attendance system. After scanning their tag or fingerprint in the device, the students tag number and fingerprint is compared with stored value. The students will be marked as present when the stored value is matched with the scanning value or else it will be marked as absent. Finally, the attendance of the students will be updated in the website.

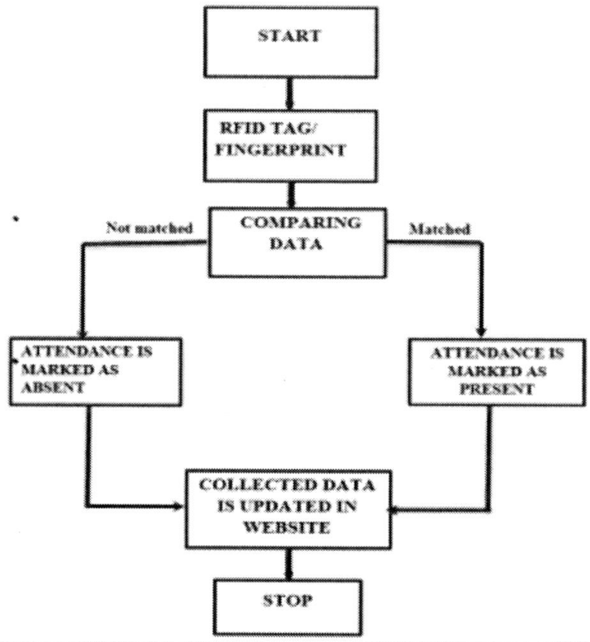

Figure 2. Flow diagram of the proposed attendance management system

The process flow of the designed device setup is given below as vivid steps:

- The RFID Reader is fixed in the device which is used to scan the RFID Tag of the students.
- In Addition, Fingerprint sensor is available in the device which is used to scan the fingerprint of the student.
- Either scanning their Tag to the Reader or scanning their fingerprint to the fingerprint module, the attendance of the students is marked as present.
- After the scanning process is completed, the request is sent to the server to update the results to the corresponding website which is under the control of the admin.
- The attendance of the students is updated on the website with corresponding date and time.
- The attendance report of the student can be exported as a spreadsheet.

The designed system can switch over between two modes – RFID and fingerprint seamlessly and hence provide a fail proof authentic attendance system as compared to the existing models. Raspberry pi can double as a hosting platform for the web UI.

Figure 3. Hardware prototype

3. Results

Webpage creation and Online hosting is performed on https://somee.com. As a free .NET Hosting. It provides a storage capacity of 150Mb and the monthly transfer of 5Gb. **Figure 4.** shows the designed login page for the attendance portal. On successful login, the user is navigated to the registration page as shown in **Figure 5**. The basic details of the student may be fed in this page. The option is also made available to make bulk registration of students by the competent authority.

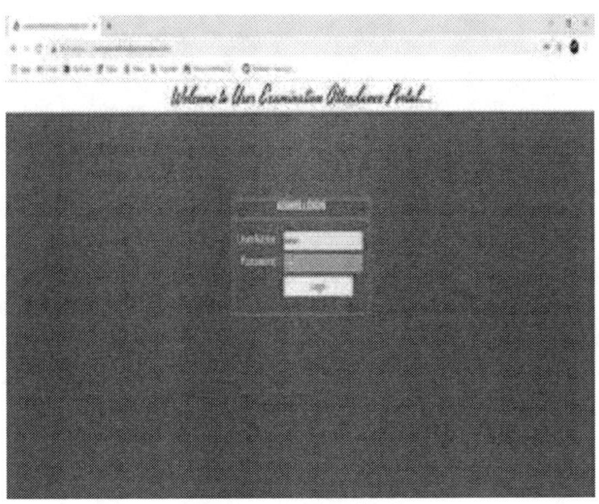

Figure 4. Login Page.

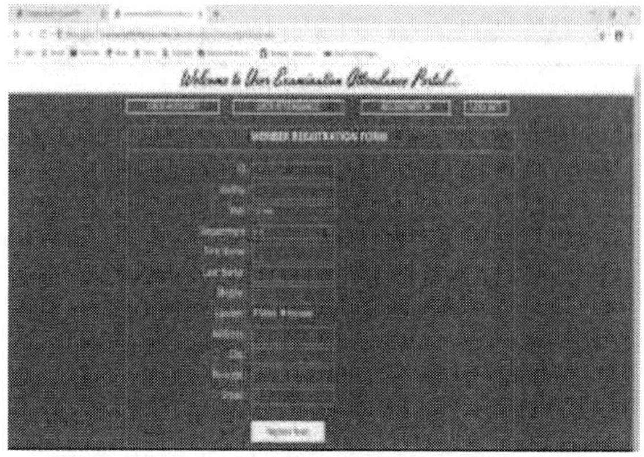

Figure 5. Registration page.

Registration page allows each student to update his basic details. Provision for biometric storage and cloud access is built into the system.

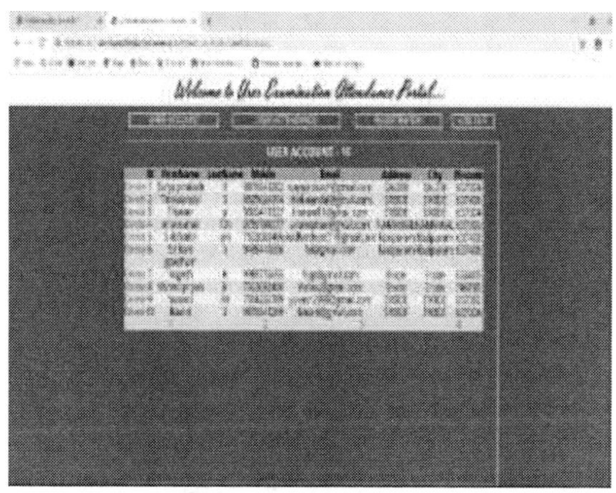

Figure 6. User Account Session

The administrator is provided with the privileges to view and edit the details of each user as shown in Fig. 6. Varying levels of access is provided depending on the user or administrator privileges. Interactive GUI enables updation of user account with ease. Bulk upload of data feature is also provided.

Figure 7. Attendance Details page

As and when a fingerprint scan or an RFID tag is read, attendance field for the particular student gets updated in the central server. This updation is made semi-autonomous with the faculty concerned having to authenticate the process. This provides a two-level authentication keeping the issue of malpractices to a minimum.

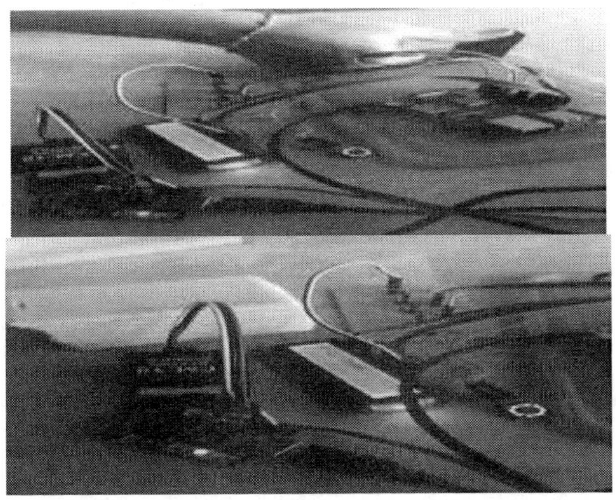

Figure 8. Mode Selection – Fingerprint and RFID

As discussed earlier, the designed system is a hybrid structure with two authentication modes namely RFID tag and a biometric scan (fingerprint). This hybrid nature of authentication ensures that the right student takes up the exam and there is very little space for forgery. Moreover, the provision of authentication/ approval by the faculty ensures impersonation is kept to the minimum or null. Fig 8 shows the two modes of access from the designed system.

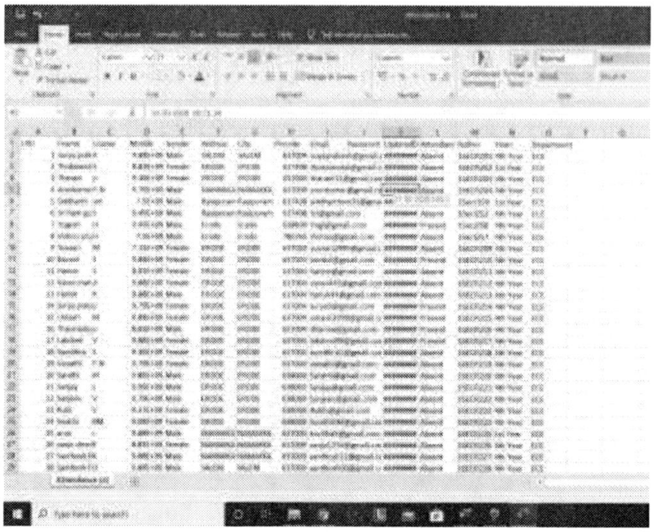

Figure 9. Spreadsheet Creation from the Attendance Webpage

Finally, for documentation and such purposes, the attendance details can be exported as a sheet in .xls format. Facility to include multiple formats is also being worked out. Report generation of any day and any session can be accessed with ease by the administrator.

4. Conclusions

The designed system offers many advantages as against the existing manual attendance system seen in exam halls. Though various authentication schemes are proposed in literature, they remain as papers and no tangible product/ solution is seen in real world. Our design is portable and a cost-effective solution. The system helps in avoiding the human errors and proxy punching. It provides persistent and protected storage. It has put an end to the tedious process of maintaining a log book for attendance and at the same time it has saved 10 to 15 minutes of time that is been previously spent on taking attendance during the exam time. This system can be implemented instead of traditional attendance system. It can be implemented in both schools and colleges as an effective means of taking attendance which leads to less time consumption and cost effective.

Improvements on the GUI and android based app are to be carried out to make the system attractive and acceptable to the user. Application of security protocols also needs to be worked out.

References

[1] Gupta, Pallavi Vermal Namit. "Fingerprint based Student Attendance system using GSM." International Journal of Science and Research (IJSR) ISSN (Online) (2013): 2319-7064.
[2] Talaviya, Gunjan, Rahul Ramteke, and A. K. Shete. "Wireless fingerprint based college attendance system using Zigbee technology." International Journal of Engineering and Advanced Technology (IJEAT) 2.3 (2013): 201-203
[3] Yadav, Ravishankar, and Sumita Nainan. "Design of RFID based student attendance system with notification to parents using GSM." International Journal of Engineering 3.2 (2014).
[4] Kamaraju, Maddu, and Penta Anil Kumar. "Wireless fingerprint attendance management system." 2015 IEEE International Conference on Electrical, Computer and Communication Technologies (ICECCT). IEEE, 2015
[5] Sudha, K. Lakshmi, et al. "Barcode based student attendance system." International Journal of

Computer Applications 119.2 (2015).
[6] Potadar, Mahesh P., et al. "Biometric Attendance Recording and Communication System." International journal of innovations in engineering and technology (IJIET) 5.2 (2015): 230-234
[7] Saheed, Y. K., et al. "Attendance management system using barcode identification on students' identity cards." The Pacific Journal of Science and Technology 17.2 (2016): 224-230.
[8] Joshi, Rakhi, V. V. Shete, and S. B. Somani. "Android based smart learning and attendance management system." International Journal of Advanced Research in Computer and Communication Engineering 4.6 (2015): 256-260.
[9] Nath, Asoke, and Aashijit Mukhopadhyay. "A Novel Student Attendance Tracking System Using Android Mobile Phone." (2017).
[10] Noor, Siti Aisah Mohd, et al. "Android-based attendance management system." 2015 IEEE Conference on Systems, Process and Control (ICSPC). IEEE, 2015.
[11] Buddhiwant, Aakruti, et al. "Smart Attendance Application." International Journal of Engineering and Management Research (IJEMR) 7.2 (2017): 221-224.
[12] Kumar, B. Dinesh, and S. Kareemulla. "Smart Mobile Attendance System for Employees Using QR Scanner." Asian Journal of Applied Science and Technology (AJAST) 1.5 (2017): 35-39.
[13] Srinidhi, M. B., and Romil Roy. "A web enabled secured system for attendance monitoring and real time location tracking using Biometric and Radio Frequency Identification (RFID) technology." 2015 International Conference on Computer Communication and Informatics (ICCCI). IEEE, 2015.
[14] Jacob, John, et al. "Mobile attendance using near field communication and one-time password." 2015 International Conference on Green Computing and Internet of Things (ICGCIoT). IEEE, 2015.
[15] Al Sheikh, Rahmah, et al. "Developing And Implementing A Barcode Based Student Attendance System." International Research Journal of Engineering and Technology (IRJET) Volume 6 (2019).
[16] Bhalla, Vishal, et al. "Bluetooth based attendance management system." International Journal of Innovations in Engineering and Technology (IJIET) 3.1 (2013): 227-233
[17] Kumar, Jay, et al. "Automatic attendance monitoring and tracking system using Bluetooth and face identification." International Journal of Advanced Research in Electronics and Communication Engineering 5.4 (2016): 1166-1170
[18] Tiwari, Aditi S., et al. "Optimized design of student attendance system using rfid." International Conference on Machine Learning, Electrical and Mechanical Engineering. 2014.
[19] Idachaba, F. E., and O. O. Oni. "Development of a Bluetooth-Based Contactless Automated Attendance Capture System for Large Halls and Auditoriums." (2015): 47-50.
[20] Lodha, Riya, et al. "Bluetooth smart based attendance management system." Procedia Computer Science 45 (2015): 524-527
[21] Patel, Unnati A., and S. Priya. "Development of a student attendance management system using RFID and face recognition: a review." International Journal of Advance Research in Computer Science and Management Studies 2.8 (2014): 109-119.
[22] Singhal, Zatin, and Rajneesh Kumar Gujral. "Anytime anywhere-remote monitoring of attendance system based on RFID using GSM network." International Journal of Computer Applications 39.3 (2012): 37-41
[23] Arulogun, O. T., et al. "RFID-based students attendance management system." International Journal of Scientific & Engineering Research 4.2 (2013): 1-9.
[24] Chiagozie, Ononiwu G., and Okorafor G. Nwaji. "Radio frequency identification (RFID) based attendance system with automatic door unit." Academic Research International 2.2 (2012): 168.
[25] Arbain, Norakmar, et al. "LAS: Web-based laboratory attendance system by integrating RFID-ARDUINO technology." 2014 2nd International Conference on Electrical, Electronics and System Engineering (ICEESE). IEEE, 2014.
[26] Kurniali, S. "The development of a web-based attendance system with RFID for higher

education institution in binus university." EPJ Web of Conferences. Vol. 68. EDP Sciences, 2014.

[27] Yuru, Zhang, Chen Delong, and Tan Liping. "The research and application of college student attendance system based on RFID technology." International Journal of Control and Automation 6.2 (2013): 273-282.

[28] Yadav, Ravishankar, and Sumita Nainan. "Design of RFID based student attendance system with notification to parents using GSM." International Journal of Engineering 3.2 (2014).

[29] Mekala V, et al. "Face Recognition Based Attendance System." International Journal of Innovative Technology and Exploring Engineering, Vol.8, Issue-12, (2019) 520-525.

[30] Vibin Mammen Vinod, et al. "A Comprehensive Study On Academic And Industry Authentication And Attendance Systems" International Journal Of Scientific & Technology Research Volume 9, Issue 03, (2020) 5426-5432.

[31] Silvia, A. Cathrine Joe, M. Mekala M. Dhevaki, and E. Mohanapriya. "Multi-Face Authentication-Based Digital Image Processing Automatic Classroom Attendance Management System." (2018).

[32] Wagh, Priyanka, et al. "Attendance system based on face recognition using eigen face and PCA algorithms." 2015 International Conference on Green Computing and Internet of Things (ICGCIoT). IEEE, 2015.

[33] Khatun, Amena, et al. "Design and implementation of iris recognition based attendance management system." 2015 International Conference on Electrical Engineering and Information Communication Technology (ICEEICT). IEEE, 2015.

[34] Sayanekar, Pratik, et al. "Customized NFC enabled ID card for Attendance and Transaction using Face Recognition." International Research Journal of Engineering and Technology 3.9 (2016): 1366-1368.

[35] Kuriakose, R. B., and H. J. Vermaak. "Developing a Java based RFID application to automate student attendance monitoring." 2015 Pattern Recognition Association of South Africa and Robotics and Mechatronics International Conference (PRASA-RobMech). IEEE, 2015.

[36] Jaikumar, K., et al. "Fingerprint based student attendance system with SMS alert to parents." Int. J. Res. Eng. Technol 4.2 (2015): 293-297.

[37] Ya'acob, Norsuzila, et al. "RFID lab management system using Arduino microcontroller approach associate with webpage." Journal of Scientific Research and Development 3.2 (2016): 92-97.

[38] Zainal, Nur Izzati, et al. "Design and development of portable classroom attendance system based on Arduino and fingerprint biometric." The 5th International Conference on Information and Communication Technology for The Muslim World (ICT4M). IEEE, 2014.

[39] Ayeleso, E. C., et al. "Identity card authentication system using qr code and smartphone." International Journal of Science, Engineering & Environmental Technology (IJOSEET) 2.9 (2017): 61-67.

[40] Saparkhojayev, Nurbek, and Selim Guvercin. "Attendance Control System based on RFID-technology." International Journal of Computer Science Issues (IJCSI) 9.3 (2012): 227.

[41] Sunehra, Dhiraj, and V. Surender Goud. "Attendance recording and consolidation system using Arduino and Raspberry Pi." 2016 International Conference on Signal Processing, Communication, Power and Embedded System (SCOPES). IEEE, 2016.

[42] Walia, Hitesh, and Neelu Jain. "Fingerprint Based Attendance Systems-A Review." International Research Journal of Engineering and Technology 3.5 (2016): 1166-1171.

[43] Kohalli, Shivanand C., et al. "Smart Wireless Attendance System." International Journal of Computer Sciences and Engineering 4.10 (2016): 131-137.

[44] Shakil, Md, and Rabindra Nath Nandi. "Attendance management system for industrial worker using finger print scanner." Global Journal of Computer Science and Technology (2013).

Performance Investigation of PLC Hardware for Portable Two-Wheeler Dynamometer Test rig

Shubham Maniar[1*], Ashish Umbarkar[2], Omkar Morbale[2]

[1]Department of Mechanical Engineering, MIT School of Engineering, MIT ADT University Loni, Pune, India.
[2]R&D Department SAJ Test Plant, Pune, India.

E-mail: 1299shubham@gmail.com

Abstract. The dynamometer is an instrument used for torque, force and velocity measurements. In a two-wheeler dynamometer frame, that is the easiest thing to use. So, until now, the testing has been held in the same place, but doing something new is making the whole system portable and being able to carry it anywhere. Therefore, PLC should be used to measure the parameters for input purposes, as many types of PLC are available. The machine is then connected to SCADA for an additional point. In order to monitor the working of the two-wheeler chassis dynamometer, the sensor is also used in the device and can changed the parameters.
The entire dynamometer will be controlled using PLC and HMI (human-machine interface). Like the Sensor Actuators, it can integrate the device with the hardware. I decided to go with Mitsubishi PLC for this research because it has all the specifications I need and the cost that makes this a cost-effective system is lower. The effects of all the parameters are displayed on the HMI panel. For the sensors of the test rig, the simulation was developed and 3 percent of errors were identified compared to the results of the simulation with the data sheet.

Keywords: Dynamometer, Portable Chassis Dynamometer, PLC, SCADA.

1.Introduction

The quantity of two-wheeler is much greater in our country as we live in a developing country. In India, the number of two-wheeler is around 70 per cent of India's total vehicle. Of this 70 percent, about 55 percent of two vehicles come under the 100cc to 350cc range. These two-wheelers are mainly used daily for the purposes of daily commuting. The condition of Indian roads is not as decent as we all realize, so these two-wheelers have a lot of wear and tear. And there is a fair decrease in output to these vehicles over a period of time.

In the chassis dynamometer system, the dynamometer functions as an energy-absorption unit. This chassis dynamometer was designed to measure the condition of the vehicle in order to carry out various basic tests, such as performance tests between the standard air filter and the K&N air filter, acceleration tests between two speed points, maximum speed tests, tests of different gear speeds, load tests, and tests of fuel consumption [2]. Measurement unit for dynamometer, mechanical force or electricity, transmitted by rotary shaft. As power is the combination of torque (turning force) and angular velocity, torque measurement devices are basically all dynamometer power measurement devices. The shaft's velocity is measured singly. A flexible metal ring that bends when a force is applied in such a way that the amount of bending tends to collapse as a measure of the force applied is used in the instruments for force measurement, and a hydraulic 'load cell' that calculates compressive loads in terms of fluid pressure. PLC is a system that monitors input devices constantly to make decisions based on personalized PLC programming to manage the state of a machinery's output devices. Mitsubishi FX series PLC programming port contact protocol. Presentation of a portion of the LabVIEW-based communication software. The serial communication is introduced between the PLC and the Industrial PC [3]. In order to increase productivity in everyday life, temperature measurement and control are very important in the industrial and agricultural sectors. Traditional temperature control techniques, which not only waste a lot of workforce and financial resources, but also cause faults and accidents due to incorrect temperature measurements, are needed for manual practice. Therefore, real-time wireless temperature monitoring is useful for timely inspection and security of control products, as well as saving the labor force [1].

2. Field Survey

Table 1. Field Survey

Parameter	Value	Two-Wheeler
Max Torque	28.00 Nm	Royal Enfield model classic 350
Max Power	29.91 Kw	KTM model 250
Maximum Wheelbase ratio	1412 mm	Hero X-Pulse
Maximum weight	195 kg	Royal Enfield thunderbird 350
Maximum speed	9000 rpm	KTM model 250

3. Methodology

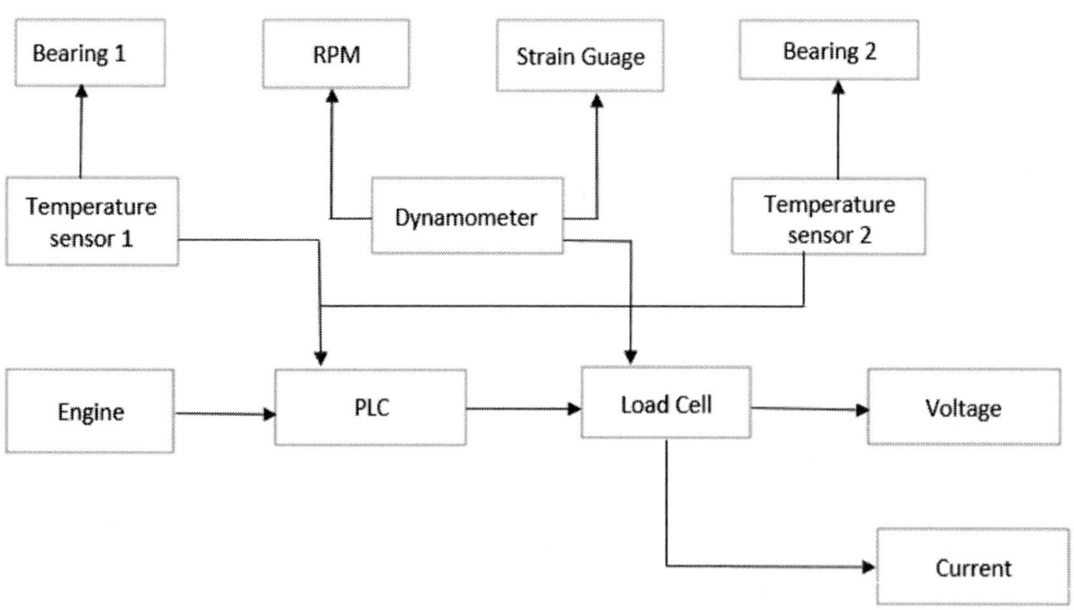

Pre requirement of Mitsubishi PLC

- Mitsubishi Provides its Programming Language with its PLC Hardware
- Use of any Other Programming Language can be Work with Mitsubishi PLC
- It has Low Cost Compared to Other PLC

Using of Mitsubishi PLC (fx5u-32mr)

Compatible with the I Q programmable logic controller (PLC), the platform (Mitsubishi FX5U series) has 32 digital inputs (type 24VDC sink / source) and 32 relay / transistor outputs. Link MELSEC (3C/4C Frames) and the communication protocol to MELSOFT. It has program capacity of 64k steps, and 32 built-in inputs. By offering a wide variety of new and existing add-on options, the FX5U continues the FX tradition of complete flexibility. Integrated standard Ethernet functions, I / O analogue, data logging, position control, security, communication, and networking are further enhanced. This will allow users to recognize more powerful systems, but overall fewer components, saving time and money [4].

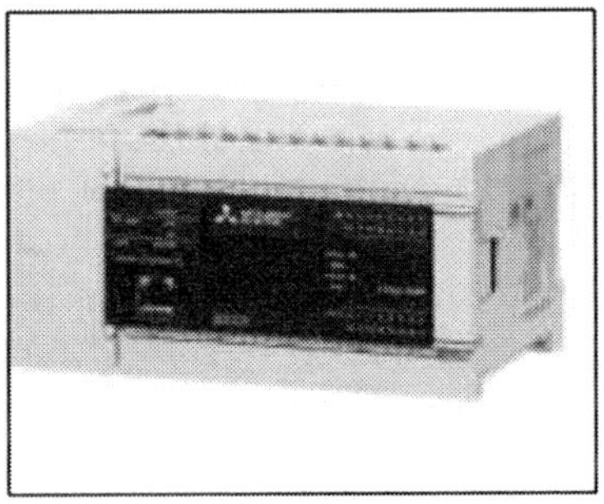

Figure 1. Mitsubishi fx5u-32mr PLC [4]

Parker Drive 514C

The 514C industrial environment controller is designed and should be mounted in a closed box that provides controller and user protection. At the terminals cited, the controller should be permanently grounded. The 514C controller is built to track permanent magnet or DC Shunt cutter speeds. All 4 operating quadrants will have control of the motor speed. The controllers are equipped to work within the range of 110 Vac to 415 Vac at 50 or 60 Hz from a single-phase AC power supply. For internal power generation and principal supply contactor sequencing, auxiliary supply is required. A linear, closed loop system with feedback from either a tach generator or armature voltage, with a selectable feedback source to turn, regulates the DC engine rpm. A current loop within the speed loop also guarantees the addition of regulated current levels to the engine, with programmable switches being able to scale the actual levels for motor protection, a stall detection circuit is given and the current will be streamed within the engine range of 55 to 60 seconds. At the event of a short circuit, the safety of the controller is provided by an instant overcurrent trip circuit that overrides the power. [5].

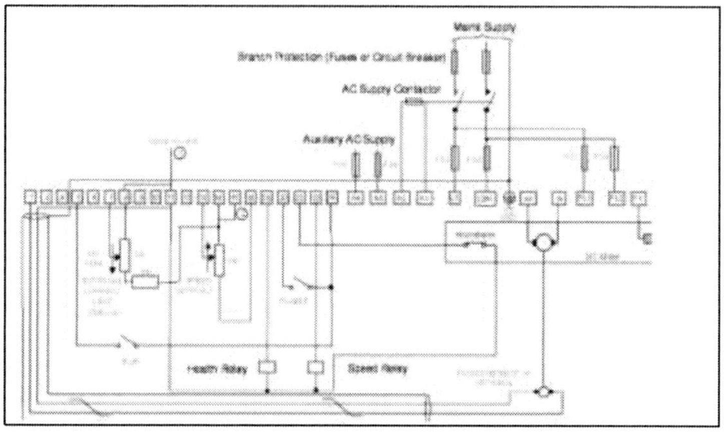

Figure 2. Circuit Diagram of Parker Drive 514C [5]

Types of sensor used in dynamometer

Load cell

A load cell is a device that transforms a force or charge into a measurable output. Load cells can come in various types, including hydraulic, pneumatic, strain gauge, piezoelectric, and capacitance, but the subject of this handbook will be strain gauge load cells. The most popular strain gauge load cells are classified as a device which transforms a force or load into an electrical equivalent signal or a digitized load value. General load cells are ideal for periodic static force measurement applications consisting of measuring devices, structural testing devices, and material testing. [6]

Table 2. Specification of Load cell

Supply Voltage	110-500V +10% user selectable
Auxiliary supply	110/120 or 220/240V +10% user selectable. Single phase 50-60Hz +10%
Ambient	0-40°C - Altitude: up to 1000m without derating
Overload	150% for 60 seconds

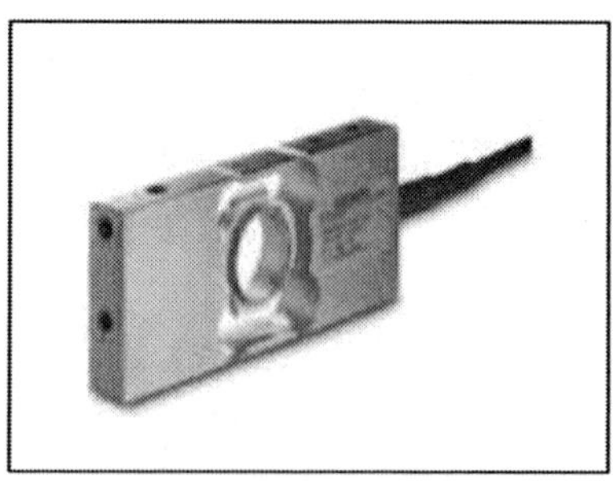

Figure 3. Load cell [6]

Pre requirements of RTD PT 100 Sensor

- As per the standard bearing No. 6209 SKF selection the temperature required maximum 120° C where the RTD PT 100 measure temperature range between 0- 200° C

- It gives an accurate reading compare to others sensors.
- The Important key point of the sensor is it contains a probe type sensor so it doesn't require special mounting.

RTD PT-100 Temperature sensor

A conductor is a substance in which the electrons are less bound to the outer orbit of the atom. If the energy of the material is increased, the atoms will move (vibrate) more and more by heating up, for example, and at some point, these electrons can leave their orbit and float freely inside the space between the atoms. The higher the temperature, the quicker the atoms move, and the harder it is for the electrons to move through the space between the atoms, so there is less space between them. [8].

Table 3. Specification of RTD-PT100

Temperature range	−60 to +500 °C
Accuracy	±0.3 °C

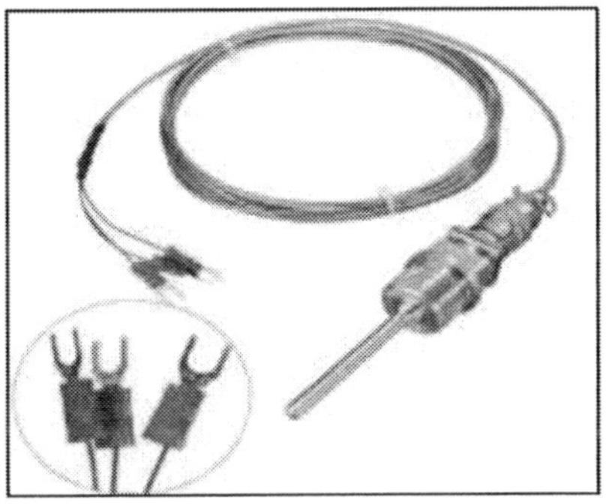

Fig 4. RTD PT-100

Pre requirement of Strain Gauge

As per the Field Survey it has been concluded that the maximum torque is 28Nm.

Strain Gauge

A strain gauge is made of very fine wire, or foil, placed and fixed to a flexible backrest in a grid pattern. If there is a change in the shape of the strain gauge, its electrical resistance can change. The wire or foil in the strain gauge is designed in such a way that a linear change in the resistance occurs when force is applied in one direction. The tension force is extended by a strain gauge, allowing it to become thinner and wider, leading to resistance increases. The same is true for compression power. The pressure gauge compresses, becomes thicker and shorter, and resistance decreases. The strain gauge is mounted on a flexible backrest that allows a load cell to be easily added, representing the minute change to be measured.[7]

Table 4. Specification of Strain Gauge

Measurable Strain	2 to 4% maximum
Temperature Range	−30°C to +80°C

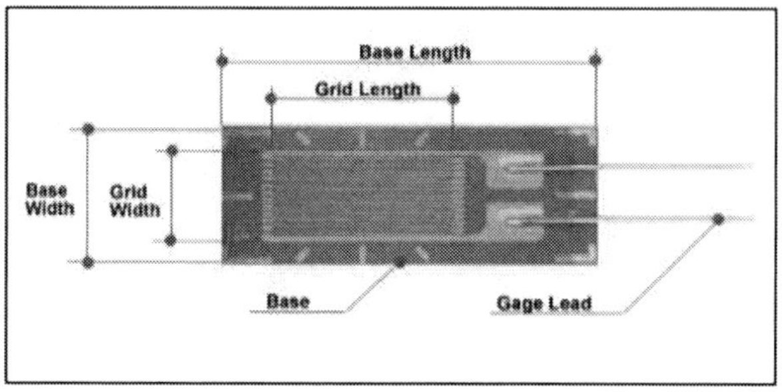

Fig 5. Strain Gauge

Pre requirement of Magnetic Pulse Pickup

The rotor of the dynamometer has a gear of 24dp which makes easy to use pulse pickup of measuring the RPM. It can also withstand higher temperature up to 113°C.

Pulse Pickup

A pulse pick sensor mounted on the case defines the velocity, which detects a 60-tooth wheel located on the coupling core. All electrical signals are routed to a terminal box mounted on the bedside table which provides the dynamometer control system with the interface

Table 5. Specification of Pulse Pickup

Gear Pitch	24dp
DC Resistance	1200 ohms
Operating Temperature	73°C to 113°C

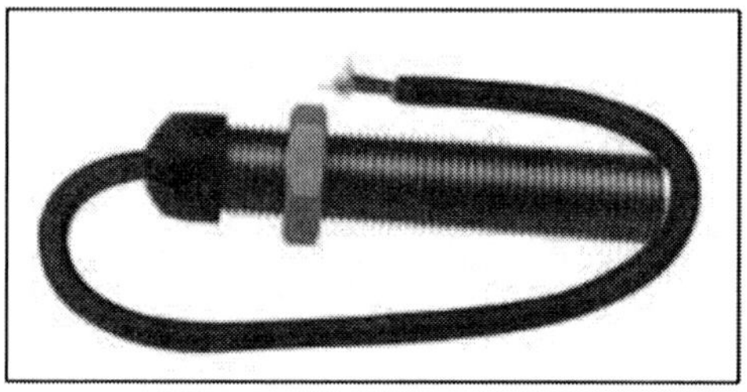

Fig 6. Pulse Pickup

4. Result & Validation

Simulation with the sensor

For the Simulation with the sensors I have used the Arduino Board for input purpose because PLC is not available in the proteus software.

RTD PT-100 or Temperature resistance detectors, are instruments used for temperature measurement. These sensors are among the most accurate sensors available which cover wide temperature ranges.

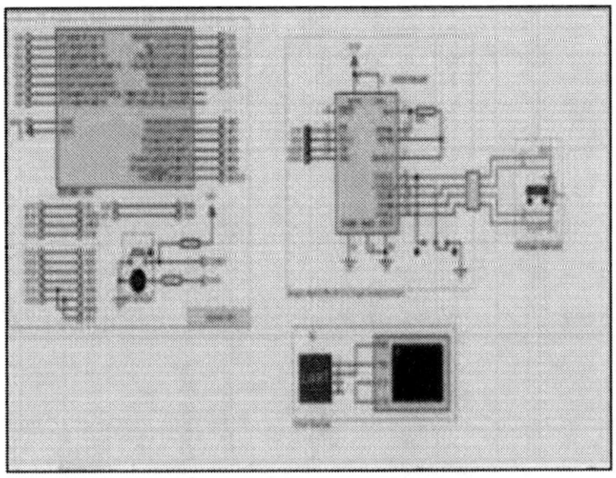

Fig 7. Simulation of RTD-PT100 Sensor

- Grove-UART- Grove UART is a model of Grove Software Module which is customizable. Pin 1 is the RX line (used by the base unit to receive data, so it's an input) where the TX line is Pin 2 (used by the base unit to send the Grove module data).
- Connector-An electrical connector is a system used to connect electrical conductors and build an electrical circuit.
- Arduino Uno- Inputs-light on a sensor, finger on a button-and transform to output-can be read by Arduino boards-activate a motor, switch on an LED, and publish something online.
- Capacitor-Capacitor is a simple storage system for holding and releasing electrical charges, as the circuit requires.
- DRDY Pin- This pin is used for advanced uses where you are asking the sensor to start reading and then wait until this pin is big.
- SDO- For the input sent from Max31865 to the processor, this is Serial Data Out.
- SDI- This is a serial data input for the data sent to Max31865 from the processor.

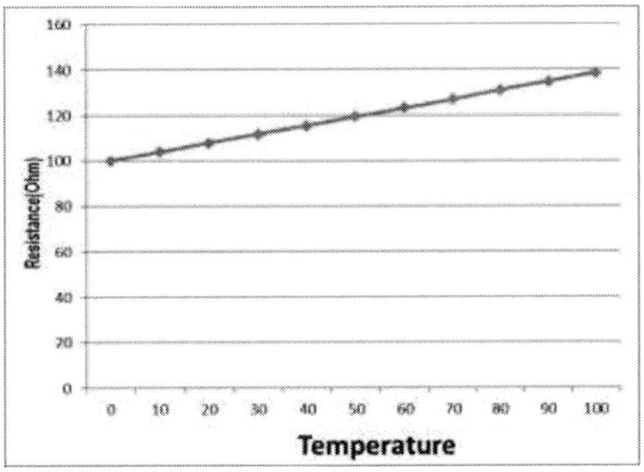

Fig 8. Graph of Temperature with resistance

Load Cell- Load cell, are the instruments used for measurement of weight carried out by the system.

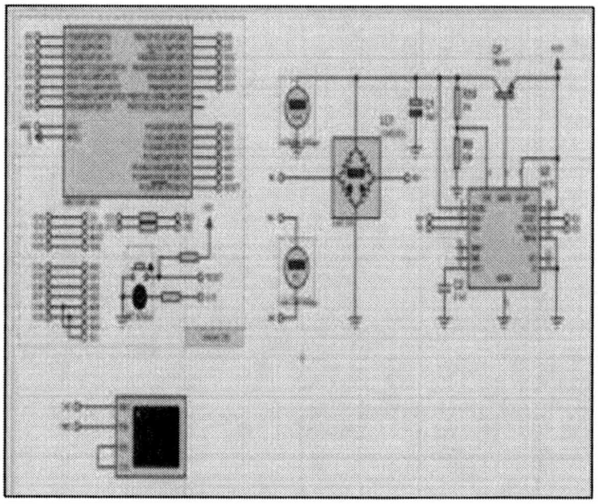

Fig 9. Simulation of Load cell

- HX711 is a 24bit ADC which amplifies and digitally converts the output of the load cell. The Arduino is then fed with this amplified value. The load cell senses the weight and gives an analogue electrical voltage to the HX711 Load Amplifier Module.
- 2N4403- It is a type of general-purpose PNP Bipolar junction transistor used in low-power switching and amplifier applications where holes and electrons are transferred.
- Capacitor-It is an electronic passive component, with two terminals. A condenser is an instrument that stores electrical energy in an electrical field.
- Resistor-A resistor is a passive two-terminal electrical component which as an element of a circuit implements electrical resistance. In electronic circuits, resistors are used to reduce current flow, modify signal levels, separate voltages, bias active elements and terminate transmission lines, among other applications.
- Arduino Uno- Inputs-light on a sensor, finger on a button-and transform to output-can be read by Arduino boards-activate a motor, switch on an LED, and publish something online.

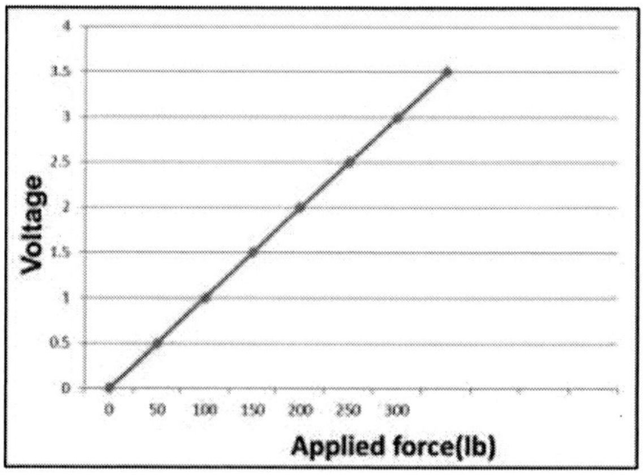

Fig 10. Graph of Force with Voltage

Strain Gauge- The strain gauge is mounted on a flexible backrest that makes it simple to apply to a load cell, representing the minute changes that need to be measured.

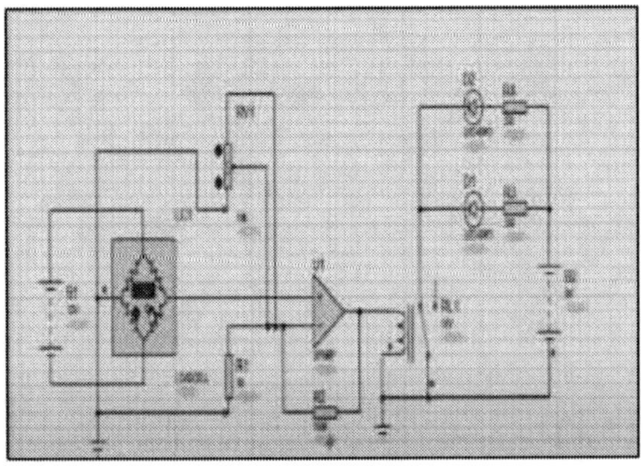

Fig 11. Simulation with Strain Gauge

- O-PAMP- The Operational Amplifier is a DC-coupled high-gain electronic voltage amplifier with a distinct input and typically a single-end output.
- POT-HG- It is an instrument called as potentiometer used for measuring the voltage. Potentiometers are typically used on audio equipment to control electrical devices such as volume controls. Potentiometers operated by a mechanism can be used as transducer of position,
- Relay- It is used when a circuit can be operated by a low-power signal. The signal coming from one source to another also works to move.
- Resistor-A resistor is a passive two-terminal electrical component which as an element of a circuit implements electrical resistance. In electronic circuits, resistors are used to reduce current flow, modify signal levels, separate voltages, bias active elements and terminate transmission lines, among other applications.

5. Conclusion

The two-wheeler chassis dynamometer is a test rig which is used to test the two-wheeler's performances. In previously used dynamometer was equipped with the Microcontroller for collecting performance data. But due to high costs of microcontroller it's not affordable to use in Portable Two- Wheeler chassis dynamometer. Hence in the present study the PLC & SCADA system is implemented for a better output and more precious outcome. PLC allows us to expand the setup by having a greater number of I/O Ports.

The system will work under any conditions, such as extreme heat, extreme mechanical vibrations, electrical noise when in cooperated with a two-wheeler dynamometer test setup. The sensors will allow us to read all the parameters of the vehicle and enable us to solve this problem. The processing speed increasing by 20%.

The simulation was built for the sensors of the test rig and 3% of errors were found compared to the outcome of the simulation with the data sheet.

References

[1] Yu Qian, Zhongming Luo, Zhuofu Liu, Hailu Zhao, Chenyang Li, Yang Song, Desheng Nan, Jingwei Wei, Application of RTD sensor in the real time measurement and wireless transmission,

Harbin University of Science and Technology, 2014 Fourth International Conference on Instrumentation and Measurement, Computer, Communication and Control.

[2] Shiyu Zhou, Jianjun Shi, Department of Mechanical Engineering, University of Michigan, Ann Arbor, MI 48109, USA. "Active Balancing and Vibration Control of Rotating Machinery: A Survey" *The Shock and Vibration Digest, Vol. 33, No. 4, July 2001 361-371-2001 Sage Publications.*

[3] Sun Jian, Lu Peng, Fu Yaqiong, Liu Ruixiang., The Implementation and Application of Programming Port Communication between Industry PC and Mitsubishi FX Series PLC, *Proceedings of 2008 3rd International Conference on Intelligent System and Knowledge Engineering, 2008 IEEE.*

[4] Manual of Mitsubishi PLC

[5] Manual of Parker Drive-HA463296 Issue 8

[6] Siemens SIWAREX Load Cell Operating *Instruction-Edition 06/2009.*

[7] Dan Mihai stefanescu, Strain Gauges and Wheatstone Bridges – Basic Instrumentation and New Applications for Electrical Measurement of Non-Electrical Quantities, *8th International Multi-Conference on Systems, Signals & Devices.*

[8] Edval J. P. Santos, Senior Member IEEE, and Isabela B. Vasconcelos, Member IEEE, RTD-based Smart Temperature Sensor: Process Development and Circuit Design, *26th INTERNATIONAL CONFERENCE ON MICROELECTRONICS (MIEL 2008).*

Low Powered Self-Testable ALU

M. Nagharjun[1*], **V. Ravi**[1]
[1]School of Electronics Engineering(SENSE), Vellore Institute of Technology, Chennai

Email: nagharjun.m2018@vitstudent.ac.in

Abstract-The Arithmetic and Logic Unit(ALU) is a Significant Segment of a CPU as it does all the arithmetic and logical operations and computations required by all the processes running inside the CPU. Being such an important part inside the processor, It requires more power and thus it would be efficient and feasible if it can run on low power and also can test for any faults in the circuit itself. This paper completely focuses on building a Low Powered ALU with the implementation of Built-In Self-Test(BIST) mechanism for efficient arithmetic and logical operations. In the design, power-efficient circuits such as Wallace Tree Multiplier and Carry Look-Ahead Adder is being used. The created design is synthesized and simulated on the Xilinx Vivado software tool and implemented on Digilent Xilinx Basys 3 Artix-7 (XC7A35T-1CPG236C) Field Programmable Gate Array (FPGA). The overall power consumption of the design is 75mW and the ALU has self-testability function.

Keywords: Arithmetic and Logic Unit, Xilinx, Wallace Tree Multiplier, Carry Look-Ahead Adder, Built-In Self-Test, Field Programmable Gate Array.

1. Introduction

As day by day, more and more transistors are embedded in the Integrated Circuits, and power dissipation plays an important role in any type of circuit produced. Looking into the Arithmetic and Logical Unit which is a fundamental part of the Central Processing Unit, needs to be optimized to dissipate lower power. The Field Programmable Gate Arrays(FPGA's) are reprogrammable, flexible and versatile hardware device which can be programmed to any digital circuit using Hardware Description Languages like Verilog or VHDL and has many applications. Due to these advantages, the Arithmetic and Logic Unit has been programmed and implemented through FPGA. The main function of the ALU is to perform bitwise Arithmetic and Logical operations assigned to the ALU by the processor. In the Fetch – Decode – Execute process of the CPU, the ALU performs the Execute process i.e., the required computations for many different processes running in the device.

2. Problem Statement

As ALU is one of the main part of the processes done by the CPU, it faces many issues. The main concern is Power Dissipation and Consumption. It hinders the efficiency of the circuit. Whenever a circuit is needed to be efficient, it requires higher performance which automatically leads to higher power being consumed and dissipated because of the higher number of many different complex operations to be done in a short amount of time. The amount of power dissipating is directly proportional to the amount of heat released. Every similar transistor has a same Junction Temperature value where, if the temperature due to the heat released by the transistors crosses the Junction Temperature, the transistor may get damaged and can cause circuit failure. Due to this, the ability of the CPU to reach higher clock rates has been restricted. The other major problem is Area Consumption of the circuit. As the modern circuits needs new and upgraded functionalities, the density of transistors in the circuit increases. So efficient use of area is required. Another problem is Faults present in the circuit which leads to the production of wrong outputs and computations. This may also be a case due to higher power being dissipated and due to transistor aging which disrupts the logic of the circuit.

3. Literature Survey

Many different types of methods have been proposed to address this problem statement and provided, have been efficient. Some of the effective methods have been discussed.

Jamuna S et al., [1] proposed the BIST logic design for ALU on FPGA. This mainly focused of the detection of faults in the circuit by using BIST architecture with 94.2% code coverage. T Esther Rani et al., [2] proposed a method which focusses on reducing the power of the ALU by 70% compared to traditional design by using low power 1-bit full adders with 10 transistors. Typically, a 1-bit full adder requires a total of 16 transistors which has been reduced here to 10 to get an effective and efficient result. AnkitMitra et al., [3] proposed the Implementation and Design of Low Power 16-bit ALU with Clock Gating Technique where, whenever the circuit is not in use, the clock signals are frozen and power is saved. N Sureka et al., [4] proposed An Efficient High Speed Wallace Tree Multiplier. The focuses on building a low powered and efficient Wallace tree multiplier by using Carry Save Adder to reduce the latency which is 44.4% faster than the traditional Wallace tree multiplier.

4. Methodology

As discussed in the previous topics, the ALU being such a major part inside the processor consumes and dissipates high power. Power Consumption and Dissipation can be reduced if we reduce the delay in the circuit as it performs operations in a lesser amount of time which decreases the overall power consumption. One of the delays that can be reduced is the propagation delay. The circuits for these algorithms are complex but it is power efficient. Now as the ALU is power efficient, another concern is the generation of faulty outputs. This is most commonly due to Stuck at Faults where a part of the circuit is stuck at either 0 or 1. Taking this problem into consideration, a Built-In Self-Test mechanism is being introduced where the circuit tests itself for any faults.

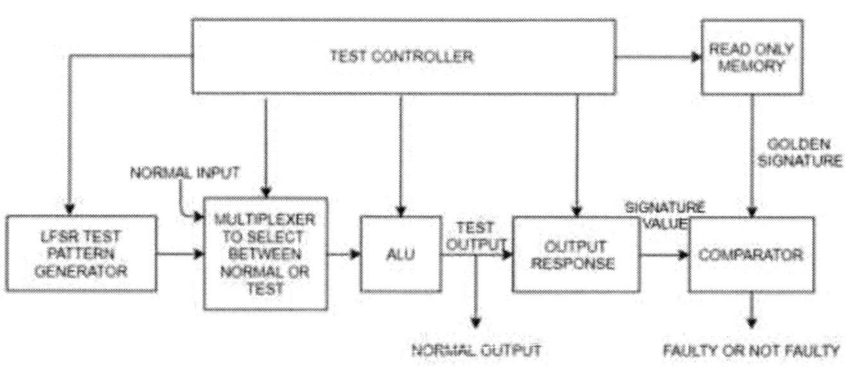

Figure 1 : Total system Architecture

The final architecture of the system is shown in the Figure 1. The TEST CONTROLLER is the main block which holds all the sub blocks together. The MULTIPLEXER TO SELECT BETWEEN NORMAL OR TEST block is used to select between a Normal ALU and Testing ALU. In the case of Normal ALU, the input is directly fed into the system. In the case of Testing ALU, the input is fed by the LFST TEST PATTERN GENERATOR block which consists of one or two LFSR's depending on the type of operation. Next comes the ALU block where the main arithmetic or logical operation takes place. In the case of a Normal ALU, the output is shown without undergoing any further actions. In the case of Testable ALU, the output is fed to the OUTPUT RESPONSE block which is the MISR and it is compared against the GOLDEN SIGNATURE stored inside the READ ONLY MEMORY block which contains the correct outputs, in the COMPARATOR block. The output of the comparator block checks whether the output is FAULTY OR NOT FAULTY.

4.1. Arithmetic and Logic Unit

The ALU deals with both Arithmetic and Logical operations. The Arithmetic part performs Addition, Subtraction, Multiplication, Division, Increment, and Decrement. The Logical part performs Logic Gate related operations and Shifting of bits. Each and every Arithmetic and Logical operation has a fixed opcode and an operation can be called upon by inputting the opcode for the required operation. The CPU fetches the instruction from memory and supplies it to the Instruction Register which decodes the instruction into opcodes and operands. According to the opcodes, the operation is performed using operands.

Table 1 : Opcodes for ALU Operation

OPCODES	**ALU OPERATION**
0000	A + B
0001	A - B
0010	A * B
0011	A / B
0100	A + 1
0101	A - 1
0110	A AND B
0111	A OR B
1000	A XOR B
1001	NOT A
1010	A NAND B
1011	A NOR B
1100	SHIFT RIGHT
1101	SHIFT LEFT
1110	GREATER COMPARE
1111	EQUAL COMPARE

Table 1 shows the various Arithmetic and Logical operations performed by the ALU and its control bit for calling the operations. For example, if 1011 is the OPCODE, then the function of the ALU is to perform the NOR operation between the given two operands and if 0000 is selected, the operation of the ALU is to perform addition(+) operation between the given two inputs.

4.1.1. Carry Look-Ahead Adder.

The carry look-ahead adder is also called a fast adder. Unlike the conventional ripple carry adder, the carry look-ahead adder reduces the propagation delay which is caused due to the waiting time for the ith full adder the get the i-1th full adder carry, by using a complex architecture. The carry look-ahead adder has 2 signals, Carry Propagate(P) and Carry Generate (G). Each and every bit of the sum and carry has been calculated in parallel using these given equations whereas in the conventional adder, each block has to wait until the carry of the previous block to arrive. Compared to other conventional and frequently used adders, the Carry Look Ahead Adder consumes both less power and low area[9].

Carry Propagate
Pi = Ai XOR Bi

Carry Generate
Gi = Ai AND Bi

Sum
Si = Pi XOR Ci

Carry
Ci+1 = Gi OR (Pi AND Ci)

The Sum and Carry are calculated using the above equations.

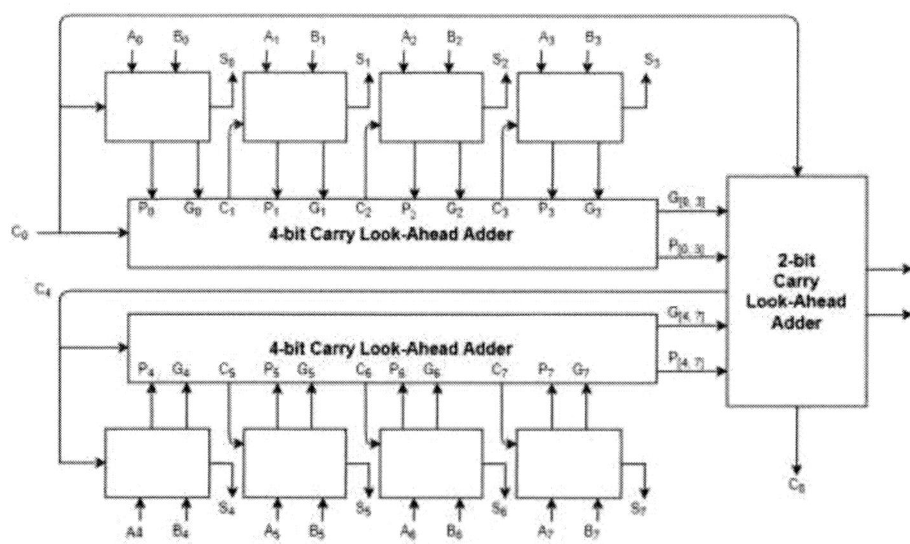

Figure 2 : Carry Look-ahead Adder Architecture

Figure 2 shows the circuit of carry look-ahead adder. This architecture consists of two 4-bit Carry Look-Ahead Adders, one 2-bit Carry Look-Ahead Adder and blocks to calculate Carry Generate(G_i) and Carry Propagate(P_i) using the equations given above.

4.1.2. *Wallace Tree Multiplier*

It is a hardware implementation of an effective and efficient digital circuit which is used for multiplication. This circuit also reduces the propagation delay to a great extent. As each and every bit of both of the operands are multiplied, partial products are being generated. The output is produced by reducing the partial products using half and full adders in parallel until we get a 2-layered partial product, and finally either using half adders or carry look-ahead adder to produce the final output. This architecture is used due to its ability to perform operations in low power and also has faster-switching speed. Compared with the conventional array multiplier, the order of the delay comes down from O(n) to O(log n) in the Wallace tree multiplier. Comparing with widely used other multipliers like Booth Multiplier, Array Multiplier, the Wallace Tree Multiplier consumes less power and low area[10].

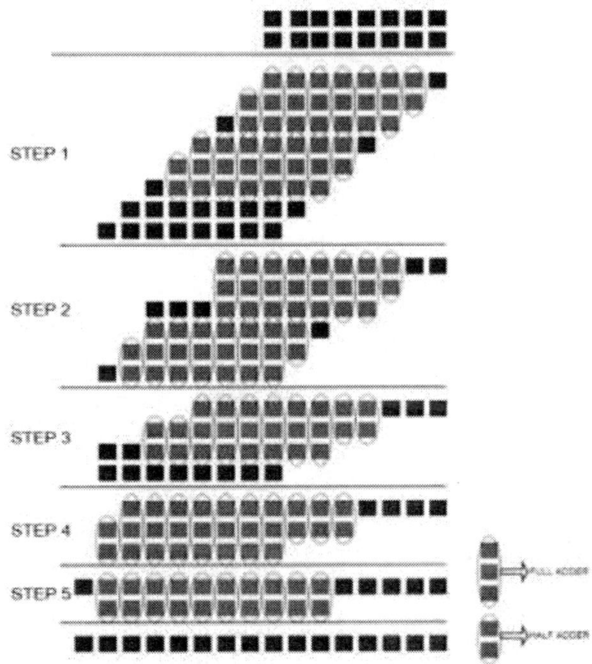

Figure 3 : Steps for Wallace Tree Multiplier

The figure 3 shows the operation of an 8-bit Wallace tree multiplier. Each of the Squares represent the bits. The Black squares represent the bits that are static and does not need to undergo any operation. The Grey squares represent the bits which are going to undergo either Full Adder or Half Adder operation. At Step 1, AND operation is performed between each and every bits of the two 8-bit numbers. At Step 2, Step3, Step 4 and Step 5, Full Adder operations and Half Adder operations are performed simultaneously and sequentially. The final 16-bit output is being generated.

4.2. *Built In Self Test*

BIST is a self-testing mechanism for digital IC's where random test vectors generated by Linear Feedback Shift Register(LFSR) are applied to the circuit under test under the control of the BIST controller. The output responses are compacted by the signature analyzer which is the MISR in the OUTPUT RESPONSE to produce a Signature Response Value which is further compared with the Golden Signature Value which is stored inside the ROM or some memory. If the Signature Response Value and Golden Signature are the same, then the output is not faulty. If the Signature Response Value and Golden Signature are different, the output is faulty. In this way, we can detect any faults present internally in the circuit.

4.2.1. LFSR

This hardware circuit is based on a shift register with feedback. This circuit generates pseudo-random patterns for testing the circuit. To produce 8-bit patterns, we use 8 D-Flipflops. The points where the feedback are taken are called taps and it affects the generation of pseudo random patterns.

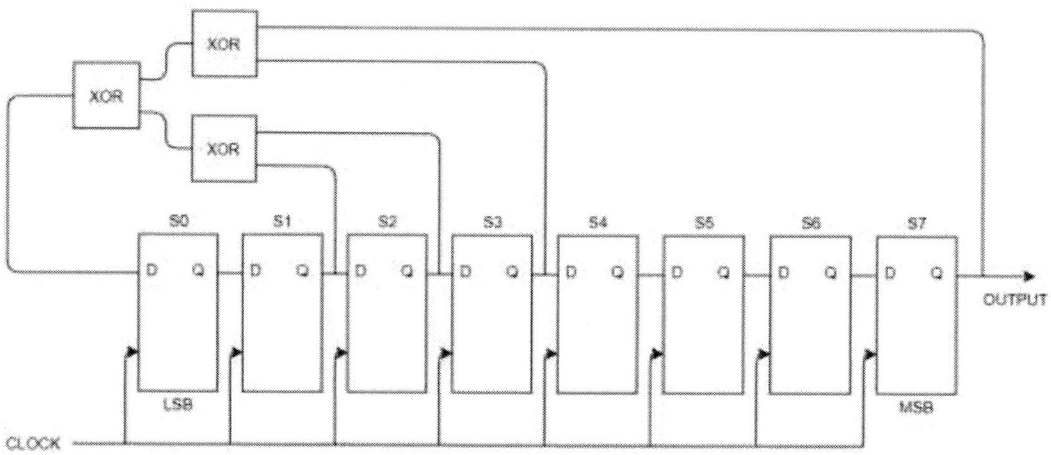

Figure 4 : LFSR Architecture

The above diagram depicts the architecture of the Many to One Linear Feedback Shift Register used. The taps are made at 1st, 2nd, 3rd and 7th points to produce a 8-bit pseudo random number.

4.2.2. Output Response or MISR

We cannot compare all of the outputs produced by the ALU, so we need to compact the outputs. Multiple Input Signature Register(MISR) is used to compact the outputs produced by the ALU. MISR is a hybrid of LFSR which is used for the compaction of the output responses of the ALU to produce a the Syndrome also called as the Signature Response which is compared with the Golden Signature.

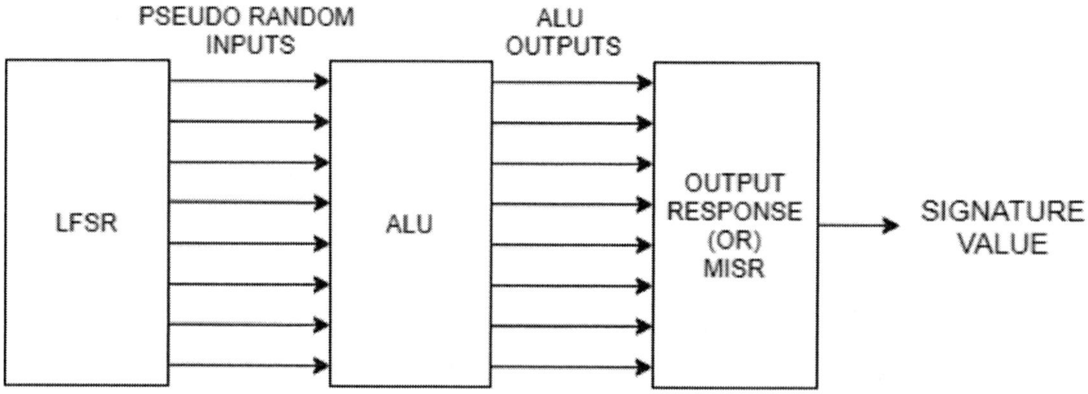

Figure 5 : MISR Working

In figure 5, the LFSR produces many pseudo random patterns and supplies to the ALU. The ALU produces many outputs which are supplied as inputs to the OUTPUT RESPONSE which compacts the multiple inputs and produces a single Signature Response value.

5. Implementation and Result

The whole implementation of the design is done in Basys 3 Artix 7 FPGA board. The complete RTL schematic and architecture of the design is given.

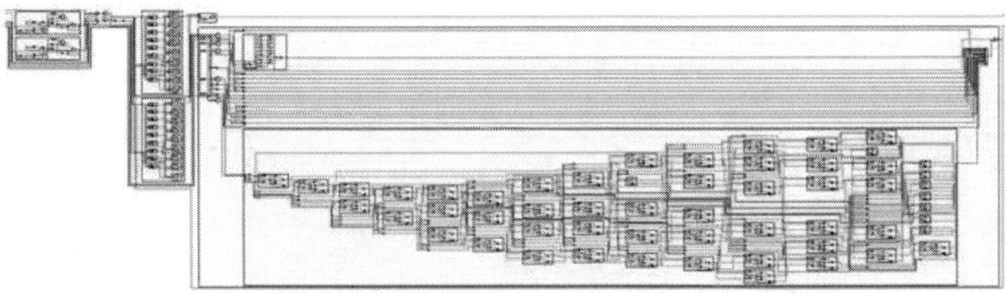

Figure 6 : RTL Design

Figure 6 shows the total view of the RTL architecture of the design.

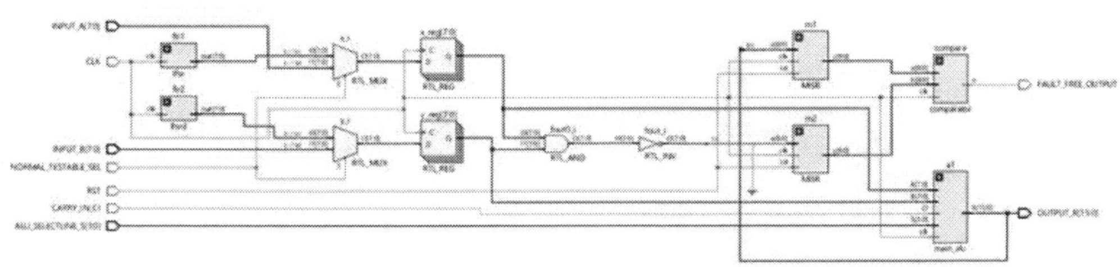

Figure 7 : RTL Design Simplified

The RTL design in Figure 7 has 2 LFSR's, 2 MISR's, 2 Multiplexers, 2 registers x and y to store the values of A and B if the ALU functions normally or the test patterns in the case of testing, 1 ALU block which has Wallace Tree Multiplier and Carry Look Ahead Adder, memory element with the signature response value and a comparator to check whether the circuit is faulty. When the NORMAL_TESTABLE_SEL line has logic 1, then the testing part will be done and when NORMAL_TESTABLE_SEL line has logic 0, then the normal ALU will function.

5.1. Power

The overall power consumed by the design is 75mW. It consists of 72mW in Static Power and 3mW in Dynamic Power.

TOTAL ON-CHIP POWER	
Static Power (96%)	0.072W
Dynamic Power (4%)	TOTAL: 0.003W Clocks: <0.001W Signals: <0.001W Logic: <0.001W I/O: 0.002W

Table 2 : Overall Power Consumption

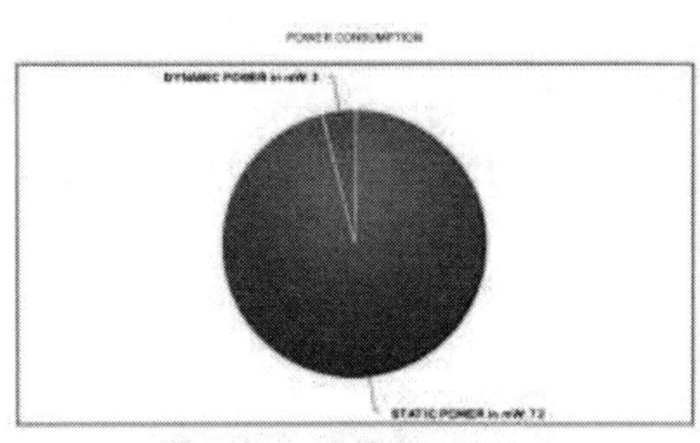

Figure 8 : Overall Power Consumption

IOP Conf. Series: Materials Science and Engineering 1012 (2021) 012048 doi:10.1088/1757-899X/1012/1/012048

5.2. Area
The total area occupied by the design is shown below.

LUT's	FLIP FLOPS
286	120

Table 3 : Overall Area Consumption

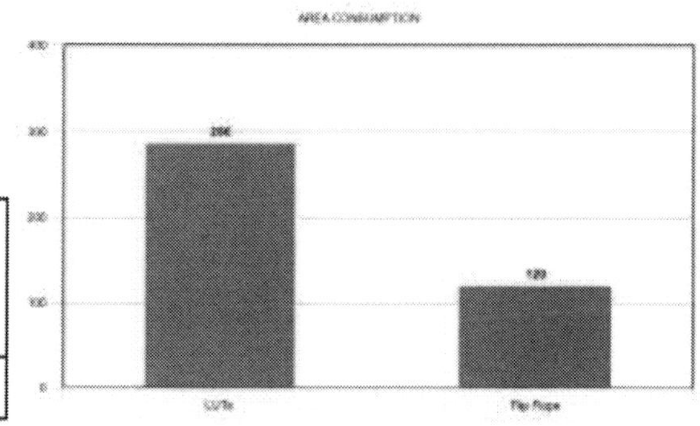

Figure 9 : Overall Area Consumption

6. Simulation and Output
The entire simulation of the proposed design is done in Vivado. There are in total 16 different arithmetic and logical operations that can be simulated and here 5 of the important and frequently used simulations are done. Along with this, the Self-Testable ability of the ALU is also shown through the simulation.

6.1. Normal ALU Mode

Multiplication operation is performed where,
NORMAL_TESTABLE_SEL = 0 indicating the ALU is in normal mode
INPUT_A = 60, INPUT_B = 80 indicating the operands
ALU_SELECTLINE_S = 0010 indicating the opcode for multiplication operation
OUTPUT_R = 4800 indicating the output of multiplication

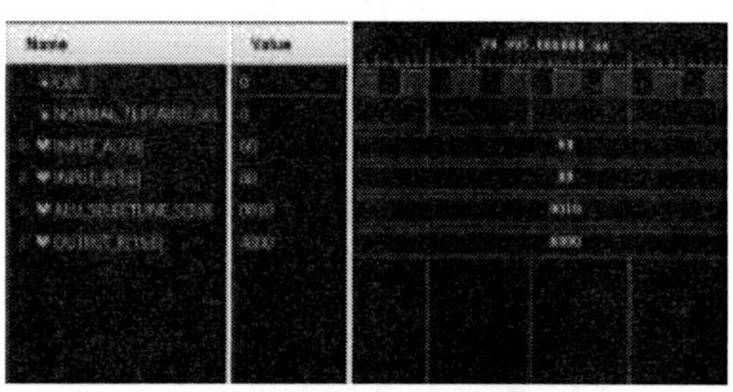

Figure 10 : Multiplication Simulation

Addition operation is performed where ,
NORMAL_TESTABLE_SEL = 0 indicating the ALU is in normal mode
INPUT_A = 60, INPUT_B = 80 indicating the operands
ALU_SELECTLINE_S = 0000 indicating the opcode for addition operation
OUTPUT_R = 140 indicating the output of addition

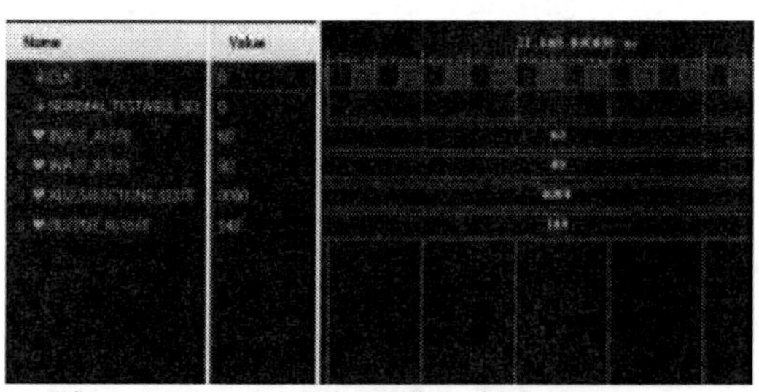

Figure 11 : Addition Simulation

OR Gate operation is performed
where,
NORMAL_TESTABLE_SEL = 0
indicating the ALU is in normal mode
INPUT_A = 00111100,
INPUT_B = 01010000 indicating
the operands
ALU_SELECTLINE_S = 0111 indicating
the opcode for OR Gate operation
OUTPUT_R = 0000000001111100
indicating the output of OR Gate

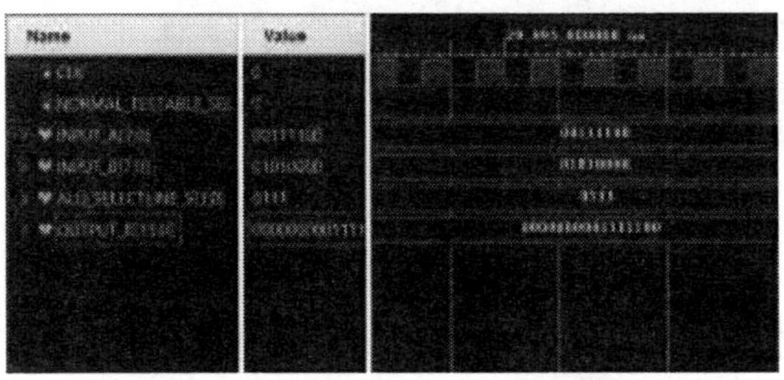

Figure 12 : OR Gate Simulation

Greater than operation is performed
where ,
NORMAL_TESTABLE_SEL = 0
indicating the ALU is in normal mode
INPUT_A = 80, INPUT_B = 60 indicating
the operands
ALU_SELECTLINE_S = 1110 indicating
the opcode for Greater than operation
As 80 > 60, OUTPUT_R = 1 indicating
the output of Greater than

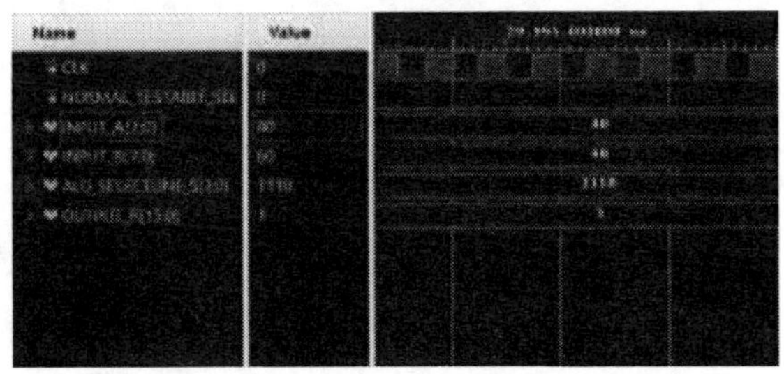

Figure 13 : Greater Than Simulation

Shift Left operation is performed
where,
NORMAL_TESTABLE_SEL = 0
indicating the ALU is in normal mode
INPUT_A = 00111100 indicating
the operand
ALU_SELECTLINE_S = 1101 indicating
the opcode for Shift Left operation
OUTPUT_R = 0000000001111000
indicating the output of Shift Left

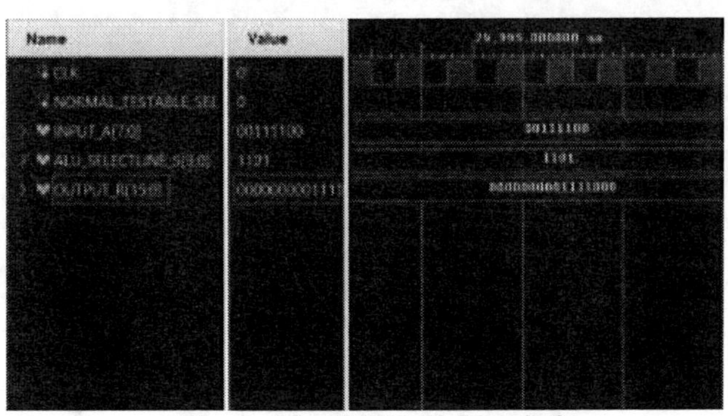

Figure 14 : Shift Left Simulation

6.2. Testable ALU Mode
Here NAND gate is used for testing whose opcode is 1010.

6.2.1. Non-Faulty
The NAND of the two of the LFSR generated pseudo random patterns are compacted and the signature response value produced is SIGNATURE_OUTPUT and ROM_OUTPUT is the memory containing the Golden Signature. Both the SIGNATURE_OUTPUT and ROM_OUTPUT are compared and as both are same, the fault free output is logic 1.

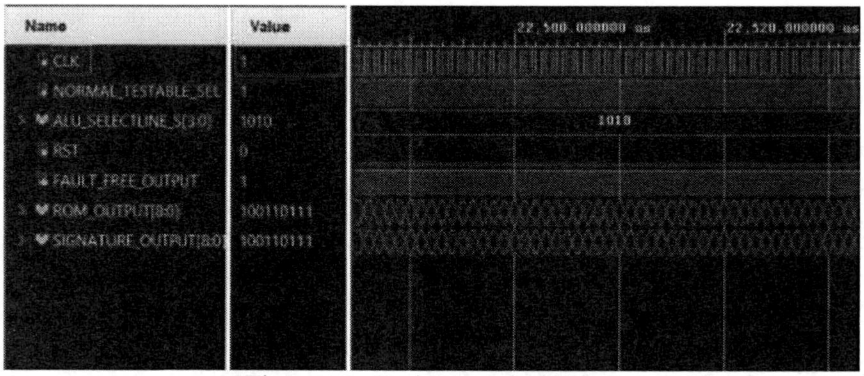

Figure 15 : Fault Free Simulation

6.2.2. Faulty
The NAND of the two of the LFSR generated pseudo random patterns are compacted and the signature response value produced is SIGNATURE_OUTPUT and ROM_OUTPUT is the memory containing the Golden Signature. Both the SIGNATURE_OUTPUT and ROM_OUTPUT are compared and as both are not the same, the fault free output is logic 0.

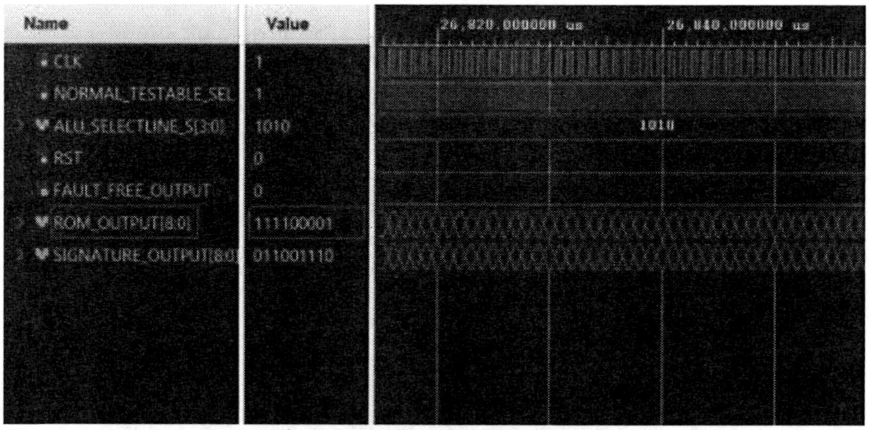

Figure 16 : Faulty Simulation

7. Result and Performance Analysis

The total power consumed by the ALU is 75mW with 72mW in Static power and 3mW in Dynamic power.

In comparison with the paper Jamuna S et al.,[1] which uses BIST Mechanism, consumes 1.061W, there has been a significant reduction in the power consumption in the proposed design with 75mW.

In comparison with the work by R Dhanabal et al.,[5], the combined power consumption for all of the ALU operations is 229.7mW, and the proposed design consumes much less power.

Now comparing the Dynamic Power which is 3mW with self-testable functionality in the proposed design with the work of B Pandey et al.,[8] which proposes Clock Gating technique to reduce power, consumes 3mW in Dynamic Power with clock gate on 100MHz.

Table 4: Performance Comparison

DESIGN	STATIC POWER	DYNAMIC POWER	TOTAL POWER
Jamuna S et al., [1]	-	-	75mW
Dhanabal R et al., [5]	-	-	229.7mW
B Pandey et al., [8]	-	3mW	-
Proposed Design	72mW	3mW	75mW

8. Conclusion and Future Scope

It is very much required that the design of the ALU should be low powered. In this work, a low powered, fast, and testable ALU has been implemented. The ALU can be used for normal purposes and can detect if any fault is present or not. The design was targeted and implemented on Basys 3 Artix 7 FPGA and the overall power consumption is found out to be 75mW. The design has been simulated and is giving positive results.

One of the improvements which can be done is implementing the pipelining strategy to get more efficient results with less delay. Deep Learning algorithms can be implemented to improve the performance of the design by learning the frequently used operations.

References

[1] J. S, D. P, K. Shashikala and K. K. K, "Design and Implementation of BIST logic for ALU on FPGA," 2018 International Conference on Networking, Embedded and Wireless Systems (ICNEWS), Bangalore, India, 2018, pp. 1-5.Another reference

[2] T. E. Rani, M. A. Rani and R. Rao, "AREA optimized low power arithmetic and logic unit," 2011 3rd International Conference on Electronics Computer Technology, Kanyakumari, 2011, pp. 224-228.

[3] AnkitMitra., "Design and implementation of low power 16 bit ALU with clock gating"

International Journal of Advanced Research in Computer Engineering & Technology (IJARCET) Volume 2, Issue 6, June 2013, ISSN: 2278 – 1323

[4] N. Sureka, R. Porselvi and K. Kumuthapriya, "An efficient high speed Wallace tree multiplier," *2013 International Conference on Information Communication and Embedded Systems (ICICES)*, Chennai, 2013, pp. 1023-1026.

[5] R, Dhanabal & Bharathi, V. & Salim, Saira & Thomas, Bincy & Soman, Hyma & Sahoo, Sarat Kumar. (2013). Design of 16-bit low power ALU-DBGPU. Intern. J. Eng. Technol.. 5. 2172-2180.

[6] Kishore, P., Sridevi, P.V., Babulu, K., & Chandra, K. (2015). A Novel Low power and Area Efficient Carry-Lookahead Adder Using MOD-GDI Technique.

[7] T. Thubrikar, S. Kakde, S. Gaidhani, S. Kamble and N. Shah, "Design and implementation of low power test pattern generator using low transitions LFSR," *2017 International Conference on Communication and Signal Processing (ICCSP)*, Chennai, 2017, pp. 0467-0471.

[8] B. Pandey, J. Yadav, M. Pattanaik and N. Rajoria, "Clock gating based energy efficient ALU design and implementation on FPGA," 2013 International Conference on Energy Efficient Technologies for Sustainability, Nagercoil, 2013, pp. 93-97, doi: 10.1109/ICEETS.2013.6533362.

[9] A. K. Panigrahi, S. Patra, M. Agrawal and S. Satapathy, "Design and Implementation of a high speed 4bit ALU using BASYS3 FPGA Board," 2019 Innovations in Power and Advanced Computing Technologies (i-PACT), Vellore, India, 2019, pp. 1-6.

[10] B. Harish, K. Sivani and M. S. S. Rukmini, "Design and Performance Comparison among Various types of Adder Topologies," 2019 3rd International Conference on Computing Methodologies and Communication (ICCMC), Erode, India, 2019, pp. 725-730, doi: 10.1109/ICCMC.2019.8819779

[11] U. Mandal, R. Banerjee and R. Mishra, "A Comparative Study of Arithmetic Multipliers," 2018 2nd International Conference on Electronics, Materials Engineering & Nano-Technology (IEMENTech), Kolkata, 2018, pp. 1-5, doi: 10.1109/IEMENTECH.2018.8465371.

[12] B. Lamba and A. Sharma, "A review paper on different multipliers based on their different performance parameters," 2018 2nd International Conference on Inventive Systems and Control (ICISC), Coimbatore, 2018, pp. 324-327, doi: 10.1109/ICISC.2018.8399088.

[13] S. Abraham, S. Kaur and S. Singh, "Study of various high speed multipliers," 2015 International Conference on Computer Communication and Informatics (ICCCI), Coimbatore, 2015, pp. 1-5, doi: 10.1109/ICCCI.2015.7218139.

[14] B. Koyada, N. Meghana, M. O. Jaleel and P. R. Jeripotula, "A comparative study on adders," 2017 International Conference on Wireless Communications, Signal Processing and Networking (WiSPNET), Chennai, 2017, pp. 2226-2230, doi: 10.1109/WiSPNET.2017.8300155.

[15] J. Saini, S. Agarwal and A. Kansal, "Performance, analysis and comparison of digital adders," 2015 International Conference on Advances in Computer Engineering and Applications,

Ghaziabad, 2015, pp. 80-83, doi: 10.1109/ICACEA.2015.7164650.

[16] K. B. Jaiswal, Nithish Kumar V, P. Seshadri and Lakshminarayanan G, "Low power wallace tree multiplier using modified full adder," 2015 3rd International Conference on Signal Processing, Communication and Networking (ICSCN), Chennai, 2015, pp. 1-4, doi: 10.1109/ICSCN.2015.7219880.

[17] Shahzad Asif, Yinan Kong, "Low-Area Wallace Multiplier", VLSI Design, vol. 2014, Article ID 343960, 6 pages, 2014. https://doi.org/10.1155/2014/343960

[18] N. Ravindran and R. M. Lourde, "An optimum VLSI design of a 16-BIT ALU," 2015 International Conference on Information and Communication Technology Research (ICTRC), Abu Dhabi, 2015, pp. 52-55, doi: 10.1109/ICTRC.2015.7156419.

[19] A. Deeptha, D. Muthanna, M. Dhrithi, M. Pratiksha and B. S. Kariyappa, "Design and optimization of 8 bit ALU using reversible logic," 2016 IEEE International Conference on Recent Trends in Electronics, Information & Communication Technology (RTEICT), Bangalore, 2016, pp. 1632-1636, doi: 10.1109/RTEICT.2016.7808109.

[20] Dhanumjaya K., Kiran Kumar G., Giriprasad M.N., Raja Reddy M. (2010) Design and Modeling of Power Efficient, High Performance 32-bit ALU through Advanced HDL Synthesis. In: Das V.V., Vijaykumar R. (eds) Information and Communication Technologies. ICT 2010. Communications in Computer and Information Science, vol 101. Springer, Berlin, Heidelberg. https://doi.org/10.1007/978-3-642-15766-0_3

[21] Shinde J.R., Sharma S., Dash L. (2020) An Optimization Design Approach for Arithmetic Logic Unit. In: Gunjan V., Garcia Diaz V., Cardona M., Solanki V., Sunitha K. (eds) ICICCT 2019 – System Reliability, Quality Control, Safety, Maintenance and Management. ICICCT 2019. Springer, Singapore. https://doi.org/10.1007/978-981-13-8461-5_81

[22] Shylashree, N., Venkatesh, B., Saurab, T.M. et al. Design and analysis of high-speed 8-bit ALU using 18 nm FinFET technology. Microsyst Technol 25, 2349–2359 (2019). https://doi.org/10.1007/s00542-018-4112-y

[23] Gary K. Yeap., "Practical Low Power Digital VLSI Design" 1998 Springer US Kluwer Academic Publishers

[24] Computer Architecture. A Quantitative Approach. Fourth Edition. John L. Hennessy. Stanford University. David A. Patterson. University of California at Berkeley.

[25] Modern Digital Electronics · R P Jain. Tata McGraw-Hill Education, Jun 1, 2003

Automated System for Lavatory Disinfection

Sneha Joseph[1*], Paul A Thaikattil[2], T. Suresh[1], Seethalakshmi. E[1]

[1]Department of Electronics and Communication Engineering R.M.K. Engineering College, Kavaraipettai, Tamilnadu, India,
[2]Festo India Private Ltd. India

E-mail: sjh.ece@rmkec.ac.in

Abstract- Instead of that conventional method of chemical (sanitizer spray) disinfection, this paper introduces a novel idea to fight the attack of microbes causing infections and diseases from the source like lavatory and specially its seat. Here we utilize the germicidal power of ultra violet light ray's .i.e. Ultra Violet C range of the light spectrum against the viruses (SARS-CoV-2), bacteria (E. coli) and other microbes. This system of disinfection constitutes the UV LEDs placed on the arm above the toilet seat. The range of wavelength is from 200nm to 280nm. Implementation of this technique ensures safety, by human free room with the automatic motion control using Arduino, motor and door locking gadgets. UVC light is very advantageous over other methods as it is eco-friendly, chemical free, non-corrosive, and no hazardous waste is produced and also inactivates viruses by attacking the DNA strands in the microbes ensuring that the replication does not happen further stops spreading of contagious diseases leading to pandemic. Usage of toilet paper and other methods used conventionally could come to an end. Thus an effective idea is derived for the hygiene establishment.

Keywords: UVC, SARS-CoV-2, non-corrosive, DNA, germicidal, lavatory, automation.

1. Introduction

During this COVID-19 pandemic year, we must have a cover over us to protect from the spreading of the diseases. But by reducing the consumption of environment resources; such as cutting of trees for papers and production of hazardous waste residues. All these are replaced by UVC for germicidal purposes.

Since the past years, we all are facing many infective diseases; these infections come from many sources, including contaminated hands and unclean surfaces. However these could be decreased by doing serious sanitizing efforts, and by a continuous effort to reduce these infections, healthcare facilities have turned to several disinfection methods. Ultraviolet light spectrum is a form of light that a human eye cannot see or sense. That is present in electromagnetic spectrum in between X-rays and visible light. In the sun's ultraviolet light emission; much of it is assimilated by ozone layer. The unique characteristic of this light is that a specific range of its wavelengths, those between 200 and 300nms and are categorized as germicidal which means they are able of inactivating microorganisms, such as bacteria, viruses and protozoa. This ability has allowed large adoption of the Ultra Violet rays spectrum as an environmentally friendlier, chemical-free, and highly efficient way to disinfect and guard water against harmful microbes. In Ultraviolet spectrum in the Ultra Violet-C band is called to be UV-C germicidal spectrum, which is utilized in large scale for purification of the equipment for reuse. And thus creating sterile environments. They are also used in the food and water industries to deactivate microbes and germs.

Unlike the chemical approaches of disinfection, Ultra Violet provides spontaneous and efficient deactivation of microbes by using a physical procedure. The Ultra Violet radiation has depicted efficacy against the pathogens, including that are responsible for hepatitis, typhoid, polio, cholera and other parasitic, viral and bacterial diseases. In addition, UV light (maybe only UV or combined with hydrogen peroxide .i.e. photo sensitizer) can decontaminate the chemicals pollutants such as insecticides, pest repellents, the industrial solvents and waste from pharmaceuticals through the process known as UV-oxidation. Microorganisms like bacteria and protozoan are destructed by utilizing Ultra Violet spectrum of light as a result of which there is damage in the nucleic acid. The enormous energy produced by the short wavelength Ultraviolet Violet energy, basically 254 nm, is captivated by cell's DNA strands and RNA strands. This engrossing of Ultra Violet energy creates new bonds between the nearby nucleotides, creating multiple bonds or else dimers. Dimerization of nearby molecules, specifically thymine takes place. The most vulgar photochemical damage will occur. Formation of numerous dimers of thymine in the DNA's of virus and bacteria prevents production of replicas and reduce the capability to infect.

2. Background

Contemporarily, many procedures are presently there for sterilization in pharmaceutical and medical field products like these chemical treatment, thermal treatment, ultraviolet radiative treatment and plasma radiation [1]. Such as in the COVID-19, pandemic period, Personal protective equipment used are as facemask respirators, protective isolation garments, goggles for eye protection and also face shields that an all are utilized to attain the promised safety of the health care workers. There are use and throw items and reusable wearable's also available in the commercial market. Reusable garments can be worn for several times. The sterilization or disinfection procedure includes washing, parching or sterilization. Thus process, of using dry steam at 1210C for 30 minutes are usually used process of sterilization [2]. The United States Centers for Disease Control gives recommendation for cleaning of face protection and eye guards using a cloth to clean and also use it with saturated with hospital solution EPA for disinfection or disinfectant wipe prior to the reuse [3].

In the chemical sterilization process, the chemical ingredients such as formaldehyde, glutaraldehyde and hydrogen peroxide are being used as sterilizing agent to inactivate viruses like SARS-CoV-2 also [4]-[6].Effectiveness and efficacy of UV disinfection is influenced due to the intensity of the light and its exposure time [7], [8]. From the wide range of Ultra Violet spectrum of light; Here we use the Ultra Violet Range from 210 to 365 nm specified UV LEDs [9].

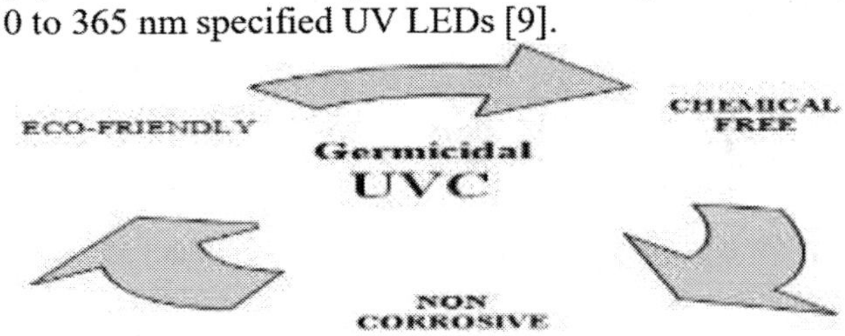

Fig. 1 Major advantages of the Germicidal UV

3. Proposed Method

In Here we the property of log reduction which means logarithmic reduction, where it will be defined as the predicted degree of the dosage needed for the specific disinfection or sterilization. The logarithmic range of reduction is the percent of physically sterilized, inactivated microorganisms by this radiation procedure. By example, in log one reduction we can get a reduction in the pathogen that is to be deactivated reduced to 10% as in per the infection level before Ultra Violet exposure and disinfection. Microorganism number is also reduced by a count 10 or 1 log. And, the log 2 reduction will have around 1% of microbe that is by the factor 100 and thus goes on.
UV C dosage and response comparison defines at which rate particular microbes is inactivated by applying specific dosage of UV exposure. This computation is given as either by the amount by which microbes disinfected or the remaining proportion as a UV dose function. Response of UV radiation is computed by the following mathematical representation;

Log inactivation = log10 (No/N)

And;

> No = infectious microorganisms concentration before the exposure to Ultra Violet radiation.

> N = infectious microorganisms concentration after the radiation exposure of UV radiation.

Each microbe category has a unique curve for dose-response. This curve describes how every microorganism responds to different dosage of UV radiation of light.

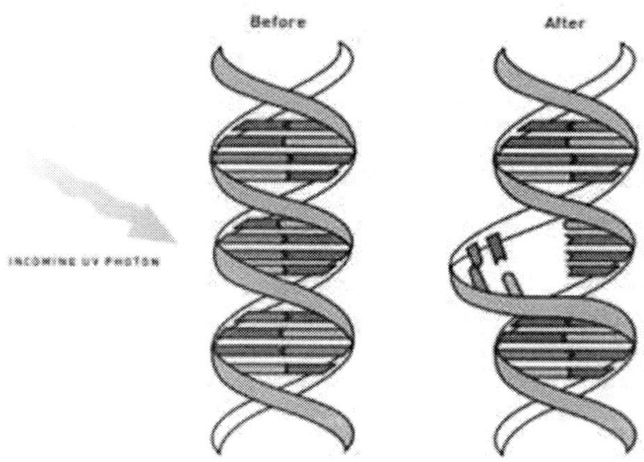

Fig. 2 DNA strand deactivation process

Log 3 reduction is chosen here with respect to the following parameters:

1. Bulb Power

The bulb we use in the system is UV-C LEDs of wattage 5W.

2. Distance

The distance between the light source and the surface is 5cm.

3. Duration

The duration for the UV exposure is less than 5mins of time.

4. Shadowing

Here the shadowing is negligible for log 3 reduction since the distance is 5cm. We use multiple LEDS from different direction in order to avoid the effect of shadowing.

Log 3 reduction ensures 99.9% reduction in germs and microbes. Requires an exposure of 6020J/m^2 to attain Log 3 reduction. The table below gives the exposure time for a distance of 5cm for different UVC LEDs power ratings.

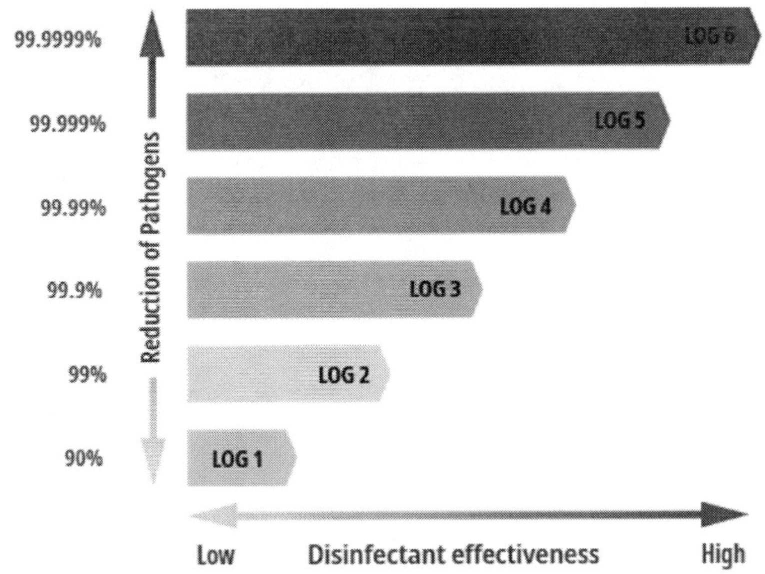

Fig. 3 The effectiveness of Log reduction in percentage.

5cm	Log3 (99.9%) (6020)	Log4 (99.99%) (10000)
5W	5m	10m
15W	2m	4m
25W	1m	2m
35W	55s	90s

Fig. 4 The comparison of log 3 and 4 reduction with a distance of 5cm.

This disinfection technique cleanses both bowl and the seat of the lavatory overall. The proposed system consists of UV LEDs embedded on the arm lid over the toilet seat; along this the motor is connected with the arm and to the controller box. Different parts of the system are briefed below:

3.1. UVC LEDs

These LEDS emits UVC light which is invisible to human eyes, but inactivates the microbes (eg.COVID-19). UVC light has much stronger germicidal properties than both UVA and UVB. UV C band is strongly sucked by RNA and DNA strand bases in the nucleus leading to structural molecular damage through a photo dimerization process. Thus results in virus deactivation, so that they are no longer will be able to multiply or replicate. The emission wavelength of them is 260- 285 nm and beam angle of 130 Deg. Forward Bias of max 9V.

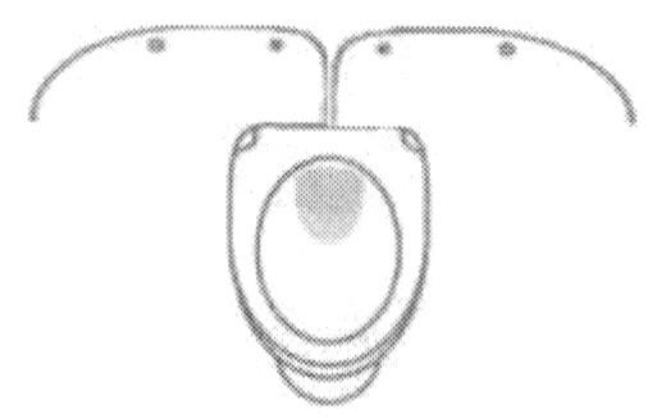

Fig. 5 Open position of the lid while UVC LEDS OFF.

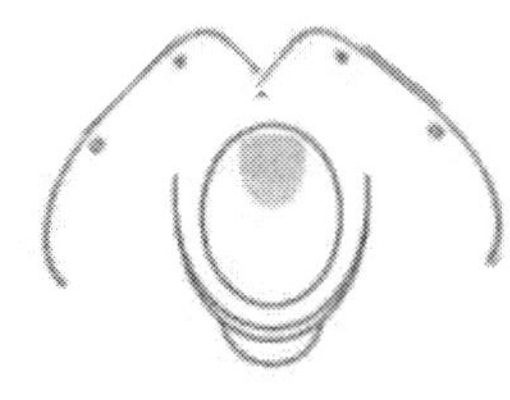

Fig.6 Intermediate Position of the lid

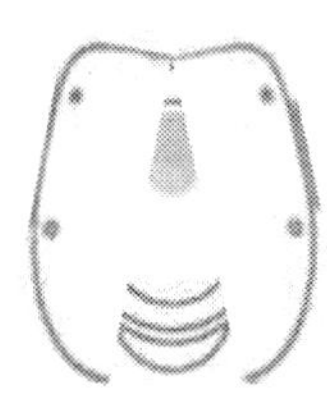

Fig. 7 Closed Position of the lid while the UVC operation is done.

3.2. Closet lid actuation

PVC lid of the closet will open sideways actuated by stepper motor powered gear system. *The Ultra Violet C band LEDs will be spread throughout the lid partition to have any even light distribution.*

3.3. Control box

This consists of Arduino module that controls the motor actuation and auto door lock mechanism taking the relevant feedbacks from the sensors.

3.4. Electrical door lock

Solenoid operated door locking mechanism to avoid UVC subjection during disinfection process ensuring fool proving and safety.

3.5. Sensors and indicators

1. Inner door knob with occupancy indication.
2. Limit switch to detect door knob actuation.
3. Limit switch to determine solenoid lock position.
4. Limit switch to ensure door close.
5. Limit switch to ensure toilet lid position

3.6. Auto door close

Self-closing spring door hinge for auto door close is fixed.

4. Experiments

Different microorganisms have different sustainability to Ultra Violet light radiation; few microbes in this world are much affected, whereas other microbe requires more time and dosage UV C exposure for exact inactivation. An appropriate UV dosage is for the optimum deactivation of the targeted organisms. The UV dosage is to be calculated using the following mathematical representation equation:

Dosage = Ultra Violet light Intensity (E) x Exposure time (t)

Or otherwise, UV Dosage = E x T,

Where;

>Dosage is got in joules per meter square (J/m^2) and also in milli joules/centimeter square (mJ/cm^2).

> Ultra Violet Intensity is to be measured in terms of milli watts/centimeter square (mW/cm^2).

>Time of Subjection is measured in terms of seconds(s).

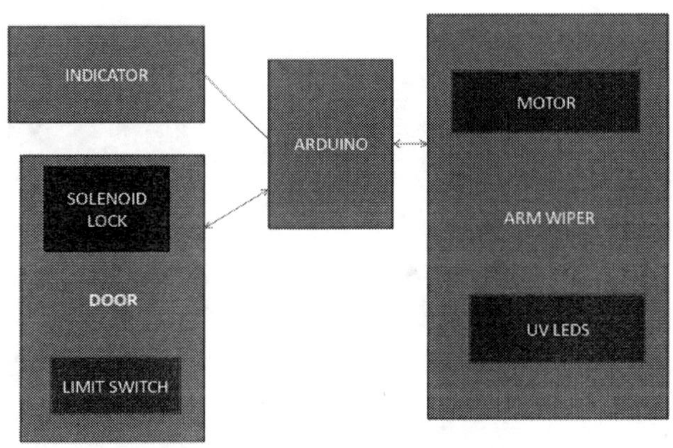

Fig. 8 Block Diagram of automated lavatory.

5. Results

This novel system for automated lavatory was simulated and modeled using Tinkercad simulation software. The following figures are the output of the proposal.

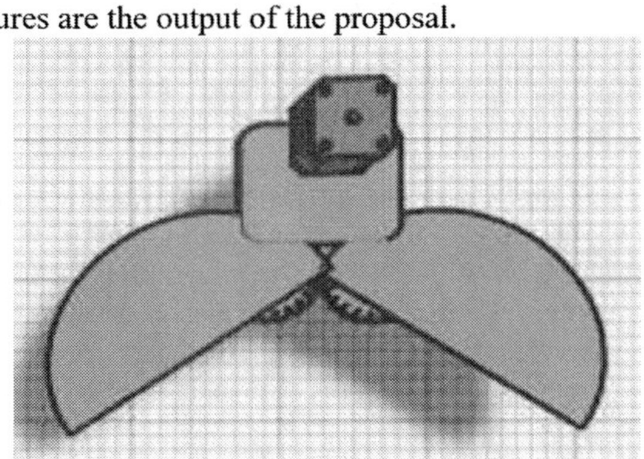

Fig. 9 Toilet seat cover with UV LED attached designed in the CAD

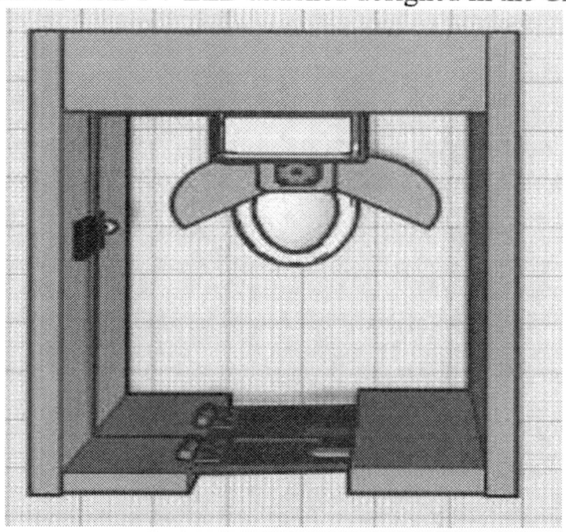

Fig. 10 Automated Lavatory with all the functions

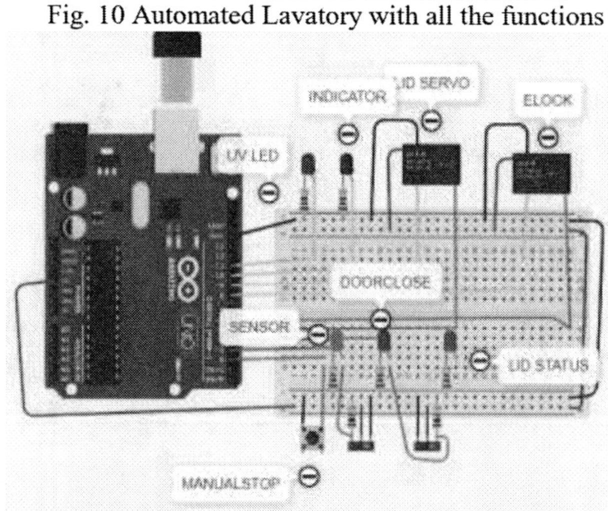

Fig. 11 Simulated Circuit Setup of the proposed idea

6. Discussions

Ultra Violet LEDs enables new application such as water, surface and air decontamination and inactivation. And also in the bio agent detection, identification process, UV curing and biomedical analytical instrumentation UV is used. UV radiation inactivates viruses present on the upper surface which will be directly prone to radiation of Ultra Violet spectrum. Conventionally mercury lamps which are used as source of Ultra Violet radiation; but compared to UV C LEDs, lamps made of mercury does not have a rapid switching ON or OFF features. And they usually require a warming up and cooling down process that consumes energy and time both. Overall the lifetime hours of these mercury lamps reduces if they are more often switched on or off. These LED devices have low noise but high modulation in frequency, flexible type of form factor, space and the power spectral distribution, high internal, quantum's efficiency and there is an ability to achieve high efficacy. This is compared favourably with conventional resources, e.g lamps of mercury.

Yes the components such as the lid and arm is only exposed to water which is water resistant and other parts are placed in the casing to protect from water. After the installation process is over, the system is user friendly and similar to the conventional system. On emergency there is E-stop switch which can be used. Of course there is water saving as we use the germicidal power of UV -C light spectrum for disinfection instead of the conventional method using fluids and water for cleaning purpose. Thus water and toilet paper saving is ensured here.

7. Conclusions

In this scenario of the pandemic caused due to a small virus, it is important and essential to break the chain spread by strict sanitization measure especially in public places and the lavatories in public places which have the highest risk of exposure to the pandemic causing virus. The practical and cost effective method can be to utilize the sterilizing property of UV rays in the light spectrum. This can easily and safely implement in sanitizing lavatories in public places by the above discussed method. The total process of disinfection will take 3 minutes only. Therefore this is quick and efficient method.

References

[1] A. J. Jinia et al., "Review of Sterilization Techniques for Medical and Personal Protective Equipment Contaminated With SARS-CoV-2," in *IEEE Access*, vol. 8, pp. 111347-111354, 2020, doi: 10.1109/ACCESS.2020.3002886.

[2] H. M. Zins and m. Howard, "Reusable medical textiles," in Handbook of Medical Textiles. Sawston, U.K.: Woodhead, 2011, pp. 80-105.

[3] U.S Centers for Disease Control and Prevention. (2020). Strategies for Optimizing the Supply of Eye Protection. *CDC Website*. Accessed: May 28, 2020. [Online]. Available: https://www.cdc.gov/coronavirus/2019-ncov/hcp/ppe-strategy/eye-protection.html.

[4] S. N. Rudnick,J.J. McDevitt, M. W. First, and J.D.Spengler,"Inactivating influenza viruses on surfaces using hydogen peroxide or triethylene glycol at low vapor concentrations, " *Amer. J Infection Control*, vol.37, no.10, pp. 813-819, Dec. 2009, doi: 10.1016/j.ajic.2009.06.007.

[5] R.A. Heckert, M. Best, L. T. Jordan, G.C. Dulac, D. L. Eddington, and W.G. Sterritt, "Efficacy of vaporized hydrogen peroxide against exotic animal viruses.," *Appl. Environ. Microbiol.*, vol.63, no. 10, pp. 3916-3918, 1997, doi:10.1128/aem.63.10.3916.3918.1997.

[6] L. Hall, J. A. Otter, J. Chewins, and N.L. Wengenack, "Use of hydogen peroxide vapor for deactivation of mycobacterium tuberculosis in a biological safety cabinet and a room," *J. Clin. Microbiol.*, vol. 45, no. 3, pp. 810-815, Mar. 2007, doi: 10.1128/JCM.01797-06.

[7] A. Laroussi, F. C. Dobbs, Z. Wei, A. Doblin, L. Ball, K. Moreira, F. F. Dyer, and J. P. Richardson, "Effects of excimer UV radiation on microorganisms," in Proc. *Conf. Rec. IEEE Pulsed Power Plasma Sci.*,

[8] K. F. McDonald, R. D. Curry, and P. J. Hancock, "Comparison of pulsed and CW ultraviolet light sources to inactivate bacterial spores on surfaces," *IEEE Trans. Plasma Sci.*, vol. 30, no. 5,pp.1986–1989, Oct. 2002.

[9] M. S. Shur and R. Gaska, "Deep-Ultraviolet Light-Emitting Diodes," in *IEEE Transactions on Electron Devices*, vol. 57, no. 1, pp. 12-25, Jan. 2010, doi: 10.1109/TED.2009.2033768.

Study on Human Fingers Perception over Textured Surfaces for the Textural Applications in Haptics Technology

Dadi Ravikanth[1*], P. Hariharan[2]

[1]Department of Mechatronics Engineering, Manipal Institute of Technology, Manipal Academy of Higher Education, Manipal - 576104.
[2]Department of Manufacturing Engineering, College of Engineering Guindy, Anna University Chennai – 600025.

E-mail: dadi.ravikanth@manipal.edu

Abstract. Haptics technology uses skin as a source for feedback and helps in the perception of the stimulus. There are various types of actuators or displays used in haptics perception, like electro-tactile, vibrotactile, thermal coil and physical pins. These actuators require larger neural density for being used as an application. Hairy and Non-Hairy skin, its neural density pattern must be observed for perception memory applications. The textural applications in haptics technology are one of the kinds which use skin and its perception levels. The paper discusses the behaviour of fingertips and fingertip's skin behaviour pattern towards textural applications in tactile feedback. The experiment in the paper uses psychophysical assumptions and single-blinded method of perception over six human participants under three environmental condition, and 10 stimuli which are used in everyday activity. The experimental result focuses on the perception time and perception memory of human fingertips, which will a comparative data with real-time haptic feedback. The experimental methods discussed in the paper addresses low-cost perception experiment for a preliminary study of haptics textural applications.

Keywords: Haptics, Psychophysics, Textural Applications, Human Perception.

1. Introduction

Haptics is the study of science and behaviour of skin, and the applications which use skin and its property with mechanical constrains through embedded protocols is termed as Haptics Technology [9]. The concepts like virtual reality, mixed reality, teleoperated and sections of human-machine interfacing has a shadow of human perception, and haptic interaction. The most common haptic feedback is perceived using vibrotactile displays/actuators [4]. The other unexplored devices are electro-tactile displays, physical pins, thermal displays, and mixed displays [5][6]. The human body is covered with 98.6% with skin and it is the most effective sensing part compared to other human sensors due to its perception memory. The skin is classified dermis and epidermis and holds millions of mechanoreceptors. Mechanoreceptors are cells which are responsible for differential human skin sensation, like pressure, pain, stiffness, and force [2]. The receptors are classified based on the depth of penetration from the outer surface of human skin, as rapid responsive and slow responsive [8].

IOP Conf. Series: Materials Science and Engineering 1012 (2021) 012050 doi:10.1088/1757-899X/1012/1/012050

The use of haptic devices and actuators are widely used in engineering applications like entertainment, automobile safety systems, avionics, rehabilitation and specially challenged human physical interaction, without perceiving skin behaviour. The behaviour of human skin changes from hairy skin to non-hairy skin. Non-hairy skin is very less responsive towards physical stimulus than hairy skin.

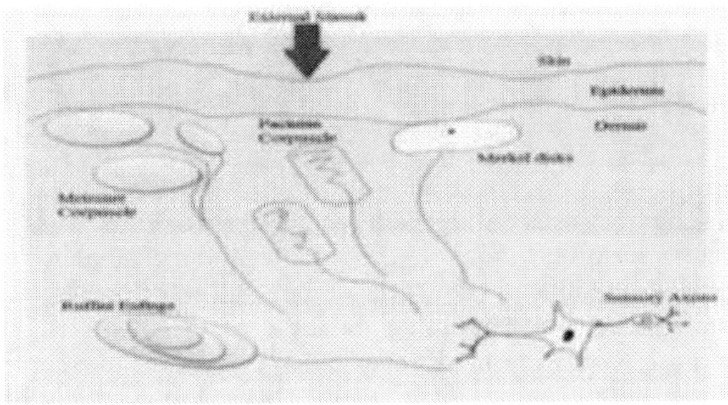

Figure 1: The classification of skin and its receptors representation [1].

Places like palm, feet, forehead, tongue, and parts of the face are very active towards a mild change in external stimuli. The external stimuli responses may change with the noise and noise in the skin is generated through sweat, humidity, physical work or occupation, and temperature. In haptics technology, the majority of the applications use human hand for physical interaction with machines or virtual reality applications. The property of skin or any human sensor is judge using psychophysical experiments. The psychophysical experiments discuss the responsive time of skin towards any stimuli and give the behaviour of in terms of just noticeable difference (JND). The ratio of sensory responses to its upper and lower limits is given as JND. There are many methods to calculate JND, but the best method was coined by Weber–Fechner law. The paper discusses the responses of human fingers towards possible external stimulus in three different environmental conditions to check the response time to check the psychophysics of skin and its sensitivity [3]. The experimental results are validated with the data taken from 1080 samples using descriptive statistics.

2. Experimentation

In this paper, the psychophysical experiment was carried out over six human male subjects and ten material subjects with single-blinded technique over three different environmental conditions. The conditions were taken to check the response time irrespective of correctness. The selection of participants was done using the finger scales measurement. Six participants with age less than 30 years old and not more than 25 years old were subjected for the experiment, assuming the range as an average [7]. The ink impressions of each participant were taken and using digital imaging microscope the selection criteria were taken into account. The figure (2) shows the best outer cutaneous surface and distorted cutaneous surface. The method is a very basic understanding of human skin sensitivity and skin thickness based on the physical work engaged in everyday activity. Six participants were selected out of thirty-two participants in random profession.

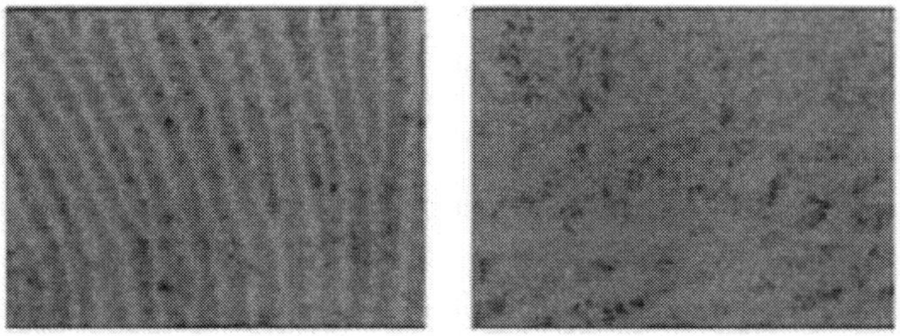

Figure 2: Finger ridges seen in the video microscope measuring system: Left is the best selective & Right one is distortive data.

The experimental setup for surface roughness recognition was done using a test bench setup. Two models were identified to perform the experiment, circular and rectangular. A trial run was performed for recognition. The circular setup had better reachability and was selected for the experiment, shown in figure (3). The experiment used ten stimuli for the experiment. Metal, Brick, Cardboard, Cloth, Glass, Paper, Rough Paper, Rough Plastic, Rubber and Wood are the most common stimuli that have a reference in perceptual memory and are quite commonly used in everyday life. Each stimulus was 4 cm X 4 cm in surface dimension with variable thickness. The categorization of stimuli was noted using roughness values. The roughness value can be found from contact and non-contact type metrology setup. In the experiment, the roughness was noted using the non-contact type roughness measuring setup. The stimulus was placed in equal distance from each other and 50 cm away from participant, shown in figure (4).

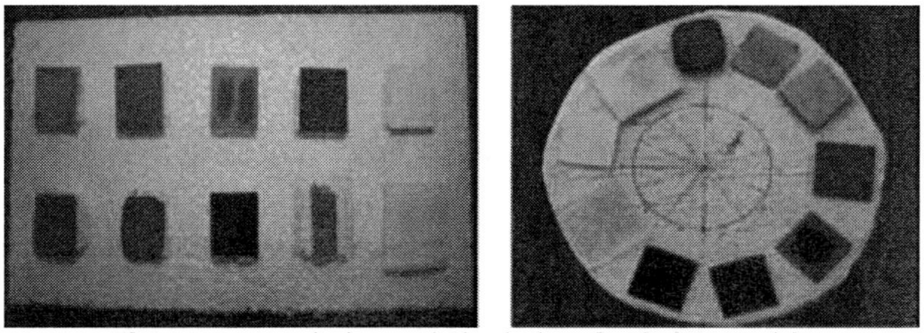

Figure 3: Rectangular base and Circular base for placing materials

Six participants were subjected to the ten stimuli and recognition test was conducted using the single-blinded experimental method. Ten stimuli were subjected over six fingers (5 fingers + Index finger repeated the second time) over three environmental conditions. Air-conditioned environment (20°C to 24°C), room temperature and open environment were the three environmental conditions.

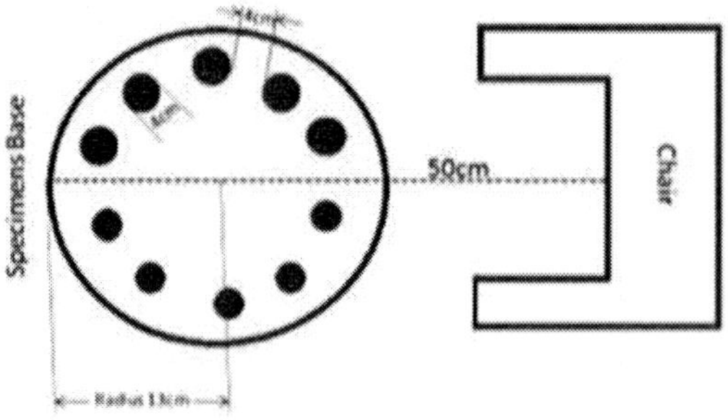

Figure 4: Dimension and orientation of experimental setup

The conditions are not controlled environments and are subject to fluctuations. The three conditions were taken to make sure maximum samples of data are available to see the behavior of the fingers towards textural perception. Each participant is given 10 stimuli one after another over left hand and asked to identify over test bench using the right hand. The time taken for perception is noted irrespective of appropriateness. The process is repeated over three environmental conditions generating 1080 samples (Six Participants, Six Fingers, Three environmental conditions and Ten stimuli) over each participant.

3. Results

The results are explained taking descriptive statistics as a key root and understanding the best suitable conditions. The results are discussed individually for each environmental condition.

1. 20 °C -24°C temperature condition:

In this condition, all the finger did detect the stimuli and has very less point of error in identifying the stimuli. The maximum time taken to detect a particular stimulus was 8.375 seconds, whereas the minimum time of detecting was 0.546, with a standard deviation from 1.351 to 2.087. The statistical observation is shown in table (1). For the first trail thumb was used and it took total time for perseverance as 53.568 seconds (which is the cumulative addition of the average values of each finger). This time is the sum of the entire stimulus over a single finger for all the participants. With the increase in the number of trails, the standard deviation and total time for perseverance tend to decrease due to the perceptual memory and repeatability. The total correct detected stimulus was close to 93%.

Table 1: Descriptive Statistics of 20 °C -24°C temperature condition (Average values of each finger over 10 materials).

Fingers	MEAN	Standard Deviation (σ)	Total Time	Minimum Detected time	Median	Maximum Detected time
	(Seconds)					
Thumb	5.356	2.087	53.568	1.850	5.475	8.375
Index	4.026	2.059	40.261	1.210	4.090	7.261
Middle	3.406	1.871	34.061	0.715	3.665	5.931
Ring	2.911	1.690	29.115	0.558	3.167	5.413
Little	2.834	1.836	28.341	0.576	2.982	6.053
Index Re	2.285	1.351	22.855	0.546	2.497	4.433

2. Room temperature condition:

In this condition, total time duration taken to detect was 108.306 seconds (Which is the cumulative addition of the average values of each finger) which had a maximum duration of 29.206 seconds. The minimum time taken to detect one particular stimulus is 0.323 seconds. Just like the previous condition, the time duration reduced with a particular rate under multiple iterations which are represented in table (2). The correct detected stimulus was 68% which is 1.36 times lesser than 20-24°C temperature condition.

Table 2: Descriptive Statistics of room temperature condition (Average values of each finger over 10 materials).

Fingers	MEAN	Standard Deviation (σ)	Total Time	Minimum Detected time	Median	Maximum Detected time
	(Seconds)					
Thumb	10.830	8.653	108.306	1.791	9.375	29.206
Index	6.850	5.410	68.504	0.94	5.867	18.076
Middle	5.398	4.366	53.981	0.84	4.582	13.958
Ring	4.876	4.428	48.764	0.351	3.511	13.796
Little	3.935	3.188	39.352	0.573	3.045	10.623
Index Re	3.420	3.406	34.206	0.323	2.648	11.068

3. Open environmental condition:

Unlike the above two conditions, here the trend did change, and an uneven pattern was found in detecting the entire stimulus. The total average time taken was 323.524 seconds, which is three times more than room temperature first iteration and six times more than 20-24°C temperature condition, Table (3). There was a reduction in minimum detected time under multiple iterations which were observed in higher roughness values like cloth and brick. The standard deviation that explains the change of duration occurring between iterations remains irregular and thus bringing out the correct detected stimulus to 11% which is far lesser than the above conditions.

Table 3: Descriptive Statistics of open environmental condition (Average values of each finger over 10 materials).

Fingers	MEAN	Standard Deviation (σ)	Total Time	Minimum Detected time	Median	Maximum Detected time
	(Seconds)					
Thumb	32.361	11.503	323.524	15.276	31.968	49.201
Index	30.840	10.155	308.405	13.523	31.788	46.255
Middle	28.735	8.552	287.356	13.272	29.112	41.855
Ring	27.709	11.145	277.093	9.783	27.494	44.516
Little	29.651	12.421	296.956	8.487	29.569	50.058
Index Re	25.088	9.225	251.793	8.581	25.305	39.334

4. Discussion

The air-conditioned condition had rapid response compared to room temperature. The maximum time duration in both the conditions has shown a decrease in response repeatability as shown in figure (5). The response time in perception in controlled conditions shows very less time compared to room temperature and perception memory is quick in controlled condition.

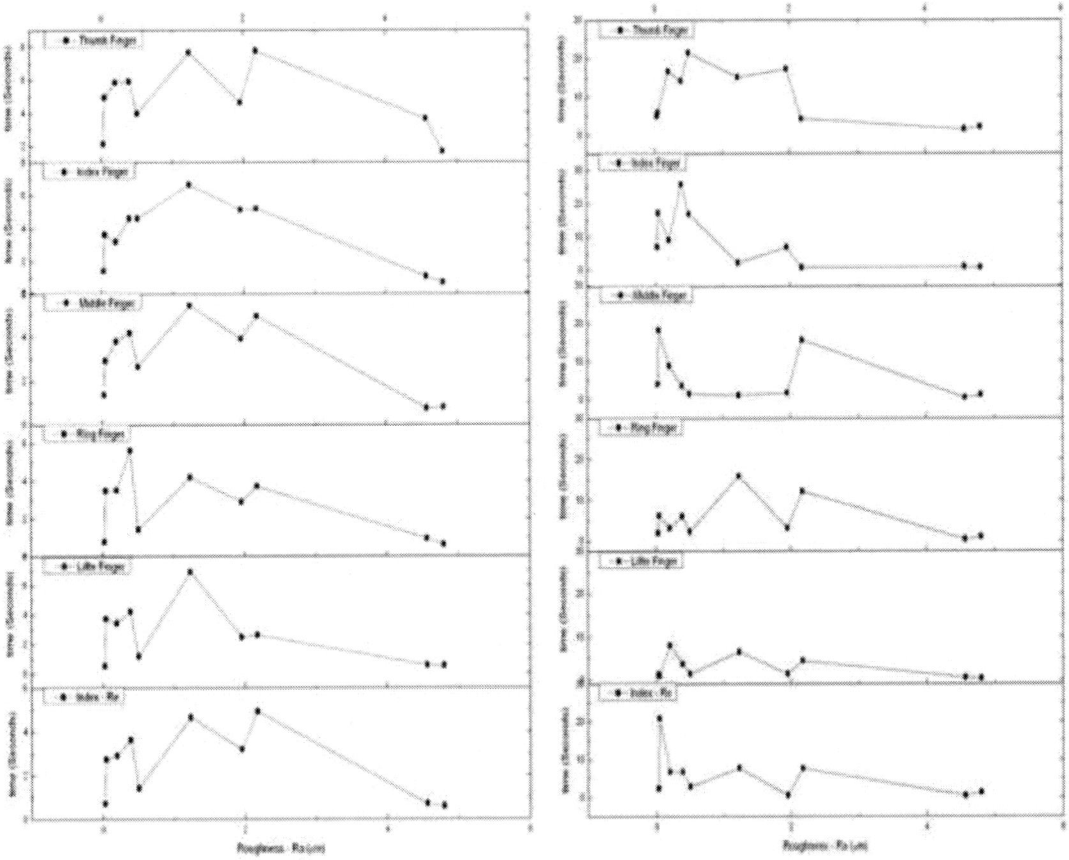

(a) (b)

Figure 5: The responsive graph of human fingers towards the stimulus. (a) left: 20 °C -24°C temperature condition and (b) right: Room temperature condition.

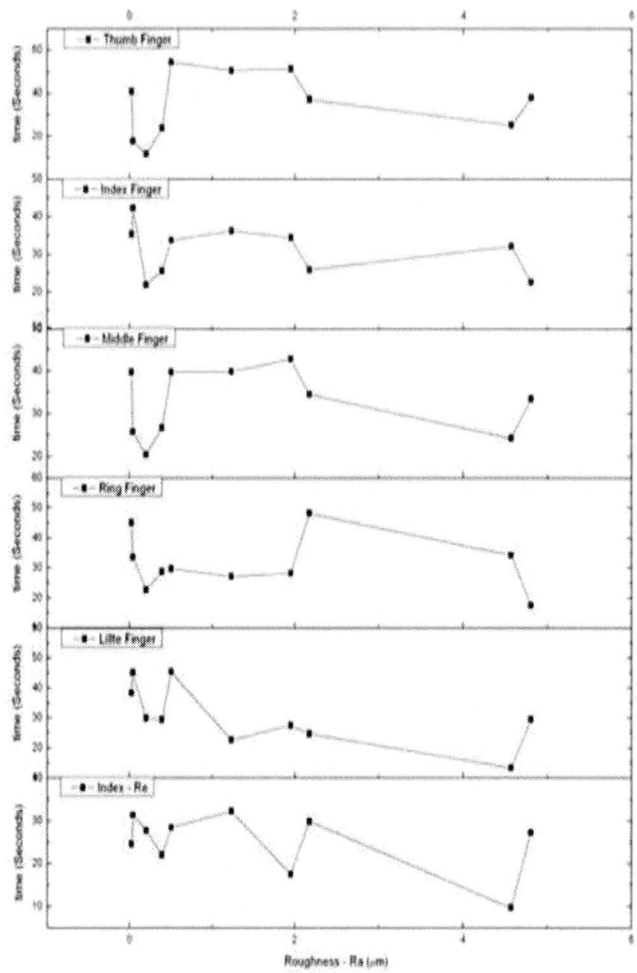

Figure 6: The response graph of the human fingertip in an open environment.

While comparing the results with open environment condition the perception duration was very high, stating the noise insertion in the perception, figure (6). Comparing the total time duration, over repetitive experiments on fingertips the perception time has reduced drastically proving the concept of perception memory. The use of index finger in haptics textural applications can be very useful for rapid recognition. The index finger and 20-24°C temperature condition were identified as an ideal combination for haptics textural applications for preliminary study. A similar experiment can be conducted over any haptic display to identify the best-fit frequencies and intensities for textural applications.

5. Conclusion

The psychophysics experiments have less explored to confirm the use of skin in Haptic applications. The experimental setup discussed comprises of limited stimuli in application but can give required assessment and perception point for haptic application. The use human fingertips due to its dexterity creates or generates multiple applications like gaming, self-driving safety system, user feedback in medical teleoperation and other elite stages of engineering. Performing psychophysics experiments. The psychophysics experiment can vary the perception data and parameters based on source. The source can be fingertips like the work discussed in the paper or can be with respect to stimuli. A researcher can select any parameter for designing application but for rendering haptics model, the stimulus individual response is required. The proposed work is time saving and can be used as a data set for deep learning algorithm and generate more data value to see the optimum frequency vs intensity for the roughness.

References

[1] Dadi, R.; Hariharan, P. Design of Electrocutaneous Tactile Display over Human Fingertip for Textural Applications in Space Manufacturing Feedback. Procedia Computer Science 2018. 20(3). 684 – 690.
[2] Bensmaia, S.; Hollins, M. Pacinian representation of fine surface texture. Attention, Perception and psychophysics 2015.67(5).842 – 854.
[3] Jones, L.; Tan, H.Z. Applications of Psychophysical Techniques to Haptic Research. IEEE Transaction on Haptics 2013. 6(3). 268 – 284.
[4] Hiroyuki, K. Electro-tactile Display: Principle and Hardware. Pervasive Haptics: Science, Design, and Application. Springer. Japan 2016. pp. 79 – 96.
[5] Ksjimoto, H.; Suzuki, M.; Kanno, Y. Hamsatouch: Tactile vision substitution with smartphone and electro-tactile display. Human Factors in Computing Systems 2014. 1273 – 1278.
[6] Keller, T.; Kuhn, A. Electrodes for transcutaneous (surface) electrical stimulation. Journal of Automatic Control 2008. 18(2). 35 – 45.
[7] Reuter, E.; Voelcker-Rechage, C.; Vieluf, S.; Godde, B. Touch perception throughout working life: effects of age and expertise. Experimental Brain Research 2012. 216(2). 287 – 297.
[8] Gerling G.J., Lesniak D.R., Wan L. Validating a Population Model of Tactile Mechanotransduction of Slowly Adapting Type I Afferents at Levels of Skin Mechanics, Single-Unit Response and Psychophysics. IEEE Transition on Haptics 2013. 7(2). 216 – 228.
[9] Hannaford, B.; Okamura, A. M. Haptics. In: Siciliano B., Khatib O. (eds). Springer Handbook of Robotics 2016. 1063 – 1084.

Semi-Autonomous Parking system for Automatic Transmission vehicles

S. S. Ohol[1], S. Kaware[1*]

[1]Department of Mechanical Engineering, Department of Automotive Technology, College of Engineering, Pune [COEP], Wellesley Raod, Shivajinagar, Pune- 411 005.

E-mail: sameerkawre5@gmail.com

Abstract. This paper proposes an automated perpendicular parking strategy for a car which is to be designed. The design considers general cases of perpendicular parking for a rectangular body within a rectangular space. The system works in phases named: The Scanning phase, the parking environment is detected by ultrasonic sensors mounted on the extreme ends of body and a parking position and maneuvering path is produced if the space is enough. Then in the positioning phase, the model positions itself and get aligned with the parking space avoiding potential collisions. Finally, in maneuvering phase, the model moves to the parking position in the parking space in a specified path, which requires clutch, brake control, backward and forward maneuvers depending on the dimensions of the parking space. Based on the characteristics, a collision-free path is planned about the surroundings. The strategy is to be integrated into an automated parking system, and implemented in the car, showing capable of safe parallel parking in tight situations and as automated parking device to help vehicle drivers.

Keywords: Semi-Autonomous Parking, Automated Manual Transmission, Automatic Transmission, Ultrasonic Parking sensors, Parking Algorithm.

.

1. Introduction

Autonomous parking is an autonomous maneuvering system which can perform angled, parallel, or perpendicular parking of any automobile. Aim of this system is to enhance the safety and comfortability of driving in such constrained space where attention, skill and experience required in controlling the vehicle is high.

Parking of a car requires space width greater than car body's actual width because driver need to open the doors to come out of the vehicle after parking it. Also, space concern is increased due to increasing number of vehicles used. Even for small space, wrong judgement of driver may result in damage to car body. Hence, attempts are made to make 'Parking' an autonomous task.

The motion can be made smooth and clean using sensors to avoid any collision by processor-controlled motion of vehicle within the available space. The 'Parking maneuver' is achieved by simultaneous and coordinated control of steering and speed, keeping constant check on actual situation of surrounding.

2. Methodology

The method by which the system works as follows:
1. Attach proximity sensors at extreme locations of vehicle to cover complete body ensuring no body-part is outside the envelope of sensors.
2. Collect distance data to determine appropriate parking location by the processor.
3. Control whole maneuvering process (propulsion and braking) till desired position is obtained.
4. Continuous check on proximity to ensure zero possibility of collision.

The longitudinal speed control system for the containing speed within specified safe value is developed. The expected speed tracking can be done by the speed data gained from vehicle itself. The speed data can be obtained by the control system to analyze the condition and throttle and brake control.

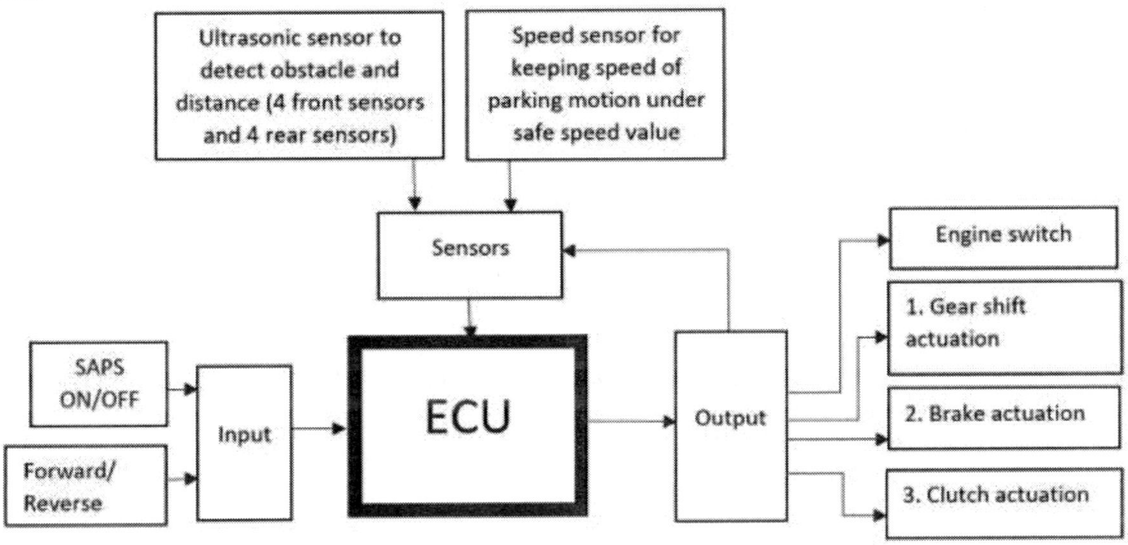

Figure 1. Block diagram of the ECU system.

3. Location of Sensors

The system features ultrasonic sensors to measure the distances to nearby objects. This is to help reduction in system cost by use of already available resources.
Working of Rear sensors and Front sensors can be activated when respective gear is selected, and other side can be deactivated. This is to be done for avoiding unnecessary warnings when there is no motion towards reverse direction when Drive gear is in action, and vice versa.
Ultrasonic sensors are to be positioned as shown in figure:

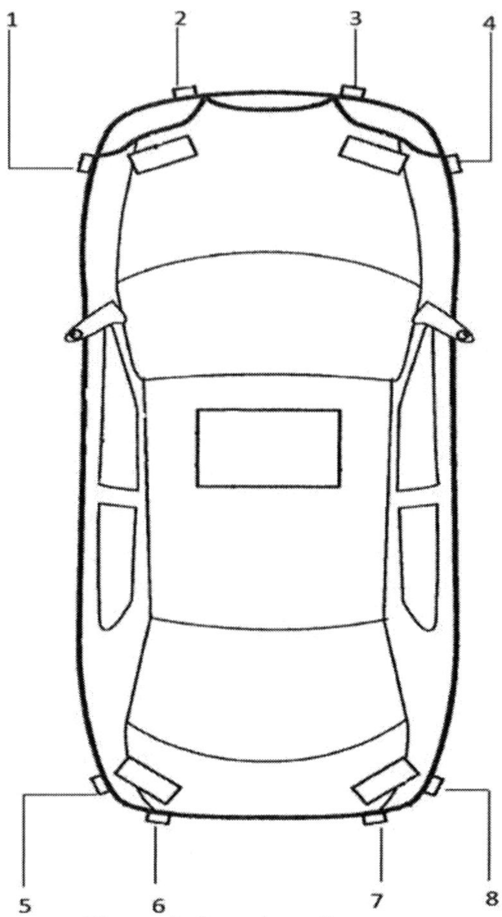

Figure 2. Location of sensors.

4. Development of Algorithm of Automated Parking

The Algorithm was developed to trace the following working method:

Distance data collected from ultrasonic sensors attached to the body of car is to be used to determine relative location of car with respect to obstacles and target parking location.

b. Then a path is to be determined to reach the target location. After this, parking process is to be initiated and moving the car to get the desired target location.

The operational flowchart for the system designed is as follows:

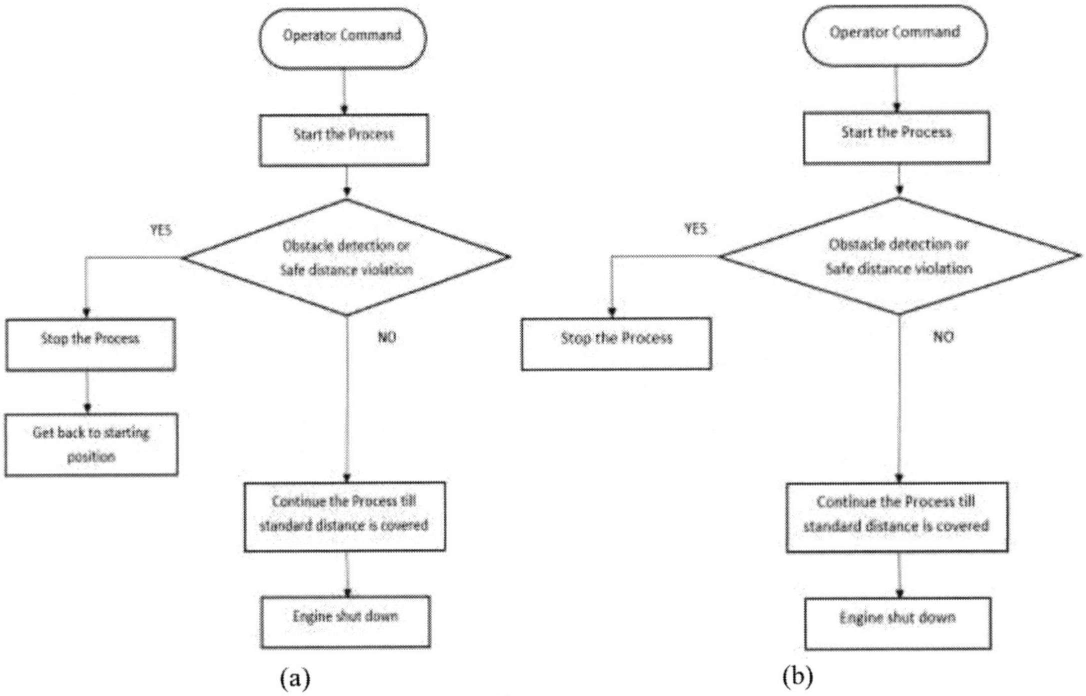

Figure 3. System flowchart for (a) PARK command and (b) CALL-OUT command.

The Algorithm developed is as follows:

- Step 1: START
- Step 2: Read command from Operator (PARK or CALL_OUT)
- Step 3: Initialize system with DISTANCE = 0, VELOCITY = 0, STANDERD_DISTANCE = 4 meters
- Step 4: Start Engine
- Step 6: Shift gear according to received command (Forward motion – Gear D, Reverse motion – Gear R)
- Step 7: Clutch engagement and Motion initiation
- Step 8: Read VELOCITY and check VELOCITY <= 10 kmph. If VELOCITY > 10 kmph, control clutch actuation and maintain value of VELOCITY
- Step 8: Read DISTANCE while in motion and move until DISTANCE = STANDERD_DISTANCE in given direction
- Step 9: Read distance data from all 8 ultrasonic sensors during motion and check for SAFE_DISTANCE >= 0.2 meters
  or When SAFE_DISTANCE < 0.2 meters
  If task is PARK
- Step 9.1: Stop motion in ongoing direction. Give warning notification to operator and get back to the initial location by moving distance DISTANCE.
  If task is CALL_OUT
- Step 9.2: Stop motion in ongoing direction and give warning notification to operator.
- Step 10: If DISTANCE = STANDERD_DISTANCE, Stop the motion and Notify the operator about the completed task.
- Step 11: Shut down the Engine
- Step 12: STOP

5. Simulation Scenario

For simulation of this Autonomous system, environmental assumptions are set for standard working scenario of system. The assumptions are:

1. Working area is horizontal which is free from any irregularity.
2. Parking space is surrounded as shown.

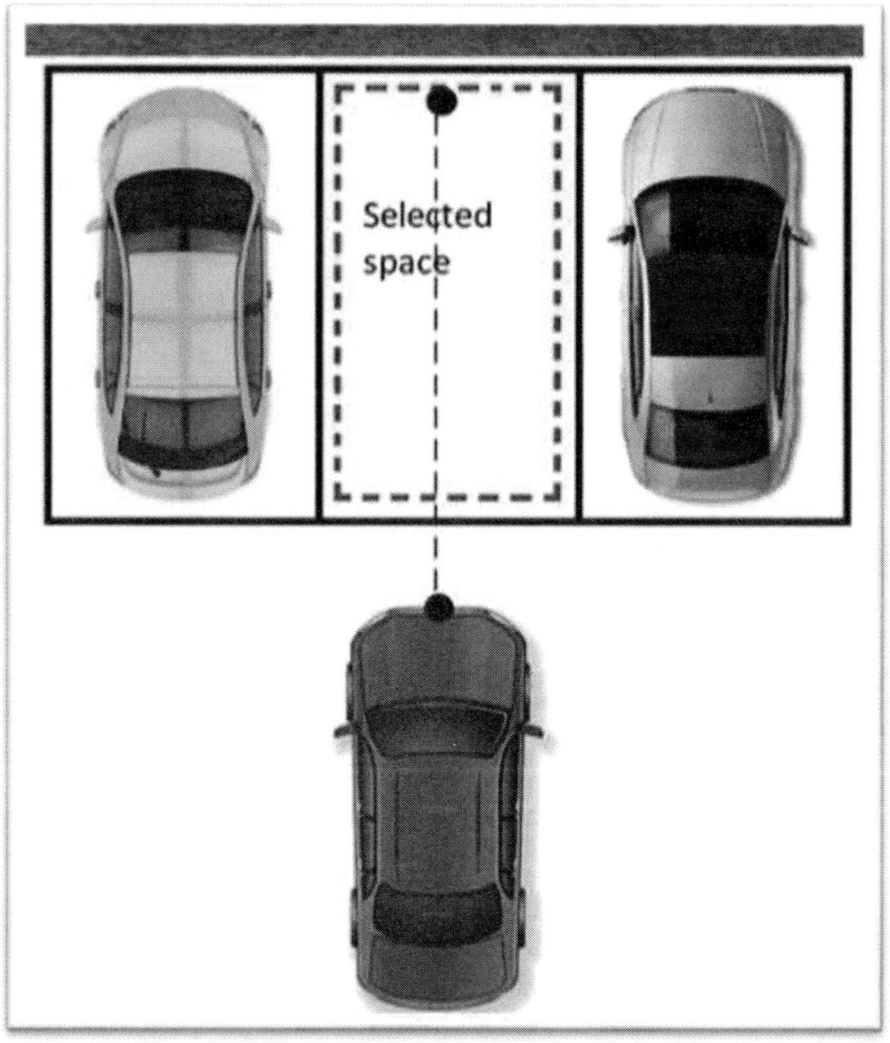

Figure 4. Target parking space.

Initial state for parking function to initiate is:

1. Engine is off.
2. Brakes are free.
3. Gear position is Neutral.
4. Steering wheel is CENTRE position.

Distance constraints are as follows:

1. Safe distance on Left and Right side is 20 cm.
2. Safe distance for Front and Rear area is 15 cm.
3. Standard distance to be covered in Parking function is 4 meters (Assuming that Ego Car body length is 3.5 meters)

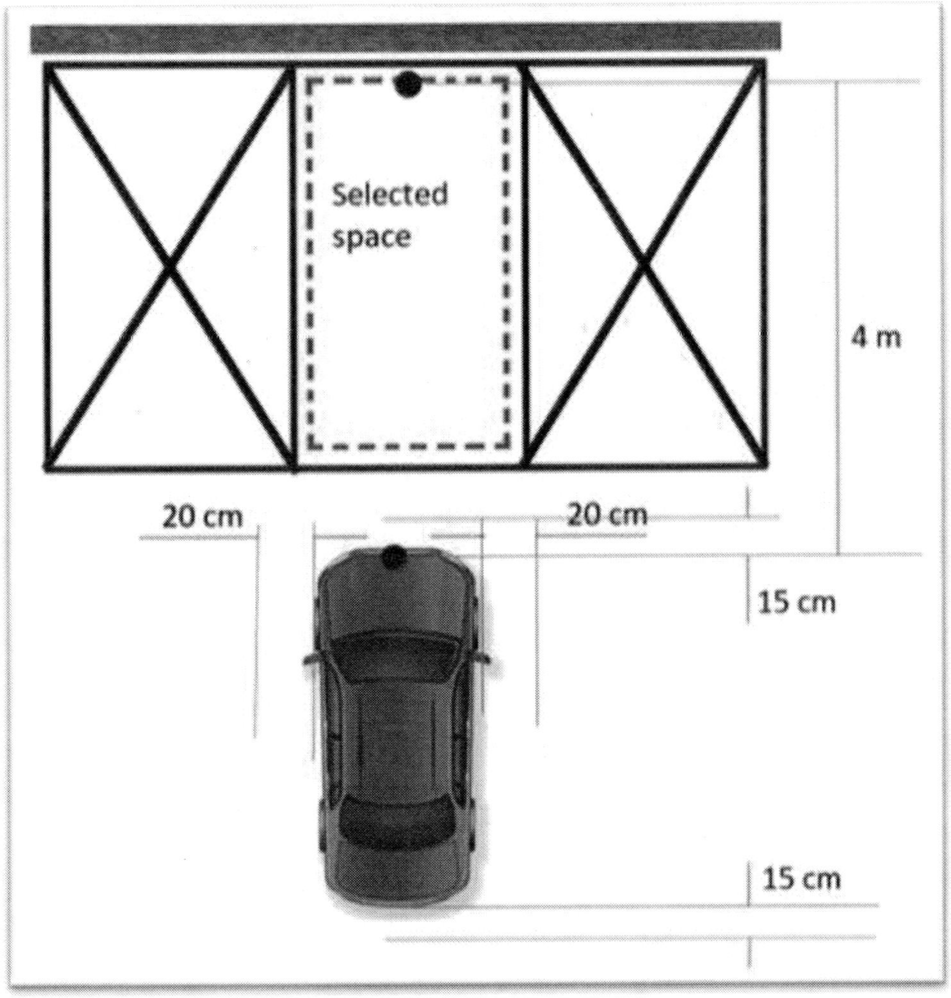

Figure 5. Positions and distances.

Vehicle model is to be prepared as follows:
Ego car: Rectangular body
Length =3.5 meters
Width=1.7 meters
y = Global Y position of the car point
v = Speed of the car
d = Distance measured from initial position of motion
T = Throttle position (Positive for Acceleration)
Manipulative variable
G = Gear position (D or R)

6. Simulation Process

The simulation is done on MATLAB SIMULINK to examine the prepared control method. The overview of the system is shown in following figure:

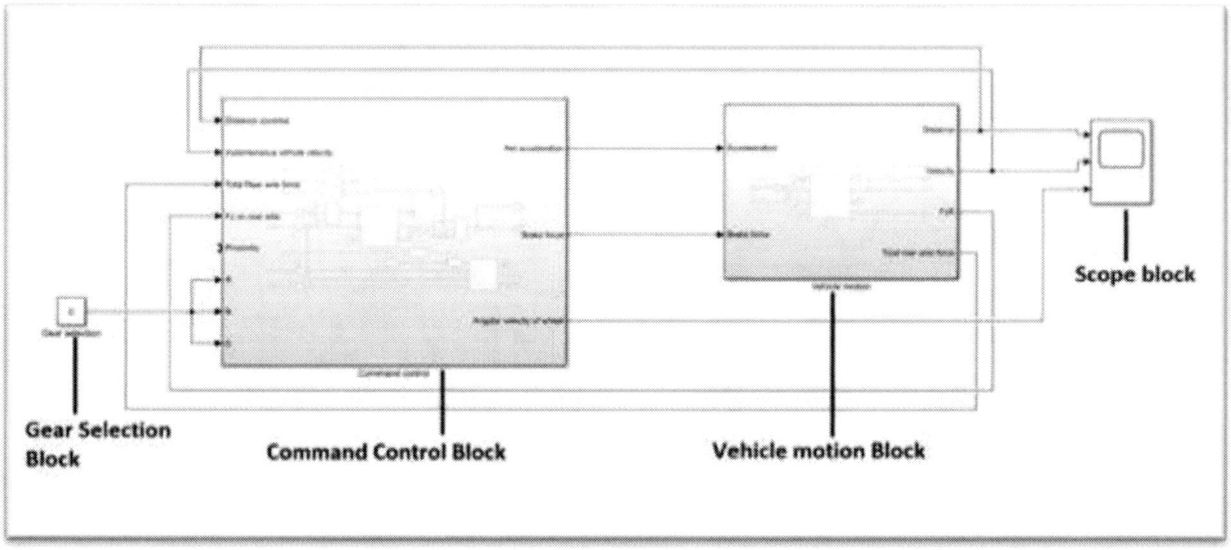

Figure 6. Overview of Simulation system.

The simulation process comprises Vehicle motion block and Command control block. The Vehicle motion block takes Acceleration and brake force as input and gives Distance traveled, Instantaneous velocity and Rear axle forces as output. This works as a last block of simulation Process. The Command control block have inputs from User as Front Parking, Reverse parking, or Callout; Nearby object distance data from Proximity sensors; and feedback data from Vehicle motion block which is used in controlling the motion giving Net acceleration, Net brake force as output.

Each of them is explained as follows:

1. Command Control block:

This block is going to control all actions of vehicle according to surrounding conditions and given command. The block is developed so far as given below:

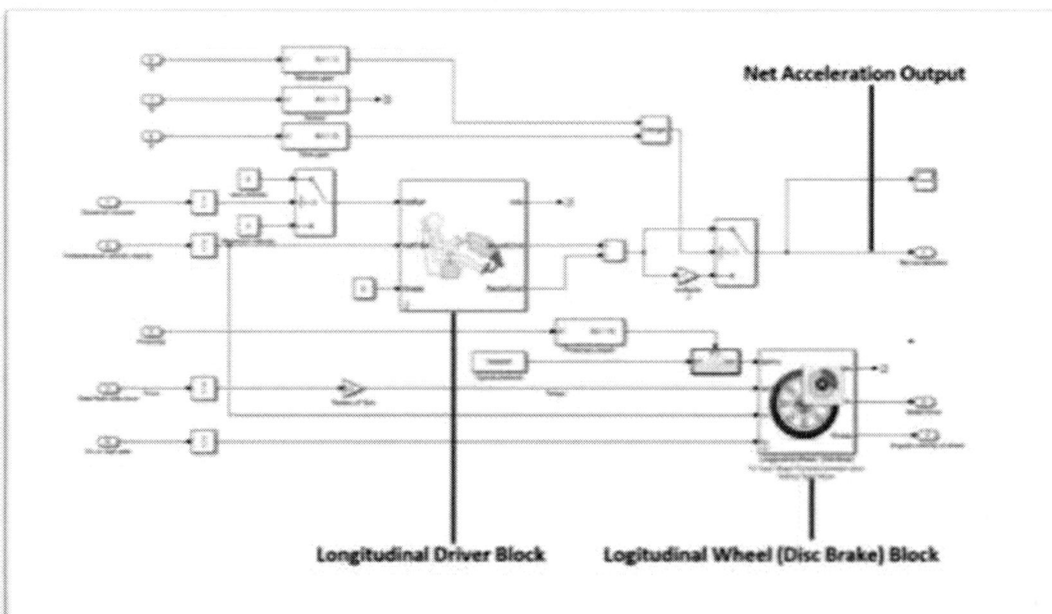

Figure 7. Command Control Block interior.

2. Vehicle motion block:

This block receives the control signal from Command control block such Acceleration and Brake force and performs motion of vehicle. This block will simulate the vehicle motion and gives output as Distance covered, Velocity of motion and Rear axle force. These outputs are further used by Command control block for keeping vehicle motion parameters in check, whereas the Distance and Velocity shows the behavior of test subject. The block interior is as shown below:

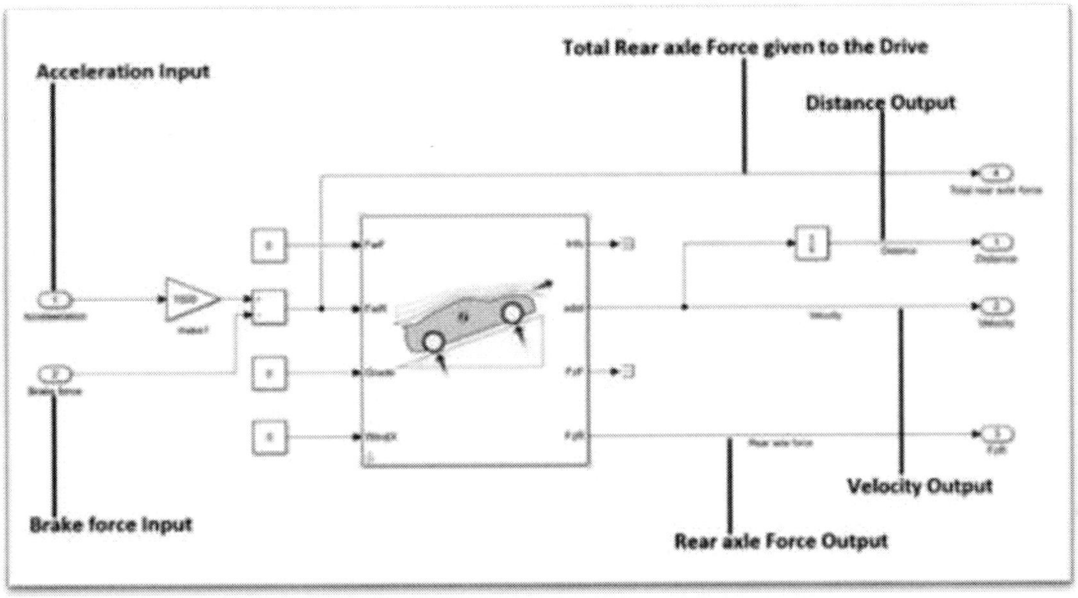

Figure 8. Vehicle motion Block interior.

7. Simulation Results

The Simulation process was conducted on three different cases:

1. Forward Parking:

a. In this case, the car engages DRIVE gear and commences the motion towards parking space. The initial distance is set as Zero. As the motion progress, the distance counters starts and measures to 4 meters. After that, the motion ceases and Car stops. This case is shown in following graph:

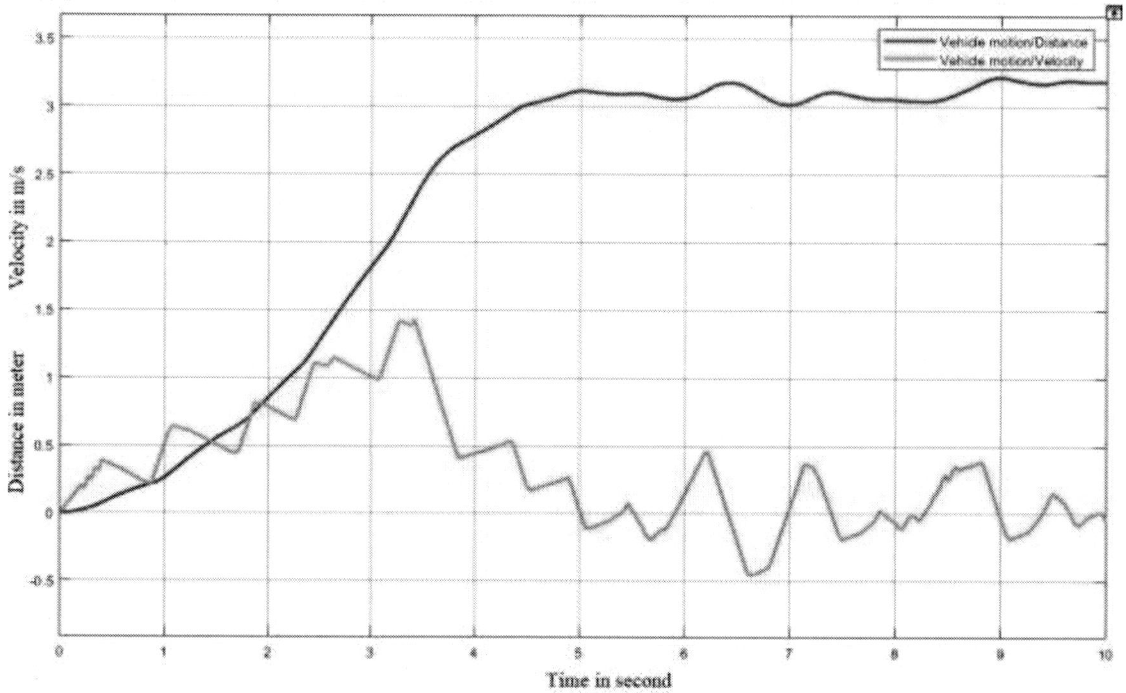

Figure 9. Forward Parking motion parameters behaviour

The Red line indicates the distance covered by vehicle during the process. The Green line shows the velocity of motion. As the car reaches the maximum distance (Here it is 4 meters), the velocity reaches gradually to zero.

b. In case there is any obstacle in the motion path (here, at 2m ahead), the car will stop before that by controlling throttle and brake functions and will get back.

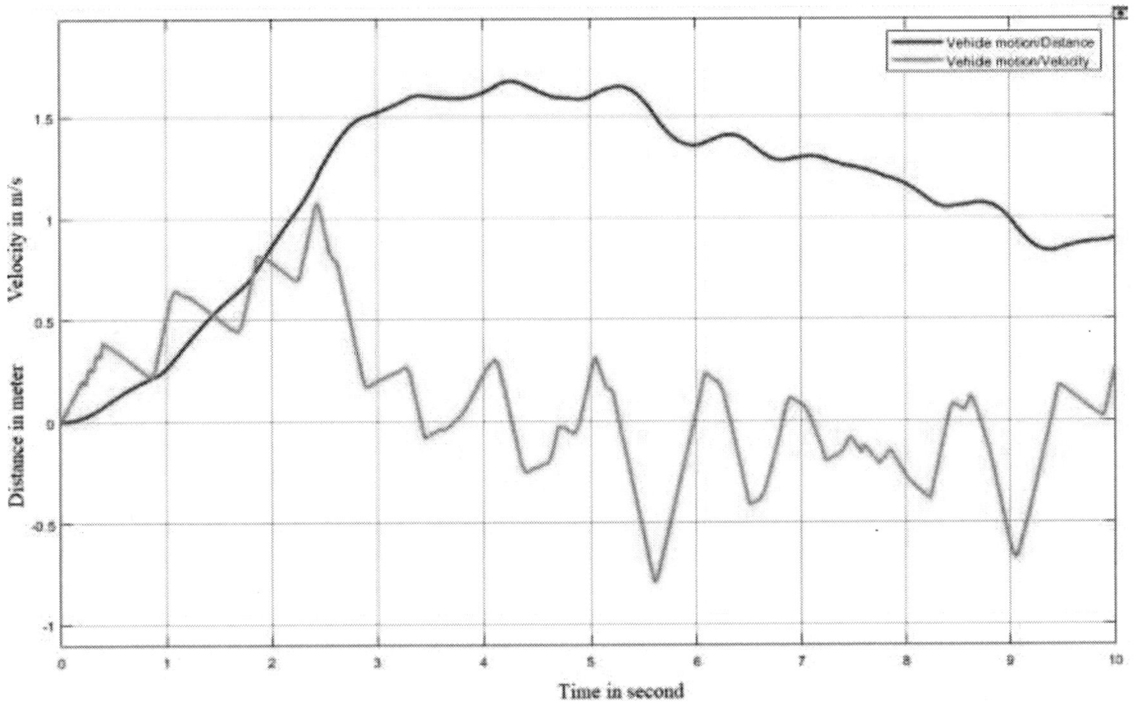

Figure 10. Forward Parking Obstacle detection behaviour

c. During Retrieval, the motion behaviour of car is as shown below:

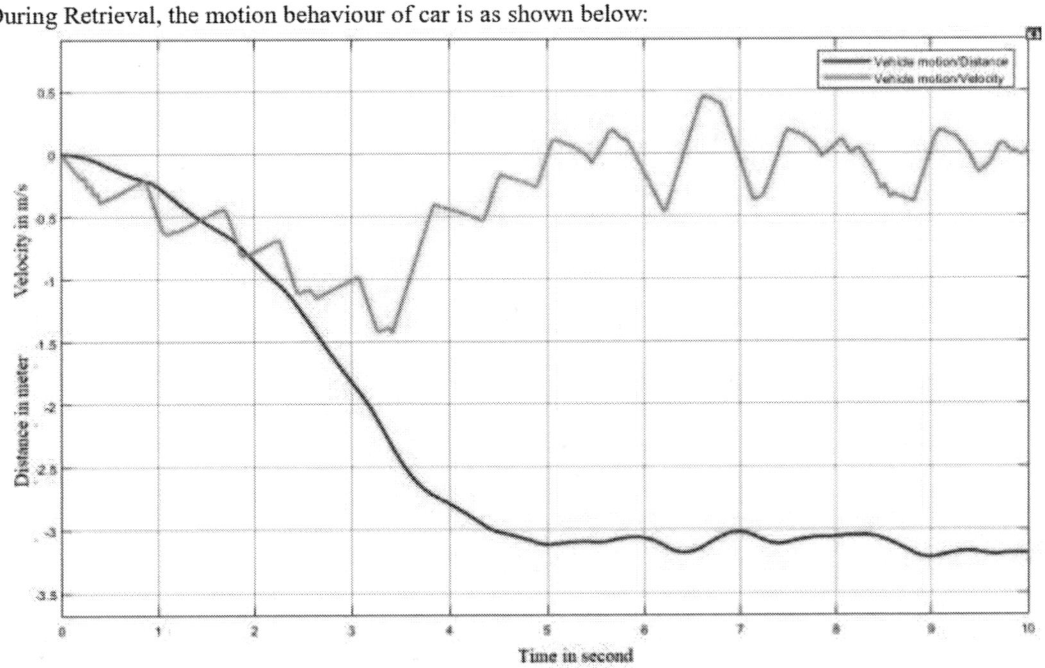

Figure 11. Forward Parked Retrieval motion parameters behaviour

2. Reverse Parking:

a. During this case, the process starts with shifting gear REVERSE. Then, the initial distance is set as Zero by the counter. As the motion progress, the distance counters starts and measures to 4 meters. After that, the motion ceases and Car stops. This case is shown in following graph:

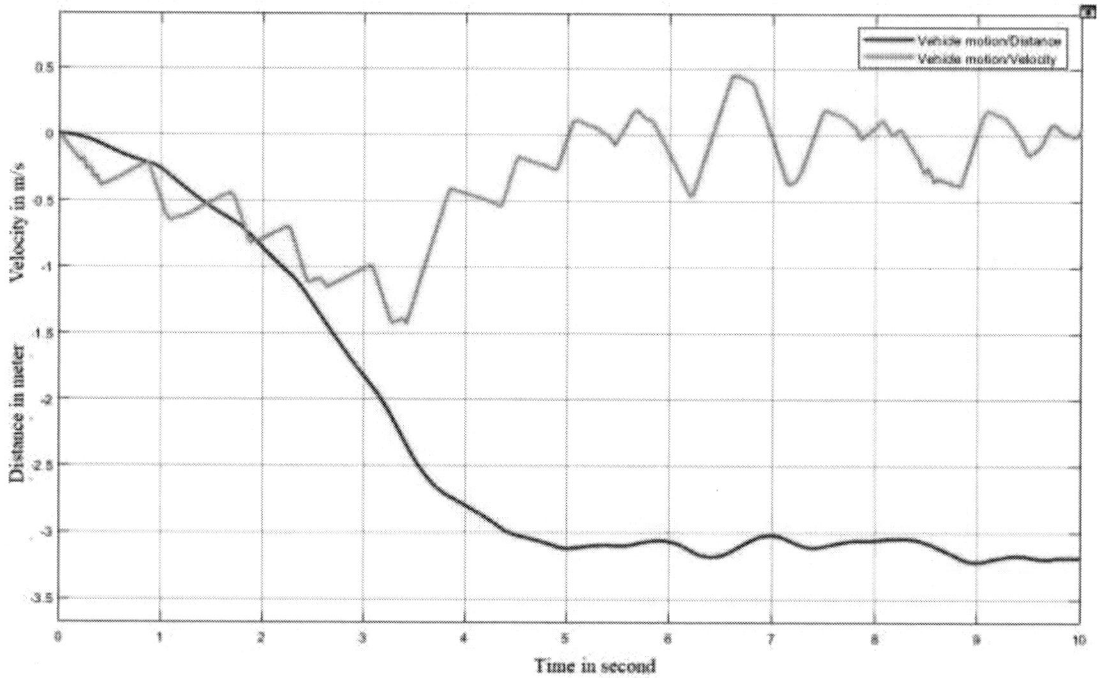

Figure 12. Reverse Parking motion parameters behaviour

b. In case there is any obstacle in the motion path (here, at 1 meter behind), the car will stop before that by controlling throttle and brake functions and will get back to initial position.

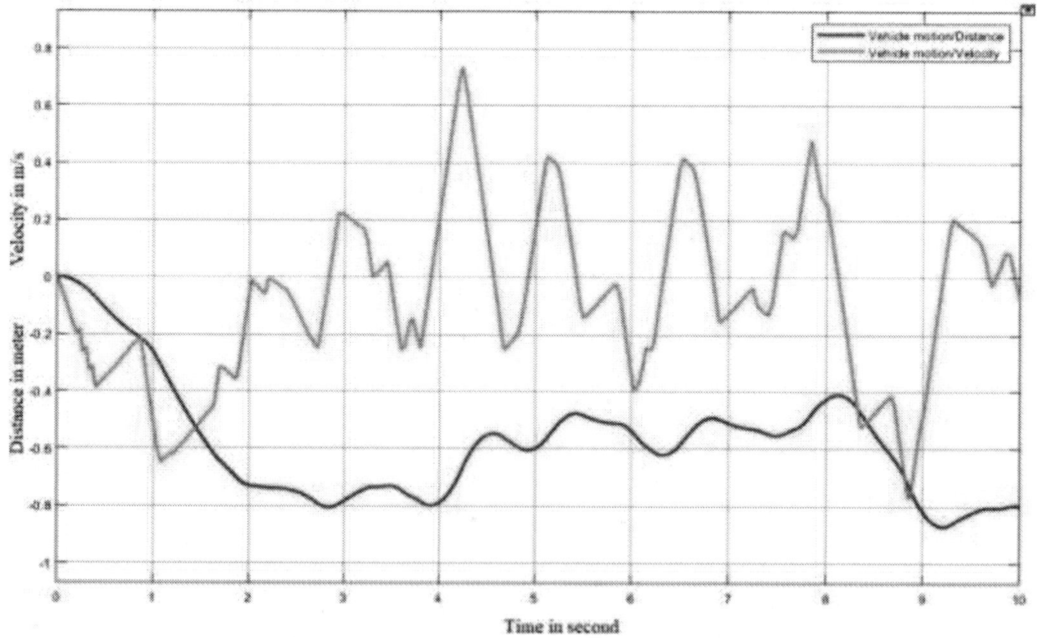

Figure 13. Reverse Parking obstacle detection behaviour

c. During Retrieval, motion characteristics are as shown below:

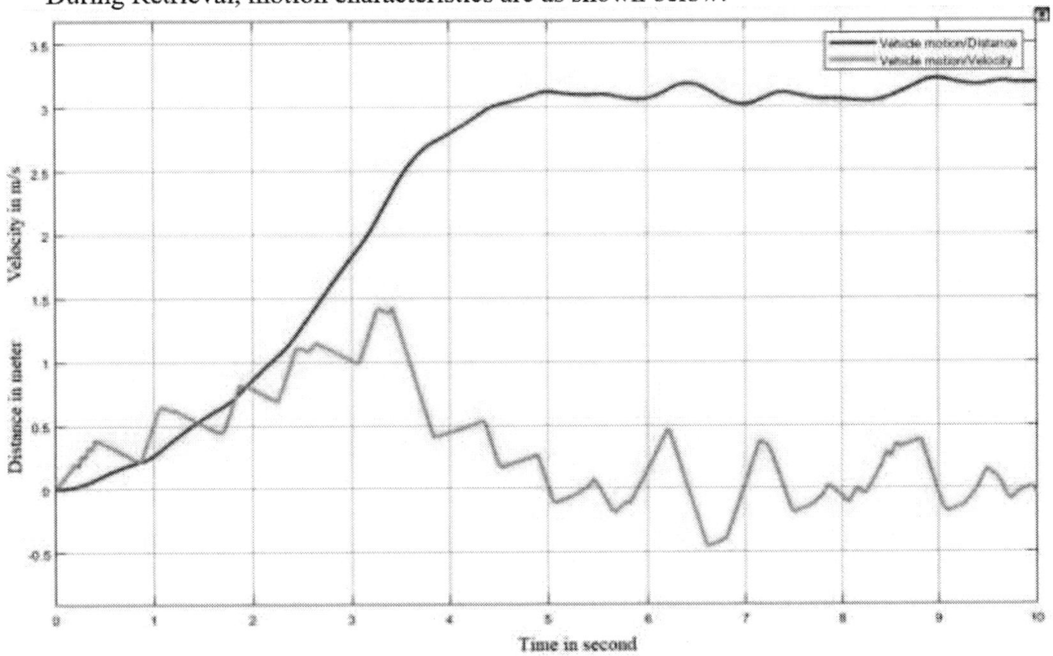

Figure 14. Reverse Parked Retrieval motion parameters behavior

8. Conclusions

The system architecture for Semi-Autonomous Parking System is thoroughly studied with the help of Literature Review and its possible new development have been proposed in this report.

The concepts rely on cheap sensors that are already used in conventional cars, along with a processor and communication device for user command. The designed algorithm and procedure for motion control is simulated on MATLAB SIMULINK. The proposed system starts by manual switch command for Forward or Reverse Parking. The proximity sensors around the body activates and gives distance data to the ECU of Parking Controller. The vehicle moves obeying given command with specified motion velocity and Distance covered. Any obstacle detected will be treated according to the task. During this, any possible collision between Car and surrounding objects is avoided by continuous check on distance from the nearest objects. If any obstacle is detected at any point of process within decided action sequence and distance parameter, the car will stop the motion in the direction and will:

- STOP and get back to initial position when working on Parking function and
- STOP and stay in the same location when CALL-OUT function is running.

As the PARK function or CALL-OUT function is completed, the vehicle will stop and will notify the user about completion. After that, system will switch off and the system will transfer control to the user. The method is capable of parking the car in straight line in Forward and Backward parking position by input commands given by the user. The Distance movement control is within 75% accuracy.

References

[1] US Patent US008862321B2 (2014)
[2] US Patent US009656691B2 (2017)
[3] K. Jiang (School of Manufacturing & Mechanical Engineering, The University of Birmingham B 15 2TT UK) and L. D. Seneviratne (Department of Mechanical Engineering, King's College

London WC2R 2LS UK) "A sensor guided autonomous parking system for nonholonomic mobile robots" IEEE (1999)

[4] M. Wada, K. Yoon, H. Hashimoto, S. Matsuda, "Development of advanced parking assistance system using human guidance," in Advanced Intelligent Mechatronics, pp. 997–1002. IEEE/ASME International Conference (1999)

[5] F. Gomez-Bravo, F. Cuesta, A. Ollero, "Parallel and diagonal parking in nonholonomic autonomous vehicles," Engineering Applications of Artificial Intelligence, vol. 14, no. 4, pp. 419 – 434, (2001).

[6] T.-H. Hsu, J.-F. Liu, P.-N. Yu, W.-S. Lee, J.-S. Hsu, "Development of an automatic parking system for vehicle," pp. 1–6 in Vehicle Power and Propulsion Conference, VPPC '08. IEEE, (2008)

[7] Plamen Petrov, Fawzi Nashashibi, Member, IEEE, and Mohamed Marouf "Path Planning and Steering Control for an Automatic Perpendicular Parking Assist System"(2015)

[8] Emese Szadeczky-Kardoss and Balint Kiss, Dept. of Control Engineering and Information technology Budapest University of Technology and Economics, Hungary "Path planning and Tracking control for an Automatic Parking Assist System" (2008)

[9] Chrysler Auto www.chrysler.com/bmo.pacifica.2019.html#/build/exterior/40001/CUC201905RUCT53B/2DD/ERC,DFH,Z1A,SDC,TYK,WPA,AL,UAV,X3,PXR, 27D,APA

[10] Autonomous Parking Market: Global Industry Analysis 2012 – 2016 and Opportunity Assessment; 2017 – 2027 www.futuremarketinsights.com/reports/autonomous-parking-market

[11] Paromtchik, Igor; Laugier, Christian "Automatic Parallel Parking and Returning to Traffic". Video Proceedings of the IEEE International Conference on Robotics and Automation. Leuven, Belgium. (1998)

[12] Paromtchik, Igor "Planning Control Commands to Assist in Car Maneuvers" (PDF). Proceedings of the IEEE International Conference on Advanced Robotics. Coimbra, Portugal. pp. 1308–1313 (2003).

[13] Grabianowski, Ed "How Self-Parking Cars Work" (2006).

[14] Viswanath K. Reddy and Swaroop Laxmeshwar (M.S.R.S.A.S) "Design and Development of Low-Cost Automatic Parking Assistance System" (2014)

[15] Warren T. Good, 8035 S.W. 8th Ave., Portland, Oreg. 97219; Jacob E. Wilhauer, Jr., 3325 SW. 44th, Portland, Oreg. 97221 "US Patent on Automatic parallel Parking System" (1988)

[16] Jianfeng Wang, Weihua Li, Jun Li, Yiqun Liu, Baoyu Song, and Haibo Gao "Modeling a Driver's Directional and Longitudinal Speed Control Based on Racing Track Features" (2018)

[17] Mihai Chirca and Guillaume Martin, Roland Chapuis and Christophe Debain, and Roland Lenain "Autonomous Valet Parking System Architecture" (2015)

[18] Jagadhambal. K [M.E], Arunkumar.B "Autonomous Self Parking and Self Retrieving Vehicle With GPS Less Keyfob for Vehicle Tracking" (2013)

[19] Shidian Ma, Haobin Jiang, Mu Han, Ju Xie, and Chenxu Li "Research on Automatic Parking Systems Based on Parking Scene Recognition" (2017)

[20] Insop Song, Keith Gowan, Jason Nery, Henrick Han, Tony Sheng, Howard Li, Fakhreddine Karray "Intelligent Parking System Design Using FPGA" (2006)

[21] Volkswagen's 10-year evolution of Park Assist; heading toward trained parking and higher levels of autonomy www.greencarcongress.com/2016/11/20161126-volkswagen

[22] Simulation and Testing Importance by SAE https://www.sae.org/news/2019/06/av-simulation-testing- trends#:~:text=The%20key%20technology%20for%20getting,to%20market% 20will%20be%20simulation.&text=Automakers%20and%20Tier%201s%20have,driving%20to%20work%20in%20collaboration.

Obstacle Avoidance Algorithms: A Review

Talabattula Sai Abhishek[1], Daniel Schilberg[2], Arockia Selvakumar Arockia Doss[1*]

[1]School of Mechanical Engineering (SMEC), Vellore Instittute of Technology, Chennai – 600127, Tamil Nadu, India
[2] Mechatronics and Research, Bochum University of Applied Sciences, Am Hochschulcampus 1, 44801 Bochum, Germany.

Email:arockia.selvakumr@vit.ac.in

Abstract. It is only with time that the efficiency or the effectiveness of the algorithms for obstacle avoidance gets better, and experiences of any kind can be inferred for the betterment of the knowledge on this domain. For a mobile robot navigating its way from starting point to an ending point while traversing through deterrents, needs to divide the problem into sub problems. It fundamentally involves, sensory data interpretation, choosing apt algorithm based on the objective function, and configuring the mobile robot accordingly to attain desired output. In this paper, few essential classifications for obstacle avoidance and robot navigation algorithms are discussed. Importance of the hardware aspect of the robot is undeniable. A set of algorithms were classified into 2 main classifications which are further divided into sub classifications in an arranged and concise manner. This information can be used to develop a suitable model for the given problem. Understanding fundamental ideas or strategies would allow in developing a novel extended strategy, although few specialized strategies are saturated, nevertheless can still be used as valuable alternatives. These alternatives may involve algorithms which have paramount potential, per se which are interestingly similar to the functioning of a brain, nature-inspired, etc.

1. Introduction

Mobile robots are increasingly used in automated industrial environments. There are additionally different applications like planet investigation, transportation, reconnaissance, landmine location, and so on. In every one of these applications, all together that the mobile robots play out their undertakings, collision free path planning is a pre-imperative. The designed robots are to travel self-rulingly in different static and non-static situations. Navigation is a fundamental errand in the field of robotics, which can be categorized into 2 sorts: Local navigation and Global navigation. Global navigation: Prior information on condition ought to be accessible. Local Navigation: Robots are to choose or control its movement and direction self-rulingly utilizing prepared sensors in a hurry. Navigation of a portable robot includes view of condition, localization and guide building, perception and way arranging and movement control. To accomplish this, few Traditional and Heuristic techniques can be utilized. This paper tends to these techniques and the extension in this exploration field while checking on characterized research papers with depictions of Flowcharts at necessary topics of discussion.

2. Obstacle Avoidance Strategies

In the Cell-decomposition strategy, complete environment around the robot is isolated into small sections, some are deterrent-free, and the rest are busy with impediments. The limits are cultivated amongst cells. This strategy is useful for ideal way finding in an old style approach; however the way is chosen by the matrix goal of the earth. The Sampling based strategies are proficient for high dimensional issues but solution is probabilistically acquired. In this technique the pace of the assembly is thoroughly relied on the utilization of nearby organizer. In traditional techniques the principle detriment is that it endures the neighbourhood trap condition which is wasteful to actualize. Accordingly, as of late analysts are persuaded towards the heuristic methodologies and it requires some investment and simpler than old style strategies. In heuristic methodology, the fuzzy logic strategy essentially relies on the "Assuming THEN" rules. The all out number of fuzzy guidelines for structuring a logic framework is settled on the quantity of info boundaries. By help of these principles surrounding the objective methodology, deterrent shirking and the way making arrangements for robot must be chosen.

Artificial Neural Network is extremely popular in the context of the heuristic methodology technique. At the point when we acquaint ANN with way arranging issue mostly three boundaries are to be approved: 1) consolidation of tactile information; 2) deterrent avoidance; 3) trajectory planning. In any case, in way arranging (path planning) issues, ANN is tedious strategy. The way arranging issues for versatile robot utilizing Neuro-Fuzzy strategy talked about by numerous specialists.

2.1. *Classical or Traditional Methods*

2.1.1. Cell Decomposition. The fundamental thought behind cell decomposition is to segregate between geometric regions or cells that are free and zones that are involved by obstructions. A significant part of the cell decomposition strategy is the position of limits between cells. In [1], Chenghui Cai and Silvia Ferrari developed a methodology which was made for planning the detecting system of a mechanical sensor sent to organize various fixed targets arranged in an obstruction populated workspace. Existing way organizing techniques are not honestly fitting to robots whose fundamental objective is to aggregate sensor estimations using a restricted field of view. The system manufactures an availability graph with perception cells that is pruned and changed into a choice tree from which a perfect identifying technique can be processed. The ampleness of the perfect distinguishing systems got by this procedure is appeared through a mine-chasing application. Numerical tests show that these methodologies outflank most limited path, complete inclusion, arbitrary, and matrix search techniques.

In [2] Probabilistic Cell Decomposition for Path Planning, it is examined by Frank Lingelbach that the working behind cell decomposition techniques is to parcel into disjoint sets, called cells. These cells structure the hubs of a non-coordinated network graph. The idea of cell decomposition is joined with probabilistic sampling to acquire a technique called Probabilistic Cell Decomposition. As for the situation for nodes, two hubs are associated by an edge if and just if the relating cells are neighbouring. Cell markings are made dependent on the rules in the calculation with following being the significant parts of the calculation: Cell shape, Graph search, Probabilistic Sampling, Cell Splitting, Local Path Planning. The PCD organizer is utilized contrasting the presentation of the diverse sampling plans. The calculation can without much of a stretch be modified for various types of robots or situations, yielding upgrades effectively.

2.1.2. Potential Field Methods. The potential field strategy is to comprehensively be use for self-sufficient mobile robot capable of path planning with the help of its gainful logical examination. As opposed to various strategies, robot development planning subject to an APF can consider the issues of crash avoiding and direction estimation at the same time. The system is to drive the robot ultimately to the designated objective while protecting it from hammering into obstacles present in the workspace. If the potential field power isn't clearly chipped away at the robots, all the issues can be conveniently disentangled. The strategy that fabricates a model to re-enact reliable state heat move with variable warm conductivity to deal with this issue is reasonable.

In [3] , the gathering proposed a novel artificial potential field strategy reliant on gravity chain that interfaces and the key part is that, in this technique the potential field power doesn't directly chip away at the robot, and the gravity chain just expects an overseeing occupation to the robot. Thus, this new potential field technique deals with the issue of close by minima. If gravity factors are set into the above condition, we will get a great deal of centres and these centres contain the path determined for the robot to traverse through. In any case, the Trajectory/Path won't for the most part be no-where close to an inventive brain in terms of smoothness of path. So in the proposed methodology the robot will reliably run the edge between two concentrations as it can makes locus smoother.

In [4], another arrangement is presented by F. Arambula and M. Padilla for self-sufficient course of a mobile robot, considering enhanced artificial potential fields along with a hereditary calculation. In customary artificial potential field strategies, the robot is pulled in by the objective position, and excused by a couple of obstacles. Use of a single interest point can incite catch conditions where the strategy can't make the resultant force expected to keep up vital good ways from tremendous obstacles. Regardless, with various right hand interest centres, the robot can avoid solidly isolated obstacles. Likewise, using hereditary calculation, builds up a gathering of exercises, for instance, MoveForward and TurnLeft, using variable length chromosomes. The plan of the perfect potential field is thus constrained by a hereditary calculation.

The robot is addressed as a particle influenced by a scalar potential field as a segment of engaging and appalling potentials. It by then moves in the game plan space showed as two-dimensional matrix, where each cell inside arrangement space can be included by the robot, the objective or one of the block cells. Advancement of the potential fields for robot course is an irksome issue generally on account of the gigantic number of elements included and the non-direct nature of the objective work. Hereditary calculations have displayed their abilities to progress huge, complex limits, dodging close by minima through the simultaneous evaluation of various concentrations in the chase space. The arrangement engages the mobile robot, outfitted with sonar sensors, to investigate through dark, high thickness and obstruction dispersals. Moderate objective centres have been used here to coordinate the robot through lobby corners and through the passage of the reasonable room.

In [5], Priyanka S. and her team went after direction planning which is a prime technique for course assessment. Looking at based calculations will allow for progress to the destination while avoiding obstacles. An Enhanced Artificial Potential Field makes the directions for adaptability of mobile robots and simultaneously guarantees the practicality and movement of the direction. Concentrating on the issue that the conventional APF can't change in accordance with the stunning direction planning and fall a prey to the close by perfect game plan as analyzed previously and E-APF strategy is manifested for Wheeled Mobile Robot course estimation. The directions that are made by the WMR are geometrically conceivable. Decisions and desires are made by utilizing the model worked than cautiously holding fast to the static direction codes. The point of view of direction planning is to explore a briefest division path under explicit restrictions in a given circumstance with crash free development. The water back and forth movement stream method was contributed to deal with issues of getting captured into close by minima circumstances.

2.2. Heuristic Methods.

2.2.1. Fuzzy Logic Controller Technique

In [6], Tang Sai Hong and his group discussed about the working of Fuzzy Logic controller in detail. To begin with, a route framework is separated into 2 layers viz, High-Level global arranging and Low-level responsive control. The information on workspace or environment is accessible by the robot is entirely or incompletely equipped. Since it isn't equipped for altering the movement course in nearness of unanticipated or moving hindrances, it neglects to arrive at target.

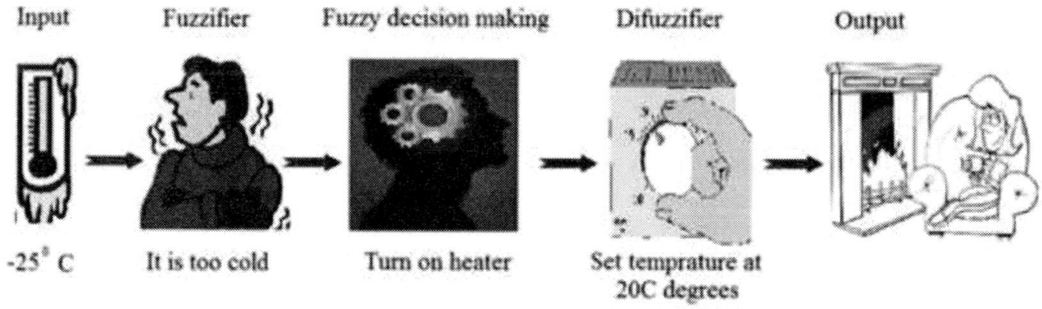

Fig1. Example of Fuzzy Control System

Control orders are produced dependent on the discernment activity design. In this way, it has a brisk reaction in responding to unanticipated impediments and vulnerabilities which alter the movement course. Fuzzy control frameworks are information based frameworks based on an assortment of Fuzzy 'If-Then' principle.. The centrality and adequacy of the Fuzzy Logic (FL) in taking care of the route issue is then introduced. It uses this heuristic information for achievement of a philosophy to create observation activity based techniques for mobile robot's route. The sides of the robot were considered as yields for the chief layered-fuzzy system. Over the long haul, the last yields of the controller are the immediate speed and the turning pace of the robot. The sensors arrangement is depicted in **Fig.2**.

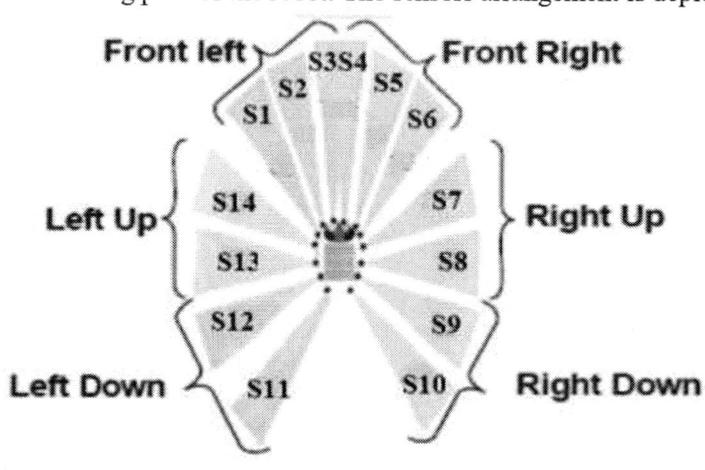

Fig.2 Infrared Sensor Arrangement

To improve the complete presentation of a route framework, complex route undertakings are separated into various more straightforward and littler subsystems (practices) which is known as conduct based framework. In a conduct based framework, every conduct gets specific tangible data and changes them into the predefined reaction as shown in **Fig.1**. The typical fuzzy logic controller configuration includes: Initialization, Fuzzification, Inference, Difuzzification. The traversal of the robot through the obstacles in the sample environment is shown in **Fig.3**.

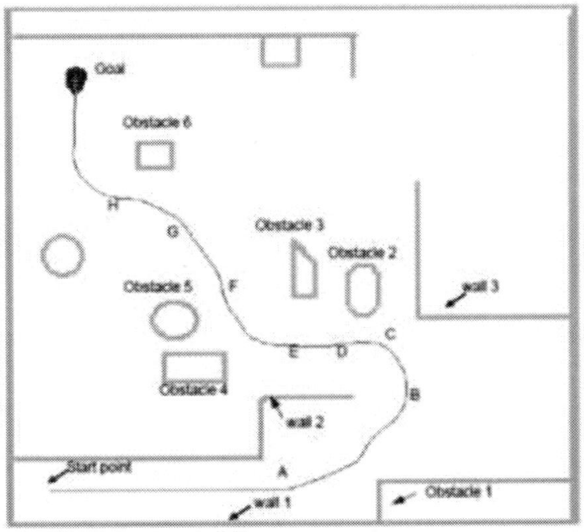

Fig.3 Navigation of the robot in a sample environment

In [7], C.G.Rusu and group built up a fuzzy obstruction evasion framework for a self-ruling mobile robot utilizing IR discovery sensors. Additionally, actualized control design for conduct based mobile robot was introduced. The mobile robot can connect with an obscure environment utilizing a responsive methodology controlled by tangible data. A standard based fuzzy controller with receptive conduct was executed and tried on a two wheels mobile robot outfitted with infrared sensors to perform impact free route. The control issue for the two-wheel mobile robots is the means to autonomously control the left-wheeled engine and right-wheeled engine. A little two wheels mobile robot was considered for the analyses. The robot was dependent on online sensor data procured while route is traversed through. A methodology called conduct based or sensor-based techniques are proposed with a usage of fuzzy based responsive control. The essential structure of a fuzzy logic controller (FLC) comprises of three theoretical segments:

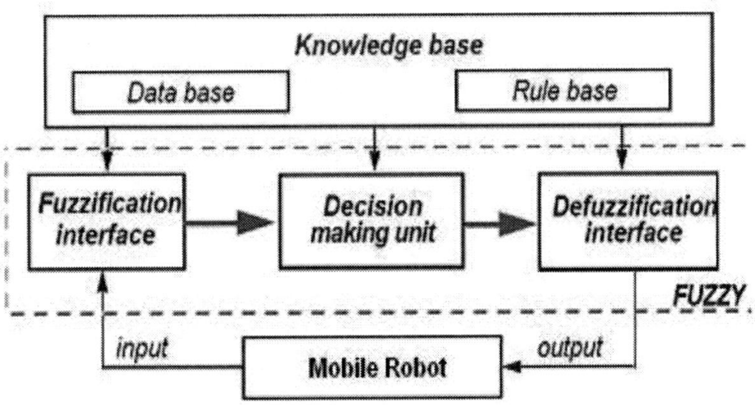

Fig4. Configuration of a Fuzzy Logic Controller

1) Fuzzification of the info yield factors;
2) Rule base that contains a lot of fuzzy principles;

3) Reasoning instrument that plays out the inference method on the principles and offered realities to determine a sensible yield.

With regards to mobile robot control, a fuzzy logic based framework has the preferred position that it permits instinctive nature of sensor-based route and can without much of a stretch change etymological data into control signs. The basic configuration of a fuzzy logic controller is shown in **Fig.4**.

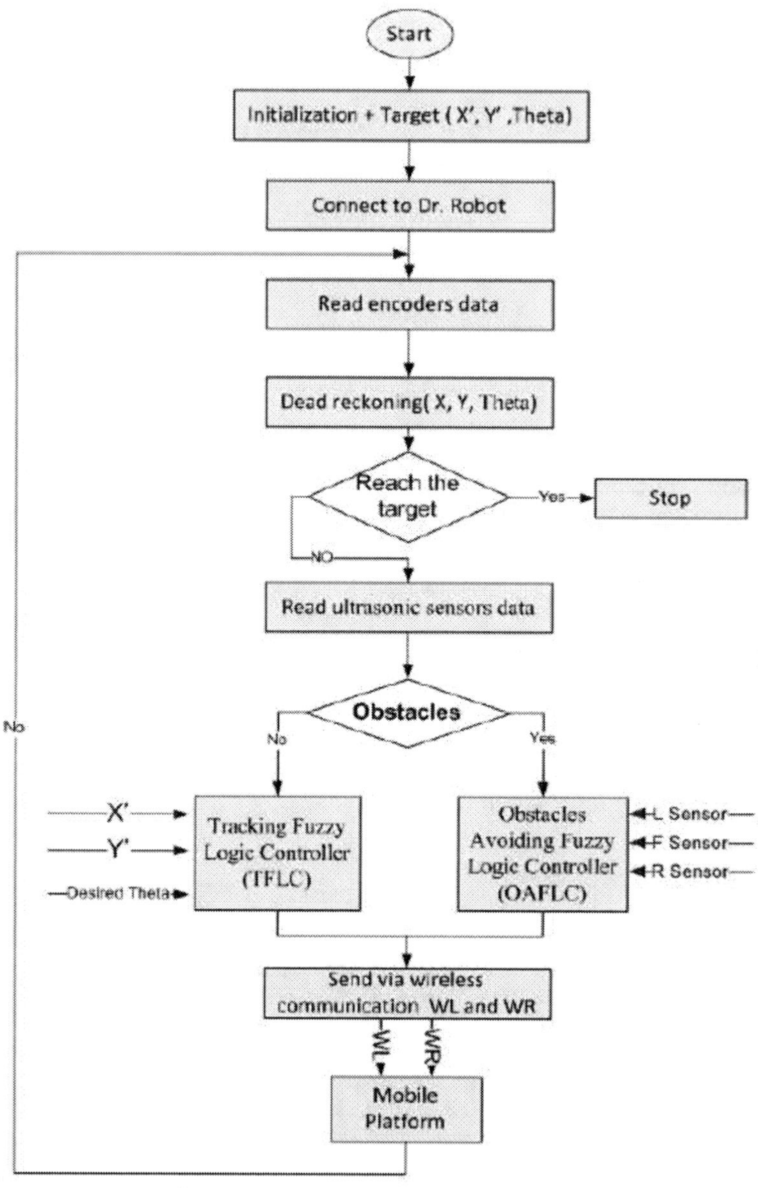

Fig.5 Flow chart of the Fuzzy Logic Algorithm

In [8], an online route procedure for WMR subjected to obscure unique environment utilizing FL strategies was researched by Mohammed and his group. The fuzzy logic control needn't bother with the numerical model of the procedure. TFLC and OAFLC are en-joined to ultimately reach the destination without colliding with deterrents. Dead-reckoning is utilized to assess the position and direction of a WMR during movement. Dead-reckoning strategy depends on the past known position and direction at given time to figure out the next position and direction. Agreeing the exploratory outcomes, the augmentation to a lot of mobile robots should effectively be possible. It is better understood in a flow chart of fuzzy logic algorithm as in **Fig5**.

2.2.2. Neural Network Technique. Deep Learning exploits the various levelled models roused by human cerebrum structures. Deep learning has effectively, in any event to some degree, understood a few primer discernment issues for robots, much the same as what a human cerebrum can do. Deep learning, up till now, has only sometimes been utilized for making decisions and controlling. In [9], Indoor impediment shirking was followed to exhibit the viability of various levelled structure that melds a CNN with a choice procedure by Lie Tai and group. It's significantly minimal network structure utilizes raw depth pictures as input or inference, creates control orders as network yield. Raw sensor estimations were considered as contribution for the development of the network, and the robot control orders are straightforwardly produced utilizing the various levelled network.

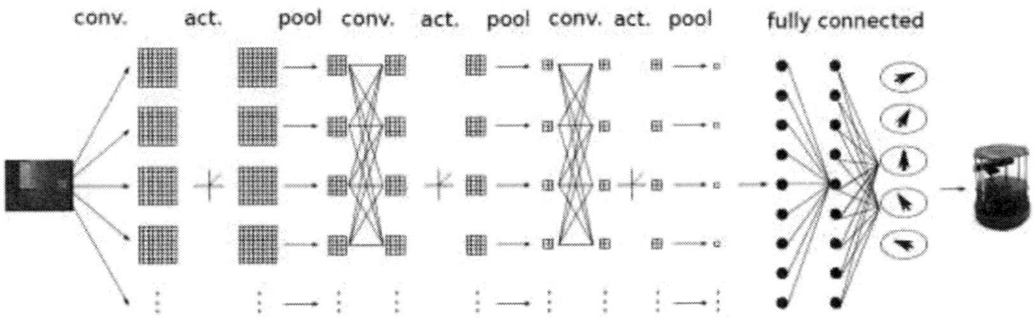

Fig.6 The proposed model combines CNN with fully-connected neural network for robot control.

Depth pictures are taken as the main contribution of the network. The Depth maps are taken as the inputs for the network. A convoluted network is utilized to empower a versatile robot to gain proficiency with a component extraction methodology. Despite the fact that the model is prepared in an administered manner, the utilization of CNN maintains a strategic distance from the estimation of hand-made highlights, which permits the operator better adjust to various types of situations. Not at all similar to geometric models which manage separation data, the model is prepared by accepting human directions as references, and it exceptionally emulates a cerebrum in the comparative manner to settle on a choice. In the proposed model, just the immediate sensor input is required and deterrent evasion task is accomplished consequently.

2.2.3. Neuro-fuzzy Technique. In [10], a neuro-fuzzy methodology is introduced so as to direct a portable robot by G.N.Marichal and group. By specifying required amount of fuzzy set rules by acknowledging the possible situations the robot can be subjected to. The neuro-fuzzy framework is made up of a 3 layered neural system, which is ready to separate consequently a lot of fuzzy rules after a preparation stage. So as to consider the conduct of these systems, two portable robots have been created. Ultrasonic sensors have been utilized with the intention to estimate the distances between the incoming deterrents and the robot. So as to arrive at an objective an electronic compass has likewise been consolidated.

The deliberate distances are related with some fuzzy sets as indicated by a participation work. Indeed, it can be given into 3 classes of distance: far-away, medium and close by. Consequently, a few sets fuzzy-rules these given rules determine appropriate speeds to the robot's motors as per the need. Similarly, 3 sets were developed to determine robots motor speed: slow, medium and fast. Besides, a value must be determined by the membership function. In this way, a defuzzification system must be applied in solicitation to gain the particular characteristics for the robot engine speeds. The supposed essential layer or data layer contains a couple of centre points, where every centre point comprises of radial basis neuron. There are a couple of boundaries in the neuro-fuzzy structure which choose the association between the data sources and the yields.

2.2.4. Genetic Algorithm Technique. The principle objective of way arranging (path planning) is to decide the ideal conceivable way from the initial instance and characterized objective situation in the given time. In [11], a way arranging strategy by using genetic algorithm (GA) is introduced by Samadi and Othman. The enhanced way of computing length and effective cost is produced by GA advancement. This encompasses a worldwide way arranging technique with hexagonal network map displaying. Interestingly, robot creators and computer game designers utilize various strategies for taking care of the way arranging issue so as to discover the upgraded existing way in the map. The way arranging algorithms are likewise isolated into two classifications, which are nearby and worldwide algorithms. The nearby algorithms create way while the robot travels through nature and can deliver another way dependent on the ecological changes. While worldwide algorithms need to have past information about the earth; along these lines, nature ought to be static.

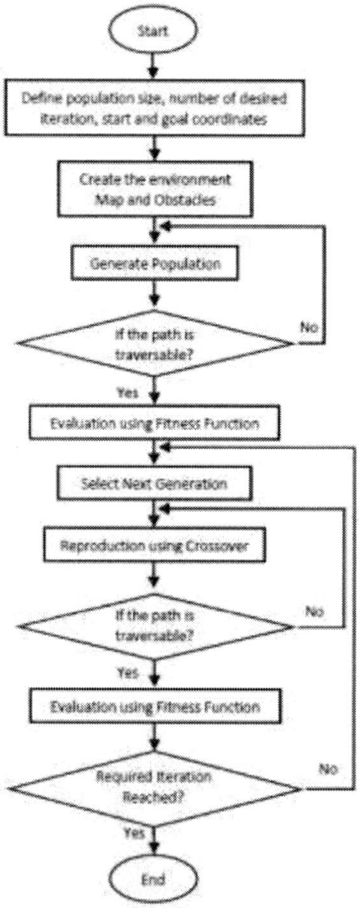

Fig.7 The Flow chart of the proposed algorithm

Numerous analysts have been taking a shot at way arranging field utilizing GA to create the ideal way and exploiting GAs advancement capacity. GA is an intrinsically equal inquiry strategy and has the ability to scan for the ideal way in a given domain. The significant issue in robot route is enhanced guide which ought to be produced by certain rules, for example, cost, time, separation and vitality.

Among these parameters, time and separation are the most significant models for the analysts. The software and equipment condition is depicted and the underlying qualities for the populace with the related outcomes are shown and broke down. In all the tests the underlying populace was fixed to concentrate on the impacts of cycle changes on the improved created way by the proposed algorithm. The GA has a decent ability to improve the way arranging algorithm and create an advanced way regarding way length and cost of the way. The utilization of unique choice and hybrid capacities had the option to diminish the GA count time and prompted better way arranging algorithm. The workflow of the proposed algorithm is shown in **Fig.7**.

2.2.5 Ant Colony Optimization Technique. In bio-persuaded heuristics perfect models have been used extensively in a couple of districts of planning fields. One of these calculations is Ant Colony Optimization (ACO). The ACO (Ant Colony Optimization) calculation is a smoothing out strategy reliant on swarm knowledge. In [12], Michael Brand and his team managed Abstract-Path masterminding. The goal is to attain the most restricted and crash free course leading to the objective.

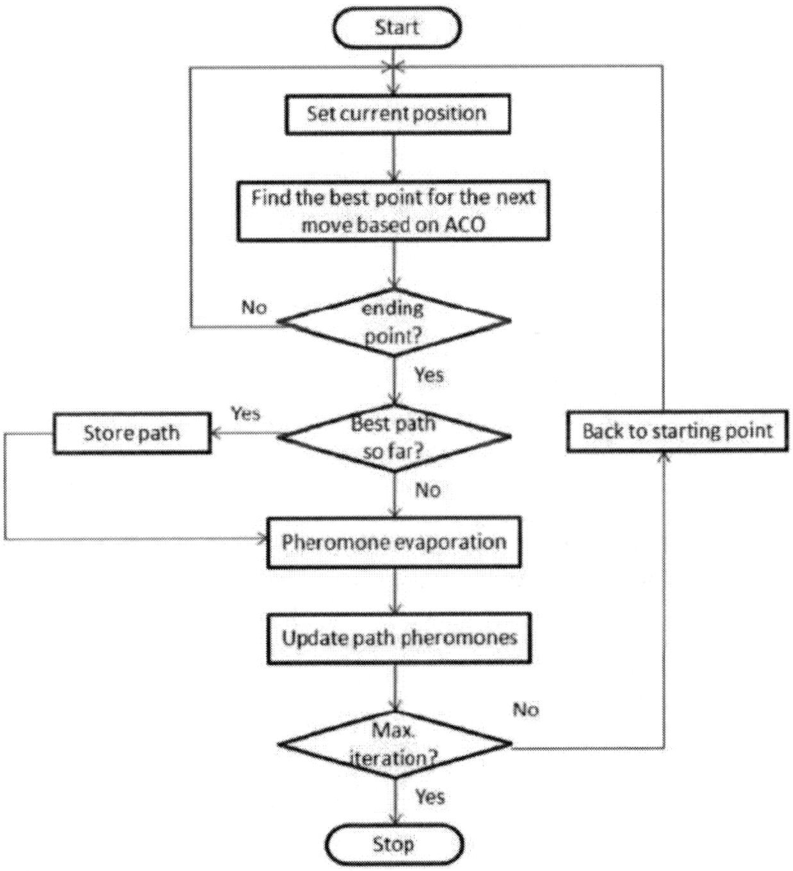

Fig.8 Computational Flow chart of ACO

Two re-instatement plans (i.e., the local introduction and the global introduction) are examined and thought about. Since we will probably locate the most brief way between the beginning and completion positions, the all out way length is picked to be the expense or prize related with every conceivable arrangement. It is better understood by illustrations of flow charts such as in **Fig.8** and **Fig.9**.

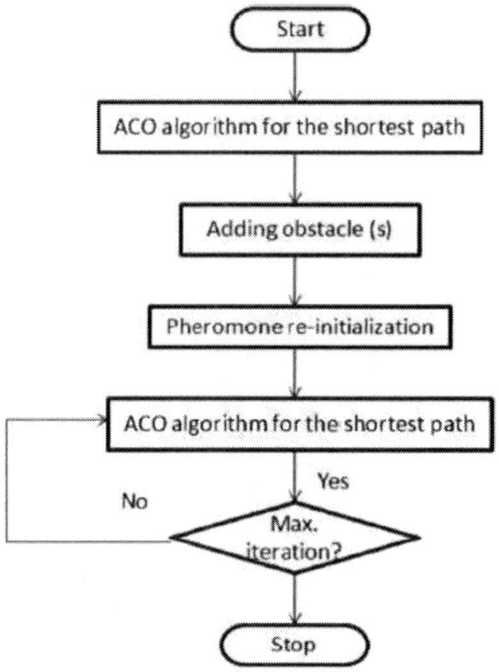

Fig.9 Flow chart of ACO in a dynamic environment

APF and ACO were applied independently as worldwide way orchestrating and local-way masterminding procedures which were presented by Hao Mei and gathering in [13]. Ants leave chemical compound known as "pheromone", on the course on which they traverse. The amount of pheromone is in reverse degree to the length of the way. Ants can choose dependent on the amount of pheromone present on the course meaning, that the quantity present is increased by increase in visits. This methodology will be gone on until all the ants select the most constrained way. Pheromone delivered by ant colony improvement was in like manner used to shield artificial potential field from getting local least. APF regularly gets into local least with no issue. Neural-framework can pick up from existing data and the data depiction is especially inconvenient. GA is a transformative calculation, and prepared to decide structure smoothing out issues. ACO is good for the mix headway issues, for instance, way orchestrating, anyway it can't be applied in novel condition. The team proposes a way masterminding approach in an exceptional condition that consolidates the worldwide coordinator and local coordinator. The portrayal of the ACO used in this paper is according to the accompanying:
1) Initialize the pheromone as per the strategy expressed previously.
2) Construct a course arrangement from the beginning stage to the objective point dependent on the pheromone data.
3) Update the quantity of pheromone.
4) Exit and yield the ideal arrangement if the arrangements have united or come to the up furthest reaches of emphasis, in any case hop to Step 2.
Results of re-enactment show that by blending ACO and APF algorithms, global-ideal and ongoing snag evasion can both be fulfilled.

2.2.6 Particle Swarm Optimization. In [14], Adaptive PSO was discussed. In adaptable robot course, the robot needs to plot its own particular way to reach the destination without hitting the deterrents while taking into account of previous behaviours. On the off chance that the model proceeds with the least path in minimal time. During following the most concise way, robot in like manner needs to manage the obstruction position. This most extreme separation can be settled using the referenced separation recipe. The obstructions are scattered arbitrarily in nature. At first the developed robot is no-where near the obstruction and with knowledge on the deterrent. Exactly when the robot enters as far as possible territory of the obstacle, the APSO calculation is started to check an appropriate situation for robot. Boundaries like time elapsed and length of path wandered out by the robot to arrive at objective with various kinds of ecological diversions. It can be concluded that in APSO, the robot is capable to feasibly compute least path and time in comparison with standard PSO.

2.2.7 Machine Learning or Predictive Approach. For the portable robots that can coexist with humans, it is important to find some kind of harmony among security and proficiency. It is of prime importance to make a forecast about the human's location. In [15], Hamasaki and group, introduced an estimating arrangement of person on foot development tendency, and expectation arrangement of passerby development. They led the expectation try different things with genuine watched information, and demonstrated that framework can add to the adjusting security and productivity of working versatile robot. So as to work versatile robots in a domain coexisting with humans, it is important to evade deterrents to avoid interference with humans. It's not usual for versatile robots having to maintain a strategic distance from collision in a situation that people can keep away from each other without any problem.

There is consistently an exchange off among wellbeing and effectiveness, the proficiency is diminished when the robot moves gradually for the purpose of security, and the security is decreased when the robot moves rapidly for the purpose of productivity. If there is a chance of crash. With the acknowledgement of the progress of human's advancement can be applied as the difference in the speed. While assuming human as a molecule, we can get a virtual drive from the difference in the speed. Sensors like, Laser Range Finder (LRF) can be used to create a database of humans' motion-captures. The ecological sensor viewed for motion of the human for certain hours in intervals of times. In the process of trajectory estimation, the watched or experienced information acts as the prime factor for forecast. We need to think about the blunder of estimation. Particularly the blunder of velocity, that will incredibly impact the outcome. The flow chart of proposed algorithm is shown in Fig.10.

In [16], a formulation which incorporates, collision expectation and evasion for portable robots, in presence of moving snags is introduced by K.J.Kyriakopoulos and G.N.Sardis. The calculated route should interface the underlying and target centres without hammering into the fixed things while fulfilling certain objectives. The calculation is ought to follow an administrative perspective while guaranteeing that the robot is progressing towards the destination on keeping up a key good ways from impacts with those moving impediments with which crash is anticipated reliant on material information. The basic idea is to adjust the speed en route. By this it can be inferred that, the adaptable robot may be constrained with time in computing best outcome with experiences from past. In [17], the issue of seeing ecological trails using a solitary monocular picture obtained from the POV of a robot traversing, was introduced by Alessandro and group. All in all, by working on the entire picture without a moment's delay, our framework/system yields the principle direction of the path in contrasted with the view direction. Qualitative and quantitative outcomes registered on an enormous genuine world datasets show that the methodology outperforms the accuracy of humans which were subjected to same image classification task.

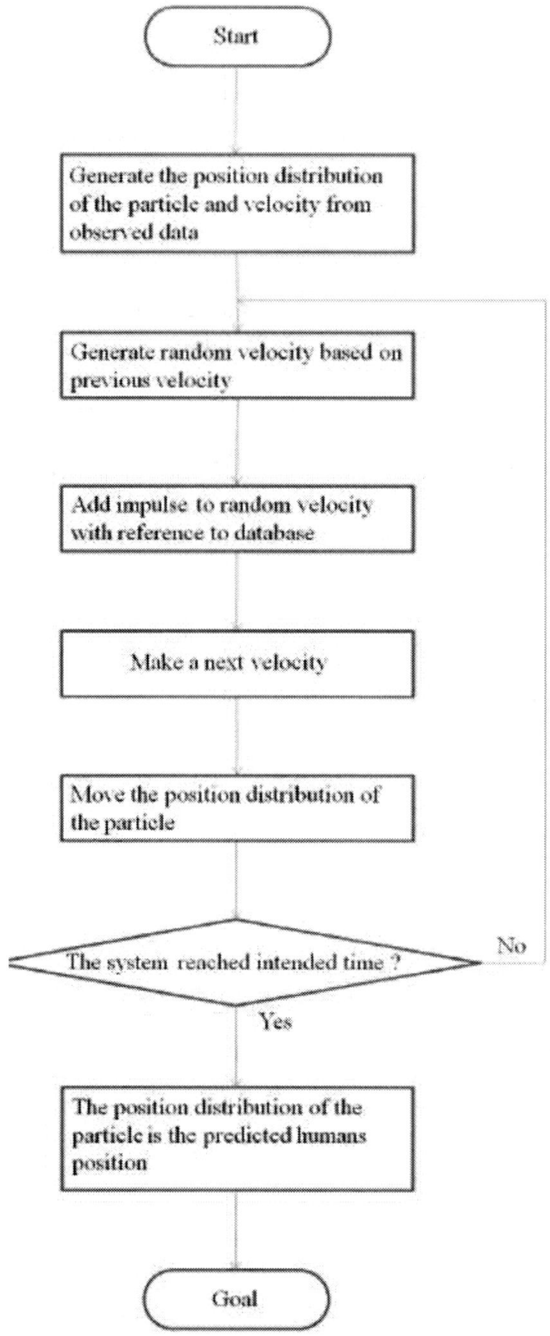

Fig.10 Flow chart of proposed algorithm

Following a man-made path, (for example, those normally crossed by explorers or mountain-bikers) is a difficult and for the most part unsolved assignment. So as to effectively follow a trail a robot has to clearly have sight of the path to begin with to obtain further information on trajectory. Seeing genuine path in these conditions is a very difficult and fascinating example recognition issue. A few previous works managing trail discernment/perception tackled a division issue, i.e., planned for figuring out which zones of the input information of picture compares to the picture of the path. So as to unravel this assignment, one needs to explicitly characterize which visual highlights portray a path. The saliency of picture captured is utilized to map the trail. The work followed a different methodology and cast the path discernment issue as a picture classification task. With a supervised-AI dependent on Deep Neural Networks (DNNs), a best in class deep learning strategy that works legitimately on the picture's crude pixel values. With this, highlights are found out straightforwardly from the information. DNN is used as a picture classifier to receive the system architecture.

In [18], D.Nakhaeinia and group, portray a novel receptive deterrent evasion strategy to aid versatile robot to compute route in a dynamic condition. The methodology is dependent on "situated-activity paradigm" and "separate and overcome" strategies to drive the robot..The Virtual Semi-Circles (VSC) way arranging technique is partitioned to 4 modules: division, assessment, decision, movement generation. Comparing to 6 sensors course of action the robot work space is partitioned to six sub spaces. The subspaces are: Right (R), Front-right (FR), Right-front (RF), Front-left (FL), Left-front (LF) and Left (L) as shown in Fig.11; which speak to the target bearing and hindrances position. three semi-circles are accepted around the robot with range of 1m, 2m, and 3m from its inside which separate the robot's work space to three areas. These subspaces speak to the snags distance from the robot:
1) N (Near): Inside of the semi-hover with sweep of 1m;
2) M (Middle): The existing hole between the two semi-hovers with sweep of 1m and 2m;
3) F (Far): The existing hole between two semi-hovers with sweep of 2m and 3m.
The robots subspaces can be visualised as in **Fig.11**.

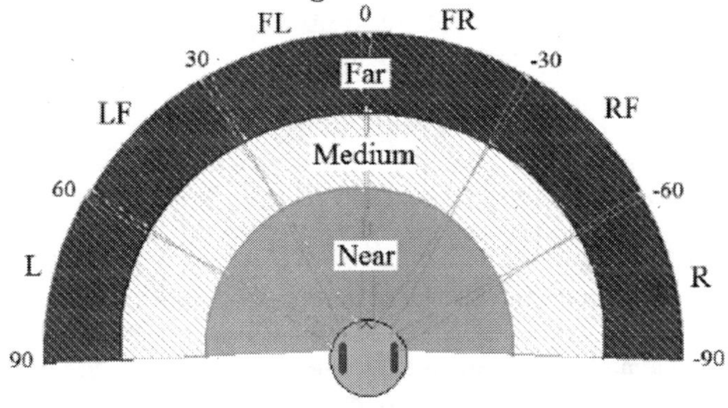

Fig.11 The robot's subspaces

Therefore, the robot work space is separated to 18 districts totally. In each progression of the robot development, the tactile information is updated and demonstrates the robot's status to the snags. Therefore, the exact impediment positions and free deterrent spaces are acquired relating to the sonar readings. To build the exactness of the robot manoeuvrability, the subspaces can be changed (expanded/diminished) as indicated by the robot size and condition characteristics with changing sweep of the semi-circles or rakish divisions. Each locale gets a value of 1, 2 or 3 comparing to the obstructions position (dim districts) from the robot. Therefore, the locales with higher value are considered as protected area which the areas ought to have the value of multiple. In each control period, the districts are updated dependent on the tangible information to identify the robot's situation. At the same time, the districts values are extricated from the impediments position and free snag regions to help the robot in decision making for next activity. Therefore, the following stage is to choose which safe area improves the route way and reaches the robot to the objective without any collision in the most limited time. The robot needs to settle on a decision to pick appropriate way toward the objective among existing safe areas. First the situations are arranged as following:
1) Target in safe district: this happens when safe locale and target have the same course.

2) Target in different locale: this happens when safe districts and target have different heading. The nearest sheltered district to the objective has the most noteworthy priority in picking a way. Therefore, the priority is with the protected locale which has the same bearing with the objective. In any case, if there are more than one safe district when target is in different area, the closest sheltered locale to the objective course is the best way. When an appropriate way chose by the decision module, the robot vanquishes the free spaces toward the objective in the way. The movement commands depend on the protected district heading and value. The VSC approach differs from the existing techniques in the utilization of straightforward algorithm with high productivity, incorporating different modules which don't require huge memory and calculation. Be that as it may, the methodology cannot acquire the shortest way/path. In [19], A system for visual route with synchronous impediment evasion is introduced by Andrea Cherubini and Francois Chaumette. The methodology blends strategies from APF, repetition along with visual surveying. The impediments are demonstrated by occupancy framework/grid. In the system that has been created, the camera must recognize the highlights important for route, while the range scanner detects the deterrents before the robot. An appropriate function can be used to achieve smooth the transits.

2.3. Comparisons/Combinations. In [20], Nikos and gathering proposed a model where an electrostatic PF worldwide way coordinator is represented with a 2 layered fuzzy method (using sonar sensor data) of reasoning enlistment engine and completed for constant convenient robot course in a 2-D dynamic condition. The essential layer of the fuzzy engine consists of sensor readings into a fuzzy variable, giving data about potential crashes. The sensor readings are culmination of commitments to the fundamental layer. The combination of EPF and FL allows the potential field to structure the ways while keeping up a vital good ways from impact with all obstacles. The PF utilization doesn't move toward the fuzzy reasoning, and thusly an alternate control must be utilized. In [21], Tewolde and Sheng talk about, the issue of hardware path getting together with respect to splash shaping structures. Shower framing is a creating performing process used in the amassing industry to fabricate things, for instance, body segments of automobiles and planes. A couple of issues ought to be comprehended before a computerized instrument orchestrating structure can be grasped for usage of manufacturing plant floor. It is very well deduced that ACO outmatches GA in path orchestrating similar to execution. Course and impediment avoidance are important issues for the powerful use of a self-sufficient convenient robot. To allow the robot to move between its present and last plans with no impact inside the overall condition, development orchestrating needs a ton of treatment. Accordingly, to create collision free way it ought to have appropriate movement arranging just as impediment shirking plan.

In [22], Sourav Dutta discusses about the impediment shirking of a wheeled versatile robot in organized condition by utilizing PSO based neuro-fuzzy methodology. Here, a 3 layered neural system utilizing PSO as a self-learning algorithm to decide the ideal trajectory without any collisions. To produce without collision way of a vehicle like robot during its route among a few moving hindrances, it ought to have legitimate movement arranging just as deterrent shirking plans. Both logical like potential field strategy just as diagram based methods have been utilized to take care of the route issues of robots including static deterrents. In any case, each and every such system may not be reasonable for on-line utilization in light of their characteristic computational multifaceted nature and impediments. Since ANN can get acquainted with the circumstances, various specialists have adequately with application of feed-forward. In this work, a PSO calculation is used for learning inspiration driving the neuro-fuzzy structure. Sensors are used to get the separation between the robot and the obstruction. Contributions of the neuro-fuzzy framework depend on the yield information acquired from sensors. In [23], Mario Volf and group thought about the adequacy through advantages and disadvantages of 2 techniques important for robot path computation. By this, physical movement arranging issue into geometrical way arranging issue. The robot is to be considered as a point defined in a proper environment or space in presence of deterrents. When the robot doesn't have prior information on the snags, yet it facilitates during movement execution. Continuous proficiency is desired, instead of reaching the destination. A graphical technique using information about workspace is used to obtain an unoccupied way.

2.4. *Importance of Sensors/Scanners.* In [24], Byoung-Suk Choi and gathering proposed a novel calculation that improves the localization by interweaving an RFID system with an ultrasonic sensor structure to obtain desired output. This results in coping up with the draw backs of distance with help of each other. In order to confirm the proposed calculation, two examinations were coordinated. The calculations are issued for range estimation of the RFID. Another goal of this paper is to find an arrangement to effectively lessen weakness of Indoor RFID system. The weakness in RFID structure estimations was lessened by the LEC strategy. Exact and convincing estimation was then possible. In [25], Guoqiang Fu and gathering managed to utilize Triangulation Laser Scanner for Deterrent Detection. The purpose of the work is to display the chance of expelling central data from the robot natural components by strategies for little, direct, low-power, and negligible exertion requesting devices, which, likewise, can be scaled down in order to get ready a lot humbler robots. To perform harmless examination of faulty conditions, self-ruling flexible robots frequently use run sensors to help course and planning. Sorted out light allows the robot to definitely measure the geometric features of enveloping articles, along these lines, adding to the choice creation method of the robot when it achieves tasks, or even more only for progressively secure course and examination.

In [26], Andrea Cherubini and team worked on self-sufficient Visual Navigation. Hindrance avoidance has customarily been taken care of by two systems: the deliberative approach, by and large comprising of a development coordinator, and the responsive philosophy, considering the instantaneous distinguished data. It consolidates a responsive arm based technique to guarantee way following, hindrance bypassing, and crash evading by deceleration. The work relied upon two speculations: All things are firm. The development is essentially translational over minute time breaks. Block speeds are spoken to inside a visual course plot. To assess the hindrance speeds, Kalman-based onlooker was organized. At that point, the speeds to anticipate potential crashes among robot and obstructions were utilized. The results show that, by anticipating the hindrance relocations inside the up-and-comer extremities, the robot direct is increasingly secure and smoother hence, higher speeds can be accomplished. In [27], Joydeep Biswas and gathering worked on robot localization using depth cameras. The sheer volume of data made by significance cameras gives a test to process constantly and explicitly when used for indoor convenient robot localization and course. The Fast Sampling Plane Filtering (FSPF) calculation tests irregular neighbourhoods along with usage of RANSAC. In [28], the fuzzy method of reasoning calculation which Mohamed Karim and gathering corrected the robot position with an Extended Kalman Filter. The robot needs to differentiate its uproarious position and the ground truth. Undoubtedly, the fuzzy systems rely upon the possibility of fuzzy sets with every segment of the set being weighted. This capacity relies upon the assurance of a particular number of components and the choice of its structure. We join this conduct with the position remedy utilizing EKF algorithm.

In [29], Santos Ortiz and his group introduced the advancement of an observation framework for indoor conditions to permit autonomous route for reconnaissance portable robots. The framework is made by two sections. The initial segment is a responsive route framework wherein a portable robot moves staying away from hindrances in condition, utilizing the distance kinetic sensor. The second piece of this framework utilizes an ANN to perceive various designs of nature. In, robotics, ANN is utilized in many ways to perform its best way utilizing kinetic sensor to locate the robot in any given location. Furthermore, the two subsystems (Reactive Anti-Collision System AND Corridors Recognition utilizing artificial neural Networks) will be coordinated as components of a total reconnaissance framework or surveying system. In [30], Jeongdae Kim and his group introduced some starter consequences of the recognition of moving deterrents (especially strolling people) by the utilization of a solitary camera joined to a versatile robot. At the point when a camera moves, straightforward moving article recognition procedures for a fixed camera, for example, foundation deduction or picture differencing, cannot be utilized.

A picture is initially separated into little squares, and then the movement of each square is looked by contrasting two back to back pictures. If the movement between coordinating squares is significantly enormous, the square in the present picture is classified as having a place with moving articles. The strategy is verified by the indoor route trials of a robot. The plan of a round cluster of ultrasonic sensors around the body of a versatile robot is now one of the most famous strategies for obstruction evasion. As of late presented time-of-flight (TOF) cameras have numerous advantages for the autonomous route of a portable robot. In comparison to laser scanner, TOF camera furnishes 3D information with an ongoing casing rate and is a lot quicker. The camera alignment and information pre-preparing are important to get steady estimation. The Block-Based Motion Estimation (BBME) technique has been broadly utilized especially in video coding. Video pictures have worldly excess, which is identified with the spatial similarity that transiently associated outlines have. Since the square based coordinating has a normal algorithmic structure, it is conceivable to ideally actualize the strategy for the continuous handling of the two marginally different pictures of sound system cameras. At the point when individuals are at a farther distance, their movement in successive pictures shows up more slow, and they may not be found as moving items. In this way, the algorithm can be applied all the more effectively to objects at nearer distances to the camera (i.e., robot).

In [31], N.K.Philip and his group dealt with the issue of, object discovery and Obstacle Avoidance for portable robot utilizing Stereo camera. The objective of their examination is to build up a constant deterrent identification and obstruction shirking for autonomous route of versatile robots utilizing a sound system camera in an unstructured domain. Overall, position data is gotten from GPS and range sensors. All in all, to summarize the approach:

1) Arrangement of deterrents in the vicinity;

2) Kinematic deterrent creation AND obstruction by projection;

3) Potential function comparing to each obstruction.

Finally, deterrent avoidance using potential field calculation re-enacted close by sound framework model, 3D entertainment and stance estimation calculation. In [32], Saurav Kumar and gathering presented after Sensor Fusion of Laser and Stereo vision camera. LRF can be used to create 2D guides of the situation. The 2D framework acquired in the wake of pruning and downsizing 3D point cloud is combined with the 2D map produced by the laser to make an occupancy matrix. Primary spotlight on lessening the computational weight engaged with combination process and precise location and evasion of impediments in the environment of a versatile robot. In [33], G.Benet and gathering worked with the help of IR sensors for path estimation. IR sensors are better in response time than US sensors. All things considered, the standard application of IR sensors is impact avoidance as opposed to dynamic range distinguishing.

2.5. *Team of Robots.* In [34], Hamed Rezaee and FarzanehAbdollahi talk about the issue of arrangement *command on* a gathering of adaptable robots reliant on the virtual structures. In the virtual structure, each robot is considered as an electric charge. In the proposed method, every convenient robot finds its situation in the development independently. They proposed a methodology for keeping up a key distance from obstructions reliant on the social structure. In this technique, when a flexible robot gravitates towards its goal reliant on social structures, a rotational potential field is applied to lead the versatile robot to avoid the obstacle, without arranging in local least positions. The distances between the flexible robots coordinated independently to arrive at a standard polygon development, and the arrangement was kept up while it was moving. In [35], Dimitra Panagou and Vijay Kumar address arrangements of non-holonomic flexible robots. For instance, adaptable robots working inside usually need access to worldwide situating estimations. The distinctions of the proposed approach differentiated and other vision-set up development control frameworks depend upon the nonappearance of correspondence, the idea of restricted FOV. The last determination is the rule distinction of this current paper differentiated and different papers on keeping up perceivability issues incorporating robots with field-of-view requirements.

In [36], Mohamed A. Kamel and Youmin Zhang investigate the development control and hindrance evading of various differentially decided WMRs considering the pioneer aficionado strategy. MPC is applied to the linearized model to attain control inputs. An obstacle avoidance calculation is also embedded to the control system to allow the individual robots to avoid impediments on the go. Development relies upon a pioneer disciple procedure, and the control calculation is executed in a decentralized manner. In the wake of getting the desired linearized model uses linear model predictive control strategy to improve the controller plan for such WMR nonlinear structures. The standard advantages of the proposed procedure is the decentralization, its capacity to recognize the obstructions, and update of control contributions. In [37], Ignacio Mas and Christopher Kitts, managed the cluster space control procedure progresses improving particulars and checking of the development of adaptable multi-robot system. The controller produces bunch orders that convert into substantial robot level developments. This strategy conceptualizes the n-robot structure determined as a component of gathering attributes, for instance, position, bearing, and orientation. These changes license bunch orders to be changed over to robot-explicit orders. Our speculation is that such correspondence improves ease of use by offering control. Holonomic robots are those that can absolutely turn about its own pivot without float of its point of convergence of gravity from a lone point. Non-holonomic robots are dynamically like vehicles that need to dislodge their point of convergence of gravity in order to turn.

2.6. *Scope in the Field/Relevant Research Fields.* In [38], Adrian Green and group proposed a novel technique for hindrance avoidance. The oddity of this methodology isn't just the distance parameter but also the relative angles of the robot and obstruction in the environment. In this way, we isolate the whole design region into 2 sub-regions viz., sheltered and snag evasion region. The exchanging floor comprises of direction, characterized through the relative demeanour amongst the robots heading towards the snag. Evading crash is one of the chief research fields in mobile robotics. Different researchers around the globe are working on static and dynamic crash avoidance calculations. One such arrangement of rules is the Optimal Reciprocal Collision Avoidance that proposals with numerous robots moving in a joint territory without exacting impact, that still with out imparting and without brought together preparing. Along these lines, Ali Khan and his team toiled on Collaborative ORCA in [39], This arrangement of rules is successful in taking care of crash avoidance. Nonetheless, the problem of halt frequently appears while the robots need to explore through thickly packed conditions in joint zone. The design is to move robots in a little joint region and accomplish crash avoidance without managing gridlocks. For this thought process, this convention is extended into Collaborative Optimal Reciprocal Collision Avoidance to cure the issue of stops. The convention is enlivened from rules and settles the problem of halts adequately. The convention transformed into inspected upon Player-Stage basically based recreation where gridlocks were stayed away from accurately.

The ORCA algorithm makes certain that each one the robots are moving in a collision unfastened state. In role-based totally ORCA that is completed with the aid of sensing of the surroundings by using the sensory network and retaining tune of it continuously while it moves and reacts to its surroundings. The movements taken are in my opinion computed for each robotic on its personal without communicating to different robots or significant coordination. Yet ORCA makes certain that the robots are in motion without any collision. Each robotic avoids collision on its personal with out the know-how of the opposite robot's reasoning and moves. The simplest supply of know-how is the sensory network via which every robotic personally obtains the information of the environment. Collaborative Optimal Reciprocal Collision Avoidance (C-ORCA) is very simple and is stimulated through the traffic policies and ordinary using. According to C-ORCA whenever one-of-a-kind robots are on a path of collision the robots take use of the guideline set stimulated by using visitors laws as opposed to choosing the most effective and finest path. Thus, similarly to ORCA based totally reasoning based totally on the PCS idea, a site visitors rule precedence is used to pick out the direction of action.

In [40], Chaochao Chen and Paul Richardson, present a dynamic recurrent neuro-fuzzy system (DRNFS) with brief memory for obstruction avoidance of mobile robots in obscure conditions. The boundary optimization is discovered through the arranged subsidiary calculation, and the shape re-arrangements is finished by utilizing the rugged strategies similitude degree. Accordingly, the problematic path can be noticeable. In this manner, a unique route framework that has the usefulness of memory carport is anticipated. The (DRNFS) gadget is a powerful planning machine that can manage time-fluctuating difficulties. With the potential of briefly putting away realities, the DRNFS has been executed effectively inside the estimation, personality, forecast and control of entangled time-different powerful machine. The boundaries, we prescribe a measure alluded to as the ragged strategies comparability degree to assess the level of the similitude of two fuzzy arrangements. At the point when the confirmation of the distinction of fuzzy guidelines is seen as littler than a predefined edge, the 2 fuzzy guidelines are expelled and another fuzzy principle is made. So as to connote unmistakably that the proposed machine can make a chief route determination the utilization of the capacities of memory stockpiling and handling, the DRNFS inside the machine is changed through a fuzzy standard based controller to play out the hindrance avoidance conduct.

The data from both the predominant and the most recent past are recruited inside the DRNFS to make it practical that the robot can maintain a strategic distance from confinements simply like the human main thrust with short memory. The DRNFS is remembered for a conduct based thoroughly plan to harvest robot self supporting route with various parts: global objective chasing and conduct co-appointment. The outcomes show that the propelled device with brief memory can make the robot arrive at the reason positions simultaneously as effectively taking off various styles of limits, and can likewise accomplish extra greatest snag averting courses in certain inevitabilities in contrast with the fuzzy principle based absolutely contraption. In [41], Peggy Wang and Caitlin Hogan compare a path making plans model for the Stanford Doggo, a quadruped robot specific in its compliant gait and ability to jump variable heights. Implementation of a couple of commonplace noisy sensors utilized in robotics in addition to particle filtering, permits for an extremely sensible version of navigation. The world is modelled as a Hidden Markov Model, with every tile having a perception opportunity of getting a dynamic impediment, together with some other vehicle. Additionally, it had already protected a form of graphical interface, which is useful for each debugging and example of the work. This grid-primarily based model is a useful tool in growing and trying out self sustaining modes of motion for jumping-enabled robots such as the Stanford Doggo quadruped. The model is fairly reflective of the robotic's bodily and navigational properties, despite the fact that the real robot presently lacks most of the sensors currently used for course planning on this version (as the robot is not self sufficient). The manner which sensor fusion was carried out for the version supports the modular addition or removal of any other sensors, which is a critical characteristic for any checking out environment to have.

In [42], Seyyed Mohammad and his gathering did path optimization based Riccati equation approach-based (SDRE) for absolutely best control way to deal with a robot. The robot in each time steps wishes to choose standards which incorporates speed, security, environmental factors, and distance in acknowledge to characterized objectives after which figure the correct control approach. Additionally, getting records related with the environmental factors to keep away from confinements, do the first in class steering, and become mindful of nature is important. The robot should astutely see and act utilizing adequate calculations to control required control and route issues. In this paper, sharp route of a cellular robot in an environment with sure fixed hindrances (perceived to the robotic) and best directing through Riccati condition contingent upon SDRE is thought of. This methodology permits the robot to do the best heading making arrangements in static situations. The idea of expanded linearization and the SDRE controller for nonlinear most alluring principles are provided, at that point the extra phases of opportunity outfitted through the non-uniqueness of the SDC definition is investigated. The correct idea for each robotic's course making arrangements is to make a most appropriate mix calculation with regards to the novel shape of each robotic all together that each arrangement of rules might be ensured the constraints of the contrary calculation.

In [43], Zihao Yuan and group, proposes a dynamic algorithm for accurate mobile robot navigation through a portable incorporated system ready with a Pioneer P3-DX robot and an array of sonar sensors. The proposed set of rules can dynamically calculate the most efficient path to reach the vacation spot, and travel the route with the aid of averting any obstacle in its course. The robot calls for specific commands to successfully navigate indoors. While navigating, the robotic need to also avoid barriers as detected by way of the sonars. The proposed technique makes use of easy trigonometric functions to calculate a shortest direction from the present day area to its destination. Once the space from the destination and magnitude of rotation has been calculated, the robot rotates for that reason, and begins shifting closer to the vacant spot. During this manner, the robotic constantly video display units its surroundings through sonars to detect any objects in its route. If the sonars detect no objects, the robotic dynamically estimates its current location, calculates the new distance and perspective of rotation, and reaches its destination with an excessive precision.

In [44], Aamir Ahmad and his group, adapt to the difficulty of agreeable localization and objective observing with a group of moving robots. They form the problem as a least squares minimization inconvenience and show that this difficulty can be accurately understood the use of meagre optimization strategies. Static tourist spots at respected position are utilized to layout a typical reference body for the robots and the objectives. Investigations done the utilization of a rigid of genuine robots show higher precision in contrast with a Kalman get out. The technique for multi-robotic moving milestone graph optimization (MMG-O) incorporates the resulting steps:

1) Create a graph speaking to postures of the robots and positions and speeds of the objectives (hubs) and the perceptions made through the robots (edges).

2) Stack all the perceptions on the whole to make an single non-linear least squares blunder/error function.

3) Use non-linear least squares solver to compute the minimum of this blunder function.

The methodology was tested in a robot football situation and contrasted it with an EKF-based methodology. The results show that our strategy closes in expanded exactness in the estimation and to an advanced versatility in inevitabilities wherein a higher wide assortment of robots is required. In [45], Patrick Benavidez and Mo Jamshidi introduced the structure for the route and objective following machine for a cell robot. Many oversee plans had been introduced for route what's more, target following for use on cellular robots. Equal preparing of route and other important activities permit a robot to perform obstruction avoidance and objective following in genuine time, leaving long calculations to different operators in a contraption. One of the guideline inconveniences in shading photograph preparing is the assortment of hues. A shading territory known as rg-chromaticity is utilized to expel the light force from the commitment of the shading. With rg-chromaticity, essential and auxiliary colorings can be remoted into one of kind limits around dark red-green chromaticity territories. With the shade portioned picture, applicants of items recognized to be positive shading can be detached in a photograph outline and tried further. Arranged handling of the force photograph includes finding the safe area and the planning a course the utilization of the profundity picture.

In [46], D. Ibrahim and his group provided take a shot at Navigation of cell robots inside the nearness of boundaries. The robot will explore amongst obstructions without hitting them and arrive at the necessary objective point. The created robot route programming program incorporates the executions of four calculations. The fuzzy calculation anyway effectively dodged the boundaries making choices of where to interchange bearing since it moved toward the hindrances. The fuzzy calculation forestalled the hindrances through creation choices where to change course since it moved toward the impediments. Fuzzy standard based absolutely with fuzzy derivation system is actualized to find turning viewpoint of the robot. The ragged calculation may be a direct result of its likeness to the astute human thinking and dynamic methodology. In [47], Hongjiu Yang and gathering present a particular control plot for a couple of issues on observing and impediment avoidance of a wheeled cell robotic with non-holonomic requirement. An all-encompassing nation spectator is brought to evaluate obscure unsettling influences and speed data of the wheeled cell robot. A nonlinear controller is intended to procure observing objective and obstacle avoidance in entangled situations. The proposed nonlinear oversee plot for a wheeled cellular robotic on the elements stage has been structured without the conviction on the consul speed checking with non-holonomic imperative.

In [48], Dilyana Budakova and group, referenced the most extreme fundamental ways to deal with complex developments, considering the area of the incorporating static contraptions. The acknowledgment of articles and limits inside the environmental factors where in people and robots cooperate; situating; Proper correspondence can take zone while applied on multi-specialist, notwithstanding human-robotic exchange situations. The computerized road network ought to be soaked with records to be advantageous and powerful. Time to Collision and furthermore the amazing/clashing practices are typically utilized measurements to assess the crash possibility. Powerful development oversee calculations are basis for the triumphant self supporting activity of the cell robots. They choose the correct moves to be taken essentially dependent on the state of the robot.

The Dynamical Motor Primitive (DMP) technique utilizes dynamic frameworks to speak to oversee rules. DMP is generally executed inside the impersonation figuring out how to analyze different engine gifts alongside motion. The measurable demonstrating techniques are each other way of encoding strategy thus. For instance, a constant concealed Markov variant is utilized for encoding the teacher's exhibit moves in the so-known as Mimesis Model. Exercises of the robotic with any strategy, including picked up from exhibit, may moreover now and again have negative generally speaking execution. The blend of movement planning calculations, path making arrangements to watch, taking off obstructions, move inclination, creating the incredible direction, equal impact avoidance procedures, sport hypothesis calculations for limiting opportunity to crash, decision making and situation mastery, chance appraisal, vehicle correspondence brings about the chance of realizing a self continuing automobile.

In [49], TaeseokJin proposes a route set of rules for a mobile robotic, which is keenly searching the point region in obscure powerful conditions utilizing an ultrasonic sensor. Rather than the utilization of "sensor combination" technique which creates the direction of a robot dependent on the environmental factors form and tactile records, "order combination" approach is utilized to administer the robot movements. It is set up as a chain of importance of fuzzy principle bases which permits conveyance of insight among specific reason fuzzy-practices. The hindrance avoidance and direction making arrangements are two uncommon subjects, which makes the control shape of seeking after - objective together mind boggling. To show the presentation of proposed approach, genuine investigations are finished. The exploratory outcomes show that the cell robotic can explore to the reason point securely under obscure situations and furthermore can abstain from moving limits independently. A creation plans procedure for non-holonomic cellular controllers that utilizes simple and constant capacities comprising of polynomials is advanced by utilizing Evangelos Papadopoulos and IoannisPoulakakis in [50]. The methodology decouples kinematically, the controller from the stage by building acceptable paths that drive it to a last setup and is basically founded on planning the non-holonomic limitation to a space wherein it might be inconsequentially fulfilled. The technique allows in for direct authority over the stage direction. The advanced change additionally maps Cartesian region limits to changed over ones and takes into account impediment avoidance by developing the request for the polynomials that are utilized in making arrangements directions.

Mobile controller structures, alongside a cellular stage outfitted with controllers, are of top notch importance fundamentally because of their ability to arrive at targets which are in any case open air of the controller achieve. Applications for such structures have large amounts of mining, creation, ranger service, planetary investigation and the military. A basic problem in cell robots is development making arrangements which is engaged with getting open circle controls that steer the stage from an underlying realm to a last one, without disregarding the non-holonomic limitations. The determined paths need to fulfil the non-holonomic imperative of the mobile stage, must be able to convince the gadget faraway from constraints, which may likewise exist in its workspace, and must be computationally less expensive to process. A change that maps the non-holonomic imperative to a region where it can be inconsequentially happy changed into utilized. The resulting paths and directions are smooth in light of the idea of the guide and to utilize smooth polynomials. The properties of the proposed change might be applied for the situation confinements exist inside the workspace of the robotic: It maps Cartesian space snags to groups of changed over ones. It is demonstrated that hindrance avoidance can be practiced through developing the request for the polynomials which may be used in making arrangements directions. Encasing standard deterrents in basic shapes, for example, ovals or circles is encouraged for calculation of the additional boundaries required.

3. Conclusion

As the movement/motion of the robot is dynamic, henceforth beyond question, heuristic methodology is selected. Nature-inspired algorithms for deterrent evasion and navigation have had a great deal of significance in research ground. Exploratory outcomes show the viability of ACO-based and GA-based techniques, and the nature of arrangements that can be accomplished. There is a compromise on the nature of arrangement and time of execution of calculations of ACO. With expanding multifaceted nature of the issue, the ACO-based strategy delivers preferred quality arrangements over the GA-based technique, at an expense of higher execution time. It very well may be gathered that ACO outmatches GA in path planning as far as execution, although GA weighs in factors such as safety, distance, etc. On the other hand, a blend of heuristic strategies can be utilized to accomplish the ideal yield.

References

[1] Chenghui Cai and Silvia Ferrari 2009 Information-Driven Sensor Path Planning by Apporximate Cell Decomposition **39** 3 672-689.
[2] Frank Lingelbach 2004 Path Planning using Probabilistic Cell Decomposition **1** 1-6.
[3] Lei Tang, Songyi Dian, Gangxu Gu, Kunli Zhou, Suihe Wang, Xinghuan Feng 2010 A Novel Potential Field Method for Obstacle Avoidance and Path Planning of mobile robot **1** 633-637.

[4] F.ArambulaCosio and M.A.Padilla Castaneda 2003 Autonomous Robot Navigation using Adaptive Potential Fields **40** 9 1141-1156.
[5] Priyanka Sudhakara, Velappa Ganapathy, B.Priyadharshini, KarthikaSundaran 2018 Obstacle Avoidance and Navigation Planning of a Wheeled Mobile Robot Using Amended Artificial Potential Field Method **133** 998-1004
[6] Tang Sai Hong, Danial Nakhaeinia and Babak Karasfi 2012 Application of Fuzzy Logic in Mobile Robot Navigation **1** 21-36
[7] C.G.Rusu, I.T.Birou, E.Szöke 2010 Fuzzy Based Obstacle Avoidance System for Autonomous Mobile Robot **1** 1-6
[8] Mohammed Faisal, RamdaneHedjar, Mansour Al Sulaimanand Khalid Al-Mutib 2013 Fuzzy Logic Navigation and Obstacle Avoidance in an Unknown Dynamic Environment **10** 37 1-7
[9] Lei Tai, Shaohua Li, Ming Liu 2016 A Deep-Network Solution Towards Model-less Obstacle Avoidance **1** 1-6
[10] G.N.Marichal, L.Acosta, L.Moreno, J.A.M-endez, J.J.Rodrigo, M.Sigut 2001 Obstacle avoidance for a mobile robot - A Neuro-Fuzzy approach **124** 2 171-179
[11] Masoud Samadi, MohdFauzi Othman 2013 Global Path Planning for Autonomous Mobile Robot using Genetic Algorithm **1** 726-730
[12] Michael Brand, Michael Masuda, Nicole Wehner, Xiao-Hua Yu 2010 Ant Colony Optimization Algorithm for Robot Path Planning **3** 436-440
[13] Hao Mei, Yantao Tian, Linan Zu 2006 A Hybrid Ant Colony Optimization Algoritm for Path Planning of Robot in Dynamic Environment" International Journal of Information Technology **12** 3 78-88
[14] Harshal S. Dewang, Prases K. Mohanty, Shubhasri Kundu 2018 A Robust Path Planning For Mobile Robot Using Smart Particle Swarm Optimization **133** 290-297
[15] Shunsuke Hamasaki, Yusuke Tamura, Atsushi Yamashita and Hajime Asama 2011 Prediction of Human's Movement for Collision Avoidance of Mobile Robot **1** 1634-1638
[16] K.J.Kyriakopoulos and G.N.Saridis 1993 An Integrated Collision Prediction andAvoidance Scheme for Mobile Robots inNon-stationary Environments" **29** 2 309-322
[17] Giusti and Alessandro 2015 A machine learning approach to visual perception of forest trails for mobile robots **1** 2 661-667
[18] Tang S.H. and C.K.Ang 2013 A Reactive Collision Avoidance Approach for Mobile Robot in Dynamic Environments" Journal of Automation and Control Engineering **1** 1 16-20
[19] Andrea Cherubini and Francois Chaumette 2010 A Redundancy-Based Approach for Obstacle Avoidancein Mobile Robot Navigation **1** 8-15
[20] Nikos C. Tsourveloudis, Kimon P. Valavanis, and Timothy Hebert 2001 Autonomous Vehicle Navigation Utilizing Electrostatic Potential Fields and Fuzzy Logic **17** 4 490-497
[21] Girma S. Tewolde and Weihua Sheng 2008 Robot Path Integration in Manufacturing Processes - Genetic Algorithm Versus Ant Colony Optimization **38** 2 278-287
[22] Sourav Dutta 2010 Obstacle Avoidance of mobile robot using PSO basedNeuro Fuzzy Technique Engineering **02** 02 301-304
[23] DusanGlavaski, Mario Volf, Mirjana Bonkovic 2009 Robot motion planning using exact cell decomposition and potential field methods **1** 126-131
[24] Byoung-Suk Choi, Joon-Woo Lee and Ju-Jang Lee 2011 A Hierarchical Algorithm for Indoor Mobile Robot Localization Using RFID Sensor Fusion **58** 6 2226-2235
[25] Guoqiang Fu, Paolo Corradi, Arianna Menciassi, and Paolo Dario 2011 An Integrated Triangulation Laser Scanner for Obstacle Detection of Miniature Mobile Robots in Indoor Environment **16** 4 778-783
[26] Andrea Cherubini, Fabien Spindler, Francois Chaumette 2014 Autonomous Visual Navigation and Laser-based Moving Obstacle Avoidance **15** 5 2101-2110
[27] Joydeep Biswas and Manuela M. Veloso 2012 Depth Camera Based Indoor Mobile Robot Localization and Navigation **32** 14 1-7

[28] Mohamed Karim Ettouhami and Younes Raoui 2013 Human-Robot Collision Avoidance with RFID Sensors Using Fuzzy Logic and Extended Kalman Filter **7** 52 2555-2567

[29] Diogo Santos, Marcos Gomes and Daniel Oliva 2012 Mobile Robots Navigation in Indoor Environments Using Kinect Sensor **1** 1-6

[30] Jeongdae Kim and Yongtae Do 2012 Moving obstacle avoidance of a mobile robot using a single camera **41** 911-916

[31] R.Lagisetty, N.K.Philip, R.Padhi and M.S.Bhat 2013 Object Detection and Obstacle Avoidance for Mobile Robot using Stereo Camera **1** 605-610

[32] Saurav Kumar, Daya Gupta and Sakshi Yadav 2010 Sensor Fusion of Laser & Stereo Vision Camera for Depth Estimation and Obstacle Avoidance **1** 26 22-27

[33] G. Benet, F. Blanes, J.E.Simó, P. Pérez 2002 Using infrared sensors for distance measurement in mobile robots **40** 4 255-266

[34] Hamed Rezaee and FarzanehAbdollahi 2014 A Decentralized Cooperative Control Scheme With Obstacle Avoidance for a Team of Mobile Robots **61** 1 347-354

[35] Dimitra Panagou and Vijay Kumar 2014 Cooperative Visibility Maintenance forLeader–Follower Formations inObstacle Environments **30** 4 831-844

[36] Mohamed A. Kamel and Youmin Zhang 2015 Decentralized Leader-Follower Formation Control with ObstacleAvoidance of Multiple Unicycle Mobile Robots **1** 406-411

[37] Ignacio Mas and Christopher Kitts 2012 Obstacle Avoidance Policiesfor Cluster Space Controlof Non-Holonomic Multi-Robot Systems **17** 99 1-12

[38] JingfuJin, Adrian Green and Nicholas Gans 2014 A Stable Switched-System Approach to Obstacle Avoidance for Mobile Robots in SE(2) **1** 1533-1539

[39] Shehryar Ali Khan, Yasar Ayaz, Mohsin Jamil, Syed Omer Gillani, Muhammad Naveed, Ahmed Hussain Qureshi and Khawaja Fahad Iqbal 2015 Collaborative Optimal Reciprocal Collision Avoidance for Mobile Robots **8** 8 203-212

[40] Chaochao Chen and Paul Richardson 2010 Mobile robot obstacle avoidance using short memory - a dynamic recurrent neuro-fuzzy approach **34** 2 1-17

[41] Peggy (Yuchun) Wang and Caitlin Hogan 2019 Path Planning with Dynamic Obstacle Avoidance for a Jumping-Enabled Robot **1** 1-8

[42] Seyyed Mohammad Hosseini Rostami, Arun Kumar Sangaiah, Jin Wang and Hye-jin Kim 2018 Real-time obstacle avoidance of mobilerobots using state-dependent Riccati equation approach **1** 1-13

[43] Zihao Yuan, Kevin Laubhan, Kumar Yelamarthi 2014 A Efficient and Dynamic Algorithm forAccurate Mobile Robot Navigation **1** 1-6

[44] Aamir Ahmad, Gian Diego Tipaldi, Pedro Lima and Wolfram Burgard 2013 Cooperative Robot Localization and Target Trackingbased on Least Squares Minimization **1** 5696-5701

[45] Patrick Benavidez and Mo Jamshidi 2011 Mobile Robot Navigation and Target TrackingSystem **1** 299-304

[46] R. Abiyev, D. Ibrahim , B. Erin 2010 Navigation of mobile robots in the presence of obstacles **41** 10 1179-1186

[47] Hongjiu Yang, Xiaozhao Fan, Peng Shi and Changchun Hua 2016 Nonlinear control for tracking and obstacleavoidance of a wheeled mobile robot withnonholonomic constraint **24** 2 1-6

[48] DilyanaBudakova, Galya Pavlova, RoumenTrifonov and Ivan Chavdarov 2019 Obstacle avoidance algorithms for mobile robots **1** 78-83

[49] TaeseokJin 2012 Obstacle Avoidance of Mobile Robot Basedon Behavior Hierarchy by Fuzzy Logic **12** 3 245-249

[50] Evangelos Papadopoulos and IoannisPoulakakis 2001 and Obstacle Avoidance for Mobile Robots **4** 3967-3972

Hybrid Microgrid: Energy Management under Different Loading Condition using Typhoon HIL's Real-time Digital Simulator

Ruchi Verma[1*], Hemant Kumar Verma[1], Priyanka Verma[1]

[1]Electrical Engineering, Department (PSE), Shri Shankaracharya Group of Institution, Junwani, Bhilai, Chhattisgarh, India.

E-mail: ruchiverma150196@gmail.com

Abstract. This paper presents real time simulation of hybrid microgrid consist AC microgrid along with DC microgrid and utility grid, where developed Hill Climbing algorithm for even power transfer between AC microgrid plus DC microgrid ensuring stable and reliable system operation under different loading conditions. Also generalized function generated to deal with the power flow from utility grid to AC microgrid and DC microgrid. Typhoon HIL control centre used for the accomplishment of hybrid micro grid. The proposed system is developed in the Schematic Editor. Hybrid microgrids are simulated using the real-time emulator HIL402 for real-time simulation within the SCADA environment.

Keywords: Hybrid micro grid, power flow management, hill climbing, interlinking converter, real-time simulation, Typhoon HIL.

1. Introduction

Due to concern of global warming phenomenon and integration of environmental issue and carbon emission lead to progressive amalgamation of distributed renewable resources in power networks. This area rapid growth increases interest in microgrid concept. Microgrid is group of electrical sources such as distributed energy resources, loads and storage connected with utility system grid system to supply power in rural areas. It operates as controllable system which can function in grid tied mode in addition to Iceland mode [2], [3].

Microgrids can classify in three types AC, DC and hybrid micro grid according to its type of connection. Among all three types, hybrid microgrid is most convenient as it consist advantage of AC as well as DC microgrid. Hybrid microgrid incorporates AC plus DC distributed units, storages and both AC and DC loads, for power conversion, incorporates interlinking converters. According to connection to the bus Hybrid micro grid can categorize in three types hybrid AC coupled microgrid, hybrid DC coupled microgrid and hybrid AC DC microgrid. Hybrid AC DC microgrid consist advantage of AC coupled and DC coupled microgrid together. **Figure.1** demonstrates the classic hybrid AC DC microgrid block diagram [4], [5].

Hardware-In-the-Loop (HIL) system is laboratory technologies which allow real-time simulation within the software to a power converter or undersized electrical power system, at the same time it can interfaced with real-world hardware. A powerful phase of technique is a power system be-designed and tested in software atmosphere and interfered with any real laboratory hardware. This accumulates time as well as money on the choice of hardware components in addition to the structure from scratch.

The merging of ICs improves system reliability and power quality [7], [9]. Uses of ICs are responsible for voltage and frequency regulation in whichever services. Endow with active-reactive power control to reach appropriate power division in together grids tied and untied mode of operation with optimized operating cost [5].

The power allotment among each connected system need a proper management of energy according to load demand. Voltage and frequency stability in microgrids in AC DCs should be maintained whether its grid tied or island mode. Therefore, a finest power control must be performed to offer sufficient power sharing among devices. Various literatures have been classified control strategies in the primary, secondary and tertiary. Primary control level concluded droop based techniques are most preferable for both operating modes [10], [11]. Secondary level classified into two strategies centralized and decentralized. For low scale micro grids centralized strategies are more useful, whereas decentralized ones are suitable for multiple users with large scale system. In tertiary level strategy centralized approaches are more common than decentralized ones [10]. In [1] discussed a new hill climbing power flow algorithm in which the verdict to prevent the perturbations is based on condition $f_{pu} \approx V_{dc\_pu}$ to stop the perturbation. This facilitates proportional power sharing in AC also DC sub grids.

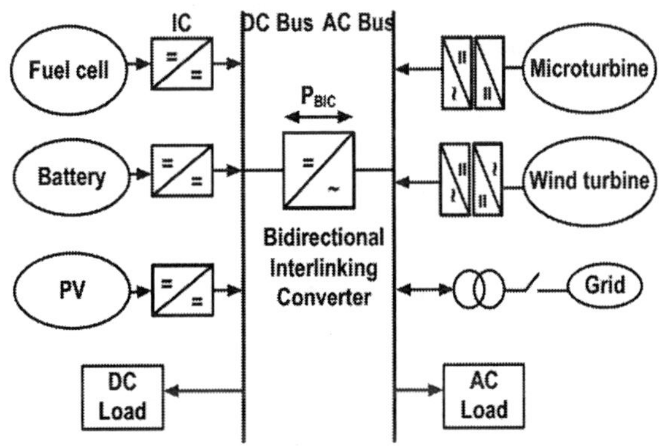

Figure.1 Classic block diagram of hybrid AC-DC micro grid.

Proportional integral (PI) controllers in hybrid microgrids are extensively used in the old days; several papers have discussed drawbacks of PI controllers. In [15] compared various adaptive controlling methods like fuzzy logic, bacterial search algorithm etc. also stated that change in operating point is a critical issue in micro grids which can be handle by using intelligent optimization algorithms and adaptive control techniques which increase the interest for adaptive control methods. Some literature shows comparison between perturb and observe algorithm and hill climbing algorithm, and due to straightforwardness, easy implementation and low down cost with faster response than P&O; hill climbing algorithm is more suitable [16], [17]. In [12],[14] elaborated about controlling hybrid AC/DC microgrid which involves energy storage and renewable energy with pulsed loads. Explained about modelling of the AC/DC Hybrid Microgrid.

A quicker manner to certify research proposals, is expected to be complementary related to microgrid modules and requires further control approaches to operate under an envelope command. Thus, the environment near the real world stands as an important obligation to build up, test and make lawful solutions for such purpose. To provide the system with each reimbursement related to micro grid approach. Therefore real-time simulation can help to establish an adequate test-bed to accelerate researches and innovations.

In literature [6] presented the real-time simulation technologies as well as their relativity to smart-grid approaches and the progress of real-time simulators, industrial as well as educational backgrounds are outlined and its current confronts are pointed. Also stated characteristics of Typhoon HIL, in which numerous processors operate in corresponding to match the marked stage at simulation runs and a host computer is placed to organize the model off-line; then compiled and loaded onto the target platform also for monitoring the results of the simulation.

[8] Presents a master / slave communication-based power management system to include PV and energy storage systems, meant for a real-world grid-interactive hybrid microgrid. For the implementation of the microgrid Typhoon HIL Control Centre is in employment with SCADA interface and the power management system.

On basis of Electrical Engineering Related tools and modelling advantage of Typhoon HIL's Virtual HIL device over other Simulation Software are as follows [18]:

It uses its own model editor developed for real time simulation and no need of third party software so no complication of integration with third party tool.

Typhoon uses FPGAs as the computational platform to achieve the precision in time from 500 nano second to 20 micro second. There is no need to develop first offline model and then retransforming the same in to real time.

Controller board programmed using any third party tool like MATLAB, PSCAD, ETAB, PSIM etc. can be interfaced directly with HIL Simulator. Hence closed loop testing platform can be created.

In the proposed work, a hybrid micro grid system is proposed which contain both AC and DC sources connected with utility grid through ICs. The system serves to both AC and DC load demand. For power flow control in ICs for sub grids a new hill climbing algorithm is applied. The power sharing under different loading conditions managed by applying generalized function. The software model is developed in Typhoon HIL schematic editor inside the SCADA environment and interfaced with Real–time emulator HIL402. The remains are structured as follow. Section II give explanation modelling and configuration of the system. The power flow of the system discussed in section III. The real time simulation with case study in session IV. Result and discussion presented in section V. session VI concluded the presented work.

2. Hybrid Microgrid Modelling and Configuration

Here considering a micro grid system which consist AC and DC sub grids both interlinked to AC main grids. There are ICs are placed for power division among the grids and sub grids. The hybrid microgrid configuration show in **Figure.2** A DC sub grid contains DC source and DC loads connected with AC sub grid containing AC source and AC load through bidirectional converter. Both AC plus DC sub grid also connected to AC main grid through ICs as shown in **Figure. 2**.

The electrical diagram of system exposed in **Figure. 3** Diagram consist main grid bus V_{abc_main} and I_{abc_main} are main grid voltage and current respectively. There are three interlinking converters are connected, first IC is regular conventional 6 bridge converter connected between utility grid along with DC sub grid named as IC_1 in diagram. Second converter IC_2 is AC-DC-AC five pulse converter, connected between DC sub grid and AC sub grid. There is control circuit box which controls the entire interlinking converters. The last one is IC_3 connected between AC sub grid and main utility grid, which is a 3×3 Matrix converter or 5 pulses AC-DC-AC converter. The current direction changes according to loading conditions.

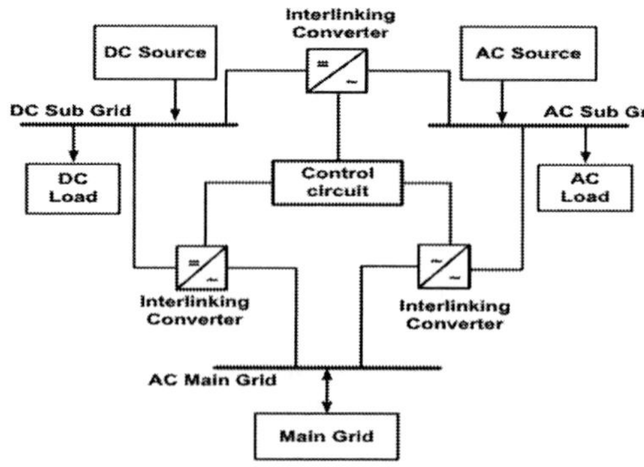

Figure. 2 Block diagram of inter connection of hybrid micro grid.

The last one is IC_3 connected between AC sub grid and main utility grid, which is a 3×3 Matrix converter or 5 pulses AC-DC-AC converter. The current direction changes according to loading conditions.

DC source droop characteristic specified as follow:

$$V_{dc} = V_{dc\_max} - n_{dc}P_{dc} \quad (1)$$

Where n_{dc} is droop coefficient, V_{dc_max} is maximum value of DC source, P_{dc} output power from DC source. AC droop characteristic for voltage also frequency is:

$$V_m = V_{m\_max} - n_{acv}Q_{ac} \quad (2)$$

$$f = f_{max} - n_{acf}P_{ac} \quad (3)$$

Where $n_{acv}$is voltage and n_{acf} is a frequency droop coefficient respectively, V_{m_max} is maximum voltage from AC source and f_{max} maximum permissible frequency. P_{ac} And Q_{ac} are active along with reactive powers of ac source respectively.

The interlinking converter (IC_{DC-AC2}) model is mathematically expressed as:

$$V_{a2\_IC} = m\frac{V_{dc}}{2}\sin(\theta + \delta) \quad (4)$$

$$V_{b2\_IC} = m\frac{V_{dc}}{2}\sin(\theta + \delta - \frac{2\pi}{3}) \quad (5)$$

$$V_{b2\_IC} = m\frac{V_{dc}}{2}\sin(\theta + \delta + \frac{2\pi}{3}) \quad (6)$$

The voltage of three phases a,b,c are expressed as equation (4) & (6). WhereV_{a2_IC}, V_{b2_IC} and V_{c2_IC} are IC voltages and δ, θ and m are power angle, instantaneous angle and modulation index respectively.

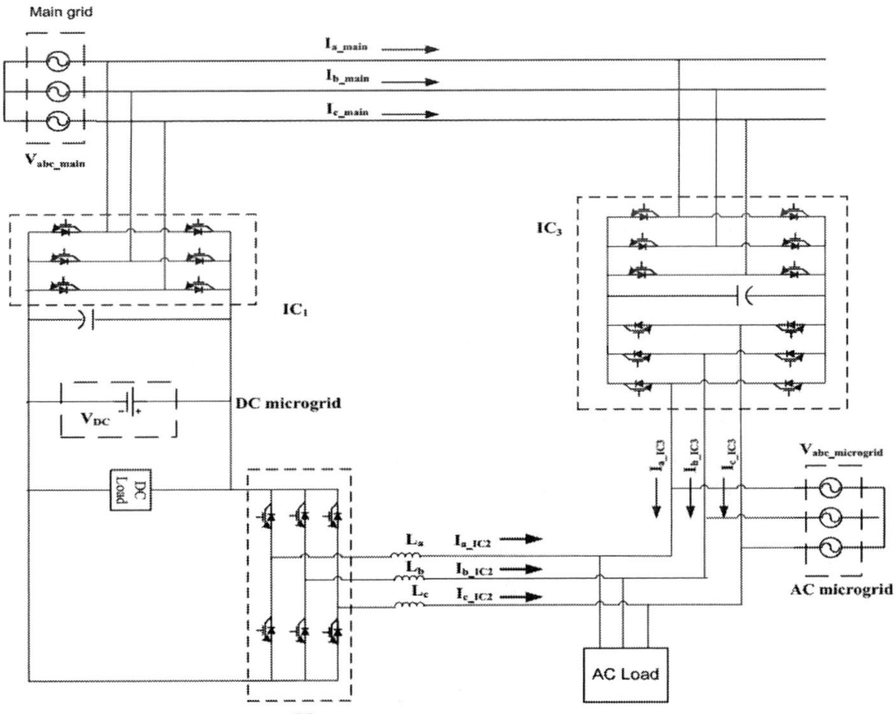

Figure. 3 Electrical diagram of hybrid micro grid system model.

In IC_{DC-AC2}, hill climbing power flow algorithm is applied for controlling action of IC as shown in flow chart in **Figure. 4**. Fast subroutines quickly determine instantaneous values of power angle and modulation index. And also called when the system observe any change in values. The HC subroutine performs perturbation in power angle.

After finishing perturbation, control transferred to Steady state where values of frequency as well as voltage are monitored. The parameters of main grid voltage, dc voltage from DC sub grid and power transferred across sub grids.

In Fast sub routine using feature scaling which is procedure to regularize variables within certain range, value of frequency and DC voltage converted into per unit values. In algorithm value of frequency and DC voltage ranged between -1 to 1.

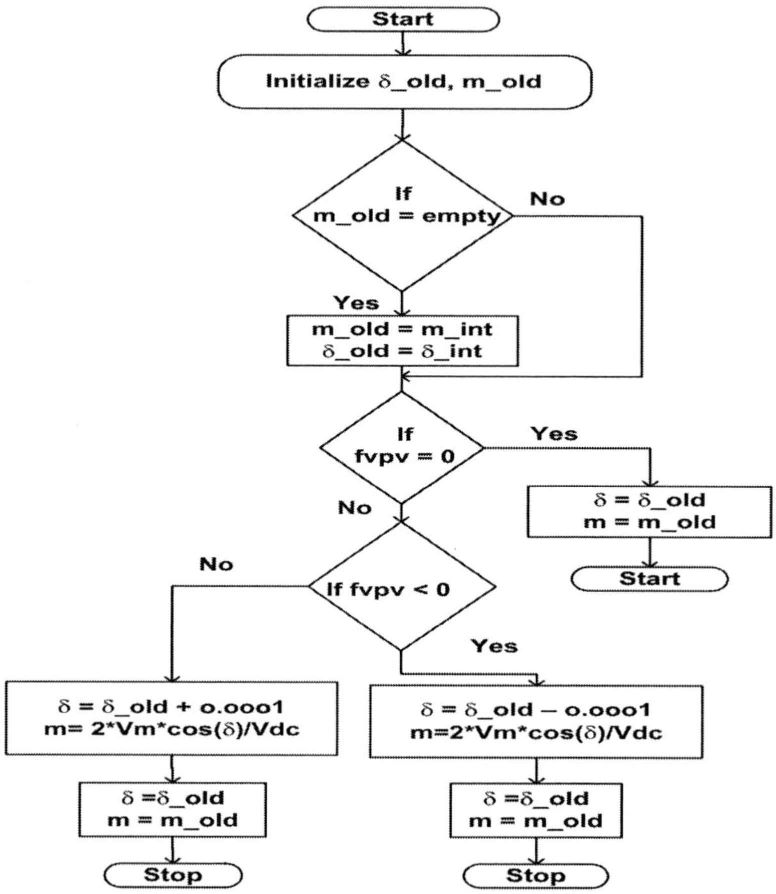

Figure. 4 Flow chart used for Hill Climbing algorithm

For the other ICs there is a MATLAB function generated for power sharing among AC sub grid, AC main grid and DC sub grid explained in section 3.

3. Methodology Power Flow Analysis of Hybrid Microgrid

In the energy system the power management is very important task to make available power according to system load demand without any time delay. In the proposed system the MATLAB function designed to grant power to each load and keep power balance in system. The whole process for IC control shown as flow chart in **Figure. 5**. The flow chart represents different loading conditions and control response of interlinking converter according to the loading conditions.

Function for main grid through which the power sharing is controlled is explained further. The activation function have a condition when input power for IC is greater than zero equation (11) then power from main grid transferred to DC micro grid. The amount of power will be calculated by equation (12). From equation (11), (12) and (13) P_I , P_{main} , P_{ac} , P_{dc} denotes power transfer from IC, main grid, AC micro grid and DC micro grid respectively. AC_{load} and DC_{Load} denotes AC and AC power demand.

$$P_I > 0 \quad (11)$$

$$P_{main} = DC_{load} - P_{dc} - P_I \tag{12}$$

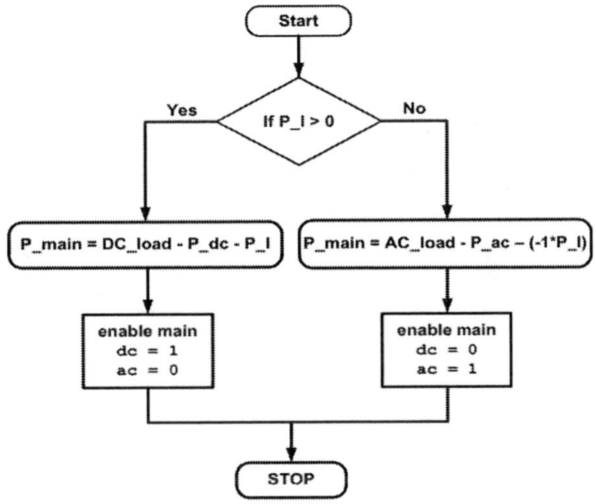

Figure. 5 Flow chart of power flow in hybrid micro grid.

This will enable power transfer as of main grid to DC micro grid. When condition as equation (11) is not satisfied then the amount of power transfer will be calculated from equation (13).

$$P_{main} = AC_{load} - P_{AC} - (-1 \times P_I) \tag{13}$$

This will enable power transfer from main grid to AC micro grid. When IC shows zero power it means load supplied from micro grids. There is no need to flow power from main grid. It means the power is flowing in normal condition. In this way the power flow control is performed.

Real-time Simulation and Case study

HIL simulation

The proposed model embraces the emulator hardware and application software that supports a diversity of model configurations. As result of circuit compiler the circuit parameters could be easily modified. To illustrate the model to be simulated in real time circuit compiler consist of a schematic editor and software to facilitate by compiling the model description into instructions for the processor. In simulation the hybrid micro grid average model is designed in schematic editor. In the proposed model the value of AC source rated 10 kW with frequency range between 49Hz to 51Hz. The DC source value rated at 10 kW along with DC voltage ranging between 590V to 615V. The interlink converter ranged up to 5 kW it means it can share power only under the range. There are some cases simulation studies carried out with different loading conditions in SCADA environment.

Real time implementation

The system hardware is simulated in real time HIL platform. The model of the hybrid microgrid is simulated using an HIL402 emulator. **Figure.6** shows the experimental setup.

In each one switching period system has to perform the following tasks:

- Hardware is connected to the computer where the hybrid microgrid model is developed in Typhoon HIL software/ schematic editor. Hardware HIL402 measures the input signals according to it creates real time environment for microgrid system.
- The system works in real time environment. The hardware consist bidirectional programme for input and output. After running the model output signals will show in SCADA panel.
- The signals shown in SCADA are real time simulation results. From SCADA panel the parameters can be changed and vary as per requirement.

IOP Conf. Series: Materials Science and Engineering 1012 (2021) 012053 doi:10.1088/1757-899X/1012/1/012053

Figure. 6 Experimental hardware setup

HIL experimental results

In hardware part model is connected to HIL402 emulator for real time simulation. Typhoon HIL402 environment permit for extraordinary control of active as well as reactive power flow, current harmonics along with grid voltage sources with arbitrary magnitude, phase and frequency.

Case 1

At case 1 all the loads are constant. Where AC and DC loads are rated up to 5 kW as exposed in **Figure. 7**. In **Figure. 7(a)** and **7(b)** shows AC and DC load demand. **Figure. 7(c)** shows micro grid power P_{ac}, **Figure. 7(d)** shows DC micro grid power P_{dc} and **Figure. 7(e)** shows main grid power P_{ac_main} respectively, **Figure. 7(f)** shows power output of IC.

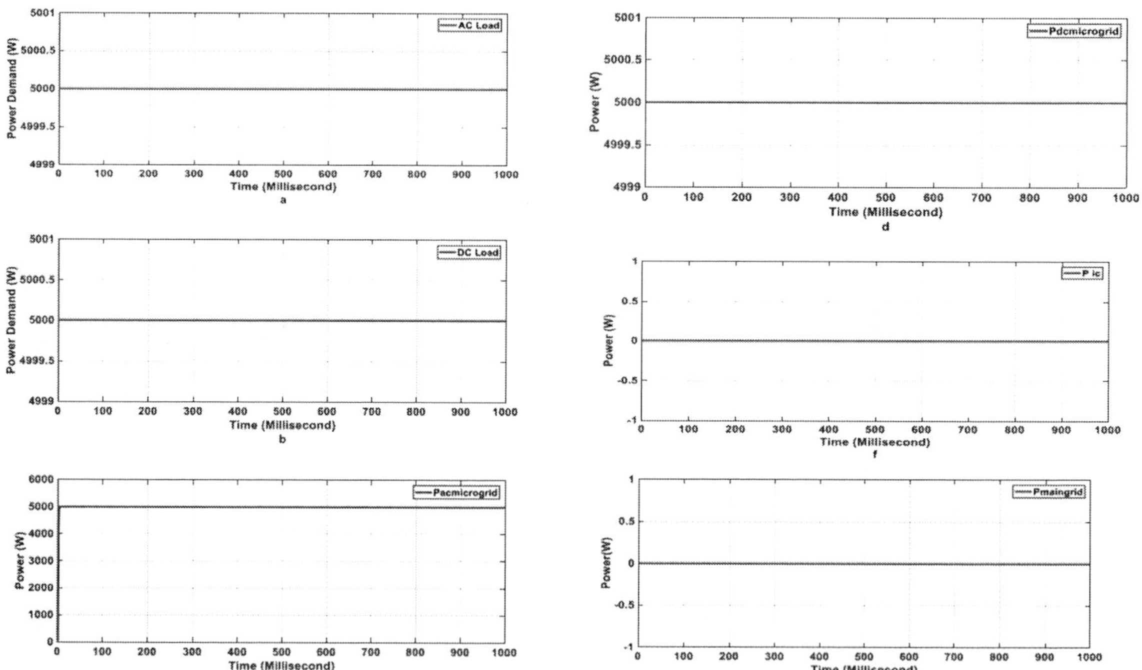

Figure.7 Output values with constant load (a) AC load and (b) DC load, (c) AC microgrid power, (d) DC microgrid power, (e) main grid power and (f) power transfer trough IC.

Figure. 7 shows event in 1 when both loads are constant rated at 5KW. AC micro grid, DC micro grid both generating 5KW power and completing load demand by own generation, therefore no power supplying from main grid and IC, both shows 0 power.

Case 2

Inside case 2 AC load is 15 kW and DC load is 5 kW as shown in **Figure. 8**. In **Figure.8(a)** show AC and (b) DC load demand. (c) Shows power output from AC micro grid power P_{ac} 9 kW, DC micro grid power P_{dc} 10 kW and main grid power P_{ac_main} 1 kW. Power output through IC is 5 kW and –ve sign shows power transferring from DC microgrid to AC load.

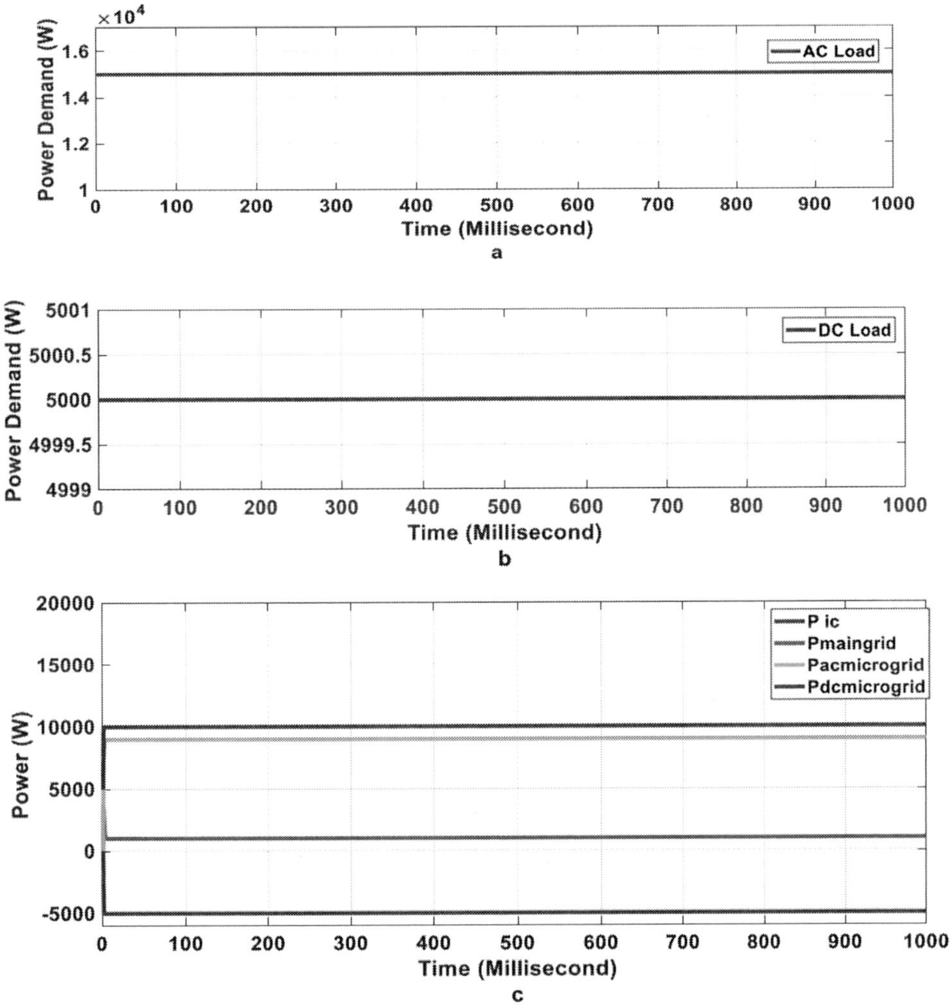

Figure. 8. Output values with variable load (a) AC load, (b) DC load, (c) AC microgrid power, DC microgrid power, main grid power and power transfer trough IC.

Case 3

In case 3 AC load is 15 kW and DC load is 10 kW as show in **Figure. 9**. In **Figure. 9(a)** shows AC and **(b)** DC load demand. (bc shows power output from AC micro grid power P_{ac} 9 kW, DC micro grid power P_{dc} 10 kW and main grid power P_{ac_main} 5 kW. Power output through IC is 5 kW and –ve sign shows power transfering from DC microgrid to AC load.

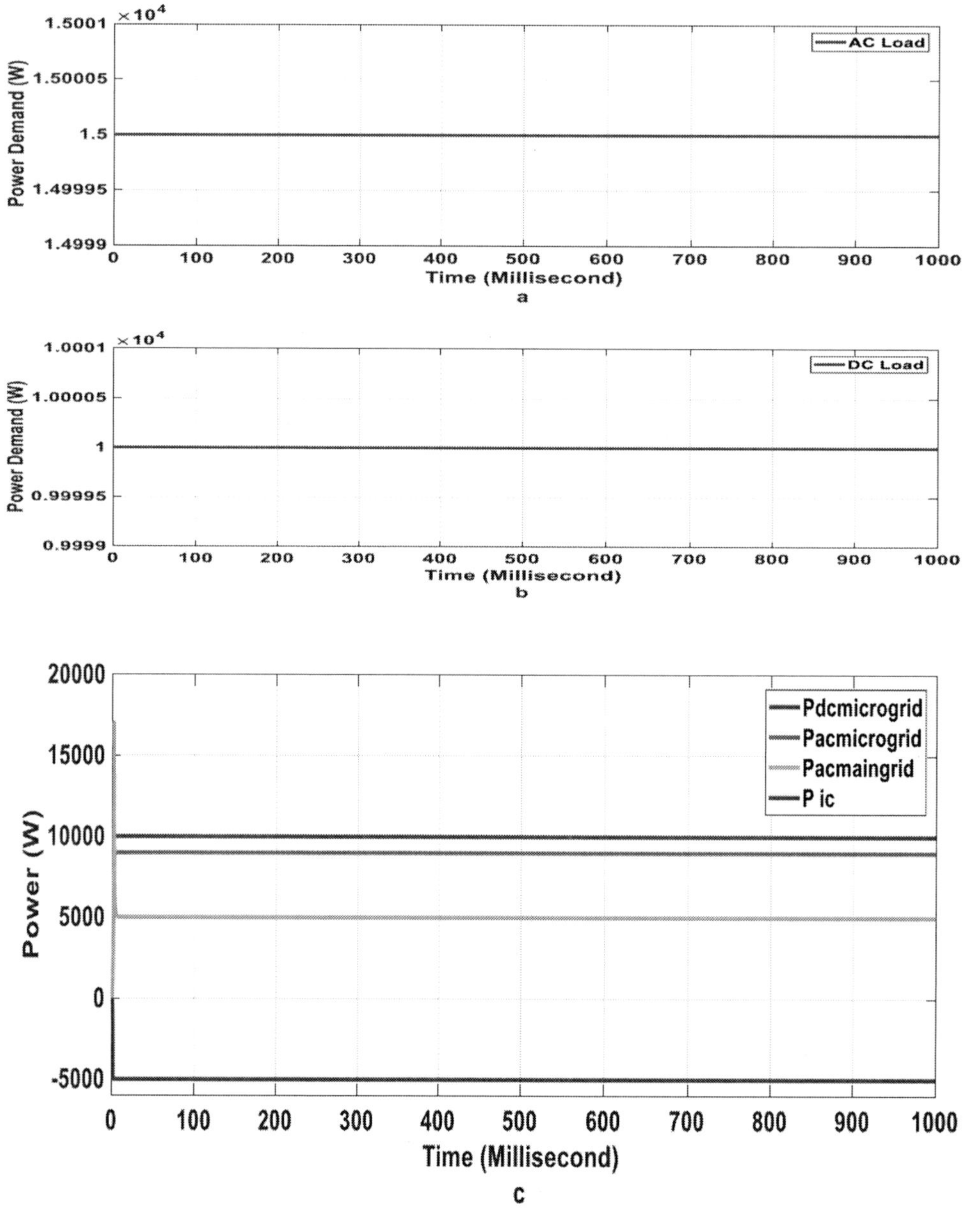

Figure. 9 Output values (a) AC load and (b) DC load, (c) AC microgrid power, DC microgrid power, main grid power and power transfer trough IC.

Case 4

In case 4 AC load is 10 kW and DC load is 5 kW as revealed in **Figure. 10**. In **Figure. 10(a)** shows AC and (b) DC load demand (c) Shows power output from AC micro grid power P_{ac} 5 kW, DC micro grid power P_{dc} 10 kW and main grid power P_{ac_main} 0 kW. Power output through IC is 5 kW and –ve sign shows power transferring from DC microgrid to AC load.

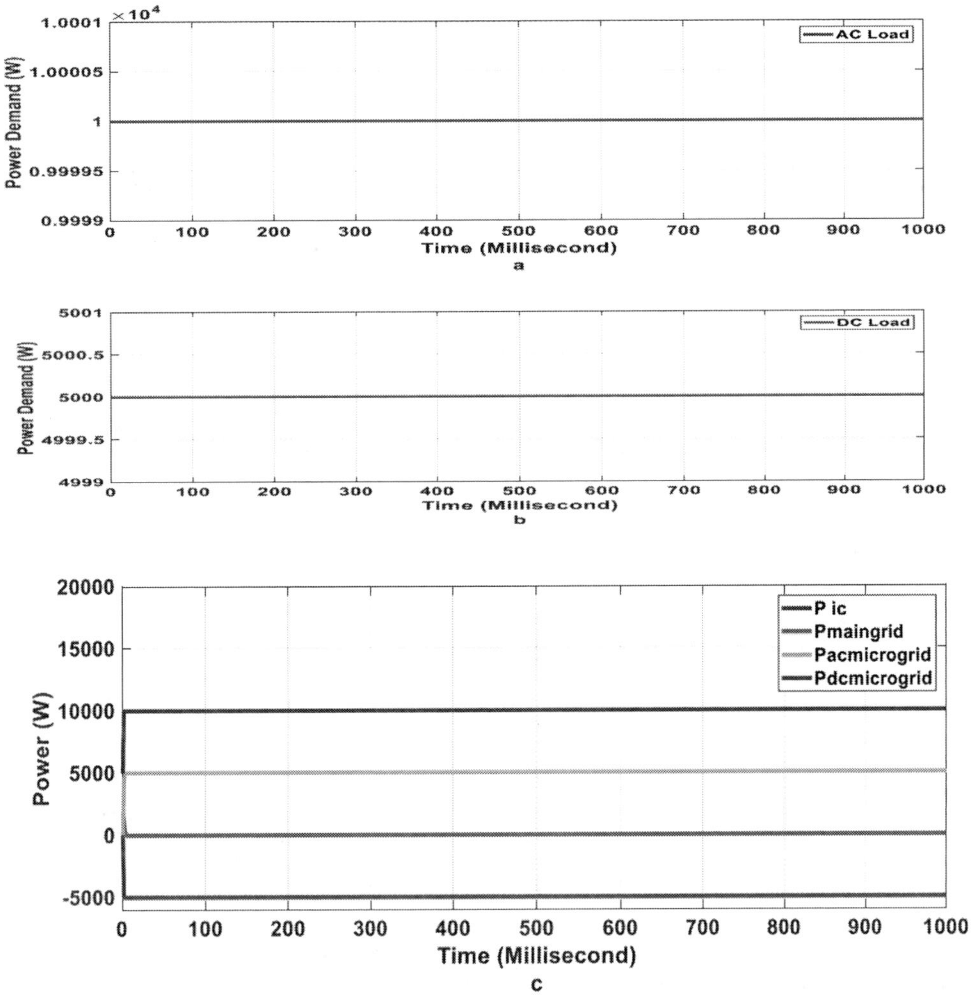

Figure.10 Output values (a) AC load and (b) DC load, (c) AC microgrid power, DC microgrid power, main grid power and power transfer trough IC.

Case 5

In case 5 AC load is 5 kW with DC load is 15 kW as exposed in **Figure. 11**. In **Figure. 11(a)** and **11(b)** shows AC and DC load demand. **Figure.**11**(c)** shows power output from AC micro grid power P_{ac} 10 kW, DC micro grid power P_{dc} 10 kW and main grid power P_{ac_main} 0 kW. Power output through IC is 5 kW and –ve sign shows power transfering from AC microgrid to DC load.

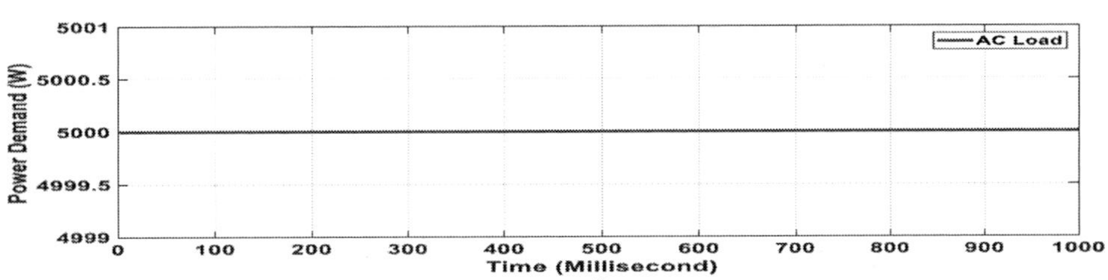

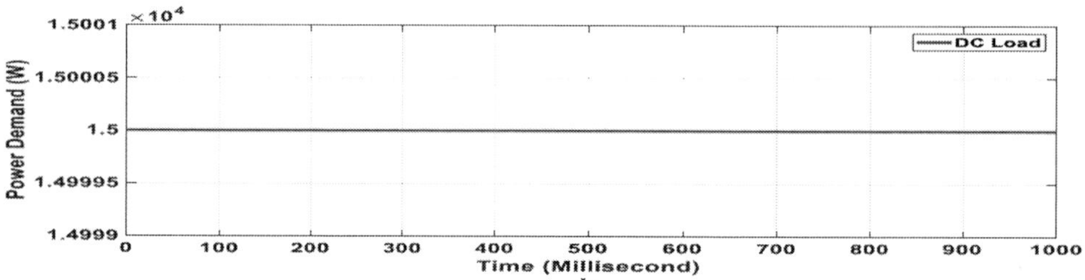

b

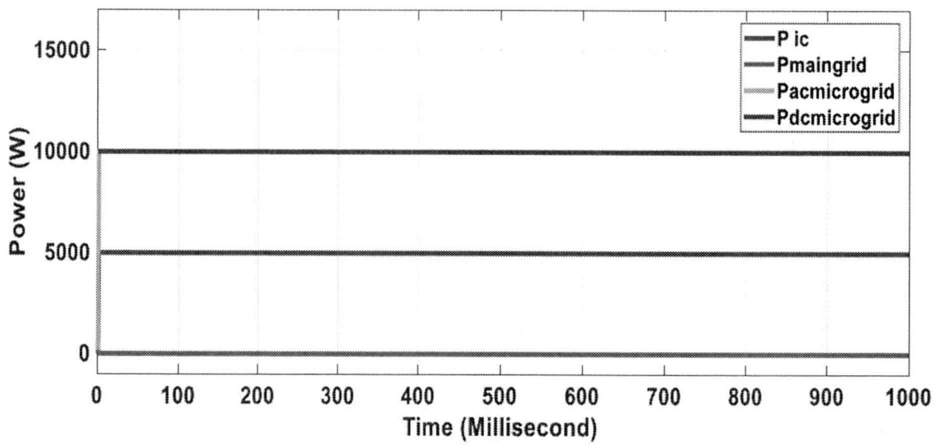

c

Figure.11 Output values (a) AC load and (b) DC load, (c) AC microgrid power, DC microgrid power, main grid power and power transfer trough IC.

TABLE I

EVENTS DURING DIFFERENT CASES

Case	**Loading Conditions**		**Change in power**			
	AC Load (KW)	***DC Load (KW)***	***P_{dc} (KW)***	***P_{ac} (KW)***	***P_{ac_main} (KW)***	***P_I (KW)***
1	5	5	5	5	0	0
2	15	5	10	9	1	-5
3	15	10	10	9	5	-5
4	10	5	10	5	0	-5
5	5	15	10	10	0	5

Table 1 shows events during different loading cases. The power management according to load is successfully done by three sources AC, DC micro grid and main utility grid. All ICs performing synchronously as per load demand weather its constant or variable.

4. Result and discussion

The purpose of presented paper is to perform real time power management within hybrid micro grid via hill climbing algorithm. In this system consist more complex structure and there is another function for power flow the whole system is developed in real time environment using HIL402 hardware. In our system power flow manage among AC micro grid, DC micro grid, main grid with AC load, DC load, different loading conditions which is more complex. Now challenge is hill climbing algorithm has to work to manage power among three sub grid system and it is performing that. Hence, this paper shows the accessibility of hill climbing with large, complex system and also synchronized work with other functions also incorporates benefits of Typhoon HIL real time emulation.

5. Conclusion

In this paper, in real time environment a coordinated control of hybrid micro grid is presented. There are two sub grid systems AC grid and DC grid system, called as sub grid system. Both connected with AC main grid called as utility grid. Both sub grids are interlinked with interlinking converter where Hill climbing control algorithm is applied for proportional power division in AC along with DC sub grids. To control power flow from utility grid a generalized function is generated. To investigate and verify the system a generalized Typhoon HIL software model is developed in HIL SCADA environment. To show the system power flow different loading conditions are made and the simulated model is emulated in HIL402 and results are observed which validated the performance of the control strategy.

References

[1] O. Khan, S. Acharya, M. Al Hosani and M. S. El Moursi. 2018. Hill Climbing Power Flow Algorithm for Hybrid DC/AC Microgrids. *IEEE Transactions on Power Electronics*, Volume: **33**, 5532 - 5537.

[2] Ander Ordono, Eneko Unamuno and Jon Andoni Barrena. 2019. Interlinking converters and their contribution to primary regulation: a Review. *Electrical Power and Energy Systems* ***111***, 44-57.

[3] Saroja Kanti Sahoo, A. K. Sinha and N. K. Kishore. 2018. Control Techniques in AC, DC, and Hybrid AC-DC Microgrid: A Review. *IEEE Journal of Emerging and Selected Topics in Power Electronics, Volume:* ***6***, 738 – 759.

[4] Eneko Unamuno and Jon Andoni Barrena. 2015. Hybrid ac/dc microgrids—Part I: Review and classification of topologies. *Renewable and Sustainable Energy Reviews* ***52***, 1251-1259.

[5] Quan Zhou, Mohammad Shahidehpour and Zhiyi Li. 2019. Two-Layer Control Scheme for Maintaining the Frequency/Voltage and the Optimal Economic Operation of Hybrid AC/DC Microgrids. *IEEE Transactions on Power Systems, Volume:* ***34***, 64 - 75.

[6] Luis Ibarra and Antonio Rosales. 2017. Overview of Real-Time Simulation as a Supporting Effort to Smart-Grid Attainment. *Energies, 10*(6),817.

[7] Lihu Jia, Yongqiang Zhu and Shaofei Du. 2018. Analysis of the Transition between Multiple Operational Modes for Hybrid AC/DC Microgrids. *CSEE Journal Of Power And Energy Systems, Volume:* ***4***, 49 – 57.

[8] Hooman Ekhteraei Toosi, Adel Merabet.2019. Central Power Management System for Hybrid PV/Battery AC-Bus Microgrid Using Typhoon HIL. *IEEE 28th International Symposium on Industrial Electronics (ISIE), Canada,* 10.1109/ISIE.2019.8781277.

[9] Ajay Gupta, Suryanarayana Doolla and Kishore Chatterjee. 2017. Hybrid AC-DC Microgrid: Systematic Evaluation of Control Strategies. *IEEE Transactions on Smart Grid, Volume:* ***9,*** 3830 - 3843.

[10] Eneko Unamuno and Jon And oni Barrena. 2015. Hybrid ac/dc microgrids-Part II: Review and classification of control strategies. *Renewable and Sustainable Energy Reviews* ***52,*** 1123-1134.

[11] Josep M. Guerrero, Juan C. Vasquez and José Matas. 2011. Hierarchical Control of Droop-Controlled AC and DC Microgrids—A General Approach Toward Standardization. Ieee *Transactions On Industrial Electronics, Volume:* ***58,*** 158 - 172.

[12] Tan Ma, Mehmet Hazar Cintuglu and Osama Mohammed. 2015. Control of Hybrid AC/DC Microgrid Involving Energy Storage, Renewable Energy and Pulsed Loads", *IEEE Industry Applications Society Annual Meeting*, 10.1109/IAS.2015.7356857.

[13] Peng Wang, Chi Jin and Dexuan Zhu. 2014. Distributed Control for Autonomous Operation of a Three-Port AC/DC/DS Hybrid Microgrid.*IEEE Transactions on Industrial Electronics, Volume:* ***62,*** 1279 - 1290.

[14] Hao Zheng, Hongwei Ma and Kaiqi Ma. 2017. Modeling and Analysis of the AC / DC Hybrid Microgrid with Bidirectional Power Flow Controller. *China International Electrical and Energy Conference (CIEEC)*, 25-27.

[15] Magdi S. Mahmoud and Nezar M. Alyazidi. 2017. Adaptive intelligent techniques for microgrid control systems: A survey. *International Journal of Electrical Power & Energy Systems*, Volume **90**, 292-305.

[16] Sabah B. Siad and Ahmad Malkawi. 2019. Nonlinear control of a DC MicroGrid for the integration of distributed generation based on different time scales. *International Journal of Electrical Power & Energy Systems, Volume* ***111***, 93-100.

[17] Fangrui Liu, Yong Kang, and Yu Zhang .2008.Comparison of P&O and Hill Climbing MPPT Methods for Grid-Connected PV Converter. *3rd IEEE Conference on Industrial Electronics and Applications, Singapore,* 10.1109/ICIEA.2008.4582626.

[18] Ruchi Verma and Hemant Kumar Verma. 2019. Hybrid Microgrid: Energy Management under Different Loading Condition. *I-Manager's Journal on Power Systems Engineering; Nagercoil* Vol. **7**, Iss. 3, 36-44.

[19] R. Swarnkar, H. K. Verma and R. N. Patel. 2019. Comparative Analysis of Isolated and Non-Isolated Bi-Directional DC-DC Converters for DC Microgrid. *3rd International Conference on Recent Developments in Control, Automation & Power Engineering (RDCAPE), NOIDA, India,* 2019, pp. 557-562.

[20] Microgrid Testbed. Available online: https://www.typhoonhil.com /doc/brochures/Typhoon-HIL-and-SPS-Microgrid-Testbed.pdf.

PSO Based Selective Harmonic Elimination PWM for Cascaded H-Bridge Multilevel Inverter

M.S.Sivagama Sundari[1*], D.Periyasamy[2]

[1*]Department of Electrical and Electronics Engineering, Amrita College of Engineering and Technology, Erachakulam, Nagercoil – 629901
[2]Department of Electrical and Electronics Engineering, Imayam College of Engineering , Tiruchirappalli – 621206

E-mail: mssivagamasundari@gmail.com

Abstract. A Single phase Cascaded H-bridge multilevel inverter topology fed by Photovoltaic source for the reduction of harmonics is proposed in this paper. Selective Harmonic Elimination technique is employed for elimination of 5th, 7th, 11th and 13th lower order harmonics and the particle swarm optimization algorithm is employed for estimating the switching angles. As a novelty, in order to maintain the power balance among the various sources, a cyclic selection procedure is adopted that over a period of operation also the sources equally contribute for the output power. MATLAB software is used for optimization and the results show the effectiveness of the proposed algorithm. The performance of the Particle Swarm Optimization Algorithm has been compared against the performance of the Genetic Algorithm.

Keywords: multilevel inverter, cascaded h-bridge multilevel inverter, total harmonic distortion, selective harmonic elimination-pulse width modulation, particle swarm optimization

1. Introduction

In recent years, electrical energy consumption is increasing rapidly due to the increasing demand of energy in the world. As a result of this, numerous fossil fuel sources and other resources which have great contribution in global warming due to greenhouse gases emission are largely consumed. Due to these negative impacts caused by these resources, a rapid progress in finding an alternative and renewable energy has attained a significant interest in the area of researches to eliminate the shortage of fossil fuels and reduce the global warming concern. In the last three decades, renewable energy has become a challenging field and many researchers have made renewable energy the main focus in order to create new sustainable, natural abundance and environmental friendly nature energy resources. This interest of environmentally friendly power will proceed in future in light of the fact that the creation of sustainable power sources are required to fulfill 20% and half of the absolute energy needs in 2020 and 2050 separately. In 2015, worldwide last energy utilization which originated from environmentally friendly power sources is around 22.8% and the portion of environmentally friendly power sources incorporating hydro power in power age is about 28%. Because of the chance of giving energy less reliance on the petroleum products, various types of renewable energy resources such as solar, wind, tidal, waves, geothermal heat, hydropower, especially solar photovoltaic conversion have gained increased acceptance and growth in recent times[1]. Significant advantages of PV panels include clean and reliable energy production and suitability for distributed generation. In addition, the prices for Photovoltaic modules are drastically decreasing. In fact, it is expected that the PV global installed capacity can reach upto 3000GW, approximately 11% of the global electricity production by 2050. [2]

An inverter is an important component used to convert DC power into AC power from the PV panels in the PV system which is to be given to the grid. PV inverters types have been introduced and explained [3-6]. The cascaded H-bridge multilevel inverter needed a separate DC source for individual H-bridge and have been used for high power and high voltage applications. Also this topology is suitable for medium and large grid-connected PV systems [7-10]. In addition to that, isolated dc sources in the multilevel inverter circuit make their own voltage control. At the same time, the cascaded h-bridge multilevel inverter has given great care due to its simplicity of control and modularity and as a suitable candidate for the forthcoming generation of reliable grid-connected PV system. In solar photovoltaic fed fifteen level Cascaded Multilevel Inverter(CMLI) [11], an optimized harmonic stepped waveform technique is introduced to eliminate lower order harmonics. In eleven level inverter[12], optimum relevance has been found out between switching angles and dc voltages to achieve minimum total harmonic distortion.

In this paper, the PSO algorithm is employed to minimize lower order harmonics and fundamental component is to be satisfied. The performance of the PSO Algorithm has been compared against the performance of the Genetic Algorithm. The novelty uses an automatic sequential scheme such that the total energy delivered the state of charge of the batteries and the terminal voltages are all equal in all the DC sources

2. Cascaded H-bridge multilevel Inverter topology

The block diagram of proposed topology contains PV source from solar panel, boost converter, multilevel inverter, battery and a filter is shown in **figure.1.** Multi-string technology with PV modules is employed. The PV array string is linked together to a boost converter with maximum power point tracking. The power produced by the PV modules is continuously changing due to weather conditions, since the irradiance is not constant during the whole day. By using the Perturb and Observe P & O algorithm, maximum power is extracted from the PV array. The eleven level inverter output is AC voltage which is linked together with the load.

.

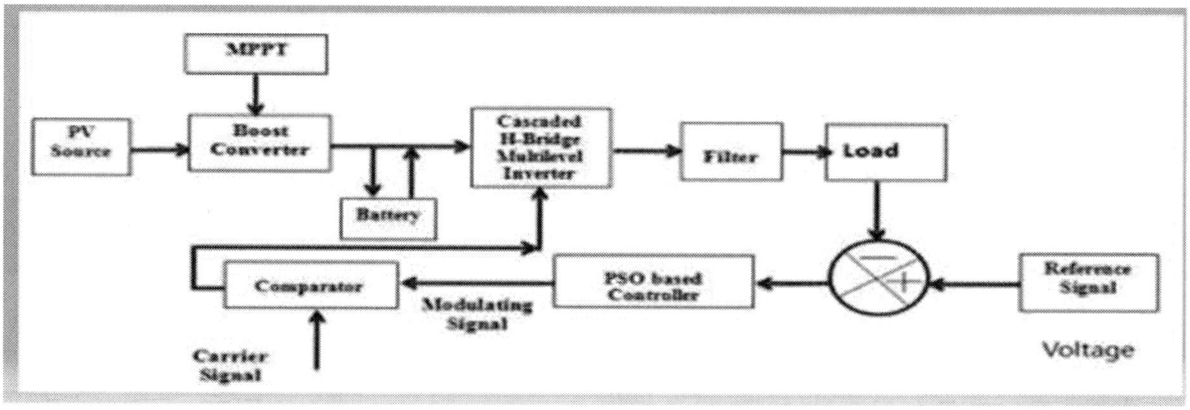

Figure 1. Block diagram of the proposed topology.

The proposed topology is shown in **figure.2.** Five H-bridge inverters are connected in series to form a single-phase 11-level inverter. The H-bridge inverter output voltage gives vHT with 5 levels by providing PV panels with equal dc link voltage.

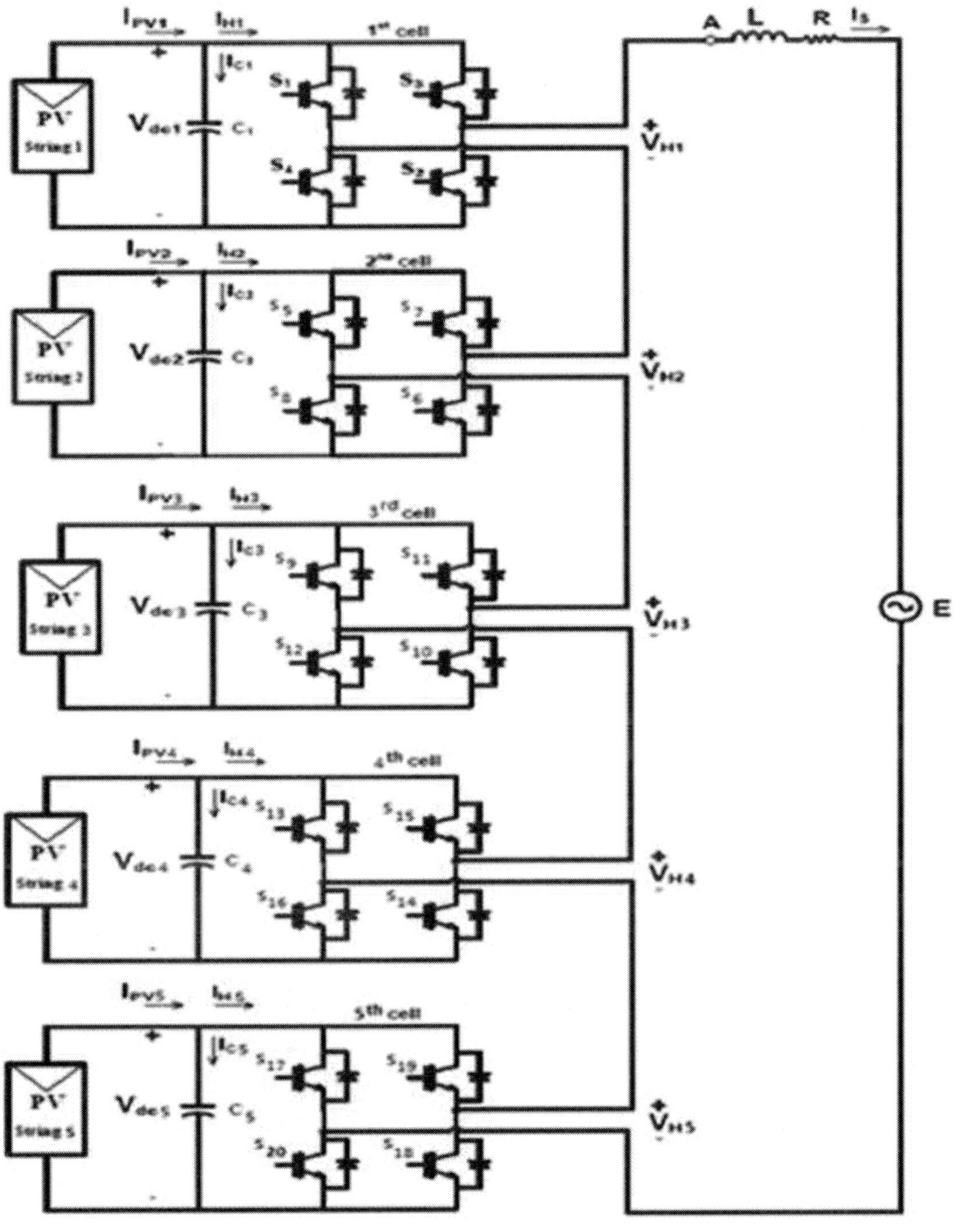

Figure 2. Solar fed Cascaded h-bridge inverter topology

3. Particle Swarm Optimization Algorithm

The particle swarm optimization (PSO) algorithm was employed for estimating the switching angles θ1, θ2, θ3, θ4 and θ5 which will result to the lower order harmonics 5th, 7th, 11th and 13th elimination in a multilevel inverter. An objective function uses the set of five trigonometric transcendental equations for harmonics elimination.

4. Selective Harmonic Elimination Pulse Width Modulation

To eliminate 5th, 7th, 11th and 13th lower order harmonics, mathematical methods are used in SHE PWM technique. This technique is employed to find out the switching angles such as θ_1, θ_2, θ_3,..., θ_N so that amplitude of fundamental voltage is satisfied and (N-1) odd harmonics can be minimized. The harmonic components in the multilevel waveform are as follows:

* amplitude of dc component is equal to zero
* amplitude of fundamental component, n = 1, and odd harmonics amplitude are as follows:

$$h_1 = 4Vdc/\pi \cos\theta_k \; s_k=1 \text{ , and}$$

$$h_n = 4Vdc/n\pi \cos n\theta_k \; s_k=1$$

* amplitude of all even harmonics are equal to zero

In the quarter-wave waveform of symmetrical multilevel inverter, only odd harmonics is to be minimized. To control the amplitude of fundamental component and to minimize 5th, 7th, 11th and 13th lower order harmonics, switching angles θ_1, θ_2, θ_3, θ_3, θ_4 and θ_5 are calculated for a particular modulation index to reduce the harmonics and the nonlinear transcendental equations must be calculated.

$$\cos\theta_1 + \cos\theta_2 + \cos\theta_3 + \cos\theta_4 + \cos\theta_5 = 3\pi M/4 \quad (1)$$

$$\cos 5\theta_1 + \cos 5\theta_2 + \cos 5\theta_3 + \cos 5\theta_4 + \cos 5\theta_5 = 0 \quad (2)$$

$$\cos 7\theta_1 + \cos 7\theta_2 + \cos 7\theta_3 + \cos 7\theta_4 + \cos 7\theta_5 = 0 \quad (3)$$

$$\cos 11\theta_1 + \cos 11\theta_2 + \cos 11\theta_3 + \cos 11\theta_4 + \cos 11\theta_5 = 0 \quad (4)$$

$$\cos 13\theta_1 + \cos 13\theta_2 + \cos 13\theta_3 + \cos 13\theta_4 + \cos 13\theta_5 = 0 \quad (5)$$

The modulation index is given by

$$M = h_1 sV \quad (6)$$

Where, h_1 is the amplitude of fundamental component. By changing the modulation index from equation (6), the amplitude of fundamental component is controlled.

5. Power balance between the H-Bridge Inverters

In a cascaded h-bridge multilevel inverter operated with a number of isolated PV sources, with step modulation there is a possibility that the individual PV sources and the corresponding inverter units are not equally loaded. This may cause the sources that deliver more power to get more burdened. In order to avoid this situation a cyclic change over system was employed. Also the five H-bridge units were cyclically shifted periodically over every cycle so that fairly uniform loading could be affected on each inverter unit and the corresponding power supply.

6. Simulation Studies

The performance of the SHE PWM based cascaded h-bridge inverter topology with isolated PV dc sources is determined through MATLAB/SIMULINK software. The simulation elements and the parameters are given in Table 1

The simulink model of the proposed system is shown in **figure.3.** The simulation circuit contains five H-bridges and PWM block contain parameters as pulse width period, phase delay and amplitude. The obtained output voltage is eleven level stepped waveform and reduces the harmonics.

Table 1. Parameters of the Cascaded H-Bridge Inverter

Parameters	**Values**
No. of H-Bridge levels	5
No. of Switches	20
PV source voltage for individual H- bridge	20 V
Fundamental frequency	50Hz
Load resistor	100 Ohm
Load Inductor	40mH

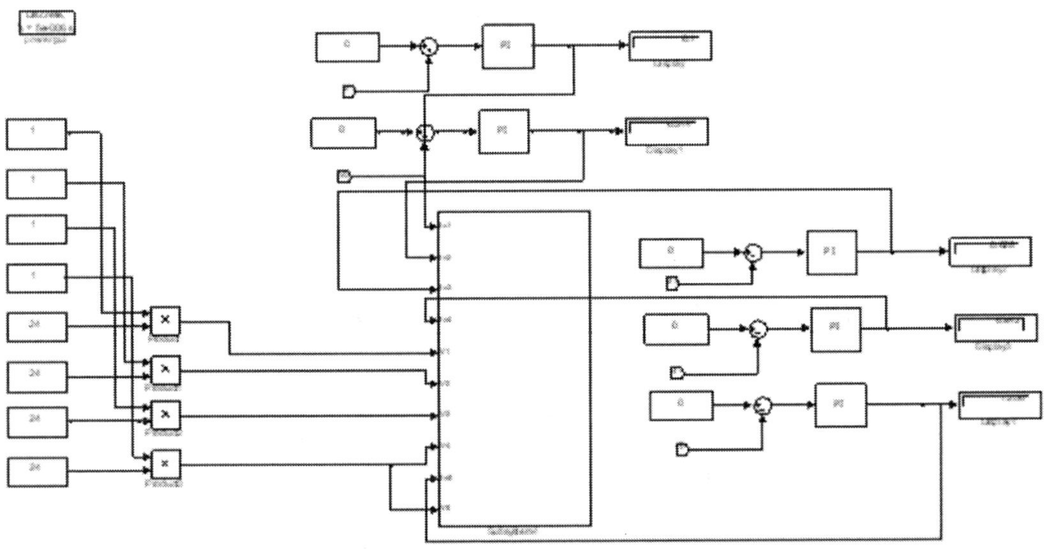

Figure 3. Simulation model of cascaded h-bridge eleven level inverter topology

The results are taken for modulation index M = 0.71 in MATLAB software . The Output voltage and current waveform of Cascaded h-bridge eleven level inverter are shown in **figure. 4 & 5.**

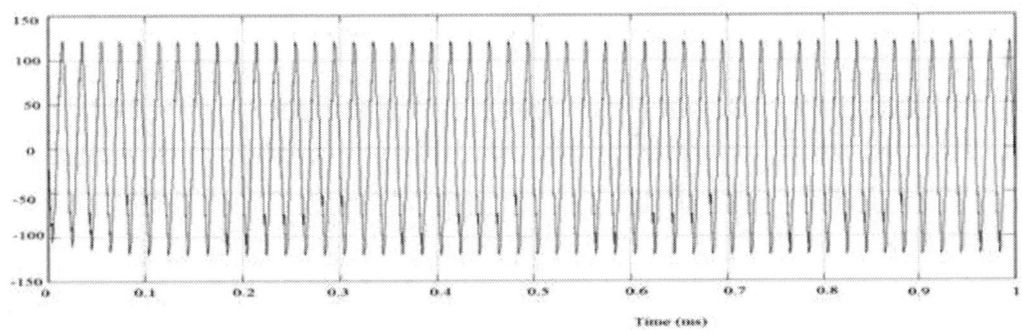

Figure 4. Output voltage waveform

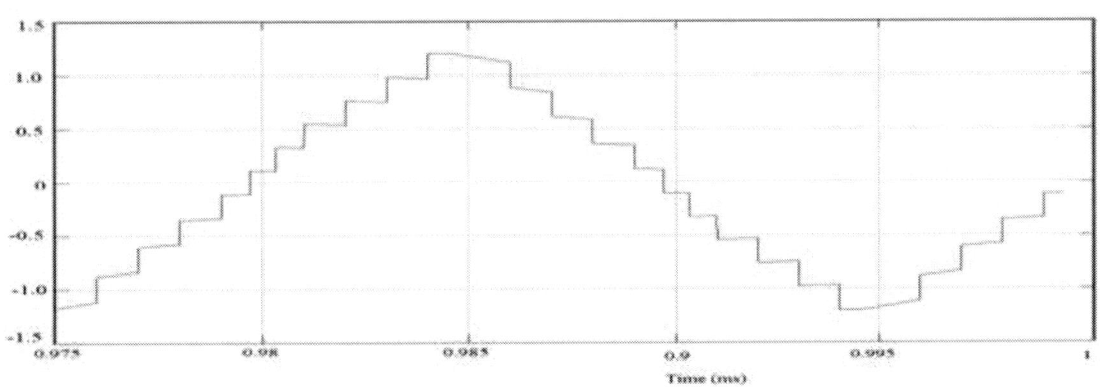

Figure 5. Output current waveform

THD is the summation of all consonant parts of the current or voltage waveform. The quantity of music to be diminished is equivalent to the quantity of exchanging points - 1 are acquired.

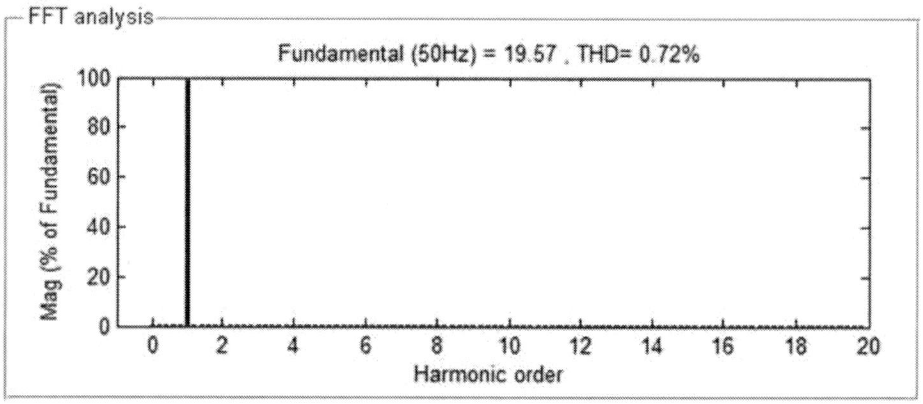

Figure 6. Harmonic spectrum of the inverter output current

From the above harmonic analysis shown in **figure.6**, lower order harmonics such as 5th, 7th, 11th, and 13th magnitudes are very small. Furthermore, the solar panel parameters are listed in Table 2.

Table 2. Parameters of the Solar Panels

	Parameters	Values
DC Side	MPP voltage	17.7V
	MPP current	6.5A
	Open circuit voltage	22.2V
	Short circuit current	7.97A
	Temperature	25°C
	Solar Irradiance	1000W/m^2
AC Side	Rated grid voltage	230V
	Rated grid current	0.5A
	Switching frequency	20Khz
	Inductor filter	0.9µH

Table 3. Comparison of output current THD

Modulation Index M	Switching angles in terms of radians					THD(%)	
	θ_1	θ_2	θ_3	θ_4	θ_5	Simulation	Theoretical
0.51	0.79	0.89	1.06	1.26	1.34	6.54	6.02
0.63	0.60	0.66	0.87	1.03	1.12	4.22	4.08
0.67	0.44	0.56	0.75	0.98	1.08	2.25	2.05
0.71	0.25	0.42	0.66	0.94	1.02	0.72	0.62

The comparison of output current THD for various modulation indices are indicated in table 3. The performance of the PSO Algorithm has been compared against the performance of the Genetic Algorithm. For the same modulation index the results obtained in PSO are compared with that obtained with GA in table 4.

Table 4. Comparison of output current THD (proposed algorithm with existing algorithm

Modulation Index M	THD (%) PSO		THD (%) GA	
	Simulation	Theoretical	Simulation	Theoretical
0.51	6.54	6.02	7.78	7.12
0.63	4.22	4.08	6.04	5.25
0.67	2.25	2.05	4.68	4.01
0.71	0.72	0.62	3.02	2.55

The versatility of the proposed logic is that it can be easily extended for multi-level inverters of any size. For example the same procedure can be used if the proposed idea is design a 15 level inverter or more or a 9 level inverter or less. The only issue will be to formulate the objective function with the required number of variables that influence the removal of harmonics. If the requirement is a 9 level inverter then the number of H bridges will be 4 and 4 switching angles will have to be determined. For the purpose of eliminating three harmonics and to ensure the magnitude of the fundamental to the required level four equations will be formulated to describe the SHE system. Then these four equations will be used to form the objective function to find the four switching angles.

Similarly a 15 level inverter can be constructed using 7 numbers of H bridges and 7 number of switching angles will have to estimate. An objective function incorporating the seven switching angles as the variables to be estimated will be formed.

7. Conclusions

In this paper, elimination of fifth, seventh, eleventh and thirteenth lower request music utilizing SHEPWM system is examined. To tackle the SHE issue and to discover the exchanging points, PSO algorithm is employed with equal dc sources in cascaded H-bridge multilevel inverters. Using an automatic sequential scheme, with the equally distributed fall of state of charge of the batteries, the batteries would exhibit the same terminal voltage thus the switching angle estimation would not be affected. Simulation results are introduced for an eleven-level H-connect inverter to affirm the specific hypothetical results.. The selected area of work opens up new avenues for further research. The different directions of research include usage of unequal DC sources for the cascaded H bridges, asymmetrical operation of the bridges with voltage ratios in the order of 1:3:9:81 or 1:2:4:8 etc. for minimal switch operation, enhancement of levels of the inverter and implementation of much modern search algorithms. Besides since most of the applications require real time control of modulation index the avenues for research in this direction are also open. Proposed topology produces high quality output voltage and gives better current total harmonic distortion.

References

[1] J. M. Carrasco, L. G. Franquelo, J. T. Bialasiewicz, E. Galvan, R. C. Portillo Guisado, M. A. M. Prats, J. I. Leon, and N.Moreno-Alfonso. 2006 Power-electronic systems for the grid integration of renewable energy sources: A survey. IEEE Transactions on Industrial Electronics **53(4)**1002–1016.

[2] 2011 SEPA Utility Solar Rankings [Online]. Available: www.solarelectricpower.org

[3] S. B. Kjaer, J. K. Pedersen, and F. Blaabjerg. 2005 A review of single phase grid connected inverters for photovoltaic modules.IEEE Transactions on Industrial Applications **41(5)** 1292–1306.

[4] J. M. A. Myrzik and M. Calais. 2003String and module integrated inverters for single-phase grid connected photovoltaic systems—A review in Proceedings IEEE Bologna Power Tech Conference **2** 8.

[5] S. Daher, J. Schmid, and F. L.M. Antunes. 2008 Multilevel inverter topologies for stand-alone PV systems. IEEE Transactions on Industrial Electronics **55 (7)** 2703—2712.

[6] R. Mechouma, B. Azoui, and M. Chaabane.2012 Three-phase grid connected inverter for photovoltaic systems, a review in Renewable Energies and Vehicular Technology (REVET), First International Conference 37-42.
[7] L. M. Tolbert, F. Z. Peng. 2000 Multilevel converters as a utility interface for renewable energy systems in IEEE Power Engineering Society Summer Meeting, Seattle, Washington 1271-1274.
[8] H. Ertl, J. Kolar, and F. Zach. 2002 A novel multicellular DC-AC converter for applications in renewable energy systems. IEEE Transactions on Industrial Electronics **49(5)**1048-1057
[9] F. Filho, Y. Cao, and L. M. Tolbert 2010 11-level cascaded H-bridge grid-tied inverter interface with solar panels in IEEE Procedings Applied Power Electronics Conference and Exposition (APEC) 968-972.
[10] C. D. Townsend, T. J. Summers, and R. E. Betz. 2012 Control and modulation scheme for a cascaded H-bridge multi-level converter in large scale photovoltaic systems in IEEE Proceedings Energy ConversionCongress and Exposition (ECCE) 3707 – 3714.
[11] S.Albert Alexander,Manigandan Thathan 2015 Optimal Harmonic Stepped Waveform Technique for Solar Fed Cascaded Multilevel Inverter. Journal of Electrical Engoneering Technology **10** 742-751
[12] Firouzkouhi 2019 Control of Cascaded H-bridge Multilevel inverter based on optimum regulation of switching angles, with FPGA implementation European journal of electrical and computer engineering **3(3)** 1-5.

Challenges in p-type Oxide Based Thin Film Transistor

Keka Mukhopadhyaya[1,2*], Srividya P[1]

[1]Department of ECE, RV College of Engineering, Bengaluru, Karnataka, INDIA
[2]Department of EEE, CMR Institute of Technology, Bengaluru, Karnataka, INDIA

*E-mail: keka.m@cmrit.ac.in

Abstract- p-type Thin Film Transistor (TFT) exhibits performance characteristics which is inferior to its n-type counterpart, which in turn limited its scope of wide application mainly in the domain of transparent and flexible electronics. This paper thoroughly reviews the basic operation of TFT, different parameters used to evaluate performance. A comparison of performance analysis of different oxide based p-type TFT viz. copper oxide, tin oxide and nickel oxide are done against different relevant parameters.

Keywords: p-type TFT, n-type TFT, Copper Oxide, Nickel Oxide, Tin Oxide.

1. Introduction

Journey of TFT started in the year of 1925, but main development started from the year of 1970s. Scientists looking for an alternative for crystalline Silicon, made TFT relevant for wide use in the logic integrated circuit, because of its low cost and large area application possibilities. Complementary metal oxide semiconductor (CMOS) is a major feature in analog and as well as in digital electronic systems because of its low power consumption. Though oxide based TFT is one of the potential candidates for transparent electronic application for flexible display device but the most important challenges faced by researchers is unavailability of high performance p-type semiconductor material which can work with its n-type counterpart with equal efficiency. The reason behind the poor device characteristics can be attributed to low hole mobility. Other alternative of oxide based TFT is p-type organic TFT but most of the device exhibits poor stability along with low hole mobility.

2. Device structure of TFT and its operation

TFTs are field-effect devices with three terminals, their principle of working is nearly same as MOSFETs with three terminals as source (S), drain (D) and gate (G). On applying a suitable voltage in the gate electrode, an accumulation layer is formed near to the semiconductor/dielectric interface which regulates the current between S and D. But the substrate materials and manufacturing temperatures of MOSFET and TFTs differ quite a lot [1,2]. Various materials can be used as an active layer in TFTs, each are marked with their own advantages and disadvantages. For example, amorphous are having very low mobility in spite of advantage of deposition over large surfaces at low temperature. Polycrystalline silicon TFTs are extremely suitable for high speed application but processing speed quite high. Nano crystalline Silicon TFTs combine advantages of both amorphous and polycrystalline TFTs as they consist of Si crystalline embedded inside the amorphous silicon. In recent past oxide based TFT has emerged as one of the potential candidates specially in the area of solar cell based technology, solid- state based sensors application, display panels devices, smart windows, transparent and flexible electronic devices etc.

TFT can be Top Gate (TG) and Bottom Gate (BG) structure based upon the position of gate electrode. Further classification of TFT can be coplanar (C) and staggered (S) configuration based upon the relative positions of TFT. Each device structure of TFT has its own advantages and disadvantages, based upon material used along with its manufacturing method and as well as the working environment of the TFT devices. In TG structures, there is no overlapping area between the source/drain and the gate electrode, which results in minimum parasitic capacitance and mainly suitable for high speed application [2,3]. Not only that the TG structure is mainly desirable when a semiconductor material needs high temperature for fabrication which may affect the layers previously deposited. Additional advantages for TG, insulator and electrode act as a passivation layer, which gives structural protection of the semiconductor layer from external damage, as well as decreases fabrication steps thereby reduced cost. But TG structure of TFT is not preferable for liquid crystal display (LCD) based application, as the backlight unit cannot shield light to reach the semiconductors because of absence of metal electrode. Therefore, improvement of stability of oxide based semiconductor for LCD application is required. Because of simplicity in fabrication of BG structures, they are widely used.

Other than these conventional TFTs there are many improved structures fabricated and analyzed over the time. These are to fulfill the recent demand of large scale integration and for which nanoscaled channel length has become essentials. Reduction of driving voltage, without any compromisation with driving capability at the output end and as well as reduction of channel length are a few contradictory criteria. The vertical TFT demonstrated with these objectives showed a good potential. Fabrication steps included multilayer stacking of source, dielectric and drain. Another recent structure drew attention to all is double gate (DG) structure of TFT, inclusion of an additional gate improves controlabilty. Yi-Hsuan Hsiaoet. A et. al. [4] modeled variability caused because of randomly distributed grain boundary and also due to trap location for a vertical gate, three dimensional NAND Flash memory utilizing double gate TFT configuration. Double gate and vertical gate structure reported with better mobility, sub threshold swing and on-off current ratio[5] and stability [6]. Chong E et. al. has focused in their research to attain a high performance Si- In-Zn-O based TFT [7], processing temperature kept below 200° C and introduced a buried channel structure where a low resistance layer was embedded inside the channel layer which ultimately increases the on current of the TFT and field effect mobility. In the domain of flexible electronic device organic semiconductors and polymer dielectrics got their huge applications. Keeping this in mind Jang J et. al. [8] fabricated cylindrical gate TFT, which reduces the size and hence high packing density is achieved. For this purpose, the group used photocurable polyvinyl cinnamate (PVCN) as gate dielectric for its high bending stability. A smooth metallic fiber used as cylindrical substrate and polymer dielectric is deposited by dip coating method. The steps followed after this are UV irradiation for cross-link and thermal heating which made the coating uniform and smooth. The device fabricated with PVCN dielectric exhibited superior quality than PVP dielectric made organic TFT.

3. TFT Device Performance Parameters

Like any other field effect device, performance analysis of TFT is done with the help of static characteristics namely output and transfer characteristics. This analysis is also applicable for p- type TFT. The output characteristics can be divided into two distinct parts based upon voltage level of drain to source voltage, V_{DS}. When the level of voltage V_{DS} is less, resistance of channel is mainly manipulated by applying the gate to source voltage, V_{GS}. This is because, accumulated charge is distributed uniformly along the channel. Therefore, field effect between source and drain can be modeled as variable resistance which can be controlled by V_{GS} and linear increment of current happens with V_{DS}. Saturation region is marked by the condition when V_{DS} is greater than the V_{GS}. For a particular value of V_{GS}, current practically remain constant in this region. The two distinct regions discussed is followed the following current equations [9,10]:

$$I_{DS} = -\mu_{FE} C_{ox} \frac{W}{L} [(V_{GS} - V_{TH})V_{DS} \ -\frac{1}{2} V_{DS}^2] \tag{1}$$

$$I_{DS} = -\mu_{sat} C_{ox} \frac{W}{2L} (V_{GS} - V_{TH})^2 \tag{2}$$

But satisfactory performance analysis is not possible with the help of output and transfer characteristics. Hence, few important parameters are defined, for example: mobility, on/off current ratio, turn on voltage, threshold voltage and sub-threshold swing which give us more elaborate idea about performance of TFT. These parameters can be obtained from transfer characteristics. These parameters are mobility, on/off ratio of current , turn-on voltage, threshold voltage and sub-threshold swing.

Figure 1. Sequence of solution based fabrication

4. p- type oxide TFT deposition process

High performance n-type oxide based TFT has already taken a major role in the application of large area display device. Therefore, automatically research focus moves towards finding a p- type oxide based TFT. As complementary device technology is the only solution for less consumption of power in both analog and as well as digital technologies. But the main hindrance for achieving a high performance p type TFT is its very low mobility. This is because of electronic configuration of metallic oxide. In most of the metal oxide semiconductor the conduction band minima (CBM) are spatially spread on s orbital of metal and valence band maxima (VBM) is localized 2p orbital of oxygen. But carriers derived from valence band are having much lower mobility than carriers generated from conduction band minima. Due to the large effective mass of hole compared to electron, hole exhibits a hopping conduction and, therefore, have reduced mobility. [11,12]. To overcome this and obtaining a better performance TFT, hybridized orbitals of O 2p with metal cations orbital is one of the best solutions. Accordingly, number of oxide materials of p-type is analysed, having common molecular formula CuAO2 where, 'A' can be boron, aluminium, galium or indium. None of these metals showed a potential as a p-type candidate, however, addition of another copper, Cu2O has showed a good potential as p-type candidate. Here VBM is the result of hybridization of 2p orbital of oxygen with 3d orbital of copper [13]. Another approach to find out the p-type material is by hybridization of VBM with 2p orbital of oxygen and s orbital of metallic cations. Outcome of this approach is the formation of PbO, Bi2O and SnO based materials. But PbO is n-type and Bi2O3 based oxide exhibited a very low mobility. SnO based TFT can turn out to be a suitable p- type candidate. In NiO, formation of VBM occurred through hybridization of 2p orbital of oxygen with partially filled 3d of Ni and this combination turned out as another good option for fabricating p-type TFT. In the subsequent section, discussion will be done about aforementioned three p-type oxide based TFT viz. oxide of copper, tin and nickel.

5. Deposition techniques and performance analysis of Copper Oxide thin films

Copper Oxide TFT can be fabricated by different methods, having great impact on electrical, structural and optical properties.

5.1 *Thermal Oxidation (TO)*

It is one of the simple and conventional methods for fabrication with the additional advantage of low cost. Heating of copper sheet is done between 200°C to 1000°C for several hours which results in formation of CuO and Cu_2O. Relative amount of these two phases depend upon partial pressure of oxygen, oxidation temperature and time. As low temperature formation of Cu_2O is more prominent, whereas at high temperature transition from Cu_2O to CuO takes place [14].

Table1. Different sputtering technique of copper oxide TFT

Partial Pressure of O_2	Sputtering Temp/ Annealing Temp (°C)	Sputtering Pressure (Torr)	Sputtering Power (Watt)	Electrical Resistivity, (Ohm cm)	Ref
24	200	6.3	DC-60	32.7	[17]
9	RT / 200	4.5	RF-50	104	[18,19]
4	RT	6	RF-200	3.85	[20]
16	RT	1.5	RF-80	20.8	[21]

5.2 *Vacuum-based physical techniques*

Sputtering is used as a deposition technique for thin films of a material onto a substrate, using gaseous plasma. Reactive magnetron sputtering is one of the popular techniques for oxide film fabrication. Key feature of sputtering is accurate control in stoichiometry and the deposition rate is quite high at low temperature. It can be classified into two types: Direct current and Radio frequency sputtering. Properties of the film deposited depend upon the parameters like partial pressure of oxygen, flow rate of gas, pressure in chamber and temperature of substrate.

5.3 *Pulsed Laser Deposition (PLD)*

It is a physical vapor deposition (PVD) method using pulsed laser beam of high power within a vacuum chamber and therefore to strike a target material for deposition. Growth of Cu_2O layer is also possible by pulsed laser deposition [15,16]. This technique needs high process temperature and giving a high hole mobility ranging from 30-107 cm_2/Vs. Maximum field effect mobility is reported by [16]. The Table 1 is showing the sputtering parameters of Cu_2O deposited thin film.

5.4 *Solution-based wet-chemical techniques*

Wet techniques are simpler, faster and cheaper and at the same time can produce good quality films. Researchers have worked utilizing this technique by using different starting material for example Copper (II) acetate, copper (II) chloride [22,23], or copper (II) sulphate etc. But copper acetate came out as most promising material because of high solubility and low sensitivity to moisture when prepared from acetate group [24,25]. Many activities were done in this research field with different starting materials and conditions for coating and analysis. The Figure 1 is showing the flow chart for the process and Table 2 is showing the performance analysis for different fabrication condition.

6. Deposition techniques and performance analysis of Tin oxide TFT

Research on p-type TFT using SnO started in the year of 2008 by Ogo et al [31]. They have grown. For this purpose, they have grown SnO epitaxially on yittria stabilized zirconia (YSZ) substrate by PLD method at a temperature 575°C [32]. But ratio of Ion/ Ioff was not satisfactory because of high hole density. PLD requires high temperature processing, to overcome that RFMS method is introduced by Fortunato et al [33].

Table 2. Performance analysis of copper oxide based p-type TFT for different fabrication methods

Channel Layer	Structure /Method	Substrate /Dielectric	Threshold Voltage(V)	Mobility ($cm^2V^{-1}s^{-1}$)	I_{ON}/I_{OFF}	SS	Ref
Cu_2O	C-TG/ PLD	Si/HfON	-0.8	-	$3x10^6$	0.18	[26]
CuxO	S- BG/ ALD	Si/SiOX	-1.9	5.6	$1.8x10^5$	0.75	[27]
CuxO	S-BG/ SC	Si/A12O3	26	0.32	$5x10^4$	1.1	[28]
CuOx	S-BG/ SC	Si/SiO2	-	$2.8x10^{-3}$	$4.9x10^4$	0.7	[29]
CuO	C- BG/ DBP	Si/ nanocomposite	-	0.01	105	1.2	[30]

Here sputtering process is followed by annealing process at 200°C temperature under normal atmospheric condition. At this temperature transition happens from amorphous structure to polycrystalline structure, mixture of α-Sn and β-SnOX. This shows p-type conductivity, when partial pressure of oxygen is maintained between 5% to 15% [33]. A p-type SnO with double gate TFT structure is proposed by Zhong et al. in 2015 [34]. The device can perform switching operation in double gated mode and as well as single gated mode also. The device performance remarkably improves under double gate mode w.r.t on current and mobility compare to bottom gate or top gate operation. The reason behind this is because of accumulation of charge carriers in the middle of channel which reduces surface scattering considerably. Not only that, transfer characteristics under single mode can be tuned by the biased voltage in the other gate. Chen et al. in their paper in 2017 [35] discussed about treatment of SnO channel by fluorine plasma, this improved the device performance with respect to Ion/Ioff, field effect mobility and sub threshold swing and extremely small magnitude of off current. Enhanced performance of this treated plasma can be attributed to reduction of channel roughness, passivation of interface trap and oxygen vacancies. Modelling and simulation of plasma fluorination is also done by Rajshekar et al. [36]. Table 3 summarized the performance analysis of tin oxide based p-type TFT.

7. Deposition techniques and performance analysis of Nickle Oxide TFT

In 2008, Shimotani et. al. studied NiO based p type TFT using electric double layer(EDL) [39] and the performance analysis of TFT made it evident that future study in this domain might be quite helpful. In 2013, Jiang et al. fabricated a p-type nanocrystal NiO-based TFT by thermal oxidation method which exhibited higher performance parameters [40]. In the report of Lin et al. p-type solution-processed NiOx TFT is mentioned with a significantly higher statistics when they introduced Sn dopant [41]. Doping of Sn induces more amorphous phase. The improvements in TFT performances mainly due to shift of Fermi level into the midgap. Soon afterwards, in 2017, Liu et al. proposed Cu doped NiO thin films of p type [42]. In 2018, a p-type nickel oxide TFT used aluminum oxide as the gate dielectric and inkjet printing (IJP) technique [43], obtaining a hole mobility of 0.78 $cm^2 V^{-1}s^{-1}$, a threshold voltage of −0.6 V, a sub_threshold swing of 1.37 V/decade and an on-off current ratio of 5.3×10^4. Table 4 summarized the performance analysis of nickel oxide based p-type TFT.

Table 3. Performance analysis of tin oxide based p-type TFT

Processing Method of SnO TFT	Threshold Voltage(V)	Mobility ($cm^2V^{-1} s^{-1}$)	I_{ON}/I_{OFF}	Ref
Grown Epitaxially on YSZ substrate by Plasma Laser Deposition at 575 ^{0}C	4.8	1.3	102	[32]
RFMS method, Annealing Temperature 200^{0}C	-5	1.2	103	[33]
Double Gated structure	-	6.54	$> 10^5$	[34]
Fluorine plasma treatment employed on SnO Channel	-	2.13	$> 10^6$	[35]
Vacuum thermal evaporation method along with thermal annealing and oxygen plasma treatment at low cost	-	5.59	50	[37]
Low Temperature Processing	-10	4.6	$7x10^4$	[38]

Table 4. Performance analysis of nickel oxide based p type TFT

Channel Layer	Structure / Method	Substrate /Dielectric	Threshold Voltage(V)	Mobility ($cm^2V^{-1} s^{-1}$)	I_{ON}/I_{OFF}	SS	Ref
NiO	TG/ -	NiO/EDL	/	$1.6x10^{-4}$	130	/	[39]
NiO	S-BG/ TO	Si/SiO2	-11.4	5.2	$2.2x10^{3}$	3.91	[40]
Sn:NiOx	C-BG /SC	Glass/AlOx	-1.44	0.97	10^{6}	0.24	[41]
Cu:NiOx	S-BG/ SCS	Glass/ZrO2	0.45	1.53	$3x10^{4}$	0.13	[42]
NiOx	S-BG/ IJP	Si/ Al2O3	-0.6	0.78	$5.3x10^{4}$	1.37	[43]

8. Conclusions

In past few decades progress of TFT is enormous. This paper analyzed and put forward work of researchers using Copper oxide, Tin Oxide and Nickel oxide for developing p-type TFT. It can be concluded that, the performance of p type TFT has improved a lot, but still not satisfactory, if mobility is compared with the n-type counterpart. Therefore, more research is needed in the fabrication technique and deposition methods to improve its functionality in the display device application. Organic TFT is also another potential candidate to develop p-type TFT. It has been observed in general, cost of fabrication for oxide based TFT is less compared to polycrystalline TFT. TFT is primarily used as a large area display device along with that several researches are ongoing to evaluate in sensing devices in the biochemical domain, especially where flexible or foldable substrates are used. Areas of further research should be for reduction of noise and improvement of stability.

References

[1] Wang Z., Nayak P.K., Caraveo-Frescas J.A., Alshareef H.N. 2016. Recent Developments In P-Type Oxide Semiconductor Materials And Devices Adv. Mater., **28**(20) 3831-3892.
[2] Petti L., Münzenrieder N., Vogt C., Faber H., Büthe L., Cantarella G., Bottacchi F., Anthopoulos T.D., Tröster G. 2016. Metal Oxide Semiconductor Thin-Film Transistors For Flexible Electronics Appl. Phys. Rev **3**(2) 021303.
[3] Choi J.Y., Lee S.Y. 2017. Comprehensive Review On The Development Of High Mobility In Oxide Thin Film Transistors J. Korean Phys. Soc. **71**(9) 516-527.
[4] Yi-Hsuan Hsiao, Hang-Ting Lue, Wei-Chen Chen, Chih-Ping Chen, Kuo-Ping Chang, Yen-Hao Shih, Bing-Yue Tsui, and Chih-Yuan Lu. 2012. Modeling the Variability Caused by Random Grain Boundary and Trap-location Induced Asymmetrical Read Behavior for a Tight-pitch Vertical Gate 3D NAND Flash Memory Using Double-Gate Thin-Film Transistor (TFT) Device, in Proceedings of the IEEE International Electron Devices Meeting (IEDM), San Francisco, CA, USA, 609–612.
[5] Münzenrieder N., Zysset C., Petti L., Kinkeldei T., Salvatore G.A., Tröster G. 2013. Flexible Double Gate A-IGZO TFT Fabricated On Free Standing Polyimide Foil Solid-State Electron. **84** 198-204.
[6] Son K.S., Jung J.S., Lee K.H., Kim T.S., Park J.S., Park K.C., Kwon J.Y., Koo B., Lee S.Y. 2010. Highly Stable Double-Gate Ga-In-Zn-O Thinfilm Transistor IEEE Electron Device Lett. **31**(8) 812-814.
[7] Chong E., Kim B., Lee S.Y. 2012. Reduction Of Channel Resistance In Amorphous Oxide Thin-Film Transistors With Buried Layer in IOP Conf. Series: Mater. Sci. Eng. **34** 012005.
[8] Jang J., Nam S., Hwang J., Park J.J., Im J., Park C.E., Kim J.M. 2012. Photocurable Polymer Gate Dielectrics For Cylindrical Organic Field Effect Transistors With High Bending Stability J.

Mater. Chem. **22**(3) 1054-1060.

[9] Hong D., Yerubandi G., Chiang H.Q., Spiegelberg M.C., Wager J.F. 2008. Electrical Modeling Of Thin- Film Transistors Crit. Rev. Solid. State **33**(2) 101-132.

[10] Pierret R.F. 1996. Semiconductor Device Fundamentals Addison- Wesley, Boston, MA, USA.

[11] Raebiger H., Lany S., Zunger A. 2007. Origins Of The P-Type Nature And Cation Deficiency In Cu2O And Related Materials Phys.l Rev. **76**(4) 045209.

[12] Togo A., Oba F., Tanaka I., Tatsumi K. 2006 First-Principles Calculations Of Native Defects In Tin Monoxide Phys. Rev. **74**(19) 195128.

[13] Hiramatsu H., Ueda K., Ohta H., Hirano M., Kamiya T., Hosono H. 2003. Degenerate p-Type Conductivity In Wide-Gap Lacuos1−Xsex (X=0-1) Epitaxial Films Appl. Phys. Lett. **82**(7) 1048-1050.

[14] A.H.Jayatissa, K.Guo, A.C.Jayasuriya. 2009. Fabrication Of Cuprous And Cupric Oxide Thin Films By Heat Treatment Appl. Surf. Sci. **255**(23) 9474–9479.

[15] K.Matsuzaki, K.Nomura, H.Yanagi, T.Kamiya, M.Hirano,H. Hosono. 2009. Effects Of Post-Annealing On (110) Cu 2 O Epitaxial Films And Origin Of Low Mobility In Cu 2 O Thin-Film Transistor Phys. Status solidi A, **206**(9) 2192–2197.

[16] X.A.Zou, G.J.Fang, L.Y.Yuan, M.Y.Li, W.J.Guan, X.Z.Zhao 2010. Top- Gate Low-Threshold Voltage P-Cu2O Thin-Film Transistor Grown On Sio2/Si Substrate Using A High-K Hfon Gate Dielectric IEEE Electron Device Lett. **31**(8) 827–829.

[17] C.-L.Chu, H.-C.Lu,C.- Y.Lo, C.Y.Lai, Y.H.Wang 2009. Physical Properties Of Copper Oxide Thin Films Prepared By Dc Reactive Magnetron Sputtering Under Different Oxygen Partial Pressures Phys. Condens. Matter, , **404**(23-24) 4831–4834.

[18] E.Fortunato, V.Figueiredo, P.Barquinha, E.Elamurugu, R. Barros, Gonçalves,S. C.K. Park, C.S. Hwang, R.Martins. 2010. Thin-Film Transistors Based On P-Type Cu2O Thin Films Produced At Room Temperature Appl. Phys. Lett. **96**(19) 192102.

[19] V.Figueiredo, E.Elangovan, R.Barros, J.V.Pinto, T.Busani, R.Martins, E.Fortunato. 2012. p-Type CuxO Films Deposited at Room Temperature for Thin-Film Transistors J.Disp.Technol. **8**(1) 41–47.

[20] T.H.Darma, A.A.Ogwu, F.Placido 2011. Effects Of Sputtering Pressure On Properties Of Copper Oxide Thin Films Prepared By Rf Magnetron Sputtering Mater. Technol. **26**(1) 28–31.

[21] Z.Q.Yao, S.L.Liu, L.Zhang, B.He, A.Kumar, X.Jiang, W.J.Zhang, G.Shao 2012. Room Temperature Fabrication Of P-Channel Cu2O Thin-Film Transistors On Flexible Polyethylene Terephthalate Substrates Appl. Phys. Lett. **101**(4) 042114.

[22] S.Y.Kim, C.H.Ahn, J.H.Lee, Y.H.Kwon, S.Hwang, J.Y.Lee, H.K.Cho. 2013. p-Channel Oxide Thin Film Transistors Using Solution-Processed Copper Oxide ACS Appl. Mater. Interfaces **5**(7) 2417– 2421.

[23] P.Pattanasattayavong, S.Thomas, G. Adamopoulos, M.A. McLachlan, T.D. Anthopoulos. 2013. p-Channel Thin-Film Transistors Based On Spray-Coated Cu_2O Films Appl. Phys. Lett. **102**(16) 163505.

[24] A.Y.Oral, E.Menşur, M.H.Aslan, E.Başaran. 2004. The Preparation Of Copper(II) Oxide Thin Films And The Study Of Their Microstructures And Optical Properties Mater. Chem. Phys. **83**(1) 140–144.

[25] L.Armelao, D.Barreca, M.Bertapelle, G.Bottaro, C.Sada, E.Tondello 2003. A Sol–Gel Approach To Nanophasic Copper Oxide Thin Films Thin Solid Films **442**(1-2) 48–52.

[26] Zou X., Fang G., Yuan L., Li M., Guan W., Zhao X. 2010. Top-Gate Low Threshold Voltage P-Cu2O Thin-Film Transistor Grown On Sio2/Si Substrate Using A High- Hfon Gate Dielectric IEEE Electron DeviceLett. **31**(8) 827-829.

[27] Maeng W., Lee S.H., Kwon J.D., Park J., Park J.S. 2016. Atomic Layer Deposited P-Type Copper Oxide Thin Films And The Associated Thin Film Transistor Properties Ceram. Int. **42**(4) 5517-5522.

[28] Liu A., Nie S., Liu G., Zhu H., Zhu C., Shin B., Fortunato E.,Martins R., Shan F. 2017. In Situ

One-Step Synthesis Of P-Type Copper Oxide For Low-Temperature, Solution-Processed Thin-Film Transistors J. Mater. Chem. C **5**(10) 2524-2530.

[29] Jung T.S., Lee H., Park S.P., Kim H.J., Lee J.H., Kim D., Kim H.J. 2018. Enhancement Of Switching Characteristic For P-Type Oxide Semiconductors Using Hypochlorous Acid ACS Appl.Mater. Interfaces **10**(38) 32337-32343.

[30] Reker J., Meyers T., Vidor F.F., Hilleringmann U. 2019. Inorganic p-Channel Thin-Film Transistors Using Cuo Nanoparticles in Proc. SPIE **11043** 1104312.

[31] Ogo Y., Hiramatsu H., Nomura K., Yanagi H., Kamiya T., Hirano M., Hosono H. 2008. p-Channel Thin- Film Transistor Using P-Type Oxide Semiconductor SnO Appl. Phys. Lett. **93**(3) 032113.

[32] Fortunato E., Barros R., Barquinha P., Figueiredo V., Park S.H.K., Hwang C.S., Martins R. 2010. Transparent P-Type Snox Thin Film Transistors Produced By Reactive Rf Magnetron Sputtering Followed By Low Temperature Annealing Appl.Phys. Lett. **97**(5) 052105.

[33] Fortunato E., Martins R. 2011. Where science fiction meets reality? With oxide semiconductors! Rapid Research Letter **5**(9) 336-339.

[34] Zhong C.W., Lin H.C., Liu K.C., Huang T.Y. 2015. Improving Electrical Performances Of P-Type Sno Thin-Film Transistors Using Doublegated Structure IEEE Electron Device Lett. **36**(10) 1053-1055.

[35] Chen P.C., Chiu Y.C., Zheng Z.W., Cheng C.H., Wu Y.H. 2017. Influence Of Plasma Fluorination On P- Type Channel Tin-Oxide Thin Film Transistors J. Alloys Comp. **707** 162-166.

[36] Rajshekar K., Hsu H.H., Kumar K.U.M., Sathyanarayanan P., Velmurugan V., Cheng C.H., Kannadassan D. 2019. Effect Of Plasma Fluorination In P-Type Sno Tfts: Experiments, Modeling, And Simulation IEEE T. Electron. Dev. **66**(3) 1314-1321.

[37] Lee H.N., Song B.J., Park J.C. 2014. Fabrication Of p-Channel Amorphous Tin Oxide Thin-Film Transistors Using A Thermal Evaporation Process J. Display Technol. **10**(4) 288-292.

[38] Barros R., Saji K.J., Waerenborgh J.C., Barquinha P., Pereira L., Carlos E.,Martins R., Fortunato E. 2019. Role Of Structure And Composition On The Performances Of P-Type Tin Oxide Thin-Film Transistors Processed At Low-Temperatures Nanomaterials **9**(3) 320.

[39] Shimotani H., Suzuki H., Ueno K., Kawasaki M., Iwasa Y. 2008 .p-Type Field-Effect Transistor Of Nio With Electric Double-Layer Gating Appl. Phys. Lett. **92**(24) 242107.

[40] Jiang J., Wang X., Zhang Q., Li J., Zhang X.X. 2013. Thermal Oxidation Of Ni Films For P-Type Thin-Film Transistors Phys. Chem. Chem. Phys. **15** 6875-6878

[41] Lin T., Li X., Jang J. 2016. High Performance P-Type Niox Thin-Film Transistor By Sn Doping, Appl. Phys. Lett. **108**(23) 233503.

[42] Liu A., Zhu H., Guo Z., Meng Y., Liu G., Fortunato E., Martins R., Shan F. 2017. Solution Combustion Synthesis: Low-Temperature Processing For P-Type Cu:Nio Thin Films For Transparent Electronics Adv. Mater. **29**(34) 1701599.

[43] Hu H., Zhu J., Chen M., Guo T., Li F. 2018. Inkjet-printed p-type Nickel Oxide Thin-Film Transistor Appl. Surf. Sci. **441** 295-302.

Acknowledgement
We express our gratitude to Visvesvaraya Technological University, Jnana Sangama, Belagavi, India for supporting this research work.

Single stage Boost Cascaded Multilevel Inverter Based on Switched Inductor Structure

P.Sriramalakshmi[1*], Rahul Singh[2], Sneha[3]

[1]School of Electrical Engineering, Vellore Institute of Technology, Chennai – 600173.
[2]Engineer - security intelligence-analytics-assurance-l2.1, Wipro Ltd.
[3]Software Engineer, Wipro Ltd

E-mail: sriramaakshmi.p@vit.ac.in

Abstract. This paper presents the experimental investigations of Switched Inductor cell (SL-FLCHBqSBI) based Five-Level Cascaded H-Bridge quasi Switched Boost Inverter for Renewable Energy (RE) applications. The SL-FLCHBqSBI topology is based on the ripple input current Switched Inductor Boost Inverter (SLBI) structure which is available in the literature. In this work, a stiff DC voltage of 36 V is taken as input to deliver a power of 100 W across a resistive load is designed. The control signals for the active devices are generated using Phase-Shifted sinusoidal Pulse Width Modulation (PS-PWM) modulation technique. The SL-FLCHBqSBI topology provides a higher voltage gain of 2.83 compared to the conventional topologies for a low duty ratio. The simulation is performed in MATLAB/Simulink environment and the simulation results are presented. A hardware set up is implemented and experiments are carried out to check the consistency of the theoretical concepts with the working of the proposed topology in practical.

Keywords: Voltage Source Inverter; Phase Shifted triangular carrier PWM; Shoot Through; quasi Switched Boost Inverter (qSBI); Switched Inductor Boost Inverter (SLBI)

1. Introduction

The Voltage-Source Inverter (VSI) [1- 2] converts the DC voltage to the AC voltage which is always less than the DC input voltage. A DC-DC boost converter is connected with a traditional inverter module to achieve a maximum ac output voltage. This results in a two stage power conversion system. This circuit yields low efficiency but the cost is high. A conventional two stage Multi Level Inverter (MLI) is required to boost and invert the low DC voltage availed from the RE sources for medium voltage applications. The Multi-Level Inverters (MLIs) are featured with good quality voltage, reduced Total Harmonic Distortion (THD), reduced voltage stress, better Electro Magnetic Interference (EMI) and reduced switching losses [3-9]. The suitable arrangement of DC voltage links of MLI considerably delivers the high output voltage with improved power quality [7-9].

MLIs are categorized into three as follows:- Flying Capacitor (FC), Neutral Point Clamped Inverter (NPC) and Cascaded H bridge Inverter (CHB) inverter. Compared to CHB type, the remaining two types need more number of switching devices. CHB inverters find major applications in industry due to their high reliability, modular features and easy controllability compared to NPC and FC [1], [7-9]. More number of identical modules is connected in cascade to increase the output voltage levels of CHB which in turn reduces the filter size at the load terminals. However, the inverted voltage obtained from the traditional CHB offers the always less than the input dc voltage. To eliminate this issue, Z source inverter and its derived circuits are suggested by the researchers. These converter circuits are able to boost the dc voltage using shoot through pulses generated by modified PWM techniques. Hence it produces a boosted and inverted voltage in single stage of power conversion [10]-[25]. Only limited voltage gain is possible with ZSI/qZSI converter circuits.

Various ZSI based MLI circuits are proposed which possess the advantages of both MLI and ZSI circuits. In ZSI/qZSI based CHB [26-32], each module produces boosted DC link voltage with duty ratio control. The single phase qZSI based CHB is presented in [29] with the details of closed loop control. But the impedance network of Z source based circuits increase the cost and weight of the system. Switched Boost Inverter and its derived topologies are presented to overcome the problems with the ZSI based in [33-35]. The class of quasi SBI is available to overcome the drawbacks of SBI like discontinuous input current, higher voltage stress across the capacitor [36]. A comparison between qZSI and qSBI topologies is presented in [37]. Due to their advantages, qSBI based CHB topology is proposed in [39]. A three phase cascaded five level qSBI is discussed in [40]. To achieve a high boost factor in fuel cells and photovoltaic based applications, switched-inductor (SL) [38], switched capacitor (SC) [41] and transformer structures [42] are used along with the switched boost network topology. A five level cascaded switched-capacitor (SC) based qSBI is presented in [43]. To overcome the shortcomings of SBI, SLBI topology is introduced.

In this research article, the experimental investigation of a five-level SL based cascaded H-Bridge qSBI presented and it is named as SL-FLCHBqSBI topology in this article. This new topology is based on the ripple-input current SLBI topology [38]. Two identical SLBIs modules are cascaded to obtain the five level ac output voltages. The overall load voltage of the cascaded inverter is superior to one SLBI module. In addition, an improved harmonic spectrum of output voltage is obtained.

2. Five Level CHB qSBI based on switched inductor (SL-FLCHBqSBI)

Inspired from the literature and as a preliminary study, modeling and simulation of a mobile smart waste bin in webots is presented in this section. This section also describes the use of different sensors, actuators, and controllers associated with the modeling of the mobile smart bin.

2.1 Circuit topology

Figure. 1 illustrates a five- level switched inductor based cascaded H-bridge qSBI. Two identical SLBI modules A and B [38] are connected in cascade to obtain a five level voltage at the inverter terminals. The source voltages V_{g1}, V_{g2} are taken as 24 V each.

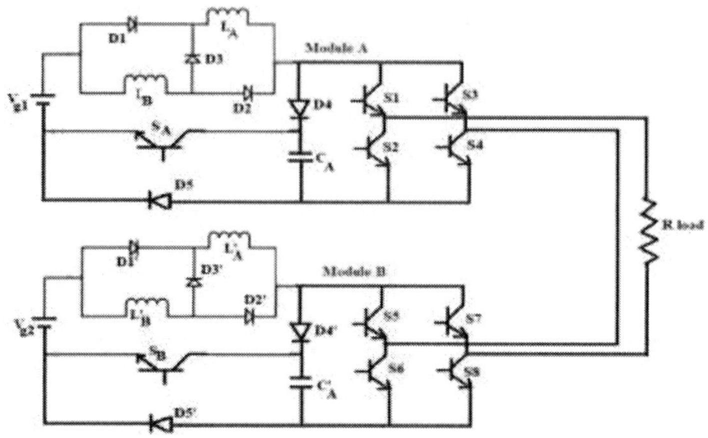

Figure 1. Five level cascaded H bridge qSBI based on switched inductor structure

2.2 *Operating Principle of SL-FLCHBqSBI*

Two identical modules of ripple current SLBI topology explained in [38], is cascaded here to form a five level CHB qSBI based on SLBI. It can operate in two stages i.e. shoot-through stage and non-shoot-through stage. Both stages of operations are depicted in the **Figure. 2** and **Figure. 3**.

2.2.1 Shoot- through stage

Figure. 2 shows the shoot through stage of module A of SL-FLCHBqSBI topology. The capacitor is initially assumed to be fully charged. The diodes D4 and D5 are reverse biased. The boost network switch S_A conducts along with the inverter leg switches S_1 and S_2 which causes the shoot through state in the circuit. During this stage, capacitor C_A charges the inductors L_A and L_B.

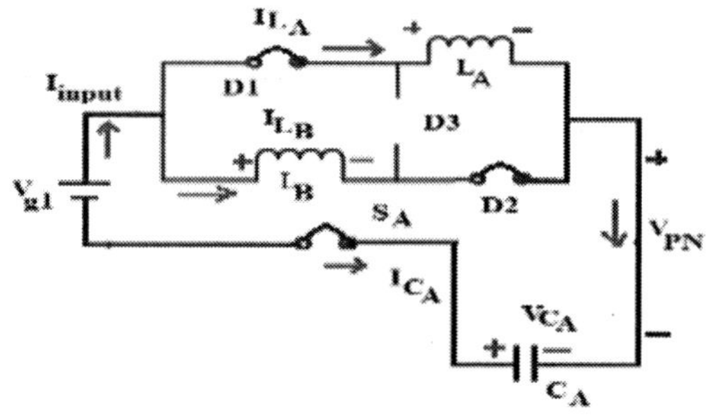

Figure 2. Shoot through stage

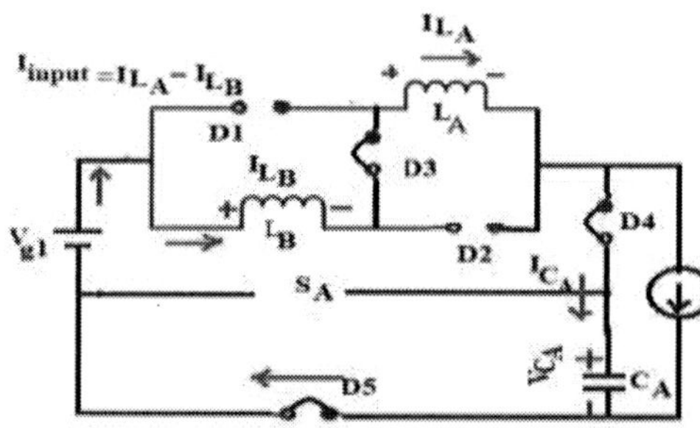

Figure 3. Non shoot through stage

2.2.2 ***Non-shoot through stage***

Figure. 3 shows the non-shoot through stage of module A. In this stage, inductors discharge through the capacitors. The diodes D4 and D5 are forward biased whereas diodes D1 and D2 are reverse biased. The switch S_A is turned off and the load current flows as depicted in **Figure.3**.

2.2.3 *Steady state analysis*

The voltage across the inductors and current through the capacitors during shoot- through stage are derived as,

$$L_A \frac{dI_{L_A}}{dt} = L_B \frac{dI_{L_B}}{dt} = V_{g1} + V_{C_A} \tag{1}$$

$$C_A \frac{dV_{C_A}}{dt} = -I_{L_A} - I_{L_B} = -I_{input} \tag{2}$$

The inductor voltage and capacitor current values during non-shoot through state [38] are derived as,

$$L_A \frac{dI_{L_A}}{dt} = V_{g1} - V_{C_A} - V_{L_{B\_non}} \tag{3}$$

$$L_A \frac{dI_{L_A}}{dt} = V_{L_{B\_non}} \tag{4}$$

$$C_A \frac{dI_{C_A}}{dt} = I_{L_A} - I_{PN} = I_{L_B} - I_{PN} = I_{input} - I_{PN} \tag{5}$$

Using equations (1) to (5), inductor voltage V_{LB_non} during the non-shoot through state is obtained as,

$$V_{LB\_non} = -\frac{D}{1-D}(V_{g1} + V_{C_A}) \tag{6}$$

Similarly, current through inductor (L_A) and voltage across the capacitor (C_A), it gives,

$$V_{C_A} = -\frac{1+D}{1-3D} V_{g1} \tag{7}$$

$$I_{input} = I_{L_A} = I_{L_B} = \frac{1-D}{1-3D} I_{PN} \tag{8}$$

$$I_{input} = (1+D) I_L \tag{9}$$

The DC link voltage (V_{PN}) across the inverter of the module A during non-shoot through state is derives as,

$$V_{PN} = V_{C_A} = \frac{1+D}{1-3D} V_{g1} = BV_{g1} \tag{10}$$

The module A [38] of the proposed SL-FLCHBqSBI provides the boost factor (B) as,

$$B = \frac{1+D}{1-3D} \tag{11}$$

3. Design equations for passive elements

The value of inductors L_A and L_B is given by,

$$L_A = \frac{V_{g1} + V_{C_A}}{I_{L_A}} DT_s \tag{12}$$

$$L_B = \frac{V_{g1} + V_{C_A}}{I_{L_B}} DT_s \tag{13}$$

The value of capacitance C_A, C_B are given by

$$C_A = \frac{I_{input}}{V_{C_A}} DT_s \quad ; \quad C_B = \frac{I_{input}}{V_{C_B}} DT_s \tag{14}$$

4. Simulation Results

For the simulation analysis the topology is designed for 200 W power rating. Both the modules are supplied with the input voltage of 24 V separately to obtain a five level ac voltage across the resistive load. The expressions mentioned in (12)-(14) are used to design the passive elements. The phase shifted triangular carrier PWM is used to produce the firing pulses for the switching devices [39].The design values used in the MATLAB simulations are presented in Table 1.

Table 1 Design specifications

Parameters	Attributes
Source Voltage (V_g)	24 V (each module)
Capacitors (C_A & C_B)	470 µF
Load Resistance (R_L)	100 Ω
Shoot through duty ratio (D)	0.26
Inductors (L_A & L_B)	1 mH
Switching Frequency (f_S)	10 KHz
Modulation Index (M)	0.74

The simulation results obtained using MATLAB/SIMULINK platform is shown in **Figure 6**. With the V_{g1} and V_{g2} of 24 V, the inverter generates the inductor currents (I_{LA}, I_{LB}) of 5 A as depicted in **Figure 4(a).** The diodes D1 and D2 are in series with the inductors and hence the same current flows through the diodes as shown in **Figure. 4(a).** The diode currents D3, D4 are depicted in **Figure. 4(b).** The switch voltage stress S_A in the boost network of module A and source current waveform are depicted in **Figure. 4(b). Figure. 4(c)** depicts the voltage across the capacitors C_A and C_B, DC link voltages (V_{PN1} and V_{PN2}) of modules A and B which are obtained the same as 120 V.

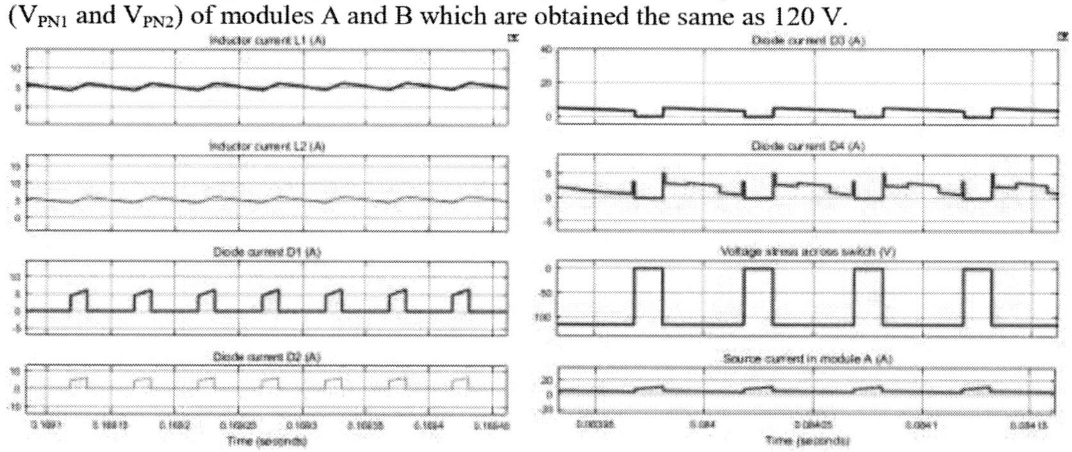

(a) (b)

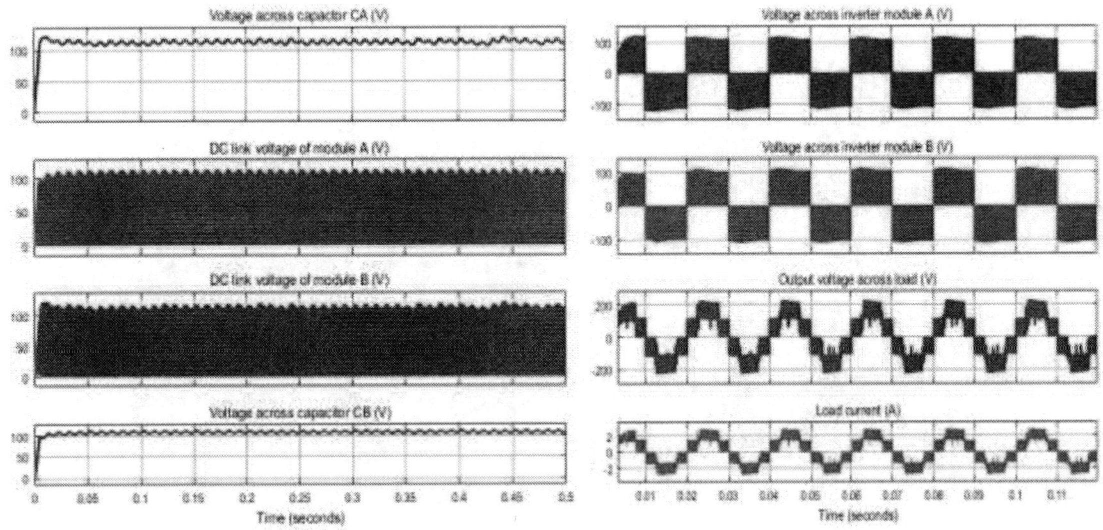

(c) (d)

Figure 4. Simulation results of five level CHB qSBI based on switched inductor
(a) Inductor currents (I_{LA}, I_{LB}) and diode currents (I_{D1}, I_{D2})
(b) Diode currents (I_{D3}, I_{D4}), switch stress (S_A) and source current
(c) Capacitor stress (C_A, C_B), DC link voltages (V_{PN1}, V_{PN2}) of module A and module B
(d) Output voltage across inverter module A, B, load voltage and load current

Figure 4(d). depicts the voltage across the inverter module A and B of 120 V and five level peak ac load voltage of 240 V. The five level peak load current is obtained as 2.25 A as in **Figure 4(d).**

5. Experimental results

A hardware set up of 100 W is developed to experiment the high boost capability of SL-FLCHBqSBI is depicted in **Figure 5**. The hardware design parameters are presented in Table.2. **Figure. 6(a)** to **Figure 6(g)** exhibit the hardware results of the SL-FLCHBqSBI topology. The switching pulses in **Figure 6(a)** and Figure 6(b) are generated in FPGA platform.

Table 2. Hardware Design Parameters

Parameters	Specification
Input Voltage	18 V(each module)
Output Power	100 W
Resistor	100 Ω
Capacitors (C_A & C_B)	100 μF
Modulation Index (M)	0.74
Power IGBT- H20R1203	1200 V,20 A
Shoot-through Mosfet- IRFP250N	200 V, 30 A
Inductors (L_A & L_B)	1 mH
Driver Circuit- TLP250	20 V, 1.5 A
Switching Frequency (f_S)	10 kHz
Shoot through duty ratio (D)	0.26
Power Diode- LT6221	100 V,20 A

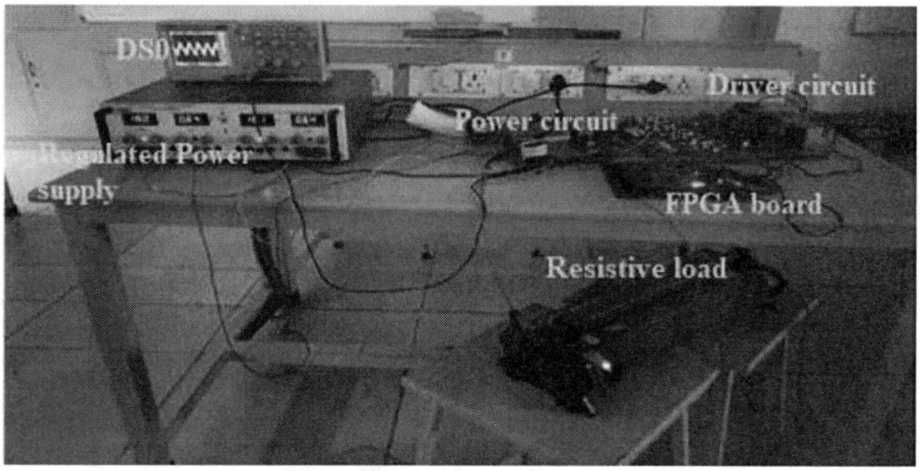

Figure 5 Experimental setup

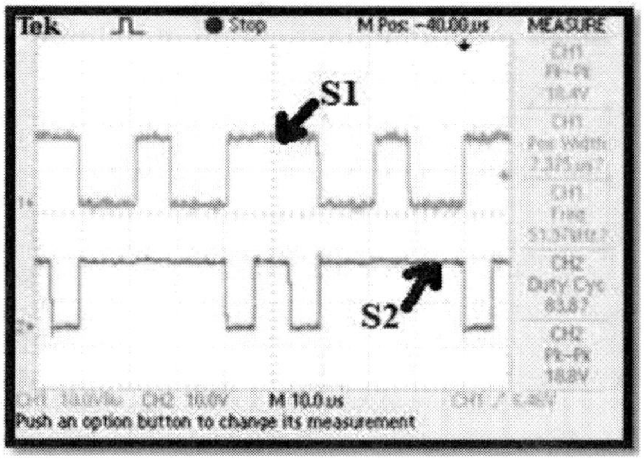

(a)

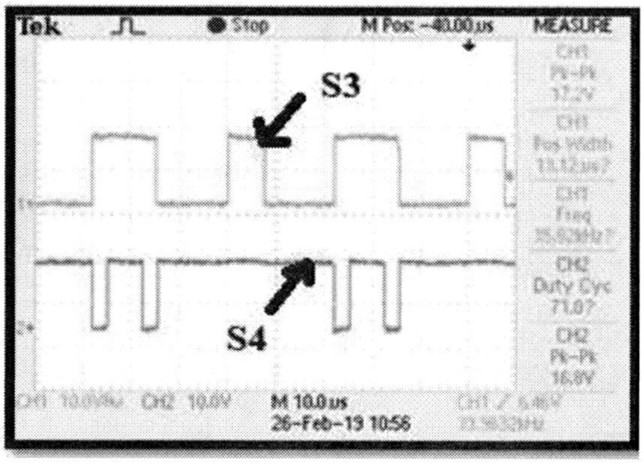

(b)

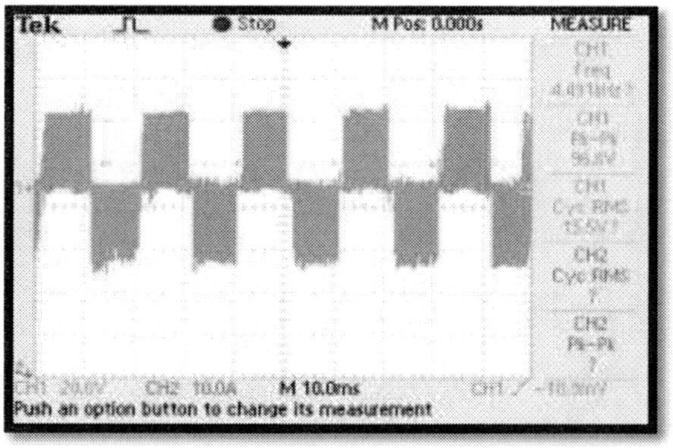

(c)

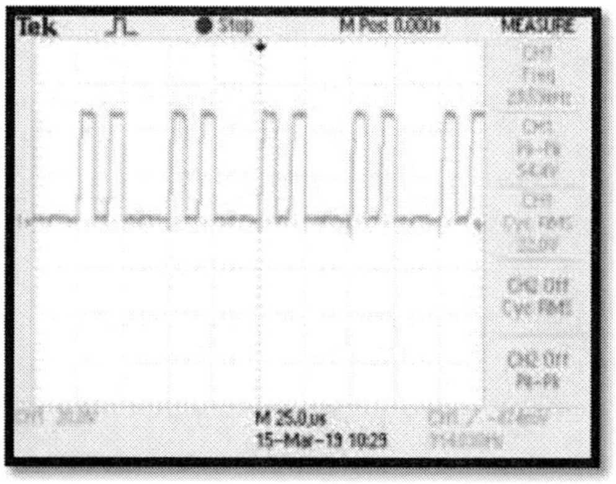

(d)

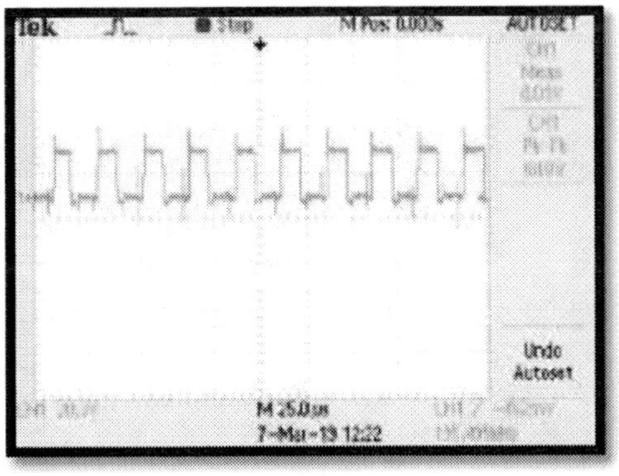

(e)

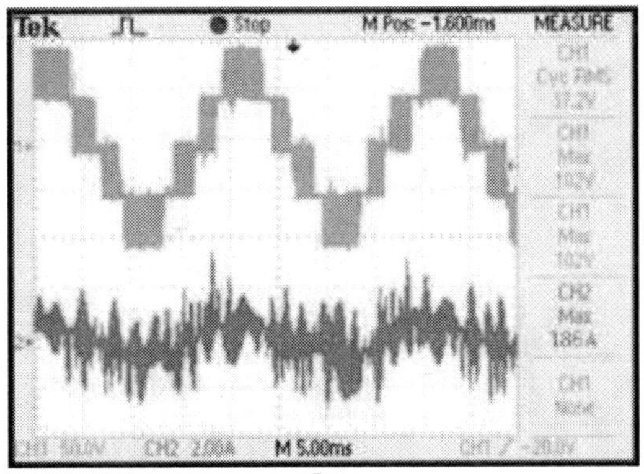

(f)

Figure 6. Experimental results of SL-FLCHBqSBI topology

(a) Gate pulse to switch S1 and S2
(b) Gate pulse to switch S3 and S4
(c) Inverter output voltage
(d) Inverter switch stress
(e) Peak DC- link voltage
(f) Load voltage and load current waveforms

Figure.6(a) and **Figure. 6(b)** represents the gate pulses given to module A of the inverter through the gate driver circuit. **Figure. 6(c)** shows the output voltage waveform of one module of SL-FLCHBqSBI. The value of 48.4V peak is obtained across the load. **Figure. 6(d)** shows the inverter switch stress of SL-FLCHBqSBI. The value of 54.4V peak voltage is obtained across the inverter switch. **Figure. 6(e)** shows the DC-link voltage of module A of SL-FLCHBqSBI. The value of 60V peak is obtained across the DC-link. The output voltage obtained across the load of SL-FLCHBqSBI is depicted in **Figure. 6(f).** The values of 102V peak output voltage and 1.86 A output current is obtained across the load.

6. Performance Comparison

Table 3 presents the comparison of passive elements and switch count among the cascaded SL-FLCHBqSBI and classical cascaded MLI circuits. The cascaded SL based qSBI uses four inductors and 10 switches. In addition, it has 2 capacitors and 10 diodes and. It has two reduced number of capacitors, same number of inductors and passive devices, two additional active devices compared to cascaded qZSI. The SL-FLCHBqSBI has two less capacitors; four less diodes compared to SCCFqSBI keeping the switch count the same. The switching losses are high compared to qZSI topology [39]. But, the conduction loss is significantly less.

Table 3 Circuit elements comparison of existing MLIs with SL-FLCHBqSBI

Components	Cascaded DC-DC boosted VSI [13]	Cascaded qZSI [39]	Cascaded qSBI [30]	SC-CFqSBI[43]	SL-FLCHBqSBI (Proposed)
Switches	10	8	10	10	10
Capacitors	2	4	2	4	2
Diodes	10	10	12	14	10
Inductors	2	4	2	2	4

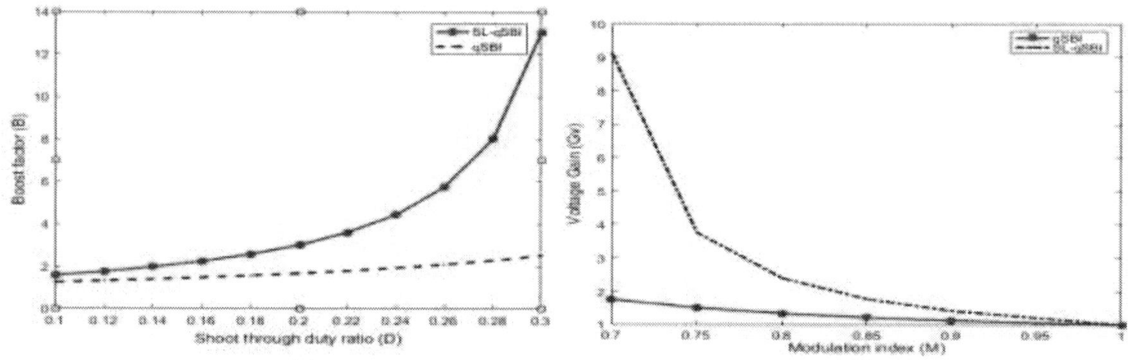

(a) (b)

Figure7. Performance comparison of SL-FLCHBqSBI topology

(a) D Vs B

(b) (M) Vs (B)

Figure. 7(a) depicts the comparison between boost factor (B) and duty ratio (D). It is understood that a small duty ratio of SL-qSBI offers a high boost (B). The variation between voltage gain (G) and modulation index (M) of the conventional cascaded qSBI and SL-FLCHBqSBIis presented in **Figure. 7(b).**It is clear that the SL-FLCHBqSBI provides a high voltage gain conventional CHBqSBI topology at the same modulation index.

7. Conclusion

In this research article experimental investigation of single phase cascaded five levels quasi qSBI based on switched inductor structure is discussed. The working principle of the proposed circuit and steady state analysis of a five level cascaded H-bridge qSBI based on switched inductor are detailed in this article. From the experimental investigations, it is obvious that SL-FLCHBqSBI offers a five level maximum ac voltage of 102V with the source voltage of 36V. A voltage gain of 2.83 is achieved at reduced shoot through duty ratio. Higher efficiency of 94.86% is achieved. Hence the topology is suitable for photo voltaic applications where a low voltage is required to provide a high boosted ac voltage.

References

[1] Chen, S., Lai, Y.M., Tan, S.C., Tse, C.K. 2009, Fast response low harmonic distortion control scheme for voltage source inverters, IET Power Electron, **2(**5) 574–584

[2] Dasgupta, S., Sahoo, S.K., Panda, S.K. 2011, Single-phase inverter control techniques for interfacing renewable energy sources with microgrid–part I: parallel-connected inverter topology with active and reactive power flow control along with grid current shaping, IEEE Trans. Power Electron, **26**(3) 717–731

[3] J.Rodriguez, L. Jih-Sheng, P.Fang Zheng, 2002, Multilevel inverters: a survey of topologies, controls, and applications, IEEE Trans. Ind. Electron, **49**(4) 724–738

[4] A.Nabae, I.Takahashi, H. Akagi ,1981, A new neutral-point-clamped PWM inverter", IEEE Trans. Ind. Appl,**17**(5) 518–523

[5] K.K.Gupta, S.Jain, 2012 ,Topology for multilevel inverters to attain maximum number of levels from given DC sources, IET Power Electron, **5**(4) 435–446

[6] T.A. Meynard, H.Foch,1992, Multi-level choppers for high voltage applications, **2**(1) 45–50

[7] P. Fang Zheng, L.Jih-Sheng, L., J.W.McKeever, 1996, A multilevel voltage source inverter with separate DC sources for static VAR generation, IEEE Trans. Ind. Appl, **32**(5) 1130–1138

[8] R.S Alishah, D. Nazarpour, S.H Hosseini, 2015,Reduction of power electronic elements in multilevel converters using a new cascade structure, IEEE Trans. Ind. Electron, **62**(1) 256–269

[9] M.T Khosroshahi, 2014, Crisscross cascade multilevel inverter with reduction in number of

components, IET Power Electron, **7**(12) 2914–2924
[10] Peng, F.Z, 2003, Z-source inverter, IEEE Trans. Ind. Appl., **39**(2) 504–510
[11] Tang, Y., Xie, S., Ding, J.,2014, Pulsewidth modulation of Z-source inverters with minimum inductor current ripple, IEEE Trans. Power Electron., **61**(1) 98– 106
[12] Liu, J.B., Hu, J.G., Xu, L.Y, 2007, Dynamic modeling and analysis of Z-source converter-derivation of AC small signal model and design-oriented analysis, IEEE Trans. Power Electron., **22**(5) 1786–1796
[13] Liu, Y., Ge, B., Abu-Rub, H., Peng, F.Z, 2014, Modelling and controller design of quasi-Z-source inverter with battery-based photovoltaic power system, IET Power Electron., **7**(7) 1665–1674
[14] Sun, D., Ge, B., Yan, X., et al.,2014, Modeling, impedance design, and efficiency analysis of quasi-Z source module in cascade five level photovoltaic power system, IEEE Trans. Ind. Electron., **61**(11) 6108–6117
[15] Liu, H., Liu, P., Zhang, Y.,2012, Design and digital implementation of voltage and current mode control for the quasi-Z-source converters, IET Power Electron, **6**(5) 990–998
[16] Ellabban, O., Mierlo, J.V., Lataire, P, 2012, A DSP-based dual-loop peak DC-link voltage control strategy of the Z-source inverter, IEEE Trans. Power Electron., **27**(9) 4088–4097
[17] Hanif, M., Basu, M., Gaughan, K., 2011, Understanding the operation of a Z-source inverter for photovoltaic application with a design example, IET Power Electron., **4**(3) 278–287
[18] Calzo, G.L., Lidozzi, A., Solero, L., Crescimbini, F.,2012, Three-phase Z-source power supply design, IET Power Electron.,**5**(8) 1270–1281
[19] Rajaei, A.H., Mohamadian, M., Dehghan, S.M., Yazdian A.,2010, Single-phase induction motor drive system using z-source inverter, IET Electr. Power Appl., **4**(1) 17–25.
[20] Y. Tang, S.-J. Xie, C. H. Zhang, Z.-G. Xu, 2009, Improved Z-source inverter with reduced capacitor voltage stress and soft-start capability, IEEE Trans. Power Electron., **24**(2) 409–415
[21] M. K. Nguyen, Y. C. Lim, J. H. Choi, Y. O. Choi,2016, Trans-switched boost inverters, IET Power Electron.,**9**(5), 1065–1073
[22] M. Zhu, K. Yu, F. L. Luo, 2010, Switched inductor Z-source inverter, IEEE Trans. Power Electron., **25**(8) 2150–2158
[23] M. K. Nguyen, Y. C. Lim, and G. B. Cho, 2011, Switched-inductor quasi-Z source inverter, IEEE Trans. Power Electron., **26**(11) 3183– 3191
[24] K. Deng, J. Zheng, J. Mei, 2014, Novel switched-inductor quasi-Z-source inverter, J. Power Electron.,**14**(1) 11–21
[25] Rymarski, Z., Bernacki, K.,2014, Influence of Z-source output impedance on dynamic properties of single-phase voltage source inverters for uninterrupted power supply, IET Power Electron., **7**(8) 1978–1988
[26] Liu, Y., Ge, B., Abu-Rub, H., Peng, F.Z., 2014, Phase-shifted pulse-width-amplitude modulation for quasi-Z-source cascade five level inverter-based photovoltaic power system, IET Power Electron.,**7**(6) 1444–1456
[27] M.D. Seyed, M. Mustafa, G.Reza, 2012, Analysis and carrier-based modulation of Z-source NPC inverters, in International Journal of Electronics, **99**(8) 1075- 1099.
[28] Poh Chiang Loh, F.Blaabjerg, Chow Pang Wong, 2007, Comparative Evaluation of Pulse width Modulation Strategies for Z-Source Neutral-Point-Clamped Inverter, IEEE Transactions on Power Electronics, **22**(3) 1005-1013
[29] Y. Zhou, L. Liu, H. Li, 2013, A high- performance photovoltaic module-integrated converter (MIC) based on cascaded quasi-Z-source inverters (qZSI) using eGaNFETs, IEEE Trans. power Election, **28**(6) 2727 – 2738
[30] D. Sun, B. Ge, X. Yan, D. Bi, H.Y. Zhang Liu , H. Abu, L. Ben, F. Z. Feng, 2014, Modeling, impedance-design, and efficiency analysis of quasi-Z source module in cascaded multilevel photovoltaic power system, IEEE Trans. Ind. Election, **61**(11) 6108 – 6117
[31] Banaei, M.R., Dehghanzadeh, A.R., Salary, E., Khounjahan, H., Alizadeh, R, 2012, Z-source-

based fivelevel inverter with reduction of switches, IET Power Electron, **5**(3) 385–392

[32] Gao, F., Loh, P.C., Blaabjerg, F., Teodorescu, R., Vilathgamuwa, D.M, 2010, Five-level Z-source diode-clamped inverter, IET Power Electron, **3**(4) 500–510

[33] Upadhyay.S., Ravindranath.A., Mishra.S., Joshi.A, 2010, A Switched-Boost Topology for Renewable Power Application, IEEE, IPEC, 758-762

[34] S. Mishra.S., Adda.R., Joshi.A, 2012, Inverse Watkins–Johnson topology- based inverter, IEEE Trans. Power Electron, **27**(3) 1066–1070

[35] Ravindranath.A., Mishra.S., Joshi.S, 2013, Analysis and PWM control of switched boost inverter, IEEE Trans. Ind. Electron, **60**(12) 5593–5602

[36] M. K. Nguyen, T. V. Le, S. J. Park ,Y. C. Lim, 2015, A class of quasi-switched boost inverters, IEEE Trans. Ind.Electron, **62**(3) 1526 – 1536

[37] M. K. Nguyen, Y. C. Lim, S. J. Park, 2015, A comparison between single-phase quasi-Z-source and quasi-switched boost inverters, IEEE Trans. Ind. Electron **62**(10) 6336 – 6344

[38] M. K. Nguyen, T. V. Le, S. J. Park, Y. C. Lim, J. Y. Yoo, 2015, A class of high boost inverters based on switched-inductor structure, IET Power Electron., **8**(5) 750–759

[39] Tan-Tai Tran, Minh-Khai Nguyen, 2017, Cascaded five-level quasi-switched-boost inverter for single-phase grid-connected system, IET Power Electron.,**10**(14) 1896-1903

[40] V.T. Tran, M.K. Nguyen, M.H. Yoo, Y.O. Choi and G.B.Cho, 2017, A three- phase cascaded H-bridge quasi switched boost inverter for renewable energy, 20th International Conference on Electrical Machines and Systems (ICEMS),1-5

[41] Minh-Khai Nguyen, Truong-Duy Duong, Young-Cheol Lim, Yi-Gon Kim, 2018, Switched-Capacitor Quasi-Switched Boost Inverters, IEEE Trans. Ind Electron, **65**(6) 5105-5113

[42] D.T. Do and M.K. Nguyen, 2018, Three-Level Quasi-Switched Boost T-Type Inverter: Analysis, PWM Control, and Verification, IEEE transactions on industrial electronics, **65**(10) 8320 – 8329

[43] P.Sriramalakshmi, V.T. Sreedevi, 2018, A single phase cascaded five level Quasi Switched Boost Inverter based on switched capacitor structure, IEEE International Conference on Power Electronics, Drives and Energy Systems (PEDES) 1-6

Design and Simulation of Magnetic Levitation System

M Maheedhar[1], T.Deepa[1*]
[1]School of Electrical Engineering, Vellore Institute of Technology, Chennai, TamilNadu -600127.

E-mail: deepa.t@vit.ac.in

Abstract. This paper deals with the designing of magnetic levitation methodology depends on finite element method simulation. Sequences of simulations were performing, by changing the turn's number of the coil and by varying the current passing through coil to construct the coil capable of generating the field of 0.6016T. From the simulation result it was found that by varying the number of turns of the coil and current passing through coil, magnetic field generation also varied to levitate the body.

Keywords: Levitation, magnetic field, current, turns density.

1. Introduction

This magnetic levitation system is an advanced technology used in transportation system and in industrial areas [2]. The advantages of this system are dirt free, frictionless these advantages leads to the evolution of magnetic levitation system [9].

Due to frictionless, in transport system the vehicle can high speeds. Main aim of this paper was to suggest the maglev system modelling for levitating the body of certain mass. The FEMM (Finite Element Method Magnetics) simulation allowed explaining an optimal coil to levitate the body of certain mass. Hence the simulation will helpful to develop a methodology for deigning the magnetic levitation coil to a known mass.

To attain good results with less cost different geometries and various types of materials were simulated. This software is an 2D open source package, it solves the problems of both magnetostatic and AC problem [3]. This maglev is used for measuring the densities of samples in industries without knowing the mass and volume of the sample [4]. In axial maglev system by varying the distance between magnets the sensitivity of the system also can vary [5].

For triangular arrangement of high temperature superconductor bulks have the good performance [1]. The six degree of freedom maglev system is simple in view of communication and number of coils [2]. Any conductor which carries current is enclosed by the path must flow through surface once [7]. This FEM software is advantageous tool which can be used for formulation and analysis of problems of engineering [8].

FEM is applied to the problems of electromagnetism by using basic concepts of calculus [8]. In MAGLEV transport system the Linear-quadratic regulator cause more stability and uniform response compared to other controllers [9].

2. Methodology

In FEMM software [6] the simulation for coil was performed. The objective of simulation is to develop a methodology of magnetic levitation coil design. Simulation will helpful to design perfect coil for hanging the body of known mass without any mechanical support. The magnetic field produce by the coil can be varied along varying the turn's number and electric current flowing through that coil. Thus by varying the number of turns of the coil and current passing through the coil most suitable coil can design for levitating the body of known mass.

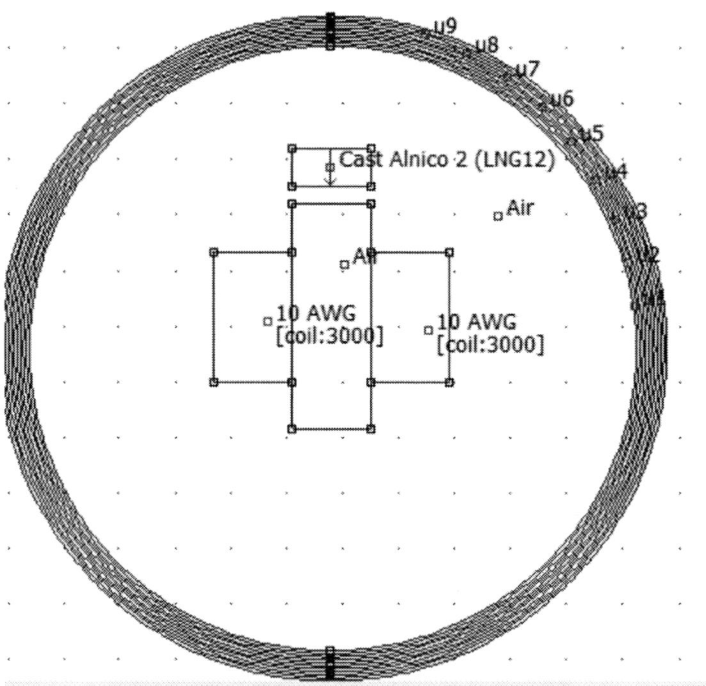

Figure 1. Representation of coil.

Figure 1 shows the representation of designed coil. For designing the MAGLEV system, Neodymium magnet and coil with 3000 turns and with the gauge of 10 AWG, and air is used as a medium. At constant number of turns the current flowing through the coil is varied as 5A, 5.5A, 6A, 6.A, 7A, 7.5A and at constant current number of turns also varied as 500, 1000, 1500, 2000, 2500, 3000.

Figure 2 shows the mesh analysis of the system; this mesh analysis is an integral part in engineering simulation which makes the complicated geometries into simple. This mesh analysis effects the accuracy, convergence, and speed of simulation

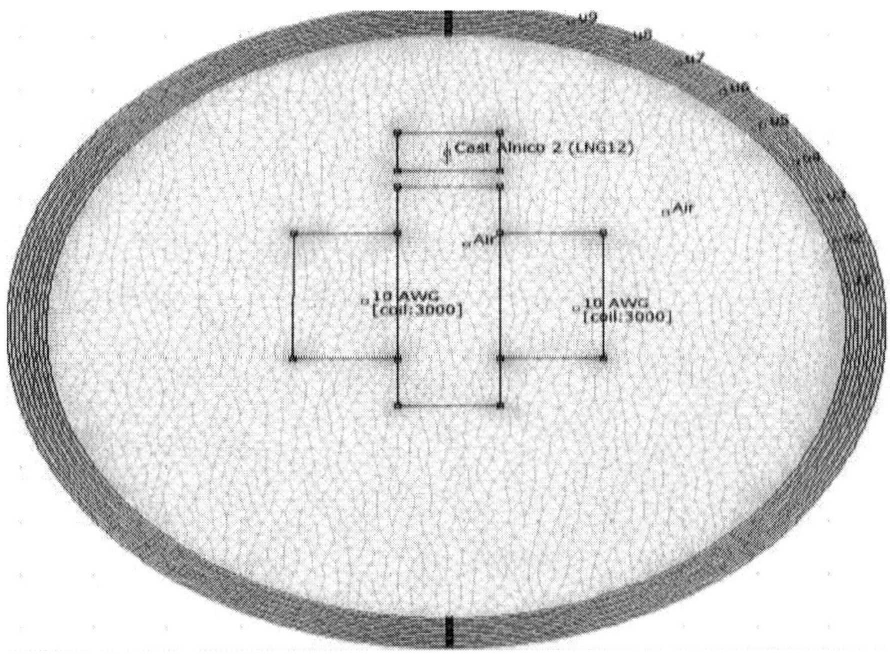

Figure 2. Mesh Analysis

3. Results and Discussion

Initially coil with 500 turns with the medium of air were simulated. This below table 1 and 2 shows the variation of magnetic field by variation of turn's number of the coil and current flow through coil respectively. As the turn's number of the coil was varied from 500, 1000, 1500, 2000, 2500, 3000 magnetic field varies from 0.007376 T to 0.270092 T. Similarly by varying the current flowing through the coil from 5A to 7A with the interval of 0.5A magnetic field varies from 0.270092 T to 0.6099T.

Table 1. Variation of magnetic field density with coil turns

No. of turns of the coil	Magnetic flux density (T)
500	0.007376
1000	0.029787
1500	0.06726
2000	0.119801
2500	0.1874
3000	0.270092

Table 2. Variation of magnetic field by variation of current

Coil Current (A)	Magnetic flux density (T)
5	0.270092
5.5	0.32681
6	0.38893
6.5	0.4564
7	0.52937
7.5	0.6099

4. Conclusions

From the simulation results it was concluded that by increasing the turn's number of the coil and current flow through coil magnetic field produced by the coil also increases. In some systems where space is main constraint, varying the turn's number of the coil is not possible in such cases the magnetic field produced by the coil is varied only by varying the current flowing through the coil.

References

[1] Liang G. Aug 2015. Study of the maglev performance of the side-mounted high-temperature superconductor maglev rotating system. IEEE Trans. Appl. Supercond 25(4).
[2] Zhu Haiyue, Teo Tat Joo, Pang Chee Khiang. 2017 Design and modeling of a six-degree-of freedom magnetically levitated positioner using square coils and 1-D Halbach arrays.IEEE Trans. Ind. Electron. 64(1):440–50.
[3] Nan-Ho Kim, 2015 Introduction to Nonlinear Finite Element Analysis. 1st ed. Springer, Softcover reprint of the original.
[4] Qiu-Hua Gao, Wen-Bo Li, 2020 A centrifugal magnetic levitation approach for high-reliability density Measurement, Sensors & Actuators: B. Chemical volume 304 127362.
[5] Chengqian Zhang, Peng Zhao, Daofan Tang, Neng Xia, 2020 Axial Magnetic Levitation: A High-sensitive and Maneuverable Density-based Analysis Device, Sensors & Actuators: B. Chemical volume 304 127362.
[6] David Meeker, Finite Element Method Magnetics. http://www.femm.info/wiki/HomePage. Internet.
[7] Hayt WH, Buck JA 2019. Engineering Electromagnetics. New York, NY: McGraw-Hill Education;.
[8] Cardoso José Roberto. 2016 Electromagnetics through the Finite Element Method: A Simplified Approach Using Maxwell's Equations. São Paulo: CRC Press. 198 p.
[9] Yaseen Mundher HA, Abd Haider J. March 2018 Modeling and control for a magnetic levitation system based on SIMLAB platform in real time. Results Phys. [s.l.], 8. Elsevier BV p. 153–9.

Embedding Complete Bipartite Graph into Quadrilateral Snake

Leo Joshwa. P [1*], A. Berin Greeni[1]
[1]School of Advanced Sciences, Vellore Institute of Technology, Chennai – 600127.

E-mail: joseleowa18@gmail.com

Abstract. This paper deals with the technique that maps a guest graph into a host graph, known as Graph embedding. This tool is used in the simulation of interconnection networks. The exact wirelength of embedding complete bipartite graphs into quadrilateral snakes is considered as a research problem in this paper.

Keywords: complete bipartite graph, quadrilateral snakes, wirelength.

1. Introduction

The computing architecture utilizes interconnection network as the communication between the processor cores and memory modules. Thus an interconnection network is nothing but a graph. Thus the process of data exchange between parallel computers is made feasible in a computer network. Distinctive feature of graph embedding is of major concern in the design of the interconnection network. It is a strategy where the guest graph D maps to a host graph F [1]. Numerous applications are likely to be modeled as graph embedding problem. For instance, one architecture can be simulated into another by modelling as an embedding problem [2]. Dilation, congestion and wirelength are important measure of an embedding which we compute in this paper for embedding complete bipartite graph into quadrilateral snakes.

2. Basic concepts

Definition 2.1. [1]
An one - one mapping $\mathcal{E}$ of the vertex set of a graph D into the vertex set of a graph F is called an embedding if every edges (x, y) in D is mapped to a path between $\mathcal{E}(x)$ and $\mathcal{E}(y)$ in F.

Definition 2.2. [1]
Let $\mathcal{E} : D \rightarrow F$ be an embedding. For $e \in E(F)$, the number of path in $\{ p_{\mathcal{E}}(u, v): p_{\mathcal{E}}(u, v)$ is a path between $\mathcal{E}(u)$ and $\mathcal{E}(v)$ in F for $(u, v) \in E(F)$ that contains $e \}$ is called the congestion on e with respect to $\mathcal{E}$ and is defined by $c_{\mathcal{E}}(e)$. For $S \subseteq E(M)$, we define $c_{\mathcal{E}}(S)=\sum_{v \in S} c_{\mathcal{E}}(v)$.

Definition 2.3. [1]
Let $\mathcal{E} : D \rightarrow F$ be an embedding. Then the congestion of $\mathcal{E} : D \rightarrow F$ is given by $c_{\mathcal{E}}(D,F) = max\, c_{\mathcal{E}}(e)$. The congestion problem is to find $\min_{\mathcal{E}} c_{\mathcal{E}}(D,F)$ where the minimum is taken over all embedding $\mathcal{E} : D \rightarrow F$.

Definition 2.4. [3]
The wirelength of an embedding $\mathcal{E}: D \rightarrow F$ is given by $WL_{\mathcal{E}}(D, F) = \sum_{e\in E(F)} c_{\mathcal{E}}(e)$. The wirelength of embedding D into F denoted as $WL(D, F) = \min_{\mathcal{E}} WL_{\mathcal{E}}(D,F)$ where the minimum is taken over all embedding $\mathcal{E}: D \rightarrow F$

The wirelength problem [1, 3] of a graph D into F is to determine $WL_{\mathcal{E}}(D, F)$. The wirelength problem is NP-complete[4].

The wirelength problem of various architectures such as hypercube into grid [3], complete bipartite graph into grid[5], complete bipartite graph into necklace graph[6] and complete bipartite graph into sibling trees[7] are studied.

Definition 2.5. [8]
A subgraph K of a graph D is said to be convex if all the shortest paths between any two vertices of K lie in K. An edge cut M of D is said to be a convex cut if $D\backslash M$ splits into two components, each of which is convex.

Lemma 2.6. (Congestion Lemma) [9]
For an embedding $\mathcal{E}$ of a graph D into F, let M be a convex edge cut of F such that $F\backslash M$ splits into components F_1 and F_2 and let $D_1 = \mathcal{E}^{-1}(F_1)$ and $D_2 = \mathcal{E}^{-1}(F_2)$. Suppose D_1 and D_2 are maximal subgraph of D and $P_{\mathcal{E}}(u,v)$ with $u \in D_1$ and $v\in D_2$ contains exactly one edge in M for every $(u, v) \in E(D)$, Further

$$c_{\mathcal{E}}(M) = \sum_{v\in V(D_1)} d_G(v) - 2|E(D_1)|$$
$$= \sum_{v\in V(D_2)} d_G(v) - 2|E(D_2)|.$$

Lemma 2.7. (k - Partition Lemma) [3, 10]
Let $\mathcal{E}: D \rightarrow F$ be an embedding. Let $E^k(F)$ represent the collection of edges of F, where each edge of F is repeated exactly k times. Let $\{X_1, X_2, \ldots, X_p\}$ be a partition of $E^k(F)$ such that each X_i is an edge cut of F. Then the wirelength of embedding D into F with respect to $\mathcal{E}$ is given by

$$WL_{\mathcal{E}}(D, F) = \frac{1}{k} \sum_{i=1}^{p} c_{\mathcal{E}}(X_i).$$

3 Quadrilateral Snakes

Definition 3.1. [11]
Complete bipartite graph denoted as $K_{m,n}$, is a special type of bipartite graph whose every vertex of one set X is connected to every vertex of other set Y, with $|X| = m$ and $|Y| = n$.

Lemma 3.2. [5] $K_{\lceil \frac{l}{2} \rceil, \lfloor \frac{l}{2} \rfloor}$ is the maximum subgraph of $K_{m,n}$ on l vertices.

Even-odd labeling of $K_{m,n}$

If (X, Y) is the partition of $V(K_{m,n})$, with $|X| = m$ and $|Y| = n$, $m \geq n$. The vertices of X and Y are labeled as $1, 3, \ldots, 2m-1$ and $2, 4, \ldots, 2n$ respectively.

Definition 3.3. [12]
A quadrilateral snake $Q(n)$ is obtained from a path $v_1, v_2, \ldots v_{n+1}$ by joining vertices v_i and v_{i+1} to new vertices x_i and y_i respectively and adding edges $x_i\, y_i$ for $1 \leq i \leq n$ See **Figure 1**.

Remark 3.4. The quadrilateral snake $Q(n)$ has $3n+1$ vertices and $4n$ edges.

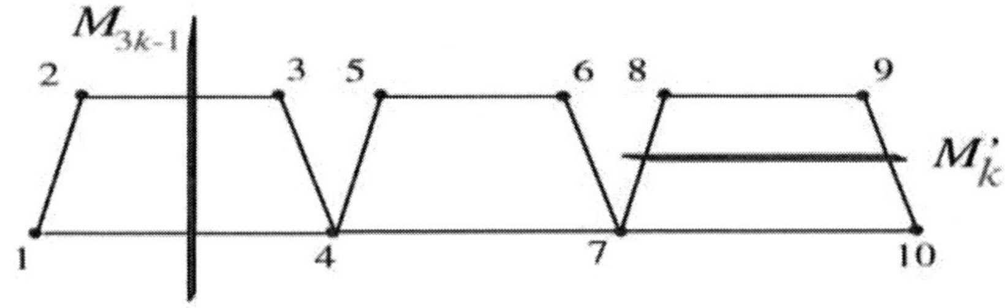

Figure 1: Edge cut of Quadrilateral snake $Q(3)$

Algorithm A

Input : The complete bipartite graph $K_{\lceil\frac{3n+1}{2}\rceil,\lfloor\frac{3n+1}{2}\rfloor}$ and the quadrilateral snake $Q(n)$.

Algorithm : Using Even-odd labelling, label $V\left(K_{\lceil\frac{3n+1}{2}\rceil,\lfloor\frac{3n+1}{2}\rfloor}\right)$. Label $V(Q(n))$ as follows: the vertex v_k is labeled as $3k-2$ for $1 \leq k \leq n$ and u_i, w_i as $3k-1$, $3k$ respectively for $k = 1, 2, \ldots, n-1$.

Output : An embedding $\mathcal{E}: K_{\lceil\frac{3n+1}{2}\rceil,\lfloor\frac{3n+1}{2}\rfloor} \rightarrow Q(n)$ given by $\mathcal{E}(y) = y$ gives the wirelength minimum.

Proof of correctness : For $1 \leq k \leq n$, let $M_{3k-1} = \{(3k-2, 3k+1), (3k-1, 3k)\}$ be the edge cut of the quadrilateral snake $Q(n)$. The edge cut M_{3k-1}, $1 \leq k \leq n$ disconnects $Q(n)$ into components: F_{k1}^{3k-1}, F_{k2}^{3k-1} Clearly $V(F_{k1}^{3k-1})$ is labeled consecutively. See **Figure 1**. By Lemma 3.2, $D_{k1}^{3k-1} = \mathcal{E}^{-1}(F_{k1}^{3k-1})$ and $D_{k2}^{3k-1} = \mathcal{E}^{-1}(F_{k2}^{3k-1})$ are maximum subgraph of $K_{\lceil\frac{3n+1}{2}\rceil,\lfloor\frac{3n+1}{2}\rfloor}$. Thus the edge cut M_{3k-1}, $1 \leq k \leq n$ satisfies conditions (1) and (2) of the Congestion Lemma. Clearly for $k = 1, 2, \ldots, n$ $C_{\mathcal{E}}(M_{3k-1})$ is minimum. Similarly, for $1 \leq k \leq n$, let $M_k' = \{(3k-1, 3k), (3k-2, 3k+1)\}$ be the edge cut of the quadrilateral snake $Q(n)$. The edge cut M_k', $1 \leq k \leq n$ disconnects $Q(n)$ into components : F_{k1}', F_{k2}', Clearly $V(F_{k1}')$ is labeled consecutively. See **Figure 1**. By Lemma 3.2, $D_{k1}' = \mathcal{E}^{-1}(F_{k1}')$ and $D_{k2}' = \mathcal{E}^{-1}(F_{k2}')$ are maximum subgraph of $K_{\lceil\frac{3n+1}{2}\rceil,\lfloor\frac{3n+1}{2}\rfloor}$. Thus the edge cut M_k', $1 \leq k \leq n$ satisfies conditions (1) and (2) of the Congestion Lemma. Clearly for $k = 1, 2, \ldots, n$, $C_{\mathcal{E}}(M_k')$ is minimum. Therefore, using Partition lemma the wirelength is obtained.

Theorem 3.5.
The wirelength of the complete bipartite graph $K_{\lceil\frac{3n+1}{2}\rceil,\lfloor\frac{3n+1}{2}\rfloor}$ denoted as D and the quadrilateral snake $Q(n)$, denoted by F is given by

$$WL(D,F)=\begin{cases}\frac{1}{2}\left[\sum_{k=1,3}^{n}[(3k-1)(3(n-k)+2)]+\sum_{k=2,4}^{n-1}9k(n+1-k)-(3n+1)\right]+n(3n-1): & n \text{ is odd}\\ \frac{1}{2}\left[\sum_{k=1,3}^{n}[(3k-1)(3(n-k)+2)]+\sum_{k=2,4}^{n-1}9k(n+1-k)-(3n+2)\right]+n(3n-1): & n \text{ is even}\end{cases}$$

Proof. Following the Algorithm A, we split the proof into two cases.

Case 1 (n is odd) By Congestion Lemma,
$C_{\mathcal{E}}(M_{3k-1})=\binom{3n+1}{2}(3k-1)-2I_G(3k-1)$ for $1\leq k\leq n$ and $C_{\mathcal{E}}(M'_k)=\binom{3n+1}{2}(2)-2=3n-1$, for $1\leq k\leq n$. Clearly,

$$C_{\mathcal{E}}(M_{3k-1})=\begin{cases}\frac{1}{2}[(3k-1)(3(n-1)] & : k \text{ is odd}\\ \frac{1}{2}[9k\,(n+1-k)-(3n+2)]: & k \text{ is even}\end{cases}$$

Using Partition Lemma we obtain,

$$WL(D,F)=\frac{1}{2}\left[\sum_{k=1}^{n}C_{\mathcal{E}}(M_{3k-1})\right]+\sum_{k=1}^{n-1}C_{\mathcal{E}}(M'_k)$$

$$WL(D,F)=\frac{1}{2}\left[\sum_{k=1,3}^{n}[(3k-1)(3(n-k)+2)]+\sum_{k=2,4}^{n-1}9k(n+1-k)-(3n+1)\right]+n(3n-1)$$

Case 2 (n is even) By Congestion Lemma,
$C_{\mathcal{E}}(M_{3k-1})=\binom{3n+1}{2}(3k-1)-2I_G(3k-1)$ for $1\leq k\leq n$ and $C_{\mathcal{E}}(M'_k)=\binom{3n+1}{2}(2)-2=3n-1$, for $1\leq k\leq n$
Clearly,

$$C_{\mathcal{E}}(M_{3k-1})=\begin{cases}\frac{1}{2}[(3k-1)(3(n-1)] & : k \text{ is odd}\\ \frac{1}{2}[9k\,(n+1-k)-(3n+2)]: & k \text{ is even}\end{cases}$$

Using Partition Lemma we obtain,

$$WL(D,F)=\frac{1}{2}\left[\sum_{k=1}^{n}EC_f(M_{3k-1})\right]+\sum_{k=1}^{n-1}EC_f(M'_k)$$

$$WL(D,F)=\frac{1}{2}\left[\sum_{k=1,3}^{n}[(3k-1)(3(n-k)+2)]+\sum_{k=2,4}^{n-1}9k(n+1-k)-(3n+1)\right]+n(3n-1).$$

Definition 3.6. [12]
The quadrilateral snake $pQ(n)$ has n blocks of quadrilateral, where each block contains p number of quadrilaterals, with one edge common. When $n = 2$, it is called double quadrilateral.

Remark 3.7. The quadrilateral snake $pQ(n)$ has $(2pn + n) + 1$ vertices and $(3pn + n)$ edges.

Algorithm B

Input : $K_{\lceil\frac{(2pn+n)+1}{2}\rceil,\lfloor\frac{(2pn+n)+1}{2}\rfloor}$, the complete bipartite graph and $pQ(n)$, $p>1$, the quadrilateral snake.

Algorithm : Using Even-odd labelling, label $V\left(K_{\lceil\frac{(2pn+n)+1}{2}\rceil,\lfloor\frac{(2pn+n)+1}{2}\rfloor}\right)$. Label $V(pQ(n))$ as follows:
Let L_i denote the path induced by the vertices in $Q(n)$, $1\leq i\leq 3n+1$. Sequentially, label the quadrilateral snake L_i, $1\leq i\leq p$ of $pQ(n)$ with numbers $1, 2, \ldots p(2n+1)+1$ alternately from left to right and from right to left, beginning with L_1.

Output : An embedding $\mathcal{E} : K_{\left\lceil\frac{(2pn+n)+1}{2}\right\rceil,\left\lfloor\frac{(2pn+n)+1}{2}\right\rfloor} \longrightarrow pQ(n),\ p > 1$ given by $\mathcal{E}(y) = y$ gives wirelength minimum. See **Figure 2**.

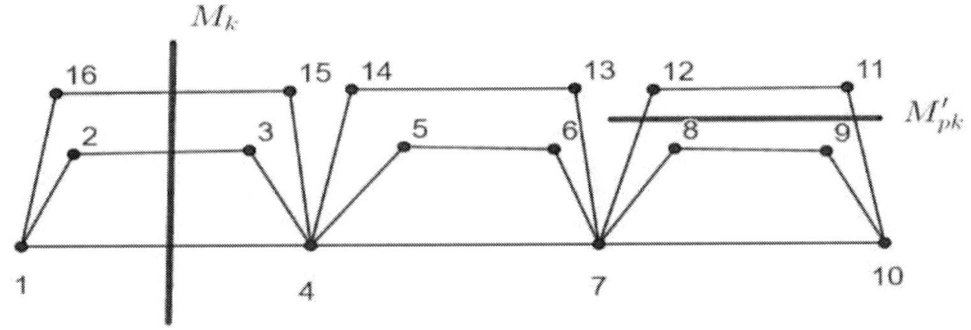

Figure 2: Edge cut of Quadrilateral snake $2Q(3)$.

Proof of correctness : Let M_k, $1 \le k \le n$ be the edge cut sets which cuts the k-blocks of the quadrilateral snake $pQ(n)$ such that M_k disconnects $pQ(n)$ into two components F_{k1} and F_{k2} where $V(F_{k1})$ contains $k + p(2k - 1) + 1$ vertices with atmost equal number of odd and even labels. By Lemma 3.2, $D_{k1} = \mathcal{E}^{-1}(F_{k1})$ and $D_{k2} = \mathcal{E}^{-1}(F_{k2})$ are maximum subgraph of $K_{\left\lceil\frac{(2pn+n)+1}{2}\right\rceil,\left\lfloor\frac{(2pn+n)+1}{2}\right\rfloor}$. Thus the edge cut M_k, $1 \le k \le n$ satisfies conditions (1) and (2) of the Congestion Lemma. Therefore $C_{\mathcal{E}}(M_k)$ is minimum for $k = 1, 2, ..., n$. For $1 \le k \le n$, $p > 1$, let M'_{pk} be the edge cut which cuts the quadrilateral snake $pQ(n)$ horizontally in each block as algorithm A. Clearly for $k = 1, 2, \ldots, n$, $C_{\mathcal{E}}(M'_{pk})$ is minimum. Therefore, using Partition lemma the wirelength is obtained.

Theorem 3.7.
The embedding of complete bipartite graph $K_{\left\lceil\frac{(2pn+n)+1}{2}\right\rceil,\left\lfloor\frac{(2pn+n)+1}{2}\right\rfloor}$ into the quadrilateral snake $pQ(n)$, $p > 1$ yields minimum wirelength.

4 Conclusion
The wirelength of embedding complete bipartite graph into quadrilateral snake is studied in this paper.

References
[1] S.L. Bezrukov, J.D. Chavez, L.H. Harper, M. R¨ottger and U.P. Schroeder, 1998. Embedding of hypercubes into grids, Mathematical Foundations of Computer Science, Lecture Notes in Computer Science, **1450**, 693 – 701.
[2] S.L. Bezrukov, J.D. Chavez, L.H. Harper, M. R¨ottger and U.P. Schroeder, 2000. The congestion of n-cube layout on a rectangular grid, Discrete Mathematics, **213**, 13 – 19.
[3] P. Manuel, I. Rajasingh, B. Rajan and H. Mercy, 2009. Exact wirelength of hypercube on a grid, Discrete Applied Mathematics, **157**(7), 1486 – 1495.
[4] M.R. Garey and D.S. Johnson, 1979. Computers and intractability: a guide to the theory of NP completeness, Freeman, San Francisco, California.
[5] A.B. Greeni and I. Rajasingh, 2017. Embedding complete bipartite graph into Grid with optimum congestion and Wirelength, International Journal of Network and Virtual

Organisation, **17**, 64 – 75.

[6] A.B. Greeni, 2020. Embedding complete bipartite graphs into necklace graphs, Procedia Computer Science, **172**, 199-203.

[7] A.B. Greeni, 2020. Embedding Complete Bipartite Graph into Sibling Tree with Optimum Wirelenth, Journal of Combinatorial Mathematics and Combinatorial Computing, **112**, 115-126.

[8] S.L. Bezrukov, 1999. Edge isoperimetric problems on graphs, Graph Theory and Combinatorial Biology, Bolyai Soc. Math. Stud. 7, L. Lovasz, A. Gyarfas, G.O.H. Katona, A. Recski, L. Szekely eds., Budapest, 157 – 197.

[9] M. Miller, R.S. Rajan, N. Parthiban and I. Rajasingh, 2015. Minimum linear arrangement of incomplete hypercubes, The Computer Journal, **58**(2), 331 – 337.

[10] M. Arockiaraj, P. Manuel, I. Rajasingh and B. Rajan, 2011. Wirelength of 1-fault hamiltonian graphs into wheels and fans, Information Processing Letters, **111**, 921–925.

[11] J.M. Xu, 2001. Topological structure and analysis of interconnection networks, Kluwer Academic Publishers, Boston.

[12] T.Manjula, R.Rajeswari, 2016. Dominator coloring of Quadrilateral snake, Triangle snake and Barbell graph, in Second International Conference on Science Technology Engineering and Management (ICONSTEM), Chennai, 115-119, doi: 10.1109/ICONSTEM.2016. 7560934.

A Similarity Measure for Text Document Using Term Cardinality

S. Sumathi[1], **G. Hannah Grace**[2]

[1]School of Advanced Sciences, Division of Mathematics, Vellore Institute of Technology, Chennai, Tamilnadu, India.

E-mail: sumathi.manivannan04@gmail.com

Abstract- With enormous development of digital technology, data is being generated at rapid rate with various application domains. Data has to be extracted or filtered to find useful information. A basic concept for these tasks and applications are the distance measures to effectively determine how similar two objects are. In this paper, a novel similarity measure for clustering text documents is proposed using the cardinality of the terms in the documents. The bench mark algorithm k-medoids is used for clustering task. The results obtained from the proposed distance measure are compared with other standard distance measures like Manhattan, Euclidean distance measure. Dunn Index is used to analyze the cluster validation of the results obtained from the distance measure.

Keywords: Document Clustering, k-Medoids, Similarity/Distance Measure, Dunn Index.

1. Introduction

In past few years there has been a sudden increase in the volume of text documents used on internet. Thus, there is a need for efficient methods for understanding the documents by the machine itself. To address this problem, document similarity plays a vital role for machines to process natural language. A few applications of documents similarity are document clustering, document summarization, document categorization and query-based search [1].

Text documents can be represented as vector space model. In this model, each document is represented as a vector, in which the components are each term of the document. Usually, raw text documents are not preferred for clustering text. Hence, the document has to be pre-processed before clustering task. This preprocessed text document is called a corpus and is used as an input document for clustering. The steps involved in pre-processing are tokenization, stop-word removal, converting to upper case to lower case and stemming [2]. The corpus is now transformed to term document matrix by using the raw count of the terms present in the document. This term document matrix forms the base of the document clustering task. This matrix can be used with several known weighting scheme. The most popular weighting scheme is the term frequency and inverse document frequency (tf-idf) [3]. The flowchart of proposed approach of this paper is shown in **figure 1**.

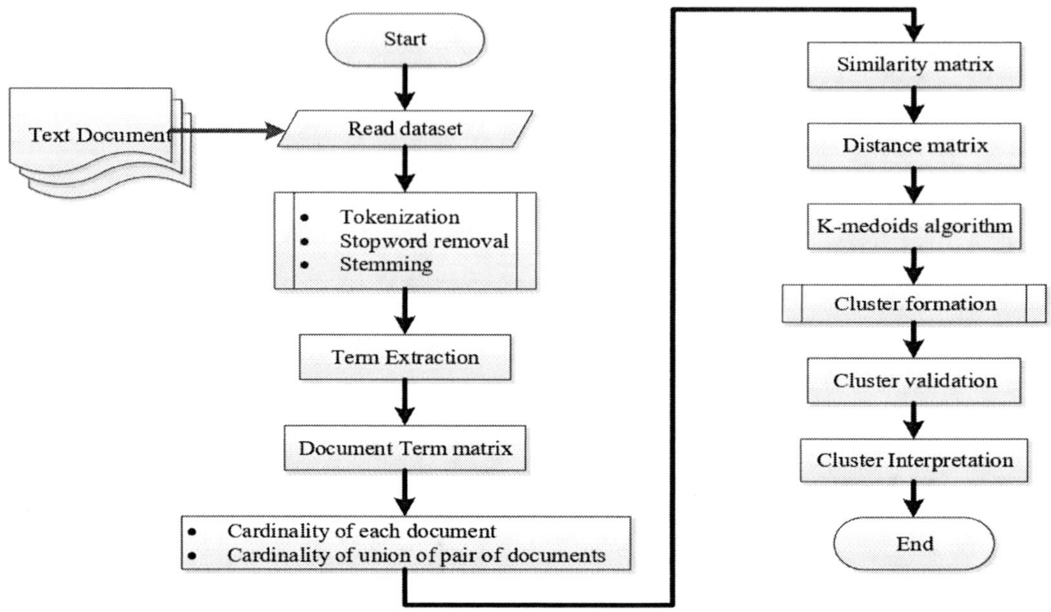

Figure 1. Flowchart of proposed approach

2. Distance Measures

Distance measure is a numerical value which describes how dissimilar two objects are. The similarity measure is another notion of distance measure i.e., how similar two objects are. The similarity value ranges from 0 (no similarity) to 1 (most similarity) [4]. The distance is given as 1 minus the similarity value. The most similar objects have the least distance implying that the similarity measure is high and vice-versa.

2.1. Definition

A metric on a set X is a function called the distance function or simply distance, if there exist a map $d: X \times X \rightarrow \mathbb{R}$ (where $\mathbb{R}$ is the set of real numbers) such that for all x, y, z in X, the function satisfy the following axioms:

(1) $d(x,y) \geq 0$ (non-negativity, or separation axiom)
(2) $d(x,y) = 0$ if and only if $x = y$ (coincidence axiom)
(3) $d(x,y) = d(y,x)$ (symmetry axiom)
(4) $d(x,z) \leq d(x,y) + d(y,z)$ (Triangle inequality)

2.2. Euclidean Distance

The Euclidean distance between two data points is defined by the square root of the sum of the squared differences of the all co-ordinates of the data points. For x, y in n-dimensional space, the Euclidean distance d is defined as:

$$d(x,y) = \sqrt{\sum_{i=1}^{n} (x_i - y_i)^2}$$

It is the most common metric for numerical data [5].

2.3. Manhattan Distance

The Manhattan distance between two data points is defined by the absolute value of sum of the differences of the all co-ordinates of the data points. For x, y in n-dimensional space, the Manhattan distance, d is defined as:

$$d(x,y) = \sum_{i=1}^{n} |x_i - y_i|$$

This is also called taxicab distance [6].

3. k-Medoids

Clustering is an unsupervised learning, where the objects of similar patterns are grouped together to form clusters. The clusters obtained can be used to interpret and group similar objects. k-Medoids is a partition approach, in which the objects are partitioned into k clusters. The initial medoid is chosen randomly. The clusters are based on minimum distance. The centroid in k-means is the mean of the objects within the cluster whereas in k-medoids the centroid is taken as an object itself. In general, k-Medoids is used to cluster the types of data in which the mean of objects is not defined or unavailable [7]. The flowchart of k-medoids algorithm is shown below in **figure 2.**

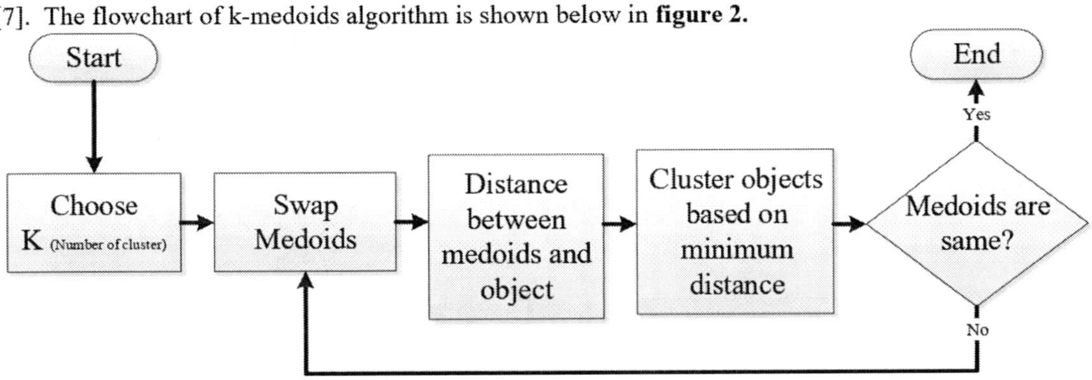

Figure 2. Flowchart of k-medoids algorithm

4. Dunn Index

The Dunn Index (DI) is a validation index used to analyze the compactness of the cluster obtained from the clustering task. For any cluster partition V of the dataset X with k number of partition (namely $C_1 \dots C_k$), the Dunn index is given by,

$$D^{(V)} = \frac{\min\limits_{1 \le i \le j \le k, i \ne j} \delta(C_i, C_j)}{\max\limits_{1 \le i \le k} \Delta_k}$$

where $\delta(C_i, C_j)$ is the distance between the two different cluster C_i, C_j (inter-cluster distance) and Δ_k is the maximum intra-cluster distance of the k-th cluster [8]. In general, a higher value of Dunn index suggests a better cluster quality.

5. Proposed Method

Clustering of text document is a challenging task. Distance measure plays a major role in clustering algorithm. Though, geometric distance measure like Euclidean, Manhattan are handy, the clusters obtained are based on the structure produced by the compactness of the cluster. In this paper, we have proposed a similarity measure based on term cardinality of the document. After pre-processing the documents, the terms of each document is represented as a set and the proposed similarity measure makes use of the cardinality of the document. The cardinality of the document is the number of terms involved in the document.

The similarity measure between each pair of documents A and B is given by,

$$\mathrm{Sim}(A,B) = \frac{\mathcal{C}_A \cdot \mathcal{C}_B}{\mathcal{C}_{A \cup B}^2}$$

where $\mathcal{C}_A$ is the cardinality of terms in document A and $\mathcal{C}_{A \cup B}$ is the cardinality of the union of terms in document A and document B.

Using the above similarity measure the distance between two documents A and B is given by

$$\mathrm{dist}(A,B) = 1 - \mathrm{Sim}(A,B)$$
$$= 1 - \frac{\mathcal{C}_A.\mathcal{C}_B}{\mathcal{C}^2_{A\cup B}}$$

The proposed distance measure satisfies the three conditions of metric.

1. As $\mathcal{C}^2_{A\cup B} \geq \mathcal{C}_A.\mathcal{C}_B$ and $\mathcal{C}_{A\cup B} > 0$, for all A, B it follows that $\mathrm{dist}(A,B) \geq 0$. Hence axiom 1 is satisfied.
2. Notice that, $\mathrm{Sim}(A,A) = 1$, as the cardinality of union of same documents equals to A. Thus, $\mathrm{dist}(A,B) = 0$ whenever $A = B$. Hence axiom 2 is satisfied.
3. It is clear that $\mathrm{Sim}(A,B) = \mathrm{Sim}(B,A)$. It follows that $\mathrm{dist}(A,B) = \mathrm{dist}(B,A)$. Hence axiom 3 is satisfied.
4. Let A, B, C be any three documents. We shall consider the following cases.

Case (i)

Let, $A \subseteq C$ and $B \subseteq C$ then,

$$dist(A,C) + dist(C,B) = \left(1 - \frac{\mathcal{C}_A.\mathcal{C}_C}{\mathcal{C}^2_{A\cup C}}\right) + \left(1 - \frac{\mathcal{C}_C.\mathcal{C}_B}{\mathcal{C}^2_{C\cup B}}\right)$$

$$\geq 2 - \left(\frac{\mathcal{C}_A.\mathcal{C}_{A\cup B}}{\mathcal{C}^2_{A\cup B}} + \frac{\mathcal{C}_{A\cup B}.\mathcal{C}_B}{\mathcal{C}^2_{A\cup B}}\right) \quad (\text{Since } \mathcal{C}_{A\cup B} \leq \mathcal{C}_C)$$

$$\geq 2 - 2\left(\frac{\mathcal{C}_A.\mathcal{C}_B}{\mathcal{C}^2_{A\cup B}}\right)$$

$$= 2\left(1 - \frac{\mathcal{C}_A.\mathcal{C}_B}{\mathcal{C}^2_{A\cup B}}\right) = 2 * dist(A,B) \geq dist(A,B)$$

Case (ii)

Suppose, $C \subseteq A$ and $C \subseteq B$ then for any elements in A or B but not in C will only increase $dist(A,C) + dist(C,B)$, but not $dist(A,B)$. If $A = B = C$, then $0 + 0 \geq 0$. Thus, the triangular inequality holds for all cases.

Hence, our proposed measure is a metric.

Thus, the similarity measure between two documents is given by the ratio of product of cardinalities of terms to the square of cardinality of the union of terms.

6. Dataset

The documents used for clustering task are downloaded from the BBC sport website [9]. The number of documents in the BBC sports dataset is 737 which come under five classes namely athletics, cricket, football, rugby, tennis. This document was preprocessed and reduced to a document term matrix which consists of 2042 terms.

7. Results and Discussion

In this paper, we have illustrated the proposed measure with a toy dataset which consist of 6 documents and 8 terms [10]. The raw count is tabulated below in **Table 1**.

Table 1. Term-Document Matrix

	cluster	cluto	Distance	document	evaluate	high	measure	use
D1	1	1	0	0	0	1	0	0
D2	1	1	0	0	0	0	0	0
D3	1	0	0	1	0	1	0	0
D4	1	0	1	2	0	0	0	0
D5	1	0	0	1	1	0	2	1
D6	0	0	2	1	1	0	1	1

As an example, $D_1 = \{cluster, cluto, high\}$ and $D_2 = \{cluster, cluto\}$.

Thus $D_1 \cup D_2 = \{cluster, cluto, high\}$.

Here, $\mathcal{C}_{D_1} = 3,\ \mathcal{C}_{D_2} = 2$ and $\mathcal{C}_{D_1 \cup D_2} = 3$

Thus, $Sim(D_1, D_2) = \frac{3.2}{3^2} = \frac{2}{3} = 0.667$.

The similarity matrix obtained for the above dataset is given in **table 2**.

Table 2. Similarity matrix

	D1	D2	D3	D4	D5	D6
D1	1	0.667	0.562	0.36	0.306	0.234
D2	0.667	1	0.375	0.5	0.278	0.2041
D3	0.562	0.375	1	0.5625	0.4167	0.306
D4	0.36	0.5	0.5625	1	0.4167	0.4167
D5	0.306	0.278	0.4167	0.4167	1	0.694
D6	0.234	0.2041	0.306	0.4167	0.694	1

The k-medoids algorithm (for k=2, 3, 4) employed with Euclidean, Manhattan and our proposed measure are used for clustering task. The clusters obtained are validated using Dunn Index. The Dunn Index obtained for Euclidean, Manhattan and our proposed measure is evaluated and it is shown in **table 3**.

Table 3. k-medoids Algorithm using 3 different metrics assessed by Dunn Index

	DISTANCE METRICS		
Number of Cluster	Euclidean	Manhattan	Proposed Metric
k=2	0.6123	1	0.7
k=3	0.7071	0.8	1.0011

Further, the proposed measure is analyzed using documents downloaded from the BBC sport website. As the original document consist of 5 class labels, we have analyzed the results for $k = 5$ using k-Medoids algorithm. The clustering results are validated using Dunn Index. The value obtained using Dunn Index is given in **table 4**.

Table 4. k-medoids algorithm using 3 different metrics assessed by Dunn Index for BBC Sports dataset

	DISTANCE METRICS		
Number of Cluster	Euclidean	Manhattan	Proposed Metric
k=5	0.07952198	0.04494382	**0.207456**

From the above table, we can see that the cluster obtained from the proposed measure is of good quality when compared to both Euclidean and Manhattan.

8. Conclusion

In this paper, a novel distance measure for text document clustering is proposed. The BBC Sports dataset used is pre-processed and a document term matrix is obtained. This matrix is employed with bench mark algorithm k-medoids for clustering task. The results obtained from the proposed distance measure are compared with other standard distance measures like Manhattan, Euclidean distance measure. Dunn Index is used to analyze the cluster validation of the results obtained from the distance measure. We find the proposed measure perform well for sparse matrix. Our future work is to extend this work for weighted terms and also to employ other clustering algorithm with the proposed measure.

References

[1] Michael W. Berry and Jacob Kogan, Text Mining: Applications and Theory, John Wiley & Sons. Ltd, (2010).
[2] Ronen Feldman and James Sanger, The Text Mining Handbook - Advanced Approaches in Analyzing Unstructured Data, Cambridge University Press, (2007).
[3] Charu Aggarwal, ChengXiang Zhai, Mining Text Data (Springer) Ed. (2012).
[4] Santini, S., Jain, R.: Similarity Measures. IEEE Trans. Pattern Analysis and Machine Intelligence 21(9), 871–883 (1999)
[5] Wang H, Wang H, Wang W, Wang W, Yang H, Yang H, et al. Clustering by pattern similarity in large data sets. 2002 ACM SIGMOD international conference on Management of Data. New York, New York, USA: ACM Press; 2002. p. 394.
[6] Xu, R., & Wunsch, D. (2005). Survey of clustering algorithms. IEEE Transactions on neural networks, 16(3), 645-678.
[7] Saxena, Amit, et al. "A review of clustering techniques and developments." Neurocomputing 267 (2017): 664-681.
[8] Legány, Csaba, Sándor Juhász, and Attila Babos. "Cluster validity measurement techniques." Proceedings of the 5th WSEAS international conference on artificial intelligence, knowledge engineering and data bases. World Scientific and Engineering Academy and Society (WSEAS) Stevens Point, Wisconsin, USA, 2006.
[9] D. Greene and P. Cunningham. "Practical Solutions to the Problem of Diagonal Dominance in Kernel Document Clustering", Proc. ICML 2006.
[10] Grace, G. Hannah, and Kalyani Desikan. "Document clustering using a new similarity measure based on energy of a bipartite graph." Indian Journal of Science and Technology 9 (2010): 40.

IOP Conf. Series: Materials Science and Engineering 1012 (2021) 012060 doi:10.1088/1757-899X/1012/1/012060

Parallelisable Regular Languages

Mohana. N[1*], Kalyani Desikan[1], V. Rajkumar Dare[2]

[1]School of Advanced Sciences, Vellore Institute of Technology, Chennai.
[2]Department of Mathematics, Madras Christian College, Tambaram, Chennai.

E-mail: mohana.n@vit.ac.in

Abstract. In this paper, we have defined Parallelisable strings and languages in order to reduce the recognition time of strings. Parallelisable languages have been introduced and their recognition has been studied through parallelisable finite automaton. We have restricted our study to languages that comprise of words that contain a common subword. By introducing parallelism on the subword we have been able to reduce the recognition time. Also, we have studied the properties of parallel regular languages.

Keywords: Formal language theory, Parallelisable languages, Parallelisable regular languages, Parallelisable finite automaton.

1. Introduction

Parallel programming has some advantages that make it suitable as a solution approach for certain types of computing problems that can be best solved using multiprocessors. Nevertheless, the disadvantages of parallel programming must be taken into consideration before embarking on this challenging activity.

The main advantage of parallel programming is the efficient execution of code. Parallel programming saves time by allowing applications to be executed in a shorter duration of time. Since the code is executed efficiently, parallel programming often scales with the size of the problem and therefore, larger problems can be solved. Parallel programming provides concurrency, particularly, multiple actions being performed simultaneously.

Parallel programming surpasses the limits imposed by sequential computing [1,5,10,11]. Sequential programming is often constrained by physical and practical factors. This limits its capability to build faster sequential computers. For example, the speed of a sequential computer is dependent on the speed at which data moves through its hardware. The bandwidth of a physical medium imposes constraints on the transmissions that are made through it. (for example, the speed of light or the transmission limit of copper wire).

The process of dividing a sequential program into processes that can be run in parallel is called parallelisation. Analysing data dependencies is complex in a sequential program, and hence, it becomes difficult to parallelise efficiently. There are two possible approaches for the parallelisation of a program. The first one is to augment the sequential language such that the programmer supplies information to indicate which parts of the program are to be executed concurrently. The other approach is to explicitly write the code for each processor and mechanically identify the parallelisable

elements of a program, and consequently write compliers that automatically transform a sequential program into one that can be efficiently run on a parallel machine.
In this paper, section 2 provides the basic definitions and notations to define Parallelisable languages. Also, the mathematical definition of Parallelisable strings and languages with examples. Section 3 includes finite representation in terms of finite automaton (FA) for Parallelisable languages and their properties.

2. Preliminaries

Consider Σ a finite, non-empty collection of alphabets. A finite sequence of symbols called a string u over Σ. Length of u is given by |u| and defined as the number of symbols u. An empty string is denoted by ϵ or λ which has zero length. Concatenation of two strings v and w is v.w or vw.
v=xuy is a string for some substrings x and y. The substring u is called a prefix of v if x= ϵ. Similarly, u is a suffix of v if y= ϵ.
For example, consider a string x=aabb then the substrings are $\{$a, aa, aab, abb, bb, ab, b$\}$. The collection ϵ, a, aa, aab, aabb are prefixes of the string x and the set of suffixes of x is {aabb, abb, bb, b, ϵ.
A language L is a subset of $\Sigma^*$and the collection of all strings of length n over Σ is denoted as $\Sigma^n$where Σ^0= ϵ.

Definition 2.1
Two main definitions of regular languages:
- A language L is regular if there exists a DFA or NFA M such that L(M)=L
- A language L is regular if there exists a regular expression 'r' such that L(r)=L

Example: 2.2
Consider the language$L = \{(ab)^n : n \geq 0\}$. L is regular since it is accepted by the Deterministic finite automaton Figure.1 and Non-deterministic finite automaton Figure 2.
Also, L can be written as a regular expression $(ab)^*$.

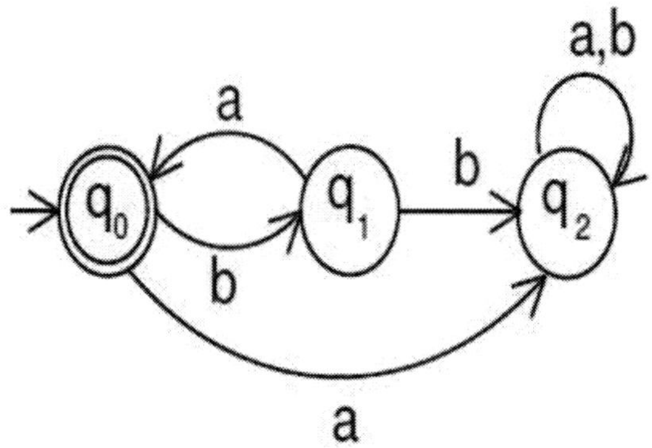

Figure. 1 Deterministic Finite Automaton

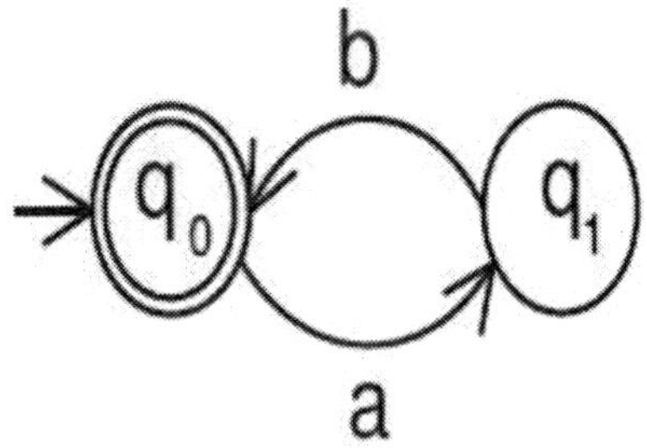

Figure 2. Non-Deterministic Finite Automaton

2.1 Parallelisable strings and Parallel languages

Parallelisation happens when we have more than one input. Parallelisation of any string of length more than one can be defined as follows:

Definition 2.3
If w=w1w2...wn, wi $\epsilon\Sigma$ and $w\epsilon\Sigma^{>1}$then
P(w)=w1||w2||...||wn
where || is an operator that is used to split or divide the string w into atleast two substrings.

Parallelisation is a process by which a string is broken down into a finite number of alphabets. It is used to speed up the recognition of a string. It is obvious that, if $w\epsilon\Sigma \cup \{\epsilon\}$ then parallelisation cannot happen on w.
If a string of length n takes n units of time to complete through in a sequential process then parallelisation enables the string to be completed in less than n units of time.
A machine that reads a string in one unit of time is more interesting where in a string of length n is parallelized n times.
Example: 2.4
Let $\Sigma = \{a, b\}$, $x\epsilon\Sigma^*$and x=abaab then the parallelisable string P(x) can be written as,
P(x) = a||b||a||a||b.

Definition: 2.5
Let $S_u = \{x = vuw\epsilon\Sigma^*: u = u1u2 \ldots un$ is a substring of x, |u|>1} then
$P(S_u) = \{xP(u)y: P(u) = u1||u2|| \ldots ||un\}$ where $x, y\epsilon\Sigma^*$ and ui$\epsilon\Sigma$
If L=$\cup S_u$ then L is a parallelisable language.
Example: 2.6
If L=S_{aba}, where aba can be a prefix, suffix or infix of L, then P(L)={x(a||b||a)y: $x, y\epsilon\Sigma^*$} where a||b||a is prefix if x== ϵ, suffix if y= ϵ and infix if $x, y\epsilon\Sigma^+$
Example: 2.7
If L={aa, bb}then L is not a Parallelisable language. Since there is no common subword u.
Example: 2.8
L={a,b} is not a Parallelisable language.There is no subword of length greater than 1 in L Therefore, parallelisation does not exist on L.
Definition: 2.9
Let L1, L2 $\subseteq \Sigma^*$ be two parallelisable languages. Then, L1||L2={x||y: xϵL1, yϵL2}
Definition: 2.10
If L1=Σ^* u1, L2=u2 Σ^* then L1 L2=S_{u1u2}, P(L1 L2)=P(S_{u1u2}) where u1, u2 $\epsilon\Sigma^{>1}$

Example: 2.11
The collection of all strings over {0, 1}which contains 11 as a substring is a regular language.

$L=\{u \in \{0,1\}^*: 11 \text{ is a substring of } u\}$
$=\{v11w: v,w \in \{0,1\}^*\}$
$=\{0,1\}^*11\{0,1\}^*$
$P(L)=\{0,1\}^*1\|1\{0,1\}^*$

Hence, the corresponding regular expression of L is $\{0,1\}^*11\{0,1\}^*$ and that of P(L) is $\{0,1\}^*1\|1\{0,1\}^*$.

Example: 2.12

The collection of all strings over $\Sigma=\{a, b\}$ that contains ba or bb as a substring is a regular language.

$L=\{u: \text{ba is a substring of } u\} \cup \{u: \text{bb is a substring of } u\}$
$=\{vbaw: v,w \in \Sigma^*\} \cup \{xbby: x,y \in \Sigma^*\}$
$=\Sigma^*\{ba\}\Sigma^* \cup \Sigma^*\{bb\}\Sigma^*$
$P(L)=\Sigma^*\{b\|a\}\Sigma^* \cup \Sigma^*\{b\|b\}\Sigma^*$

Hence, the corresponding regular expression of L can be expressed as (a+b)*ba(a+b)*+(a+b)*bb(a+b)* and that of P(L) is (a+b)*(b||a)(a+b)*+(a+b)*(b||b)(a+b)*.

2.2 Parallelisable Finite automaton

Finite automata are main tools to understand regular languages better as language accepting devices. A finite automaton **A**=(Q, Σ, δ, q0, E) is a quintuple. Here Σ is the finite alphabet, Q is the finite collection of states, q0 is an initial (start) state and E⊂Q is the collection of end states and the function δ is given by δ: $Q \times (\Sigma \cup \epsilon\}) \rightarrow 2^Q$

We now introduce, the transition called Parallelisable transition δp for Parallelisable Finite automaton from the transition of Branching automaton [2] which is equivalent to an extended transition in Finite automaton as follows:

$$\delta p \subseteq \delta^* \times \delta join \times \delta^* \times \delta fork \times \delta^*$$

where $\delta^*: Q \times \Sigma^* \rightarrow Q$, $\delta fork \subseteq Q \times Mn(Q)$, $\delta \subseteq Q \times \Sigma \times Q$ and $\delta join \subseteq Mn(Q) \times Q$ and Mn(Q) is a set of non-empty, non-singleton multistates as defined in branching automaton [2] but multistates are of the form $\{q_{i1}, q_{i2},..., q_{in}\}$ for some i.

For example, consider the multistates are created between the states q_{i} and q_{j} for some i and j for the string of length n then $Mn(Q)=\{q_{i1}, q_{i2},...,q_{in}\}$ in fork transition and $M_{n}(Q)=\{q_{j1}, q_{j2},...,q_{jn}\}$ will be in join transition.
Figure 3 represents the graphical representation (state diagram) of fork and join transitions respectively.

Figure 3. Fork and Join Transition

Formally, a Parallelisable Finite automaton **Ap**=(Q, Σ , δp, q0, E) is a quintuple. Here Σ is the finite collection of input alphabets, Q is the finite collection of states, δp is as defined earlier, q0 belongs to Q and E is a subset of Q.

IOP Conf. Series: Materials Science and Engineering 1012 (2021) 012060 doi:10.1088/1757-899X/1012/1/012060

A string w $\epsilon\Sigma^*$ is accepted by a Parallelisable Finite automaton **Ap** if δp ((q0,w))∩E ≠ ϕ
The collection of all strings recognised by the Parallelisable Finite automaton is the language accepted by **Ap** and is written as L(**Ap**)={ w $\epsilon\Sigma^*$| δp ((q0,w) ϵ E.

Example: 2.13
If L=S_{aba}, where aba is a substring of L then P(L)={x(a||b||a)y:x,y $\epsilon\Sigma^*$} Figure 4 depicts P(L).

Example: 2.14
The language L over Σ ={0,1} that contains 01 as a substring is regular.
That is, L={x01y : x,y $\epsilon\Sigma^*$} is regular and P(L)={x(0||1)y : x,y $\epsilon\Sigma^*$}$ is parallel regular. Refer Figure 5.

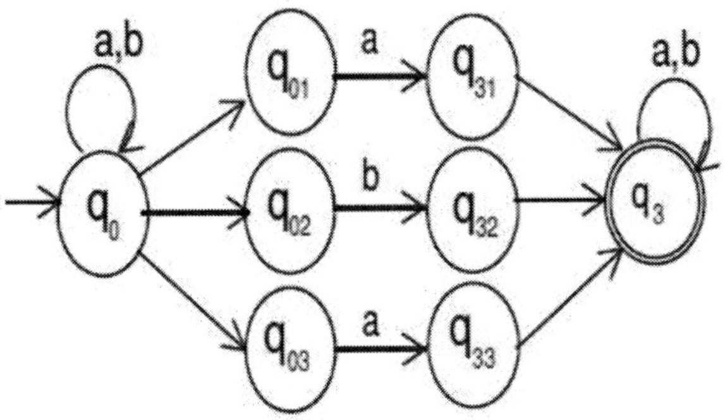

Figure 4. Parallelisable finite automaton P(L)={x(a||b||a)y:x,y $\epsilon\Sigma^*$}

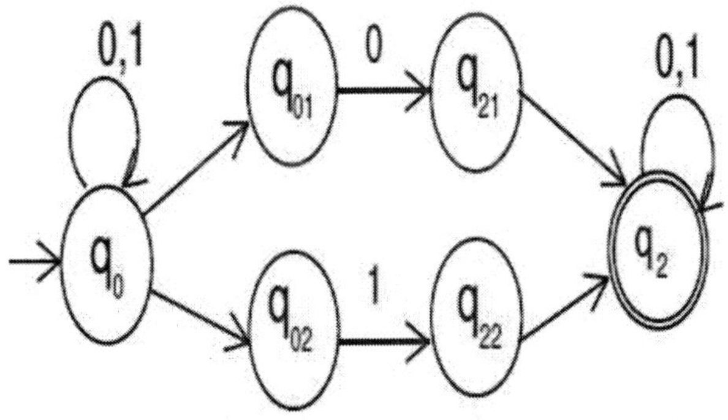

Figure 5. Parallelisable finite automaton P(L)={x(0||1)y:x,y $\epsilon\Sigma^*$}

2.3 Finite Automaton vs Parallelisable Finite Automaton
We now illustrate how a common substring is recognised in a finite automaton and its corresponding parallelisable finite automaton.
Let Σ ={a,b}, L=S_{aaa} where S_{aaa}={x(aaa)y : x,y $\epsilon\Sigma^*$}. Transitions for the parallelisable substring aaa on FA and Parallelisable Finite Automaton are given below:

$$\begin{aligned}\delta^*(q0, aaa) &= \delta\ (\delta^*(q0,aa),a)\\ &=\delta\ (\delta\ (\delta^*(q0,a),a),a)\\ &=\delta\ (\delta\ (\delta\ (\delta^*(q0, \epsilon),a),a),a)\\ &= \delta\ (\delta\ (\delta\ (q0,a),a),a)\\ &= \delta\ (\delta\ (q1,a),a)\\ &= \delta\ (q2,a)\end{aligned}$$

$= q3$

where δ *(q0,x) corresponds to the transitions of FA in Figure 6.

Given below is the corresponding transition in a parallelisable finite automaton in Figure 7

$$\begin{aligned}\delta p(q0, aaa) &= \delta join\ \delta\ \delta fork(q\_\{0\}, aaa)\\ &= \delta\ join\ \delta\ (\{q\_\{01\}, q\_\{02\}, q\_\{03\}\}, aaa)\\ &= \delta\ join\ \{\ \delta\ (q\_\{31\},a), \delta\ (q\_\{32\},a), \delta\ (q\_\{33\},a)\}\\ &= \delta\ join\{q\_\{31\}, q\_\{32\}, q\_\{33\}\}\\ &= q\_\{3\}\end{aligned}$$

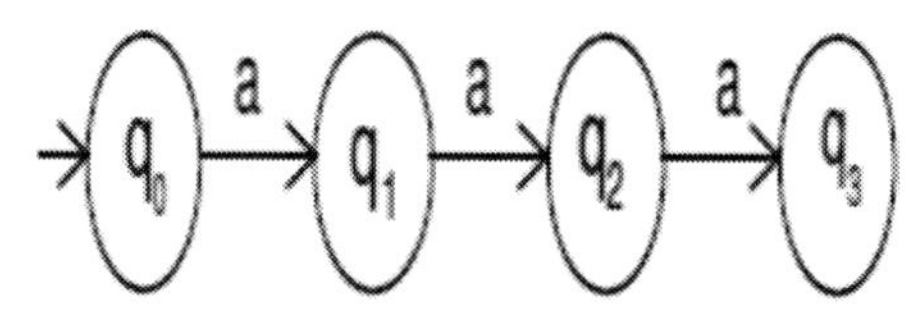

Figure 6. Finite automaton for a string aaa

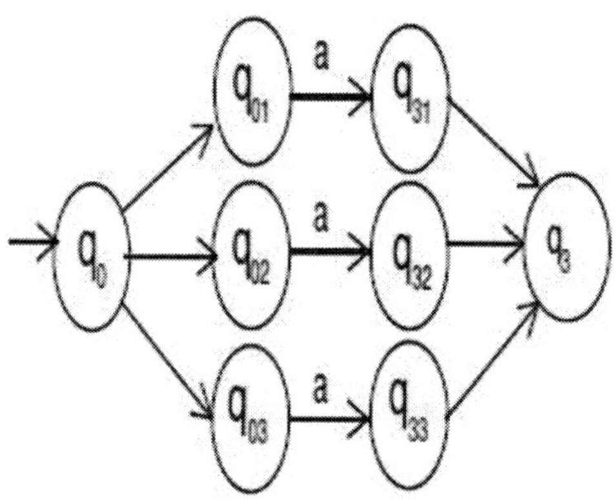

Figure 7. Parallelisable Finite automaton for a parallelizable string a||a||a

Lemma: 2.15

A parallelisable language L is recognised by a FA **A** in |w| units of time, where |w| is the length of the string w ϵ L.

Proof:

If L=S_u then L is a parallelisable language, where $S_u=\{w=xuy: u \epsilon \Sigma^*$ is a substring of w and $x,y \epsilon\Sigma^*\}$.
Consider **A** =(Q, Σ , δ, q0, E) a FA and wϵS_u then δ*(q0,w)ϵ E.

$$L(\mathbf{A})=\{w\epsilon S_u: \delta\ ^*(q0,w)\epsilon\ E\}$$

Assume that, δ (q,a)=p takes one unit of time for transition where q,p ϵQ and a$\epsilon\Sigma$ then δ^*(q,w)=p takes |w| units of time for transition where w ϵ $\Sigma^*$and q,pϵ Q.

Hence, a parallelisable language L takes |w|=|x|+|u|+|y| units of time in finite automaton.

Lemma: 2.16

For every finite automaton **A** which accepts a parallelisable language in |x| units of time, there exists a Parallelisable Finite Automaton **Ap** which accepts the string x in less than |x| time units.

Proof:

Let **A**=(Q, Σ , δ, q0, F) be an finite automaton which accepts a parallelisable language.
Construct **Ap** =(Qp, Σ , δp, q0, F).

$$\delta p \subseteq \delta^* \text{ X } \delta join \text{ X } \delta^* \text{ X } \delta fork \text{ X } \delta^*$$

where Qp=Mn(Q) ∪ Q.
δ^*: Q X Σ *→Q, δfork ⊆ Q X Mn(Q), δjoin ⊆ Mn(Q)X Q
and Mn(Q) is a set of non-empty, non-singleton multistates as defined in section 2.2.

We, now, prove by induction on the size of a parallelisable string u=u1u2, P(u)=u1||u2 and u1, u2ϵΣ. Consider a parallelisable language L=S_u where u= u1u2.
By the above lemma, L is recognized by a finite automaton **A** in |x| units of time where u is a subword of x.
We now define a parallelisable finite automaton for accepting L=S_u where u=u1u2 by parallelising u as P(u)=u1||u2 as given below.
δp(q_{i},u) =δjoin $\delta^*\delta$fork(q_{i}, u_{1}||u2)
=δjoin δ^*({q_{i1}, q_{i2}}, u1||u2)
=δjoin (δ (q_{i1}, u1), δ (q_{i2}, u2))
=δjoin{q_{j1}, q_{j2}}
=q_{j}
The above construction gives only the transition of parallelising substring. Prefix and suffix of the strings from L can be recognized by usual finite automaton.
The substring u=u1u2 takes two units of time in **A**. But, this can be recognized in one unit of time by **Ap.**
That is, δp(q_{i},x)=q_{j} if and only if δ^*(q_{i},x)=q_{j} for some i and j.
We prove that for all x ϵ$\Sigma^{>1}$. Since parallelisation happens only on the strings of length more than one.
δp(q_{0},x)=q_{j} if and only if δ^*(q_{0},x)=q_{j} for some j.
This proves the result, because
xϵL(**A**) ⇔ δ^*(q_{0},x)∩F≠ ϕ
⇔ δp(q_{0},x)ϵ
⇔ xϵ L(**Ap**)
If |u|=2 then δ^*(q_{i},u)=q_{j} also δp(q_{i},u)=q_{j} for some i and j with δfork(q_{i})={q_{i1}, q_{i2}} and δ{join}({q_{j1}, q_{j2}})=q_{j} so that the statement is true in this case. Assume that the statement of the lemma is true for strings whose cardinality is n which takes n units of time in **A** and one unit of time **Ap**.
Let u ϵ$\Sigma^{>1}$with |u|=n and a ϵΣ
By inductive hypothesis,
δp(q_{i},ua)=q_{j} if and only if δ *(q_{i},ua)=q_{j} for some i and j and
δfork(q_{i})={q_{i1}, q_{i2},...,q_{in},q_{i(n+1)}}and δ join({q_{j1}, q_{j2},...q_{jn}, q_{j(n+1)}})=q_{j}
Moreover, the string ua takes (n+1) units of time in **A** and one unit of time **Ap**.
We generalize the above induction that any parallelisable substring u of length greater than 1 can be recognized in one unit of time in **Ap**.
Also, a string x belongs to L is recognised by **Ap** in |x|-|u|+1 units of time which less than |x|.
Hence, by induction, the statement is true.

Lemma: 2.17
For every Parallelisale Finite Automaton **Ap** there exists a Finite Automaton **A** such that L(**Ap**)=L(**A**).
Proof:
If x is a parallelisable string then δ (q,x) is equivalent to some δ *(q,x). If P(x)=x1||x2||...||xn is accepted by **Ap** then x=x1x2...xn is a string which is accepted by **A**.

Theorem: 2.18
If L is parallelisable regular then L is regular.
Proof:

If $L=\cup S_u$ then L is parallelisable where $S_u=\{x=vuw\epsilon\Sigma^*: u=u1u2...un \text{ is a substring of } x, |u|>1\}$ is recognised by parallelisable finite automaton. Then by lemma 2.17, S_u is accepted by finite automaton. Since S_u is regular, $L=\cup S_u$ also regular. Hence, proved

Theorem: 2.19

If L is a regular then L is may not be parallelisable regular.

Proof:

We prove the theorem by example.

Consider a regular language $L=\{(ab)^n\}:n\geq 0\}$. L={a, ab, abb, abbb, abbbb,...}. Here L is not parallelisable. Since it contains a string 'a' of length one which cannot be parallelised. Hence, L is not parallelisable and therefore L is not parallelisable regular.

Theorem: 2.20

If L is a non-regular language then L is not parallelisable regular.

Proof:

We prove the statement by giving argument. By lemma 2.15 and 2.16, for every parallelisable language we can construct finite automaton and parallelisable finite automaton. It means that if a finite automaton accepts parallelisable language then there exists a parallelisable finite automaton. Non-regular languages did not have finite automata for their recognition, so that we cannot have a parallelisable finite automaton to accept the language. Hence proved.

Theorem: 2.21

The set of parallelisable regular languages is a subclass of regular languages.

Proof:

Result is obvious by theorem 2.18, 2.19 and 2.20.

3. Conclusion

In this paper, we have defined parallelisable strings and languages from Σ^*. Finite representation of parallelisable languages has been provided through parallel regular expression and parallelisable finite automaton. Parallelisable regular language has been introduced and studied their properties.

References:

[1] J. Hopcroft and J. Ullman, Introduction to Automata Theory, Languages and Computation, Addison-Wesley, Reading, MA,1979.
[2] K. Lodaya and P. Weil, Series parallel posets: algebra, automata and languages, in Proc. STACS (Paris 98), LNCS 1373 (M. Morvan, C. Meinel, D. Krob, eds.) (1998) 555-565.
[3] K. Lodaya and P. Weil, Series-Parallel languages and the bounded-width property, Theoretical Computer Science, 237 (2000) 347-380.
[4] D. Kuske, Infinite series-parallel posets: logic and languages, ICALP 2000, Lecture Notes in Computer Sceince Vol. 1853, Springer-Verlag, 2000, 648-662.
[5] W. Thomas, Languages, automata and logic, in: G. Rozenberg, A. Salomaa (Eds.), Handbook of Formal Languages, Springer, Berlin, 1997, 389-455.
[6] Ivan M. Havel, Finite Branching Automata, Kybernetika, 10 (1974) 281—302
[7] Nicolas Bedon, Logic and Branching automata, Logical methods in computer science, 11 (2015) 1—38
[8] K. Lodaya, P. Weil, A Kleene iteration for parallelism, Foundations of Software Technology and Theoretical Computer Science, (1998) 355-366.
[9] K. Lodaya, P. Weil, Rationality in algebras with a series operation, Information and Computation,(2001) 269-293.
[10] D. Beauquier and J.E. Pin, Languages and Scanners, Theoretical Computer Science, 84 (1991) 3-21.
[11] Eilenberg. S, Automata, Languages and Machines, Vol. A (1974) Academic Press.

Optimization of Harmonious coloring problem in Uniform theta graph and torus Network

Franklin Thamil Selvi.M.S[1*], A.Amutha[2]

[1] Department of Mathematics, Sathyabama Institute Of Science and Technology, India.
[2]Department of mathematics, The American College, Madurai, India.

E-mail: thamizanand@gmail.com

Abstract- Coloring of graphs has been extraordinarily widened areas of investigation. A coloring of a graph can be depicted by a capacity that maps pieces of a graph into some plan of numbers commonly called labels or hues, all together that some property is satisfied. In this article we deal with the harmonious chromatic number of central graph of generalized uniform theta graph and torus network. A coloring $c: V \rightarrow \mathbb{N}$ of the nodes is a harmonious coloring if and only if the coloring is proper and for each and every line $(e_1, e_2) \in E$ the line color $(c(e_1), c(e_2))$ is distinctive that is it emerge solely once.

Keywords: Harmonious coloring, Central graph, Uniform theta graph, Torus Network.

1. Introduction

Graph coloring rise ordinarily in various convenient conditions where it is needed to fragment a great deal of things into social occasion with the goal that the people of each get-together are normally amazing as shown by some measure. An imaginative method of coloring a graph richly is harmonious coloring. A coloring $c: V \rightarrow \mathbb{N}$ of the nodes is a harmonious coloring if and only if the coloring is proper and for each and every line $(e_1, e_2) \in E$ the line color $(c(e_1), c(e_2))$ is distinctive that is it emerge solely once. In the year 1982 Frank Harray and Plantholt first posted the above way of coloring. Nevertheless the finest part is delivered by Lee Hopcroft and Krishnamoorty[5]. Lu has acquired evaluations for harmonious chromatic number of a meager families of graphs[6,8]. He has additionally found the bounds for binary and trinary tree[7]. The lower bound of harmonious chromatic number might be considered as an edge accepting an interesting shading pair[1,4]. The greatest part of this coloring is concern with estimation results[2]. So instigator get down to shot at central, line and total of some known graphs[3,4]. In this article we are indeed interested in determining the minimum number of colors used in coloring the central graph of uniform theta graph and torus network harmoniously. Thus a central graph is the subdivision of each edge by a node exactly once and joining all the non adjacent nodes.

2. Main Results

In this section we discuss the minimum number of colors used in central graph of Uniform theta graph and Torus network which is colored harmoniously.

2.1. *Uniform Theta Graph*

A generalised theta graph $\theta(s_1, s_2, \ldots\ldots, s_\alpha)$ consist of a pair of end nodes joined by α internally disjoint paths of lengths $s_1, s_2, \ldots\ldots, s_\alpha \geq 1$. These end nodes are called north pole (N) and south pole (S) and the paths are called longitudes. A generalized theta graph in which all the paths are of same length is called a uniform theta graph denoted by $\theta(n, \alpha)$[1]. Hence n denotes the number of internal vertices in each longitude.

Let $l_1, l_2, \ldots\ldots, l_\alpha$ denote the longitudes of the uniform theta graph $\theta(n, \alpha)$. The number of vertices and edges in $\theta(n, \alpha)$ are $\alpha n + 2$ and $\alpha(n + 1)$ respectively. Its diameter is $n + 1$.

We utilize a specific naming of nodes of the Uniform theta graph in request to define an appropriate coloring. First let n be odd. We partition the vertex set into two sections called the upper and lower parts. The nodes v_i's represent the upper half nodes and the nodes w_i's for the lower half nodes. The nodes in the upper half are named from left to right beginning from the centre of the longitudes and listed towards the North Pole. Likewise in the lower half are named from left to right beginning from the centre and listed towards the South Pole. Let $V_1 = \{v_1, v_2, \ldots\ldots, v_{\lceil \frac{n}{2} \rceil \alpha}\}$ and $V_2 = \{w_1, w_2, \ldots\ldots, w_{\lfloor \frac{n}{2} \rfloor \alpha}\}$ then the node set is $V = V_1 \cup V_2 \cup \{N, S\}$. See Figure 1, a uniform theta graph with $n = 3$ and $\alpha = 6$.

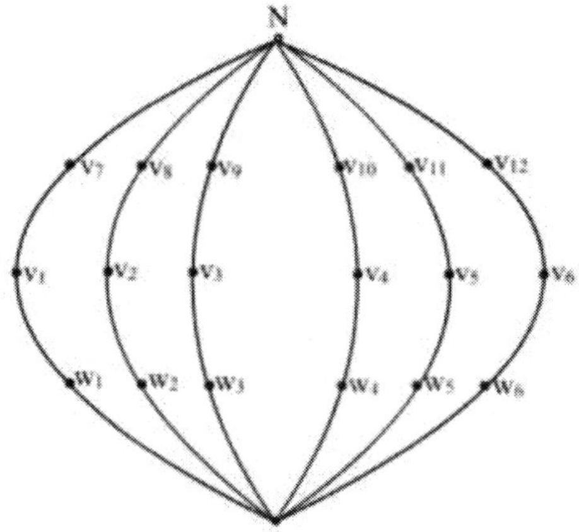

Figure 1: A uniform theta graph G with $n = 3$ and $\alpha = 6$

2.2. *Theorem*

Let $G = \theta(n, \alpha)$ then $\chi_h\{C[\theta(n, \alpha)]\} = \Delta\{C[\theta(n, \alpha)]\} + \Delta[\theta(n, \alpha)] + 1$.

Proof

Let the node set of G be $V = V_1 \cup V_2 \cup \{N, S\}$ where $V_1 = \{v_1, v_2, \ldots\ldots, v_{\lceil \frac{n}{2} \rceil \alpha}\}$ and $V_2 = \{w_1, w_2, \ldots\ldots, w_{\lfloor \frac{n}{2} \rfloor \alpha}\}$ by the annotation of central graph let u_i be the nodes of subdivision of the lines of $\theta(n, \alpha)$. Fix colors to the nodes $V = V_1 \cup V_2 \cup \{N, S\}$ as $c_i : 1 \leq i \leq \alpha n + 2$ which is in turn is the maximum degree of central graph of uniform theta graph with the addition of one, that is $\Delta\{C[\theta(n, \alpha)]\} + 1$. The balance $\alpha(n + 1)$ nodes are colored with $c_{\alpha n+3}, c_{\alpha n+4}, c_{\alpha n+5}, \ldots, c_{\alpha n+\alpha+2}$ in anticlockwise direction which is in turn is the maximum degree of G that is $\Delta[\theta(n, \alpha)] = \alpha$. Thus the above defined coloring is harmoniously colored with least number of colors. Hence $\chi_h\{C[\theta(n, \alpha)]\} = \alpha n + \alpha + 2$, that is $\chi_h\{C[M(s, t)]\} = \Delta\{C[\theta(n, \alpha)]\} + \Delta[\theta(n, \alpha)] + 1$.

2.3. Torus Network

The torus is one of the well known geographies for interconnecting processors to manufacture elite multi computers. In mathematics, a torus is a donut moulded surface of insurgency created by rotating a hover about a hub coplanar with the circle. It is a work with fold over connections. Torus are bipartite if and just if all side lengths are even. They are Hamiltonian, regular and vertex symmetric.
A one-dimensional torus is just a cycle or a ring. A two dimensional torus network contains st nodes, organized in two dimensions with s, t nodes per dimension. A $s \times t$ torus is denoted by $TR(s,t)$.
An $s \times t$ torus $TR(s,t)$ is defined as a graph with vertex set $V = \{(j,k): 1 \leq j \leq s, 1 \leq k \leq t\}$ and edge set $E = \{((j_1, k_1), (j_2, k_2)): (j_2 = (j_1 + 1)(mods) \Lambda k_1 = k_2) V (j_1 = j_2 \Lambda k_2 = (k_1 + 1)(modt)\}$
The number of nodes in $TR(s,t)$ is st and the number of edges is $2st$. See Figure 2 a torus network with $s = 4$ and $t = 5$

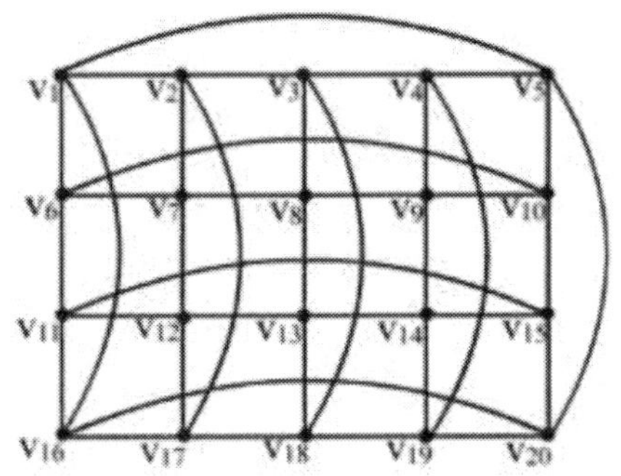

Figure 2: Torus $TR(4,5)$

2.4. Theorem

Let $G = TR(s,t)$ then $\chi_h\{C[TR(s,t)]\} = \Delta\{C[TR(s,t)]\} + \Delta[TR(s,t)] + 1$.

Proof

Let the node set of G be $V = \{v_{n(j-1)+k}: j = 1,2,3, \dots \dots, s \; and \; k = 1,2,3, \dots \dots, t\}$ and by the annotation of central graph let w_i be the nodes of subdivision of the line of $TR(s,t)$. Fix colors to the nodes $V = \{v_{n(j-1)+k}: j = 1,2,3, \dots \dots, s \; and \; k = 1,2,3, \dots \dots, t\}$ as $c_i : 1 \leq i \leq st$ which is in turn is the maximum degree of central graph of extended mesh with the addition of one that is $\Delta\{C[TR(s,t)]\} + 1$. The balance $2st$ nodes are colored with c_{st+1}, c_{st+2}, c_{st+3}, c_{st+4} colors in anticlockwise direction which is in turn is the maximum degree of G that is $\Delta[TR(s,t)] = 4$. Obviously the above defined coloring is harmoniously colored with least number of colors. Hence $\chi_h\{C[TR(s,t)]\} = st + 4$, that is $\chi_h\{C[TR(s,t)]\} = \Delta\{C[TR(s,t)]\} + \Delta[TR(s,t)] + 1$.

3. Conclusion

In most of the coloring problems optimization is a challenge on certain classes of graphs. In this work, we have optimized the harmonious coloring number of centralgraph of uniform theta graph and Torus Network.

Our future work is to investigate the performance of the other classes of graphs such as line, total middle of some interconnection networks.

References

[1] Bharati Rajan, Indra Rajasingh, Amutha A and Paul Manuel, Minimum Tree Spanner Problem for Generalized Theta Graphs, Third National Conference Mathematical and

ComputationalModels (NCMCM-2005), pp. 441-447,2005.

[2] K. J. Edwards and C. J. H. McDiarmid, "New upper bounds on harmonious colorings", Journal of Graph Theory, Vol.18, pp. 257-267, 1994.

[3] Franklin Thamil Selvi.M.S and A.Amutha, "A study on harmonious coloring of snake derived networks", Journal Of Engineering Science and Technology, Vol.11, No.12, pp.1736-1743, 2016.

[4] Franklin Thamill Selvi.M.S and A.Amutha, "A study on harmonious chromatic number of total graph of central graph of generalized Petersen graph", Journal of Ambient Intelligence and Humanized Computing, https://doi.org/10.1007/s12652-020-01697-6, 2020.

[5] E. Hopcroft and M. S. Krishnamoorthy, "On the harmonious coloring of graphs", SIAM Journal on Algebraic and Discrete Methods, Vol.4, pp. 306-311, 1983.

[6] Zhikang Lu, "On an upper bound for the harmonious chromatic number of a graph", Journal of Graph Theory,Vol.15, pp. 345-347, 1991.

[7] Zhikang Lu, "The harmonious chromatic number of a complete binary and trinary tree", Discrete Mathematics, Vol.1181993, pp.165-172, 1993.

[8] Zhikang Lu, "Estimates of the harmonious chromatic numbers of some classes of graphs", Journal of Systems Science and Mathematical Sciences,Vol.13, pp. 218-223, 1993.

Problematizing Hybridity and Third Place in *The Immigrant*

Caro Velma J[1*], Alan G[2]

[1]School of Social Sciences and Languages, Vellore Institute of Technology, Chennai - 600127.

E-mail: carovelma.j2019@vitstudent.ac.in

Abstract- The study purveys the hybrid nature of the first generation of settlers in The Immigrant. Manju Kapur has imbibed the bourgeois characteristics in the characters like Ananda, Dr Sharma and Nina. Identity crisis, a quest for self-identity and a contrast in Indian and Canadian style of living is predominantly evidenced in the novel, which leads to the idea of third place. This paper significantly identifies an unquenchable need for third places in the immigrants. The findings are of two parts: the first part identifies the hybrid qualities of the characters and the second part of the paper spots the need for third places in the characters. Further study correlates the characteristics of the Third Places with the places found in the novel by stating those as a source of elevation from the clutches of westernisation. The amalgamation of Hybridity and 'Third Places' theories dissect the novel whereby justifies the title of the paper.

Keywords: Third places, Hybridity, Immigrant, Postcolonial

1. Introduction

Hybridity is an inevitable and vastly used sub-theory under postcolonial theory. Gina Wisker in her book Key Concepts in Postcolonial Literature comments on Hybridity as, "it is both widely used and disputed, referring to the creation of new transcultural forms within the space produced by colonisation where people, indigenous, immigrated, settled, colonising and colonised, live and move"[18]. The immigrants accommodate these qualities as essential survival needs to amalgamate with the society, due to the supremacy of western ideas. In terms of creating a high impact by adapting high class identity immigrants relinquish their cultural root. Some immigrants strive to withhold their cultural identity, that results in hybridity, while most of them lose it and regaining that identity remains void. In this realm of life, it is essential for human beings to have social bonding and express their self through emotions. This processed syndrome of the social beings is being fulfilled at some social places, which are termed as 'Third Places' by Ray Oldenburg, in the work The Great Good Place (1998). The idea of Third places is looked up as centers for maintaining psychological well-being of people who suffer from loneliness, who feel detached from society, who seek peace or feel devoid of joy. These places are, "known for their friendliness and qualities that support sociability and place attachment" [9]. Third place serves as a helping aid for the immigrants in relating themselves with the society. The paper problematizes the concept of Third places and its relevance with the hybrid immigrant characters present in the novel. The paper also reflects the hybrid state of the characters and their essential need for third places, in order to recreate their identity and guard their mental well-being.

2. Theorising Hybridity in *The Immigrant*

Manju Kapur, the author of the novel has experienced a livelihood at Canada while she was studying in Dalhousie University, which reflects the concept of hybridity in her work The Immigrant. This is specified by Dr. Shanthi in her thesis stating that, "Hybridity reflects a conscious discontinuity that comes from immigration and displacement of writer experience" [14]. The plot revolves around Nina, an unmarried middle aged lecturer at Delhi University from a middle class vegetarian family, living with her widowed mother. The story unfolds when Nina reaches her early thirties and is hitched to an NRI dentist from Canada, which increases her discomfort and tension towards the future, relationships, identity, career and marriage. Being married and settling in a new place, especially miles apart from her mother and motherland was not easy for Nina. She regrets her decision of agreeing to marry Ananda, an NRI but eventually ends marrying him.

While travelling to Canada by flight, she muses upon the literary works like Prothalamion, The Waste Land, Heart of Darkness and many other works, as they picturize the places she travels over, like the River Thames in England on her way to Heathrow. She tries to establish a sense of connectivity with the things around her. This shows her love for English and her craving for western lands without any guilt of disowning her culture or mother tongue. In the words of John McLeod, hybridity is described "as a way of thinking beyond exclusionary, fixed, binary notions of identity based on ideas of rootedness and cultural, racial and national purity" [8]. Nina, at her first act of migration, steps beyond these notions by readily associating herself with the western land and having passion for English language and literary works, as she was an English professor.

She represents herself as a foster daughter of that place as she proudly utters these lines, "O Thames, I come from far away shores to greet you, receive me your foster daughter, who loves your literature and language better than her own" [6] when she enters Heathrow. The place substantiates her yearning for literature and living experience at western lands, as she was once living in Brussels with her father in her childhood. Her love for literature gives her satisfaction and she experiences everything while walking around the streets of Heathrow. She also states, "So near, yet so far" [6], which represents her state of being close with her desired dream land but is far from its culture and away from her home. Her mixed emotions reflect the state of her getting uprooted from her social bondage, family, tradition and culture.

When she enters Heathrow she witnesses different shades of grey sky, grey planes and grey tarmac which soothes her, but in the later part the grey environment at Canada gives her a void, dim, dull and gloomy presence. She says, "Day after day, nothing but grey. It is so depressing." [6], which reflects her change of mind and attitude. The colour grey plays an indicating role of relating Nina's emotions with the greyness around her. "Everything is very strange" [6], "At home one was never really lonely" [6] and "I feel so lost here" [6] are statements denoting Nina's loss of belongingness, identity and craving for home. As an immigrant Ananda feels exuberant after meeting Nina, as he once again gains his culture through her presence but in turn gives her the Western grey experiences. At Canada, staying at home, sleeping and doing away with households was her only major routine.

In the words of Homi K. Bhabha, "The idea of hybridity […] attempts to stress the mutuality of cultures" [1] and in accordance with this statement, Nina is addressed as a mix of the East and the West as her behaviour, responsibilities and love for her mother reflects Indianness and her tastes and manner of speaking, thinking and interests represent the other. Nina's hybrid behaviour is expressed where she has passion for English literature and the cold climate but at the same time she defends Indian society against Ananda's accuses on Indian culture and praises for Canadian practices. Nina also regrets the western stereotypical minds as she was blocked at the immigration due to her skin tone, and writes an imaginary note to Anand stating, "I am the wrong colour, I come from the wrong place" [6]. The judgemental gazes of western eyes made her unstable and miserable.

Ananda respects and abides by the Indian tradition while he is in India but seems varied in Canada. He insists the Gellers to be there for his wedding as they were the only family like people he had in Canada. In the novel familial bond is found lubricant yet liminal. Individuality is given importance as in the statement of Gary stating, "One can't have family breathing down one's neck" [6]. Ananda is comfortable with the idea yet earns for a family, as he had no family. In the later part, he feels odd when Nina tries to find some kind of ties with Dr Sharma and his family. He also consoles Nina, saying that it is not a practice of Canadians being dependent on relatives or even visiting them often. Ananda at his early immigrant life experiences harsh hybrid realities, which transforms him into a complete hybrid cultured person. Supporting Ananda's character John McLeod in the Beginning Postcolonialism says that, "Hybrid identities [...] remain perpetually in motion, pursuing errant and unpredictable routes, open to change" [8].

"Another hybrid Diwali over" [6] is a notable exhausting statement made by Ananda on the first year of his immigrant life at Canada. The settlers celebrate Diwali inside closed halls, wearing sarees and dhotis, frustrates Ananda as he knows the reality. Ananda had celebrated every religious ceremony with his family at varied temples with special food varieties and new clothes, but this celebration of showing off the traditional practices of India to the next generation, made Ananda hopeless and disappointing. Also, the conversation between Dr Sharma and his nephew Ananda reflects the cold reality of immigrants hooked up in a delusion of being culturally rooted by celebrating a ritual once in a year, despite being miles apart. Depending on the words of Bill Ashcroft "the space of hybridity itself is the space in which cultural meanings and identities always contain the traces of other meanings and identities" [1], the victimization of Dr Sharma who couldn't stay three weeks in India, but behaves totally Indian on a particular day can be justified. He eats vegetarian and drapes his women in sarees but behaves totally Western on the other days, probes Ananda to mention his uncle as a fraud. On the contrary, Nina is found to be missing all the cultural and religious practices, yet is unable to practice those in a different and far away western land. In this way Nina can also be addressed as fraud as she has only emotions towards her cultural practices but no action taken towards that.

Ananda living in Canada, voluntarily starts adapting western culture and his first step closer was by eating fish. Ananda's action of "taking his first bite of a fillet soaked in lemon and tartar sauce, asking his mother's forgiveness, but feeling liberated" [6] represents that it was not a big struggle for him in adapting himself against the norms of his cultural vegetarianism. Ananda's only connect to the culture were his parents and as he has already lost them, he did not worry much about losing his culture, but in the case of Nina it was different. Nina as a child loves to fly abroad, but now stumbles to cope with Canadian culture. She resists taking meat for a long course of period, preferring only dal and veggies. Every nuances and segments of adaptation consumed a huge amount of time and a lot of thinking from the past to the present. Thus hybridity "can be compared to this space of deferral and possibility (thus a culture's difference is never simple and static but ambivalent, changing, and always open to further possible interpretation)" [1].

In the case of food, aroma and taste evokes emotion and connection.

> ...the smells caught her mind and shut her mouth. Turmeric, yellow turning into brown as it bubbled in hot oil, red chillies that crackled as they roasted, onions and garlic that turned pink then brown, releasing sweet sharp smells, tomatoes that became soupy as they were swished around, cumin and coriander that gave out pungent flavours, these smells and imagined sights travelled across the world from north India to eastern Canada to kick her sharply in the stomach. [6]

The appetizing words used in these lines describe the emotion and the bonding between Indian food and the Indian himself, deep in one's thoughts. Nina relishes Indian food for its flavoured aromas and varied colours at the Taj Mahal restaurant, when compared to her habitual raw salads, breads, rice or dal at Halifax.

Over some comfortable years of friendship with Gary, Ananda moved in with Gary's family; as a paying guest in their basement. Their way of living and lifestyle inspired Ananda. Gary became

IOP Conf. Series: Materials Science and Engineering 1012 (2021) 012062 doi:10.1088/1757-899X/1012/1/012062

Ananda's best pal and he wanted Ananda to move on with his life, so he introduced him to Sue. Ananda failed in his relationship, as he found uncomfortable with Canadian women. Ananda witnessed his inability to love a white woman was due to his connection with Indian ethics and values. As Indianness preaches pre/post marital affairs as a taboo, the 1st part of Ananda's life reflects such tradition but completely turns out the other way as he gets into an affair with Mandy in the later part. Bhabha sees hybridity "as part of the tendency of discourse analysis to de-historicize and de-locate cultures from their temporal, spatial, geographical and linguistic contexts [8]. Ananda represents it and in this case he is viewed as a hybrid character. He also never allows Nina to do all the households and it seemed uneasy for Nina, for such acts was looked upon as the basic duty of an Indian wife. He used to do all these before her arrival and he continues to do so as Canadians were self-reliant. Even Nina's act of showing her love through cooking food for him was found a big task loaded upon her in the eyes of Ananda. Even at some special evenings if Nina cooks special dishes for him, Ananda in turn does the vessels that night. This may seem to be an act of love and care to a certain point, but Ananda's actions clearly reflect the western ideas imbibed in him.

After few months of Indian culture followed in a western land, Ananda suggests western clothes for Nina as she was like an exhibit to others in her bright coloured Indian clothes. Nina refused but later had no choice as the climate became colder for her salwars or silk sarees failed to resist the piercing cold. Nina felt so awful of her appearance in jeans and shirt. Nina stresses that she is feeling uncomfortable but Ananda compliments on it. This cultural conflict is well defined by Zizek stating, "The hybrid nature of societies [...] implies that [...] uncertainties is not something that only occurs between individuals of different traditions or cultures, they also characterize much interaction between individuals within the same nation or ethnic collective" [4]. According to these words, Ananda, an Indian by birth indirectly insists Nina to adapt western clothing culture. The way the author describes Nina's appearance in western clothes clarifies us of her standing out of place. The novel itself comments on the adaptation of clothing as a dilemma faced by immigrants.

> Women who are not used to wearing Western clothes find themselves in a dilemma. If they focus on integration, convenience and conformity they have to sacrifice habit, style and self-perception. The choice is hard, and in Nina's case it took months to wear down her resistance. [6]

Nearing the end of the novel, Nina completely adapts the western style of clothing which helped her fit in Canadian atmosphere. "As immigrants fly across oceans they shed their old clothing, because clothes maketh the man and new ones help ease the transition" [6]. Not only the clothes but also the culturally constructed virtual vows were broken. Nina ends up doing a course in Library science at the University and falls in love with Anton. She claims an extramarital affair with Anton due to her husband's inability to satisfy her sexually and is not found guilty. She legalised it stating, "Judging from the evidence and the sexual therapy centres, every citizen in North America regarded good sex as their inalienable right. It was her right too" [6]. The lines indifferently prove her cultural up rootedness, but in the course of time culture plays its innate role when she realises her immorality leading to cultural glitch and turns away from Anton. As John McLeod says, "Hybrid identities are never total and complete in them" [8] Nina innocently gets stuck between whirlpool of cultures. A state of void vacuum space develops between the married couple as she discovers Ananda's affair and her mother's death adds to it. The plot subsides on Nina's decision to move out of Ananda's house and make her own living as a graduate. She decides on owning some space for her. Nina's longing for familial bonding is never accomplished till the end of the novel for she abandons Ananda and travels abroad envisaging her career. Her quest for being rooted in any culture leads her in discovering Third places.

3. Problematizing Third Places in *The Immigrant*

The Great Good Place originally published in the year 1998, introduces the term 'Third place', which was coined by an urban sociologist Ray Oldenburg. He characterises third place as "a place of refuge other than the home or workplace where people can regularly visit and communicate with friends, neighbours, co-workers, and even strangers" [9]. As a sociological concept, Third place has gained attention and positive recognition in various fields. An article titled "Every Community Deserves a Third Place" published in The New York Times states that "the 'third place' (home and work being the first and second) that neighbourhoods need is to thrive and, in many cases, survive" [3]. In encountering and victimizing a hybrid state of living, the characters eventually develop an unquenchable thirst for third places and venture into them, that interests them or according to its availability. The immigrants in an unknown land go in search of third places and some readily adapt to any third place environment irrespective of their interests, for if anyone proffers a place due to its vicinity and easy accessibility.

As the quote denotes, "The ingredients are exclusive and emergent and [...] change with the shifting patterns of life style" [12] each immigrant's need differs in time and varied environment. In this novel, the Indians settled in Canada suffer non-socialisation and loneliness and venture into community gathering places, where they are not judged. The immigrants depend on some third places where they can completely rely on, as their identity is hidden in such community gatherings. The Third places are determined according to its eight specified characteristics, listed by various scholarly authors at varied time periods. One among them is Jeffrey Wimmer, who has compiled it together and has listed all the eight characteristics in one of the chapters of the book titled, Multiplayer: the Social Aspects of Digital Gaming.

Neutral Ground is the first listed characteristics, which states that, "...individuals are free to come and go as they please" [17] to a third place. The fourth characteristic, Accessibility and Accommodation pair up with it and represents the library visited by Nina, as a third place. Library accommodates everyone irrespective of any classifications and is not financially or politically bound. Nina personally decides on visiting a nearby library and spends a lot of time there. She cherishes reading books and becomes a regular visitor. She also gets employed at the library and enjoys staying there escaping her loneliness. "The mood is playful" [15] in a third place as Nina reflects her playfulness in helping children at the library to read story books.

Nina probes into markets and nearby places spending time in shopping when she feels lonely at home, which is noted as the first place. Her eagerness to hang out with Sue and later the plight she feels at that juncture reflects her need for third places. It is worse if the only place to share one's space is home and adding to this the only person Nina can speak with is her husband. On the other hand, Ananda shares his space at home and also at work place, which is termed as the second place by Oldenburg. These central characters choose Indian restaurants as third place to relish their shared identity as Indians. They also attend parties at their friends and relatives. These places "are a home away from home. There are 'regulars' who find the atmosphere comfortable enough to 'root' them there" [5], like Ananda and Nina who revive Indianness by visiting these places.

On deciding to join Library Management at the University, Nina stops visiting the library often. At college she is introduced to a feminist group by a friend, where meetings are held regularly on Fridays. Nina finds a place for her to share her discomforts with others and feels socially connected. Conversations are the mainly focused activity at these groups. Levelling is a Third place characteristic where "an individual's rank and status in the workplace or society at large are of no import" [15], which Nina gains at the feminist group. She spots her comfort zone at the feminist group and marks it as a third place that prevailed for a shorter lifespan. Towards the end of the novel Nina starts travelling to various libraries located at different places where she found pleasure as it was also related to her studies.

Sue introduces Nina to La leche league, a club of married women, for helping her out of her stressful infertile contemplations. Unlike the feminist group, Nina feels more comfortable and opens up her worries and queries with the other members of the group. They had women from different walks of life, who take turns to empty their cups and remains non-judgmental and never comments or ask intrusive questions. Nina felt "a plain, homely, non-pretentious, accepting quality to the place" [10], as one of a distinct characteristics of third places.

Another character to be highlighted is Dr. Sharma, an immigrant settled long before Ananda migrated to Canada. Influenced by western ideas he represents himself as a Canadian except for a particular day, Diwali. He is one of the members of the Indian club at Canada that organises a closed room party for Indians on Diwali. As "...third place is welcoming and comfortable, is visited by regulars, and is a place to meet old friends and make new ones" [9], this party where the Indian immigrants gather for celebration serves as a third place for Dr Sharma.

Third Places help people, more particularly as the paper pictures the immigrants, to share their emotions and express their inner self as transparent as it is in a community of unknown, non-judgmental people. As proposed by Oldenburg, Third places are "the people's own remedy for stress, loneliness, and alienation" [11]. Every human being prefers such places to feel free from biased thoughts and stereotypical ideas. Ananda presents Mandy's house as his third place for he spends most of his leisure time there after spending a resentful amount of time at his first and second places i.e., his house and clinic. He has an extra marital affair with Mandy when Nina is out of town, busy with her studies. This can be considered a third place to a certain extent as he feels satisfied and it extends only for a short period. This can also be related with Nina's dark room experiences, an extramarital affair with Anton which was also a short term relationship. The sexual encounter with Anton made her feel "autonomous, independent [...] strange that the sex did not make her feel guilty" [6]. She utters the statement "She had lived" [6] which projects Anton's room as a third place, where Nina seeks to visit again, but restrains as it was against her cultural norms.

Metro, an online journal suggests that "We all need a third place" [16], in one of its article titled "Why everyone needs a 'third place' in their life" written by Henry Oliver. Third Place is viewed as a basic need for modern people living in a technological world to connect themselves with human beings and emotions rather distancing from machines. In an immigrants life it is more inevitable. Every third places that has been noticed has a sense of rootedness and a spiritual regeneration in the immigrant characters, which is also a characteristic nature of third places. The characters exhibit recreated optimistic behaviours while spending time at third places. If the statement of Dr. Markoç, "social sharing of people with other people is a human need" [7] is true, then there is a predominant and vital need of a wooden plank called Third Places for every immigrant, who sinks in the Western oceans.

4. Conclusion

The paper brings out the hybrid state of the immigrant characters in the novel and flashes their need for Third places to comfort themselves in the errands of westernization. The settlers caught in dilemma between the 'Hybridity' and 'Third places', strive to withstand their cultural identity. Most of the immigrants lose the battle of correlation with their culture against Western influence and domination. In such a state of flux, the immigrants thrive for third places and help themselves out of lonely devoid state. Third places retract them from the clutches of cognition that induces earning for their indigenous land. Survival and sustainability in a western land becomes liable with a prolonged dependence over third places.

As stated by Oldenburg (1998), "It is a fortunate aspect of the third place that its capacity to serve the human need for communion" [11] encourages communal living in a western land, where individualism and one's self is considered prominent. This recreates a small sociocultural environment for the immigrants to lead a life not much different from that of an Indian society. The findings of the paper conclude with a positive note featuring the stability of culture in the immigrants escaping identity crisis and loneliness in alliance with Third places.

References

[1] Ashcroft, B., Griffiths, G., & Tiffin, H. (2007). Post-Colonial Studies: The Key Concepts (2nd ed.). Taylor & Francis. https://doi.org/10.4324/9780203933473
[2] Butler, S. M., & Diaz, C. 2016. "Third places" as community builders. Brookings. https://www.brookings.edu/blog/up-front/2016/09/14/third-places-as-community-builders/
[3] Every Community Deserves a Third Place. 2014. The New York Times. https://www.nytimes.com/roomfordebate/2014/04/13/the-pros-and-cons-of-gentrification/every-community-deserves-a-third-place
[4] Fahlander, F. (2007). THIRD SPACE ENCOUNTERS: HYBRIDITY , MIMICRY AND INTERSTITIAL PRACTICE. In P. Cornell & F. Fahlander (Eds.), Encounters | Materialities | Confrontations: Archaeologies of Social Space and Interaction (pp. 15–41). Cambridge Scholars Press. https://www.academia.edu/187363/Third_Space_Encounters_Hybridity_Mimicry_and_Interstitial_Practice
[5] Jeffres, L. W., Bracken, C. C., Jian, G., & Casey, M. F. 2009. The Impact of Third Places on Community Quality of Life. Applied Research in Quality of Life, 4(4), 333–345. https://doi.org/10.1007/s11482-009-9084-8
[6] Kapur, M. 2010. The Immigrant. Random House India.
[7] Markoç, İ. 2019. Twitter in the context of Oldenburg's Third Place Theory. IBAD Sosyal Bilimler Dergisi, 79–89. https://doi.org/10.21733/ibad.610335
[8] McLeod, J. 2011. Beginning Postcolonialism. Viva Books.
[9] Mehta, V., & Bosson, J. K. 2009. Third Places and the Social Life of Streets. Environment and Behavior, 42(6), 779–805. https://doi.org/10.1177/0013916509344677
[10] Myers, P., & Sleeman, M. 2012. There's No Place Like Home? In Going Home: Essays, Articles, and Stories in Honour of the Andersons (pp. 33–40). lulu.com. https://books.google.co.in/books?id=ja_1AwAAQBAJ&pg=PA37&redir_esc=y#v=onepage&q&f=false
[11] Oldenburg, R. 1999. The great good place: Ray Oldenburg: Free Download, Borrow, and Streaming: Internet Archive. https://archive.org/details/greatgoodplaceca00olde_2
[12] Oldenburg, R., & Brissett, D. 1982. The third place. Qualitative Sociology. https://link.springer.com/article/10.1007/BF00986754?error=cookies_not_supported&code=9ed87dac-39b0-442f-92d0-1997f1cb3573#citeas
[13] Purnell, D. 2019. Public Parks: Third Places or Places Eliciting Moral Panic? Qualitative Inquiry, 25(6), 531–534. https://doi.org/10.1177/1077800418806612
[14] Shanthi, D. 2000. Hybridity and Postcolonialism (Thesis Chapter). Shodhganga. https://webcache.googleusercontent.com/search?q=cache:CsrvX-SUb9kJ:https://shodhganga.inflibnet.ac.in/bitstream/10603/164402/5/05_chapter%25202.pdf+&cd=11&hl=en&ct=clnk&gl=in
[15] Steinkuehler, C. A., & Williams, D. 2006. Where Everybody Knows Your (Screen) Name: Online Games as 'Third Places'. Journal of Computer-Mediated Communication, 11(4), 885–909. https://doi.org/10.1111/j.1083-6101.2006.00300.x
[16] Why everyone needs a 'third place' in their life. 2019. Metro. https://www.metromag.co.nz/society/society-society/why-everyone-needs-a-third-place-metro-

top-50-cafes
[17] Wimmer, J. 2014. There Is No Place Like Home: The Potential of Commercial Online Gaming Platforms to Become Third Places. In T. Quandt & S. Kroger (Eds.), Multiplayer: The Social Aspects of Digital Gaming (Routledge Studies in European Communication Research and Education) (1st ed., pp. 111–122). Routledge. https://books.google.co.in/books?id=zd_eAQAAQBAJ&pg=PA114&redir_esc=y#v=onepage&q&f=true
[18] Wisker, G. 2007. Key Concepts in Postcolonial Literature (Key Concepts: Literature). https://www.google.co.in/books/edition/Key_Concepts_in_Postcolonial_Literature/rphMDwAAQBAJ?hl=en

A Study on Finite Difference Explicit Numerical Method of Symmetric Hyperbolic Partial Differential Equation

P. Sowmiya[1], G.K. Revathi[2*], M.Sakthipriya[3]

[1]Shri Sakthikailassh Women's college, Salem-636 003.Tamilnadu, India.
[2]Division of Mathematics, School of advanced sciences, Vellore Institute of Technology, Chennai 127, Tamilnadu, India.
[3]Vaigai arts and science women's college, Valappady, Salem-636111, Tamilnadu, India.

E-mail: gk_revathi@yahoo.co.in

Abstract. Explicit method approaches numerical analysis for obtaining numerical approximation to the solution in time dependent partial differential equation and ordinary differential equation which is required in computer application and physical processes. Mainly explicit methods are used to calculate the system at current time. Explicit method are very simple but not stable. Mainly in explicit numerical methods, parameters are calculated based on previous levels and also explicit solutions are independent to other values. For example the unsteady heat equation can be solved explicitly. The present paper is concerned with explicit numerical method of symmetric hyperbolic partial differential equation by using constant indices and also we proved the monotonical preserving values and also discussed the symmetrical hyperbolic partial differential equation with constant coefficients.

Keywords: explicit, differential, numerical, symmetric, monotone

1. Introduction

A differential equation are those which involves partial derivates of a dependent variable with more than one independent variable which is called partial differential equation.A partial differential equation which is a mathematical equations involves two or more independent variables and partial derivatives of unknown function with respect to the independent variables. Partial differential equation can be solved analytically. The use of finite differential method is to approximate the solution to approximate the solution to a problem, one must discretize the problem's domain. This is usually done by dividing the domain into a uniform grid. This means that finite difference methods produce sets of discrete numerical approximation to the derivative often in a time stepping manner. This can be overcome by using the explicit numerical method of symmetric hyperbolic partial differential equation, which has been discussed in our paper.Explicit partial differential equation occurs when nodal points do not have any dependent variable. If it contains n nodal points then each equation can be solved explicitly. The unstudy heat equation can be solved explicitly. The terms of hyperbolic equation of partial differential equation has been discussed earlier in [1-5]. Mainly symmetric hyperbolic systems are very important because they form a subclass of strongly hyperbolic system in which the initial value problems are well posed in partial difference equation. An attempt has been made to introduce a new criteria to make it possible to change the constant indices to prove the monotones values and symmetric hyperbolic with constant coefficients.

2. Basic definition

2.1. Definition

The solution of the conservation law by means of an *explicit numerical method*

$$y_i^{n+1} = H\left(y_{i-k}^n, \cdots, y_{i+k}^n;\ \Delta x_i, \Delta t^{n+\frac{1}{2}}\right) \tag{1}$$

on a mesh with cell widths, cell centres and timesteps by $x_i = \frac{1}{2}\left(x_{i-\frac{1}{2}}, x_{i+\frac{1}{2}}\right)$. The pointwise error in the method is $e_i^n = y_i^n - y(x_i, t^n)$

2.2. Definition

An explicit numerical method (1) is *monotone* iff it preserves inequalities between sets of numerical results

1. For all y_i^n, for all v_i^n
2. If for all i, $y_i^n \le v_i^n$
3. Then for all i, $y_i^{n+1} = H\left(y_{i-k}^n, \cdots, y_{i+k}^n;\ \Delta x_i, \Delta t^{n+\frac{1}{2}}\right)$

$$y_i^{n+1} \le H\left(v_{i-k}^n, \cdots, v_{i+k}^n;\ \Delta x_i, \Delta t^{n+\frac{1}{2}}\right)$$

$$y_i^{n+1} \le v_i^{n+1}$$

2.3. Definition

Two mesh function y_i^n and v_i^n are L_1- contracting iff for all $m \ge n \ge 0$,

$$\|y^m - v^m\|_1 \le \|y^n - v^n\|_1 \tag{2}$$

These mesh functions are *total variation diminishing* (TVD) iff for all $m \ge n \ge 0$,

$$TV(y^m) \le TV(y^n) \tag{3}$$

These mesh functions are *monotonicity- preserving* iff

$$\begin{aligned} &for\ all\ i, y_i^n \le y_{i+1}^n \Rightarrow for\ all\ m > n, \forall i, y_i^m \le y_{i+1}^m \\ &for\ all\ i, y_i^n \ge y_{i+1}^n \Rightarrow for\ all\ m > n, \forall i, y_i^m \ge y_{i+1}^m \end{aligned} \tag{4}$$

2.4. Definition

The total variation TV of a function $w(x)$ is defined as

$$TV(w) = \lim_{\epsilon \to 0} \sup \frac{1}{\epsilon} \int_{-\infty}^{\infty} |w(x+\epsilon) - w(x)| dx$$

The *total variation* in space has the form

$$TV(w) = \sum_{i=-\infty}^{\infty} |w_{i+1} - w_i| \tag{5}$$

2.5. Definition

First order *symmetric hyperbolic* system

$$D_t y = \sum A_i D_i y \tag{6}$$

where $y = y(x,t)$ is a vector variable. A_i are symmetric matrices depends smoothly on x but not on t, D_i denote partial differentiation with respect to $x_i, \forall i = 1, \cdots, k$.

2.6. Definition

The *difference scheme* is of the form

$$y_h^{n+1} = \sum C_i\, y_{h+i}^n \tag{7}$$

where C_i are matrices that are function of $x = h\delta$, h is scalar index and the integer i ranges over a set of a finite number of indices.

2.7. Definition

The equation (6) takes the form

$$D_t y = A D_x y \tag{8}$$

gives $D_x D_t y = A D_{xx}^2 y$, $D_{tt}^2 y = A D_t D_x y = A^2 D_{xx}^2 y$.

Setting into (7) gives

$$y_h^n + D_t y\epsilon + \frac{1}{2} D_{tt}^2 y\epsilon^2 + O(\delta^3) = \sum C_i\, y_h^n \sum iC_i D_x y\delta + \frac{1}{2}\sum i^2 C_i D_{xx}^2 y\delta^2 + O(\delta^3)$$

Equating the coefficient of $y, D_x y, D_{xx}^2 y$ on both sides, it gives

$$\sum C_i = I, \quad \sum iC_i = \lambda A, \quad \sum i^2 C_i\, \lambda^2 A^2 \tag{9}$$

where $\epsilon = \lambda\delta$.

These equations for C_i should be solvable if we take all C_i but C_{-1}, C_0, C_1 equal to zero.

Now,

$$C_{-1} = \frac{1}{2}(\lambda^2 A^2 - \lambda A), \quad C_0 = I - \lambda^2 A^2, \quad C_1 = \frac{1}{2}(\lambda^2 A^2 + \lambda A) \tag{10}$$

is the unique solution of equation. Thus resulting difference scheme is called the *Lax-Wendroff scheme* (LW scheme)

$$y_h^{n+1} = C_{-1} y_{h-1}^n + C_0 y_h^n + C_1 y_{h+1}^n \tag{11}$$

The CFL condition for the convergence of the scheme

$$\lambda^{-1} \geq \sigma_{max}, \quad -\lambda^{-1} \geq \sigma_{min}$$

where σ_{max} and σ_{min} denote the largest and smallest eigenvalues of symmetric matrix A.

2.8. Definition

Symbol of the difference scheme (7) to be

$$C(\theta) = \sum C_i\, e^{ji\theta} \tag{12}$$

where C_i are matrices.

2.9. Definition

Non-local difference operator of the form

$$W_\delta = \sum W_i T^i \tag{13}$$

where W_i are not zero for $|i| > N$, but tend to zero rapidly as $|j| \to \infty$, we get

$$W(\xi) = \sum W_i\, e^{ji\xi} \tag{14}$$

is a *smooth periodic* function of ξ.

1. Note

Denote $\tilde{y}(\xi)$ the Fourier transform of y. Then

$$\widetilde{W_\delta y} = W(\delta\xi)\tilde{y}(\xi) \tag{15}$$

2.10. Definition

C_δ- *difference* operator $C_\delta y = v$ of the form

$$v_h = \sum C_i\, y_{h+i}$$

The symbol of C_δ is denoted by $C(\xi)$

$$C(\xi) = \sum C_i\, e^{ji\xi} \tag{16}$$

.

3. Main Results

3.1 Theorem

Consider the discrete value y_i^n as

$$y_i^n = \tilde{y}(i\Delta x, n\Delta t) + O(\Delta t^2) + O(\Delta x^2) + O(\Delta t\Delta x)$$

which satisfy the explicit upwind difference scheme

$$y_i^{n+1} = y_i^n - \frac{c\Delta t}{\Delta x}[y_i^n - y_{i-1}^n] \tag{17}$$

where $\tilde{y}$ is twice continuously differentiable in x and t and it satisfies a modified equation of the form

$$\partial_t\tilde{y} + \partial_x c\tilde{y} = e \equiv O(\Delta t) + O(\Delta x) \tag{18}$$

Then the modified e satisfies

$$e = \frac{c\Delta x}{2}\left(1 - \frac{c\Delta t}{\Delta x}\right)\partial_{xx}^2\tilde{y} + O(\Delta t) + O(\Delta x)$$

Proof:

Given, $\tilde{y}$ is twice continuously differentiable

$$y_i^{n+1} = \tilde{y}(x, t + \Delta t) + O(\Delta t^2) + O(\Delta x^2) + O(\Delta t\Delta x)$$

$$y_i^{n+1} = \tilde{y}(x, t) + \partial_t\tilde{y}\Delta t + \frac{1}{2}\partial_{tt}^2\tilde{y}\Delta t^2 + O(\Delta t^2) + O(\Delta x^2) + O(\Delta t\Delta x)$$

And

$$y_{i-1}^n = \tilde{y}(x - \Delta x, t) + O(\Delta t^2) + O(\Delta x^2) + O(\Delta t\Delta x)$$

$$y_{i-1}^n = \tilde{y}(x, t) - \partial_x\tilde{y}\Delta x + \frac{1}{2}\partial_{xx}^2\tilde{y}\Delta x^2 + O(\Delta t^2) + O(\Delta x^2) + O(\Delta t\Delta x)$$

Since $\tilde{y}$ satisfies modified equation (18)

$$\partial_t\tilde{y} = e - \partial_x c\tilde{y} \tag{19}$$

$$\partial_{tt}^2\tilde{y} = \partial_t[e - \partial_x c\tilde{y}]$$

$$\partial_{tt}^2\tilde{y} = \partial_t e - c\partial_x\partial_t\tilde{y}$$

Using (19), we get

$$\partial_{tt}^2\tilde{y} = \partial_t e - c\partial_x[e - \partial_x c\tilde{y}]$$

$$\partial_{tt}^2\tilde{y} = \partial_t e - c\partial_x e + c\partial_{xx}^2 c\tilde{y}$$

$$\partial_{tt}^2\tilde{y} = \partial_t e - c\partial_x e + c^2\partial_{xx}^2\tilde{y} \tag{20}$$

Substitute the approximation into explicit upwind difference scheme, we have (17)

$$y_i^{n+1} - y_i^n = -\frac{c\Delta t}{\Delta x}[y_i^n - y_{i-1}^n]$$

$$\frac{y_i^{n+1} - y_i^n}{\Delta t} = -\frac{c}{\Delta x}[y_i^n - y_{i-1}^n]$$

$$\frac{y_i^{n+1} - y_i^n}{\Delta t} + c\frac{y_i^n - y_{i-1}^n}{\Delta x} = 0 \tag{21}$$

Take,

$$y_i^{n+1} - y_i^n = \tilde{y}(x,t) + \partial_t\tilde{y}\Delta t + \frac{1}{2}\partial_{tt}^2\tilde{y}\Delta t^2 + O(\Delta t^2) + O(\Delta x^2) + O(\Delta t\Delta x)$$
$$-\tilde{y}(x,t) - O(\Delta t^2) - O(\Delta x^2) - O(\Delta t\Delta x)$$
$$y_i^{n+1} - y_i^n = \partial_t\tilde{y}\Delta t + \frac{1}{2}\partial_{tt}^2\tilde{y}\Delta t^2 \quad (22)$$

And

$$y_i^n - y_{i-1}^n = \tilde{y}(x,t) + O(\Delta t^2) + O(\Delta x^2) + O(\Delta t\Delta x) - \tilde{y}(x,t) + \partial_x\tilde{y}\Delta x$$
$$-\frac{1}{2}\partial_{xx}^2\tilde{y}\Delta x^2 - O(\Delta t^2) - O(\Delta x^2) - O(\Delta t\Delta x)$$
$$y_i^n - y_{i-1}^n = \partial_x\tilde{y}\Delta x - \frac{1}{2}\partial_{xx}^2\tilde{y}\Delta x^2 \quad (23)$$

Substitute (22) and (23) in (21), it becomes

$$0 = \frac{\partial_t\tilde{y}\Delta t + \frac{1}{2}\partial_{tt}^2\tilde{y}\Delta t^2}{\Delta t} + c\frac{\partial_x\tilde{y}\Delta x - \frac{1}{2}\partial_{xx}^2\tilde{y}\Delta x^2}{\Delta x}$$
$$0 = \partial_t\tilde{y} + \partial_{tt}^2\tilde{y}\frac{\Delta t}{2} + O(\Delta t) + c\left[\partial_x\tilde{y} - \partial_{xx}^2\tilde{y}\frac{\Delta x}{2}\right] + O(\Delta x)$$
$$0 = \partial_t\tilde{y} + c\partial_x\tilde{y} - c\partial_{xx}^2\tilde{y}\frac{\Delta x}{2} + \partial_{tt}^2\tilde{y}\frac{\Delta t}{2} + O(\Delta t) + O(\Delta x)$$

By (20), we get

$$0 = \partial_t\tilde{y} + c\partial_x\tilde{y} - \frac{c\Delta x}{2}\partial_{xx}^2\tilde{y} + \frac{\Delta t}{2}[\partial_t e - c\partial_x e + c^2\partial_{xx}^2\tilde{y}] + O(\Delta t) + O(\Delta x)$$
$$0 = \partial_t\tilde{y} + c\partial_x\tilde{y} - \frac{c\Delta x}{2}\partial_{xx}^2\tilde{y} + \frac{\Delta t}{2}\partial_t e - \frac{c\Delta t}{2}\partial_x e + \frac{c^2\Delta t}{2}\partial_{xx}^2\tilde{y} + O(\Delta t) + O(\Delta x)$$
$$0 = \partial_t\tilde{y} + \partial_x c\tilde{y} - \frac{c\Delta x}{2}\left(1 - \frac{c\Delta t}{\Delta x}\right)\partial_{xx}^2\tilde{y} + \frac{\Delta t}{2}[\partial_t e - c\partial_x e] + O(\Delta t) + O(\Delta x)$$

Since e is some arbitrary, we have

$$\partial_t\tilde{y} + \partial_x c\tilde{y} - \frac{c\Delta x}{2}\left(1 - \frac{c\Delta t}{\Delta x}\right)\partial_{xx}^2\tilde{y} + O(\Delta t) + O(\Delta x) = 0$$
$$\partial_t\tilde{y} + \partial_x c\tilde{y} = \frac{c\Delta x}{2}\left(1 - \frac{c\Delta t}{\Delta x}\right)\partial_{xx}^2\tilde{y} + O(\Delta t) + O(\Delta x)$$

By given (18),

$$\therefore e = \frac{c\Delta x}{2}\left(1 - \frac{c\Delta t}{\Delta x}\right)\partial_{xx}^2\tilde{y} + O(\Delta t) + O(\Delta x)$$

Hence the proof.

3.2 Theorem

Consider $H(w_{-k}, \cdots, w_k; \Delta x, \Delta t)$ is differentiable in each of its w_l arguments for $-k \leq l \leq k$. The solution is

$$y_j^{n+1} = H\left(y_{j-k}^n, \cdots, y_{j+k}^n;\ \Delta x_j, \Delta t^{n+\frac{1}{2}}\right)$$

is a monotone scheme iff for all $-k \leq l \leq k, \partial_{w_l}H \geq 0$

Proof:

Given any initial data y_i^0 and $\epsilon > 0$.

Let $v_i^0 = y_i^0 + \delta_{ij}\epsilon$

Claim: $\partial_{w_l}H \geq 0$

Since the scheme is monotone

For all i, it is true

$$v_i^0 \geq y_i^0$$

We have,

$$H\left(y_{i-k}^n, \cdots, y_{i+k}^n;\ \Delta x_i, \Delta t^{n+\frac{1}{2}}\right) = y_i^1 \leq v_i^1 = H\left(v_{i-k}^n, \cdots, v_{i+k}^n;\ \Delta x_i, \Delta t^{n+\frac{1}{2}}\right)$$

$$y_i^1 \leq v_i^1$$

Take $i + l = j$ for $-k \leq l \leq k$, we get

$$0 \leq v_i^1 - y_i^1$$

$$0 \leq \frac{1}{\epsilon}\left[H\left(v_{i-k}^n, \cdots, v_{i+k}^n;\ \Delta x_i, \Delta t^{n+\frac{1}{2}}\right) - H\left(y_{i-k}^n, \cdots, y_{i+k}^n;\ \Delta x_i, \Delta t^{n+\frac{1}{2}}\right)\right]$$

$$0 \leq \frac{1}{\epsilon}\left[H\left(y_{i-k}^n, \cdots, y_j^n + \epsilon, \cdots, y_{i+k}^n;\ \Delta x_i, \Delta t^{n+\frac{1}{2}}\right) - H\left(y_{i-k}^n, \cdots, y_j^n, \cdots, y_{i+k}^n;\ \Delta x_i, \Delta t^{n+\frac{1}{2}}\right)\right]$$

$$\partial_{w_l} H\left(y_{i-k}^n, \cdots, y_{i+k}^n;\ \Delta x_i, \Delta t^{n+\frac{1}{2}}\right) \to 0 \ as\ \epsilon \to 0$$

$$\therefore \partial_{w_l} H \geq 0$$

Conversely,

There is a vector w on the line segment between u and v.

Claim: y_j^{n+1} is a monotone scheme

Thus, we have

$$H\left(v_{i-k}^n, \cdots, v_{i+k}^n;\ \Delta x_i, \Delta t^{n+\frac{1}{2}}\right) = H\left(y_{i-k}^n, \cdots, y_{i+k}^n;\ \Delta x_i, \Delta t^{n+\frac{1}{2}}\right) + \sum_{l=-k}^{k} \partial_{w_{i+l}} H\left(w_{i-k}, \cdots, w_{i+k}; \Delta x_i, \Delta t^{n+\frac{1}{2}}\right)\left(v_{i+l}^n - y_{i+l}^n\right)$$

Since the partial derivatives of H are all nonnegative .

For all y_j^n, v_j^n, For all j such that

$$y_j^n \leq v_j^n$$

For all j such that

$$y_j^{n+1} \leq v_j^{n+1}$$

$\therefore y_j^{n+1}$ is a monotone scheme

Hence its claimed.

3.3 Lemma

The explicit numerical method (1) on a uniform mesh such that H does not depend explicitly on x or t. If the numerical solution is L_1- contracting, then the scheme is TVD.

Proof:

Claim: The scheme is TVD

It is enough to show $m = n + 1$ such that $TV(y^m) \leq TV(y^n)$

Given,

$$y_i^{n+1} = H\left(y_{i-k}^n, \cdots, y_{i+k}^n;\ \Delta x_i, \Delta t^{n+\frac{1}{2}}\right)$$

Here, $m = n + 1$

$$y_i^m = H\left(y_{i-k}^n, \cdots, y_{i+k}^n;\ \Delta x_i, \Delta t^{n+\frac{1}{2}}\right)$$

So,

$$v_i^m = y_{i-1}^m = H\left(y_{i-k-1}^n, \cdots, y_{i+k-1}^n;\ \Delta x_i, \Delta t^{n+\frac{1}{2}}\right)$$

Since the solution is L_1- contracting, we have

$$TV(y^m) = TV(y^{n+1})$$

Using (5), we get

$$TV(y^m) = \sum_{i=-\infty}^{\infty} |y_i^{n+1} - y_{i-1}^{n+1}|$$

By $v_i^m = y_{i-1}^m$,

$$TV(y^m) = \sum_{i=-\infty}^{\infty} |y_i^{n+1} - v_i^{n+1}|$$

We know that,

$$\|y^{n+1} - v^{n+1}\|_1 = \sum_i |y_i^{n+1} - v_i^{n+1}| \Delta x_i$$

From this equality, it becomes

$$TV(y^m) = \frac{1}{\Delta x} \|y^{n+1} - v^{n+1}\|_1$$

Using (2)

$$TV(y^m) \leq \frac{1}{\Delta x} \|y^n - v^n\|_1$$

$$TV(y^m) \leq \sum_{i=-\infty}^{\infty} |y_i^n - v_i^n|$$

$$TV(y^m) \leq \sum_{i=-\infty}^{\infty} |y_i^n - y_{i-1}^n|$$

$$\therefore TV(y^m) \leq TV(y^n)$$

The scheme is TVD.

The proof is completed.

3.4 Lemma

Consider the numerical scheme

$$y_i^{n+1} = y_i^n - \frac{\Delta t^{n+\frac{1}{2}}}{\Delta x_i} [\tilde{f}(y_{i-k-1}^n, \cdots, y_{i+k}^n) - \tilde{f}(y_{i-k}^n, \cdots, y_{i+k-1}^n)]$$

If the numerical solution is total variation diminishing and it is constant for large spatial indices, there exists $y_-, y_+, n > 0, M > 0, \ \forall\, i \leq -M$ such that $y_i^n = y_-$ and $\forall\, i \leq M$ such that $y_i^n = y_+$. Then the scheme is monotonicity preserving.

Proof:

Claim: The scheme is monotonicity preserving.

It is enough to show for $m = n + 1$ such that $y_i^m \leq y_{i+1}^m, \ \forall i$.

Take $i \geq M + k$

$$y_i^{n+1} = y_i^n - \frac{\Delta t^{n+\frac{1}{2}}}{\Delta x_i} [\tilde{f}(y_{i-k-1}^n, \cdots, y_{i+k}^n) - \tilde{f}(y_{i-k}^n, \cdots, y_{i+k-1}^n)]$$

$$y_i^{n+1} = y_+ - \frac{\Delta t^{n+\frac{1}{2}}}{\Delta x_i} [\tilde{f}(y_+, \cdots, y_+) - \tilde{f}(y_+, \cdots, y_+)]$$

$$y_i^{n+1} = y_+$$

Take $i \leq -M - k$

$$y_i^{n+1} = y_i^n - \frac{\Delta t^{n+\frac{1}{2}}}{\Delta x_i} [\tilde{f}(y_{i-k-1}^n, \cdots, y_{i+k}^n) - \tilde{f}(y_{i-k}^n, \cdots, y_{i+k-1}^n)]$$

$$y_i^{n+1} = y_- - \frac{\Delta t^{n+\frac{1}{2}}}{\Delta x_i}[\tilde{f}(y_-,\cdots,y_-) - \tilde{f}(y_-,\cdots,y_-)]$$

$y_i^{n+1} = y_-$

Since y_i^n is non-decresing.

$$TV(y^{n+1}) \leq TV(y^n)$$

Now,

$$y_+ - y_- = \sum_{i=-\infty}^{\infty} [y_{i+1}^{n+1} - y_i^{n+1}]$$

$$y_+ - y_- \leq \sum_{i=-\infty}^{\infty} |y_{i+1}^{n+1} - y_i^{n+1}|$$

$$y_+ - y_- \leq TV(y^{n+1})$$

$$y_+ - y_- \leq TV(y^n)$$

$$y_+ - y_- \leq \sum_{i=-\infty}^{\infty} |y_{i+1}^n - y_i^n|$$

$$y_+ - y_- = \sum_{i=-\infty}^{\infty} [y_{i+1}^n - y_i^n]$$

Its follows that,

$$\sum_{i=-\infty}^{\infty} [y_{i+1}^{n+1} - y_i^{n+1}] = \sum_{i=-\infty}^{\infty} [y_{i+1}^n - y_i^n]$$

$$y_{i+1}^{n+1} - y_i^{n+1} \geq 0 \;\; \forall i$$

$$y_{i+1}^{n+1} \geq y_i^{n+1}$$

$$y_i^{n+1} \leq y_{i+1}^{n+1}$$

$$\therefore y_i^m \leq y_{i+1}^m \;\; \forall i$$

The scheme is monotonicity preserving.

Hence proved.

3. 5Lemma

Consider the difference scheme

$$y_i^{n+1} = y_i^n + D_{i+\frac{1}{2}}\Delta y_{i+\frac{1}{2}}^n - C_{i-\frac{1}{2}}\Delta y_{i-\frac{1}{2}}^n$$

and if $\Delta y_{i+\frac{1}{2}}^n = y_i^{n+1} - y_i^n$ such that for all $i, 0 \leq C_{i+\frac{1}{2}}$, $0 \leq D_{i+\frac{1}{2}}$ and $C_{i+\frac{1}{2}} + D_{i+\frac{1}{2}} \leq 1$. Then the scheme is TVD.

Proof:

Claim: The scheme is TVD.

It is enough to show $m = n + 1$ such that $TV(y^m) \leq TV(y^n)$

Take $m = n + 1$

$$TV(y^m) = TV(y^{n+1})$$

Using (5)

$$TV(y^m) = \sum_{i=-\infty}^{\infty} |y_i^{n+1} - y_{i-1}^{n+1}|$$

$$TV(y^m) = \sum_i \left| y_{i+1}^n + D_{i+\frac{3}{2}} \Delta y_{i+\frac{3}{2}}^n - C_{i+\frac{1}{2}} \Delta y_{i+\frac{1}{2}}^n - y_i^n - D_{i+\frac{1}{2}} \Delta y_{i+\frac{1}{2}}^n + C_{i-\frac{1}{2}} \Delta y_{i-\frac{1}{2}}^n \right|$$

$$TV(y^m) = \sum_i \left| D_{i+\frac{3}{2}} \Delta y_{i+\frac{3}{2}}^n + [\, y_{i+1}^n - y_i^n] - C_{i+\frac{1}{2}} \Delta y_{i+\frac{1}{2}}^n - D_{i+\frac{1}{2}} \Delta y_{i+\frac{1}{2}}^n + C_{i-\frac{1}{2}} \Delta y_{i-\frac{1}{2}}^n \right|$$

$$TV(y^m) = \sum_i \left| D_{i+\frac{3}{2}} \Delta y_{i+\frac{3}{2}}^n + \Delta y_{i+\frac{1}{2}}^n - C_{i+\frac{1}{2}} \Delta y_{i+\frac{1}{2}}^n - D_{i+\frac{1}{2}} \Delta y_{i+\frac{1}{2}}^n + C_{i-\frac{1}{2}} \Delta y_{i-\frac{1}{2}}^n \right|$$

$$TV(y^m) = \sum_i \left| D_{i+\frac{3}{2}} \Delta y_{i+\frac{3}{2}}^n + \left[1 - C_{i+\frac{1}{2}} - D_{i+\frac{1}{2}}\right] \Delta y_{i+\frac{1}{2}}^n + C_{i-\frac{1}{2}} \Delta y_{i-\frac{1}{2}}^n \right|$$

$$TV(y^m) \le \sum_i D_{i+\frac{3}{2}} \left| \Delta y_{i+\frac{3}{2}}^n \right| + \sum_i \left[1 - C_{i+\frac{1}{2}} - D_{i+\frac{1}{2}}\right] \left| \Delta y_{i+\frac{1}{2}}^n \right| + \sum_i C_{i-\frac{1}{2}} \left| \Delta y_{i-\frac{1}{2}}^n \right|$$

$$TV(y^m) \le \sum_i \left[D_{i+\frac{3}{2}} + \left[1 - C_{i+\frac{1}{2}} - D_{i+\frac{1}{2}}\right] + C_{i-\frac{1}{2}} \right] \left| \Delta y_{i+\frac{1}{2}}^n \right|$$

$$TV(y^m) \le \sum_i \left| \Delta y_{i+\frac{1}{2}}^n \right|$$

$$TV(y^m) \le \sum_i |\, y_{i+1}^n - y_i^n|$$

$$TV(y^m) \le TV(y^n)$$

The scheme is TVD.
Hence the proof.

3. 6Theorem

The difference scheme (6) with constant coefficient is stable if the norm of the power of its symbol $C(\theta)$ defined by (12) are uniformly bounded

$$|C^n(\theta)| \le k, \quad for\ some\ k, \forall \theta, n \qquad (24)$$

Proof:
The norm of y^n,

$$\|y^n\|_\delta^2 = \delta \int |C(\theta)^n\, \tilde{y}^o(\theta)|^2 \bar{d}\theta$$

Using (24)

$$\|y^n\|_\delta^2 \le \delta \int |k\, \tilde{y}^o(\theta)|^2 \bar{d}\theta$$

$$\|y^n\|_\delta^2 \le \delta k^2 \int |\, \tilde{y}^o(\theta)|^2 \bar{d}\theta$$

Parseval relation is

$$\|y^n\|^2 = \delta \int |\, \tilde{y}^n(\theta)|^2 \bar{d}\theta$$

By this relation, it becomes

$$\|y^n\|_\delta^2 \le k^2 \|y^0\|_\delta^2$$

It expressed in terms of the operator C_δ linking y^{n+1} to y^n. The nth power norm of C_δ is $\le k, \forall n$.

∴ The scheme (6) is stable.

IOP Conf. Series: Materials Science and Engineering 1012 (2021) 012063 doi:10.1088/1757-899X/1012/1/012063

Hence proved.

2. Note
If $|C(\theta)| \leq 1, \forall\theta$, then (24) holds with $k = 1$.

3.7Theorem

The LW scheme (10) and (11) with constant coefficient is stable provided that the CFL condition is satisfied.

Proof:

Take $\lambda A = A$, the CFL condition that the eigenvalues of A lies between -1 and 1. Compute the symbol by LW scheme

$$C(\theta) = C_{-1}e^{-i\theta} + C_0 e^0 + C_1 e^{i\theta} \tag{25}$$

Using (10),

$$C(\theta) = \frac{1}{2}(A^2 - A)e^{-i\theta} + (I - A^2)\cdot 1 + \frac{1}{2}(A^2 + A)e^{i\theta}$$

$$C(\theta) = \frac{1}{2}A^2e^{-i\theta} - \frac{1}{2}Ae^{-i\theta} + I - A^2 + \frac{1}{2}A^2e^{i\theta} + \frac{1}{2}Ae^{i\theta}$$

$$C(\theta) = I - A^2 + \frac{1}{2}A^2(cos\theta - isin\theta) + \frac{1}{2}A^2(cos\theta + isin\theta) - \frac{1}{2}A(cos\theta - isin\theta) + \frac{1}{2}A(cos\theta + isin\theta)$$

$$C(\theta) = I - A^2 + 2\left(\frac{1}{2}A^2cos\theta\right) + 2\left(\frac{1}{2}A\ isin\theta\right)$$

$$C(\theta) = I - A^2 + A^2cos\theta + A\ isin\theta$$

$$C(\theta) = I + A^2(cos\theta - 1) + A\ isin\theta$$

Similarly,

$$C^*(\theta) = I + A^2(cos\theta - 1) - A\ isin\theta$$

Now,

$$C^*(\theta)C(\theta) = (I + A^2(cos\theta - 1) - A\ isin\theta) + (I + A^2(cos\theta - 1) + A\ isin\theta)$$

$$C^*(\theta)C(\theta) = \left(I + A^2(cos\theta - 1)\right)^2 - (A\ isin\theta)^2$$

$$C^*(\theta)C(\theta) = I + A^4(cos\theta - 1)^2 + 2A^2(cos\theta - 1) - (-A^2sin^2\theta)$$

$$C^*(\theta)C(\theta) = I + A^4(cos\theta - 1)^2 + 2A^2(cos\theta - 1) + (A^2sin^2\theta)$$

$$C^*(\theta)C(\theta) = I + A^4(cos\theta - 1)^2 + A^2(2(cos\theta - 1) + sin^2\theta)$$

$$C^*(\theta)C(\theta) = I + A^4(cos\theta - 1)^2 - A^2(cos\theta - 1)^2$$

$$C^*(\theta)C(\theta) = I + (A^4 - A^2)(cos\theta - 1)^2$$

The eigenvalues of A are ≤ 1 in absolute value, $A^4 - A^2$ is negative, such as

$$C^*(\theta)C(\theta) \leq I \tag{26}$$

For any vector w, then

$$|C(\theta)w|^2 = (C(\theta)w, C(\theta)w)$$

$$|C(\theta)w|^2 = (w, C^*(\theta)C(\theta)w)$$

$$|C(\theta)w|^2 = (w, Iw)\ [by\ (26)]$$

$$|C(\theta)w|^2 = (w, w)$$

$$|C(\theta)w|^2 = |w|^2$$

The equation (26) we have,

$$C^*(\theta)C(\theta) \leq I$$

$$|C(\theta)|^2 \leq I$$

$$|C(\theta)| \leq 1, \forall\theta$$

From the note: 2,

∴ The scheme is stable.

The proof is completed.

3.8 Theorem

The symbol of a difference scheme with constant coefficient $C(\xi)$, there exists a smooth periodic function $W(\xi)$ whose values are symmetric positive definite matrices that satisfies

$$C^*(\xi)W(\xi)C(\xi) \leq W(\xi) \quad (27)$$

Then the scheme is stable.

Proof:

The norm of y,

$$\|y\|_W^2 = \int W(\delta\xi)\tilde{y} \cdot \tilde{y} d\xi$$

Take y as $v = C_\delta y$, we get

$$\|v\|_W^2 = \int W(\delta\xi)\tilde{v}(\xi) \cdot \tilde{v}(\xi) d\xi$$

Using (15) as $v = C_\delta y$

$$\tilde{v} = \widetilde{C_\delta y} = C(\delta\xi)\tilde{y}(\xi)$$

Now,

$$\|v\|_W^2 = \int W(\delta\xi)C(\delta\xi)\tilde{y}(\xi) \cdot C(\delta\xi)\tilde{y}(\xi) d\xi$$

$$\|v\|_W^2 = \int C^*(\delta\xi)W(\delta\xi)C(\delta\xi)\tilde{y}(\xi) \cdot \tilde{y}(\xi) d\xi$$

By given (26), it becomes

$$\|v\|_W^2 \leq \int W(\delta\xi)\tilde{y}(\xi) \cdot \tilde{y}(\xi) d\xi$$

$$\|v\|_W^2 \leq \|y\|_W^2$$

Taking square root on both sides, we get

$$\|v\|_W \leq \|y\|_W$$

$$\|C_\delta y\|_W \leq \|y\|_W$$

For any power of C_δ is ≤ 1,

$$\|C_\delta^n y\|_W \leq \|y\|_W$$

We know that,

$$C_1\|y\| \leq \|y\|_W \leq C_2\|y\|$$

Here,

$$\|C_\delta^n y\| \leq C_1^{-1}\|C_\delta^n y\|_W$$

$$\|C_\delta^n y\| \leq C_1^{-1}\|y\|_W$$

$$\|C_\delta^n y\| \leq C_1^{-1}C_2\|y\|$$

∴ This shows inequality.

Hence proved.

4. Finding

We develop the convergence of numerical method by explicit numerical method and approximation is to discretize the variables x and t by replacing derivatives by difference quotients, then in an symmetric system of conserved hyperbolic equation. Hence we find out the solution by using monotonically preserving values.

5. Conclusions

Numerical partial differential equation is the large area of study. This includes the components in the area of applications, mathematics, science and computer. Sometimes differential equations are very difficult to solve analytically or models which are needed for computer simulations. In these cases, finite differences are used to solve the equations instead of analytical ones. Thus we use the explicit numerical method by using monotone values and symmetric hyperbolic equation with constant coefficients. Further higher studies can be extended to finite Implicit numerical method of symmetric hyperbolic partial differential equation using variable coefficients.

References

[1] John. A. Trangenstein, ``Numerical solution of Hyperbolic partial differential equation", Cambridge university press, New York, 2009.
[2] Ami Harten, ``High resolution schemes for Hyperbolic conservation laws", Journal of computational physics 135, 260-278, 1997.
[3] Peter D. Lax, ``Hyperbolic partial differential equation", New York university, New York.
[4] Amiya Kumar Pani, "Lecture notes on finite element method for elliptic problems", Indian Institute of Technology, Bombay, 2012.
[5] Dietrich Braess, "Finite element – Theory, Fast solvers and Application in Elasticity Theory", Cambridge university press, New York, 2007.
[6] Susanne C. Brenner, L. Ridgway Scott, "Texts in Applied Mathematics: The mathematical Theory of finite element methods", Third edition, Springer publication, 2008.
[7] Volker John, "Numerical methods for partial differential equation", Summer semester, 2013.
[8] Christian Clason, "Finite Element methods lecture notes" arXiv: 1709.08618v1[math.NA] 25 sep 2017.

The Study of Advanced Scientific Experiments in *Avatar*- A Postcolonial Perspective

P. Vijayasekaran[1*], G. Alan[2]

[1]School of Social Sciences and Languages, Vellore Institute of Technology, Chennai-600127.

E-mail: vijay.sekarmotoe@gmail.com

Abstract: Scientific advancement is used for the welfare of many people across the world. Even in the times of pandemic, the advancement of science has paved way for digital communication. But in this paper, an analysis is going to be made on how the humans had misused the scientific advancement to conquer a planet called Pandora in the movie *Avatar* to destroy the inhabitants of the planet who are referred to as Na'vi and to plunder a valuable natural mineral called unobtanium. This movie is an indirect adaptation of science versus nature, the argument to be made in this research paper is that nature is the base for everything, but people are deconstructing it with the scientific advancements, it is arguably like placing our own lives into misery. The question is that do we really need science for a happy life but the indigenous people in Pandora lived happily before the invasion of science in their planet, this is going to be further analysed in this research paper. The title emphasizes on the scientific advancement because this paper aims to discuss the impact of the advanced science that is depicted in the movie.

Keywords: science, colonization, postcolonialism, nature, environment.

1. Introduction

The scientific advancements in the DNA research, as well as in the field of hi-tech weapon system, also in the space research, all the advanced types of scientific experiments are portrayed in the movie, but the impact of these advancements are further going to be analysed in this paper. The impact of science can be seen in various fields, but colonization is a process which is done through various modes and for different motives. The scientific effect of colonization among the people in a new planet is expressed in the movie which can also be a sign of warning for human beings.

2. Methodology

This paper proposes to study the colonial effects on nature. The methods used to identify the problems in colonization are, firstly an analysis about the scientific advancements when it is used by the humans for colonization and the negative implication is discussed, the second thing is the conflict between nature and science because of the humans and the importance of nature for human beings for now and of course for the future generation. These aspects are going to be analysed effectively.

IOP Conf. Series: Materials Science and Engineering 1012 (2021) 012064 doi:10.1088/1757-899X/1012/1/012064

3. Science experiments by the RDA

In this movie *Avatar*, science is taken as a tool for colonization, this can be argued by various examples, in the beginning of the movie Jake Sully is taken to the planet of Pandora as a replacement for his brother who was dead. The RDA has implemented a human into a body of Avatar. This is one of the major scientific advancements because making a robot is different from creating an alike of a living being. A blue coloured, seven-feet tall 'Avatar' body is created for the sake of entering into the planet of Pandora, this is a new level in colonization, because so far people have invaded as reformers and traders but entering a new planet with the appearance of the native inhabitant Na'vi is a terrific scientific advancement which is portrayed in the movie. Though it can be considered as a positive technological improvement, as by providing a new body for a disabled person who can walk and run with the new 'Avatar' body, but the real motive of Quaritch is cruel, as he wants to destroy the native inhabitants to plunder the unobtanium. After accomplishing the mission, Colonel Quaritch promises Jake to bring back his walking ability which is also merely based on business motive.

Many questions can be raised that if science is mandatory for a happy life? And if so, the native inhabitants are quite happy without any scientific advancement. But Jake is accepted as a leader in Pandora, this is indifferent in their custom i.e., to accept an outsider as a leader. The indigenous people respected the rituals and customs of their ancestors. It can be certainly argued that they valued nature over science.

The RDA is an administration formed by the humans to conquer the planet Pandora, in fact they had all the technical equipment and a much developed technology eventhough they are in hunt for the wealth of the other planet. The important aspect which is to be taken into account is that the movie happens in future and this makes us to wonder that if this is the future of human beings. The movie makes a strong impact that we humans will deplete the natural resources and we would misuse the science. The colonial impact is based on capitalist motive, they invaded Pandora not to develop them but to capture their land and wealth, like what the whites did to the black people, in the words of Frantz Fanon in *Black Skin White Masks* (2008), "The Negro problem does not resolve itself into the problem of Negroes living among white men but rather of Negroes exploited, enslaved, despised by a colonialist, capitalist society that is only accidentally white (p. 157)" [2]. These are some of the colonial comparisons that can be made between the whites and the RDA but the RDA had gone to the extent of colonizing the other planet with the usage of hi-tech science and technology.

Jake, in the beginning of the movie watches a television footage and says that these are the cloned versions of Bengal tiger which was extinct over many years, it gives us a glimpse about the future, and the people in the future are portrayed in a scene with masks and we have to remember that the movie was released in 2009. Such a huge prediction is made by the director which shows us the importance of nature and our natural resources, it gives us a hint that humans are diluting everything, after the extinction of the species humans are creating it again with the immense science development, if a new hybrid can be produced with the usage of science then why we cannot save the same existing species with the use of science, this is a deep question which we need to find an answer before it is too late. In the words of Richard Gunderman in his article "H.G. Wells and Orwell on Whether Science Can Save Humanity",

> Wells' enthusiasm for science had political implications. Having contemplated in his novels the self-destruction of mankind, Wells believed that humanity's best hope lay in the creation of a single world government overseen by scientists and engineers. Human beings, he argued, need to set aside religion and nationalism and put their faith in the power of scientifically trained, rational experts [3].

H.G. Wells in his novels had portrayed the impact of science on modern man, he implied all kinds of scientific advancements in human lives and he insisted that science should be used productively. It can be connected with Avatar, because most of the scenes reflect the cruel reality that is man depleting all the natural resources with the help of science and trying to colonize another planet which may happen in future as depicted in the movie. The impact of science should be positive for all living creatures and that is the respect that human can contribute towards nature.The warcrafts, nuclear weapons shown in the movie are very advanced and such kinds of weapons are invented by the humans for the protection, but in this movie, the RDA had used it for the negative purpose, they broke the customs and traditions of the Na'vi without showing any mercy. The hometree of the Na'vi is destroyed as the RDA did not show any sentimental value for the inhabitants. Meanwhile Jake is sent as an agent to spy on them. In the words of Ania Loomba (2005) in *Colonialism/Postcolonialism*,

> Colonisers differed in their modes of interacting with the local populations, thus producing variable racial discourses and identities. For example, the Spanish in America and the Portuguese in India settled down in the lands they colonised, adopted local manners and intermarried in a way that the English derided. Eventually, inter-marriages and concubinage blurred racial distinctions and created a population which acted as a strong base for colonial rule (p. 96) [4].

As per the words of Ania Loomba the colonizers implemented various techniques to colonize the natives. In *Avatar*, Jake Sully visits the land of Pandora as a 'Na'vi' in the artificially created body of Na'vi called 'Avatar', later he falls in love with one of the indigenous girls called Neytiri. Initially Jake was sent to Pandora to know more about the Na'vi but later he turns against the RDA, moreover he is considered as a leader of the Na'vi. The important difference between the Na'vi and RDA is that the power of the former is nature and the later is science.

4. Nature vs Science

The climax of the movie is nothing but the indirect war between nature and science. The problem with the RDA is that they did not value any culture of the Na'vi, one of the most important of customs of the Na'vi is that they seek permission from their goddess Eywa, the god of nature, for any event from any ritual to selecting the life partner the Na'vi had proceeded only after getting the blessing of goddess Eywa. Another significant factor about the Na'vi is that they respected all the living beings of Pandora. For instance, in a particular scene Neytiri saves Jake from the attack of an animal by killing it and later she says that it was the fault of Jake and she repents for killing it, this is something hardly we can find with human beings. They valued nature in each and every aspect that they only used the dragons for flying, only after the acceptance of the dragons. They valued each and every living being of Pandora.

The indigenous people of Omaticaya tribe referred RDA as 'Skypeople' because of the ships they used to enter into Pandora. The term 'Skypeople' denoted a kind of superiority but the bitter truth is that they did not enter there for a better purpose. In the words of Lynn Jr. et.al.(1996), in the *Ecocriticism Reader*, "What people do about their ecology depends on what they think about themselves in relation to things around them. Human ecology is deeply conditioned by beliefs about our nature and destiny-that is, by religion" (p. 9) [5]. The concern for humans towards nature is very low when we compare it with that of the Na'vi. Colonel Quaritch is merciless that he kills all the creatures with his advanced weapons with the help of his military. But the basic difference is that the natives are very respectful towards nature unlike humans who have exploited all the natural resources. Even with science the team of RDA could not make a positive impact over the human mankind. The character of Jake Sully is well drafted by the director, this is because eventhough Quaritch offered him to bring back is mobility, Jake refused and stood by the side of the Na'vi, because he felt connected towards the naturalistic environment of Pandora. Moreover, in the final scene of the movie Jake is permanently transformed as a Na'vi with the help of the goddess Eywa, this could be taken as if the science could only provide the body but the souls are only connected by the nature which is ever superior. In this movie, the negative impact of the science has depleted the natural wealth of the earth and if they succeed in colonizing Pandora, the future of Pandora will be the same. They will destroy all the natural wealth of it and they will conquer another planet and that too with the help of science.

5. Colonial dominance

The colonial dominance of the humans can be seen via the character of Colonel Quaritch. This is because of the desire for wealth. Fanon in his *The Wretched of the Earth*, expressed the impact of colonizers over the colonizers, he wrote about the sufferings of the negro people who were under such bad circumstances because of the white invaders, this happened because of the power [6]. The whites called themselves as superior and imposed their culture, in the same way the team of RDA with their advanced technology claimed that they are superior to the Na'vi and they had a plan to construct many buildings over their land. They are not really there to help the Na'vi but only to take away all the unobtanium from them. There are several countries around the world where people are living with hunger and famine and this could turn out to be the fate of the whole world without any resource in the future as it is portrayed in the movie. From man colonizing the man to humans colonizing the alien may happen in the near future if we are unaware about our future. Our future is uncertain that we should apply the knowledge and science for the welfare of the people. The important aspect of the movie is that the difference shown by the director, the colonial mindset of human beings, he did not portray humans as glorious or noble rather he showed the wicked attitude of the humans who are ready to destroy another planet to safeguard themselves. Colonialism can be done through various modes, it may be through business, language or any other aspect but this movie has taught us that how science can have an impact on another planet. We have seen several movies on how the science has turned up against the human beings for instance we can give the example of the movie *Hollow man* (2000), which depicted the misusage of science. But *Avatar* shows us that how the impact of science can put the people of another planet into misery. Saving the other planet is a different story but in this movie they have failed to save the earth itself. There is no means of attacking the other planet after depleting our own planet. This will lead to a war between two worlds as depicted in the climax of the movie. The argument that can be raised here is rather than using the science to conquer another planet we may use the science to save our home earth. In the words of Richard Gunderman about Wells,"While championing science, Wells recognized that scientific progress could also lead to human misery" [3]. Science can be used for the welfare of the people but if we misuse it, this will lead us into misery.
Modern day colonialism has changed the attitude of human minds. The countries are finding advanced nuclear weapons for the protection of their countries, it is pretty good thing from protection perspective but what will the fate of the innocent lives, in this movie, a Na'vi says that if a war happens they will lose lot of people, this is the mindset of an ordinary man. Colonel Quaritch is ruthless about the lives of the Na'vi, with the usage of giant robots and hi-tech warships he is in a rage to destroy the planet of Pandora which is a kind of colonialism not even to colonize but to kill them and plunder their wealth. This is a fascistic approach of human beings which is expressed by John Marmysz in his "Rooting for the Fascists in James Cameron's Avatar" states that the fascistic mindset of the human beings is expressed in this movie, he lauds about the making of the movie but he expresses his concern that many people are fascinated about this fascistic mindset which is a negative attitude [8].To put it in general, the world has changed now, people are involved in theft because of hunger and unemployment, this movie has shown our future, if this situation prevails we would be robbing and begging another planet for our survival, this is a saddest truth in our contemporary world.

6. Conclusion

This paper is an attempt to analyse the pros and cons of science through the movie *Avatar*, and to indicate that how science has been used as a tool to colonize Pandora. To analyse this movie from a postcolonial perspective is not only the major theme of this paper, but also the aftermath of the postcolonial effect is deeply analysed, moreover the harm after colonizing the natural environment is exposed in the movie. The impact of the movie is vast, because the movie itself 3-D animated and they have spent crores to achieve such a kind of movie. It shows us the importance of science in our modern day world. Regarding the theme of the movie, it is a sign of our future, how humans will be in future. It maybe a science fiction 'movie', but the content expressed in this movie is highly connected with our modern lifestyle. The argument is that the Na'vi (natives) are quite happy without science but desperately they struggled in the final scene to protect their planet. As it is a movie we can accept the victory of Na'vi over the advanced RDA, but in reality it would be very difficult to fight against such powerful weapons. We have to accept the fact that science has made our lives easier but it should be redirected in the right path, a clear vision is needed for science and that should be meant only for productive needs. The movie itself has shown many positive factors of science, for instance blessing the disabled Jake Sully with the walking ability is a gift of science but the negatives are also clearly indicated in the movie, the Na'vi is a representation of other living creatures in the world, the duty of the human beings is to protect nature with the helping hand of science and not to fight against it.

References

[1] Avatar. 2009. Directed by James Cameron, performances by Sam Worthington andZoe Saldana,Twentieth Century Fox.
[2] Fanon, Frantz. Black Skin, White Masks.2008.2nd ed., Pluto Press
[3] Gunderman, Richard (2020, October 11), H.G. Wells and Orwell on Whether Science Can Save Humanity.https://fee.org/articles/hg-wells-and-orwell-on-whether-science-can-save-humanity/.
[4] Loomba, Ania., Colonialism/Postcolonialism. 2005. 2nd ed., Routledge, Taylor and Francis group, London and New York.
[5] White, JR., Lynn, Fromm, Harold., The Historical Roots of Our Ecologic Crisis, From Transcendence to Obsolescence: A Route Map. The Ecocriticism Reader: Landmarks in Literary Ecology. 1996. edited by Cheryll Glotfelty and Harold Fromm, The University of Georgia Press Athens and London.
[6] Fanon, Frantz. The Wretched of the Earth.1963. Grove Widenfeld, New York.
[7] Martinez, Falquina Silvia, The Pandora Effect: James Cameron's Avatar and a Trauma Studies Perspective, ATLANTIS Journal of the Spanish Association of Anglo-American Studies.**36**(2).115-131.
[8] Marmysz, John, Rooting for the Fascists in James Cameron's Avatar, 2012.Film and Philosophy, edited by Dan Shaw. Ethics and Existentialism II,**16** special interest edition.101-120.
[9] Bartels A., Eckstein L., Waller N., Wiemann D. (2019).Postcolonialism And Ecology. In:Postcolonial Literatures in English". J.B. Metzler, Stuttgart.109-129.
[10] Bennett, Bruce, Loving the Alien:Indigenous Protest and Neo-Colonial Violence in James Cameron's Avatar. (2014) .in KatarzynaMarciniak, Imogen Tyler, eds.,Immigrant Protest: Politics, Aesthetics and Everyday Dissent, State University of New York Press. 97-119.

Embedding of Circulant Networks into Wheels

Stalin Mary[1], **Indra Rajasingh**[1*]

[1] Mathematics Division, School of Advanced Sciences, Vellore Institute of Technology, Chennai – 600127.

E-mail: indrarajasingh@yahoo.com

Abstract- Circulant graphs have been studied over many years because of its vast applications in telecommunication industry. We find the minimum wirelengths of embedding the circulant graph $G(n;\ \pm\{1,2\})$ and $G(n;\ \pm\{1,2,\ldots,j\})$ into variation of wheel graphs on n vertices in this paper.

Keywords: Embedding, Congestion, Circulant graph, Wheel graph

1. Introduction

Graph is a mathematical model represented by points and lines joining certain pairs of points. These points are addressed as vertices or nodes and the lines are addressed as edges or links. Graphs occur naturally in several real world applications such as transportation network, electronic circuit, social network, communication network etc.,[1]. Networks are graphs whose vertices represent modules and lines represent routes that convey or transmit messaages. Links also represent transfer of data. An one-one function $\mathcal{E}$ of a graph G into a graph H is called an embedding of G into H if every edge (x, y) in G is mapped to a path between $\mathcal{E}(x)$ and $\mathcal{E}(y)$ in H [2]. Circulant networks have optimal routing capabilities and optimal fault tolerance [3, 4, 5, 6]. This paved way for using circulant networks as telecommunication networks [4]. They are also used in arrangement of binary codes [7]. The undirected circulant graphs are used in ILLIAC type computers [8]. A wheel graph W_n consists of $n-1$ number of vertices in the outer cycle which are all connected to a vertex called its hub. In wheel network, information flows to and from a single person. It is a communication network where the employees get information fast. Very often, circuit layout designs include wheel networks [15]. Gear graphs and Helm graphs are obtained from wheel graphs [8]. The minimum wirelengths of embedding $G(n;\ \pm\{1,2\})$ and $G(n;\ \pm\{1,2,\ldots,j\})$ into wheels, one-point union of wheels and union of wheels with cut edges are obtained in this paper.

2. Preliminaries

Definition 2.1. **Maximum Subgraph Problem** [9]: For a given graph G on n nodes and a given integer $k, k \leq n$, finding a subgraph on k nodes with number of edges greater than or equal to any other subgraph of G on k vertices, is called Maximum Subgraph Problem.

Definition 2.2. [10] : Let G and H be finite graphs with n vertices. An one-one mapping $f: V(G) \rightarrow V(H)$ inducing a one-one mapping $P_f: E(G) \rightarrow \{P_f(u,v): P_f(u,v)$ is a path between $f(u)$ and $f(v)$ in H for $(u,v) \in E(G)\}$ is called an embedding of G into H.

Definition 2.3. Let $f: G \rightarrow H$ be an embedding. For $e \in E(H)$, the number of paths in $\{P_f(u,v)$: $P_f(u,v)$ is a path between $f(u)$ and $f(v)$ in H for $(u,v) \in E(G)\}$ that contain e is called the congestion on e with respect to f and is denoted by $c_f(e)$. For $S \subseteq E(H)$, we define $c_f(S) = \sum_{e \in S} c_f(e)$.

Definition 2.4. [12] A subgraph K of a graph G is said to be convex if all the shortest paths between any two vertices of K lie in K. An edge cut S of G is said to be a convex cut if $G \backslash S$ splits into two components, each of which is convex.

Definition 2.5. Let $\mathcal{E}: G \rightarrow H$ be an embedding. Then, the expression $\sum_{e \in E(H)} c_{\mathcal{E}}(e)$ is called the wirelength of embedding G into H with respect to $\mathcal{E}$ and is denoted by $WL_{\mathcal{E}}(G,H)$.
$\min_{\mathcal{E}} WL_{\mathcal{E}}(G,H)$ is called the wirelength of embedding G into H and is denoted by $WL(G,H)$. The wirelength problem [11] of G into H is to determine $WL(G,H)$.

Definition 2.6. **Congestion Lemma** [12] For an embedding $\mathcal{E}$ of a graph G into H, let S be a convex edge cut of H such that $H \setminus S$ splits into components H_1 and H_2 and let $G_1 = \mathcal{E}^{-1}(H_1)$ and $G_2 = \mathcal{E}^{-1}(H_2)$. Suppose G_1 and G_2 are maximal subgraphs of G and $P_{\mathcal{E}}(u,v)$ with $u \in G_1$ and $v \in G_2$ contains exactly one edge in S, for every $(u,v) \in E(G)$, then

$$\begin{aligned} c_{\mathcal{E}}(S) &= \sum_{v \in V(G_1)} deg_G(v) - 2|E(G_1)| \\ &= \sum_{v \in V(G_2)} deg_G(v) - 2|E(G_2)| \end{aligned}$$

Lemma 2.7. **r-Partition Lemma** [13] Let $\mathcal{E}: G \rightarrow H$ be an embedding. Let $E^r(H)$ be the multiset in which each edge of H is repeated r times. Let $\{R_1, R_2, \ldots, R_t\}$ be a partition of $E^r(H)$ where each R_i is an edge cut of H. Then

$$WL_{\mathcal{E}}(G,H) = \frac{1}{r}\sum_{i=1}^{t} c_{\mathcal{E}}(R_i)$$

Definition 2.8. **Circulant graph** [14] Let $K \subseteq \left\{1,2,\ldots,\left\lfloor \frac{n}{2} \right\rfloor\right\}$, $n \geq 3$. A graph G with vertex set $V = \{0,1,\ldots,n-1\}$ and the edge set $E = \{(i,j): |j-i| \equiv k(mod\, n), k \in K\}$ is called a circulant graph and is denoted by $G(n; \pm K)$.

Lemma 2.9. [15] Any set of r consecutive vertices on the outer cycle of $G(n; \pm 1)$, $1 \leq r \leq n$ induces a subgraph of $G(n; \pm K)$ with maximum number of edges.

Lemma 2.10. [15] A maximum subgraph on r vertices of $G(n; \pm K)$, where $K = \left\{1,2,\ldots,\left\lfloor \frac{n}{2} \right\rfloor\right\}$, $1 \leq j < \left\lfloor \frac{n}{2} \right\rfloor, 1 \leq r \leq n,\ n \geq 3$ has the number of edges as,

$$I_G(r) = \begin{cases} \frac{r(r-1)}{2}; & r \le j+1 \\ rj - \frac{j(j+1)}{2}; & j+1 < r \le n-j \\ \frac{1}{2}((n-r)^2 + (4j+1)r - (2j+1)n); & n-j < r \le n \end{cases}$$

Definition 2.11. **Wheel graph** [16] Consider a cycle on $n-1$ nodes. Take a new node and join it to each of the $n-1$ nodes on the cycle. The resultant graph is called a wheel graph and is denoted by W_n. The new vertex of degree $n-1$ is called the hub of W_n.

3. Embedding of $G(n;\ \pm\{1,2\})$ into wheel W_n

Embedding Algorithm $G(n; \pm\{1,2\}, W_n)$
Input: $G(n; \pm\{1,2\}), n \ge 5$ and W_n.
Algorithm:

i. The nodes of $G(n; \pm 1)$ are given labels $0, 1, 2, \dots, n-1$ in the clockwise sense.
ii. The hub vertex is given label 0 and the cycle vertices are given labels $1, 2, \dots, n-1$ in the clockwise sense. See Figure 1.

Output: An embedding $\mathcal{E}(x) = x$ have minimum wirelength.
Proof of Correctness:
Let $S_i = \{(i,\ i+1), (i+2,\ i+3)\},\ 0 \le i \le n-2$, where all the numbers are taken modulo n. Each S_i is an edge cut of W_n. $\{S_i\}, 1 \le i \le n-1$ is a partition of $E^2(W_n)$. $W_n \backslash S_i$ has two components H_{i_1} and H_{i_2}. For $0 \le i \le n-1$, since each S_i is a cut induced by 4 consecutive labels, G_{i_1} and G_{i_2} are also induced by consecutive labels. Hence, they are maximum subgraphs in $G(n; \pm\{1,2\})$. Again, since $\{S_i\}_{0 \le i \le n-1}$ is a partition of $E^2(W_n)$, by Congestion Lemma and 2-Partition Lemma, we see that the wirelength is minimum.

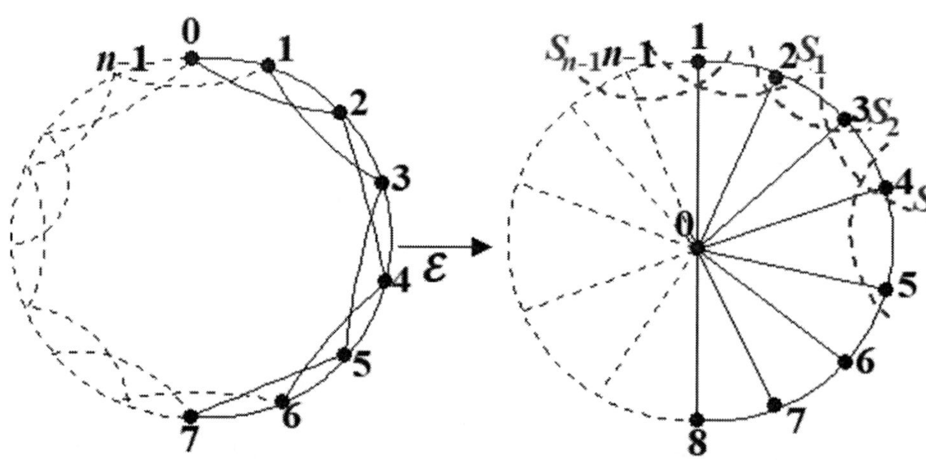

Figure 1. Embedding of $G(n;\ \pm\{1,2\})$ into wheel W_n.

Theorem 3.1. The wirelength of $G(n; \pm\{1,2\})$ into W_n, $n \ge 5$ is given by $WL(G(n; \pm\{1,2\}), W_n) = 3n - 1$.

Proof. Let G be the circulant graph $G(n;\pm\{1,2\})$ and H be the wheel graph with $n-1$ vertices in the outer cycle. By congestion lemma and 2-partition lemma,

$$WL(G(n;\pm\{1,2\}),W_n)=\frac{1}{2}\sum_{i=0}^{n-1}c_{\mathcal{E}}(S_i)$$
$$=\frac{n-1}{2}[4(2)-2(1)]$$
$$=3(n-1)$$

Therefore, the minimum wirelength of embedding $G(n;\pm\{1,2\})$ into W_n, $n\geq 5$ is $3(n-1)$.

Proceeding along the same line, we arrive at the following general result.

Theorem 3.2. The wirelength of embedding $G(n;\pm\{1,2,\ldots,j\})$ into W_n is

$$WL(G(n;\pm\{1,2,..,j\}),W_n)=(r-1)(n-1).$$

4. Embedding of $G(n;\ \pm\{1,2\})$ into one-point union of wheels

Definition 4.1. Let $W_{n_1},W_{n_2},\ldots,W_{n_k}$ be wheels such that their hubs are merged as one single vertex. The resultant graph is called an one-point union of wheels denoted as $W_{n_1}\wedge W_{n_2}\wedge\ldots\wedge W_{n_k}$. See Figure 2.

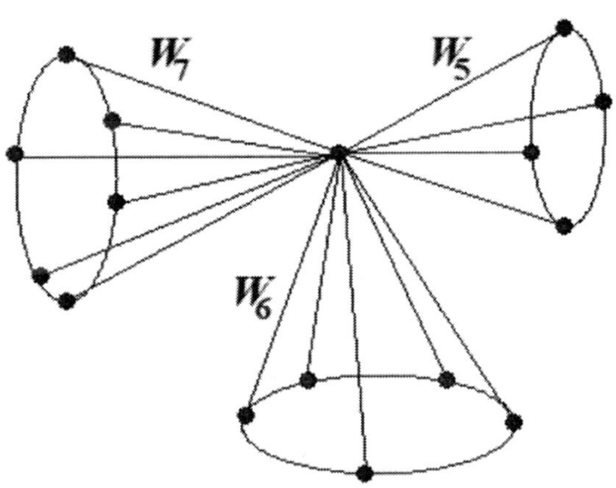

Figure 2. $W_5\wedge W_6\wedge W_7$.

Theorem 4.1. Let $G(n;\ \pm\{1,2\})$ be the circulant graph and $\wedge_{i=1}^{k}W_{n_i}$ be the one-point union of k wheels and let $n=n_1+n_2+\cdots+n_k+1$. Then

$$WL\left(G,\bigwedge_{i=1}^{k}W_{n_i}\right)=3(n-k)$$

Proof. Edge cuts similar to those in the Proof of Correctness of the embedding algorithm for W_n, and the 2-partition lemma yeild

$$WL\left(G, \bigwedge_{i=1}^{k} W_{n_i}\right) = 3(n_1 - 1) + 3(n_2 - 1) + \cdots + 3(n_k - 1)$$

$$= 3(n - k)$$

5. Embedding of $G(n;\ \pm\{1,2\})$ into union of wheels with cut edges

Definition 5.1. Let $W_{n_1}, W_{n_2}, \ldots, W_{n_k}$ be wheels with $v_1, v_2, \cdots, v_k$ as hub vertices. Let $e_1, e_2, \cdots, e_{k-1}$ be edges such that $e_i = (v_i, v_{i+1}), 1 \leq i \leq k-1$. See Figure 3. The resultant graph is denoted by

$$H = \bigvee_{i=1}^{k} W_{n_i}$$

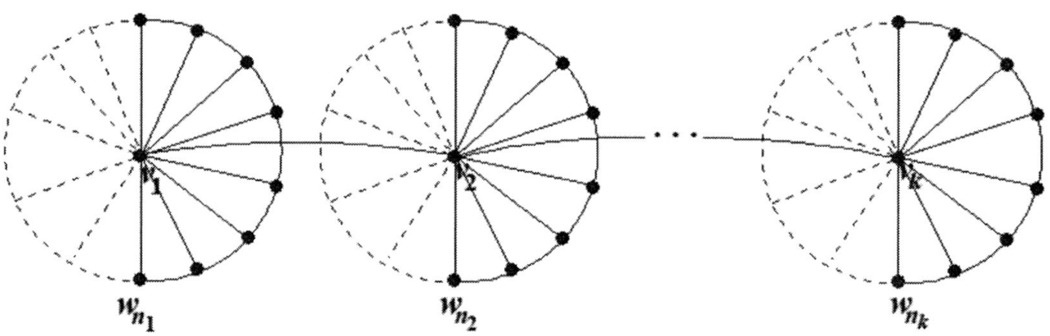

Figure 3. Wheels with cut edges.

Theorem 5.1. Let $G(n;\ \pm\{1,2\})$ be the circulant graph and let $H = \bigvee_{i=1}^{k} W_{n_i}$, where $n_1 + n_2 + \cdots + n_k = n$. Then, $WL(G,H) = 3(n-k) + \sum_{i=1}^{k} \mathcal{E}(n_i)$, where $\mathcal{E}(n_i)$ is the expression in Theorem 2.10.

6. Conclusion

Finding wirelength of an embedding is an NP-complete problem. Hence this study becomes significant while dealing with standard interconnection networks. In this paper, the circulant network has been considered as the guest graph with a variation of wheel graphs as the host graph.

References

[1] Palash Goyal and Emilio Ferrare. 2018. Graph embedding Techniques, Applications and Performance: A Survey, Knowledge-Based Systems 151 78-94.

[2] Jasintha Quadras, Micheal Arockiaraj, Indra Rajasingh, and Arul Jeya Shalini. 2014. Wirelength of Circulant Network Into Nanostar Dendrimer, Journal of Computational and Theoretical Nanoscience 11 1223-1229.

[3] R. S. Wilkov. 1972. Analysis and design of reliable computer networks IEEE Transactions on Communications 20(3) 660-678.

[4] Francis T. Boesch, Jhing-Fa Wang. 1985. Reliable circulant networks with minimum transmission delay IEEE Transactions on Circuitsand Systems 32(12) 1286-1291.

[5] R. Beivide, E. Herrada, J. L. Balcazar, A. Arruabarrena. 1991. Optimal distance networks of low degree for parallel computers Transactions on Computers 40(10) 1109-1124.

[6] J. C. Bermond, F. Comellas, D. F. Hsu. 1995. Distributed loop computer networks: A survey Journal of Parallel Distributed Computing 24(1) 210.
[7] Maurice. Karlin1969. New binary coding results by circulants IEEE Transactions on Information Theory 15(1) 81-92.
[8] Jasintha Quadras and S. Sarah Surya.2017. Wirelength of Circulant Networks into Wheel Related Graphs Annals of Pure and Applied Mathematics 14(1) 69-75.
[9] L.H. Harper. 2004. Global Methods for Combinatorial Isoperimetric Problems, first ed., The Press Syndicate of the University of the Cambridge.
[10] S.L. Bezrukov, J.D. Chavez, L.H. Harper, M. Rttger, U.P. Schroeder. 1998. Embedding of hypercubes into grids In MFCS. 693-701.
[11] P. Manuel, I. Rajasingh, B. Rajan, H. Mercy. 2009. Exact wirelength of hypercube on a grid Discrete Applied Mathematics 157(7) 1486-1495.
[12] M. Miller, R.S. Rajan, N. Parthiban and I. Rajasingh. 2015. Minimum linear arrangement of incomplete hypercubes The Computer Journal 58 (2) 331-337.
[13] M. Arockiaraj, P. Manuel, I. Rajasingh, B. Rajan. 2011. Wirelength of 1-fault Hamiltonian graphs into wheels and fans Information Processing Letters 111(18) 921-925.
[14] J.M. Xu. 2001. Topological Structure and Analysis of Interconnection Network Kluwer Academic Publishers, Dordrecht, Boston, London.
[15] I Rajasingh, P Manuel, M Arockiaraj, B Rajan. 2013. Embeddings of circulant networks Journal of combinatorial optimization 26(1) 135-151.
[16] I.I.Javaid and S.Shokat. 2008. On the partition dimension of some wheel related graphs Journal of Prime Research in Mathematics 4 154-164.

Total and Inverse Domination Numbers of Certain Graphs

V.Shalini[1], **Indra Rajasingh**[1*]

[1] Division of Mathematics (SAS), Vellore Institute Technology, Chennai-600127, India

E-mail: indrarajasingh@yahoo.com

Abstract- For any graph G having vertex set $V(G)$ then the subset set $D \subseteq V(G)$ is known as a dominating set if every single vertex of G not belonging to D is adjoining to not less than one vertex in D. The domination number $\gamma(G)$ is the minimum number of elements contained in a minimum dominating set D of G. Any subset D in $V(G)$ is known as total domianting set if each and every vertex of V in G is adjoining to not less than one vertex of D. The set which contains minimum number of elements among all total dominating set is the minimum total dominating set and its cardinality denoted as total domination number $\gamma_t(G)$. The inverse dominating set D' is defined as that D is a minimum dominating set of G, if there exist an another dominating set say D' in $V - D$ corresponding to D and its cardinality is the inverse domination number $\gamma'(G)$. In this paper we give the total and inverse domination numbers of certain graphs.

Keywords: dominating set, total domination number, inverse dominating set and inverse domination number.

1. Introduction

A graph or a network is a mathematical model comprising of points called vertices or nodes and lines called edges or links wchich join certain pair of points. The number of vertices incident at a point is called the degree of that vertex and is denoted by $\deg_G(v)$. The notation $\Delta(G)$ and δ (G) denoted the maximum and minimum degrees of a graph G respectively. A vertex $v \in V$ is called pendant vertex or end vertex of G if $deg(v)$ =1. An edge of a graph is said to be pendant edge if one of its vertices is a pendant vertex [1].

Around 1960 the study of dominating sets in Graph theory began. According to Hedetneimi and Laskar (1990)[2], the problems in domiantion were studied around 1950's onwards, but its rate of research much more increased from the year 1970. In 1958 the concept of domination number in graphs known as "coefficient of external stability" was defined by Berge, later it should be called as domianting set by Ore in 1962[4]. The notaion $\gamma(G)$ for the domination number of graph G was used by Cockayane and Hedetneimi[5] in 1977. Also they gave an interesting survey results of time about dominating sets in graphs [6],[7].

Any subset D in $V(G)$ is known as total domianting set if each and every vertex of V in G is adjoining to not less than one vertex of D. The set which contains minimum number of elements among all total dominating set is the minimum total dominating set and its cardinality denoted as total domination number $\gamma_t(G)$. This concept was introduced by Cockayne et al. [8] which was verymuch used in fault tolerance.

The inverse dominating set D' is defined as that D is a minimum dominating set of G, if there exist an another dominating set say D' in $V - D$ corresponding to D. Among all inverse dominating set there is a set with minimum number of elements is called minimum inverse dominating set and its cardinality denoted as the inverse domination number $\gamma'(G)$ which was introduced by introduced by Kulli and Sigarkanthi [9] and it was studied by several graph theorists.

In rest of the sections we give the total and inverse domination numbers for Lollipop graph $L_{m,n}$, Fly graph $F(m,n)$ and Jellyfish graph $J(r,s)$.

2. Definitions

Definition 2.1: A lollipop graph is the graph obtained from a complete graph K_m with one vertex is attached by a path of length n and it is denoted by $L_{m,n}$.

Definition 2.2[10]: A shell graph S_n obtained by taking $(n - 3)$ concurrent chords in a cycle C_n. All the chords are coincident in one vertex is called the apex vertex.

Definition 2.3[10]: Two bow graphs with their apexes merged is called a double shell.

Definition 2.4[10]: A fly graph is a bow graph with exactly two pendant edges at the apex vertex.

Definition 2.55[10]: The jellyfish graph $J(r,s)$ is from a $4-$cycle $uvwxu$, by joining v and x with an edge and appending r pendant edges to u and s pendant edges to w.

Definition 2.6: The floor function of a real number x is the greatest integer less than or equal to x and it is denoted by $\lfloor x \rfloor$. Suppose that $n \leq x < n + 1$, where n is an integer, then $\lfloor x \rfloor = n$.

Definition 2.7: The roof or ceiling function of a real number x is the greatest integer less than or equal to x and it is denoted by $\lfloor x \rfloor$. Suppose that $n - 1 < x \leq n$, where n is an integer, then $\lceil x \rceil = n$.

3. Preliminary Results

Theorem 3.1[11]: If the graph G has p vertices and maximum degree Δ, then $\gamma(G) \geq p/(\Delta + 1)$.

Theorem 3.2[8]: Let G be a connected graph with vertices $p \geq 3$, then $\gamma_t(G) \leq 2p/3$.

Theorem 3.3[12]: For Path P_n , $n \geq 2$ then $\gamma(P_n) = \lceil n/3 \rceil$.

Theorem 3.4[13]: For Path P_n , $n \geq 2$ then

$$\gamma_t(P_n) = \begin{cases} \frac{n}{2}, & if\ n \equiv 0(mod\ 4) \\ \left\lfloor \frac{n}{2} \right\rfloor + 1, & otherwise \end{cases}$$

Theorem 3.5[8]: For any graph G with p vertices and no isolates, then $\gamma_t(G) \leq p - \Delta(G) + 1$.

Theorem 3.6[9]**:** For Path P_n , $n \geq 2$ then

$$\gamma'(P_n) = \begin{cases} \left\lceil \frac{n}{3} \right\rceil + 1, & if\ n \equiv 0(mod\ 3) \\ \left\lceil \frac{n}{3} \right\rceil, & otherwise \end{cases}$$

4. Main Results

Theorem 4.1: Let G be Lollipop graph $L_{m,n}$ then

$$\gamma_t(L_{m,n}) = 2 + \begin{cases} \frac{n-2}{2}, & if\ n-2 \equiv 0(mod4) \\ \left\lfloor \frac{n-2}{2} \right\rfloor + 1, & otherwise \end{cases}$$

Proof: Assume G be a Lollipop graph on $m+n$ nodes and $\left[\frac{m\{m-1)}{2}\right] + n$ lines where $m \geq 2, n \geq 1$. In G there is no isolated vertex. We know that for a complete graph K_m the domination number is always one.

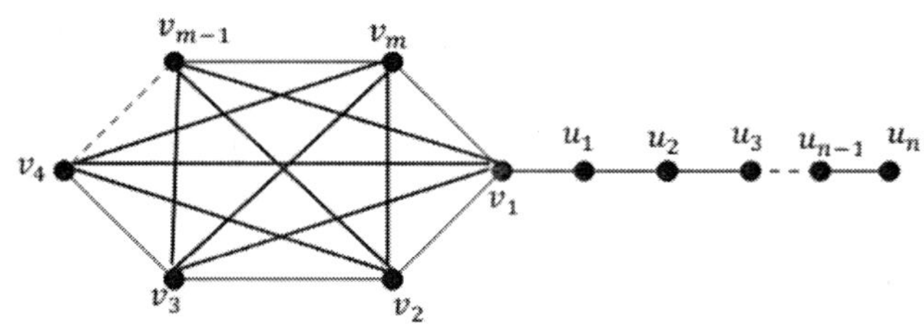

Figure 1. Lollipop Graph $L_{m,n}$

For minimizing the total dominating set we choose two consecutive vertices $\boldsymbol{v_1}$ and $\boldsymbol{u_1}$ from $\boldsymbol{G}$. Here $\boldsymbol{v_1}$ dominates $\boldsymbol{v_2, v_3 \dots v_m, u_1}$ and $\boldsymbol{u_1}$ dominates $\boldsymbol{u_2}$, therefore we need to find out the total dominating set for path $\boldsymbol{P_{n-2}}$. From the Theorem 3.4 wee know that for path $\boldsymbol{P_{n-2}}$ is

$$\boldsymbol{\gamma_t(P_{n-2}) = \begin{cases} \frac{n-2}{2}, & if\ n-2 \equiv 0(mod\ 4) \\ \left\lfloor \frac{n-2}{2} \right\rfloor + 1, & otherwise \end{cases}}$$

Thus for Lollipop graph $\boldsymbol{L_{m,n}}$ the total domination number is

$$\gamma_t(L_{m,n}) = 2 + \begin{cases} \frac{n-2}{2}, & if\ n-2 \equiv 0(mod4) \\ \left\lfloor \frac{n-2}{2} \right\rfloor + 1, & otherwise \end{cases}$$

Hence the proof.

Theorem 4.2: Let G be Lollipop graph $L_{m,n}$ then the inverse domination number is

$$\gamma'(L_{m,n}) = 1 + \begin{cases} \left\lceil \frac{n}{3} \right\rceil + 1, & if\ n \equiv 0(mod\ 3) \\ \left\lceil \frac{n}{3} \right\rceil, & otherwise \end{cases}$$

Proof: Let G be Lollipop graph on $m+n$ vertices and $\left[\frac{m(m-1)}{2}\right] + n$ edges where $m \geq 2, n \geq 1$. In G there is no isolated vertex. Also we know that for complete graph K_m the inverse domination number is one. Suppose v_1 is the dominating set for K_m , we need to choose any one of the vertex from $v_2, v_3 \dots v_m$ to dominate K_m and it is clear that $\gamma'(K_m) = 1$. Then we need to choose inverse dominating set for path P_n of vertices $u_1, u_2, u_3, \dots u_n$. From the Theorem 3.6 we know that,

$$\gamma'(P_n) = \begin{cases} \left\lceil \frac{n}{3} \right\rceil + 1, & if\ n \equiv 0(mod\ 3) \\ \left\lceil \frac{n}{3} \right\rceil, & otherwise \end{cases}$$

Therefore we have,

$$\gamma'(L_{m,n}) = \gamma'(K_m) + \gamma'(P_n)$$

Hence it is clear that the inverse domination number for Lollipop graph $L_{m,n}$ is

$$\gamma'(L_{m,n}) = 1 + \begin{cases} \left\lceil \frac{n}{3} \right\rceil + 1, & if\ n \equiv 0(mod\ 3) \\ \left\lceil \frac{n}{3} \right\rceil, & otherwise \end{cases}$$

Hence the Proof.

Theorem 4.3: Let G be Fly graph $F(m,n)$. Then the total and inverse domination numbers are

$$\gamma_t(F(m,n)) = 2$$

$$\gamma'(F(m,n)) = 2 + \left\{\left\lceil \frac{m}{3} \right\rceil + \left\lceil \frac{n}{3} \right\rceil\right\}$$

Proof: Assume the graph G as a fly graph having shells of orders m and n excluding the apex. After including one apex and two pendant vertices in G, the number of vertices of $F(m,n)$ is $m+n+3$ with $2(m+n)$ edges. The vertex u denoted as the apex of the fly graph and the vertices from top to bottom in the left and the right wing are denoted as follows $v_1, v_2, v_3 \dots v_m$ and $u_1, u_2, u_3 \dots u_n$ respectively. Clearly G has two pendent vertices which are denoted as v and w.

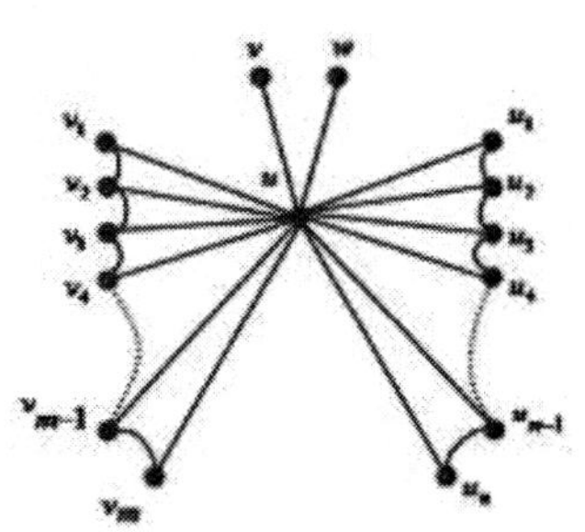

Figure 2. Fly Graph $F(m,n)$

IOP Conf. Series: Materials Science and Engineering 1012 (2021) 012066 doi:10.1088/1757-899X/1012/1/012066

The domination number for $F(m,n)$ is 1 since the apex vertex u is the minimum dominating set of G. Then by choosing any one adjacent vertex of u from the vertices of left or right wing gives the total domination number of G.ie, $\gamma_t(F(m,n)) = 2$.

For finding the inverse domination number of (m,n) , we need to choose the vertices other than the apex vertex u from G ie $(to\ choose\ vertices\ from\ left\ and\ right\ wing\ of\ G)$. The vertices $v_1, v_2, v_3 \dots v_m$ in left wing induce a path P_m and the vertices $u_1, u_2, u_3 \dots u_n$ in right wing induce a path P_n . By using the Theorem 3.3 we have $\gamma(P_m) = \left\lceil \frac{m}{3} \right\rceil$ and $\gamma(P_n) = \left\lceil \frac{n}{3} \right\rceil$. By adding the above two domination number with pendent vertices v and w gives the inverse domination number for $F(m,n)$.

Therefore the inverse domination number for $F(m,n)$ is

$$\gamma'(F(m,n)) = 2 + \{\gamma(P_m) + \gamma(P_n)\}$$

$$\gamma'(F(m,n)) = 2 + \left\{\left\lceil \frac{m}{3} \right\rceil + \left\lceil \frac{n}{3} \right\rceil\right\}.$$

Hence the proof.

Theorem 4.4: Let G be Jellyfish graph $J(r,s)$ then the total and inverse domination numbers are

$$\gamma_t(J(r,s)) = 3$$

$$\gamma'(J(r,s)) = r + s + 1$$

Proof: Assume the graph G be a jellyfish graph with $r+s+4$ nodes and $r+s+5$ lines where $r, s \geq 3$.

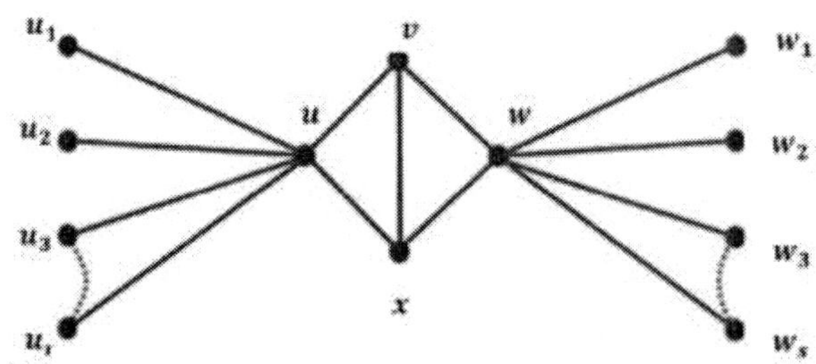

Figure 3. Jellyfish Graph $J(r,s)$

Since the vertex u and w are the minimum dominating set and domination number is 2, we need to choose either the vertex v or x to get the total dominating set for Jellyfish Graph. Thus

$\gamma_t(J(r,s)) = 3$.

For finding the inverse domination number, we have to choose the pendant vertices $u_1, u_2, u_3 \dots u_r$ and $w_1, w_2, w_3 \dots w_s$ also choosing either the vertex v or x.

Thus we have $\gamma'(J(r,s)) = r + s + 1$. Hence the proof.

IOP Conf. Series: Materials Science and Engineering 1012 (2021) 012066 doi:10.1088/1757-899X/1012/1/012066

References

[1] G. Chartrand, Lesniak. 2005. Graphs and Digraphs. fourth ed., CRC Press Boca Raton.

[2] S. T. Hedetniemi and R. C. Laskar. 1991. Topics on domination. North Holland.

[3] B. Bollob'as and E. J. Cockayne.1979. Graph-theoretic parameters concerning domination independence and irredundance J. Graph theory (3) 241-249 .

[4] O. Ore. 1962.Theory of Graphs Amer. Math. Soc. Colloquium Pub., Amer. Math. Soc., Providence. Rhode Island 38 (206).

[5] E. Cockayne and S. Hedetniemi. 1977. Towards a theory of domination in graphs Networks 247-261.

[6] T. W. Haynes, S. T. Hedetniemi and P. J. Slater. 1998. Fundamentals of domination in Graphs New York: Dekker.

[7] U. Niranjan, R. B. V. Subramanyam, V. Khanaa. 2010. Developing a Web Recommendation System Based on Closed Sequential Patterns Communications in Computer and Information Science 10(10) 171-179

[8] E. J. Cockayne, R. M. Dawe and S. T. Hedetniemi. 1980. Total domination in graphs, Networks (10) 211-219

[9] V. R. Kulli, S. C. Sigarkanti. 1991. Inverse domination in graphs, Nat. Acad.Sci. Letters (14) 473–475.

[10] A. Sugumaran and E. Jayachandran. 2019. Domination, Equitable and End Equitable Domination Numbers of Some Graphs , Journal of Computer and Mathematical Sciences, 10(3)399-409.

[11] Berge. C. 1973 .Graphs and Hyper graphs. North Holland Amsterdam.

[12] G. Chartrand and P. Zhang. 2005. Introduction to Graph Theory. McGraw-Hill Boston, Mass, USA.

[13] Kang, Cong X. 2014 .Total Domination Value in Graphs Util. Math. Arxiv.org. Journal (95) 263-279.

A study on human intelligence using Fuzzy Grey Relational Maps (FGRMs)

S. Arokiamary[1], M. Mary Mejrullo Merlin[2*]

[1]Department of Mathematics, Mother Gnanamma Women's College of Arts and Science, Varadarajanpet, Ariyalur, Tamilnadu, India.
[2]PG & Research Department of Mathematics, Holy Cross College (Autonomous), Trichirapalli, India.

E-mail: merlinprashanth@yahoo.com

Abstract- Fuzzy-grey system is an ambiguous system where unclear, incomplete and only small data is available. Fuzzy Grey relational maps model the causal relation between complex systems. In this fuzzy model, the nodes are fuzzy while the edges are grey and they represent the grey causal relationship between two disjoint sets of concepts. Intelligence is one of the most admired and coveted features of human mind for, human intelligence has played a greater role in placing human beings at the top of the evolutionary pyramid of creation. Intelligence is widely accepted as hereditary, but at the same time there are researches going on to prove that it is also influenced by the beliefs the individual holds. In this paper, the causal influence between intelligence and self-beliefs is studied using Fuzzy Grey relational maps.

Key words: Fuzzy relational maps, FGRM, beliefs, intelligence, grey system

1. Introduction

The cognitive processes in human beings such as reasoning and decision-making take place in a highly uncertain environment. Modelling such a complex system using classical methods is a challenging task. The uncertainty and hesitancy in human thinking make the information highly approximate and not absolutely reliable. The Fuzzy Grey Relational Maps reduce the ambiguity in the available data with the help of Grey System Theory. The grey numbers capture the uncertainty and hesitancy in the data provided by an individual. Identifying and assessing human intelligence is certainly a very indecisive process. In this paper, the causal relationship between different types of intelligence and the underlying beliefs is studied using the Fuzzy Grey Relational Maps.

2. Description of the Problem

Intelligence is a much-appreciated trait in every creature for, it enables them to find ways and means to a meaningful life and creative solutions in times of challenges. In particular, human intelligence has taken human life and evolution to what it is today. The world moves by the power of human intelligence. Human Intelligence has played a greater role in making this planet earth an advanced place to live. Human beings have worked so hard, since time immemorial, to reach this far. From the researches that have been carried out on intelligence, it can be inferred that beliefs are strappingly associated with human intelligence and their performance [1,2]. The strong association between beliefs and intelligence of an individual prompts to study the causal association between them.

2.1. Beliefs and Intelligence

Human intelligence is the capacity of an individual to understand, comprehend, learn, create, plan, solve problems, make decisions and so on. These abilities define a person's intelligence and gives them the power to carry on their life successfully. The capacities of individuals vary person to person to a greater extend. One individual who is good in one aspect may not do well in another. Assessing an individual's capacity is also another indecisive process where one may not be able to give an exact measure of the level or type of intelligence [3-6]. Because everyone has different opinions and ideas about intelligence [7, 8], it is difficult to come to conclusion as to what is the ability of an individual to become the fittest to survive. Human intelligence is not a fixed idea, but an evolving concept that varies with time and experience. What brings in the change in one's capacity to cope with life to survive and succeed in life attracts our attention [1, 2,9]. In this paper, an attempt is made to study the relationship between self- beliefs and different types of human intelligence.

2.2. Simulating the causal relationship between Beliefs and Intelligence

Simulating psychological constructs to study their properties and behaviours is definitely a difficult process as it involves a great deal of uncertainty and vagueness. The ambiguity in human thinking and understanding makes it even harder to analyse the mental constructs. The hesitancy and indecisiveness associated with human reasoning and decision-making processes have to be incorporated intelligently into the model. Producing new models that is enabled with capacity to measure and process the imprecise information is the need of the hour.

There are different theories on human intelligence which suggest the components, types or features of intelligence. Here, the model proposed by Harvard University neuropsychologist and educator Howard Gardener in his book Frames of Mind: The theory of Multiple Intelligences which describes eight kinds of intelligence is adopted [9]. Human belief system is not a well-defined structure, yet there are several primary beliefs which are universal and indispensable in every individual. The description of domains of beliefs and types of intelligence is given in table-1. In this study, seven areas of common beliefs and eight types of intelligences are chosen. The causal relationship between personal beliefs and types of intelligence using a fuzzy-grey mathematical model called Fuzzy Grey Relational Maps (FGRM) as given in figure-1.

Table-1: FGRM nodes and their description

Self-Beliefs (Domain Space)		**Human Intelligence (Range Space)**	
B_1	Survival	I_1	Linguistic/Verbal
B_2	Security	I_2	Logical/Mathematical
B_3	Connection	I_3	Visual/Spatial
B_4	Vulnerability	I_4	Bodily/Kinaesthetic
B_5	Judgement	I_5	Musical/Rhythmical
B_6	Responsibility	I_6	Interpersonal
B_7	Recognition	I_7	Intrapersonal
		I_8	Naturalist

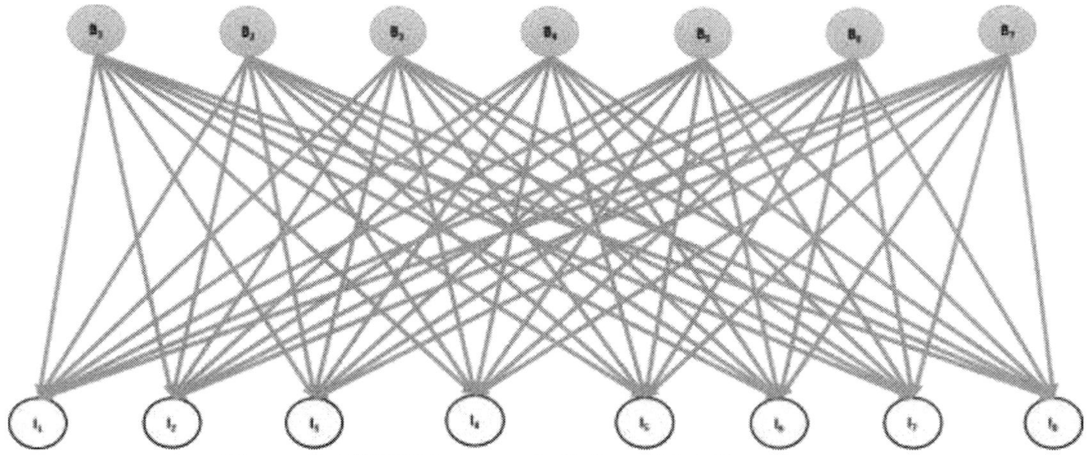

Figure-1: FGRM between beliefs and intelligence

The relationship between human beliefs and intelligence is completely grey. No one can absolutely estimate the level or identify the type of one's intelligence or one's own for that matter. The FGRM is obtained in the form of a matrix from parents and teachers who deal with children of age 3 to 6 years. The relation between the nodes of domain space and range space is given by a grey number which is an interval taking values from [-1, +1]. The relational matrix that captures the grey relationship between these two disjoint concepts is obtained from experts and the augmented matrix is calculated. The augmented relational matrix that describes the grey relation between the domain space and the range space is given in table-2.

Table-2: The augmented relational matrix $R^{Aug}(\otimes)$ based on Expert's Opinion

	I_1	I_2	I_3	I_4	I_5	I_6	I_7	I_8
B_{1+}	[0.6, 0.8]	[0.4, 0.7]	[0.8, 0.9]	[0.4, 0.7]	[0.6, 0.8]	[0.7, 0.9]	[0.3, 0.5]	[0.4, 0.6]
B_{1-}	[−0.6, −0.2]	[−0.4, −0.3]	[−0.7, −0.5]	[−0.4, −0.2]	[−0.5, −0.2]	[−0.5, −0.3]	[−0.8, −0.6]	[−0.5, −0.3]
B_{2+}	[0.4, 0.6]	[0.3, 0.7]	[0.6, 0.7]	[0.7, 0.9]	[0.5, 0.8]	[0.7, 0.9]	[0.2, 0.5]	[0.4, 0.6]
B_{2-}	[−0.5, −0.3]	[−0.5, −0.4]	[−0.5, −0.2]	[−0.6, −0.4]	[−0.6, −0.3]	[−0.4, −0.2]	[−0.8, −0.6]	[−0.7, −0.4]
B_{3+}	[0.6, 0.8]	[0.5, 0.8]	[0.5, 0.7]	[0.4, 0.7]	[0.4, 0.6]	[0.6, 0.8]	[0.3, 0.5]	[0.5, 0.8]
B_{3-}	[−0.7, −0.4]	[−0.6, −0.3]	[−0.5, −0.2]	[−0.6, −0.3]	[−0.7, −0.3]	[−0.5, −0.3]	[−0.8, −0.6]	[−0.7, −0.5]
B_{4+}	[0.6, 0.9]	[0.5, 0.8]	[0.4, 0.7]	[0.6, 0.8]	[0.5, 0.8]	[0.6, 0.9]	[0.3, 0.5]	[0.4, 0.7]
B_{4-}	[−0.7, −0.3]	[−0.7, −0.4]	[−0.6, −0.4]	[−0.7, −0.4]	[−0.8, −0.5]	[−0.5, −0.2]	[−0.8, −0.6]	[−0.6, −0.4]
B_{5+}	[0.8, 0.9]	[0.6, 0.8]	[0.5, 0.7]	[0.4, 0.7]	[0.4, 0.6]	[0.6, 0.9]	[0.3, 0.5]	[0.3, 0.6]
B_{5-}	[−0.4, −0.2]	[−0.5, −0.3]	[−0.5, −0.4]	[−0.5, −0.2]	[−0.6, −0.3]	[−0.4, −0.3]	[−0.7, −0.5]	[−0.6, −0.3]
B_{6+}	[0.7, 0.9]	[0.6, 0.8]	[0.6, 0.9]	[0.4, 0.7]	[0.5, 0.7]	[0.7, 0.9]	[0.3, 0.6]	[0.4, 0.7]
B_{6-}	[−0.6, −0.3]	[−0.8, −0.4]	[−0.7, −0.3]	[−0.6, −0.3]	[−0.5, −0.3]	[−0.5, −0.2]	[−0.8, −0.6]	[−0.6, −0.3]
B_{7+}	[0.7, 0.9]	[0.6, 0.8]	[0.5, 0.8]	[0.6, 0.8]	[0.4, 0.7]	[0.7, 0.9]	[0.5, 0.7]	[0.4, 0.6]
B_{7-}	[−0.6, −0.3]	[−0.6, −0.4]	[−0.5, −0.3]	[−0.6, −0.3]	[−0.6, −0.3]	[−0.3, −0.2]	[−0.8, −0.6]	[−0.7, −0.4]

3. Fuzzy Grey Relational Maps

Fuzzy Grey Cognitive Maps (FGCMs) was proposed by Salmeorn J.L in his research article Modelling grey uncertainty with Fuzzy Grey Cognitive Maps in 2010 [10]. The Fuzzy Grey Relational Maps (FGRMs) are constructed analogous to FGCM. Fuzzy Grey Cognitive Map (FGRM) is an extension of Fuzzy Relational Map (FRM) [11] where two uncertain systems, namely fuzzy system and grey system are combined. Like the Fuzzy system, grey system is also an ambiguous system where unclear, incomplete and only small data is available [12-21]. In FGRM, the nodes are fuzzy while the edges are grey and they represent the grey causal relationship between two disjoint concepts. FGRM are useful in studying the causal relationship between two disjoint concepts [11] while the relationship could be either, black or white or grey [10].

Vasantha Kandasamy *et al.* introduced Fuzzy Relational Maps model to study the real-world problems wherein the cause and effect relationship between two disjoint concepts is analysed in 2000. Fuzzy Relational Maps (FRM) are structurally similar to the state-of-the-art model to study the complex systems namely, Fuzzy Cognitive Maps (FCM). In FRMs the concepts (nodes) are divided into two disjoint fuzzy sets as domain space and range space. Like FCMs, the FRMs can also be extended in a similar manner to adapt to different scenarios. The greatest challenge that is encountered in formulating and functioning FRMs is assigning weight to the edges which describe the influence of one concept on another. The uncertainty and hesitancy of the experts in assigning the proper weight results in subjective valuations which may not be precise in bringing out the relationship between the two sets of nodes. Consequently, lots of efforts have gone into finding methods and techniques to suggest proper weight that would take into consideration the uncertainty and hesitancy involved in the processes of determining the weights and their inference in the complex system.

The Fuzzy Grey Relational Maps is yet another approach to study the relationship between concepts in a complex system by choosing the proper edge strength. The Fuzzy Grey Cognitive Maps (FGCM) that were first introduced by Jose L. Salmeron in 2010 [10] wherein the author describes several new ideas such as grey environment, grey values and arithmetic operation on grey numbers and whitenisation of grey numbers. Jose L. Salmeron and Elpiniki I. Papageorgiou studied Decision Support System for radiotherapy treatment planning using FGCM tool in 2012[16]. Jose L. Salmeron and Ester Gutierrez used FGCM in reliability engineering 2012[20]. Wojciech Froelich and Jose L. Salmeron analysed the evolutionary learning of FGCM in 2015 [14]. Subsequently Jose L. Salmeron and Pedro R. Palos-Sanchez studied the uncertainty propagation in FGCM With Hebbian-Like Learning Algorithms in 2017 [22]. FGCMs are used in various grey environments to study the highly uncertain systems efficiently. The FGRM tool is a combination of Grey system theory and Fuzzy Relational maps which is very much useful in highly uncertain multiple meaning environments. This tool is advantageous over regular FRMs as it includes the uncertainty and hesitancy that may arise due to shortcomings in human perception and comprehension. This fuzzy model best suited to obtain the cause and effect relationship of the problem considered for the analysis.

3.1. Dynamics of Fuzzy Grey Relational Maps

Fuzzy Grey Relational Maps (FGRM) is a fuzzy tool that models the causal relations between two disjoint set of concepts that may involve discrete, small and incomplete data [12-21, 23]. The nodes of FGRM are fuzzy and their edges are grey. The causal influence between the nodes of the domain space and the range space are given by grey weights which are interval valued numbers. The weight of each edge is given by the grey weight:

$$\otimes w_{ij} = \left[\underline{w_{ij}}, \overline{w_{ij}}\right], \forall i,j\ \underline{w_{ij}} < \overline{w_{ij}}, \left\{\underline{w_{ij}}, \overline{w_{ij}}\right\} \in [-1, +1]$$

where i and j denote the concepts in Domain Space and Range Space respectively.

The resultant inference of the one node influencing the other is given by [21]

$$\otimes C_j^{t+1} = f\left((\otimes C_j^t) + \sum_{i=1}^{N} \otimes w_{ij} \,.\otimes C_i^t \right)$$
$$= f(\otimes C^{t*}) = f\left(\left[\underline{C^{t*}}, \overline{C^{t*}}\right]\right)$$
$$= \left[f\left(\underline{C^{t*}}\right), f\left(\overline{C^{t*}}\right)\right] = \left[\underline{C^{t+1}}, \overline{C^{t+1}}\right]$$

where $f(\cdot)$ is the bipolar sigmoid function.

Since the concepts' states map in the range [-1, +1] the hyperbolic tangent function which is a bipolar sigmoid function is used. The component i of the state vector $\otimes \vec{C}_i^{t+1}$ after the inference is denoted as follows.

$$\otimes \vec{C}_i^{t+1} \in \left[\left(\frac{e^{\lambda \underline{C^{t*}}} - e^{-\lambda \underline{C^{t*}}}}{e^{\lambda \underline{C^{t*}}} + e^{-\lambda \underline{C^{t*}}}}\right), \left(\frac{e^{\lambda \overline{C^{t*}}} - e^{-\lambda \overline{C^{t*}}}}{e^{\lambda \overline{C^{t*}}} + e^{-\lambda \overline{C^{t*}}}}\right)\right]$$

The output of the function is restricted to the interval [-1,1]. Additionally, FGRM includes greyness as an uncertainty measurement and it is computed as follows. [21]

$$\emptyset(\otimes C_i) = \frac{|l(\otimes C_i)|}{l(\otimes \psi)}$$

Where $|l(\otimes C_i)|$ is the absolute value of the grey influence $\otimes C_i$ state value and $l(\otimes \psi)$ is the absolute value of the information space range denoted by $\otimes \psi$. Higher values of greyness imply higher degree of uncertainty. FGRM maps nodes' states within the interval [-1, +1] since negative values are allowed. In this case,

$$l(\otimes \psi) = f(x) = \begin{cases} 1, & if\ \{\otimes C_i, \otimes w_i\} \subseteq [0,1] \quad \forall \otimes C_i, \otimes w_i \\ 2, & if\ \{\otimes C_i, \otimes w_i\} \subseteq [-1,1] \quad \forall \otimes C_i, \otimes w_i \end{cases}$$

The grey relational matrices obtained from different experts is converted into augmented relational matrix $R^{Aug}(\otimes)$ which gives the improved approximation of all the experts who have studied the problem. The dynamics of FGRM is similar to that of FGCM which in turn is similar to FCM [12-21]. The initial vector that represents the initial uncertainty is also with grey values, i.e., $\otimes \vec{C}_0^i = [\underline{C_0^i}, \overline{C_0^i}]$. The different initial vectors interact with the augmented relational matrix $R^{Aug}(\otimes)$.

3.2. Analysis of the Problem Using FGRM model.

The choice of initial vector the output that is expected from the system. The interaction of the initial Vector with the dynamic system of causal relationship results in a stable vector. The initial vector: [0.8, 1], [0, 0], [0,0,], [0,0], [0,0], [0,0], [0,0], [0,0] is chosen to analyse the relationship between the different beliefs in the domain space and verbal intelligence in the range space.

The FGRM method takes in the uncertainty and hesitancy present in the experts' opinion for causal influences between concepts as well as within the initial vector. The resultant stable vector is also a grey number with reduced greyness compared to the original value [23]. The stable vector is converted into white number by taking appropriate values for α in the expression $C_i = \underline{C_n^i}\,\alpha + (1 - \alpha)\overline{C_n^i}$. Here, in this problem under study, equal mean weight is calculated by taking $\alpha = 0.5$. The process of finding the white numbers corresponding to the grey numbers is displayed on table-3. This table captures the causal influence of the beliefs and the verbal intelligence ($\boldsymbol{I_1}$) of the range space. The greyness value is the measure of uncertainty. If the white numbers for two causal associations are the same, then the number with less greyness is chosen.

Table-3: Calculation of stable vector and whitenisation process

	I_1	Stable Vector	Length $\|l(\otimes C_i)\|$	Greyness $\emptyset(\otimes C_i)$	Whitenisation
B_{1+}	$[0.6, 0.8]$	[0.48, 0.80]	0.32	0.16	0.64
B_{1-}	$[-0.6, -0.2]$	[-0.60, -0.16]	0.44	0.22	-0.38
B_{2+}	$[0.4, 0.6]$	[0.32, 0.60]	0.28	0.14	0.46
B_{2-}	$[-0.5, -0.3]$	[-0.50, -0.24]	0.26	0.13	-0.37
B_{3+}	$[0.6, 0.8]$	[0.48, 0.80]	0.32	0.16	0.64
B_{3-}	$[-0.7, -0.4]$	[-0.70, -0.32]	0.38	0.19	-0.51
B_{4+}	$[0.6, 0.9]$	[0.48, 0.90]	0.42	0.21	0.69
B_{4-}	$[-0.7, -0.3]$	[-0.70, -0.24]	0.46	0.23	-0.47
B_{5+}	$[0.8, 0.9]$	[0.64, 0.90]	0.26	0.13	0.77
B_{5-}	$[-0.4, -0.2]$	[-0.40, -0.16]	0.24	0.12	-0.28
B_{6+}	$[0.7, 0.9]$	[0.56, 0.90]	0.34	0.17	0.73
B_{6-}	$[-0.6, -0.3]$	[-0.60, -0.24]	0.36	0.18	-0.42
B_{7+}	$[0.7, 0.9]$	[0.56, 0.90]	0.34	0.17	0.73
B_{7-}	$[-0.6, -0.3]$	[-0.60, -0.24]	0.36	0.18	-0.42

4. Discussion

The equal mean white values of causal relationship between the personal beliefs and types of intelligence is given in table-4. From this table it is easy to compare the influence of different beliefs on different kinds of intelligence.

Table-4: Causal influences between Beliefs and Intelligence

	I_1	I_2	I_3	I_4	I_5	I_6	I_7	I_8
B_{1+}	0.64	0.51	0.77	0.51	0.64	0.73	0.37	0.46
B_{1-}	-0.38	-0.32	-0.55	-0.28	-0.33	-0.37	-0.64	-0.37
B_{2+}	0.46	0.47	0.59	0.73	0.60	0.73	0.33	0.46
B_{2-}	-0.37	-0.40	-0.33	-0.46	-0.43	-0.28	-0.64	-0.51
B_{3+}	0.64	0.60	0.55	0.51	0.46	0.64	0.37	0.60
B_{3-}	-0.51	-0.43	-0.33	-0.42	-0.47	-0.37	-0.64	-0.55
B_{4+}	0.69	0.60	0.55	0.64	0.60	0.69	0.37	0.51
B_{4-}	-0.47	-0.51	-0.46	-0.51	-0.60	-0.33	-0.64	-0.46
B_{5+}	0.77	0.64	0.55	0.51	0.46	0.69	0.37	0.42
B_{5-}	-0.28	-0.33	-0.41	-0.33	-0.42	-0.32	-0.55	-0.42
B_{6+}	0.73	0.64	0.69	0.51	0.55	0.73	0.42	0.51
B_{6-}	-0.42	-0.56	-0.47	-0.42	-0.37	-0.33	-0.64	-0.42
B_{7+}	0.73	0.64	0.60	0.64	0.51	0.73	0.55	0.46
B_{7-}	-0.42	-0.46	-0.37	-0.42	-0.42	-0.23	-0.64	-0.51

5. Conclusion

The table-4 gives the estimated causal relationship between human beliefs and types of intelligence in white numbers. From this table it is inferred that the beliefs which arise from the belief-domain $\boldsymbol{B_5}$ (Judgement) influence the linguistic/verbal intelligence $\boldsymbol{I_1}$ very much. The primitive beliefs from the belief-domains $\boldsymbol{B_1}$ (survival) and $\boldsymbol{B_2}$ (security) play a significant role in promoting logical/mathematical intelligence $\boldsymbol{I_2}$. Positive beliefs about one's own survival gives them the power to visualize and foresee things in life which in turn increases their skills related to visual/spatial intelligence $\boldsymbol{I_3}$. People who have positive beliefs about security of their life and existence develop bodily/kinaesthetic intelligence $\boldsymbol{I_4}$. Those who feel confident about the survival of their being approach life with balanced attitude and thus develop musical/rhythmical intelligence $\boldsymbol{I_5}$. All the positive beliefs have equal influence in promoting musical/rhythmical intelligence. All the positive beliefs are equally active in promoting interpersonal intelligence $\boldsymbol{I_6}$. People who have reasonable level of positive beliefs seem to feel equanimous and this improves their intrapersonal intelligence $\boldsymbol{I_7}$. Those who have strong positive beliefs emerging from the $\boldsymbol{B_3}$ (connection) domain influence a person to have naturalistic intelligence $\boldsymbol{I_8}$. The FGRM tool is very efficient in analysing the causal relationship between the disjoint sets of concepts. In this model the strength of the edges would be characterized with white, black or grey values. In order to make this system more competent we can include sophisticated training methods that would train the network to produce better and optimum results in a grey environment.

References

[1] Armstrong, T. 2014. You're smarter than you think: A kid's guide to multiple intelligences. Minneapolis, MN: Free Spirit Publishing.

[2] Armstrong, Thomas. 2009. Multiple Intelligences in the Classroom, 3rd ed. Alexandria, VA: Association for Supervision and Curriculum Development, 2009. ISBN 978-1-4166-0789-2

[3] Concepción L., Nápoles G., Bello R., Vanhoof K. 2020 On the Behavior of Fuzzy Grey Cognitive Maps. In: Bello R., Miao D., Falcon R., Nakata M., Rosete A., Ciucci D. (eds) Rough Sets. IJCRS 2020. Lecture Notes in Computer Science, vol 12179. Springer, Cham. https://doi.org/10.1007/978-3-030-52705-1_34

[4] Furnham, A., & Akanda, A. 2004. African parents' estimation of their own and their children's multiple intelligences. Current Psychology, 22(4), 281–294.

[5] Furnham, A., & Fukumoto, S. 2008. Japanese parents' estimates of their own and their children's multiple intelligences: Cultural modesty and moderate differentiation. Japanese Psychological Research, 50(2), 63–76.

[6] Furnham, A., & Wu, J. 2008. Gender differences in estimates of one's own and parental intelligence in China. Individual Differences Research, 6(1), 1–12.

[7] Sternberg, R. J. 2013. Intelligence. In D. K. Freedheim & I. B. Weiner (Eds.), Handbook of psychology: History of psychology (p. 155–176). John Wiley & Sons, Inc.

[8] Sternberg, Robert J. 1999. "The Theory of Successful Intelligence." Review of General Psychology, 3 (4), pp. 292–316, doi:10.1037/1089-2680.3.4.292.

[9] Gardner, H. 1983. Frames of mind: The theory of multiple intelligences. New York: Basic Books.

[10] Salmeron,J.L. 2010, Modelling grey uncertainty with Fuzzy Grey Cognitive Maps, Expert Systems with Applications., 37 (12), 7581-7588. https://doi.org/10.1016/j.eswa.2010.04.085

[11] Vasantha Kandasamy, W.B., and Smarandache, F. 2004 Fuzzy Relational Maps and Neutrosophic Relational Maps, HEXIS, Church Rock.

[12] Harmati I.Á., Kóczy L.T. 2020. Improvements on the Convergence and Stability of Fuzzy Grey Cognitive Maps. In: Lesot MJ. et al. (eds) Information Processing and Management of Uncertainty in Knowledge-Based Systems. IPMU 2020. Communications in Computer and

Information Science, vol 1239. Springer, Cham. https://doi.org/10.1007/978-3-030-50153-2_38

[13] Harmati, I. Á., & Kóczy, L. T. 2019. On the Convergence of Sigmoidal Fuzzy Grey Cognitive Maps, International Journal of Applied Mathematics and Computer Science, 29(3), 453-466. doi: https://doi.org/10.2478/amcs-2019-0033

[14] Papageorgiou, E. I. and J. L. Salmeron. 2012. "Learning fuzzy grey cognitive maps using nonlinear Hebbian-based approach," Int. J. Approx. Reason., 53 (1), pp. 54–65.

[15] Salmeron J.L., Papageorgiou E.I. 2014. Using Fuzzy Grey Cognitive Maps for Industrial Processes Control. In: Papageorgiou E. (eds) Fuzzy Cognitive Maps for Applied Sciences and Engineering. Intelligent Systems Reference Library, vol 54. Springer, Berlin, Heidelberg. https://doi.org/10.1007/978-3-642-39739-4_14

[16] Salmeron, J. L. and E. I. Papageorgiou. 2012. "A fuzzy grey cognitive maps-based decision support system for radiotherapy treatment planning," Knowl. Based Syst., vol. 30, no. 1, pp. 151–160, https://doi.org/10.1016/j.knosys.2012.01.008

[17] Salmeron, J. L. and E. I. Papageorgiou.2014. "Fuzzy grey cognitive maps and nonlinear Hebbian learning in process control," Appl. Intell., vol. 41, no. 1, pp. 223–234.

[18] Salmeron, J. L. and P. R. Palos-Sanchez. 2019. "Uncertainty Propagation in Fuzzy Grey Cognitive Maps With Hebbian-Like Learning Algorithms," in IEEE Transactions on Cybernetics, 49 (1), pp. 211-220, doi: 10.1109/TCYB.2017.2771387.

[19] Salmeron, J.L. 2016. "An autonomous FGCM-based system for surveillance assets coordination." The Journal of Grey System, 28(1), p. 27-35.

[20] Salmeron, J.L., Gutierrez, E. 2012. Fuzzy Grey Cognitive Maps in Reliability Engineering. Applied Soft Computing 12 (12), 3818-3824. https://doi.org/10.1016/j.asoc.2012.02.003

[21] Salmeron,J.L. 2010, Modelling grey uncertainty with Fuzzy Grey Cognitive Maps, Expert Systems with Applications., 37 (12), 7581-7588. https://doi.org/10.1016/j.eswa.2010.04.085

[22] Seyed Amin Seyed Haeri., JafarRezaei. 2019. A grey-based green supplier selection model for uncertain environments, Journal of Cleaner Production, 221, 768-784, https://doi.org/10.1016/j.jclepro.2019.02.193

[23] Zanon, L. G. and L. Cesar Ribeiro Carpinetti. 2018. "Fuzzy Cognitive Maps and Grey Systems Theory in the Supply Chain Management Context: a literature review and a research proposal," 2018 IEEE International Conference on Fuzzy Systems (FUZZ-IEEE), Rio de Janeiro, pp. 1-8, doi: 10.1109/FUZZ-IEEE.2018.8491473

Automated Detection of Diabetic Retinopathy Using Intuitionistic Fuzzy Digital Convex Envelope Segmentation Algorithm

S. Meenakshi[1*], P. Amsini[2], R. Uma Rani[2], D. Amsaveni[1]

[1]Department of Mathematics, Sri Sarada College for Women, Tamilnadu
[2]Department of Computer Science, Sri Sarada College for Women, Tamilnadu

E-mail: sylphmeenu@gmail.com

Abstract- Diabetic retinopathy is an ailment of the retinal vessels, which progresses about in most of the diabetes mellitus patients with high blood sugar levels. A framework for automated detection of disorders of the retinal vessels is proposed and examined in this article. As intuitionistic fuzzy sets are being defined through membership and nonmembership functions, they play a vital role in digital image processing to detect the disorder part more clearly. Moreover, shape is a crucial one in numerous areas such as object identification, detection, geomorphology and biology characterization. The intuitionistic fuzzy digital convex envelope is used to measure and analyze the shape of the image of the affected part. The various stages concerned in the process are image acquisition, preprocessing and segmenting the disorder region by the new proposed intuitionistic fuzzy digital convex envelope algorithm. Foremost, this automatic segmentation method reduces the manual work errors and time consuming.

Keywords: Diabetic retinopathy, intuitionistic fuzzy digital (IFD) convex set, intuitionistic fuzzy digital (IFD) convex envelope, digital image processing, preprocessing, acquisition.

1. Introduction

Medical images are at the fundamental of medical science and they are the massive source of information that needs to be used. Medical image processing has been an interdisciplinary research arena fascinating expert from Computer sciences, physics, applied mathematics, statistics, biology, engineering and medicine. Computer-aided diagnostic processing becomes a vital part of clinical routine.

Afterward Zadeh's initiation of fuzzy sets [1], numerous higher order extensions of fuzzy sets were anticipated by different researchers. Midst them Atanassov's intuitionistic fuzzy sets[2] offer a modest, flexible and dense mathematical context to deal with with the uncertainties depicting real world structures. The intuitionistic fuzzy architecture has been effectively applied to various image processing problems. A novel image enhancement method using fuzzy morphological intuitionistic reconstruction function was proposed in [3]. An efficient intuitionistic fuzzy context for image processing is established in the framework of contrast enhancement and edge detection in [4]. In [5], the thresholding technique was described with intuitionistic fuzzy sets and this tactic used intuitionistic index values by Atanassov to represent the hesitancy of the proficient in deciding which pixels belongs to the background and which belongs to the object.

In this article, a framework for automated detection of disorders of retinal vessels is proposed and examined. The various stages are concerned in proposed methods such as image acquisition, preprocessing and intuitionistic fuzzy based segmentation. Intuitionistic fuzzy sets are defined through membership and non membership functions which plays a vital role in digital image processing to detect the disorder part more clearly. This automatic method is reduced the manual work errors and time consuming. The image processing techniques are used to detect the disorders of the retinal vessels. The affected parts are detected and extracted by applying intuitionistic fuzzy set theory. This automatic method is reduced the manual work errors and time consuming.

2. Preliminaries

In this section, some definitions which play the vital role in this article are provided. Also, a brief introduction of the disease, Diabetic retinopathy is given.

2.1 Diabetic Retinopathy:

People with long standing Diabetes Mellitus can have a disorder of the retinal vessels, called Diabetic Retinopathy or Diabetic eye disease. This will damage the blood vessels in the retina and these blood vessels can swell and outflow or they can stopover the blood from passing through. In some cases, anomalous new blood vessels nurture on the retina. These disorders will tend to loss one's vision.

The following are the two foremost phases of Diabetic retinopathy:

(i) Non-Proliferative Diabetic Retinopathy: In this phase, a minor deteriorating in the small blood vessels of the retina and the percentages of the vessels may swell and outflow liquid into the neighboring retinal tissue.

(ii) Proliferative Diabetic Retinopathy: It is a progressive form of Diabetic retinopathy. This ensues when anomalous new blood vessels and scar tissue form on the exterior of the retina.

Diabetic retinopathy underwrites 4.8% of the 37 million cases of loss of sight all through the world. A recent study in city population in South India estimates prevalence of Diabetic Mellitus in adult population as high as 28% and the frequency of Diabetic retinopathy in diabetics to 18%. Each year in the US, Diabetic retinopathy records for 12% of all new cases of loss of sight. It is also the primary reason of loss of sight of people aged 20 and 64 years.

2.2 Intuitionistic fuzzy digital subset:[6]

If $A = \langle \mu_A, \nu_A \rangle$ is the intuitionistic fuzzy subset of the Euclidean plane E, The IFD subset of the lattice points in E is defined by $A_{\sim} = \langle \mu_{A\sim}, \nu_{A\sim} \rangle$ with degree of membership $\mu_{A\sim}(P) = max\{\mu_A(R) \mid R \in P^*\}$ and degree of nonmembership $\nu_{A\sim}(P) = min\{\nu_A(R) \mid R \in P^*\}$, provided $0 \leq \mu_{A\sim}(P) + \nu_{A\sim}(P) \leq 1$ for each $P \in E$.

2.3 Intuitionistic fuzzy digital convex set:[7]

An IFD set $A_{\sim} = \langle \mu_{A\sim}, \nu_{A\sim} \rangle$ of the lattice points in the Euclidean plane E is called as an IFD convex set if for any two points $P, Q \in A_{\sim}$ and $\alpha \in [0,1]$,

$$\alpha\mu_{A_\sim}(P)+(1-\alpha)\mu_{A_\sim}(Q)\geq min(\mu_{A_\sim}(P),\mu_{A_\sim}(Q))$$
$$\text{and } \alpha\nu_{A_\sim}(P)+(1-\alpha)\nu_{A_\sim}(Q)\leq max(\nu_{A_\sim}(P),\nu_{A_\sim}(Q)).$$

Notation:
C (E) denotes the set of all IFD convex sets in *E*.

2.4 Intuitionistic Fuzzy Digital Convex Envelope: [6]
If $A_\sim = \langle \mu_{A_\sim}, \nu_{A_\sim} \rangle$ is an IFD subset of the lattice points in the Euclidean plane *E*, then the IFD convex envelope of $A_\sim$ is $\mathcal{E}_{A_\sim} = \langle \mu_{\mathcal{E}_{A_\sim}}, \nu_{\mathcal{E}_{A_\sim}} \rangle$ whose membership and non membership functions are as follows,

$$\mu_{\mathcal{E}_{A_\sim}}(P)=inf\{\mu_{B_\sim}(P):\mu_{B_\sim}(P)\geq\mu_{A_\sim}(P) \text{ where } B_\sim \in \mathcal{C}(E)\}$$
$$\text{and } \nu_{\mathcal{E}_{A_\sim}}(P)=sup\{\nu_{B_\sim}(P):\nu_{B_\sim}(P)\leq\nu_{A_\sim}(P) \text{ where } B_\sim \in \mathcal{C}(E)\},$$

provided $0\leq\mu_{\mathcal{E}_{A_\sim}}(P)+\nu_{\mathcal{E}_{A_\sim}}(P)\leq 1$ and $P\in E$.

3. Methodology
In this proposed work, the concept of intuitionistic fuzzy digital convex envelope was applied to segment the affected region from the images. The aim of this work was achieved from the base of the intuitionistic fuzzy digital convex set together with the Euclidean distance based of membership and non membership values. These values are measured from the intuitionistic fuzzy digital image region. The region was extracted clearly by using this proposed algorithm.

Eye disorder image datasets were acquired through medical lab. Intuitionistic fuzzy digital convex envelope concept was applied in the image to detect the disorder part clearly. There are three main steps in this proposed work and they are as follows.

A) Image preprocessing
A digital image is basically a numerical representation of an object. After the image was acquired, the true color image f(x,y) is converted to hue, saturation and value. These values are extracted from 0 to 1. Then they normalized to [0 255].

B) Image Segmentation
Segmentation sections an image into is component regions or objects. The level of feature to which the subdivision is supported rest on the problem being resolved. The segmentation should stopover when the objects or regions of curiosity in an application have been spotted. Segmentation of nontrivial images is one of the greatest tough tasks in image processing. In this proposed work, the non membership and membership values of intuitionistic fuzzy digital convex envelope is applied to segment the disorder region more accurately compared to the existing method of fuzzy edge based segmentation.

The binary mask of all foreground objects in the input image, can be detected through the intuitionistic fuzzy digital convex envelope. The foreground defected object was detected via proposed method.

C) Measuring Image Region
After the segmentation process, each disorder region on the segmented image is measured by using the pixel values. Then calculate the standard deviation of each region in defected area through centroid, pixel values of the image and boundary box which are aided to detect the defected region.

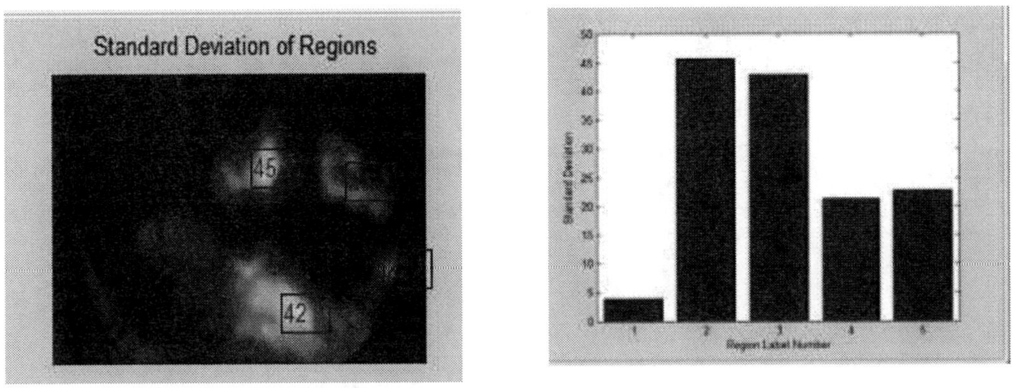

Figure 1: Standard deviation of the segmented region and plots of the values

The values of the standard deviation of the segmented regions lies between intuitionistic fuzzy convex envelope based segmentation and fuzzy edge based segmentation. The prevailing method of fuzzy edge based segmentation technique is not truthfully partitioning the disorder regions and the standard deviation region values are too low compared to the proposed algorithm.

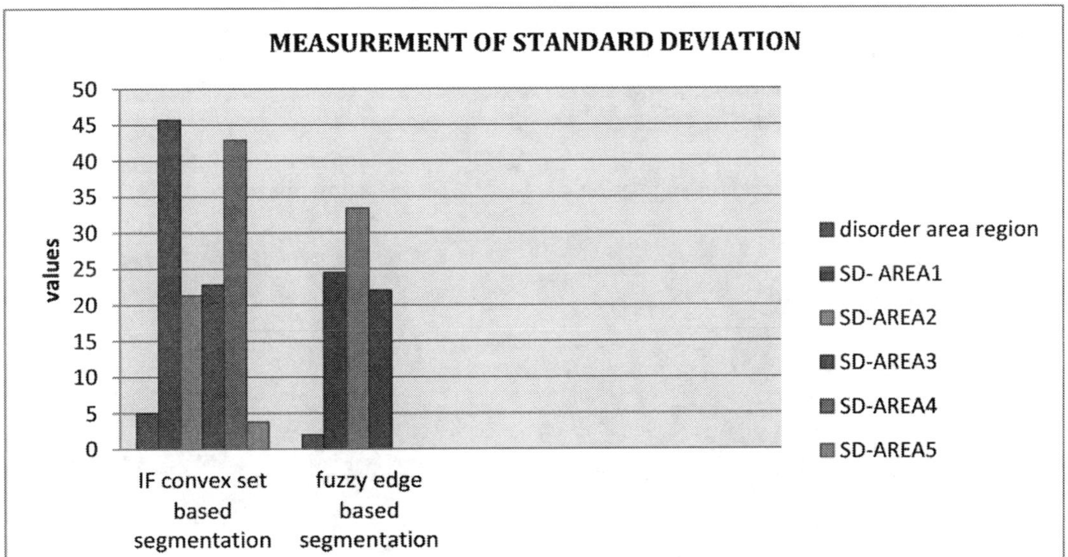

Figure 2: Differnce values between two segmentation techniques

4.Experimental Result

The proposed algorithm is developed by using matlab 2009b. Figure 3 illustrates the following.

(a) an original eye disorder image was acquired.
(b) applied the membership and non membership values of the intuitionistic fuzzy digital convex envelope to detect the disorder region
(c) computed the standard deviation between the disorder regions
(d) from the existing fuzzy edge based segmentation technique, some of the disorder regions are neglected

IOP Conf. Series: Materials Science and Engineering 1012 (2021) 012068 doi:10.1088/1757-899X/1012/1/012068

(e) segmented the regions by using fuzzy edge.

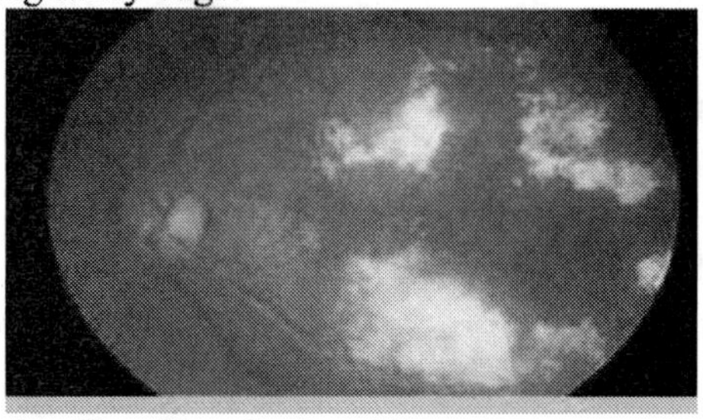

a) Eye disorder input image

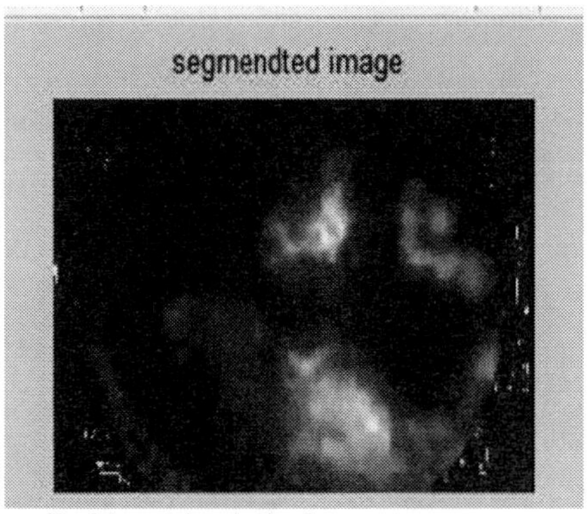

b) applied intuitionistic fuzzy digital convex envelope for segmentation

c) applied fuzzy edge segmentation

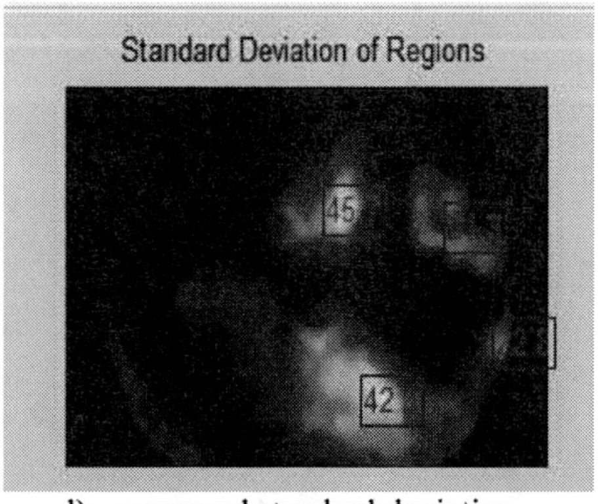

d) measured standard deviation

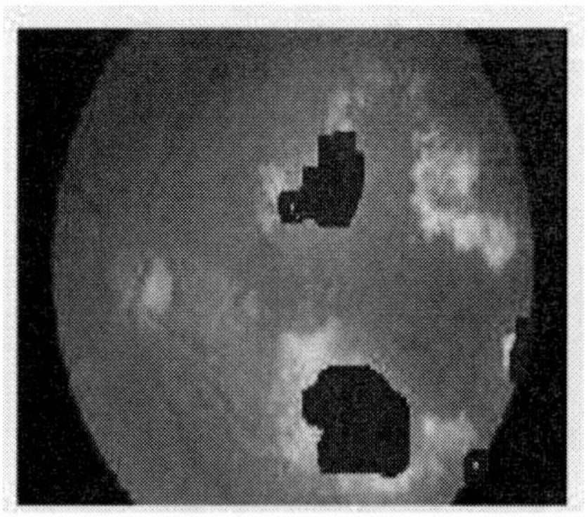

e) fuzzy segmented parts

Figure 3: Intuitionistic fuzzy digital convex envelope based convex envelope and fuzzy edge based segmentation process

5. Conclusion

Automatic detection of the eye disease is assisted to reduce the time-consuming process. In this proposed work, intuitionistic fuzzy digital convex envelope based segmentation detected the disorders accurately compared to fuzzy edge based detection. Fuzzy based segmentation excludes the disorder regions. After segmentation process, measured the standard deviation values from the region of the segmented image. The standard deviation values are defined for the segmented areas by using the proposed method. It segmented the region properly with truthful information.

References

[1] L.A. Zadeh, Fuzzy Sets. 1968. Information and Control 338-353.
[2] Krassimir T Atanassov, Intuitionistic fuzzy sets: Theory and applications. 1999. Physica - Verlag, Heidelberg, New York.
[3] Dillip Ranjan Nayak. 2016. Image enhancement using intuitionistic fuzzy reconstruction, International journal of advance electrical and electronic engineering 5(1), 2278-894.
[4] K. Ioannis, Vlachos and George D. Sergiadis, Intuitionistic fuzzy image processing. Part of the 'Studies in Fuzziness and soft computing' book series 210.
[5] Pedro Melo-Pinto, Pedro Couto, Humberto Bustince and Javier Fernandez, Image segmentation using Atanassov'situitionistic fuzzy sets. 2013. Expert systems with applications 40(1), 15-26.
[6] S. Meenakshi, J. Tamilmani, D. Amsaveni, Intuitionistic Fuzzy Digital Convexity. 2017. International Journal of Computational and Applied Mathematics 12(1), 54-63
[7] S. Meenakshi, J. Tamilmani, Intuitionistic fuzzy digital structure convex envelopes and simplexes. 2017. Journal of analysis 25(2), 167-175.

AUTHOR INDEX